BIOETHANOL PRODUCTION FROM FOOD CROPS

BIOETHANOL PRODUCTION FROM FOOD CROPS

SUSTAINABLE SOURCES, INTERVENTIONS AND CHALLENGES

Edited by

RAMESH C. RAY

ICAR-Central Tuber Crops Research Institute (Regional Centre), Bhubaneswar, India

S. RAMACHANDRAN

Birla Institute of Technology & Science, Dubai, United Arab Emirates

ELSEVIER

ACADEMIC PRESS

An imprint of Elsevier

Academic Press is an imprint of Elsevier
125 London Wall, London EC2Y 5AS, United Kingdom
525 B Street, Suite 1650, San Diego, CA 92101, United States
50 Hampshire Street, 5th Floor, Cambridge, MA 02139, United States
The Boulevard, Langford Lane, Kidlington, Oxford OX5 1GB, United Kingdom

Notices

Knowledge and best practice in this field are constantly changing. As new research and experience broaden our understanding, changes in research methods, professional practices, or medical treatment may become necessary.

Practitioners and researchers must always rely on their own experience and knowledge in evaluating and using any information, methods, compounds, or experiments described herein. In using such information or methods they should be mindful of their own safety and the safety of others, including parties for whom they have a professional responsibility.

To the fullest extent of the law, neither the Publisher nor the authors, contributors, or editors, assume any liability for any injury and/or damage to persons or property as a matter of products liability, negligence or otherwise, or from any use or operation of any methods, products, instructions, or ideas contained in the material herein.

Library of Congress Cataloging-in-Publication Data
A catalog record for this book is available from the Library of Congress

British Library Cataloguing-in-Publication Data
A catalogue record for this book is available from the British Library

ISBN: 978-0-12-813766-6

Publisher: Andre Gerhard Wolff
Acquisition Editor: Nancy Maragioglio
Editorial Project Manager: Brianna Garcia
Production Project Manager: Denny Mansingh
Designer: Victoria Pearson

Typeset by Thomson Digital

Contents

Contributors xi
Preface xiii

I
GENERAL PERSPECTIVES OF BIOETHANOL PRODUCTION TECHNOLOGIES

1. Bioethanol from Biorenewable Feedstocks: Technology, Economics, and Challenges 3

SONALI MOHAPATRA, RAMESH C. RAY, S. RAMACHANDRAN

1.1 Introduction 3
1.2 Global Scenario of Bioethanol Production 4
1.3 Renewable Feedstocks According to Their Generations 6
1.4 Biorefinery Approach 15
1.5 Biotechnology of Bioethanol Crops 17
1.6 Food Versus Fuel Debate 17
1.7 Economic Impacts of Bioethanol 18
1.8 Policy Issues 19
1.9 Bioethanol Production Technologies: Environmental Impacts and Life Cycle Assessment (LCA) 20
1.10 Conclusion and Future Perspectives 22
References 22

II
BIOETHANOL FROM FOOD CROPS

2. Disassembling the Glycomic Code of Sugarcane Cell Walls to Improve Second-Generation Bioethanol Production 31

MARCOS S. BUCKERIDGE, ADRIANA GRANDIS, EVELINE Q.P. TAVARES

2.1 Introduction 31
2.2 Sugarcane as a Source of Bioethanol 32
2.3 Sugarcane Cell Walls 32
2.4 Pretreatments and Hydrolysis and Their Impact on Covalent Linkages 35
2.5 Effect of Pretreatments on the Noncovalent Linkages of the Wall 39
2.6 Conclusions and Future Perspectives 40
References 41

3. Bioethanol Production from Corn and Wheat: Food, Fuel, and Future 45

SUJIT K. MOHANTY, MANAS R. SWAIN

3.1 Introduction 45
3.2 Corn and Wheat-Based Ethanol Production: Global Scenario 46

3.3 USA—The Global Leader in Fuel Ethanol Production Prefers Corn 50
3.4 Technological Aspects of Ethanol Production from Corn 50
3.5 Technological Aspects of Ethanol Production from Wheat 55
3.6 Socioeconomical Advantages and Food Versus Fuel Debate 56
3.7 Conclusion and Future Perspectives 57
References 58

4. Status and Perspectives in Bioethanol Production from Sugar Beet 61

CRISTINA MARZO, ANA B. DÍAZ, ILDEFONSO CARO, ANA BLANDINO

4.1 Introduction 61
4.2 Global Production Scenario of Sugar Beet 61
4.3 Sugar Beets as Raw Material 63
4.4 Sugary Juices from Sugar Beet for Bioethanol Production 64
4.5 Sugar Beet Pulp Pretreatment and Hydrolysis for Bioethanol Production 67
4.6 Ethanol Fermentation (First and Second Generation) 70
4.7 Economics and Life Cycle Assessment of Sugar Beet Ethanol 72
4.8 Future Perspectives 74
4.9 Conclusion 75
References 75

5. Sweet Sorghum for Bioethanol Production: Scope, Technology, and Economics 81

RAMESH C. RAY, KIRAN BABU UPPULURI, CHANDRASEKARAN TRILOKESH, CLAUDIA LAREO

5.1 Introduction 81
5.2 Sweet Sorghum as a Biofuel Crop 82
5.3 Processes for Conversion of Sweet Sorghum Into Bioethanol 82
5.4 Biotechnology of Sorghum Fermentation 91
5.5 Technoeconomic Feasibility and Real-Time Applications 94
5.6 Conclusions and Future Perspectives 97
References 97

6. Cassava as Feedstock for Ethanol Production: A Global Perspective 101

SANETTE MARX

6.1 Introduction 101
6.2 Cassava as Energy Crop 102
6.3 Bioethanol Production from Cassava Roots 104
6.4 Bioethanol from Cassava Wastes and Peels 106
6.5 Cassava-Based Ethanol in China 107
6.6 Cassava-Based Ethanol in Thailand 108
6.7 Cassava-Based Ethanol in Vietnam 109
6.8 Concluding Remarks 110
References 110

7. Sweet Potato as a Bioenergy Crop for Fuel Ethanol Production: Perspectives and Challenges 115

CLAUDIA LAREO, MARIO D. FERRARI

7.1 Introduction 115
7.2 General Characteristics of the Sweet Potato Crop 115
7.3 World Production and Current Uses 116
7.4 Industrial Sweet Potato 118
7.5 Chemical Composition of Sweet Potato 121
7.6 Processes for Conversion of Sweet Potato Into Bioethanol 125
7.7 Effect of the Main Variables of the Conversion Process: Solid to Liquid Ratio, Enzymes, and Temperature 134
7.8 By-Products: Use of Residual Solids for Animal Feed 136
7.9 Technoeconomic Feasibility 136
7.10 Life Cycle Analysis 140
7.11 Conclusions and Future Perspectives 143
References 144

8. Jerusalem Artichoke: An Emerging Feedstock for Bioethanol Production 149

KONSTANTINOS V. KOTSANOPOULOS, RAMESH C. RAY, SUDHANSHU S. BEHERA

8.1 Introduction 149
8.2 Characteristics of Jerusalem Artichoke 150
8.3 Characteristics of Inulin and its Yields from Jerusalem Artichoke 151

8.4 Bioethanol Production from Jerusalem Artichoke Tubers 152
8.5 Butanol 157
8.6 Technoeconomic Feasibility 158
8.7 Conclusion and Future Perspectives 158
References 158

III

BIOETHANOL FROM LIGNOCELLULOSES

9. Lignocellulosic Ethanol: Feedstocks and Bioprocessing 165

RICHA ARORA, NILESH K. SHARMA, SACHIN KUMAR, RAJESH K. SANI

9.1 Introduction 165
9.2 Lignocellulosic Bioethanol Production: an Overview 166
9.3 Feedstocks 169
9.4 Pretreatment 169
9.5 Saccharification 174
9.6 Fermentation 174
9.7 Challenges 177
9.8 Economic Assessment 178
9.9 Concluding Remarks 179
References 179

10. Bioethanol from Sugarcane Bagasse: Status and Perspectives 187

SADAT M.R. KHATTAB, TAKASHI WATANABE

10.1 Introduction 187
10.2 Why Bioethanol? 188
10.3 Why Sugarcane Bagasse? 189
10.4 Conversion Technologies of Sugarcane Bagasse 193
10.5 Evolution of Saccharomyces Cerevisiae Toward Second-Generation Bioethanol Production from Lignocellulosic Biomass 198
10.6 Conclusions and Future Prospects 206
References 207

11. Bioethanol Production from Rice- and Wheat Straw: An Overview 213

MANAS R. SWAIN, AJIT SINGH, AJAY K. SHARMA, DEEPAK K. TULI

11.1 Introduction 213
11.2 Availability of Rice Straw 214
11.3 Ethanol as Bioenergy Resources 215
11.4 Fermentation 223
11.5 Technoeconomical Feasibility of Bioethanol Production 227
11.6 Conclusion 227
References 228

12. Forest Bioresources for Bioethanol and Biodiesel Production With Emphasis on Mohua (*Madhuca latifolia* L.) Flowers and Seeds 233

SUDHANSHU S. BEHERA, RAMESH C. RAY

12.1 Introduction 233
12.2 Bioprospecting of Forest Resources for Harnessing Biofuels 234
12.3 Biorefinery from Forest Woody Biomass 234
12.4 Biorefinery from Mohua Flowers and Seeds 238
12.5 Consolidated Bioprocessing(CBP)/Engineered Microorganism 241
12.6 Comparison of Biorefinery from Forest Biomass with other Feedstocks 241
12.7 Technoeconomic Feasibility of Bioethanol Production from Forest Biomass and Bottlenecks 242
12.8 Global Biofuel Economy and Where Forest Biomass Stands 243
12.9 Conclusion and Future Prospects 243
References 243

13. Microbial Enzyme Applications in Bioethanol Producing Biorefineries: Overview 249

FRANCISCO J. RÍOS-FRÁNQUEZ, ÓSCAR A. ROJAS-REJÓN, CARLOS ESCAMILLA-ALVARADO

13.1 Introduction 249
13.2 Basics of Enzyme Technology 250

13.3 Enzyme Industrial Applications and New Insights 256
13.4 Conclusion and Future Perspectives 262
References 262

14. Application of Fungal Pretreatment in the Production of Ethanol from Crop Residues 267

THELMO A. LU-CHAU, MARÍA GARCÍA-TORREIRO, MARÍA LÓPEZ-ABELAIRAS, NATALIA A. GÓMEZ-VANEGAS, BEATRIZ GULLÓN, JUAN M. LEMA, GEMMA EIBES

14.1 Introduction 267
14.2 Mechanism Involved in the Pretreatment of Crop Residues by Ligninolytic Fungi 269
14.3 Agricultural Residues as Potential Substrate to be Fungal Pretreated 271
14.4 Development of Fungal Pretreatment 272
14.5 Parameters Affecting the Performance of Fungal Pretreatment and their Optimization 273
14.6 Strategies to Improve the Action of Fungal Pretreatment 276
14.7 Fermentation of Fungal Pretreated Crop Residues 284
14.8 Concluding Remarks 285
References 287

15. Currently Used Microbes and Advantages of Using Genetically Modified Microbes for Ethanol Production 293

BISHNU JOSHI, JARINA JOSHI, TRIBIKRAM BHATTARAI, LAKSHMAIAH SREERAMA

15.1 Introduction 293
15.2 World Fuel Ethanol Production and Challenges 294
15.3 Heterogenity in Carbohydrate Composition of Lignocellulosic Biomass 298
15.4 Inhibitors Generated During Pretreatment/Hydrolysis and their Effects on Fermenting Microorganisms 299
15.5 Engineering Bacteria for Ethanol Production 301
15.6 Conclusion and Future Perspectives 310
References 310

IV

BIOETHANOL FROM ALGAE

16. Biorefinery Approach for Ethanol Production from Bagasse 319

STAVROS E. MICHAILOS, COLIN WEBB

16.1 Introduction 319
16.2 Ethanol from Bagasse: Technological Status 321
16.3 Process Simulation as a Tool for Route Selection 324
16.4 Process Design and Modeling 325
16.5 Feasibility Analysis 331
16.6 Concluding Remarks 338
References 339

17. Biorefinery as a Promising Approach to Promote Ethanol Industry from Microalgae and Cyanobacteria 343

CARLOS E. DE FARIAS SILVA, ELENA BARBERA, ALBERTO BERTUCCO

17.1 Introduction 343
17.2 Microalgal Biomass Cultivation and Carbohydrate Production 346
17.3 Harvesting and Water/Nutrient Recycle Steps 348
17.4 Hydrolysis and Fermentation 350
17.5 Nutrients Recovery 354
17.6 Potential Bioethanol Productivity from Microalgae 356
17.7 Concluding Remarks and Future Outlook 356
References 356

18. Role of Genetic Engineering in Bioethanol Production from Algae 361

DURAIARASAN SURENDHIRAN, ABDUL RAZACK SIRAJUNNISA

18.1 Introduction 361
18.2 Algae 361
18.3 Biofuels from Algae 363
18.4 Bioethanol from Microalgae 364
18.5 Bioethanol from Macroalgae 366

18.6 Algal Polysaccharides 367
18.7 Enhancement of Carbohydrate Content for Increased Bioethanol Production 370
18.8 Prospects of Bioethanol 378
18.9 Conclusion 378
References 379

V

LIFE CYCLE ANALYSIS, ECONOMICS AND POLICY ISSUES

19. Life Cycle Assessment (LCA) of Bioethanol Produced from Different Food Crops: Economic and Environmental Impacts 385

PORITOSH ROY, ANIMESH DUTTA

19.1 Introduction 385
19.2 Life Cycle Assessment Methodology (and Dynamic Life Cycle Assessment) 387
19.3 Life Cycle Assessment Studies on Bioethanol from Food Crops 388
19.4 Land, Water, and Other Approaches in Life Cycle Assessment of Bioethanol 390
19.5 Environmental Benefits of Bioethanol 391
19.6 Economics of Bioethanol 392
19.7 Ongoing Efforts to Improve the Sustainability of Bioethanol 393
19.8 Discussion 394
19.9 Conclusions 395
References 396

20. Upgrading Comparative and Competitive Advantages for Ethanol Fuel Production from Agroindustrial Crops in Developing Countries: Mexico as a Case Study 401

NOÉ AGUILAR-RIVERA, RICARDO SERNA-LAGUNES, CHRISTIAN MICHEL-CUELLO, ARMÍN TRUJILLO-MATA

20.1 Introduction 401
20.2 Value Chain Ethanol Fuel 402
20.3 Biofuels for Developing Countries Boost Comparative Advantages 403
20.4 Land Suitability for Ethanol Production in Mexico—A Case Study 405
20.5 Final Remarks 413
References 414

21. Bioethanol in Brazil: Status, Challenges and Perspectives to Improve the Production 417

ANA K. DE SOUZA ABUD, CARLOS E. DE FARIAS SILVA

21.1 Introduction 417
21.2 Historical Background and Geographic Impositions to Sugarcane Exploitation 418
21.3 Lignocellulosic Ethanol in Brazil: Current Status, Needs, and Limitations 421
21.4 Macro and Microalgae 432
21.5 Some Alternatives to Improve the Sustainability and Energy/Environmental Balance of Brazilian Ethanol Biorefineries 433
21.6 Conclusions 439
References 439

Index 445

Contributors

Noé Aguilar-Rivera University of Veracruz, Cordoba, Mexico; University of Veracruz, Xalapa, Mexico

Richa Arora Sardar Swaran Singh National Institute of Bio-Energy, Kapurthala; School of Bioengineering and Biosciences, Lovely Professional University, Phagwara, India

Elena Barbera Department of Industrial Engineering DII, University of Padova, Padova, Italy

Sudhanshu S. Behera Department of Fisheries and Animal Resource Development, Government of Odisha, Cuttack, India

Alberto Bertucco Department of Industrial Engineering DII, University of Padova, Padova, Italy

Tribikram Bhattarai Tribhuvan University, Kirtipur, Nepal

Ana Blandino University of Cádiz, Cádiz, Spain

Marcos S. Buckeridge Laboratory of Plant Physiological Ecology (LAFIECO), Department of Botany, Institute of Biosciences, University of São Paulo, São Paulo, Brazil

Ildefonso Caro University of Cádiz, Cádiz, Spain

Ana B. Díaz University of Cádiz, Cádiz, Spain

Carlos E. de Farias Silva Department of Industrial Engineering DII, University of Padova, Padova, Italy; Institute of Chemistry and Biotechnology, Federal University of Alagoas, Maceió, Brazil; Center of Technology, Federal University of Alagoas, Maceió, Brazil

Ana K. de Souza Abud Department of Food Technology, Federal University of Sergipe, São Cristóvão, Brazil

Animesh Dutta University of Guelph, Guelph, ON, Canada

Gemma Eibes Dept. of Chemical Engineering, Institute of Technology, Universidade de Santiago de Compostela, Santiago de Compostela, Spain

Carlos Escamilla-Alvarado Universidad Autónoma de Nuevo León, San Nicolás de los Garza, Mexico

Mario D. Ferrari Instituto de Ingeniería Química, Universidad de la República, Montevideo, Uruguay

Natalia A. Gómez-Vanegas Department de Ingeniería Química, Universidad de Antioquia, Medellín, Colombia

María García-Torreiro Department of Chemical Engineering, Institute of Technology, Universidade de Santiago de Compostela, Santiago de Compostela, Spain

Adriana Grandis Laboratory of Plant Physiological Ecology (LAFIECO), Department of Botany, Institute of Biosciences, University of São Paulo, São Paulo, Brazil

Beatriz Gullón Department of Chemical Engineering, Institute of Technology, Universidade de Santiago de Compostela, Santiago de Compostela, Spain

Bishnu Joshi Tribhuvan University, Kirtipur, Nepal

Jarina Joshi Tribhuvan University, Kirtipur, Nepal

Sadat M.R. Khattab Al-Azhar University, Assiut Campus, Egypt; Research Institute for Sustainable Humanosphere (RISH), Kyoto University, Uji Campus, Kyoto, Japan

Konstantinos V. Kotsanopoulos University of Thessaly, Volos, Hellas, Greece

Sachin Kumar Sardar Swaran Singh National Institute of Bio-Energy, Kapurthala, India; South Dakota School of Mines and Technology, Rapid City, SD, United States

María López-Abelairas Department of Chemical Engineering, Institute of Technology, Universidade de Santiago de Compostela, Santiago de Compostela, Spain

Claudia Lareo University of the Republic, Montevideo, Uruguay; Instituto de Ingeniería Química, Universidad de la República, Montevideo, Uruguay

Juan M. Lema Department of Chemical Engineering, Institute of Technology, Universidade de Santiago de Compostela, Santiago de Compostela, Spain

Thelmo A. Lu-Chau Department of Chemical Engineering, Institute of Technology, Universidade de Santiago de Compostela, Santiago de Compostela, Spain

Sanette Marx North-West University, Potchefstroom, South Africa

Cristina Marzo University of Cádiz, Cádiz, Spain

Stavros E. Michailos University of Sheffield, Sheffield, United Kingdom

Christian Michel-Cuello Autonomous University of San Luis Potosí, Rioverde, Mexico

Sujit K. Mohanty Iowa State University, Ames, IA, United States

Sonali Mohapatra Department of Biotechnology, College of Engineering & Technology, Bhubaneswar, India

Francisco J. Ríos-Fránquez Tecnológico Nacional de México, Torreón, Mexico

S. Ramachandran Birla Institute of Technology & Science, Dubai, United Arab Emirates

Ramesh C. Ray ICAR-Central Tuber Crops Research Institute (Regional Centre), Bhubaneswar, India

Óscar A. Rojas-Rejón Instituto Tecnológico de Estudios Superiores de Occidente, Tlaquepaque, Mexico

Poritosh Roy University of Guelph, Guelph, ON, Canada

Rajesh K. Sani South Dakota School of Mines and Technology, Rapid City, SD, United States

Ricardo Serna-Lagunes University of Veracruz, Cordoba, Mexico; University of Veracruz, Xalapa, Mexico

Ajay K. Sharma DBT-IOC Centre for Advance Bioenergy Research Centre, Faridabad, India

Nilesh K. Sharma Sardar Swaran Singh National Institute of Bio-Energy, Kapurthala, India

Ajit Singh DBT-IOC Centre for Advance Bio-energy Research Centre, Faridabad, India

Abdul Razack Sirajunnisa Anna University, Chennai, India

Lakshmaiah Sreerama Qatar University, Doha, Qatar

Duraiarasan Surendhiran St. Joseph University, Dar es Salaam, United Republic of Tanzania

Manas R. Swain DBT-IOC Centre for Advance Bioenergy Research Centre, Faridabad, India

Eveline Q.P. Tavares Laboratory of Plant Physiological Ecology (LAFIECO), Department of Botany, Institute of Biosciences, University of São Paulo, São Paulo, Brazil

Chandrasekaran Trilokesh SASTRA Deemed University, Thanjavur, India

Armín Trujillo-Mata Technology Institute of Orizaba, Orizaba, Mexico

Deepak K. Tuli DBT-IOC Centre for Advance Bioenergy Research Centre, Faridabad, India

Kiran Babu Uppuluri SASTRA Deemed University, Thanjavur, India

Takashi Watanabe Research Institute for Sustainable Humanosphere (RISH), Kyoto University, Uji Campus, Kyoto, Japan

Colin Webb University of Manchester, Manchester, United Kingdom

Preface

In less than two decades, world biofuel production has increased more than four times, from less than 20 billion litres in 2001 to over 88 billion litres approximately, in 2017. However, the relationship between biofuel and food security is always challenging and debatable, because it is at the intersection of some major global issues: energy, food, land, and water uses, greenhouse gas emission, environment, income generation, and overall development. Among various biofuels, bioethanol, and biodiesel are the foremost commodities of global demand and economic concern. In this book, various aspects of bioethanol production technology both from food and nonfood crops, covering first (sugar and starchy crops), second (lignocelluloses), and third (algae) generation technologies, related public policies and their impact on land, water, environment and food security have been discussed in the form of 21 chapters. The United States and Brazil are two leading countries in production and usage of bioethanol, particularly in the transport sector, sharing 58% and 27% of total global production, respectively. At present, many more countries (<50) have adopted bioethanol policies to blend ethanol with gasoline for transport purpose with other concomitant approaches for a reduction in greenhouse emission and urban pollution, and not to rely heavily on food crops or land use for food production. While the intriguing theme of this book is on food versus fuel debate, the focus is mostly on futuristic policies, particularly on lignocellulosic ethanol. The book is divided into five sections: Section I: General perspectives of bioethanol production technologies and various feedstocks used; Section II: Bioethanol from food crops (sugarcane, corn, wheat, cassava, sweet potato, sugarbeet, and sweet sorghum); Section III: Bioethanol from lignocelluloses, Section IV: Bioethanol from algae, and Section V: Life cycle analysis, economics, and policy issues. The chapters have been written by eminent researchers in these fields of research/study from leading bioethanol producing countries including USA, Brazil, Canada, Mexico, India, and others. The editors are immensely thankful to all the authors for their prompt response in accepting our invitations and timely delivery of the quality manuscripts.

Ramesh C. Ray
S. Ramachandran

SECTION I

GENERAL PERSPECTIVES OF BIOETHANOL PRODUCTION TECHNOLOGIES

1 *Bioethanol From Biorenewable Feedstocks: Technology, Economics, and Challenges* 3

CHAPTER

1

Bioethanol From Biorenewable Feedstocks: Technology, Economics, and Challenges

Sonali Mohapatra, Ramesh C. Ray**, S. Ramachandran*[†]

*Department of Biotechnology, College of Engineering & Technology, Bhubaneswar, India
**ICAR-Central Tuber Crops Research Institute, Bhubaneswar, India
[†]Birla Institute of Technology & Science, Dubai, United Arab Emirates

ABBREVIATIONS

1G	First generation
2G	Second generation
3G	Third generation
4G	Fourth generation
ADPGase	ADP glucose pyrophosphorylase
CBP	Consolidated bioprocessing
CCS	Carbon Capture and Storage
COMT	3-O-methyltransferase
DOE/NETL	US Department of Energy's National Energy Technology Laboratory
FAO	Food Agricultural Organization of United Nations
FFBs	Fresh fruit-bunches
GHG	Greenhouse gas
IEA	International Energy Agency
LCA	Life cycle assessment
MRLE	Mineral rich liquid extract
SCB	Sugarcane baggase
SPRs	Sweet potato residues
TAES	University's Agricultural Experiment Station
USAF	US Air Force

1.1 INTRODUCTION

The fast depletion of the fossil fuel reserves and severe environmental concerns has necessitated the demand for alternative energy and has sparked an exponential motivation to return to a biobased economy (Demirbas, 2009; Mohapatra et al., 2016). With the high consumption rate of these nonrenewable fuels it is expected that these fossil fuel reserves will be depleted within the next 40–50 years (Chen et al., 2016). Further, the greenhouse gas (GHG) emissions contributed by the burning of fossil fuels have been stated to be the foremost culprit for global warming leading to urban pollution, variation in climatic conditions and steady rise in sea level.

Bioethanol Production From Food Crops
http://dx.doi.org/10.1016/B978-0-12-813766-6.00001-1

To address these challenges, using eco-friendly, biodegradable, and economical alternatives, such as biofuel, specifically bioethanol can be an ideal option (Balat, 2009; Demirbas, 2009). Nevertheless, economically developed countries like the USA, Brazil, China, Canada, and several EU member states have already attempted to increase their dependence on bioethanol, with the former three countries are highly successful in their attempts. This can be well exemplified by the production of bioethanol being expected to reach to an approximate of 97,800 million gal in 2017 as compared to 25,754 million gal in the previous year (GRFA, 2017). Initially, sugar crops and grain-based feedstocks were used for bioethanol production. However, the large-scale commercialization of grain-based ethanol industry has been restricted because of the grains' competition for fuel ethanol and food applications. Consequently, the development of "non-food ethanol" has been promoted from lesser-used food crops like cassava, sweet sorghum, Jerusalem artichoke, and others (Mussatto et al., 2010).

The feedstock is one of the vital areas of research for bioethanol production as the constituents of the biomass play an important role in the overall ethanol yield. Considering this fact, the analysis of bioethanol yield from different feedstocks becomes important with focus on the biochemical composition, availability, transportation, processing cost, and overall, their efficiency as fossil fuel alternative. Literature studies reveal the utilization of diverse biomass by different research groups across the world. These feedstocks for bioethanol production have been divided into different generations (Gs) depending on their composition and the technology used.

The first-generation (1G) feedstock is primarily food crops, which have high sugar and starch contents. Following it is the second-generation (2G) feedstock that contains cellulose as the primary source of saccharide with lower amounts of pentose sugars in the form of hemicellulose. Third-generation (3G) feedstock like algae, microalgae, and cyanobacteria were previously considered for bioethanol but researchers have now realized its potential more for biodiesel rather than bioethanol (Brennan and Owende, 2010; Thatoi et al., 2016). The advantage of fourth-generation (4G) biofuels is their capability in capturing and storing CO_2 + biomass materials, which have absorbed CO_2 while growing, are converted into fuel using the same processes as 2G biofuels. This process differs from second- and third-generation production because in this, at all stages of production, the carbon dioxide is captured using processes, such as oxy-fuel combustion.

Although a substantial research towards the economical production of biofuel has been implemented, the maturity of bioethanol production in commercial scale has only been achieved from 1G feedstock, to date. However, dependence on 1G feedstock for bioethanol cannot be a long-term thought as these being food crops can lead to food crises for the ever-increasing human population. Therefore, it becomes rational to discuss the economic aspects that are related to bioethanol production from the four generations of the biofuel feedstocks with a focus on the biomass composition and the technology employed for ethanol production. This overview is thus expected to mark a thoughtful insight toward future bioethanol production strategies in commercial scale.

1.2 GLOBAL SCENARIO OF BIOETHANOL PRODUCTION

From a human perspective, the world is dependent on fossil fuels for its primary energy supply. In 2014, the energy production of the world was 13,805.44 million tons of oil equivalent (Mtoe), a more than 100% increase compared to 1973 consumption of 6,213.69 Mtoe. The percentage-wise shares of energy production were: coal, 28.8%; crude oil, 31.2%; natural gas, 21.2%; nuclear power, 4.8%; hydroenergy, 2.4%; biofuels and waste, 10.2%, and other sources (includes geothermal, solar, wind, heat, and electricity trade), 1.3% (IEA, 2016). While the consuming part remained high, statistics show that only 3.0% of global energy consumption is supplied from renewable sources, which are expected to rise from 20% in the present scenario to 80% by 2050.

It is also observed that among the oil consumers, the transport sector virtually dominates with 61.5% of the total consumption. The line has to be changed to, Therefore if fossil fuels are to be replaced, the need for renewable alternatives is imperative in transportation sector. The increasing interest in the production of sustainable renewable sources like bioethanol is exemplified by increase. In the production of global biofuel production from the year 2001 to 2017, as given in Fig. 1.1. A steady increase in the production rate have been observed over the years, with an annual production of only 4874 million gal produced in the year 2001 to an anticipated production of 27,737 million gal in 2017. However, the production of bioethanol was not consistent from the year 2010–12 and the annual growth rate decreased from 22,861 to 22,715 million gal. Nevertheless, a sharp increase in the production to 23,429 million gal in the preceding year (2013) was an encouraging sustenance for the bioethanol global market (REN, 2013).

The USA and Brazil are the largest ethanol producing countries, with the former having produced nearly 15 billion gal in 2015 alone. Together, they produce 85% of the world's bioethanol. The vast majority of US ethanol is produced from corn, while Brazil primarily uses sugarcane. Brazil's bioethanol output reached a streak high of 98.3 billion L in 2016 and is expected to be 100 billion L in 2017 (Sapp, 2017). Many other countries also produce ethanol besides the United States and Brazil, albeit at a lower production scale. For example, the Republic of China started producing bioethanol in the year 2001, using corn as raw material. In 2007, grain-based feedstock was used in four ethanol plants, and their production was about 1.4 million metric tons that was gradually replaced with nonfood crops, such as cassava

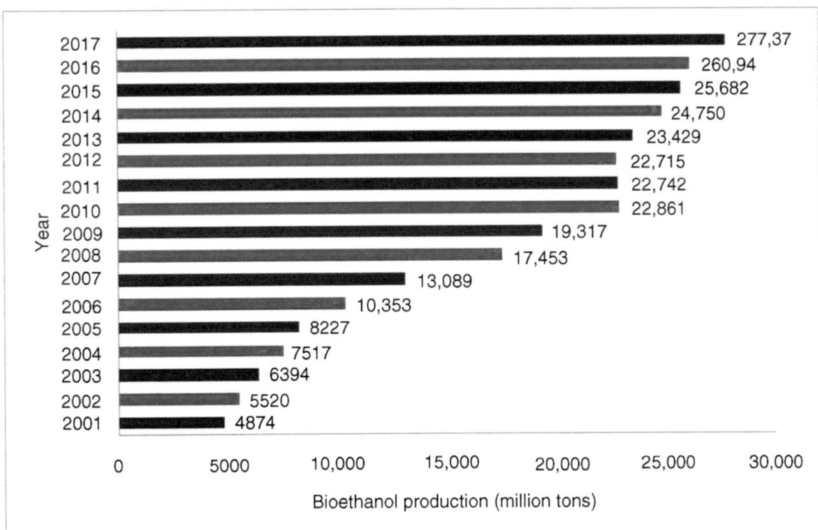

FIGURE 1.1 **World bioethanol production scenario in million tons from the year 2001–2017. 2017* is the expected bioethanol production globally.** *Source: F.O. Lichts' World Ethanol & Biofuels Report, 15 (19).*

and sorghum (Mussatto et al., 2010). At present, China has established a 5000 ton/year sweet sorghum ethanol demonstration plant with the support of the National High-Tech Program, and a 400,000 ton/year cassava ethanol project has been under development since 2005. Similarly, the Thailand government has encouraged production and use of bioethanol in transport sector, from cane molasses and cassava. In 2007, there were seven ethanol plants with a total installed capacity of 955,000 L/day, comprising 130,000 L/day cassava ethanol and 825,000 L/day molasses ethanol (Silalertruksa and Gheewala, 2011).

However, currently, most industrial scale production of ethanol belongs to the 1G biofuels, although the technology to produce 2G ethanol does exist and successfully commercialized in a few countries like the USA and Brazil. The main obstacles for implementation of 2G ethanol plant is due to high technological risks, production costs, and political/policy risks with low potential returns.

1.3 RENEWABLE FEEDSTOCKS ACCORDING TO THEIR GENERATIONS

As mentioned earlier, bioethanol can be produced from any of the four generations. The compositions of the feedstock (first, second, third, and fourth generations) and their monosaccharide/polysaccharide structures are as shown in Fig 1.2. Thus, this section is mainly focused on the in-depth studies of bioethanol production from each of the feedstocks used in the particular generation.

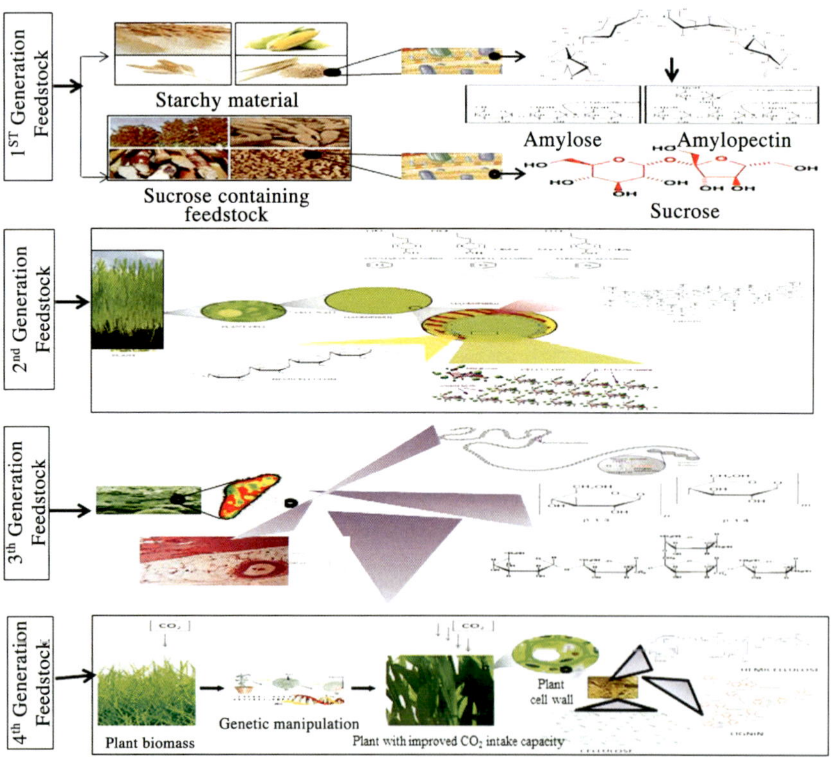

FIGURE 1.2 **Different generation of bioethanol based feedstocks and their cell wall compositions.**

1.3.1 First-Generation Feedstock

First-generation (1G) biofuels are biofuels produced primarily from food crops, such as grains, sugar cane, and tuber crops. Literature review indicates that bioethanol is produced mostly with sugarcane (Brazil) and maize (USA) followed by wheat (Canada), sugar beet, and sorghum (EU countries). Owing to its higher ethanol yield, maize accounts for 67% of the global bioethanol supply (Rulli et al., 2016). However, in terms of crop biomass production sugarcane retains the highest contributor to bioethanol production and the least consumer of water as compared to maize and wheat (Gerbens-Leenes et al., 2009, 2012). Overall, the United States produces 40 billion L of bioethanol from corn/wheat, while Brazil accounts for 25 billion L from sugarcane, China (3 billion L from corn/cassava/rice), Canada (2 billion L from corn/wheat), India (1 billion L from sugarcane/molasses), France (1 billion L from wheat/sugarcane/sugar beet), Germany (750 million L from wheat/sugarcane/sugar beet), and Australia (500 million L from sugarcane) are the remaining countries producing significant bioethanol (http://biofuel.org.uk/major-producers-by-region.html).This accounts to the fact that the use of different resources like water, land, and food equivalent are major factors determining the type of bioethanol crop produced by a country. As the 1G feedstocks have been a major source of bioethanol production, it is vital to study some details about these sugary and starchy feedstocks. The industrial process of bioethanol production and the theoretical and practical bioethanol yield are as shown in Fig. 1.3 and Table 1.1, respectively.

1.3.1.1 Sugar-Containing Feedstock

The sucrose-containing bioethanol feedstocks are mostly grown in Brazil, Germany, France, and India. They mostly include sugarcane, sugar beet, and sweet sorghum with yields of 62–74 tons/ha, 54–111 tons/ha and 50–62 tons/ha, respectively. Sugarcane molasses or black straps, are also interesting sugar containing feedstocks that are widely used for bioethanol production. The high concentrations of sucrose (around 31%) and inverted sugar (around 15%) make the dilution of the substrate compulsory, prior to fermentation. The dilution enables the optimum growth of the microorganisms along with higher fermentation yields. The conventional technique of bioethanol production from sugarcane molasses is by anaerobic fermentation. Nevertheless, aerobic fermentation of the sugarcane molasses using baker's yeast has also been reported to produce high ethanol yields of 0.669 g/g (Jayusab et al., 2016). In a similar work, 0.6 g/g ethanol was obtained from sugarcane molasses using a different species of baker's yeast (Muruaga et al., 2016). Bioethanol production from cane molasses (diluted to 15% sugar w/v) was studied using immobilized *Zymomonas mobilis* MTCC 92 entrapped in luffa (*Luffa cylindrica*) sponge discs and Ca-alginate gel beads. At the end of 96 h fermentation, the final ethanol concentrations were 58.7 ± 0.09 and 59.1 ± 0.08 g/kg molasses with luffa and Ca-alginate entrapped *Z. mobilis* cells, respectively, exhibiting 83.25 ± 0.03 and 84.6 ± 0.02% sugar conversion (Behera et al., 2012). In another study, the ethanol yields were 64.67 ± 0.016 and 65.21 ± 0.030 g/kg molasses, with luffa and Ca-alginate entrapped *Saccharomyces cerevisiae* cells exhibiting 89.90% ± 0.008% and 91.86% ± 0.072% sugar conversion, respectively (Behera and Ray, 2012).

Sugar beets and sweet sorghum molasses like sugarcane molasses are also promising sources for bioethanol production with approximately 53.0% and 56.0% of sugar, respectively. It is reported that aqueous sugars extracted from 1 kg sugar beet, can produce an ethanol yield of 0.07 kg (Santek et al., 2010).

1.3.1.2 Starch-Containing Feedstock

The major starch-based feedstocks that are used for bioethanol production include corn, wheat, and tuber crops like cassava, sweet potato, yam, and aroids. Corn-based bioethanol

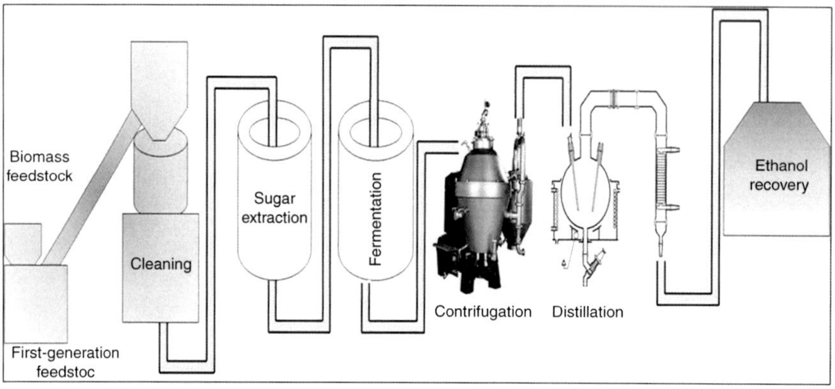

FIGURE 1.3 **Industrial process of bioethanol production from first-generation feedstock.**

TABLE 1.1 Composition, Biomass Yield, and Ethanol Production From Some of the Commonly Used Starchy and Sugary Crops for Bioethanol Production

			Ethanol yield			
Biomass	**Composition (starch/ sugar) (%)**	**Biomass yield (million tons/ha)**	**Practical (g/L)**	**Theoretical (g/L)**	**Rate of ethanol production (million liters)**	**References**
Sweet potato	10–20	1,577,300	38–45	89	270.23	Lareo et al. (2013)
Cassava	20–25	288.4	42.4	106	330.76	Kanagaraj (201 3)
Sugar beet	25–30	188.38	39.40	45.72	251.30	Salazar-Ordóñez et al. (2013)
Potato	15–20	364.8	34.9	42.30	290.65	Lareo et al. (2013)
Yam	15–30	68.1	21.78	32.30	235.535	Akponah and Akpomie (2011)
Sugarcane	12–16	1,830			402.300	Cardona et al. (2010)
Molasses	30–40	51	31.8	35.7	280	Gasmalla et al. (2012)
Bagasse	20–30	363.132	36.5	41.06	290	Roni et al. (2014)

production is dependent on the corn variety and the quality of corn used as the substrate. Recent data exhibits that 258 corn varieties have been utilized to date for bioethanol production and interestingly the starch content and saccharification efficiency are inversely related to each other in corn bioethanol production (Gumienna et al., 2016). Likewise the kernel composition, hardness of the endosperm, soil quality of the area in which the corn is planted, and the presence or absence of mycotoxins determine the corn variety. Depending on the variety of corn, the ethanol yields range from 3% to 23%, with higher bioethanol yields observed in the kernels, which have high free sugar content (Singh, 2012). Another important feedstock for bioethanol production is wheat, which was reported to replace barley for bioethanol production 30 years ago

(Muktham et al., 2016). Bioethanol produced from wheat is reported to have less GHG emissions as compared to gasoline. This was evident from a study done by Belboom et al. (2015) who reported that the GHG emissions can be reduced by 42.5%–61.2% by consumption of 1 MJ (mega joule) bioethanol produced from wheat instead of 1 MJ gasoline.

Roots and tubers (i.e., cassava, sweet potato, Jerusalem artichoke) are the underground storage organs of the tuberous plants with high concentrations of starch (25%–35%, fresh weight basis), which are suitable feedstock for 1G bioethanol production (Hoover, 2001). These crops seem to have greater potential for ethanol production than corn grains, provided economical harvesting, on-farm processing, and cost-effective techniques for conversion of starch to ethanol are developed (Ray and Swain, 2011; Thatoi et al., 2016; Wheals et al., 1999). Tuber crops like cassava with their high starch yield per hectare (36.3 tons/ha/annum) and availability of raw material all year round are promising feedstock for bioethanol production (Behera and Ray, 2014). Even though ethanol production from cassava was successfully commercialized in many countries (i.e., China, Thailand), exploration of optimum slurry concentration, enzyme load, and fermentation conditions to obtain high ethanol titre and maximum ethanol yield are some of the bottlenecks that need further research (Nguyen et al., 2014; Ray and Swain, 2011; Shanavas et al., 2011). Another area of interest that has recently attracted researchers is the use of thermus anaerobes, which support higher rates of starch/cellulose conversion to sugars and reduce cooling costs in fermentation. Recently, a thermus anaerobe, *Caloramator boliviensis*, was used at 60°C in fed-batch fermentation, which resulted in an ethanol yield of 33 g/L corresponding to 85% of the theoretical ethanol yield from saccharified cassava (Moshi et al., 2015).

Industrial sweet potatoes are bred to either increase its starch content or lowering starch liquefaction temperature (from 90°C to around 50°C) to be adaptive for ethanol production. It has been reported that some industrial sweet potatoes breeding lines (i.e., K 9807.1) developed could produce ethanol yields of 4500–6500 L/ha compared to 2800–3800 L/ha for corn (Lareo et al., 2013; Ray and Naskar, 2008; Ziska et al., 2009).

Ethanol production from food crops has some limitations because of the impact on food security and food price, while providing a bit relief on reduction of greenhouse gas emission (Balan et al., 2013). An interesting point that supports the fact that the use of food crops as commodities for biofuel may interfere with the food chain is supported by the data, which states that about 200 million people could be fed by the 1G feedstock used to meet the bioethanol demand in countries like the USA, Brazil, Canada, India, and The Netherlands (Shikida et al., 2014). A more intense study on this aspect suggests that the 1G crops used to produce one TJ of bioethanol are ample to feed 110 people (Rulli et al., 2016). Further, the water consumed by these feedstocks accounts to 3% of the global water requirements that is used for food production (Agência Nacional do Petróleo, 2015). These statistics clearly indicate the necessity of reevaluating the potential of these feedstocks for future bioethanol production. Currently there is much focus on advancement in cellulosic bioethanol production (2G) that utilizes lignocellulosic biomass.

1.3.2 Second-Generation Feedstock

Owing to the food versus fuel issues and harmful environmental impacts of large-scale production of 1G feedstocks like corn and wheat, 2G feedstocks like wood and a wide range of nonfood biomass, such as bagasse, straw, stover, stems, leaves and deoiled seed residues, and grass biomass have gained much interest in the past two decades (Mohapatra et al., 2016, 2017). This can be perceived from the statistics of the number of commercial facilities that have been currently started for 2G bioethanol productions. Statistical reports in 2016 reveal that 67 facilities have been started throughout the world for full commercial scale 2G bioethanol production out of which more than one-third have been producing ethanol in tons (US Department of Energy, 2016). Nevertheless, the US still remains as the leading 2G ethanol producer having 35% of the commercial installed capacity for the same. The processing cost, which involves pretreatment, enzymatic hydrolysis, and fermentation limits the 2G bioethanol production for the rest of the globe with pretreatment and high enzyme (cellulase) costs serving as the major constraints (Behera and Ray, 2016). Appreciatively, the recent initiative taken by National Renewable Energy Laboratory, USA, in collaboration with Novozymes and Genencor, to produce low-cost enzymes can make the commercial scale production of cellulosic ethanol more feasible (Beiter and Tian, 2015). Though several reported literatures have accounted for the economical lignocellulosic pretreatment processes, the effect of the process is seen to be substrate-dependent. Thus, it becomes inevitable to understand the effect of different pretreatment processes on different types of lignocellulosic biomass. The general process of bioethanol production from lignocellulosic biomass is demonstrated in Fig. 1.4. In a broader approach lignocellulosic biomass can be either woody or nonwoody in nature. However, to understand the nature and bioethanol production capacity from each of the biomass type an elaborate study on the feedstocks is as given in later sections.

1.3.2.1 Woody Biomass

The wood biomass often denotes to the hardwood and softwoods that are used as substrates for bioethanol production and differ in their physical properties and chemical compositions (Romaní et al., 2011). In general wood biomass is composed of nanosize cellulose microfibrils held together by hemicellulose and lignin (Alvira et al., 2010). The structure is generally formed by laying the vessels and tracheids that carry water in the middle with layering of microfibrils around them. This layer is eventually responsible for the toughness of the wood thus leading to difficulties in pretreatment. Size reduction, which is essential for increasing the surface area of the biomass, is one of the widely used pretreatment methods for these feedstocks. However, the high energy requirement of approximately 200–600 Wh/kg makes the economical prospective of ethanol production from these biomasses a challenging job (Zhu et al., 2010).

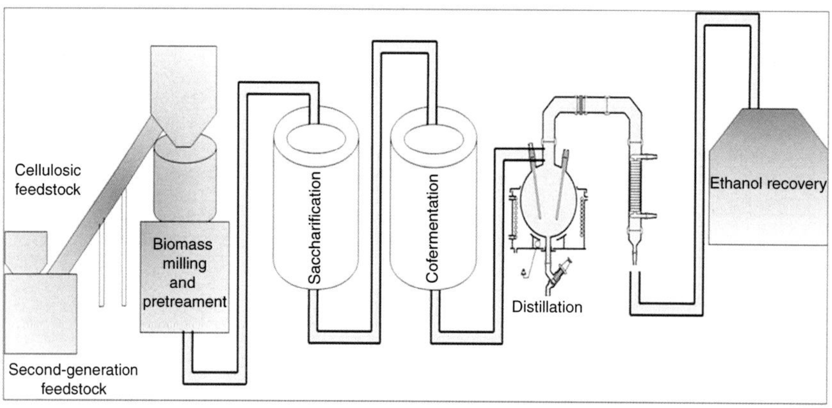

FIGURE 1.4 **Industrial process of bioethanol production from second-generation feedstock.**

1.3.2.2 Nonwoody Biomass

Nonwood biomass as compared to woody biomass has widespread availability, contains more open structures, is cheaper and easy to process, and, more importantly, requires less energy for final bioethanol production. Nonwood biomasses are broadly categorized into agricultural residues, native plants, and nonwood plant fibers. The important agricultural residues that have been explored for their bioethanol production capacity are corn stover, cassava bagasse, cereal straw, sugarcane baggase, potato peel, and oil-palm biomass. The details of the nonwoody biomass are given in the next section.

1.3.2.3 Corn Stover

The residues of the corn plant, such as the cobs, husks, leaves, and stalks that are left in the field after the corn grain is harvested are estimated to produce 80 million of ethanol gallons per year (Liew et al., 2014). Stover, being a nonfood source and a by-product of corn production, has the advantage of lower production costs. Besides, corn stover has a vital role in restocking the soil with organic matter. Nevertheless, with the appropriate safeguards, it is possible to utilize sustainable amounts of corn stover for bioethanol production. Lau and Dale (2009) obtained an ethanol concentration of 40.0 g/L from corn stover using *S. cerevisiae* as the fermenting organism.

1.3.2.4 Cassava Bagasse

Cassava is one of the starchy tuber crops grown in many countries of Asia, Latin America, and Africa. For example, in India itself 1100 cassava processing units are there, which produce 1.46 million tons of starch from 8.74 million tons of cassava. The cassava bagasse, which is the main waste (after starch extraction) commodity of the tuber crop is also rich in carbohydrate, that is, around, 30%–35% as compared to the tuber itself (Ray and Swain, 2011). Sangodoyin and Amori (2013) reported an estimated bioethanol production of 114 L from 1 ton of cassava bagasse.

1.3.2.5 Cereal Straws

Cereal grains, which are high energy–rich food for humans with 60%–70% starch produce huge quantities of by-products in the form of straws. These are the dry stalks that remain after the nutrient grain or seed has been removed. Mostly cereal crops like barley, wheat, rice, oat, corn, and sorghum are considered as important cereal crops. Calculations showcase that the estimated annual cereal straw production is about 1580.2 million tons/year from cereal crops like barley and oat from Europe, sorghum and corn from the United States, and rice along with wheat from India and China (Tye et al., 2016). The cereal straws, which consist of 33%–47% cellulose, can be possibly one of the lignocellulosic sources that can be utilized for bioethanol production.

1.3.2.6 Sugarcane Baggase (SCB)

SCB is primarily composed of lignin (20%–30%), cellulose (40%–45%), and hemicelluloses (30%–35%) (Peng et al., 2009). Because of its lower ash content (1.9%) (Li et al., 2002), SCB offers advantages over high ash contenting bagasse, such as rice straw, 14.5% (Guo et al., 2009) and wheat straw, 9.2% (Zhao and Bai, 2009). The major advantage of SCB is its immediate availability in the plant site or biorefinery site where the sugarcane juice has been extracted and processed. The integrated biorefinery approach was also evaluated and was concluded that higher production rates for ethanol was achieved in the integrated approach, rather than separate production of 1G and 2G ethanol from sugarcane bagasse plant (Behera and Ray, 2012; Furlan et al., 2013).

1.3.2.7 Sweet Potato Residues (SPRs)

Sweet potato residues (SPRs) are the biomass that are separated after extracting starch, account for more than 10% of the total dry matter of sweet potatoes. China, which is the largest producer and exporter of sweet potato with an annual production of 71 million tons worldwide, also produces 2 million tons of SPRs. SPRs were not much utilized possibly because of their high viscosity and these unutilized feedstocks plays an important role in environmental pollution. Studies on acid-catalyzed methods for release of fermentable sugars from the SPRs have shown encouraging results, but the issue of industrial wastewater discharge (Duvernay et al., 2013) and requirement of heavy investments in corrosion-resistant equipment and controlling of fermentation inhibitors remain as bottlenecks. However, recently explored enzymatic methods for utilization of SPRs have opened up new prospects for bioethanol production from sweet potato waste (Izmirlioglu and Demirci, 2012).

1.3.2.8 Oil-Palm Biomass

The oil-palm tree produces fruit bunches that are a rich source of palm oil. In 2013, the total palm-oil production was estimated to be approximately 58.3 millions with Indonesia remaining as the highest producer (53.2%), followed by Malaysia (32.9%) (MPOB, 2015). However, the processed fresh-fruit bunches (FFBs), the oil-palm fronds and trunks are left out materials that can be potential substrates for bioethanol production. Statistics show that while 40 tons/ha/annum of FBBs are produced, the amount of oil-palm fronds and trunks accounts to 10.5 and 2.8 tons/ha/annum, respectively.The total cellulose content accounts to 7.7%–14.7% for FBBs while high cellulose contents of 31.0%–32.0%, and 39.9%–41.0% are observed for the fonds and trunks, respectively. Recently, Eom et al. (2015) attempted to produce bioethanol from the oil-palm trunk utilizing both starch and cellulose degrading enzymes and obtained a high-glucose yield of 96.3% with an ethanol yield of 93.5%. Thus, these feedstocks can be used as sustainable feedstock for bioethanol production.

1.3.2.9 Native Plants

Grasses are generally considered as the native plants and the general composition of some of the industrially important grasses with their theoretical and practical bioethanol yield is given in Table 1.2. Grasses grow naturally and do not

TABLE 1.2 Composition, Biomass Yield, and Ethanol Production From Some of the Commonly Used Energy Crops for Bioethanol Production

Energy crops	Composition (%)			Biomass yield (tons/ha)	Ethanol yield		Rate of ethanol production (L/ha/year)	References
	Cellulose	Hemicellulose	Lignin		Practical (g/L)	Theoretical (g/L)		
Coastal Bermuda grass	25	37.5	6.4	600,000			10,786	Sun and Cheng (2005)
Elephant grass	22	24	24	18,000	23.4	36.4	23,700	Olukem (2015)
Moroccan grass	33–38	27–32	17–19	10,805	17.62	23.11	6,762	Semhaoui et al. (2016)
Orchard grass	32	40	4.7	74,131.61			7,672	Sun and Cheng (2005)
King grass	50	23	21	8,013	30.8	32.7	12,616	Cardonaa et al. (2016)
Switch grass	45	31	12	60,000	46.5	54.06	32,915	Keshwani and Cheng (2009)

require any special requirements for cultivation, which makes the biomass growth cost effective, as application of fertilizers and pesticides is not a necessity. Grasses are composed primarily of carbohydrate polymers (cellulose and hemicellulose) and phenolic polymers (lignin). These polysaccharides can be hydrolyzed to sugars and then fermented to ethanol. Further, the carbohydrate concentration in grasses is directly related to the bioethanol yield from biomass and the maturity of the grass is the key factor that determines its quantity in the grass. Another feature that makes grass an attractive energy crop is its potential to increase carbon storage by increasing above- and belowground biomass, specifically in C4 grasses. Among the different varieties, the most commonly used herbaceous biomass is giant miscanthus (*Miscanthus* sp.), switchgrass, Napier grass, and costal Bermuda grass. *Miscanthus* x *giganteus* is a variety of sawgrass that is capable of producing 5 to 8 times as much ethanol per acre as corn. The main feature distinguishing giant miscanthus from other biomass crops is its high lignocellulose yields with cellulose (40%–60%), hemicellulose (20%–40%), and lignin (10%–30%) contents (Brosse et al., 2012). Similarly, switchgrass (*Panicum virgatum*), which is a native warm season grass has been promoted as a model bioenergy crop because of its high bioethanol yield potential, low input requirements on marginal soils, and potential for soil carbon sequestration (Adler et al., 2006). The cellulose, hemicellulose, and lignin contents generally vary from 37% to 40%, 25% to 29%, and 18% to 25%, respectively.

Napier grass (*Pennisetum purpureum*) is a native to eastern and central Africa and has been introduced to most tropical and subtropical countries. Its high cellulosic fiber content, zero utilization of nitrogenous fertilizers, and fast-growing capability makes it an excellent cheap feedstock for ethanol production. The ability of napier grass to produce adequate biomass under limited nitrogen levels is linked to the occurrence of diazotrophic nitrogen fixing bacteria with the grass. Presence of these bacteria in soil augments the nitrogen requirement of the plant by fixing atmospheric nitrogen (Zahran, 1999). Due to its highly efficient CO_2 fixation, it is capable of producing 60 ton/ha/year of dry biomass under optimal condition and 30 t/ha/year of dry biomass under suboptimal condition. Other features that make this grass suitable for bioenergy purposes include the cellulose content of 40%–50% by weight followed by hemicelluloses and lignin, which is about 20%–40% and 10%–25%, respectively (Takara and Khana, 2015). Bermuda grass (*Conodont dactylon*) also has the advantage of good biomass yield of 14.1 to 24.2 ton/ha with high carbohydrate content (cellulose and hemicellulose) of 40%–55% and low lignin content of 20%–25% (Takara and Khana, 2015). Other grasses like cocksfoot grass, reed canary grass, big blue stem grass, and alfalfa are also documented to be potential feedstock for bioethanol production. Further, some grass varieties have also good amounts of hemicellulose present in them. These grasses can be utilized by extraction of the hemicellulose fraction for a pentose fermentation leading to bioethanol production. An example can be cited from the study done by Njoku et al. (2013) who had used the hemicellulose fraction of cocksfoot grass for bioethanol production with an ethanol yield in the range of 89–158 mL/kg of dry biomass. With minimum or nearly zero maintenance costs these grasses can be cheap sources of biomass for bioethanol production.

1.3.2.10 Natural Nonwoody Plant Fibers

1.3.2.10.1 BAST FIBER

Bast crops are a highly efficient mop crop and can grow on waste or even brackish water. Bast fibers are obtained from the outer layer of the plant fiber and in the form of fibrous bundles and comprise one-third of the weight. Similar to grasses they also have carbon dioxide absorption capacity and from 1.9 tons of carbon dioxide absorbed approximately 1 ton of cellulose is produced. Flax, ramie, kenaf, sun hemp, and industrial hemp are some of the examples of bast fiber crops, which have high cellulose content and can be utilized as bioethanol feedstock. Among these bast fiber crops European Energy Agency has identified industrial hemp, as an important sustainable potential alternative for biofuel production. (Cherney and Small, 2016). The annual production of hemp worldwide is estimated to be 0.1 million tons with cellulose content of 70.0–90.0 .The cellulose content of other bast fiber crops range from 60.0%–80.0%, 51.0%–84.0% and 68.0%–76.0% for flax, jute, and ramie, respectively (Tahir et al., 2011). However, the biofuel yield from industrial hemp and other bast fiber crops is largely unexplored.

1.3.3 Third-Generation Feedstock

Recently, algae are considered 3G feedstock and very potential candidates for bioethanol production due to their multisided beneficial aspects, such as faster-growing rates as compared to terrestrial plants, high availability, and ability to survive in harsh conditions (Chia et al., 2017; Khan et al., 2018; Silva and Bertucco, 2016). Moreover, the high lipid and carbohydrate content, high proton conversion, ability to grow in different water environments and high carbon dioxide absorption capability of algae make them a promising feedstock. The process of bioethanol production from algal biomass is shown in Fig. 1.5. Schenk et al. (2008) reported that the maximum theoretical yield for algal biomass production has been calculated at 365 tons of dry biomass/ha/year. Further, use of algal biofuels could reduce the greenhouse gas emissions from 101,000 g of CO_2 equivalent per million British thermal units (BTU) to 55,440 g. Algae are large groups of photosynthetic aquatic organisms that consist of two groups, microalgae and macroalgae, which are in unicellular form and multicellular form, respectively (Chia et al., 2017). Microalgae like dinoflagellates, green algae (*Chlorophyceae*), golden algae (*Chryosophyceae*), and diatoms (*Bacillariophyceae*) represent some of the bioethanol producing species, which are differentiated according to their protein, carbohydrate and lipid contents (Singh et al., 2010).

Similarly, macroalgae usually known as seaweeds, are classified into three main groups: brown (*Phaeophyceae*), red (*Rhodophyceae*), and green (*Chlorophyceae*). The structural cell wall of seaweeds usually consists of a matrix made up of linear sulfated galactan polymers (Yanagisawa et al., 2011). The high carbohydrate content

FIGURE 1.5 **Industrial process of bioethanol production from third-generation feedstock.**

of up to 50% of dry weight has been recently observed for some species of microalgae, that is, *Scenedesmus*, *Chlorella*, and *Chlamydomonas* (Ho et al., 2014). Likewise, seaweeds contain rich amounts of carbohydrates especially laminarin, mannan, mannitol, fucoidan, cellulose, agar, xylose, carrageenan, and alginates, which are converted into bioethanol (Hong et al., 2014).

Furthermore, both micro- and macroalgae lack lignin, making it simpler compared with terrestrial plants that lead to easy conversion of sugars into bioethanol by fermentative microorganisms (Obata et al., 2016). Though, macroalgae have high sugar content, are easy to cultivate and harvest, their conversion into bioethanol is crucial (Jambo et al., 2016) due to presence of different kinds sugars and composition, which also vary with the seasons (Obata et al., 2016). Moreover, the yield of ethanol from microalgal biomass is more than that of macroalgae due to its simpler structure composed of mainly cellulose. For example, Khambhaty et al. (2012) reported that the maximum bioethanol yield was 0.390 g/g from red seaweed species (*Kappaphycus alvarezii*) using acid hydrolysis process. However, the maximum bioethanol yield of 0.520 g/g was obtained from microalgae following the same (acid hydrolysis) process (Harun and Danquah, 2011). Even though the research on the application of algal feedstock in bioethanol is still in its naive stage, but it holds immense potential as a promising feedstock for commercial bioethanol production in near future and can help in the mitigation of global warming (Silva and Bertucco, 2016).

1.4 BIOREFINERY APPROACH

The biorefinery approach targets the optimum use of biomass for the targeted product (in the present case its bioethanol) along with value added products that can be obtained from the by-products. It is noteworthy that the progress of biorefineries can be enhanced by full exploitation of the biomass potential. The use of full spectrum of organic macromolecules (carbohydrates, oils, proteins, and lignin) and other chemical constituents of the biomass, such as antioxidants and pigments can have positive implications in biorefineries. Recently, this concept has been exemplified, where an integrated biorefinery approach was used for extraction of mineral rich liquid extract, lipid, ulvan, and cellulose from a green seaweed, *Ulva fasciata* (Trivedi et al. 2016). The cellulose was further enzymatically hydrolyzed and used for bioethanol production. The use of biorefining approach of crops for production of multiple products, such as energy, chemicals, and materials, will intensify the overall value of the biomass. Exploitation of the proteins and lignin components that are left out in the biomass after the extraction of bioethanol have also been part of active research for cost-effective productions. This concept seems promising with simultaneous bioethanol and biomethane production from sugarcane baggase along with good amount of heat generated from the extracted lignin (Rabelo et al., 2011). Similarly, high-protein residues after extraction of bioethanol from algae, which can be used as animal feed supplements, are also promising biorefinery concepts from bioethanol feedstocks (Pattarkine and Kannan, 2012). Recently, new biorefinery concepts for cellulosic alcohols have been reported stating them either to be "bolt-on" or "stand-alone" biorefineries (Chen et al., 2016). When the existing corn–grain ethanol biorefineries are corelated with other facilities, such as a sharing of feedstock, distribution supply chains, and decreased capital costs aiming at reduction in investment risk they are said to be "bolt-on" biorefineries. Contrary to this "stand-alone" biorefineries carry out all the functions by themselves. Hence, advancements in biotechnology focus on biomass of lower quality, such as grasses, waste biomass, and the recycling of waste are technologies for improved biorefinery concept.

1.4.1 Consolidated Bioprocessing (CBP) Technology for Biorefinery

CBP technology (Fig. 1.6) of bioethanol production from lignocelluloses refers to the combining of the three biological events (production of saccharolytic enzymes, hydrolysis of polysaccharides in pretreated biomass, fermentation of hexose and pentose sugars) in one bioreactor (no need for an enzyme producing reactor, like for separate hydrolysis and fermentation and simultaneous saccharification and cofermentation). Although no natural microorganism exhibits all the features desired for CBP, a number of microorganisms, bacteria, and fungi, as well as recombinant microbial strains possess some of the desirable properties (Edwards et al., 2011; Parisutham et al., 2014). The basic requirements of CBP microorganisms are: production of enzymes effectively hydrolyzing lignocellulosic heteropolymers to fermentable sugars, efficient ethanol production (titer, yield, and productivity), utilization of both hexoses and pentoses, and resistance to ethanol, fermentation inhibitors, and stressful environments (e.g., high osmotic pressure, low pH, high temperature, low nutrition capacity, fluctuating processes). To achieve these objectives, there could be two options: engineer ethanologenic yeast, that is, *S. cerevisiae*, which is a natural good ethanol producer with high tolerance to inhibitors but not effective at producing cellulose, to produce cellulases and hemicellulases or engineer a lignocellulose degraders (i.e., *Clostridium thermocellum*, *C. phytofermentans*), which produce cellulase and hemicellulase but with low ethanol production capacity, to be an efficient ethanol producer. Matano et al. (2013) developed a scheme of cell recycle batch fermentation of high solid lignocelluloses material using recombinant cellulase of yeast strain (*S. cerevisiae*) for a high yield of ethanol. Five consecutive batch fermentation of 200 g/L rice straw hydrothermally pretreated led to an average ethanol titer of 34.5 g/L. However, using recombinant yeast strain (*S. cerevisiae*) increased ethanol titer to 42.2 g/L with 86.3% of theoretical yield. There has also been a substantial progress in the development of genetic tools for free-enzyme bacterial systems, including *C. japonicas*, *C. phytofermentans*,

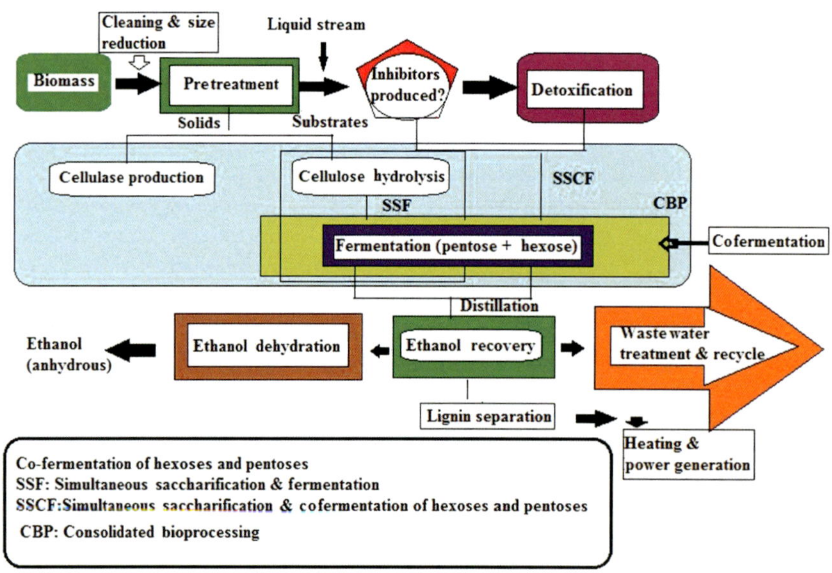

FIGURE 1.6 Consolidated bioprocessing of lignocellulose based feedstock for bioethanol production and other value-added products.

Thermoanaerobacter, and *Thermanaero-bacterium* sp. The latter, thermophillic anaerobes are a recently developed potential engineered microorganism (genetic tool) that uses a broad range of substrates, including xylan, to produce biofuel at high yield (Amore et al., 2014).

1.5 BIOTECHNOLOGY OF BIOETHANOL CROPS

With the recent reports from World Energy Council, it is anticipated that the biofuels could replace approximately 40% of all petroleum-based transport fuels by 2050. The International Energy Agency (IEA) Bioenergy Task 40 sees a far larger potential (up to 260 Ej by 2050), which would come down as a replacement of all petrofuels for transport (IEA, 2016). Economical production of bioethanol in the future will be dependent on the advances of genetic manipulation of the feedstock or the microorganisms involved in the saccharification and fermentation process. For example, scientists at the Agricultural Research Service in the United States Department of Agriculture (USDA) have worked substantially on developing the high biomass yielding of sorghum variety having ability to grow in arid soils (Msongalel et al., 2017). Scientists at Texas A and M University's Agricultural Experiment Station (TAES) have released a drought tolerant sorghum that may yield between 37 and 50 tons of dry biomass per hectare (May et al., 2016).

In case of 2G biofuels, significant advancements are made especially for engineering energy crops like switchgrass. Switchgrass genome has been modified to reduce (by over 90%) the expression of the caffeic acid 3-O-methyltransferase (COMT) gene. This modification reduced switchgrass lignin content by approximately 10% (Fu et al., 2011). Overexpression of PvMYB4, a general transcriptional repressor of the phenylpropanoid/lignin biosynthesis pathway in switchgrass can lead to very high yield ethanol production through dramatic reduction of recalcitrance (Shen et al., 2013). The research group concluded that MYB4-OX switchgrass is an excellent model system for understanding recalcitrance, and provides new germplasm for developing switchgrass cultivars as biomass feedstocks for biofuel production.

Starch is chemically composed of two types of glucan polymers, amylase and amylopectin, that are synthesized from the precursor, ADP glucose. Therefore, the regulation of ADP glucose pyrophosphorylase (ADPGase) would determine the sink strength (capacity to accumulate photosynthesis products) of starchy crops and its overexpression could result in higher starch content. Transgenic potato expressing *Escherichia coli glg*16 gene coding the bacterial ADPGase showed remarkably high starch content (60% more than the normal) in tubers. Researchers of North Carolina State University (USA) are reengineering the sweet potato to make it better suited for producing ethanol. By incorporating amylolytis genes from thermophilic bacteria from deep sea thermal vents, the group created an industrial sweet potato with double the starch content and enriched with liquefying (α-amylases) and saccharifying (amyloglucosidase) enzymes (El Sheikha and Ray, 2017; Ray et al. 2010).

1.6 FOOD VERSUS FUEL DEBATE

Using food crops for ethanol production often raises ethical and moral issues (Caniato et al. 2017). Bioethanol production is likely to compete with the food, feed, and industrial sectors, either directly, if food grains are used as the energy source, or indirectly, if bioethanol crops are cultivated on soil that is being used or would be used for food production (Galembeck, 2017; Thompson, 2012). Both effects may impact food prices and food availability if demand for the crops or for land or other inputs, such as fuels or agrochemicals, is considerably large. Bioethanol production could also reduce water availability for food production as more water would be diverted toward production of feedstock and for human and industrial consumption. However, the increases in food prices were felt temporarily in the past, because the agricultural sector had responded by increasing crop production and productivity, because of the green revolution technology, until now. An alternate strategy for bioethanol production and sustainability (availability, composition, and potential conversion yields of several feedstocks) is the economical production from lignocellulosic biomass from agricultural and forest residues, and algae. Grasses and woody plants typically have higher biomass yields/ha/annum than grains. The extent of grassland and woodland with the potential for lignocellulosic feedstock is about 1.75 billion ha worldwide (Popp et al., 2014). However, these grasslands and woodlands provide food and wood for cooking and housing to local communities, or is in use as grazing ground for livestock and barely 700 to 800 million ha of this land is suitable for economically viable lignocellulosic feedstock production (Fischer et al., 2009, 2010). Therefore, it is necessary to develop regional and national ethanol programs integrally considering all offers of lignocellulosic feedstocks available throughout a crop year to use forest and agricultural residues, agroindustrial by-products (bagasse, husks, leaves, tubercles, etc.) and plants without a specific use that generate barks, pods, fibers, leaves, and so on, in large quantities.

Hence, it seems unlikely to expect that food crops would free up substantial crop areas for planting energy (ethanol) crops. If current trends of agricultural intensification and livestock feeding efficiency growth are projected into the future, meeting global food demand might be achieved without reducing the amount of annual crop production remaining in ecosystems transforming by-products of agriculture, most of them without current use, but only in the absence of large-scale additional bioenergy production. Further, the common perception that expansion of bioenergy use will create serious competition with food and feed is not accepted by many experts. It is anticipated that more than 80% of the food/feed global future demand will be fulfilled by increment in crop productivity and developing high yield varieties and GM crops. In fact, between 1961 and 2009, global crop land grew by about 12% and agricultural production expanded by 150%, due to productivity gains (Popp et al., 2014). As a relevant outcome, the world food security situation, in general, is steadily improving as indicated by a consistent rise of average per capita food consumption and gradual reduction of malnourishment in the developing world (Goldenberg and Teixeira Coelho, 2013).

1.7 ECONOMIC IMPACTS OF BIOETHANOL

On an energy basis, bioethanol is currently more expensive to produce than gasoline in most of the countries. However, ethanol produced from sugarcane in Brazil comes close to competing with gasoline because sugarcane provides the lowest production costs (Belincanta et al., 2016), productivity is highest (6190–7500 L/ha against 3460–4020 L/ha of corn and 3400–3600 L/ha of cassava)), processing of sugarcane to sugar and further conversion to ethanol is easier, bagasse can be burnt to provide energy generation in ethanol plants and most importantly, it is favorable in terms of energetic balance. The energetic balance to convert corn into ethanol is of approximately 1:1.6, that is, for each 1 kcal of energy consumed for ethanol production, a gain of 1.6 kcal is obtained by the ethanol produced (Kim and Dale, 2005). On the other hand, the energetic balance of the ethanol production from sugarcane bagasse is 1.3 (Andreoli and De Souza, 2006). The other aspect is the expense of nonrenewable energy required to convert the sugars in the same ethanol amount. Sugarcane bagasse requires

4-fold less energy than corn, that is, 1.6 billion kcal versus 6.6 billion for corn (Andreoli and De Souza, 2006). For all these reasons, ethanol produced from corn in the USA and wheat in Canada is considerably more expensive than from sugarcane in Brazil. Ethanol from grain and sugar beet in Europe is even more expensive than those already mentioned. The production cost differences are attributed to many factors, such as costs and compositions of feedstock, transportation, capital and labor costs, scale of production, maintenance, insurance and taxes, and coproduct accounting.

Estimates of ethanol production costs from lignocelluloses is difficult because different types of feedstock as well as different production methods have been employed. In cellulosic ethanol production, the largest capital cost components are for feedstock pretreatment (17%), fermentation technology adapted (15%), and energy utilities for boilers and turbogenerators (36%) (Gupta and Verna, 2015; Solomon et al., 2007). Recently, the production of cheaper recombinant cellulase enzymes from Genencor International and Novozymes Biotech has resulted in up to 30-fold drop in cost of feedstock bioconversion into sugars for ethanol production (Mussatto et al., 2010). Several other factors, such as low-cost debt financing, integration into a biorefinery platform to increase the range of biocommodities could further lower the cellulosic ethanol production cost (Solomon et al., 2007). In the case of algal bioethanol production the prospective is not very clear. Although algae cultures have high potential yield and the ability to grow in locations unsuitable for agriculture, the biofuel production is challenged by feedstock cultivation, processing, and logistics issues as well as economic barriers, such as heavy water and energy requirements along with nitrogen and phosphorus (Silva and Bertucco, 2016).

Economic advantages of biobased ethanol industry would include value added to the feedstock, an increasing number of rural manufacturing jobs, an increased income tax, investment in plants and equipments, reduced greenhouse gases emissions, reduced country's reliance on crude oil import and market opportunities for domestic crops. In recent years, the importance of nonfood crops increased significantly. The opportunity to grow nonfood crops under the compulsory set-aside scheme is an option to increase biofuels production.

1.8 POLICY ISSUES

Policy drivers for renewable liquid biofuels have attracted high levels of financial incentives in many countries given their promise of benefits in several areas of interest to governments, including support to agricultural production, climate change, greenhouse gas emissions, trade balance, energy security, rural development, employment, and economic opportunities for developing countries (Demirbas, 2009). Similar to food security, energy security is a pertinent concern in many countries because of the fluctuating crude oil prices, foreign exchange shortage, greenhouse gas emissions, which have prompted to provide incentives to bioethanol industries. Strong demand from rapidly developing countries, especially sugarcane producers, such as China, Thailand, India, Colombia, and Mexico, and developed countries, such as the United States, Canada, and the European Union with corn and beet, is based in key lessons from Brazil experience since 1970, have supported the use of ethanol-based fuel (de Moraes et al., 2017) is adding to concerns over future food and energy prices, security and supplies with the development of state-of-the-art technology, stakeholder participation in the value chain, and biofuel mandates.

Alcohols have been used as a fuel for engines since the 19th century. Among the various alcohols, biobased ethanol is known as the most renewable eco-friendly fuel for spark-ignition engines. Ethanol has higher octane numbers, higher flame speeds, higher heat of vaporization, and broader flammability than gasoline (Balat and Balat, 2009); these properties allow for a higher compression ratio, shorter burn time and leaner burn engine, which are advantages over gasoline in an internal combustion engine (Liew et al., 2014). Bioethanol is used directly in automobiles designed to run on pure ethanol or blended with gasoline to make "gasohol." Anhydrous ethanol is required for blending with gasoline. No engine modification is required to use the ethanol blend up to 10%. World production of ethanol from sugarcane, maize, and sugar beet increased from less than 20 billion L in 2000 to more than 95 billion L in 2017 (Giovanni et al., 2010). However, this represents around 3% of global gasoline use.

For a global assessment of bioethanol production and sustainability, it's necessary to reconstruct global patterns of bioethanol trade, assess feedstock resources, plan cost-effective logistics or supply chain structure, and determine the associated displacement of water and land use in the entire value chain (Rulli et al., 2016), to prevent environmental impact and conserve existing biodiversity for future generations, as well as lowering cost of production (Blanchard et al. 2015). Further, specific policy directives are lacking in many countries to use bioethanol blending with gasoline as transport fuel. Some suggested measures are: blending mandates with gasoline, tax incentives, government purchasing policies, support for biofuel-compatible infrastructure, and conventional and emerging technologies, research, development, and innovation (including crop research, modeling of several production scenarios with different raw materials, conversion technology development and change, feedstock handling, vinasse management, and other byproducts), public education and outreach, reduction of counterproductive subsidies, investment risk reduction for next-generation ethanol production, infrastructure, facilities and creation of competitive environments, measuring the overall sustainability of biofuel and environmental protection and gradual reduction of supports as the market matures (Ribeiro et al. 2017; Viardot, 2013).

1.9 BIOETHANOL PRODUCTION TECHNOLOGIES: ENVIRONMENTAL IMPACTS AND LIFE CYCLE ASSESSMENT (LCA)

LCA (also called life cycle analysis or eco-balance) is a conceptual framework and methodology for the assessment of environmental impacts of product systems on a "cradle-to-grave" basis (Chan, 2007; Tan et al., 2002). This approach takes into account all steps of a product's life cycle: from the extraction of raw materials/natural resources, through materials processing, manufacture, distribution, use, repair and maintenance, and finally to its ultimate disposal. The entire process contributes to a wide range of impacts, such as climate change, acidification, eutrophication, the depletion of natural resources, water use, and land use. Practitioners and researchers from many domains come together in LCA to calculate indicators of the aforementioned potential environmental impacts that are linked to products—supporting the identification of opportunities for pollution prevention and reductions in resource consumption while taking the entire product life cycle into consideration (van Blottnitz and Curran, 2007). In many countries, LCA is used to support policy making of various thematic areas, such as eco-design, waste management and recycling, biorefinery of industrial commodities policy, and sustainable uses of natural resources like biofuels. In recent years, the LCA methodology has been frequently used to assess the potential economic benefits as well as ecological and environmental threats due to biofuel production (Rathore et al., 2016).

Several LCA studies have been conducted to analyze biofuel (ethanol) pathways. Wang et al. (2007) evaluated LCA emphasizing on corn and cellulosic ethanol production and concluded that ethanol had environmental benefits in terms of nonrenewable energy consumption and global warming impact. Coproduction of bioethanol with other biofuels and bioproducts is regarded as an alternative to reduce environmental impacts and to increase economic and energetic performance of biorefineries (Escamilla-Alvarado et al., 2016). Cherubini and Ulgiati (2010) analyzed a biorefinery concept using wheat straw or corn stover for bioethanol and methane production, and a complementary production of phenols; the use of both substrates had an energy return on investment (EROI) over 4. On the other hand, Melamu and Blottnitz (2011) when investigating the consequences of diverting sugarcane bagasse into 2D biofuels in the South African context instead of its incineration for electricity generation, concluded that none of the options for bioethanol production was better than bagasse for electricity production mainly because coal production/use would replace bagasse as a source of energy with worse environmental implications than the production/use of gasoline.

Enzyme production is also a big contributor to environmental impacts in lignocellulosic biorefineries. MacLean and Spatari (2009) found that process chemical and enzyme inputs would be responsible for 30%–40% of fossil energy use and 30%–35% of greenhouse gas emissions. Moreover, in the LCA performed by Nielsen et al. (2007) in which different amylases from Novozymes were evaluated, it was shown that a fungal amylase (Spirizyme plus FG) had very high environmental impacts in contrast with the other enzymes of bacterial origin (e.g., Termamyl 120 L); this difference was mainly ascribed to the longer fermentation times generally required in fungal processes. Elobeida et al. (2013) studied a coupled modeling framework: the CARD US agricultural market model and the MARKAL energy system model to capture the dynamic linkage between agricultural and energy markets that have been enhanced through the expansion of biofuel production as well as the environmental impacts resulting from the expansion. McKechnie et al. (2015) studied the impacts of process input supply chains and ongoing technology developments on the life cycle gas emission of ethanol production from corn stover in the United States. More recently, Farahani and Asoodar (2017) studied the environmental burdens and major sectors of emissions (e.g., abiotic depletion, eutrophication, acidification, global warming potential, human toxicity, ozone layer depletion, exotoxicity of freshwater, marine and terrestrial bodies, and photochemical oxidation) in molasses-based bioethanol production using the LCA methodology. Production of 1000 L molasses-based bioethanol resulted in 1322.8 kg CO_2 eq greenhouse gas emission (Farahani and Asoodar, 2017). However, comparing total greenhouse gas emissions from 1000 L bioethanol to gasoline, the net avoided greenhouse gas emission claimed were 503.2 kg CO_2 eq. Similar assessment was made on greenhouse gas performance of a Thai bioethanol system, which is inclined to decrease in the long run due to the effects from the expansion of plantation areas to satisfy the deficit of cassava and molasses (Silalertruksa and Gheewala, 2011). Bioethanol will contribute to the Thailand strategic plan on greenhouse gas mitigation in the transportation sector only if the production systems are sustainably managed, that is, coal replaced by biomass in ethanol plants, biogas recovery, and adoption of improved agricultural practices to increase crop productivity without intensification of chemical fertilizers. Achieving the year 2022 Thai government policy targets for bioethanol with recommended measures would help mitigate greenhouse gas emissions up to 4.6 Gg CO_2-eq per year.

1.10 CONCLUSION AND FUTURE PERSPECTIVES

Future bioethanol conversion from different types of biomass will depend on the conversion technologies and resultant final products. However, the biomass-based biorefinery prospective of bioethanol production will be aimed for costs of bioethanol in parity with the petro-based fossil fuels along with environmental and energy security. With this objective, it will be imperative for industrial facilities for affiliated production of other value-added products along with bioethanol. With a number of reported data from life cycle assessments highlighting on the less punitive nature of bioethanol toward greenhouse gas emissions, the replacement of the same with conventional diesel and gasoline will be a longed-for effort. Further, developing countries, including India, can be targeted for sustainable economic growth through biorefinery approach because of the presence of abundant residual biomass substrates, along with leftover agricultural and forest residues, bagasse, and so on. However, a good quantity of obstacles still act as hindrances for biomass-based biorefineries with biobased products, such as increased land use and water or environmental contamination with pesticides. From the economic prospective biorefinery technologies should be designed for less energy consumptions and ideal lignocellulosic breakdown technologies. Focus on utilization of organic biomass, such as municipal wastes, notorious weeds, drought resistant grasses, and nonedible seeds and plants along with use of genetic tools for improvement in both biomass and the microorganisms is advocated for a successful biorefinery approach.

Acknowledgments

The authors are thankful to Profs. Noé Aguilar-Rivera, Universidad Veracruzana, Facultad de Ciencias Biológicas y Agropecuarias, México, Ana Karla de Souza Abud, Food Technology Department, Federal University of Sergipe, São Cristóvão, Sergipe, Brazil, Carlos Escamilla-Alvarado, Universidad Autónoma de Nuevo León, Faculty of Chemical Sciences, Engineering of Sustainable Bioprocesses Group. Ave. Universidad S/N, 66455, San Nicolás de los Garza, N.L, Mexico, Sudhansu. S. Behera, Department of Fisheries and Animal Resource Development, Government of Odisha, India, and D. Surendhiran, Department of Biology, St. Joseph University, Dar es Salaam, United Republic of Tanzania, for critically reading the manuscript and providing valuable inputs.

References

Adler, P.R., Sanderson, M.A., Boateng, A.A., Weimer, P.J., Jung, H.J.G., 2006. Biomass yield and biofuel quality of switchgrass harvested in fall or spring. Agron. J. 98, 1518–1525.

Agência Nacional doPetróleo, 2015. Gás Natural e Biocombustíveis – ANP, Statistic on biofuel. Available from: http://www.anp.gov.br/

Akponah, E., Akpomie, O.O., 2011. Analysis of the suitability of yam, potato and cassava root peels for bioethanol production using *Saccharomyces cerevisae*. Int. Res. J. Microbiol. 2, 393–398.

Alvira, P., Tomás-Pejó, E., Ballesteros, M., Negro, M.J., 2010. Pretreatment technologies for an efficient bioethanol production process based on enzymatic hydrolysis: a review. Bioresour. Technol. 101, 4851–4861.

Amore, A., Giacobbe, S., Liguori, R., Faraco, V., 2014. The second generation ethanol production. Memor. Sci. Fisiche Natural., 113–136.

Andreoli, C., De Souza, S.P., 2006. Sugarcane: the best alternative for converting solar and fossil energy into ethanol. Revista Eco. Ener. 59, 27–33.

Balan, K.N., Valarmathi, T.N., Reddy, M.S.H., Reddy, G.A., Srinivas, J.K.M.K., 2013. Analysis of CO_2 CO and HC emission reduction in automobiles. IOP Conf. Ser. Mater. Sci. Eng. 197, 012012.

Balat, M., 2009. New biofuel production technologies. Energ. Educ. Sci. Technol. 22, 147–161.

Balat, M., Balat, H., 2009. Recent trends in global production and utilization of bio-ethanol fuel. Appl. Energy 86, 2273–2282.

Behera, S., Mohanty, R.C., Ray, R.C., 2012. Ethanol fermentation of sugarcane molasses by *Zymomonas mobilis* MTCC 92 immobilized in *Luffa cylindrica* L. sponge discs and Ca-alginate matrices. Braz. J. Microbiol. 43, 1499–1507.

Behera, S.S., Ray, R.C., 2012. Comparison of *Luffa cylindrica* L. sponge discs and Ca-alginate gel beads as immobilized matrices of *Saccharomyces cerevisiae* for bio-ethanol production from sugarcane molasses. Afr. J. Biotechnol. 11, 16172–16176.

Behera, S., Ray, R.C., 2014. Batch ethanol production from cassava (*Manihot esculenta* Crantz) flour by *Saccharomy-*

ces cerevisiae cells immobilized in calcium alginate. Ann. Microbiol. doi: 10.1007/s13213-014-0918-8.

Behera, S.S., Ray, R.C., 2016. Solid state fermentation for production of microbial cellulases: recent advances and improvement. Int. J. Biol. Macromol. 86, 656–669.

Beiter, P., Tian, T., 2015. Renewable Energy. US Department of Energy's National Renewable Energy Laboratory Data Book (NREL), Golden, CO, United States.

Belboom, S., Bodson, B., Léonard, A., 2015. Does the production of Belgian bioethanol fit with European requirements on GHG emissions? Case of wheat. Biomass Bioenergy 74, 58–64.

Belincanta, J., Alchorne, J.A., Teixeira da Silva, M., 2016. The Brazilian experience with ethanol fuel: aspects of production, use, quality and distribution logistics. Braz. J. Chem. Eng. 33, 1091–1102.

Blanchard, R., Bhattacharya, S.C., Chowdhury, M., Chowdhury, B., Biswas, K., Choudhury, B.K., 2015. A review of biofuels in India: challenges and opportunities. Presented at World Engineering Congress. Orlando, FL, United States. .

van Blottnitz, H.V., Curran, M.A., 2007. A review of assessments conducted on bioethanol as a transportation fuel from a net energy, greenhouse gas, and environmental life cycle perspective. J. Clean. Prod. 15, 607–619.

Brennan, L., Owende, P., 2010. Biofuels from microalgae-a review of technologies for production, processing, and extractions of biofuels and co-products. Renew. Sustain. Energ. Rev. 14, 557–577.

Brosse, N., Anthony, D., Xianzhi, M., Ragauskas, A., 2012. *Miscanthus*.a fast-growing crop for biofuels and chemicals production. Biofuel. Bioprod. Biorefin. 6, 580–598.

Caniato, M., Carliez, D., Thulstrup, A., 2017. Challenges and opportunities of new energy schemes for food security in humanitarian contexts: a selective review. Sustain. Energ. Technol. Assess. 22, 208–219.

Cardona, C.A., Quintero, J.A.I.C.P., 2010. Production of bioethanol from sugarcane bagasse: status and perspectives. Bioresourc. Technol. 101, 4754–4766.

Cardonaa, E., Jorge, R., Mariana, J.P, Rios, P.L., 2016. King Grass: a very promising material for the production of second generation ethanol in tropical countries. Biomass Bioenergy 95, 206–213.

Chan, A.W.C., 2007. Economic and Environmental Evaluations of Life Cycle Cost Analysis Practice: A Case Study of Michigan DOT Pavement Projects. University of Michigan, Ann Arbor, MI, United States. https://deepblue.lib.umich.edu/bitstream/handle/2027.42/50469/Wai%20Cheung%20Chan%20Master%.

Chen, M., Smith, P.M., Wolcott, M.P., 2016. U.S. biofuels industry: a critical review of opportunities and challenges. BioProducts Bus. 1, 42–59.

Cherney, J.H., Small, E., 2016. Industrial hemp in North America: production, politics and potential. Agronomy 6 (58). doi: 10.3390/agronomy6040058.

Cherubini, F., Ulgiati, S., 2010. Crop residues as raw materials for biorefinery systems: a LCA case study. Appl. Energy 87, 47–57.

Chia, S.R., Ong, H.C., Chew, K.W., Show, P.L., Phang, S-M., Ling, T.C., Nagarajan, D., Lee, D.-J., Chang, J.-S., 2017. Sustainable approaches for algae utilisation in bioenergy production. Renew. Energy, 1–15, Article in press.

Demirbas, A., 2009. Political, economic and environmental impacts of biofuels: a review. Appl. Energy 86, S108–S117.

de Moraes, M.A.F.D., Rodrigues, L., Kaplan, S., 2017. The sugarcane industry and the use of fuel ethanol in Brazil: history, challenges, and opportunities. In: Khanna, M., Scheffran, J., Zilberman, D. (Eds.), Handbook of Bioenergy Economics and Policy. Springer, New York, NY, United States, pp. 39–63.

Duvernay, W.H., Chinn, M.S., Yencho, G.C., 2013. Hydrolysis and fermentation of sweet potatoes for production of fermentable sugars and ethanol. Ind. Crop Prod. 42, 527–537.

Edwards, M.C., Henriksen, E.D., Yomano, L.P., Gardner, B.C., Sharma, L.N., Ingram, L.O., Peterson, J.D., 2011. Addition of genes for cellobiase and pectinolytic activity in *Escherichia coli* for fuel ethanol production from pectin-rich lignocellulosic biomass. Appl. Environ. Microbiol. 77, 5184–5191.

Elobeida, S., Tokgozb, R.D.T., Johnson, D.O., Kaplancl, K., Secchi, S., 2013. Integration of agricultural and energy system models for biofuel assessment. Environ. Model. Softw. 48, 1–16.

El Sheikha, A.F., Ray, R.C., 2017. Potential impacts of bioprocessing of sweet potato: review. Crit. Rev. Food Sci. Nutr. 57, 455–471.

Eom, I.Y., Yu, J.H., Chan, H., Sik, K., 2015. Efficient ethanol production from dried oil palm trunk treated by hydrothermolysis and subsequent enzymatic hydrolysis. Biotechnol. Biofuels 8, 83.

Escamilla-Alvarado, C., Poggi-Varaldo, H.M., Ponce-Noyola, M.T., 2016. Bioenergy and bioproducts from municipal organic waste as alternative to landfilling: a comparative life cycle assessment with prospective application to Mexico. Environ. Sci. Pollut. Res. 23, 1–16.

Farahani, S.S., Asoodar, M.A., 2017. Life cycle environmental impacts of bioethanol production from sugarcane molasses in Iran. Environ. Sci. Pollut. Res. Int. doi: 10.1007/s11356-017-9909-1, [Epub ahead of print].

Fischer, G., Hizsnyik, E., Prieler, S., Shah, M., van Velthuizen, N.H., 2009. Biofuels and Food Security. The OPEC Fund for International Development (OFID), Vienna, Austria, Final Report to Sponsor, March 2009.

Fischer, G., Prieler, S., van Velthuizen, H., Berndes, G., Faaij, A., Londo, M., Wit, M.de., 2010. Biofuel production potentials in Europe: sustainable use of cultivated land and pastures, Part II: land use scenarios. Biom. Bioenergy 34, 173–187.

Fu, C., Mielenz, J.R., Xiao, X., Ge, Y., Hamilton, C.Y., Rodriguez, M., et al., 2011. Genetic manipulation of lignin reduces recalcitrance and improves ethanol production from switchgrass. Proc. Natl. Acad. Sci. USA 9, 3803–3808.

Furlan, F.F., Filho, R.T., Pinto, F.H.P.B., Costa, C.B.B., Cruz, A.J.G., Giordano, R.L.C., Giordano, R.C., 2013. Bioelectricity versus bioethanol from sugarcane bagasse: is it worth being flexible? Biotechnol. Biofuels 6, 6–142. doi: 10.1186/1754-6834-6-142.

Galembeck, F., 2017. Synergy in food, energy and advanced materials production from biomass. Pure Appl. Chem. doi: 10.1515/pac-2017-0607.

Gasmalla, M.A.A., Yang, R., Nikoo, M., Man, S., 2012. Production of ethanol from Sudanese sugar cane molasses and evaluation of its quality. J. Food Process Technol. 3, 163.

Gerbens-Leenes, P.W., Hoekstra, A.Y., van der, M.H., 2009. The water footprint of bioenergy. Proc. Natl. Acad. Sci. USA 106, 10219–10223.

Gerbens-Leenes, P.W., van, L.A.R., Hoekstra, A.Y., van den, M.H., 2012. Biofuel scenarios in a water perspective: the global blue and green water footprint of road transport in 2030. Global Environ. Chang. 22, 764–775.

Giovanni, S., Banse, M., Kemfert, C., 2010. An overview of biofuel policies across the world. Energy Policy 11, 6977–6988.

Goldenberg, J., Teixeira Coelho, S., 2013. Bioenergy: how much? Environ. Res. Lett. 8. doi: 10.1088/1748-9326/8/3/031005.

GRFA (Global Renewable Fuel Alliance) Reports, 2017. Available from: http://globalrfa.org/news-media/2017-forecast-global-ethanol-production-stable.

Gumienna, M., Szwengiel, A., Lasik, M., Szambelan, K., Majchrzycki, D., Adamczyk, J., Nowak, J., Czarnecki, Z., 2016. Effect of corn grain variety on the bioethanol production efficiency. Fuel 164, 386–392.

Guo, G.L., Hsu, D.C., Chen, W.H., Chen, W.H., Hwang, W.S., 2009. Characterization of enzymatic saccharification for acid-pretreated lignocellulosic materials with different lignin composition. Enz. Microb. Technol. 45, 80–87.

Gupta, A., Verna, J.P., 2015. Sustainable bio-ethanol production from agro-residues: a review. Renew. Sust. Energ. Rev. 41, 550–567.

Harun, R., Danquah, K.M., 2011. Influence of acid pre-treatment on microalgal biomass for bioethanol production. Process Biochem 46, 304–309.

Ho, S.H., Chen, C.N.N., Lai, Y.Y., Lu, W.B., Chang, J.S., 2014. Exploring the high lipid production potential of a thermotolerant microalga using statistical optimization and semi-continuous cultivation. Bioresour. Technol. 163, 128–135.

Hong, In., Hyeon, K., Bum, J.L.S., 2014. Comparison of red, brown and green seaweeds on enzymatic saccharification process. J. Ind. Eng. Chem. 20, 2687–2691.

Hoover, R., 2001. Composition, molecular structure and physicochemical properties of tuber and root starches: a review. Carbohyd. Polym. 45, 253–267.

IEA, 2016. Energy outlook 2016. Paris, Cedex 15, France.

Izmirlioglu, G., Demirci, A., 2012. Ethanol production from waste potato mash by using *Saccharomyces cerevisiae*. Appl. Sci. 2, 738–753.

Jambo, S.A., Abdulla, R., Azhar, S.H.M., Marbawi, H., Gansau, J.A., Ravindra, P., 2016. A review on third generation bioethanol feedstock. Renew. Sust. Energ. Rev. 65, 756–769.

Jayusab, N., Mayzuhroha, A., Arindhania, S., Caroenchaic, C., 2016. Studies on bioethanol production of commercial baker's and alcohol yeast under aerated culture using sugarcane molasses as the media. Agr. Sci. Proc. 9, 493–499.

Kanagaraj, A/L R., 2013. Bioethanol production from cassava by fermentation process using saccharomyces cerevisiae. A thesis submitted University Malaysia Pahang.

Keshwani, D.R., Cheng, J.J., 2009. Switchgrass for bioethanol and other value-added applications: a review. Bioresour. Technol. 100, 1515–1523.

Khambhaty, Y., Mody, K., Gandhi, M.R., Thampy, S., Maiti, P., Brahmbhatt, K., Eswaran, K., Ghosh, P.K., 2012. *Kappaphycus alvarezii* as a source of bioethanol. Bioresour. Technol. 103, 180–185.

Khan, M.I., Lee, M.G., Shin, J.H., Kim, J.D., 2018. Pretreatment optimization of the biomass of *Microcystis aeruginosa* for efficient bioethanol production. AMB Express 7.

Kim, S., Dale, B.E., 2005. Life cycle assessment of various cropping systems utilized for producing biofuels: bioethanol and biodiesel. Biomass Bioenerg. 29, 426–439.

Lareo, C., Ferrari, M.D., Guigou, M., Fajardo, L., Larnaudie, V., Ramírez, M.B., Martínez-Garreiro, J., 2013. Evaluation of sweet potato for fuel bioethanol production: hydrolysis and fermentation. SpringerPlus 2.

Lau, M.W., Dale, B.E., 2009. Cellulosic ethanol production from AFEX-treated corn stover using *Saccharomyces cerevisiae* 424A (LNH-ST). Proc. Natl. Acad. Sci. USA 106, 1368–1373.

Li, X., Kondo, R., Sakai, R.K., 2002. Biodegradation of sugarcane bagasse with marine fungus *Phlebia* sp MG-60. J. Wood Sci. 48, 159–162.

Licht, F.O., 2017. F.O. Licht's World of Biofuels 2017 Report. World of biofuels in 2017. Available from: http://get.knect365.com/world-ethanol-biofuels/report/.

Liew, W.H., Hassim, M.H., Ng, D.K.S., 2014. Review of evolution, technology and sustainability assessment of biofuels production. J. Clean. Prod. 71, 11–29.

MacLean, H.L., Spatari, S., 2009. The contribution of enzymes and process chemicals to the life cycle of ethanol. Environ. Res. Lett. 4, 1–10.

Matano, Y., Hasunuma, T., Kondo, A., 2013. Cell recycle batch fermentation of high-solid lignocellulose using a recombinant cellulase-displaying yeast strain for high yield ethanol production in consolidated bioprocessing. Bioresour. Technol. 135, 403–409.

May, A., de Souza, V.F., de Amaral Gravina, G., Fernandes, P.G., 2016. Plant population and row spacing on biomass sorghum yield performance. Cienc. Rural 46, 46.

McKechnie, J., Pourbafrani, M., Saville, B.A., MacLean, H.L., 2015. Exploring impacts of process technology development and regional factors on life cycle greenhouse gas emissions of corn stover ethanol. Renew. Energy 76, 726–734.

Melamu, R., Blottnitz, H., 2011. Second generation biofuels a sure bet? A life cycle assessment of how things could go wrong. J. Clean. Prod. 19, 138–144.

Mohapatra, S., Dandpath, S.J., Thatoi, H., 2016. Physicochemical characterization, modelling and optimization of ultrasono-assisted acid pretreatment of two *Pennisetum* sp. using Taguchi and artificial neural networking for enhanced delignification. J. Environ. Manage. 187, 537–549.

Mohapatra, S., Pattathil, S., Thatoi, H.N., 2017. Structural and functional characterization of two *Pennisetum* sp. biomass during ultrasono-assisted alkali pretreatment and enzymatic hydrolysis for understanding the mechanism of targeted delignification and enhanced saccharification. ACS Sustain. Chem. Eng. 5, 6486–6497.

Moshi, A.P., Hosea, K.M.M., Elisante, E., Mamo, G., Mattiasson, B., 2015. High temperature simultaneous saccharification and fermentation of starch from inedible wild cassava (*Manihot glaziovii*) to bioethanol using *Caloramator boliviensis*. Bioresour. Technol. 180, 128–136.

Malaysian Palm Oil Board (MPOB), 2015. Palm Oil Economic Review & Outlook Seminar 2015. The Royale Chulan, Kuala Lumpur, Malaysia.

Msongalel, B.M., Tumbo, S.D., Kihupi, N.I., Rwehumbiza, F.B., 2017. Performance of sorghum varieties under variable rainfall in Central Tanzania. Int. Sch. Res. Notic. 2017, 2506946.

Muktham, R., Bhargava, S.K., Bankupalli, S., Ball, A.S., 2016. A review on 1st and 2nd generation bioethanol production-recent progress. J. Sustain. Bioenerg. Syst. 6, 72–92.

Muruaga, M.L., Carvalho, K.G., Domínguez, J.M., de Souza, O.R.P., Perotti, N., 2016. Isolation and characterization of *Saccharomyces* species for bioethanol production from sugarcane molasses: studies of scale up in bioreactor. Renew. Energy 85, 649–656.

Mussatto, S.I., Dragone, G., Guimaraes, P.M.R., Silva, J.P.A., Carneiro, L.M., Roberto, I.C., Vicente, A., Domingues, L., Teixeira, J.A., 2010. Technological trends, global market, and challenges of bio-ethanol production. Biotechnol. Adv. 28, 1873–1899.

Nguyen, C.N., Thanh, M.L., Son, C.K., 2014. Pilot scale simultaneous saccharification and fermentation at very high gravity of cassava flour for ethanol production. Ind. Crop Prod. 56, 160–165.

Nielsen, P.H., Oxenbøll, Wenzel, H., 2007. Cradle-to-Gate environmetnal assessment of enzyme products produced industrially in Denmark by Novozymes A/S. Int. J. LCA 12 (6), 432–438.

Njoku, I.S., Uellendahl, H., Ahrin, K.B., 2013. Comparing oxidative and dilute acid wet explosion pretreatment of Cocksfoot grass at high dry matter concentration for cellulosic ethanol production. Energ. Sci. Technol. 1, 88–98. doi: 10.1002/ese3.11.

Obata, O., Akunna, J., Bockhorn, H., Walke, G., 2016. Ethanol production from brown seaweed using non-conventional yeasts. Bioethanol 2, 134–145.

Olukem, M., 2015. Production of bioethanol from elephant grass (*Pennisetum purpureum*) stem. Thesis submitted to Ahamadu Bello University. http://kubanni.abu.edu.ng:8080/jspui/handle/123456789/6525?mode=full

Parisutham, V., Kim, T.H., Lee, S.K., 2014. Feasibilities of consolidated bioprocessing microbes: from pretreatment to biofuel production. Bioresour. Technol. 161, 431–440.

Pattarkine, V.M., Kannan, D.C., 2012. From algae to biofuel: engineering aspects. In: Gordon, R., Seckbach, J. (Eds.), The Science of Algal Fuels: Cellular Origin, Life in Extreme Habitats and Astrobiology. Springer, Dordrecht, The Netherlands, pp. 165–176.

Peng, F., Ren, J.L., Xu, F., Peng, J.B.P., Sun, R.C., 2009. Comparative study of hemicelluloses obtained by graded ethanol precipitation from sugarcane bagasse. J. Agric. Food Chem. 57, 6305–6317.

Popp, J., Lakner, Z., Harangi-Rákos, M., Fári, M., 2014. The effect of bioenergy expansion: food, energy and environment. Renew. Sustain. Energy Rev. 32, 559–578.

Rabelo, S.C., Carrere, H., Maciel, F.R., Costa, A.C., 2011. Production of bioethanol, methane and heat from sugarcane bagasse in a biorefinery concept. Bioresour. Technol. 102, 7887–7895.

Rathore, D., Nizami, A.-S., Singh, A., Pant, D., 2016. Key issues in estimating energy and greenhouse gas savings of biofuels challenges and perspectives. Biofuel Res. J. 3, 380–393.

Ray, R.C., Naskar, S.K., 2008. Bioethanol production from sweet potato (*Ipomoea batatas* L. by enzymatic liquefaction and simultaneous saccharification and fermentation (SSF) process. Dyn. Biochem. Process Biotechnol. Mol. Biol. 2, 47–49.

Ray, R.C., Naskar, S.K., Tomlins, K.I., 2010. Bio-processing of sweet potato into food, feed and bio-ethanol. In: Ray, R.C., Tomlins, K.I. (Eds.), Sweet Potato: Post-harvest Aspects in Food, Feed and Industry. Nova Science, Hauppauge, New York, United States, pp. 163–192.

Ray, R.C., Swain, M.R., 2011. Bio-ethanol, bioplastics and other fermented industrial products from cassava starch and flour. In: Colleen, M., Pace (Eds.), Cassava: Farming, Uses and Economic Impacts. Nova Science, Hauppauge, New York, United States, pp. 1–32.

REN, 2013. 21(Renewable Energy Policy Network for 21st Century) 2014. Renewables 2014 Global Status Report. Paris, France: REN21 Secretariat; 2013. p. 177.

Ribeiro, L.A., da Silva, P.P., Ribeiro, L., Dotti, F.L., 2017. Modelling the impacts of policies on advanced biofuel feedstocks diffusion. J. Clean. Prod. 142, 2471–2479.

Romaní, A., Garrote, G., López, F., Parajó, J.C., 2011. *Eucalyptus globulus* wood fractionation by autohydrolysis and organosolv delignification. Bioresour. Technol. 102, 5896–5904.

Rorke, D.C.S., Kana, E.B.G., 2017. Kinetics of bioethanol production from waste sorghum leaves using *Saccharomyces cerevisiae* BY4743. Fermentation 3 (2). doi: 10.3390/fermentation3020019.

Rulli, M.C., Bellomi, D., Cazzoli, A., Carolis, G.D., Odoricoa, P.D., 2016. The water-land-food nexus of first-generation biofuels. Sci.Rep. 6, 22521. doi: 10.1038/srep22521.

Salazar-Ordóñez, M., Pérez-Hernández, P.P., Martín-Lozano, J.M., 2013. Sugar beet for bioethanol production: an approach based on environmental agricultural outputs. Energy Policy 55, 662–668.

Sangodoyin, A.Y., Amori, A.A., 2013. Aerobic composting of cassava peels using cow dung, sewage sludge and poultry manure as supplements. Eur. Int. J. Sci. Technol. 2, 22–34.

Santek, B., Gwehenberger, G., Santek, M.I., Narodoslawsky, M., Horvat, P., 2010. Evaluation of energy demand and the sustainability of different bioethanol production processes from sugar beet. Resour. Conserv. Recy. 54, 872–877.

Sapp, M., 2017. Ethanol remains sticking point after most recent round of EU-Mercosur talks. Biofuels digest. Available from: http://www.biofuelsdigest.com/bdigest/2017/10/10/.

Schenk, P.M., Thomas, H.S.R., Stephens, E., Marx, U.C., Mussgnug, J.H., Posten, C., Kruse, O., Hankamer, B., 2008. Second generation biofuels: high-efficiency microalgae for biodiesel production. Bioenergy 1, 20–43.

Semhaoui, L., Zarguili, S.A., Rezzougl, T., Maugard, J.M.Q., et al., 2016. Bioconversion of Moroccan alfa (*Stipa tenacissima*) by thermomechanical pretreatment combined to acid or alkali spraying for ethanol production. J. Mater. Environ. Sci. 8, 2619–2631.

Shanavas, S., Padmaja, G., Moorthy, S.N., Sajeev, M.S., Sheriff, J.T., 2011. Process optimization for bioethanol production from cassava starch using novel eco-friendly enzymes. Biomass Bioenergy 35, 901 909.

Shen, H., Poovaiah, C.R., Ziebell, A., Tschaplinski, T.J., Pattathil, S., Gjersing, E., et al., 2013. Enhanced characteristics of genetically modified switchgrass (*Panicum virgatum* L.) for high biofuel production. Biotechnol. Biofuels 6, 71.

Shikida, P.F.A., Cardoso, B.F., Galante, V.A., Rahmeier, D., Bentivoglio, D., Rasetti, M., 2014. Liquid Biofuels: Emergence, Development and Prospects. Lecture Notes in Energy 27. DOI: 10.1007/978-1-4471-6482-1_2.

Silalertruksa, T., Gheewala, S.H., 2011. Environmental sustainability assessment of bio-ethanol production in Thailand. Energy 34, 1933–1946.

Silva, C.E.F., Bertucco, A., 2016. Bioethanol from microalgae and cyanobacteria: a review and tecnhological outlook. Process Biochem. 51, 1833–1842.

Singh, A., Pant, D., Korres, N.E., Nizami, A.S., Prasad, Sh., Murphy, J.D., 2010. Key issues in life cycle assessment for ethanol production from lignocellulosic biomass: challenges and perspectives. Bioresour. Technol. 101, 5003–5012.

Singh, V., 2012. Effect of corn quality on bioethanol production. Biocatal Agric. Biotecnol. 1, 353–355.

Solomon, B.D., Barnes, J.R., Halvorsen, K.E., 2007. Grain and cellulosic ethanol: History, economics, and energy policy. Biomass Bioenergy 31, 416–425.

Sun, Y., Cheng, J.J., 2005. Dilute acid pretreatment of rye straw and bermudagrass for ethanol production. Bioresour. Technology 96, 1599–1606.

Tahir, P.M., Ahmad, A.B., Saifulazry, S.O.A., Ahmad, Z., 2011. Retting process of some bast plants fibers and its effect on fiber quality: a review. Bio. Resources 6, 5260–5281.

Takara, D., Khana, K.S., 2015. Characterizing compositional changes of Napier grass at different stages of growth for biofuel and biobased products potential. Bioresour Technol. 188, 103–108.

Tan, R.R., Culaba, A.B., Purvis, M.R.I., 2002. Application of possibility theory in the life cycle inventory assessment of biofuels. Int. J. Energ. Res. 26, 737–745.

Thatoi, H., Dash, P.K., Mohapatra, S., Swain, M.R., 2016. Bioethanol production from tuber crops using fermentation technology: a review. Int. J. Sustain Energ. 35, 443–468.

Thompson, P.B., 2012. The agricultural ethics of biofuels: the food vs. fuel debate. Agriculture 2, 339–358.

Trivedi, N., Baghel, R.S., Bothwell, J., Gupta, V., Reddy, C.R.K., Lali, A.M., Jha, B., 2016. An integrated process for the extraction of fuel and chemicals from marine macroalgal biomass. Sci. Rep. 6, 30728.

Tye, Y.Y., Lee, K.T., Abdullah, W.N.W., Leh, C.P., 2016. The world availability of non-wood lignocellulosic biomass for the production of cellulosic ethanol and potential pretreatments for the enhancement of enzymatic saccharification. Renew. Sust. Energ. Rev. 60, 155–172.

US Department of Energy, 2016. Energy department announces funding for design and construction of manufacturing of biofuels. Bioproducts and Biopower, http://energy.gov/articles/energy-department- announces-funding-design-and-constructionmanufacturing-biofuels.

Viardot, E., 2013. The role of cooperatives in overcoming the barriers to adoption of renewable energy. Energy Policy 63, 756–764., 10 1016/j.enpol.2013.08.034.

Wang, M., Wu, M., Huo, H., 2007. Life-cycle energy and greenhouse gas emission impacts of different corn ethanol plant types. Environ. Res. Lett. 2, 13, 024001.

Wheals, A.E., Basso, L.C., Alves, D.M.G., Amorim, H.V, 1999. Fuel ethanol after 25 years. Trends Biotechnol. 17, 17482–17487.

Yanagisawa, M., Nakamura, K., Ariga, O., Nakasaki, K., 2011. Production of high concentrations of bioethanol from seaweeds that contain easily hydrolyzable polysaccharides. Acta Horticul. 583, 187–193.

Zahran, H.H., 1999. Rhizobium-legume symbiosis and nitrogen fixation under severe conditions and in an arid climate. Microbiol. Mol. Biol. Rev. 63, 968–989.

Zhao, X.Q., Bai, F.W., 2009. Mechanisms of yeast stress tolerance and its manipulation for efficient fuel ethanol production. J. Biotechnol. 144, 23–30.

Zhu, J.Y., Pan, Xuejun, Zalesny, Jr., R.S., 2010. Pretreatment of woody biomass for biofuel production: energy efficiency, technologies, and recalcitrance. Appl. Microbiol. Biotechnol. 87, 847–857.

Ziska, L.H., Runion, G.B., Tomecek, M., Prior, S.A., Torbet, H.A., Sicher, R., 2009. An evaluation of cassava, sweet potato and field corn as potential carbohydrate sources for bioethanol production in Alabama and Maryland. Biomass Bioenergy 33, 1503–1508.

Further Readings

Brown, R.C., 2003. Biorenewable. Resources: Engineering New Products from Agriculture. Blackwell Publishing, Ames, IA, United States, 286.

Chen, J., Bai, J., Li, L.H., Chan, C., Fang, Shuqi., 2015. Prospects for bioethanol production from macroalgae. Tr. Ren. Energy 1, 185–197.

Felipe, F., Furlan, R.T.F., Fabio, H.P.B.P., Caliane, B.B.C., Antonio, J.G.C., Raquel, L.C.G., Roberto, C.G., 2013. Bioelectricity versus bioethanol from sugarcane bagasse: is it worth being flexible? Biotechnol. Biofuel. 6, 142.

Goh, C.S., Tan, K., Lee, Tat., Bhatia, K.T.S., 2009. Bio-ethanol from lignocellulose: status, perspectives and challenges in Malaysia. Bioresour. Technol. 101, 4834–4841.

Hannon, M., Gimpel, J., Tran, M., Rasala, B., Mayfield, S., 2010. Biofuels from algae: challenges and potential. Biofuels 1, 763–784.

Khan, W., Usha, P.R., Subramanian, S., Mundaya, N.J., Rayorath, P., Hodges, D.M., Critchley, T.A., Craigie, S.J., Jeff, N., Prithivira, B., 2009. Seaweed extracts as biostimulants of plant growth develop 28, 386–399.

Mohaptra, S., Dash, P.K., Behera, S.S., Thatoi, H.N., 2016. Optimization of delignification of two *Pennisetum* grass species by NaOH pretreatment using Taguchi and ANN statistical approach. Environ. Technol. 37, 940–951.

Roni, M., Rifatunb, D.M., Satriyo, K.W., Rizal, W.A., Wheni, A.I., 2014. Alkaline pretreatment on sugarcane bagasse for bioethanol production. Energy Procedia 47, 250–254.

Sasaki, K., Tsuge, Y., Sasaki, D., Kawaguchi, H., Sazuka, T., Ogino, C., Kondo, A., 2015. Repeated ethanol production from sweet sorghum juice concentrated by membrane separation. Bioresour. Technol. 86, 351–355.

Wang, M., 2002. Energy and Greenhouse Gas Emissions Effects of Fuel Ethanol. Centre for Transportation Research, 31pp, Presentation at the Congressional Briefing Organized by the Environmental and Energy Study Institute, Washington DC, July.

Yun, G., Liu, D., Wu, T., Wu, J., Ji, X., Zhuang, L., 1998. CO_2 capture and utilization for enhanced oil recovery (EOR) and underground storage: a case study in Jilin Field. China Stud. Surf. Sci. Catal. 114, 201–206.

Zhang, L., Zhao, H., Gan, M., Jin, Y., Gao, X., Chen, Q., Guan, J., Wang, Z., 2011. Application of simultaneous saccharification and fermentation (SSF) from viscosity reducing of raw sweet potato for bioethanol production at laboratory, pilot and industrial scales. Bioresour. Technol. 102, 4573–4579.

SECTION II

BIOETHANOL FROM FOOD CROPS

2 *Disassembling the Glycomic Code of Sugarcane Cell Walls to Improve Second-Generation Bioethanol Production* 31
3 *Bioethanol Production From Corn and Wheat: Food, Fuel, and Future* 45
4 *Status and Perspectives in Bioethanol Production From Sugar Beet* 61
5 *Sweet Sorghum for Bioethanol Production: Scope, Technology, and Economics* 81
6 *Cassava as Feedstock for Ethanol Production: A Global Perspective* 101
7 *Sweet Potato as a Bioenergy Crop for Fuel Ethanol Production: Perspectives and Challenges* 115
8 *Jerusalem Artichoke: An Emerging Feedstock for Bioethanol Production* 149

CHAPTER

2

Disassembling the Glycomic Code of Sugarcane Cell Walls to Improve Second-Generation Bioethanol Production

Marcos S. Buckeridge, Adriana Grandis, Eveline Q.P. Tavares

Laboratory of Plant Physiological Ecology (LAFIECO), Department of Botany, Institute of Biosciences, University of São Paulo, São Paulo, Brazil

2.1 INTRODUCTION

In the context of sustainable development, fossil energies like oil have to be replaced by renewable energy. Bioethanol is a promising alternative. The use of bioethanol can help decrease the need for petroleum-based fuels. Bioethanol can be obtained from several different plant species, sugarcane being the most efficient (De Souza et al., 2014). Ethanol fuel from sugarcane, largely used in Brazil, has some important advantages regarding sustainability: (1) it is relatively cleaner, decreasing pollution by particulate material (Salvo et al., 2017); (2) it can be used along with forest recovery systems (Buckeridge et al., 2012); and (3) it has a great potential to mitigate carbon emissions (up to 6% of the net CO_2 emissions by 2045) and substitute up to 14% of the global oil use in the planet (Jaiswal et al., 2017).

Nowadays, bioethanol is obtained after fermentation of free mono- and disaccharides (usually glucose or sucrose) by yeast. The main sources used for bioethanol production are corn starch (glucose source) and sugarcane (sucrose source). The production of bioethanol, mainly the one made from starch in the United States, causes a competition for land and water for food production (De Souza et al., 2014). Brazil's first-generation bioethanol (based on sucrose from sugarcane) is not a direct competitor with food production due to the existence of enough land to produce food and biofuels at the same time (Jaiswal et al., 2017). However, the pressure to

Bioethanol Production From Food Crops
http://dx.doi.org/10.1016/B978-0-12-813766-6.00002-3

produce food in the world in the next decades tend to be high (FAO, 2009; Gruskin, 2012), possibly leading to such problems in Brazil as well.

An alternative measure to increase productivity (i.e., produce more bioethanol using the same or even less land) is to use lignocellulosic biomass, which decreases competition with human food production. Examples are agricultural residues (sugarcane, bagasse, and straw), forestry residues (branches, leaves, bark, etc.) or energy crops (switchgrass, elephant grass, energy-cane, etc.). Lignocellulosic biomass refers to plant biomass composed of cellulose, hemicelluloses, and lignin, these compounds being localized in plant cell walls. Indeed, cellulose can be hydrolyzed into glucose, which will be used to produce bioethanol. Bioethanol made from lignocellulosic biomass is called second-generation ethanol (Dos Santos et al., 2011).

Plant cell walls represent the most important and abundant renewable raw material on Earth. However, the access to its polysaccharides, including cellulose, is prevented by polymers localization and organization within the plant cell wall, which surrounds all plant cells. Plant cell walls exhibit a complex architecture and play numerous roles in plant growth, development, and defense against biotic and abiotic stresses (Cosgrove, 2005). So, plant cell walls are quite resistant to breakdown by mechanical or microbial actions, making the access to cellulose quite difficult. Thus, the interwoven network of polymers forms a highly organized structure—the glycomic code (De Souza et al. (2014); Buckeridge et al., 2015; Tavares and Buckeridge, 2015) that is the reason for recalcitrance to hydrolysis (Himmel et al., 2007).

2.2 SUGARCANE AS A SOURCE OF BIOETHANOL

Sugarcane (*Saccharum* sp.) is a grass species belonging to monocotyledon group, with Type II cell wall. It has been cultivated for centuries to produce sugar. Nowadays, numerous value-added products are made from sugarcane: sugar, ethanol, bioelectricity, bioplastics (Dos Santos et al., 2011). After harvesting, sugarcane is taken to the mill and crushed to extract the juice. Half of the juice is used for sugar production, and the other half is fermented and distilled for ethanol production (first generation). Around 92% of the bagasse generated after crushing is burned to produce electricity (Amorim et al., 2011). The 8% left could also be converted into ethanol, producing ca.2200 L of ethanol per hectare and reducing land use in about 29%. On the other hand, if straw were used for bioenergy production along with 50% of the bagasse, it could lead to 3700–4000 L/ha, reducing land use in 33%–38% (Soccol et al., 2010).

Bagasse and straw also represent a huge source of polysaccharides that after hydrolysis can be fermented to produce ethanol. The use of this biomass can help respond to the increase in global ethanol demand and also meet the criteria of environment protection (Buckeridge et al., 2012; Jaiswal et al., 2017). The technological challenge in making bagasse and straw good raw materials for bioethanol production is to find a way to get the highest yield for polysaccharide extraction followed by the release of free sugars that could be fermented by yeast. To do that, it is necessary to know as well as possible the structure and architecture (i.e., the glycomic code) of the sugarcane cell walls.

2.3 SUGARCANE CELL WALLS

Cell walls from sugarcane have been subjected to fractionation and the polysaccharides that compose them have been identified structurally (Fig. 2.1). Many of the covalent linkages have been identified, 14 of them being glycosidic and 8 ester linkages (M.S. Buckeridge and N.C. Carpita, A. Cambler, A.P. De Souza unpublished; De Souza et al., 2013). These findings are important for bioenergy because these covalent

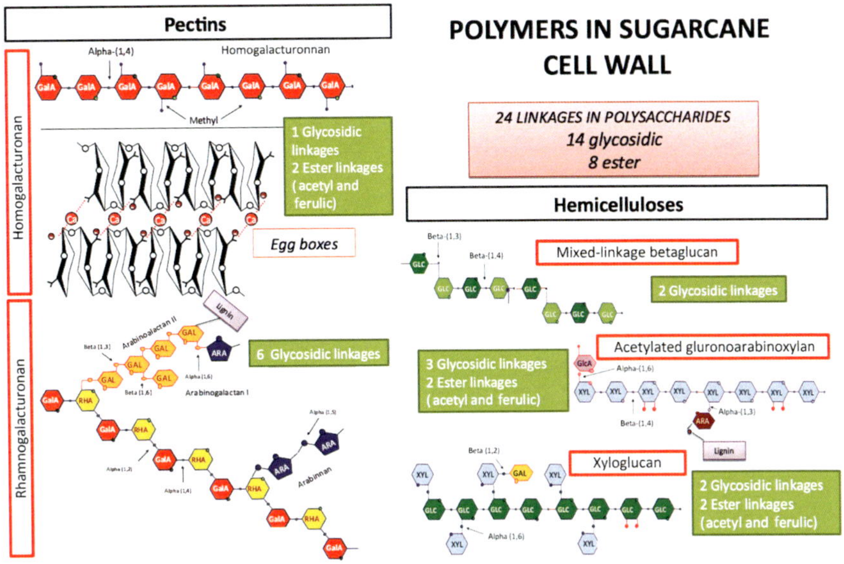

FIGURE 2.1 **Cell wall polymers in sugarcane.** The cell wall is composed of cellulose, lignin, pectins (homogalacturonan and rhamnogalacturanan I), and hemicelluloses (β-glucan, glucuronoarabinoxylan, and xyloglucan). In the green boxes, we describe the number of covalent linkages and ester linkages for each polysaccharide. The egg boxes are formed by the interaction between homogalacturonan and Ca^{++}, which induce the formation of gel-like material in regions where homogalacturonans are not methylated. *Source: Based on Buckeridge, M.S., Cavalari, A.A., Silva, G.B., 2008. Parede celular. In: Fisiologia V., Gilberto B.K. (Eds.). Guanabara Koogan, Rio de Janeiro Cap. 9:165–181.*

linkages are the main ones that have to be broken to release simple carbohydrates (monosaccharides), which could be used for fermentation and bioethanol production.

The main pectic polysaccharides detected in sugarcane cell walls are homogalacturonan and rhamonogalacturonan I (RG I) that is probably branched with linear arabinan and branched (1,3)(1,6)-linked galactans. The latter was found to be linked to lignin (A. Cambler, A.P. De Souza, and M.S. Buckeridge, unpublished results). Although pectin content is rather low in monocots (around 8% in sugarcane culms and leaves; De Souza et al., 2013), its applicability within the bioenergy context has proven to be critical (De Souza et al., 2015; Latarullo et al., 2016).

Three main types of hemicelluloses are found in sugarcane cell walls: arabinoxylans, the mixed linkage β-glucan, and xyloglucans. Arabinoxylans can be either branched (with arabinose, acetyl esters, and rarely with glucuronic acid) or unbranched (Buckeridge et al., 2015; De Souza et al., 2013). Most of the lignin in sugarcane cell walls is bound to the ferulic acid that esterifies arabinosyl residues of arabinoxylans (Buckeridge et al., 2015; A. Cambler, A.P. De Souza, and M.S. Buckeridge, unpublished results).

In current cell wall models designed for dicotyledonous species, xyloglucans are the main hemicelluloses, and their main role in the wall is to tether cellulose microfibrils (Park and Cosgrove, 2012). However, in sugarcane, xyloglucans are much less abundant than arabinoxylans (De Souza et al., 2013). Arabinoxylans may also bind to cellulose (Mikkelsen et al., 2015), but not in the same specific manner as xyloglucans. These authors found that arabinoxylans and also the mixed linkage β-glucan bind randomly to cellulose and not as specifically as happens with xyloglucans (De Lima and Buckeridge, 2001).

In the cell wall of sugarcane, cellulose is organized in the form of microfibrils (Driemeier et al., 2012; Driemeier, 2017). Microfibrils are thought to be wrapped by hemicellulosic polysaccharides, such as xyloglucans and arabinoxylans forming a macrofibrilar domain that is embedded in a matrix of pectins and structural proteins (Fig. 2.2) (Buckeridge et al., 2015, 2017). Although the polymer that binds more strongly to the surface of cellulose is xyloglucan (De Souza et al., 2013), branched arabinoxylan seems to play an important role in holding the architecture of sugarcane cell walls together. This appears to happen mainly through the crosslinking with lignin.

The mixed linkage β-glucan is an important component of the sugarcane cell wall (ca. 8%), but all experiments performed to date indicate that most of it is not strongly bound to any other components of the wall. It is possible, though, that some of the mixed linkage β-glucan could be bound nonspecifically to the surface of cellulose as suggested by Mikkelsen et al. (2015).

Altogether, the assembly of the polymers within the wall forms a composite that is held together by a range of linkage types with different strengths. At least four classes of linkages can be characterized within the wall: (1) covalent linkages, such as the glycosidic linkages among monosaccharides (which are ether linkages) and the relatively weaker ester linkages of

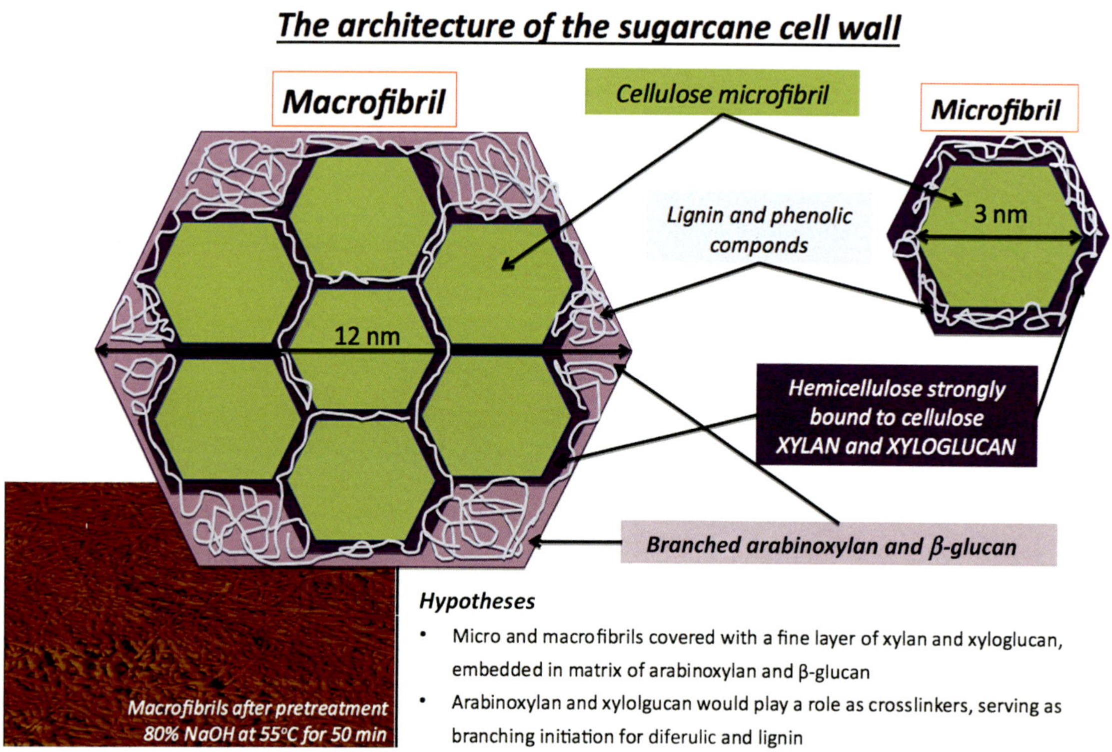

FIGURE 2.2 **The basic unit of the sugarcane cell wall architecture.** Microfibrils (top right: around 36 cellulose molecules packed in the form of a crystal structure) are assembled in bundles of seven forming a macrofibril (top left). Xyloglucan and linear xylans are thought to bind microfibrils together. The microfibrillar unit is embedded in a matrix of arabinoxylans (bound together by lignin and diferulic acids). Arabinoxylan and β-glucan are also weakly bound to themselves and cellulose through random hydrogen bond linkages. Bottom left: atomic force microscopy showing the effect of biomass pretreatment with NaOH. The macrofibrils collapse forming much larger packages of polymers. The *white bar* on the top left of the figure corresponds to 100 nm.

acetyl, methyl, and feruloyl branches to polysaccharide hydroxyls; (2) ionic linkages, such as the egg boxes formed between pectins and calcium in the middle lamella; (3) specific noncovalent linkages, such as the hydrogen bonds xyloglucans make to the surface of cellulose; and (4) nonspecific polysaccharide bonds (hydrogen bonds), such as the random attachment of the mixed linkage β-glucan and arabinoxylans to the surface of cellulose and, possibly polysaccharide–polysaccharide interactions that occur due to proximity of the polymers in the wall. Covalent, ionic, noncovalent specific, and nonspecific bonds are progressively less strong and therefore require progressively less energy to be broken. The strength of the specific (xyloglucan) binding to cellulose in comparison to the nonspecific is probably related to the number or hydroxyls involved in the binding process, which is higher in the former type of linkage.

This highly complex composite formed by linkages of different types was named the glycomic code by Buckeridge and De Souza (2014). The balance among these linkages and the strength each one contributes to maintaining the glycomic code is a key determinant of which strategies would work best for cell wall disassembly and subsequent release of free sugars for fermentation. In other words, the challenge of producing bioethanol from cell walls is the challenge of understanding how to dismantle the glycomic codes of sugarcane and other plant species.

Species that are taxonomically closer display similar cell walls' general composition. This similarity goes beyond monosaccharide composition because they lead to architectures that are quite similar and therefore similar glycomic codes. This is the case of sugarcane and miscanthus (compare De Souza et al., 2013, with De Souza et al., 2015). It can be assumed that the structural and architectural similarity among cell walls of related species may result in comparable physical–chemical properties so that the processes used in industry (pretreatments and enzyme cocktails) could be reasonably well adapted for different feedstocks.

2.4 PRETREATMENTS AND HYDROLYSIS AND THEIR IMPACT ON COVALENT LINKAGES

One of the solutions proposed for exposing polysaccharide to depolymerizing enzymes is named pretreatment (Cardona et al., 2010; Lombard et al., 2014; Soccol et al., 2010). Pretreatment processes developed to reach polysaccharides are expensive, representing a significant (ca. 20%) of the cost of the whole process (De Souza et al., 2014).

Several different types of pretreatments have been proposed and used in assays with different bioenergy feedstocks (Table 2.1). They are physical pretreatments, such as steam explosion and treatment with liquid hot water, chemical pretreatments with acid, alkali, or organosolv (supercritical CO_2). There are also biological pretreatments in which accessory enzymes are used before full hydrolysis and biological fermentation (Cardona et al., 2010; Soccol et al., 2010; Table 2.1).

The pretreatment step in the process of second-generation bioethanol production has partial effectiveness in the disassembly of the cell walls. Because of that, such treatments will mainly swell (hydrate) the walls and break a few covalent linkages (e.g., esters that bind lignin to polysaccharides and α-linked polysaccharide branches) and prepare them to be more accessible to hydrolytic enzymes. Some breakage of polymers can lead to the formation of oligosaccharides. Furthermore, in the case of sugarcane, the production of bagasse by crushing and washing with hot water can extract highly soluble polysaccharides. Pectins and β-glucans are known to be lost during the process (L.M. Azevedo, A.P. De Souza, and M.S. Buckeridge, unpublished results). Thus, although pretreatments bring the benefit of "opening" the cell

TABLE 2.1 The Effects of Different Types of Pretreatments on Biomass

Pretreatment	Technique	Mechanisms and conditions		Products	Chemical modification	Reference
Physical	Uncatalysed steam explosion	High pressure and hot water	160–270 °C	Removes hemicelluloses and modify lignin. Produces organic acids and acetic acid	Noncovalent	Zheng et al. (2009)
	Liquid hot water	Hot water	170 °C—2 h	Removes hemicelluloses, produces high concentration of xylan olygomers and polymers, organic acids and acetic acid	Noncovalent	Zheng et al. (2009); Boussarsar et al. (2009)
	Mechanical comminution	Chipping, shredding, grinding, milling are amongst the different mechanical size reduction methods	–	Unknown products	Probably noncovalent	Palmowski and Muller (1999)
	High energy radiation	Gamma rays	–	Cleaves the ß-1,4 glycosidic bonds, thus giving a larger surface area and lower crystallinity	Covalent	Takacs et al. (2000)
Chemical	Acid	HCl (1.2% v/v)	121 °C–4 h	30% of bagasse converted to reducing sugars	Covalent and noncovalent	Hernandez-Salas et al. (2009)
		Water, SO_2, and H_2SO_4	180 °C–5 min	Glucose and xylose yields on average 86.3% and 72.0%, respectively	Covalent and noncovalent	Sendelius (2005)
	Alkaline	1%–3% NaOH solution	ND	Liberates the lignin and makes cellulose and hemicelluloses available by hydrolysing uronic and acetic esteres. Swells cellulose	Covalent and noncovalent	Cardona et al. (2010)
		3% NaOH (w/v) solid to liquid ratio of 1:25	50 °C–3 h	74.9% of the hemicelluloses are hydrolyzed—xylose as predominant sugar	Covalent and noncovalent	Peng et al. (2009)
	Ammonia fiber/freeze explosion (AFEX)	1–2 kg of ammonia/kg of dry biomass	10–60 min at 60–90 °C and pressures above 3 MPa	Biomass to swell, increasing the accessible surface area, small amount of hemicellulose is solubilized in the oligosacharides and lignin distribution do not change, but the structure is rigorously altered, resulting in increase water-holding capacity and digestibility.	Mainly noncovalent	Agbor et al. (2011)
	Organosolv	Water and oxygen Supercritical CO_2 and 1-butanol-water mixture Supercritical CO_2 and ethanol–water mixture Dimethylformamide (DMF)	Operation at 200–210 °C for 150 min and 40%–60% DMF	Delignification extent: 82.7%. Solid material with nearly 83.53% of α-cellulose	Covalent and noncovalent	Rezayati-Charani et al. (2006)
	pH controlled liquid hot water	Wet oxidation low-temperature hydrolytic reaction and high-temperature oxidative reaction	Water and oxygen above 120 °C	Catalyses the transfer of hemicelluloses from the solid phase to the liquid phase. The products of hemicelluloses hydrolysis are sugar oligomers	Covalent and noncovalent	Kuo and Lee (2009)
		Wet oxidation	195 °C 15 min and alkaline pH	70% cellulose content	Covalent and noncovalent	Martín et al. (2007)

	Ionic liquids	1-allyl-3-methylimidazonium chloride (AMIMCl) and 1-butyl-3-methylimidazonium chloride (BMIMCl)	Below 100 °C	Dissolution of cellulose	Ionic liquids compete with lignocellulosic components for hydrogen bonding, thus disrupting its three dimensional network	Zhang and Lynd (2006); Zhu et al. (2006); Moulthrop et al. (2005)
Biological	Marine fungi	*Fugus phlebia* sp. MG-60	–	More than 50% of lignin in biomass was degraded and less than 10% of the holocellulose was lost.	Covalent and noncovalent	Li et al. (2002)
	Fungi	White rot fungi–Basidiomycetes	–	Capable of degrading a lignocellulose substrate	Covalent and noncovalent	Pan et al. (2005)
	Plants	Enzyme-arabinofuranosidase	–	Remove some arabinose residues from arabinoxylan	Covalent	Leite et al. (2017)
		Protein–expansins	–	Disrupts the xyloglucan-cellulose binding and increase the porosity to cell wall	Noncovalent	De Souza et al. (2013)

wall for enzyme access, it also provokes losses of different kinds.

2.4.1 Chemical Pretreatments

Biomass can be pretreated with a few types of acids. They are HCl, H_3PO_4, HNO_3, and H_2SO_4 (Cardona et al., 2010; Soccol et al., 2010). A common problem with the use of acids is the attack to covalent linkages generating diluted carbohydrate solutions that cannot be concentrated due to the high cost. Another problem is the generation of environmental hazards. Diluted sulfuric acid is the most commonly used. It usually extracts most of the polymers that are less strongly bound to others in the wall (pectins, β-glucans, and branched arabinoxylans). Depending on the concentration and temperature of the operation, covalent α-linkages (present, e.g., in the branching of arabinoxylans) could also be attacked possibly provoking debranching and increasing polysaccharide–polysaccharide interaction so that recalcitrance increases. Acid pretreatment works mainly as an extractor of many polysaccharides that could otherwise be used for hydrolysis. One of the solutions for that is to use the supernatants for further processing in the industry. However, these supernatants are highly contaminated with sulfuric acid and usually contain diluted polysaccharides. These solutions are difficult to handle and probably would require higher costs to be used for further hydrolysis and fermentation.

One of the main effects of the alkali pretreatment (usually NaOH) is the breakage of all ester linkages within the wall. Thus, all acetyls, methyls of pectins, and arabinoxylans will be solubilized and lost in the supernatant. Some of the lignin–polysaccharide linkages may be broken as well.

2.4.2 Biological Pretreatments

Several research lines have been designed to improve cell wall accessibility at a lower cost. Regarding enzyme hydrolysis, most of the studies are related to cellulose and xylans. Several enzymes have been isolated, and their structure, mode, and mechanisms of action have been unveiled (Moreira et al., 2011, 2017; Polizeli et al., 2011, 2017; Serpa and Polikarpov, 2011; Ward, 2011). Most studies use commercial cellulose as substrates, so that the interaction of it with the hemicelluloses is hardly considered (Bragatto et al., 2012). However, more recent studies have proposed and demonstrated that cell wall hydrolysis should follow a sequential action of enzymes that would follow the assembly of the glycomic code (Borin et al., 2015; De Souza et al., 2013) (Fig. 2.3).

Another proposed strategy to manipulate plant cell wall structure is through changing the biosynthetic pathways of polymers (the encryption machinery of the glycomic code; Buckeridge and De Souza, 2014) using molecular biology tools (Pauly and Keegstra, 2008; Sticklen, 2008; Tavares et al., 2015). For sugarcane, an alternative strategy has been proposed that is the use of already existent degradation mechanisms to improve hydrolysis in planta and perhaps bypass pretreatment (Buckeridge et al., 2015, 2017; Grandis et al., 2014; Leite et al., 2017; Tavares et al., 2015).

To use genetically engineered strategies so that sugarcane could be more amenable to pretreatment and hydrolysis depend on a better understanding of cell wall composition and architecture along with the modes of action of enzymes on polymers that are assembled within the wall. This knowledge has to be coupled with experiments with isolated well-known enzymes that act on the walls or isolated polymers to bring more precise information about how hydrolysis can be performed efficiently in industry.

It is also important to understand how polymers interact through weaker linkages in comparison to the covalent ones so that one would be able to know when the polymers become available for hydrolysis after pretreatment. Such models of enzyme action could be better suited

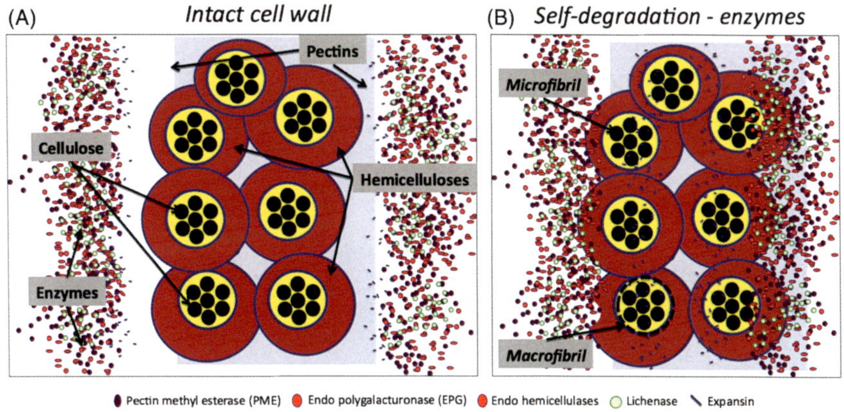

FIGURE 2.3 **Sugarcane endo enzymes from produced in sugarcane tissues that improve the pretreatment process before the polysaccharides hydrolysis.** (A) Intact cell wall architecture (Type II), with the identification of each structure. (B) The cell wall in degradation process by itself, enzymes produced from sugarcane cells act on the glycomic code disassembly and improve the access to cellulose for the fermentation process. When enzymes are added, the first protein to enter wall pores is expansin/swollenin. The enzymes follow and by the action of the pectinases first and then the hemicelluloses. *Source: Based on De Souza, A.P., Leite, D.C.C., Pattathil, S., Hahn, M.G., Buckeridge, M.S. 2013. Composition and structure of sugarcane cell wall polysaccharides: implication for second-generation bioethanol production. Bioenergy Res. 6(2):564–579. and Borin, G.P., Sanchez, C.C., de Souza, A.P., de Santana, E.S., de Souza, A.T., Leme, A.F.P. et al., 2015. Comparative secretome analysis of* Trichoderma reesei *and* Aspergillus niger *during growth on sugarcane biomass. PLoS One 10(6):e0129275. doi:10.1371/journal.pone.0129275.*

to what happens at the moment industry is processing biomass for second-generation bioethanol production.

2.5 EFFECT OF PRETREATMENTS ON THE NONCOVALENT LINKAGES OF THE WALL

The noncovalent linkages within cell walls are quite relevant for pretreatments. This is so because the use of mechanical forces combined with the elevation of temperature and the use of weak alkali or acid solutions will probably affect primarily the noncovalent bonds.

An important question that can be asked under the light of the knowledge of the glycomic code of sugarcane and other similar grasses is: What happens to the different bonds when biomass is pretreated? Table 2.1 shows an appraisal of what may happen with wall weak linkages (and some covalent linkages as well) under different pretreatments.

Alkali is also quite effective in breaking many of the polysaccharide–polysaccharide interactions, such as the egg boxes, the nonspecific polysaccharide–polysaccharide interactions, and the forces involved in the binding of xyloglucan to cellulose. This pretreatment can disassemble the glycomic code quite efficiently by separating polysaccharides that become free of ester branches and become therefore separated from lignin. Indeed, we use alkali for cell wall fractionation during our analytical procedures and can separate well different classes of polymers (De Souza et al., 2013, 2015). The alkali can be easily neutralized, but generates high concentrations of salts that become a residue in the industry and possibly elevate costs. Although alkali pretreatment could be quite effective if well controlled, one of its main nanoscopic effects on the assembled biomass (i.e., with intact glycomic code) is the secondary repackage of polymers around the micro- and macrofibrils, forming much thicker artificial "macrofibrils" (Fig. 2.2). The level of saccharification of these reassembled materials remains to be investigated. However, it can be speculated that recalcitrance might increase a little due to the secondary (artificial) interactions between hemicellulose and cellulose, which can interfere with the access of cellulose degrading enzymes to the microfibrils. On the other hand, this reassembly process can be advantageous because specific polysaccharide–polysaccharide binding (e.g., xyloglucan–cellulose and xylan–cellulose) is exchanged by nonspecific weaker bindings, probably making enzyme access easier.

Regarding thermochemical pretreatments, Sendelius (2005) found that some hydrolysis can occur that generates extraction of polysaccharides and breakage of their molecules leading to the production of oligosaccharides. Chundawat et al. (2011) studied the effects of thermochemical [ammonia fiber expansion (AFEX)] pretreatment on cell walls accounting for macroscopic and nanoscopic effects. They found that the linkages between the lignin domain and the polysaccharide domain are altered. It can be speculated that the elevation of temperature would likely break weaker linkages of the wall (specific and nonspecific polysaccharide–polysaccharide interactions) as well as break some of the covalent linkages, especially the weaker ones, such as the ester linkages and the α-linked branches.

More recently Gunawan et al. (2017) found that when AFEX pretreatment is used with corn stove, which has cell walls of the same type of sugarcane, recalcitrant residues are composed of arabinoxylans branched with glucuronic acid and pectins. It is possible that similar results would be obtained with sugarcane. However, one difference is that very little glucuronic acid is present in sugarcane arabinoxylans so that it could be less recalcitrant that maize walls. Regarding the pectins, it is more likely that this could be similar for both species since very similar pectin types are present in both materials (compare Gunawan et al., 2017, with De Souza et al., 2013).

It is important to note that the discussion on how pretreatment can affect weak forces of the cell wall of sugarcane and other biomass materials is not exhaustive. However, this item, along with Table 2.1, can help the reader to use the knowledge about the glycomic code to better understand and use other pretreatment processes.

2.6 CONCLUSIONS AND FUTURE PERSPECTIVES

The recent advances in the knowledge about the glycomic code of sugarcane cell walls are now providing means to understand how polymers are assembled. This brings another perspective about the role of different chemical forces within the cell wall composite. Mainly four different forces are involved in cell wall architecture: covalent linkages (glycosidic and ester), ionic bonds (egg boxes involving homogalacturonan assembly with calcium ions), specific polysaccharide–polysaccharide binding through hydrogen bonds (the xyloglucan cellulose domain), and nonspecific polymers binding through random interaction among polysaccharides through hydrogen bonds. Other forces and linkages may be involved that could be added to this list. One of them is the structural proteins that remain to be studied in sugarcane cell walls (Tavares and Buckeridge, 2015).

The knowledge obtained to date about the glycomic code of sugarcane walls affords to speculate about the nature of several effects that pretreatments—these processes are usually expected to separate cell wall polymers but preserving most of the covalent linkages among simple sugars—made on the walls. It is expected that by understanding the nature of the effects of pretreatments on the forces that hold together the glycomic code, one could better control pretreatment conditions so that the efficiency of second generation processes used for sugarcane and other bioenergy feedstock could be ameliorated.

Acknowledgments

Authors acknowledge financial support of the National Institute of Science and Technology of Bioethanol (FAPESP 2008/57908-6 and CNPq 574002/2008-1).

References

Agbor, V.B., Cicek, N., Sparling, R., Berlin, B., Levin, D.B., 2011. Biomass pretreatment: fundamentals toward application. Biotechnol. Adv. 29, 675–685.

Amorim, H.V., Lopes, M.L., Oliveira, J.V.C.O., Buckeridge, M.S., Goldman, G.H., 2011. Scientific challenges of bioethanol production in Brazil. Appl. Microbiol. Biotechnol. 91, 1267–1275.

Bragatto, J., Segato, F., Cota, J., Mello, D.B., Oliveira, M.M., Buckeridge, M.S., Squina, F.M., Driemeier, C., 2012. Insights on how the activity of an endoglucanase is affected by physical properties of insoluble celluloses. J. Phys. Chem. B 116, 6128–6136.

Borin, G.P., Sanchez, C.C., de Souza, A.P., de Santana, E.S., de Souza, A.T., Leme, A.F.P., et al., 2015. Comparative secretome analysis of *Trichoderma reesei* and *Aspergillus niger* during growth on sugarcane biomass. PLoS One 10 (6), e0129275.

Boussarsar, H., Rogé, B., Mathlouthi, M., 2009. Optimization of sugarcane bagasse conversion by hydrothermal treatment for the recovery of xylose. Bioresour. Technol. 100, 6537–6542.

Buckeridge, M.S., Cavalari, A.A., Silva, G.B., 2008. Parede cellular. Kerbauy, G.B. (Ed.), Fisiologia Vegetal, vol. 9, Guanabara Koogan, Rio de Janeiro, pp. 165–181.

Buckeridge, M.S., De Souza, A.P., 2014. Breaking the 'Glycomic Code' of cell wall polysaccharides may improve second-generation bioenergy production from biomass. Bioenergy Res. 7, 1065–1073.

Buckeridge, M.S., De Souza, A.P., Arundale, R.A., Anderson-Teixeira, K.J., Delucia, E., 2012. Ethanol from sugarcane in Brazil: a "midway" strategy for increasing ethanol production while maximizing environmental benefits. GCB Bioenergy 4, 119–126.

Buckeridge, M.S., Santos, W.D., Tiné, M.S., De Souza, A.P., 2015. The cell wall architecture of sugarcane and its implications to cell wall recalcitrance. In: Lam, E., Carrer, H., Silva, J.A. (Eds.), Compendium of Bioenergy Plants: Sugarcane. CRC Press, New York, NY, United States, pp. 31–50.

Buckeridge, M.S., De Souza, A.P., Tavares, E.Q.P., Cambler, A.B., 2017. Sugarcane cell wall structure and degradation: from monosaccharide analyses to the glycomic code. In: Buckeridge, M.S., De Souza, A.P. (Eds.), Advances of Basic Science for Second Generation Bioethanol From Sugarcane. Springer, New York, NY, United States, pp. 7–19.

Cardona, C.A., Quintero, J.A., Paz, I.C., 2010. Production of bioethanol from sugarcane bagasse: status and perspectives. Bioresour. Technol. 1010, 4754–4766.

Cosgrove, D.J., 2005. Growth of the plant cell wall. Nat. Rev. Mol. Cell Biol. 6, 850–861.

Chundawat, S.P.S., Donohoe, B.S., Sousa, L.C., Elder, T., Agarwal, U.P., Lu, F., Ralph, J., Himmel, M.E., Balan, V., Dale, B.E., 2011. Molti-scale visualization and characterization of lignocellulosic plant cell wall deconstruction during thermochemical pretreatment. Energy Environ. Sci. 4, 973–984.

De Lima, D.U., Buckeridge, M.S., 2001. Interaction between cellulose and storage xyloglucans: the influence of the degree of galactosylation. Carbohyd. Polym. 46, 157–163.

De Souza, A.P., Leite, D.C.C., Pattathil, S., Hahn, M.G., Buckeridge, M.S., 2013. Composition and structure of sugarcane cell wall polysaccharides: implication for second-generation bioethanol production. Bioenergy Res. 6 (2), 564–579.

De Souza, A.P., Grandis, A., Leite, D.C.C., Buckeridge, M.S., 2014. Sugarcane as a bioenergy source: history, performance, and perspectives for second-generation bioethanol. Bioenergy Res. 7, 24–35.

De Souza, A.P., Kamei, C.L.A., Torres, A.F., Pattathil, S., Hahn, M.G., Trindade, L.M., Buckeridge, M.S., 2015. How cell wall complexity influences saccharification efficiency in *Miscanthus sinensis*. J. Exp. Bot. 66, 4351–4365.

Driemeier, C., Santos, W.D., Buckeridge, M.S., 2012. Cellulose crystals in fibrovascular bundles of sugarcane culms: orientation size, distortion, and variability. Cellulose 19, 1507–1515.

Driemeier, C., 2017. Nanostructure of lignocellulose and its importance for biomass conversion into chemicals and biofuels. In: Buckeridge, M.S., De Souza, A.P. (Eds.), Advances of Basic Science for Second Generation Bioethanol From Sugarcane. Springer, New York, NY, United States, pp. 21–38.

Dos Santos, W.D., Gómez, E.O., Buckeridge, M.S., 2011. Bioenergy and the sustainable revolution. In: Buckeridge, M.S., Goldman, G.H. (Eds.), Routes to Cellulosic Ethanol. first ed. Springer, New York, NY, United States, pp. 15–26.

FAO 2009: How to Feed the World in 2050 High-Level Expert Forum, 12–13 October Rome. How to Feed the World in 2050, Food and Agriculture Organization (FAO), Rome, Italy. Available from: http://www.Fao.org/wsfs/forum2050/wsfs-Background-documents/hlef-Issues-briefs/en/.

Grandis, A., De Souza, A.P., Tavares, E.P.Q., Buckeridge, M.S., 2014. Using natural plant cell wall degradation mechanisms to improve second generation bioethanol. In: McCann, M.C., Buckeridge, M.S., Carpita, N.C. (Eds.), Plants and Bioenergy. Springer, New York, NY, United States, pp. 211–230.

Gruskin, D., 2012. Agbiotech 2.0. Nat. Biotechnol. 30 (3), 211–214.

Gunawan, C., Xue, S., Pattathil, S., Sousa, L.C., Dale, B.E., Balan, V., 2017. Comprehensive characterization of non-cellulosic recalcitrant cell wall carbohydrates in unhydrolyzed solids from AFEX-pretreated corn stover. Biotechnol. Biofuels 10, 82–95.

Hernández-Salas, J.M., Villa-Ramírez, M.S., Veloz-Rendón, J.S., Rivera-Hernández, K.N., González-César, R.A., Plascencia-Espinosa, M.A., Trejo-Estrada, S.R., 2009. Comparative hydrolysis and fermentation of sugarcane and agave bagasse. Bioresour. Technol. 100, 1238–1245.

Himmel, M.E., Ding, S.Y., Johnson, D.K., Adney, W.S., Nimlos, M.R., Brady, J.W., et al., 2007. Biomass recalcitrance: engineering plants and enzymes for biofuels production. Science 315, 804–807.

Jaiswal, D., De Souza, A.P., Larsen, S., Lebauer, D.S., Miguez, F., Sparovek, G., Bollero, G., Buckeridge, M.S., Long, S.P., 2017. Brazilian sugarcane ethanol as an expandable green alternative to crude oil use. Nat. Clim. Change 7, 788–792.

Kuo, C.H., Lee, C.K., 2009. Enhanced enzymatic hydrolysis of sugarcane bagasse by *N*-methylmorpholine-*N*-oxide pretreatment. Bioresour. Technol. 100, 866–871.

Latarullo, M.B., Tavares, E.Q.P., Maldonado, G.P., Leite, D.C.C., Buckeridge, M.S., 2016. Pectins, endopolygalacturonases and bioenergy. Front. Plant Sci. 7, 1401.

Leite, D.C.C., Grandis, A., Tavares, E.Q.P., Piovezani, A.R., Pattathil, S., Avci, U., Rossini, A., Cambler, A., De Souza, A.P., Hahn, M.G., Buckeridge, M.S., 2017. Cell wall changes during the formation of aerenchyma in sugarcane roots. Ann. Bot. 120, 693–708. doi: 10.1093/aob/mcx050.

Li, X., Kondo, R., Sakai, K., 2002. Biodegradation of sugarcane bagasse with marine fungus *Phlebia* sp. MG-60. J. Wood Sci. 48, 159–162.

Lombard, V., Golaconda Ramulu, H., Drula, E., Coutinho, P.M., Henrissat, B., 2014. The carbohydrate-active enzymes database (CAZy) in 2013. Nucleic Acids Res. 42, D490–D495.

Martín, C., Klinke, H.B., Thomsen, A.B., 2007. Wet oxidation as a pretreatment method for enhancing the enzymatic convertibility of sugarcane bagasse. Enzyme Microb. Technol. 40, 426–432.

Mikkelsen, D., Flanagan, B.M., Wilson, S.M., Bacic, A., Gidley, M.J., 2015. Interactions of arabinoxyland and (1,3)(1,4)-B-glucan with cellulose networks. Biomacromolecules 16, 1232–1239.

Moreira, L.R.S., Milanezi, N.G., Filho, E.X.F., 2011. Enzymology of plant cell wall breakdown: an update. In: Buckeridge, M.S., Goldman, G.H. (Eds.), Roots to Cellulosic Ethanol. Springer, New York, NY, United States, pp. 73–96.

Moreira, L.R.S., Corrêa, C.L., Gomes, H.A.R., Midorikawa, G.E.O., Miller, R.N.G., Filho, E.X.F., 2017. The role of fungal transcriptome analysis and side-chain hydrolyzing enzymes in sugarcane bagasse breakdown. In: Buckeridge, M.S., De Souza, A.P. (Eds.), Advances of Basic Science for Second Generation Bioethanol From Sugarcane. Springer, New York, NY, United States, pp. 81–106.

Moulthrop, J.S., Swatloski, R.P., Moyna, G., Rogers, R.D., 2005. High resolution 13C NMR studies of cellulose and cellulose oligomers in ionic liquid solutions. Roy. Soc. Chem. Commun. 12, 1557–1559.

Palmowski, L., Muller, J., 1999. Influence of the size reduction of organic waste on their anaerobic digestion. 2nd International Symposium on anaerobic digestion of solid waste. Barcelona, 15–17 June. .

Pan, X., Arato, C., Gilkes, N., Gregg, D., Mabee, W., Pye, K., Xiao, Z., Zhang, X., Saddler, J., 2005. Biorefining of softwoods using ethanol organosolve pulping: preliminary evaluation of process streams for manufacture of fuel-grade ethanol and co-products. Biotechnol. Bioeng. 90, 473–481.

Park, Y.B., Cosgrove, D.J., 2012. A revised architecture of primary cell walls based on biomechanical changes induced by substrate-specific endoglucanase. Plant Physiol. 158, 1933–1943.

Pauly, M., Keegstra, K., 2008. Cell-wall carbohydrates and their modification as a resource for biofuels. Plant J. 54, 559–568.

Peng, F., Ren, J.L., Xu, F., Bian, J., Peng, P., Sun, R.C., 2009. Comparative study of hemicelluloses obtained by graded ethanol precipitation from sugarcane bagasse. J. Agric. Food Chem. 57, 6305–6317.

Polizeli, M.L.T.M., Corrêa, E.C.P., Polizeli, A.M., Jorge, J.A., 2011. Hydrolases from microorganisms used for degradation of cell wall and bioenergy. In: Buckeridge, M.S., Goldman, G.H. (Eds.), Roots to Cellulosic Ethanol. Springer, New York, NY, United States, pp. 115–134.

Polizeli, M.L.T.M., Somera, A.F., de Lucas, R.S., Nozawa, M.S.F., Michelin, M., 2017. Enzymes involved in the biodegradation of sugarcane biomass: challenges and perspectives. In: Buckeridge, M.S., De Souza, A.P. (Eds.), Advances of Basic Science for Second Generation Bioethanol From Sugarcane. Springer, New York, NY, United States, pp. 55–79.

Rezayati-Charani, P., Mohammadi-Rovshandeh, J., Hashemi, S.J., Kazemi-Najafi, S., 2006. Influence of dimethyl formamide pulping of bagasse on pulp properties. Bioresour. Technol. 97, 2435–2442.

Sendelius, J., 2005. Steam Pretreatment Optimisation for Sugarcane Bagasse in Bioethanol Production. Master of Science Thesis. Lund University, Lund, Sweden.

Serpa, V.I., Polikarpov, I., 2011. Enzymes in bioenergy. In: Buckeridge, M.S., Goldman, G.H. (Eds.), Roots to Cellulosic Ethanol. Springer, New York, NY, United States, pp. 97–113.

Soccol, C.R., Vandenberghe, L.P.S., Medeiros, A.B.P., Karp, S.G., Buckeridge, M.S., Ramos, L.P., Pitarelo, A.P., Ferreira-Leitão, V., Gottschalk, L.M.F., Ferrara, M.A., 2010. Bioethanol from lignocelluloses: status and perspectives in Brazil. Bioresour. Technol. 101, 4820–4825.

Salvo, A., Brito, J., Artaxo, P., Geiger, F.M., 2017. Reduced ultrafine particle levels in São Paulo's atmosphere during shifts from gasoline to ethanol use. Nat. Commun. 8 (77)doi: 10.1038/s41467-017-00041-5.

Sticklen, M.B., 2008. plant genetic engineering for biofuel production: towards affordable cellulosic ethanol. Nat. Rev. Genet. 9, 433–443.

Takacs, E., Wojnarovits, L., Foldavary, C., Hargagittai, P., Borsa, J., Sajo, I., 2000. Radiat. Phys. Chem. 57, 339.

Tavares, E.Q.P., Buckeridge, M.S., 2015. Do cell walls have a code? Plant Sci. 241, 286–294.

Tavares, E.Q.P., De Souza, A.P., Buckeridge, M.S., 2015. How endogenous plant cell wall degradation mechanisms can help achieve higher efficiency in saccharification of biomass. J. Exp. Bot. 66, 4133–4143.

Ward, R., 2011. Cellulase engineering for biomass saccharification. In: Buckeridge, M.S., Goldman, G.H. (Eds.), Roots to Cellulosic Ethanol. Springer, New York, NY, United States, pp. 135–151.

Zhang, Y.H.P., Lynd, L.R.A., 2006. Functionally based model for hydrolysis of cellulose by fungal cellulase. Biotechnol. Bioeng. 94, 888–898.

Zheng, Y., Pan, Z., Zhang, R., 2009. Overview of biomass pretreatment for cellulosic ethanol production. Int. J. Agric. Biol. Eng. 2, 51–68.

Zhou, Q., Rutland, M.W., Teeri, T.T., Brumer, H., 2007. Xyloglucan in cellulose modification. Cellulose 14 (6), 625–641.

Zhu, S.D., Wu, Y.X., Chen, Q.M., Yu, N., Wang, C.W., Jin, S.W., et al., 2006. Dissolution of cellulose with ionic liquids and its application: a mini-review. Green Chem. 8, 325–327.

CHAPTER

3

Bioethanol Production From Corn and Wheat: Food, Fuel, and Future

Sujit K. Mohanty, Manas R. Swain***

*Iowa State University, Ames, IA, United States

**DBT-IOC Center for Advance Bio-energy Research, Faridabad, India

3.1 INTRODUCTION

Corn and wheat are grown and utilized not only as food and feed, but also as feedstocks for generation of renewable fuel ethanol. Production of fuel ethanol through biological fermentation of sugars extracted from sugar-rich crops (such as sugarcane) and starchy crops (such as corn and wheat) is a technically matured and commercially successful story, while those from lignocellulosic materials are still in early developmental/trail phases. For instance, global leaders in fuel ethanol production, such as the United States and Brazil produce about 15.25 billion gal (~57.7 billion L) and 7.3 billion gal (~27.6 billion L) of fuel ethanol annually primarily from starchy feedstocks, such as corn and wheat, and sugar-rich feedstock, such as sugarcane, respectively (Table 3.1). This clearly indicates the pivotal role these starchy and sugar-rich feedstocks play, particularly corn and wheat, in the global fuel ethanol production scenario.

In addition to United States and Brazil, in past few decades, many other countries around the globe are gradually emerging as global players in renewable fuel ethanol production sector by using starchy and sugar-rich feedstocks. For example, countries, such as China and Canada, are producing ~845 million gal (~3.2 billion L) and ~436 million gal (1.65 billion L) of fuel ethanol, respectively, from various starchy feedstocks, such as corn, cassava, wheat, and rice (Table 3.1), while countries, such as India, France, Germany, and Australia, are producing about 1 billion L, 1 billion L, 750 million L, and 500 million L, respectively, primarily from sugar-rich feedstock, such as sugarcane, molasses, sugar beet, and wheat (RFA, 2017). Thus, it is evident that corn and wheat are not only the top choices across the globe as the first-generation feedstocks for bioethanol production, but are also expected to remain so for decades to come. Therefore, it is becoming increasingly

Bioethanol Production From Food Crops
http://dx.doi.org/10.1016/B978-0-12-813766-6.00003-5

TABLE 3.1 World Fuel Ethanol Production by Country or Region (in Million Gallons)

Country	2007	2008	2009	2010	2011	2012	2013	2014	2015	2016
World	13,123	17,644	20,303	23,311	22,404	21,812	23,429	24,570	25,682	26,504
USA	6521	9309	10,938	13,298	13,948	13,300	13,300	14,300	14,806	15,250
Brazil	5019	6472	6578	6922	5573	5577	6267	6190	7093	7295
Europe	570	734	1040	1209	1168	1179	1371	1445	1387	1377
China	486	502	542	542	555	555	696	635	813	845
Canada	211	238	291	357	462	449	523	510	436	436
Rest of the world	315	389	914	985	698	752	1272	1490	1147	1301

Source: Data taken from Renewable Fuels Association (www.ethanolrfa.org/wp-content/uploads/2017/02/Ethanol-Industry-Outlook-2017.pdf) and the US Department of Energy Alternative Fuels Data Center (www.afdc.energy.gov/data/).

important to not only understand the global production scenario of these starchy feedstocks, but also the biotechnological processes developed so far and the socioeconomic issues involved in their utilization to fully comprehend the current global perspective of renewable fuel ethanol generation along with its future implications. This chapter primarily focuses on these aspects of corn and wheat-based fuel ethanol production.

3.2 CORN AND WHEAT-BASED ETHANOL PRODUCTION: GLOBAL SCENARIO

Global production of corn is estimated to be about 40.8 billion bushels in the year 2016–17 (Fig. 3.1). This corresponds to approximately about 1.04 billion metric tons of corn (given 1 bushel approximately equals to 56 pounds or say 1/40th of a metric ton). Among all the major

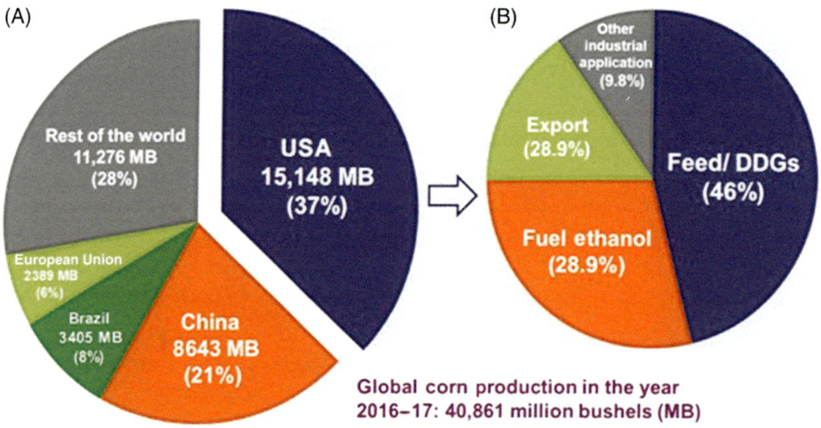

FIGURE 3.1 (A) Major corn producing countries in the world with their total production in 2016–17 in million bushels (MB) and its share (in percentage) in total global production. (B) Distribution of corn usage in USA in various application sectors. *Source: Data taken from (A) US Department of Agriculture, Foreign Agricultural Service (USDA FAS), Grain: World Markets and Trade, January 12, 2017. (B) US Department of Agriculture, Economic Research Service (USDA ERS), Feed Outlook, January 17, 2017. ProExporter Network, Crop Ending, August 31, 2017.*

corn producing countries, the United States tops the list with a major share of ~37%, which is little more than one-third of total corn production in the world, followed by China and Brazil with ~21.2% and 8.3%, respectively (Fig. 3.1).

Global corn production has sharply increased in the past few decades owing to its newly found applications in renewable liquid fuel (bioethanol) production industry (Table 3.2). Consequently, the global fuel ethanol production has also increased significantly in recent years; in fact, it has almost doubled between the years 2007 and 2016 (Table 3.2). If we particularly analyze the fuel ethanol production scenario in the United States, which produces ethanol primarily from corn, the overall ethanol production has almost tripled during these 10 years (Table 3.1). In the year 2016–17, the United States alone used about 28.9% of its total corn harvest (Fig. 3.1), which is about 4.38 billion bushels of corn, to produce over 15 billion gal of ethanol fuel (Table 3.1). The same rapid growth in ethanol production is also observed in China and other countries that primarily use corn and wheat as feedstock for ethanol production. This clearly indicates that there is not only a rapid growth in global corn production but also its utilization as feedstock in fuel ethanol industry across the globe with the United States emerging as the leader in this sector.

The United States is the top ethanol producer in the world with 15,250 million gal of fuel ethanol produced according to 2016 estimations. This account for about 58% of overall global fuel ethanol production (Fig. 3.2), which was estimated to be about 26,504 MG. Interestingly, most of this fuel ethanol comes from cornstarch as feedstock (Fig. 3.2). To be precise, almost 95% of the total fuel ethanol produced in the United States in the year 2016, which is about 14,490 MG, came from cornstarch feedstock and a minor section from wheat (3%). This implies that the United States tops the global chart not only in corn production, but also in fuel ethanol production because of its choice of using corn and wheat as feedstock for ethanol production. The availability of low-cost feedstock in sufficient quantities year-round is always a great advantage for any biotechnological process economically and the United States is a good example of that.

Brazil is the second largest producer of fuel ethanol in the world with 27% of total global

TABLE 3.2 Global Corn Production and Yield From 2015–17 (in Million Metric Tons)

	Production (million metric tons)			Yield (metric tons per hectare)		
Country/region	**2015/16**	**2016/17**	**2017/18 projected**	**2015/16**	**2016/17**	**2017/18 projected**
World	969.49	1070.51	1036.90	5.43	5.83	5.70
USA	345.51	384.78	362.09	10.57	10.96	10.72
China	224.63	219.55	215.00	5.89	5.97	6.14
Brazil	67.00	98.50	95.00	4.19	5.61	5.37
European Union	58.75	61.14	61.60	6.35	7.11	7.03
Argentina	29.00	41.00	40.00	8.29	8.37	8.16
Mexico	25.97	27.40	25.00	3.60	3.65	3.50
India	22.57	26.00	25.00	2.56	2.71	2.63
Rest of the world	196.06	212.14	213.21	—	—	—

Source: Data taken from Foreign Agricultural Services/USDA Office of Global Analysis. For a more detailed table of data see https://apps.fas.usda.gov/psdonline/circulars/production.pdf.

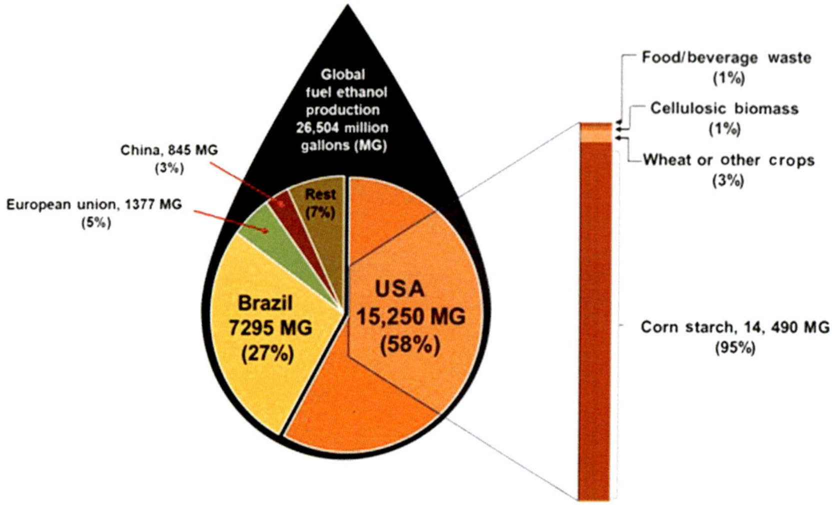

FIGURE 3.2 **Schematic showing global fuel ethanol production in 2017 and its distribution among major ethanol producing countries in the world with their total production in million gallons (MG) and share in global production (%).** *Right*: Major sources of starchy feedstocks for fuel ethanol production in USA. *Source: Data taken from Renewable Fuels Association (www.ethanolrfa.org/wp-content/uploads/2017/02/Ethanol-Industry-Outlook-2017.pdf) and the US Department of Energy Alternative Fuels Data Center (www.afdc.energy.gov/data/).*

production, which accounts to about 7295 MG of ethanol (Table 3.1). However, the major source for fuel ethanol is sugar-rich feedstock sugarcane and to a less extent from starchy crops like corn and wheat (Chum et al., 2014). Furthermore, it is interesting to note that though China shares 21% of the global corn production (Fig. 3.1) and uses corn to produce fuel ethanol, but it contributes only 3% to the global fuel ethanol production (Fig. 3.2).

3.2.1 Advantages of Using Corn as Feedstock: One Source, Many Products

Being a starchy crop with high amounts of fiber, proteins, and oils, corn has found its application not only in food and feed industry, but also in production of bulk chemicals, such as fuel ethanol, fibers, and sweeteners. A thorough analysis of the chemical composition of a shelled corn reveals that it is composed of starch (72%), fiber (9.5%), protein (9.5%), and oils (4.3%) (Mosier and Illeleji, 2006).

Thus, corn being such a widely popular and highly successful agricultural produce worldwide, and with such high content of starch, fiber, protein, and oil, it has not only found usage as food, livestock rations, and feedstock for renewable fuel ethanol production, but also in many other industrial sectors responsible for production of diverse coproducts (Sharma et al., 2016), such as:

1. Food and feed: To produce high-fructose syrup (corn syrup), beverages, alcohols, table cornstarch, sweetener, and animal feed in the form of dry distillers grains (DDGs) and salsa made from by-products.
2. Polymers: For production of polylactic acid polymers, which are sustainable versions of fibers and plastics, biodegradable styrofoam, and so on.

It is estimated that, one bushel (i.e., 56 pounds) of corn is known to provide up to 31.5 pounds of cornstarch or 33 pounds of sweeteners (high fructose corn syrup), or 22.4 pounds of polylactic acid fibers/polymers, or alternatively, one bushel of corn approximately provides 2.8 gal of ethanol along with about 17.5 pounds of DDGs via dry-milling process, or 13.5 pounds of gluten feed (with 20% protein), 2.6 pounds of gluten meal (60% protein), and 1.5 pounds of corn oil via wet-milling process (http://www.worldofcorn.com/#one-bushel-of-corn-can-provide). This rough estimate helps to quantitatively understand the significance of corn as a commercially relevant feedstock with the possibility of obtaining diverse products from a single source.

3.2.2 Global Wheat Production and Usage

Unlike corn, global production of wheat, which is estimated to be about 737.83 million metric tons in the year 2017, has almost remained stagnant over the past 3–4 years without much fluctuation (Table 3.3). The European Union tops the global list in wheat production with an average of about 150 million metric tons of wheat per year, but if we look at individual countries, China tops the list followed by India and Russia with an average of about 130, 90, and 70 million metric tons, respectively (Table 3.3).

Unlike corn, which is mostly produced in the United States and China, wheat production is well distributed across the globe and is mostly dominated by Asian and Europeans countries (Table 3.2). Plus, upon further investigation, it was found that, in the case of wheat production, the average yield is only about 3.4 metric tons/ha (Table 3.3), which is almost half of that of corn, which is about 5.7 metric tons/ha (Table 3.2). This means global wheat productivity (i.e., yield per hectare) is comparatively much lower than that of corn. This is probably the reason why, between the two agriculturally produced starchy feedstocks, corn is preferred over wheat as the feedstock of choice for fuel ethanol production in the United States and other parts of the world. Also, in many parts of the world, wheat is traditionally cultivated as a staple food primarily for human consumption and to a lesser extent as animal feed. I, and its utilization as a preferred feedstock for fuel ethanol production has not gained enough momentum the way it has happened in case of corn, particularly in last few decades (RFA, 2017).

TABLE 3.3 Global Wheat Production and Yield From 2015 to 2017 (in Million Metric Tons)

	Production (million metric tons)			Yield (metric tons per hectare)		
Country/region	**2015/16**	**2016/17**	**2017/18 projected**	**2015/16**	**2016/17**	**2017/18 projected**
World	736.97	755.00	737.83	3.27	3.39	3.34
European Union	160.48	145.70	150.00	5.98	5.33	5.63
China	130.19	128.85	130.00	5.39	5.33	5.37
India	86.53	87.00	96.00	2.75	2.88	3.13
Russia	61.04	72.53	72.00	2.39	2.69	2.64
USA	56.12	62.86	47.89	2.93	3.54	3.10
Canada	27.59	31.70	28.35	2.88	3.57	3.15
Australia	24.17	35.11	23.50	1.89	2.73	1.87
Rest of the world	190.85	191.25	190.09	–	–	–

Source: Data taken from Foreign Agricultural Services/USDA Office of Global Analysis. For a more detailed table of data see https://apps.fas.usda.gov/psdonline/circulars/production.pdf.

3.3 USA—THE GLOBAL LEADER IN FUEL ETHANOL PRODUCTION PREFERS CORN

To counteract global warming, the USA's Renewable Fuels Standards (RFS) program is planning to replace at least 20% its transportation fuel with renewable biofuels by 2022 (Schnempf and Yacobucci, 2013). Towards this goal, in 2016, the United States alone produced a record highest of 15.25 billion gal of renewable fuel ethanol by cumulative effort of 200 ethanol biorefineries operating in 28 different states. This corresponds to 58% of total amount of fuel ethanol produced globally (Fig. 3.2). The leading state among the 28 states in the United States is Iowa with 43 operating biorefineries, which produces about 4 billion gal of fuel ethanol per year, which is equal to 25.8% of total fuel ethanol produced in the United States in 2016, followed by Nebraska: 26 biorefineries with production of 2.13 billion gal (13.7%), and Illinois: 15 biorefineries with 1.75 billion gal (11.2%) (RFA, 2017). However, it is interesting to note that these states are also the major producers of corn in the United States. This indicates that US ethanol industry, which produces about 60% of global fuel ethanol, is directly dependent on agricultural production of corn (Fig. 3.1), which is about one third of total corn produced in the world (Table 3.2). Choice of corn as the feedstock for ethanol production has worked in advantage of US in becoming the global leader in fuel ethanol production.

3.4 TECHNOLOGICAL ASPECTS OF ETHANOL PRODUCTION FROM CORN

Fuel ethanol is most commonly produced by fermenting cornstarch by yeast, which converts sugars from corn kernels into ethanol. In fact, the current state of art of this well-established process is so efficient that when the best ethanol-producing yeast is used, it can turn almost all the sugar (>95%) fed to them directly into ethanol (Gulati et al., 1996). However, it is important to note that the final ethanol yield is not only dependent on the process parameters and its constraints but also on the quality and the variety of the corn or wheat grain (Singh, 2012). For example, a study by Sosulki and Sosulki (1994) suggest that the ethanol yield can vary somewhere between 3% and 23% depending on the variations in corn grain quality in terms of its kernel composition, endosperm hardness, planting location, and the presence of mycotoxins. They also found that if corn kernels contain high free sugar then it decreases enzyme consumption during saccharification thereby resulting in higher ethanol yield.

3.4.1 Corn Processing for Ethanol Production: From Farm to Fermentation

Corn undergoes many preprocessing steps before it can be considered ready for fermentation for ethanol production. However, as a basic first step postharvest, corn is shelled to remove kernels from the cob, followed by separation from impurities, such as stones and sticks by screeners or scalpers before getting stored in silos. Subsequently, a commercially well-established large-scale biotechnological process is employed (Sharma et al., 2016) via three broad steps: (1) conversion of starchy feedstocks into fermentable sugars via three major sequential unit operations namely milling, liquefaction, and enzyme-based saccharification, followed by (2) fermentation, where yeast metabolically converts these sugars into ethanol, and ultimately (3) purification, where the ethanol, thus generated, is separated out from other byproducts and impurities by distillation before it gets stored or transported to market (Fig. 3.3).

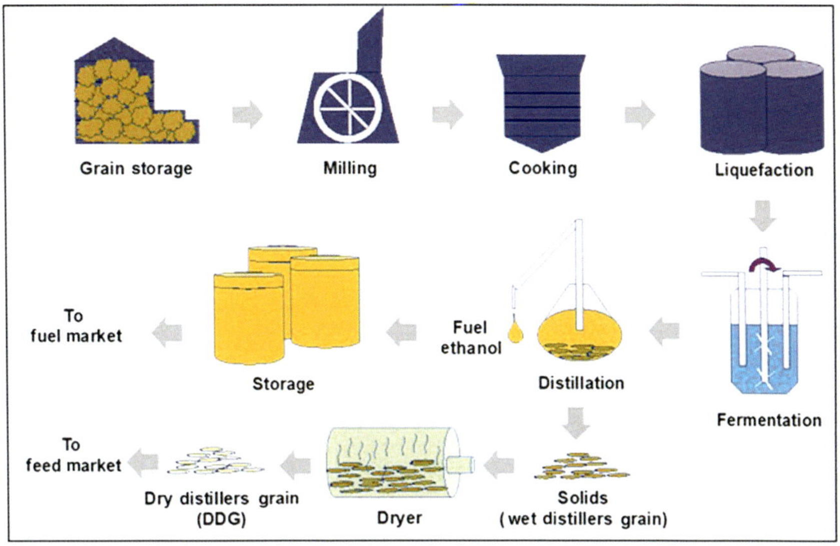

FIGURE 3.3 **A schematic diagram showing the fuel ethanol production process flow with various unit operations starting from grain storage, milling, cooking, liquefaction, saccharification, fermentation, and distillation to fuel ethanol separation and supply to market and generation of animal feed (DDGs) production.**

3.4.2 Corn Grain Milling, Liquefaction and Saccharification, and Animal Feed Generation

Milling is the very first unit operation in the biotechnological process described earlier. Based on how the grain is milled for ethanol production, this step is categorized into two methods; namely, wet milling and dry milling. Broadly defined, if the grain is first soaked in water to fractionate it into its individual components, such as starch, fiber, and germ, which are then processed separately, then it is called the wet-milling process, whereas if the whole grain and the residual components are separated at the end instead of the beginning of the process then it is called as the dry-milling process (Fig. 3.4). Thus, it is important to note that although both milling processes involve breaking down the starch present in the corn kernel into simple sugars for further processing, such as fermentation for ethanol production and distillation, the primary difference between the two milling processes is whether the entire kernel is processed as a whole (dry milling) or the corn kernel is first broken down into its individual components (i.e., germ, fiber, gluten, and starch) and then sent for processing (wet milling) (Saville et al., 2016).

Both the milling processes have their respective advantages and disadvantages (Saville et al., 2016). While the wet-milling process results in production of a number of coproducts, such as gluten feed/gluten meal, food-grade corn oil, and distillers' grains with solubles (DGS), it could separate individual components of corn grain prior to processing as described earlier, the dry-milling processes usually produce only one primary coproduct; that is, DGS, which is then used as animal feed either wet (WDGS) or dry (DDGS) (Sharma et al., 2016). However, surprisingly, in the United States, wet-milling–based ethanol production plants are far fewer in number compared to that of dry-milling–based ethanol production plants. For example,

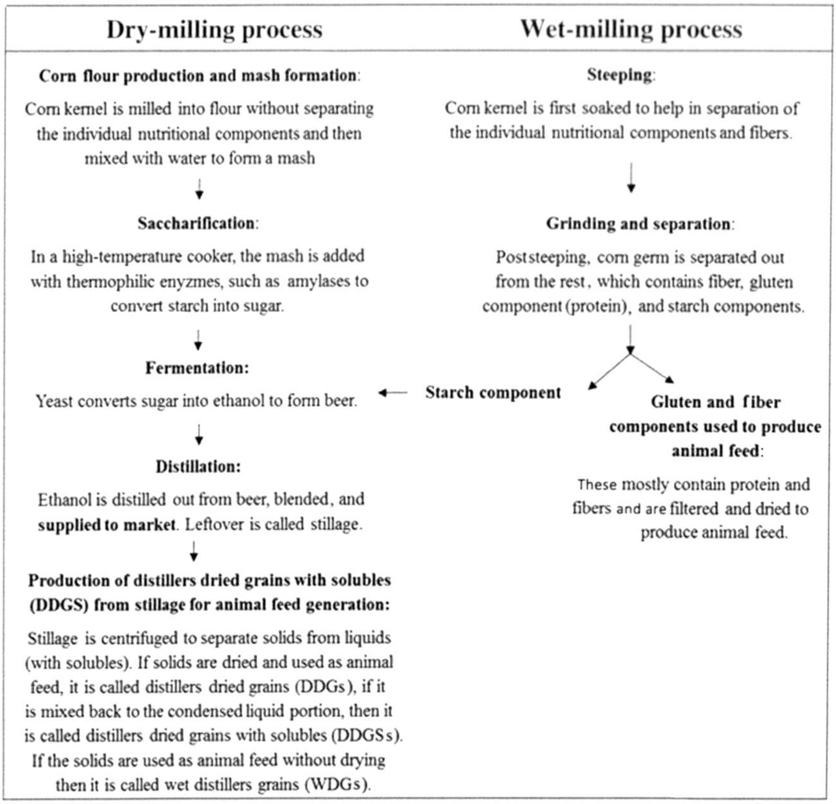

FIGURE 3.4 **A flow chart showing various steps of the dry-milling and the wet-milling process using ethanol production from corn and its coproduct (animal feed) generation.** *Source: Data taken from Saville, B.A., Griffin, W.M., MacLean, H.L., 2016. Ethanol production technologies in the US: status and future developments. In: Salles-Filho, S.L.M., Cortez, L.A.B., da Silveira, J.M.F.J., Trindade, S., da Graça Derengowski Fonseca, M. (Eds.), Global Bioethanol Evolution, Risks, and Uncertainties. Academic Press, pp. 163–180. Chapter 7; Sharma, A., Sharma, S., Verma, S., Bhargava, R., 2016. Production of biofuel (ethanol) from corn and co product evolution: a review. Int. Res. J. Eng. Technol. 3, 745–749.*

as per the data published by US Department of Agriculture in 2009, out of 180 corn-based ethanol production plants, only 11 were operating as wet-milling–based ethanol-producing plants and the rest of them, which is around 173, were operating as dry-milling–based ethanol-producing plants (RFA, 2017). In other words, about 90% of the total fuel ethanol produced from corn grain is generated via dry-milling process while the rest 10% is generated via wet-milling process (RFA, 2017). In a typical dry-milling process, enzymatic hydrolysis of cornstarch to simple fermentable sugars is accomplished via two-steps: (1) liquefaction/cooking and (2) saccharification (Mojovic et al., 2006).

Liquefaction is basically a high-temperature cooking step where with time the tightly bound grain starch becomes soft and spongy and becomes amenable to efficient enzymatic digestion. For example, a typical industrial scale liquefaction step would be cooking the ground corn mash either at 165°C for 3–5 min (very high temperature cooking), or 90–105°C for 1–3 h (Robertson et al., 2006). This step is followed by the saccharification step where enzymatic digestion of starch is initially performed at higher temperatures by incubating with thermostable alpha-amylase resulting in production of dextrins, subsequently followed by a lower temperature incubation, which is typically at about 55–60°C (Lamsal et al., 2011) or even as low as 32°C (Mojovic et al., 2006) with an enzyme called glucoamylase to further convert the dextrins into glucose. The sugary syrup, thus produced (also known as corn hydrolysate), then proceeds to the fermentation step where yeast *Saccharomyces cerevisiae* converts sugars into ethanol (Azhar et al., 2017). The ethanol is then recovered via evaporation in a distiller for ethanol separation, purification, and rectification by an well established process called distillation.

Recently, modern pretreatment methods that use techniques such as ultrasound and microwave have been developed and employed in corn grain processing to improve the glucose concentration in the sugary syrup which is obtained after liquefaction and saccharification steps (Nikolic et al., 2011). A higher glucose concentration of 6.82% and 8.48% was obtained when ultrasonic and microwaves were employed, respectively. Consequently, higher ethanol concentration of 11.15% and 13.40%, with respect to the control (untreated) samples, were also obtained from application of ultrasonic and microwaves, respectively. Thus, microwave pretreatment seems to have worked better than ultrasound. Ultimately, using this new microwave pretreatment technique, a simultaneous saccharification and fermentation (SSF) method was developed for ethanol production from corn in batch mode that resulted in a maximum ethanol concentration of 9.91% (w/w) and percentage of theoretical ethanol yield of 92.27% (Nikolic et al., 2011).

3.4.3 Microorganisms Used in Fermentation Processes for Corn-Based Ethanol Production

A myriad of microorganisms are known to date that are capable of utilizing sugars for ethanol generation (Azhar et al., 2017). Yet, *S. cerevisiae* (also widely known as brewers' yeast) is still the preferred and is one of the most widely used for commercial fuel ethanol production from sugar and starchy feedstocks (Behera et al., 2010a,b). Starting from ancient wine making to current modern day high-tech commercial-scale ethanol fermentation processes, *S. cerevisiae* has dominated this sector as the biocatalyst of choice, mainly because of the advantageous attributes it has acquired during natural evolution process over other ethanol producing microorganisms (Azhar et al., 2017; Behera et al., 2010a,b). These attributes are:

- Extremely high ethanol tolerance (Azhar et al., 2017).
- Can grow under stringent anaerobic conditions (Visser et al., 1995).
- Is least affected in the presence of oxygen (Lagunas, 1979).

The best-known strains of *S. cerevisiae* isolated so far and used industrially can convert almost ~95% of the sugar directly into ethanol (Gulati et al., 1996). *S. cerevisiae* is naturally adapted to ethanol fermentation with a very high tolerance to ethanol (150 g/L) (Claassen et al., 1999) and chemical inhibitors compared to other well-known natural ethanol producing organisms, such as *Zymomonas mobilis* (Table 3.4).

On the other hand, both *S. cerevisiae* and *Z. mobilis* have a very limited spectrum of sugar-utilizing capability for ethanol production (Table 3.4). Interestingly, neither of these naturally occurring organisms can use xylose, which is a widely available sugar from plant biomass. Fortunately, both these organisms are well studied both genetically and metabolically and so, both are complaisant to genetic modifications and consequently, engineering of their metabolic capabilities are possible. Thus, attempts have been made to overcome these above mentioned limitations of *S. cerevisiae* and *Z. mobilis* via metabolic engineering techniques (Ha et al., 2011).

TABLE 3.4 Comparative Analysis of *Saccharomyces Cerevisiae* Versus *Zymomonas mobilis*

Microorganism	Sugars utilized	Fermentation conditions	Ethanol tolerance	Other attributes	References
Saccharomyces cerevisiae	Glucose, fructose, maltose, and sucrose	Anaerobic, 30–37°C	150 g/L	Up to 95% of theoretical yield	Claassen et al. (1999)
Zymomonas mobilis	Glucose, fructose, and sucrose	Anaerobic, 30°C	100 g/L	Up to 97% of theoretical yield	Claassen et al. (1999)

Similarly, at process engineering level, options are available to either use immobilized yeast for ethanol production (Behera et al., 2010b) or use novel fermetation techniques, such as solid-state fermentation (Mohanty et al., 2009), which has shown promising results with other feedstocks. Unfortunately, they have their intrinsic drawbacks, both in terms of process economics and yeast biology, that prevents them from getting readily acceptable in industrial corn-ethanol production process (Bai et al., 2008).

Furthermore, a comparison between these two naturally ethanol producing strains (*S. cerevisiae* and *Z. mobilis*) reveals that bioprocesses with *Z. mobilis* results in higher ethanol yield and productivity (as high as 97% for theoretical yield) compared to *S. cerevisiae* (Bai et al., 2008) where it is at best 90%–95% of theoretical yield (Table 3.4). However, it is interesting to note that *S. cerevisiae* is still the preferred whole-cell biocatalyst of choice for ethanol production from corn and other starchy feedstocks (and not *Z. mobilis*) because of these two important advantages it has over *Z. mobilis*: (1) wide spectrum of sugar substrate utilizing capability (Table 3.4) and (2) the undesirability of *Z. mobilis* biomass to be used as animal feed (Bai et al., 2008).

3.4.4 Animal Feeds, an Economically Beneficial By-Product of Corn-Based Ethanol Production

In 2016, the United States, along with production of 15.2 billion gal of fuel ethanol, also produced about 42 million metric tons of high-protein animal feed from its 200-plus ethanol biorefineries operating in 28 different states (RFA, 2017), which is then fed to animals in meat and poultry (cattle, swine, and chicken), dairy, and fish industries across the globe. It is estimated that about one half of the total corn produced in the United States (46%) is used as animal feed (Fig. 3.1). In fact, it is estimated that about 30% of each bushel of corn that gets utilized in ethanol production comes back to market as animal feed in the form of DDGS, WDG, DGS, corn gluten feed, corn gluten meal, and so on (Wood et al., 2014). This puts in perspective the significance of this high-value by-product of corn-based ethanol industry and the enormous contribution it makes to the economics of this ethanol production technology, but unfortunately often overlooked or not fully appreciated in public debates, discussions, and mainstream media reports. In fact, in the past decade, 2006–16, while ethanol production in the United States has almost tripled (Table 3.1), it is also estimated that the animal feed production has also grown by a factor of eight (RFA, 2017).

Furthermore, animal feed is not the only by-product of corn-ethanol industry that has poised itself as an economically addendum, but many other by-products have also gained popularity in the market lately (Sharma et al., 2016). For example, corn distillers oil, a by-product of corn-ethanol industry has proven itself as a high value coproduct by its application in the biodiesel industry as a feedstock with an estimated market value of US$900 million. Thus, without any confusion we can easily understand the multifold economic advantages the corn-based fuel ethanol production technology has, and thus, it is not only emerging as a promising technology to meet the renewable fuel demands and mandates successfully in the years ahead, but also well-poised to address the growing global demand for food and feed.

3.5 TECHNOLOGICAL ASPECTS OF ETHANOL PRODUCTION FROM WHEAT

Apart from corn, wheat is the second most widely used starchy feedstock of choice for fuel ethanol production (Patni et al., 2013) and preferred over barley as the cereal feedstock of choice by grain distilleries for ethanol production. Recently, Terra Grains has opened North America's largest wheat-based ethanol producing facility in Belle Plaine, Saskatchewan, Canada with a production capability of 40 million gal (or say, 150 million L) of fuel ethanol and approximately 164,000 tons of DDGs annually. This facility procures about 15 million bushels of wheat annually to meet this ethanol production capacity (http://www.terragrainfuels.com/). This implies that, as per the current best-known wheat-based ethanol production technology employed commercially at industrial scale, one bushel of wheat can provide about 2.6 gal of fuel ethanol and 22 pounds of DDGs via dry-milling process as its best operating conditions (as compared to 2.8 gal of ethanol and 17.5 pounds of DDGs in case of corn; see Section 3.2.1). This means that the ethanol yield from wheat is slightly lower than that of corn while the DDGs production is comparatively higher and it also justifies why corn is preferred over wheat as the feedstock of choice for bioethanol production.

3.5.1 Grain Processing and Ethanol Production: Recent Advances

As wheat is also a starchy feedstock just like corn, the process employed to produce ethanol from wheat is almost the same as in the case of corn, starting from grain processing (milling) to fermentation and distillation (Fig. 3.3). As this is described elaborately in an earlier section, it will not be described here again to avoid redundancy. Concisely describing, wheat grains undergo five basic unit operations (1) milling (mechanical grinding to release starch components), (2) cooking and saccharification (heat, water, and enzymes alpha-amylase and glycoamylase added), (3) fermentation (by yeast where sugars are converted to ethanol), (4) distillation and rectification (ethanol separation from fermentation broth), and finally (5) drying/dehydration (Patni et al., 2013). Plus, just like in the case of corn, dry milling is also the most popular mode of operation in the case of wheat because it helps in barn separation from grain and increased starch content in the wheat flour, which ultimately results in higher ethanol titer at the end of the process. For example, Sosulki and Sosulki (1994) reported that a maximum concentration of 344–367 L ethanol per ton of wheat flour can be obtained with a maximum ethanol concentration of 15%–15.89% (v/v) when wheat flour from dry milling is used for ethanol production. In another such study, under very high gravity conditions, a maximum ethanol titer of 23.8% (v/v) was obtained, which is much higher than that obtained under normal process conditions (Thomas et al., 1993).

A typical small-scale ethanol production process is described by Patni and co-workers (2013), in which the wheat grain is cleaned and grounded to make wheat flour, which is then mixed with 250 mL of water and 0.416 gm of amylase at 70°C for 45 min followed by addition of 0.1 mL of glucoamylase at 55–60°C and incubation for 30 min. This constituted the first two steps, milling and saccharification where wheat starch is extracted and separated from barn and converted into fermentable sugars. Next about 5 g of *S. cerevisiae* (yeast) is added to the sugary syrup and left for up to 75 h in well stirred and ambient temperature condition for ethanol production which is then transferred to distillation column system to separate ethanol from fermentation mash at 80°C.

Subsequently, rectification and dehydration process virtually removes all the water from the purified alcohol resulting in up to 99.7% purity (Patni et al., 2013). However, at industrial scale more sophisticated methods are incorporated into this basic ethanol production process as described later.

Recently, various technological improvements are developed in this traditional wheat-grain to ethanol process by various wheat-based ethanol producing industries which has ultimately resulted in better process economics, which in turn makes this process more commercially viable and at par with the corn-based technology. For example, one of the largest agricultural feedstock-based ethanol producing companies, POET, LLC in South Dakota, USA, has developed and commercialized a new method called the cold-cook (BPX) technology (Saville et al., 2016) to replace the high temperature jet cooking process that is typically used during the liquefaction and saccharification step (Fig. 3.3). This novel improvement not only helps in saving thermal energy, but also helps in avoiding enzyme inactivation that typically occurs in a high-temperature jet cook operation (Saville et al., 2016). However, to compensate for the loss in efficiency that might have incurred due to this reduced temperature operation in the cold-cook technology, a more aggressive milling process is employed along with addition of various types of amylases that break down the starch more efficiently. Ultimately, the benefit of this technology is not only thermal energy saving, but also saving the activity of the enzymes added to the slurry tank, because of which the second enzyme addition to the liquefaction system is no more required (Saville et al., 2016).

3.6 SOCIOECONOMICAL ADVANTAGES AND FOOD VERSUS FUEL DEBATE

Renewable fuel (bioethanol) production from agricultural feedstocks has resulted in a positive environmental effect by significantly reducing the carbon footprint via carbon recycling. As per the estimations of the US Department of Energy, when one gallon of gasoline is replaced with an energy-equivalent amount of bioethanol the greenhouse gas emissions reduces only by ~20% (Ethanol Myths and Facts, 2013). Furthermore, a recent study by Belboom et al. (2015) reported that as high as 42.5%–61.2% of reduction in greenhouse gas emissions is possibly achieved when 1 MJ bioethanol produced from wheat is consumed instead of 1 MJ gasoline. However, despite the efficiency of corn-based ethanol production, the entire process is intrinsically limited by how much energy must be put into producing the corn in the first place. When corn is grown for producing ethanol, several cultural and agronomic practices are to be followed, such as tilling the land, sowing, and applying fertilizers, harvesting and transporting the corns to ethanol plants (Feng et al., 2010). The more energy used in these processes, the less energy and greenhouse gas emissions are saved from replacing gasoline with fuel ethanol. For example, as per USDA estimates, corn-based fuel ethanol yield only ~2.3 times the amount of energy that goes into producing it (Shapouri et al., 2010). In other words, total energy stored in corn-based fuel ethanol and its coproducts is only 2.3 times of the total energy used up during agricultural production of required feedstock (e.g., production of fertilizers, running farm equipment) as well as during the ethanol production and transportation of the final products.

Nevertheless, it has also been demonstrated that significant social and economic benefits can be achieved by incorporating renewable fuel into our economic system, primarily via generation of revenue and jobs. For example: with the United States as a case study, it is estimated that, in 2016 alone, with production of 15.25 billion gal of ethanol (Fig. 3.2) and 42 million metric tons of coproducts, 74,420 direct jobs and 264,756 indirect jobs are created, which cumulatively made US$42 billion contribution to the nation's GDP with US$23 billion in household income and US$9 billion in tax revenue

(RFA, 2017). This is an excellent example of showing how the renewable fuel industry can not only help in environment protection and climate control, but also contribute to a nation's socioeconomic welfare and development, least to mention the added advantages of energy security and energy independency that a nation gains. Due to all these advantages, many other developing/developed economics around the globe are now increasingly adopting renewable biofuel production as a replacement to traditional petrochemical-based fuel production.

3.6.1 Food Versus Fuel Debate: Food Price and Security

Although energy security concerns have pushed developed economies toward adopting sustainable green technologies, it has also raised a long-standing debate of whether corn is better used as food or as a fuel source (Tenenbaum, 2008). Recent reports have shown that utilization of food crops for renewable fuel production can come with hidden economic and environmental costs that are either not considered fully yet or aren't understood in great detail (Wise, 2012). In a recent study, these issues have been quantified and compared in terms of economics of the entire production system to determine if the beneficial effects of corn-biofuel production outweigh the costs (Richardson and Kumar, 2017). Interestingly, it was found that, utilization of corn for production of biofuel is possibly having a much higher negative environmental impact than when the same is grown and utilized for food. Similarly, bioethanol production from food crops has also been associated with negative impacts on food security, which is one of the biggest concerns raised about using food crops for producing fuel (Tenenbaum, 2008).

Furthermore, reports suggest that there isn't much land left in the United States to increase its corn production and using previously unused lands could increase greenhouse gas emissions by releasing carbon stored in them (Wise, 2012). Thus, increased use of corn for biofuel production may result in further increase in corn and food prices (Schnempf and Yacobucci, 2013). Similarly, some reports have suggested that producing ethanol from feedstocks other than corn, such as sugarcane, might require much less energy and further reduce greenhouse gas emissions (Machado et al., 2017). For instance, ethanol produced from sugarcane in Brazil yields ~8 times the energy required to produce it and lowers greenhouse gas emissions by >50% when used to replace gasoline (Chum et al., 2014). However, unlike in Brazil, the United States climate is not suitable for growing sugarcane farming, and hence, corn is likely to continue as the major feedstock in the United States in coming decades. Moreover, land acquisition for biomass/feedstock sourcing has not only been reported of people's rights and livelihoods, but also has resulted in the change of natural habitats and loss of biodiversity (Pilgrim and Harvey, 2010). Particularly, in developing countries, monitoring land acquisition is often an issue and often unreliable given the major disparities that are observed between large companies and local people (Van Derhorst and Vermeylen, 2011). Upon that, land usage also depends on different communal or cultural value sets in different geographical locations (Thornley and Gilbert, 2013), as well as the uneven distribution of climate change impacts and its local/global effects also needs to be considered in all parts of the world.

3.7 CONCLUSION AND FUTURE PERSPECTIVES

Bioethanol from corn and wheat is generally derived from starch and sugars. It is considered as the simplest source for bioethanol compare to other sources. In industrial scale, corn is considered as the primary source of the world's fuel ethanol and mostly comes from the United States. Bioethanol from wheat is not popular compared to corn as it is considered as a major staple food in many parts of the world.

But corn, with its wide range of advantages over other energy crops, is emerging as the most promising source for the bioethanol production. Agricultural practices for the corn are quite well established and its high starch content provides suitable substrates and makes the process easier to produce ethanol. Corn is also considered as the potential source to supply one fourth of US gasoline consumption. There are no direct land-use costs with the corn cultivation. So, corn can be considered as one of the most potent feedstock sources in the bioethanol production industry. However, corn and wheat have considerable drawbacks as they are both used in the food chain and at present there is a renewed search for newer alternatives for fuel feedstock.

References

Azhar, S.H.M., Abdulla, R., Jambo, S.A., Marbawi, H., Gansau, J.A., Faik, A.A.M., Rodrigues, K.F., 2017. Yeasts in sustainable bioethanol production: a review. Biochem. Biophys. Rep. 10, 52–61.

Bai, F.W., Anderson, W.A., Moo-Young, M., 2008. Ethanol fermentation technologies from sugar and starch feedstocks. Biotechnol. Adv. 26, 89–105.

Behera, S., Mohanty, R.C., Ray, R.C., 2010a. Comparative study of bio-ethanol production from mahula (*Madhucalatifolia* L.) flowers by *Saccharomyces cerevisiae* and *Zymomonas mobilis*. Appl. Energy 87, 2352–2355.

Behera, S., Mohanty, R.C., Ray, R.C., 2010b. Comparative study of bio-ethanol production from mahula (*Madhucalatifolia* L.) flowers by the immobilized cells of *Saccharomyces cerevisiae* and *Zymomonas mobilis* in calcium alginate beads. J. Sci. Ind. Res. 69, 472–475.

Belboom, S., Bodson, B., Leonard, A., 2015. Does the production of belgian bioethanol fit with European requirements on GHG emissions? Case of wheat. Biomass Bioenergy 74, 58–65.

Chum, H.L., Warner, E., Seabra, J.E.A., Macedo, I.C., 2014. A comparison of commercial ethanol production systems from Brazilian sugarcane and US corn. Biofuel Bioprod. Bioref. 8 (2), 205–223.

Claassen, P.A.M., van Lier, J.B., Lopez Contreras, A.M., van Niel, E.W.J., Sijtsma, L., Stams, A.J.M., de Vries, S.S., Weusthuis, R.A., 1999. Utilisation of biomass for the supply of energy carriers. Appl. Microbiol. Biotechnol. 52, 741–755.

Ethanol Myths and Facts, 2013. US Department of Energy: Bioenergy Technologies Office. Available from: http://www1.eere.energy.gov/bioenergy/printable_versions/ethanol_myths_facts.html.

Feng, H., Rubin, O.D., Bruce, A., Babcock, B.A., 2010. Greenhouse gas impacts of ethanol from Iowa corn: life cycle assessment versus system wide approach. Biomass Bioenergy 34, 912–921.

Gulati, M., Kohlmann, K., Ladisch, M.R., Hespell, R., Bothast, R.J., 1996. Assessment of ethanol production options for corn products. Bioresour. Technol. 58, 253–264.

Ha, S., Galazka, J.M., Kim, S.R., Choi, J., Yang, X., Seo, J., Yang, X., Seo, J., Glass, N.L., Cate, J.H.D., Jin, Y., 2011. Engineered *Saccharomyces cerevisiae* capable of simultaneous cellobiose and xylose fermentation. Proc. Natl. Acad. Sci. USA 108, 504–509.

Lagunas, R., 1979. Energetic irrelevance of aerobiosis for *S. cerevisiae* growing on sugars. Mol. Cell. Biochem. 27, 139–146.

Lamsal, B.P., Wang, H., Johnson, L.A., 2011. Effect of corn preparation methods on dry-grind ethanol production by granular starch hydrolysis and partitioning of spent beer solids. Bioresour. Technol. 102, 6680–6686.

Machado, K.S., Seleme, R., Maceno, M.M.C., Zattar, I.C., 2017. Carbon footprint in the ethanol feedstocks cultivation: agricultural CO_2 emission assessment. Agricul. Systems 157, 140–145.

Mohanty, S.K., Behera, S., Swain, M.R., Ray, R.C., 2009. Bioethanol production from mahula (*Madhucalatifolia* L) flowers by solid-state fermentation. Appl. Energy 86, 640–644.

Mojovic, L., Nikolic, S., Rakin, M., Vukasinovic, M., 2006. Production of bioethanol from corn meal hydrolyzates. Fuel 85, 1750–1755.

Mosier, N.S., Illeleji, K., 2006. How fuel ethanol is made from corn. Bioenergy: Purdue Extension. Available from: https://www.extension.purdue.edu/extmedia/id/id-328.pdf.

Nikolic, S., Mojovic , L., Rakin, M., Pejin, D., Pejin, J., 2011. Utilization of microwave and ultrasound pretreatments in the production of bioethanol from corn. Clean Technol. Environ. Policy 13, 587–594.

Patni, N., Pillai, S.G., Dwivedi, A.H., 2013. Wheat as a promising substitute of corn for bioethanol production. Procedia Eng. 51, 355–362.

Pilgrim, S., Harvey, M., 2010. Battles over biofuels in Europe: NGOs and the politics of markets. Sociol. Res. Online 15 (3), 1–16.

RFA, 2017. Ethanol Industry Outlook. Renewable Fuels Association, Washington, DC. Available from: www.ethanolrfa.org/wp-content/uploads/2017/02/Ethanol-Industry-Outlook-2017.pdf..

Richardson, M., Kumar, P., 2017. Critical zone services as environmental assessment criteria in intensively managed landscapes. Earth's Futuredoi: 10.1002/2016EF000517.

Robertson, G.H., Wong, D.W.S., Lee, C.C., Wagschal, K., Smith, M.R., Orts, W.J., 2006. Native or raw starch digestion: a key step in energy efficient biorefining of grain. J. Agric. Food Chem. 54, 353–365.

Saville, B.A., Griffin, W.M., MacLean, H.L., 2016. Ethanol production technologies in the US: status and future developments. In: Salles-Filho, Sergio LuizMonteiro, Barbosa Cortez, Luis Augusto, Jardim da Silveira, José Maria Ferreira, Trindade, S., da Graça Derengowski Fonseca, M. (Eds.), Global Bioethanol Evolution, Risks, and Uncertainties. Academic Press, Cambridge, MA, pp. 163–180, Chapter 7.

Schnempf, R., Yacobucci, B.D., 2013. Renewable fuel standard (RFS): overview and issues. Congressional Research Service. Available from: www.ifdaonline.org/IFDA/media/IFDA/GR/CRS-RFS-Overview-Issues.pdf.

Shapouri, H., Gallagher, P.W., Nefstead, W., Schwartz, R., Noe, S., Conway, R., 2010. 2008 Energy Balance for the Corn-Ethanol Industry. United States Department of Agriculture, Agricultural Economic Report Number 846.

Sharma, A., Sharma, S., Verma, S., Bhargava, R., 2016. Production of biofuel (ethanol) from corn and co product evolution: a review. Int. Res. J. Eng. Technol. 3, 745–749.

Singh, V., 2012. Effect of corn quality on bioethanol production. Biocat. Agr. Biotechnol. 1, 353–355.

Sosulki, K., Sosulki, F., 1994. Wheat as a feedstock for fuel ethanol. Appl. Biochem. Biotechnol. 45, 169–180.

Tenenbaum, D.J., 2008. Food vs Fuel: Diversion of crops could cause more hunger. Environ. Health Perspect. 116 (6), A254–A257.

Thomas, K.C., Hynes, S.H., Jones, A.M., Ingledew, W.M., 1993. Production of fuel alcohol from wheat by VHG technology. Appl. Biochem. Biotechnol. 43, 211–226.

Thornley, P., Gilbert, P., 2013. Biofuels: balance in risks and rewards. Interface Focus 3 (1), 2042–8901.

Van Derhorst, D., Vermeylen, S., 2011. Spatial scale and social impacts of biofuel production. Biomass Bioenergy 35, 2435–2443.

Visser, W., van Spronsen, E.A., Nanninga, N., Pronk, J.T., GijsKuenen, J., van Dijken, J.P., 1995. Effects of growth conditions on mitochondrial morphology in *Saccharomyces cerevisiae*. Anton. Leeuw. 67 (3), 243-L253.

Wise, T.A., 2012. US corn ethanol fuels food crisis in developing countries. Available from: http://www.aljazeera.com/indepth/opinion/2012/10/201210993632838545.html.

Wood, C., Rosentrater, K.A., Muthukumarappan, K., 2014. Techno-economic modeling of a corn based ethanol plant in 2011/2012. Ind. Crops Prod. 5, 145–155.

Further Reading

Zabed, H., Faruq, G., Sahu, J.N., Boyce, A.N., Ganesan, P., 2016. A comparative study on normal and high sugary corn genotypes for evaluating enzyme consumption during dry-grind ethanol production. Chem. Eng. J. 287, 691–703.

CHAPTER

4

Status and Perspectives in Bioethanol Production From Sugar Beet

Cristina Marzo, Ana B. Díaz, Ildefonso Caro, Ana Blandino

University of Cádiz, Cádiz, Spain

4.1 INTRODUCTION

The chemical composition of sugar beet roots makes this raw material an attractive feedstock for ethanol fermentation. In the manufacture of sugar from this crop, various intermediates, by-products, and wastes are generated, which can be used for the production of energy and other value-added products, such as biofuels. In this chapter, the current processes for the conversion of the sugary juices (raw, thin, and thick juices), molasses and sugar beet pulp into bioethanol are described and analyzed. In addition, the economics of sugar beet ethanol and the principal strategies proposed for increasing bioethanol yields are described.

4.2 GLOBAL PRODUCTION SCENARIO OF SUGAR BEET

Sugar crops are those crops cultivated primarily for the manufacture of sugar and secondarily for the production of alcohol (food and nonfood) and ethanol. There are two main sugar crops: sugar beets (*Beta vulgaris* L.) and sugarcane (*Saccharum officinale* L.). Both crops constitute the two main sources of sucrose for human consumption; approximately 80% is obtained from sugarcane, which is largely grown in tropical and subtropical countries and the remaining 20% is extracted from sugar beet, which is grown mostly in countries with moderate climates in the northern hemisphere. Seventy-one countries produce sugar from sugarcane, 43 countries produce sugar from sugar beet, and 9 countries produce sugar from both (FAO, 2009).

One considerable disadvantage facing most sugar beet–growing countries is the very short growing season. Given the relatively short period of time during which sugar beets are harvested and available for processing, a beet sugar factory operates on average only 90 days a year. The sugar beet seeds are sown in spring and harvested in the autumn or early winter. However, an advantage over sugarcane is that its cultivation requires approximately four times less water. Also, sugar

Bioethanol Production From Food Crops
http://dx.doi.org/10.1016/B978-0-12-813766-6.00004-7

content of sugar beet is about 25% higher than that found in sugarcane, but its production cost is more than twice as high as the production cost of cane sugar. Nowadays, the revenue from the sale of the by-products, that is, molasses, pulp, beet particulate matter, and carbonation lime, significantly offsets the production cost of beet sugar. Moreover, mainly due to the financial support provided to sugar industry and its use to produce ethanol fuel, sugar beet cultivation is still increasing in many countries (Erdal et al., 2007).

Sugar beet is an annual crop, propagated by the seed of the flowers. In 2014, 269,714,066 tons of sugar beet was produced worldwide, with an estimated harvested area of 4,471,580 ha and a calculated yield of 603,174 kg/ha. The largest producing countries of sugar beet in terms of tons are France, Russia, Germany, and the United States (Table 4.1). Also, these countries have the largest production areas to grow sugar beet. However, in terms of yield, Spain, Chile, and France occupy the first positions (FAOSTAT, 2017).

It becomes clear from the Table 4.1 that most of the sugar beet–producing countries are located in Europe, followed by Russian Federation and the United States. In fact, Europe is the major producer of sugar, concentrated primarily in sugar beet processing, providing nearly 80% of the world production of sugar beet. Among the 100 sugar factories that are currently operating in the European Single Market, 96 factories specialize in sugar beet processing (ezbová et al., 2013). The others are combined processors and/or specialized in cane sugar processing. Most of the sugar-producing capacities are located in Germany, France, and Poland (Maitah et al., 2016).

The Russian Federation has made a determined push into the sugar market. Its production has tripled in the past decade when it grew by almost 12% a year. However, the sugar yields are 45% lower than that of Europe. This fact is mainly due to inadequate agricultural technology (cultivars, crop management, and machinery) and constrained access to investments (FAO, 2009).

The sugar extraction rate depends on the sugar content of the sugar beet at the moment of its arrival in the processing plant. European norms define the sugar beet as marketable if it contains 14% sugar or more. For example, one ton of standard sugar beet with a sugar content of 16% processed at a sugar plant with an efficiency of 82.5% would yield 130 kg of sugar (FAO, 2009).

TABLE 4.1 Largest Producing Countries of Sugar Beet

Country	Production (tons) (Position in the ranking)	Harvested area (ha) (Position in the ranking)	Yield (hg/ha) (Position in the ranking)
France	37,844,567 [1]	407,148 [3]	929,504 [3]
Russian Federation	33,513,369 [2]	905,406 [1]	370,147 [44]
Germany	29,748,100 [3]	372,500 [4]	798,607 [9]
United States	28,381,270 [4]	463,900 [2]	611,797 [21]
Turkey	16,743,045 [5]	287,461 [6]	582,446 [25]
Ukraine	15,734,050 [6]	330,200 [5]	476,501 [32]
The Netherlands	6,821,774 [11]	75,094 [13]	908,431 [5]
Spain	3,723,309 [18]	38,414 [21]	969,258 [1]
Switzerland	1,924,260 [24]	21,040 [30]	914,572 [4]
Chile	1,732,032 [25]	18,335 [31]	944,659 [2]

4.3 SUGAR BEETS AS RAW MATERIAL

Contrary to cereals, pulses, and root crops, the main component of sugar crops is not starch but simple monosaccharides (glucose and fructose) and particularly disaccharides (sucrose or saccharose). The protein and fat contents are negligible. Sugar is the prime component that gives value to the sugar beet crop and the by-products generated in its extraction (pulp and molasses) give a benefit up to 10% of the value of the sugar.

The root of the beet, called taproot, is composed of 75% water and 25% dry matter. The dry matter comprises about 5% pulp and sugars represent 75% of the total dry matter. The pulp is insoluble in water and is composed of cellulose, hemicellulose, lignin, and pectin. The sugar content in sugar beet can vary from 12% to 20%.

The chemical composition of sugar beet roots makes this raw material an attractive feedstock for ethanol fermentation. Bioethanol is primarily produced in the world from sugarcane, sugar beet, corn, and starch, although the preparation process from sugar beets is relatively simple compared to starchy materials. In order to obtain the highest yield from sugar beet several factors must be considered, such as the breeding and variety of the roots, the location of cultivation or sodium, phosphorous, and potassium (NPK) levels of soil (Gumienna et al., 2016). The knowledge of the interdependence of those factors is decisive in the cultivation of sugar beet varieties for ethanol production and can provide a basis for directing the crop for this sector of the economy. In particular, factors stimulating the supply of fermentable sugars from growing unit should be taken into account.

Recent studies have demonstrated that the effectiveness of ethanol fermentation depends on not only the saccharose content of sugar beet, but also its relationships with other components of the dry matter. High sugar beet yield has significant effect on ethanol yield. The impurities, such as potassium, nitrogen, and sodium, have a negative effect on efficiency of the fermentation process because they could act as inhibitors of ethanologenic yeast, *Saccharomyces cerevisiae*, resulting in reduced cell growth, glucose consumption, and ethanol production. The use of high doses of fertilizers leads to an excessive accumulation of NPK, causing a high concentration of K, N, and Na in the root (Gumienna et al., 2016).

In the manufacture of sugar from sugar beet, from harvest until the final stage of the factory production process, various by-products and wastes are generated. Most of these products can be used as renewable feedstocks for the production of energy, and other value-added products, using different processes. Leaves, which are separated in the harvesting process, are partially used for animal feed. In some countries, mainly in Europe, they are also used with potato for the production of methane (Dias et al., 2012).

In the factory production process, after the operations of extraction, purification, evaporation, and crystallization the three major by-products generated are beet pulp, lime sludge, and molasses. One ton of sugar beet yields 160 kg of sugar, 500 kg of wet pulp, and 38 kg of molasses (FAO, 2009). Pulp is obtained after the beet strips diffusion operation and recovery of sugar juice. It is particularly rich in pectin (38%–62% by dry matter), hemicellulose (24%–32%) and cellulose (22%–30%), but with low lignin content (around 1%) (Kamzon et al., 2016). Each ton of sugar beet potentially yields 830 kg of sugar juice and 170 kg of wet sugar beet pulp after sucrose extraction (Rezić et al., 2013). In sugar mills, the pulp is dehydrated and granulated in form of pellets and then is sold for animal feed. These two operations constitute 30%–40% of overall energy costs of the plant and, therefore, the transformation of the pulp to bioethanol can be more profitable. As sugar beet pulp does not need to be dried for this purpose, ethanol production could be more profitable than pulp processing for animal feed. The sugar beet industry already has many built-in production advantages that favor the production of high-performance fuels at existing sugar beet processing plants. First, there are large quantities of sugar beet pulp already concentrated with no additional transportation cost. Second, energy consumption would be minimal (Donkoh et al., 2012).

Pectin can be successfully extracted from sugar beet pulp and used as food additives. As an advantage from the pectin extracted from other sources, it exhibits superior emulsifying properties and contains ample amount of soluble dietary fiber, which is crucial in functional and health foods (Huang et al., 2017). The depectinated sugar beet pulp has been investigated for the production of cellulose nanofibers (Li et al., 2014). Also, cellulose extracted from the pulp might be converted to useful derivatives by etherification, such as carboxymethyl cellulose, which is the most important water-soluble cellulose derivative, with many applications in the food, cosmetics, pharmaceuticals, and detergents industries (Toğrul and Arslan, 2003).

The juice extracted from the sugar beet contains the sugar as well as impurities (mineral salts), which are removed by adding milk of lime and carbon dioxide and then filtering. Lime sludge is the filter cake formed during filtration of calcium carbonate precipitate in the calco-carbonic purification process of sugar beet juice. It contains primarily precipitated calcium carbonate as well as minerals and organic materials removed in juice purification. This sludge stored in lagoons and disposed as solid waste is utilized for amending the soil in agriculture, as raw material in the cement production, or for the production of paper (Vaccari et al., 2005).

Finally, molasses constitutes the latest by-product generated in sugar production from sugar beet. It is the high viscous and colored runoff syrup from the crystallization step. Sugar beet molasses contain 23%–26% water, 47%–48% sugar, 9%–14% minerals (Mg, Mn, Al, Fe, and Zn), and 8%–12% nitrogenous compounds (amino acids, proteins, etc.) (Taskin et al., 2016). High sugar content in sugar beet molasses enables its use for animal feed and also represents an interesting cheap raw material source for added-value products (enzymes, lipids, ethanol, acids, yeast, etc.) by fermentation (Razmovski and Vučurović, 2012).

4.4 SUGARY JUICES FROM SUGAR BEET FOR BIOETHANOL PRODUCTION

Several by-products of the sugar beet industry, formed by sugar solutions of different types and concentrations, can be used to produce biofuels, that is, bioethanol (Anasontzis et al., 2016; Dunn and Rao, 2015; He et al., 2014), biobutanol (Jiang et al., 2015; Ndaba et al., 2015; Zheng et al., 2015), biomethane (De Vriezea et al., 2014), or biohydrogen (Abbasi and Abbasi, 2011; Das and Veziroglu, 2008). These solutions are fermented through the aerobic or anaerobic route by different microorganisms (bacteria or yeasts) to produce the mentioned biofuels. In the case of bioalcohols, the product is obtained after distillation of the fermented broth, but in the case of biogases, the product is obtained during fermentation. Likewise, these sugary juices can be used to produce many other products of interest, such as organic acids [butyric (Zigová et al., 1999), succinic (McKinlay et al., 2007), itaconic (Klement and Bachs, 2013), lactic (Hofvendahl and Hahn-Hagerdal, 2000), fumaric (Roa Engel et al., 2008), malic (Battat et al., 1991), muconic (Curran et al., 2013), triacylglycerols (Kosa and Ragauskas, 2011), or polyhydroxyalkanoates (Du et al., 2001)].

The utilization of the juices for a specific product depends on multiple factors, among them, the cost of both the by-product to be processed and the final product are limiting. Due to the bioethanol price in the market must be low, in order to be competitive with conventional fuels, it could be more profitable using sugars for other purposes. However, biofuel production from sugary juices could become an interesting alternative when the conventional fuels price goes up. Moreover, biofuels are a sustainable source of energy against nonrenewable sources.

4.4.1 Pretreatment of Sugar Beet Roots

Sugar beet roots are mechanically harvested, taking the bulb from the soil and ripping the leaves off. The detached stalks and leaves are usually kept on the field and roots are quickly transported to the extraction plants by different ways, depending on the distance (truck, train, or ship) (Panella, 2010).

The extraction plants are provided with large outdoor areas where roots can be stored in big piles once they arrive from plantations. Due to the optimal date of harvesting is concentrated into a short time; farmers have to collect the whole harvest in a few days and later the raw material is processed in the plant (Singh et al., 2014; Yarris, 1981).

Previously, bulbs are washed with a pressurized water jet to remove the sand, dust, threads, and remaining leaves. The washing sludge is decanted in order to reuse the water. Several conveyer belts lead the tubers inside the plant while the washing water is drained (Hamer et al., 1994). The first operation consists of cutting the roots into thin strips called cossettes. The cutting machinery is continuously fed by worm screws, which drag the bulbs to the hoppers. A set of knives cut the roots into small sticks (cossettes) while some trowels push them to the cutters. The form and layout of the blades determine the cossette's size, which usually ranges from 5 to 10 cm length and around 1 cm width. After the cutters, other worm screws move cossettes to the next step of the process: the sugar extraction.

4.4.2 Extraction of Juice

Juice extraction is performed in stainless steel tanks where cossettes are mixed with hot water. This operation is usually carried out in continuous and countercurrent mode. Tanks, called diffusers, are placed horizontally and can have tens of meters in length and a diameter from 0.5 to 1 m (Fares et al., 2003). At the cylinder axis, a long worm screw is placed, which turns continuously to move cossettes through the diffuser. In some diffusers, the provided screw is fixed while the cylinder rotates, causing the same effect.

Hot water (55°C–75°C) is fed to the diffuser at the cossettes output, flowing in the opposite direction to the solid, while a progressive sugar extraction is performed (Vukov and Monszpartne, 1978). Diffusers must be heat-insulated to maintain the temperature in the full length of the extractor and they are gifted with a mild slant to force gravity driven flow.

The solid–liquid ratio (SLR) in the diffusers is usually 1:2 w/w and the residence time for the solids can range from 1 to 2 h. As a result, the output stream contains a sugar concentration around 10% to 15% w/w. This sugar output stream is called "raw juice." Frequently, several food preservatives and antibacterials (e.g., sulfur dioxide) are added to this juice in order to prevent microbial contaminations (Pollach et al., 1978).

The generated pulp (95% humidity) is discharged and subsequently pressed to recover the retained juice, reducing humidity to 70%. After that, the pressed cossettes can be extruded through holes of 0.5 cm, in order to produce little cylindrical bars of cossettes of around 5-cm length and 0.5-cm diameter. These bars are dried in a continuous industrial oven to a final humidity lower than 10%, which allows their preservation. As it was previously stated, this residue is actually considered a by-product in the sugar industry, which is sold as livestock feed (Merkes et al., 2002).

4.4.3 Downstream Processing

Raw juice, besides containing sugar (principally sucrose), also includes many other substances extracted from the root, which hinder the later sugar crystallization and reduce the crystals purity. Thus the more nonsugar products are extracted, the more necessary is the application of purification steps before crystallization (Pezzi, 2011). In general, if the juice is destined to sugar crystal production, a purification step by carbonation is applied. However, this juice can be directly fermented, without any purification step, by adding a proper microbial inoculum to produce bioethanol or other bioproducts.

The most common purification treatment of raw juice is called calco-carbonic treatment (Jarski et al., 2012). It consists of two successive steps: (1) adding a water suspension of concentrated $Ca(OH)_2$ (lime); and (2) bubbling up a gas stream of CO_2. The former step raises the pH of the juice in order to precipitate many organic acids in the form of calcium salts. At these alkaline conditions, a great part of proteins and pectin are also precipitated (Sarka et al., 2015). After the lime addition, the suspension must be stirred and decanted for hours, removing the generated precipitate by filtration (Minarovicova et al., 2007). The second step consists on the pH adjustment to the original neutral value. This is achieved by the precipitation of the calcium hydroxide excess by bubbling carbon dioxide through the juice. The precipitate is now calcium carbonate and must be filtered.

After the carbonation step, a clean juice, free of nonsugar impurities, which is called "thin juice" is obtained. It is concentrated to 60% w/w sugar content by multieffect evaporators, producing a final output stream called "thick juice" (Cubero et al., 2004). Compared to thin juice it is more stable, so it can be stored for long periods. The final step of the process is performed in crystallizers, where crystallization is induced by seeding very small sugar crystals into the thick juice, while the excess of water is removed under vacuum (Vaccari et al., 1996, 2002). Crystals grow until the thermodynamic equilibrium is reached at the work conditions (Lu et al., 2017) and later they are separated of the liquid by centrifugation. The residual liquid stream is called "molasses" and it contains 50% w/w sugars and other substances, like oligosaccharides and organic acid salts. This by-product, as previously mentioned, can be diluted to be fermented for obtaining bioproducts.

Finally, crystals are dried in continuous rotary dryers and sieved in continuous vibrators sieves. Different sizes are packaged depending on the market requirements.

The diagram of the whole process to produce white sugar from sugar beet is shown in Fig. 4.1.

4.4.4 Composition and Characteristics of Sugary Juices

The raw juice, thin juice, thick juice (intermediates), and molasses (liquid by-product) can be used for bioethanol production. Raw juice has a sugar content of 10%–15% w/w and other solved substances, like proteins, organic acids, pectin, inorganic salts, depending on the extraction conditions (Koch et al., 2014). This composition makes raw juice an excellent medium for ethanol fermentation. Usually, this juice does not need pH adjustment or nutrients addition, its cost is lower than for other intermediates because fewer transforming steps are applied, and its lower content in inorganic salts favors the fermentation. However, the high risk of microbial contamination makes difficult its storage and management (Vargas-Ramirez et al., 2013).

Thin juice contains approximately the same amount of sugar than raw juice, but not other extracted substances from the root (Ruiz-Holst et al., 2003). This circumstance could become a drawback for the fermentation of thin juice to bioethanol, due to nitrogen, phosphorous, or vitamin (biotin) deficiencies. On the contrary, due to low levels of nonfermentable substances, such as polyphenols, the purification process is easier, reducing the operation cost.

As thick juice has a sugar concentration much higher than thin juice, it cannot be directly fermented and the prior dilution with water or a nutrient solution is required (Wawro et al., 2012). For this reason, it can be stored until it is required.

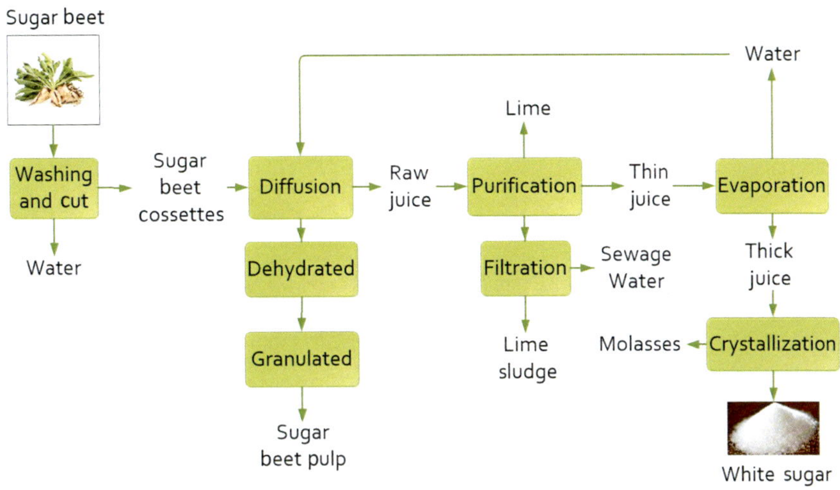

FIGURE 4.1 **Diagram of the process to produce sugars from sugar beet.**

However, the principal handicap of this juice is the higher cost compared to other juices. The high sugar concentration of molasses also requires adequate dilution to be used as raw material for ethanol production (Yeniel et al., 2006). However, the presence of high concentrations of other nonfermentable substances, like oligosaccharides (raffinose), betaine, or calcium and potassium salts, make the development of fermentation process much more difficult than that of the diluted thick juice (Behera et al., 2012). On the other hand, due to its high content of nonsugar substances and the cost associated to its recycling, the cost per unit of sugar in molasses is lower compared to the other juices. For this reason, in some cases, molasses are considered even a residue, not a by-product.

4.5 SUGAR BEET PULP PRETREATMENT AND HYDROLYSIS FOR BIOETHANOL PRODUCTION

As was previously mentioned, given sugar beet pulp composition, rich in cellulose, hemicellulose and pectin, it can be used as raw material for bioethanol production. In general, the process involves three main steps: generation of fermentable sugars (hydrolysis); fermentation of simple sugars to ethanol; and separation and purification of ethanol (Mohd Azhar et al., 2017). In order to improve sugar accessibility in the hydrolysis step, a previous pretreatment step is necessary. Fig. 4.2 shows a diagram of the steps involved in bioethanol production from sugar beet pulp.

Pretreatment and hydrolysis steps become the bottleneck of this process. Many pretreatment processes are described in literature for lignocellulosic material, but few have been applied to sugar beet pulp. It can be carried out chemically or enzymatically, though the last one is preferred to be more environmentally friendly.

4.5.1 Pretreatment of Sugar Beet Pulp

Pretreatments stand out to facilitate hydrolysis in order to obtain higher concentrations of fermentable sugars and, as a result, higher bioethanol yield (Mohd Azhar et al., 2017). There are a wide range of pretreatments with the common objective to remove structural and compositional obstacles for the hydrolysis (Buruiana et al., 2013), improving the accessibility of cellulose and hemicellulose in the hydrolysis process. In other words, they simplify the subsequent processes to produce bioethanol.

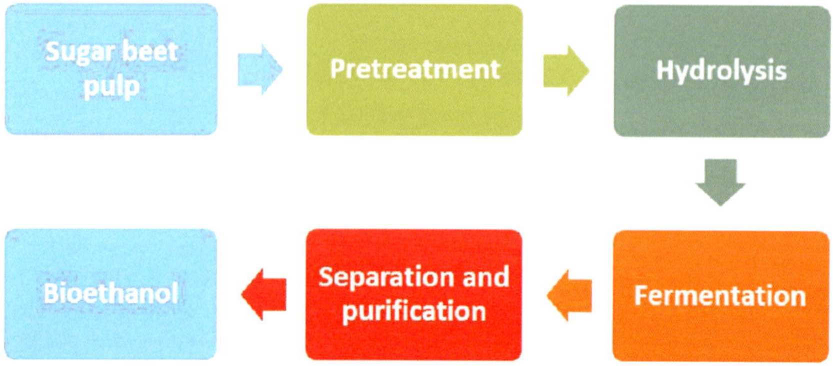

FIGURE 4.2 **Diagram of bioethanol production process from sugar beet pulp.**

In general, pretreatments produce different effects on the cellulose, hemicellulose, and lignin fractions (Alvira et al., 2010). In this way, they can be classified into physical, chemical, biological, and physicochemical, which are selected depending on the raw material. To date, few pretreatment methods have been investigated on sugar beet pulp, being the most frequently used physical (mechanical comminution), chemical (acid), and physicochemical (liquid hot water and ultrasound) methods, which are discussed in the next sections.

4.5.1.1 Mechanical Comminution

Mechanical comminution consists of the particle size reduction of the biomass and, in spite of its advantages; it is the most frequent pretreatment of sugar beet pulp. The fact of reducing solid particle size increases specific surface area and, subsequently, the accessibility of polymeric sugars to the enzymes (Alvira et al., 2010; Sun and Cheng, 2002). However, due to high-energy requirements of this pretreatment, it is not always economically feasible. There are different processes depending on the desired final size: chipping (10–30 mm), milling, or grinding (0.2–2 mm) (Sun and Cheng, 2002). Among them, the most frequently used for sugar beet pulp are around 1 mm, specifically, 0.8–1 mm (Berlowska et al., 2016), 0.3–0.5 mm (Rezić et al., 2013), 0.224–0.4 mm (Ivetić et al., 2012), and 0.08–2 mm (Micard et al., 1996).

4.5.1.2 Acid Pretreatment

Acid pretreatment is most commonly used for lignocellulosic materials, which is based on hemicellulose solubilization, improving cellulose accessibility to the enzymes (Alvira et al., 2010). This effect allows the reduction, or even elimination, of hemicellulosic enzymes needed in the hydrolysis step, reducing the final cost of bioethanol production, considering the high cost of enzymes (Zheng et al., 2013). However, during this pretreatment several compounds can be produced, such as furfural and 5-hydroxymethylfurfural (HMF), which can inhibit the fermentation process (Alvira et al., 2010; Buruiana et al., 2013; Zheng et al., 2013). In order to reduce this effect, concentrate or dilute acid can be used (Alvira et al., 2010). Compared to mechanical comminution, dilute acid is preferred for sugar beet pulp, due to its complex structure is partially broken after acid action and the inhibitory compounds are produced in a low quantity.

Several acids can be used for this pretreatment: sulfuric acid, hydrochloric acid, peracetic acid, nitric acid, or phosphoric acid. However, the most common ones employed for sugar beet pulp are sulfuric acid (Zheng et al., 2013) and hydrochloric acid (Ivetić et al., 2012).

There are two different procedures to implement the diluted acid pretreatment: at moderate temperature (T^a <160°C) or at higher temperature (T^a >160°C) (Buruiana et al., 2013). There are not many published works in literature about acid pretreatment of sugar beet pulp, but the majority of them work at moderate temperatures. After an optimization study described in literature, the best conditions seem to be sulfuric acid at 0.66 wt.% of dry biomass with a solid loading of 6 wt.% dry basis at 120°C (Zheng et al., 2013). First, the solid was heated to 100°C, temperature at which the acid was added. The pretreatment proceeds during 5 min and then was immersed in ice water to stop the reaction. After that, it was filtrated and washed to carry out the hydrolysis process. In these conditions, a hydrolysis yield of 93% was obtained (Zheng et al., 2013).

4.5.1.3 *Liquid Hot Water Pretreatment*

Liquid hot water is a pretreatment that does not employ any catalyst or chemical, being a more environmentally friendly technology (Buruiana et al., 2013). In general, this process is carried out at elevated temperature (120–200°C) for 5–20 min under pressure, to maintain water in the liquid state. The main effect produced on lignocellulosic materials is the solubilization of hemicellulose. To avoid the inhibitors formations, such as sugar decomposition, pH should be kept between 4 and 7. For sugar beet pulp, liquid hot water pretreatment is normally carried out in a lab-scale autoclave at 121°C during 20 (Díaz et al., 2017; Rezić et al., 2013; Ziemiński and Kowalska-Wentel, 2017) or 30 min (Berlowska et al., 2016).

4.5.1.4 *Ultrasound Pretreatment*

Ultrasound pretreatment consists on using an acoustic wave at high frequency in order to extract hemicelluloses, cellulose, and lignin, obtaining higher hydrolysis yields (Alvira et al., 2010). The mechanical impacts produced by the collapse of cavitation bubbles cause the opening of the solid substrate surface being more accessible to the enzymes action (Alvira et al., 2010). This pretreatment has been used for sugar beet pulp and higher yields have been obtained. Generally, it has been used as a homogenizer in continuous mode at 50% or 100% of amplitude (ultrasound power 400 W) for 5–45 min, although pretreatment time has not shown a significant influence.

4.5.2 Hydrolysis of Sugar Beet Pulp for Bioethanol Production

During the hydrolysis process, cellulose and hemicellulose present in sugar beet pulp is required to be converted into simple sugars, which are fermented to bioethanol. As it was previously mentioned, hydrolysis can be carried out chemically or enzymatically (Zabed et al., 2016).

Acid hydrolysis can be performed with diluted acid at high temperatures or with concentrate acid at low temperatures. Mostly, acids employed are sulfuric acid, hydrochloric acid, nitric acid, trifluoracetic acid, or phosphoric acid (Zabed et al., 2016). However, this type of hydrolysis provokes many problems given that for ethanol fermentation additional steps are required, such as recovery or neutralization of the acid. Other drawbacks are associated with the generation of large amounts of wastes produced and the corrosion of the equipment if concentrated acid is used (Zabed et al., 2016). This effect can be avoided by using diluted acid, but inhibitors after sugar degradation can be produced if high temperatures are applied.

On the other hand, enzymatic hydrolysis uses specific enzymes to hydrolyze the polymeric sugars. In this way, it is a more effective process, which lets it work at lower temperatures and produces fewer inhibitors. For these reasons, it is the most frequent method described in literature for this purpose.

Enzymes used for sugar beet pulp hydrolysis depend on their composition, rich in cellulose, hemicellulose, and pectin. Among them, the most important enzyme is cellulase, although supplementation with xylanases and pectinases are also used. In order to carry out the enzymatic hydrolysis, optimal work conditions of the enzymes employed are required. In this way, common temperature and pH values for cellulases are 40–50°C and 4.8–5, respectively. Table 4.2 shows a summary of the most frequent hydrolysis conditions described in literature and the type of enzyme cocktail used.

4.6 ETHANOL FERMENTATION (FIRST AND SECOND GENERATION)

As is well known, bioethanol can be produced from different raw materials classified in three types: sugars, starches, and lignocelluloses (Gumienna et al., 2014). It is what are referred to as sugars, those materials with simple sugars or disaccharides, such as saccharose, extracted with water using a simple technology, that are fermented to ethanol (first-generation bioethanol). In this case, there is no need to use additional technological operations connected with chemical or enzymatic hydrolysis, which could significantly increase the costs of biosynthesis.

Starchy and lignocellulosic materials must be previously converted into sugars by acid or enzymatic hydrolysis. Therefore, for this type of material the processing to obtain sugars to be fermented to ethanol (second-generation bioethanol) is more complicated (Leiper et al., 2006). However, these materials do not compete with food markets and show the advantage of being abundant, renewable, and inexpensive feedstocks, which makes bioethanol a more competitive alternative to fossil fuels.

As previously mentioned, given the high content of fermentable sugars in sugar beet processing by-products (molasses) and intermediates (raw, thin, and thick juice), they represent interesting raw materials for ethanol production (Tan et al., 2015). In this case, the process is relatively simple as sugar content is mainly in the form of sucrose, which can be directly fermented to ethanol (Leiper et al., 2006; Patrascu et al., 2009). In fact, in order to obtain a suitable medium for ethanol fermentation, dilution of initial sugars concentration, initial pH adjustment, and, in some cases, supplementation with micronutrients is required. In sugar concentration, the higher the initial concentration, the better ethanol yields are reached. However, it has been reported that for sugar concentrations further from 200 g/L, ethanol yield decreases given that osmotic pressure causes plasmolysis

TABLE 4.2 Summary of Sugar Beet Pulp Hydrolysis Conditions

Temperature (°C)	Solid load (w/w)	Enzymatic cocktail	pH	Agitation (r.p.m)	References
50	2%	- Celluclast 1.5L - Novozyme 188 - Pectinase	4.8	140	Zheng et al. (2013)
40	2.5%	- Celluclast 1.5L	4.8	150	Ivetic´ et al. (2012)
50	10%	- Celluclast 1.5L - Pectinase - Xylanase - B-Glucosidase	5	150	Díaz et al. (2017)
40	5%	- Celluclast 1.5L - Novozymes 43L - Viscozyme L	4.8	–	Foster et al. (2001)

of the yeast (Marx et al., 2012). In addition to this, fermentation inhibitors, such as HMF or melanoidins are contained in sugar beet processing intermediates (especially molasses), being necessary for the dilution of the fermentation media (Kawa-Rygielska et al., 2013). The optimization of the fermentation conditions (yeast strain, media supplementation, and extract concentration) and the use of appropriate cultivation techniques (fed-batch or repeated batch) can significantly increase ethanol yields, decreasing the costs of the process (Pavlecic et al., 2010).

For ethanol fermentation, *S. cerevisiae* has turned out to be a suitable microbial producer of ethanol from sugar materials (Ranković et al., 2009). This yeast ferments sucrose after its hydrolysis by invertase, which catalyzes hydrolysis of glycosidic bond between glucose and fructose (Kawa-Rygielska et al., 2013).

Given raw juice composition (described in a previous section), it can be used after pH adjustment for ethanol fermentation without nutrients additions. Moreover, it has a low price compared to the other sugar beet intermediates because fewer transforming steps are required, and its low inorganic salts content favors the fermentation (Dodić et al., 2009; Vargas-Ramirez et al., 2013). However, the high risk of microbial contamination creates difficulty for its storage and management.

Thin juice also represents a good candidate for ethanol fermentation given its high concentration of sugar; however, it does not show other extracted substances from the root (Ranković et al., 2009; Ruiz-Holst et al., 2003). This circumstance could become a drawback for the ethanol fermentation, due to nitrogen, phosphorous, or vitamin (biotin) deficiencies. On the contrary, due to low levels of nonfermentable substances, such as polyphenols, purification process is easier, reducing the operation cost. As it happens with raw juice, it shows the drawback of showing low storability and easy microbial contamination, which usually is avoided by its concentration through evaporation, which increases its production cost.

Thick juice shows a sugar concentration much higher than thin juice. For this reason, it cannot be directly fermented and the prior dilution with water or a nutrient solution is required (Wawro et al., 2012). An advantage of this property is that it can be stored until it is required.

Raw and thick juices show sufficient nutrients to support ethanol fermentation (Tan et al., 2015). However, some authors have found that ethanol production can be improved and accelerated with nutrients supplementation (Dziugan et al., 2013; Kawa-Rygielska et al., 2013).

Molasses are also used for ethanol production, being previously diluted to the required sugar concentration and pH adjustment (Dodić et al., 2009). However, given the high concentration of nonfermentable substances (described in previous section), the development of the fermentation process is much more difficult than for diluted thick juice (Behera et al., 2012). Compared to molasses, thick juice uses less water and acids for pH buffering, wastewater purification costs are reduced, and it shows easier mixing and higher levels of nutrients (Razmovski and Vućurović, 2012). In addition to this, the content of pigmenting compounds (melanoidines, caramen substances, and Fe-phenol compounds), which affects the fermenting yeast, is lower in thick juice (Vučurović et al., 2013; Vučurović and Razmovski, 2012).

Ethanol production from raw juice is higher than from thin and thick juices and molasses. Studies of Dodić concluded that impurities in raw juice compared to thin juice, only show negative effect on biomass multiplication, which is positive for this process as ethanol production starts earlier. Moreover, given the lower price of raw juice it represents a more cost-effective feedstock for bioconversion. Furthermore, considering longer process time for thin juice, thick juice, and molasses, there is more risk of cell growth inhibition.

Generally, the use of a particular juice or any other stream in the sugar industry for bioethanol production is conditioned by the cost of the raw material and the cost of the conditioning steps that are required prior the fermentation. Usually, the cheaper the raw material (per unit of sugar content), the more expensive the conditioning steps, and vice versa. Thus the optimum overall cost of first-generation bioethanol tends to be on using either raw juice or molasses.

Sugar beet pulp represents the main by-product of sugar refining industry (Gumienna et al. (2016): Zheng et al., 2013). Generally, it is sold for animal feeding at a relatively low price. However, its processing involves drying, pelletizing, and transporting, so its profitability depends on the economics of the energy and feed industries (Rorick et al., 2009). As was described in a previous section, given its composition it represents an attractive candidate to be converted into fuel ethanol (Bellido et al., 2015). Moreover, as this process is developed under aqueous conditions, the drying steps wouldn't be needed (Zheng et al., 2013). However, given that sugars in this by-product are present in complex polysaccharide structures, chemical or enzymatic hydrolysis is required, which are energy- and cost-consuming. So, for example, Gumienna et al. (2016) produced 50.7–96 cm^3 of ethanol per kg of sugar beet pulp from 75 g of sugar beet pulp soaked in a solution with $(NH_4)_2HPO_4$ at pH 5.5 (1:1 w/w). In this case, fermentations were carried out at 30°C for 72 h in stationary conditions (only one agitation per day). In this case, the enzymatic cocktail Optimash (Genencor International) with xylanase and cellulose activities was added for the hydrolysis of complex sugars. However, the main drawback of this process is the high dose of hydrolytic enzymes required to convert cellulose, hemicelluloses, and pectin into their constituent sugars, adding a significant cost to ethanol production. As a result of hemicelluloses hydrolysis, an important amount of pentoses (particularly xylose and arabinose) are also produced, which are not naturally metabolized by *S. cerevisiae* (Díaz et al., 2017; Zhong et al., 2015). Genetically modified ethanologenic bacteria, such as *Escherichia coli* KO11, *Klebsiella oxytoca* P2, *Erwinia chrysanthemi* EC 16, might be used (Zheng et al., 2013). Among these, *E. coli* KO11 turned out to be the most efficient to ferment arabinose, which also used galacturonic acid (product of pectin hydrolysis) producing 25.5 g/L ethanol.

Sugar beet pulp has been also used as support to immobilize *S. cerevisiae* to produce continuous bioethanol from thick juice without any nutrient supplementation in a packed bioreactor (Vučurović et al., 2013). Fermentation was carried out at 30°C and pH 5 for 27 days, obtaining in these conditions 54.7 g/L ethanol and a productivity of 0.498 g/g (grams of ethanol produced per gram of utilized sugar).

4.7 ECONOMICS AND LIFE CYCLE ASSESSMENT OF SUGAR BEET ETHANOL

Biofuels results from life-cycle assessments (LCA) show large variability, as a consequence of biofuels chains and methodologies diversity used to estimate inventory data. Moreover, impact categories are affected by soil types, management practices, climatic conditions, and so on (Bessou et al., 2013).

For LCA of ethanol from sugar beet, the whole process can be divided in three steps: cultivation of sugar beet, transportation, and bioethanol plant. According to a study carried out by Foteinis et al. (2011), bioethanol plant shows the highest environmental impact (62.1%), followed by sugar beet cultivation (31.9%) and transportation (23%). The use of coproducts, normally as animal feedstock, can reduce the environmental impact by 17.1%.

Bioethanol represents an attractive alternative to be used as biofuel, compared with higher petrol prices due to high crude oil prices (Deverell et al., 2009). The cost of feedstock represents the

60%–85% of the total production cost of bioethanol, which is the main factor affecting its cost and competitiveness (Kwiatkowski et al., 2006). In this sense, lignocellulosic materials are interesting feedstocks given their abundance, yield potential, and potentially better economic and greenhouse gases (GHG) efficiencies (Gray et al., 2006).

Compared to gasoline, ethanol from sugar beet has a lower environmental impact for global warming, ozone layer depletion, abiotic depletion, and photochemical oxidation categories (Bessou et al., 2013). In GHG emissions, there is a reduction of 28%–42%. Given that emitted CO_2 in the combustion of biofuels belongs to the natural cycle of carbon, it does not contribute to the global warming (Halleux et al., 2008). However, in acidification and eutrophication, sugar beet ethanol shows a higher impact given losses of reactive nitrogen in the arable field (Bessou et al., 2013).

Ethanol from sugar beet shows good water productivity if it is compared with lignocellulosic material. In fact, it is known to be relatively water efficient (Märländer et al., 2003), also in terms of water use per unit net energy produced (De Vries et al., 2010). Moreover, there is not much water available for drainage given its relatively high evapotranspiration. Compared to lignocellulosic crops, cultivation fields of sugar beet requires good quality agricultural land and it shows less net energy due to the nonconverted biomass that is not used for energy purposes (De Vries et al., 2010). In order to reduce erosion of sugar beet fields it can be used conservation tillage, requiring also less fuel for agricultural operations (Märländer et al., 2003).

Sugar beet leaves show a high concentration of nitrogen, thus they are responsible of high N_2O emissions. However, they contribute to keep soil organic carbon. GHG emissions and net energy yield could be reduced if the leaves are removed from the field and employed as raw material for anaerobic digestion and produce biogas, which could be used as transportation fuel or to generate electricity. Ecological sustainability evaluation of first generation bioethanol production from sugar beet could be significantly improved by considering the economic or energetic value of coproducts (Rozakis et al., 2013). Generally, coproducts are used internally, such as water, biogas, stillage, and nutrients that are introduced in the process, being less subjected to market forces than if they were exported outside the system boundary (Alexiades et al., 2017). So, sugar beet pulp can be used as cattle feed, thus resources used in the cultivation of dedicated fodder crops could be reduced (De Vries et al., 2014). Moreover, solids obtained after anaerobic digestion could be applied as fertilizer in the agricultural land.

In general, compared to other agricultural feedstocks, sugar beets show good environmental performance as raw material to produce ethanol in most indicators, which is the result of high yields of this crop (Muñoz et al., 2014). Moreover, slightly lower energy demand is required in this case if it is compared to starchy materials, because heat demand for sugar syrup pasteurization is lower than heat required for starch hydrolysis (Weinberg and Kaltschmitt, 2013). Sugar beet requires more energy than other raw materials for its transportation because of its high water content (75%), so it is compensated by previously mentioned factors. However, this crop is not often used because high logistic costs and low storability. The process annexed to an existing sugar refinery plant could be a promising option (Weinberg and Kaltschmitt, 2013).

The investment cost of ethanol production from sugar beets would be determined by the base capacity of the unit, the cost of equipment, the requirement for the workforce, the cost of first and auxiliary matters, electrical energy and steam, maintenance cost, and so on (Rozakis et al., 2013). In order to make ethanol from sugar beet profitable, it would be interesting to be combined with grains in order to extend the processing from 3 months to the whole year (Rozakis et al., 2013). In this way, capital and administrative charges would be extended over a longer period.

4.8 FUTURE PERSPECTIVES

The main strategies proposed for increasing bioethanol production from sugar beet juices and their by-products are focused on the fermentation stage. This is because other stages in bioethanol production are difficult to modify for several reasons: first, juices production (the extraction of juices from the sugar beet) is integrated into the sugar industry and it is a very optimized and profitable process. Therefore, it is not easy to introduce changes into the process diagram without affecting the economic success of the extraction plant. In consequence, the bioethanol industry must exploit sugary juices and other by-products from the way they are produced by the sugar industry (Hahn-Hägerdal et al., 2006). Second, regarding the separation stage of ethanol (bioethanol distillation), most of technologies currently implemented for the fermentative ethanol concentration (vacuum distillation, extractive distillation, reactive distillation, inverse osmosis, etc.) have achieved a high degree of development and efficiency, and no major improvements are expected in a short time in this field (Shihadeh et al., 2014).

Focusing, therefore, in the fermentation stage, bioethanol production can be improved in two ways: (1) by the development of the fermentative capacity of the microorganisms (Caspeta and Nielsen, 2015; Somda et al., 2011); and (2) by boosting the ethanol productivity of the process (Cardona and Sanchez, 2007; Fakruddin et al., 2012).

The principal way to improve the fermentative capacity of microorganisms is based on genetic engineering. The strategies used in this area are aimed to the development of microorganisms able to ferment simultaneously several types of sugars; to eliminate the synthesis of other metabolites that are in competition with ethanol production; and to adapt the growth and production to different hard conditions (lack of nutrients, high salts concentration, pH, or temperature extremes, etc.) (Alper et al., 2006; Sanny et al., 2010; Shaw et al., 2008; Shi et al., 2009; Wang et al., 2013; Yomano et al., 2008). Other technical option to achieve those aims is the conventional evolutionary adaptation (Fadel et al., 2013; Geddes et al., 2011; Munjal et al., 2012). However, this strategy takes much more time than genetic engineering to obtain significant results and it is not always successful. On the other hand, the use of genetically modified microorganisms in the energy industry has not as much social rejection as in the food industry.

Other strategy for increasing the fermentative capacity is the use of several microorganisms simultaneously (coculturing), where highly specialized strains in a type of sugar are used successively or simultaneously for ethanol production (Hickert et al., 2013; Magalhães-Guedes et al., 2013; Morocoima et al., 2013; Park et al., 2012). The principal handicap of this strategy is the interferences between species that could occur.

Regarding to the fermentative process implementation, there are two ways of improvement: (1) the fermentation in continuous mode, via the microorganism immobilization (Mariam et al., 2009; Razmovski and Vučurović, 2011, 2012), and (2) the fermentation in solid state mode, in the case of the solid by-products as sugar beet pulp. In order to carry out the fermentation process in continuous mode, the immobilization of the microorganism can be implemented on different types of substrates, as alginate polymers (Duarte et al., 2013), other natural solid polymers (Kirdponpattara and Phisalaphong, 2013) or some milled solid agricultural residues (Pacheco et al., 2010); and the immobilized microorganisms can be a unique strain or several types of bacteria and yeasts coimmobilized (Guo et al., 2010).

Sugar beet pulp can also be used for bioethanol production. However, the polysaccharides, the main component of this waste product, must be first hydrolyzed into more simple sugars that can be later fermented to ethanol. This process can be optimized, if both steps are carried out simultaneously by a unique microorganism in a solid state fermentation process (Moukamnerd et al., 2013; Rodriguez et al., 2010).

4.9 CONCLUSION

From the beginning of the biofuels development, most of bioethanol has been produced from crops, such as corn (North America), sugarcane (South America), or sugar beet (Europe). This is due to the high implantation of these crops in those areas, the massive exploitation of the crops in a cost-effective way, and the ease of retrieval fermentable sugars by a water extraction. However, when the use of the crops for food purposes has entered in conflict with its use for a fuel production, this strategy has been questioned. The bioethanol production is therefore moving progressively to a second-generation bioethanol (from residues) production, so that the food resources can be protected. In this sense, one of the sugar beet by-products that could have more future perspectives for bioethanol production is sugar beet pulp.

The first generation bioethanol, raw juice, is an interesting feedstock for bioethanol production, given its low purchase price and its high bioethanol yields. For this reason, the development of industrial scale plants could be highly supported.

Acknowledgments

The authors thank "Ministerio de Economía, Industria y Competitividad," "Agencia Estatal de Investigación (AEI), and "Fondo Europeo de Desarrollo Regional (FEDER)" for the financial support of this study (CTM2016-79071-R).

References

Abbasi, T., Abbasi, S.A., 2011. Renewable hydrogen: prospects and challenges. Renew. Sustain. Energy Rev. 15, 3034–3040.

Alexiades, A., Kendall, A., Winans, K.S., Kaffka, S.R., 2017. Sugar beet ethanol (*Beta vulgaris* L.): a promising low-carbon pathway for ethanol production in California. J. Clean. Prod. 172, 3907–3917.

Alper, H., Moxley, J., Nevoigt, E., Fink, G.R., Stephanopoulos, G., 2006. Engineering yeast transcription machinery for improved ethanol tolerance and production. Science 314, 1565–1568.

Alvira, P., Tomás-Pejó, E., Ballesteros, M., Negro, M.J., 2010. Pre-treatment technologies for an efficient bioethanol production process based on enzymatic hydrolysis: a review. Bioresour. Technol. 101, 4851–4861.

Anasontzis, G.E., Kourtoglou, E., Villas-Boâs, S.G., Hatzinikolaou, D.G., Christakopoulos, P., 2016. Metabolic engineering of *Fusarium oxysporum* to improve its ethanol-producing capability. Front. Microbiol. 7, 1–10.

Battat, E., Peleg, Y., Bercovitz, A., Rokem, J.S., Goldberg, I., 1991. Optimization of L-malic acid production by *Aspergillus flavus* in a stirred fermenter. Biotechnol. Bioeng. 27, 1108–1116.

Behera, S., Mohanty, R.C., Ray, R.C., 2012. Ethanol fermentation of sugarcane molasses by *Zymomonas mobilis* MTCC 92 immobilized in *Luffa cylindrica* L. sponge discs and ca-alginate matrices. Braz. J. Microbiol. 43 (4), 1499–1507.

Bellido, C., Infante, C., Coca, M., González-Benito, G., Lucas, S., García-Cubero, M.T., 2015. Efficient acetone–butanol–ethanol production by *Clostridium beijerinckii* from sugar beet pulp. Bioresour. Technol. 190, 332–338.

Berlowska, J., Pielech-Przybylska, K., Balcerek, M., Dziekonska-Kubczak, U., Patelski, P., Dziugan, P., Kregiel, D., 2016. Simultaneous saccharification and fermentation of sugar beet pulp for efficient bioethanol production. Biomed Res. Int.doi: 10.1155/2016/3154929.

Bessou, C., Lehuger, S., Gabrielle, B., Mary, B., 2013. Using a crop model to account for the effects of local factors on the LCA of sugarbeet ethanol in Picardy region, France. Int. J. Life Cycle Assess. 18, 24–36.

Buruiana, C.T., Garrote, G., Vizireanu, C., 2013. Bioethanol production from residual lignocellulosic materials: a review–Part 1. Food Technol. 37, 9–24.

Cardona, C.A., Sanchez, O.J., 2007. Fuel ethanol production: process design trends and integration opportunities. Bioresour. Technol. 98, 2415–2457.

Caspeta, L., Nielsen, J., 2015. Thermotolerant yeast strains adapted by laboratory evolution show trade-off at ancestral temperatures and pre-adaptation to other stresses. mBio 6, e00431-15.

Cubero, M.T.G., Sanz, M.C., Benito, G.G., Puerta, M.G.D., 2004. Distribution of colorants in the evaporation and first crystallization stages. Sugar Ind. 129 (9), 660–666.

Curran, K.A., Leavitt, J.M., Karim, A.S., Alper, H.S., 2013. Metabolic engineering of muconic acid production in *Saccharomyces cerevisiae*. Metab. Eng. 15 (1), 55–66.

Das, D., Veziroglu, T.N., 2008. Advances in biological hydrogen production processes. Int. J. Hydrog. Energy 33, 6046–6057.

Deverell, R., McDonnell, K., Ward, S., Devlin, G., 2009. An economic assessment of potential ethanol production pathways in Ireland. Energy Policy 37, 3993–4002.

De Vries, S.C., Van de Ven, G.W.J., Van Ittersum, M.K., Giller, K.E., 2010. Resource use efficiency and environmental performance of nine major biofuel crops, processed by first-generation conversion techniques. Biomass Bioenergy 34, 588–601.

De Vries, S.C., Van de Vena, G.W.J., Van Ittersum, M.K., 2014. First or second-generation biofuel crops in Brandenburg Germany: a model-based comparison of their production-ecological sustainability. Eur. J. Agron. 52, 166–179.

De Vriezea, J., Hennebela, T., Van den Brandea, J., Biladd, R.M., Brutonb, T.A., Vankelecomd, I.F.J., Verstraetea, W., Boona, N., 2014. Anaerobic digestion of molasses by means of a vibrating and non-vibrating submerged anaerobic membrane bioreactor. Biomass Bioenergy 68, 95–105.

Dias, M.O.S., Junqueira, T.L., Cavalett, O., Cunha, M.P., Jesus, C.D.F., Rossell, C.E.V., Filho, R.M., Bonomi, A., 2012. Integrated versus stand-alone second generation ethanol production from sugarcane bagasse and trash. Bioresour. Technol. 103, 152–161.

Díaz, A.B., Marzo, C., Caro, C., de Ory, I., Blandino, A., 2017. Valorization of exhausted sugar beet cossettes by successive hydrolysis and two fermentations for the production of bio-products. Bioresour. Technol. 225, 225–233.

Dodić, S., Popov, S., Dodic´, J., Ranković, J., Zavargo, Z., Jevtić-Mučibabić, R., 2009. Bioethanol production from thick juice as intermediate of sugar beet processing. Biomass Bioenergy 33, 822–827.

Donkoh, E., Degenstein, J., Ji, Y., 2012. Process integration and economics evaluation of sugar beet pulp conversion into ethanol. Int. J. Agric. Biol. Eng. 5, 52–61.

Du, G.C., Chen, J., Yu, J., Lun, S., 2001. Continuous production of poly-3-hydrobutyrate by *Ralstonia eutropha* in a two-stage culture system. J. Biotechnol. 88, 59–65.

Duarte, J.C., Rodrigues, J.A.R., Moran, P.J.S., Valença, G.P., Nunhez, J.R., 2013. Effect of immobilized cells in calcium alginate beads in alcoholic fermentation. AMB Express 3, 31.

Dunn, K.L., Rao, C.V., 2015. High-throughput sequencing reveals adaptation-induced mutations in pentose-fermenting strains of *Zymomonas mobilis*. Biotechnol. Bioeng. 12 (11), 2228–2240.

Dziugan, P., Balcerek, M., Pielech-Przybylska, K., Patelski, P., 2013. Evaluation of the fermentation of high gravity thick sugar beet juice worts for efficient bioethanol production. Biotechnol. Biofuels 6, 158.

Erdal, G., Esengün, K., Erdal, H., Gündüz, O., 2007. Energy use and economical analysis of sugar beet production in Tokat province of Turkey. Energy 32 (1), 35–41.

Fadel, M., Keera, A.A., Mouafi, F.E., Kahil, T., 2013. High level ethanol from sugar cane molasses by a new thermotolerant *Saccharomyces cerevisiae* strain in industrial scale. Biotechnol. Res. Int. 2013, 6.

Fakruddin, M., Abdul Quayum, M., Ahmed, M.M., Choudhury, N., 2012. Analysis of key factors affecting ethanol production by *Saccharomyces cerevisiae* IFST-072011. Biotechnology 11, 248–252.

FAO, Sugar Beet/White Sugar, 2009.

FAOSTAT. Food and Agriculture Organization of the United Nations.

Fares, K., R'zina, Q., El Amrani, M., Crepeau, M.J., Thibault, J.F., 2003. Extraction and degradation of polysaccharides from sugar beets under factory conditions. Sugar Ind. 128 (4), 243–249.

Foster, B.L., Dale, B.E., Doran-Peterson, J.B., 2001. Enzymatic hydrolysis of ammonia-treated sugar beet pulp. Appl. Biochem. Biotechnol. 91–93, 269–281.

Foteinis, S., Kouloumpis, V., Tsoutsos, T., 2011. Life cycle analysis for bioethanol production from sugar beet crops in Greece. Energy Policy 39, 4834–4841.

Geddes, C.C., Nieves, I.U., Ingram, L.O., 2011. Advances in ethanol production. Curr. Opin. Biotechnol. 22, 312–319.

Gray, K.A., Zhao, L., Emptage, M., 2006. Bioethanol. Curr. Opin. Chem. Biol. 10, 141–146.

Gumienna, M., Szambelan, K., Jele , H., Czarnecki, Z., 2014. Evaluation of ethanol fermentation parameters for bioethanol production from sugar beet pulp and juice. J. I. Brewing J. 120, 543–549.

Gumienna, M., Szwengiel, A., Szczepa ska-Alvarez, A., Szambelan, K., Lasik-Kurdy , M., Czarnecki, Z., Sitarski, A., 2016. The impact of sugar beet varieties and cultivation conditions on ethanol productivity. Biomass Bioenergy 85, 228–234.

Guo, X., Zhou, J., Xiao, D., 2010. Improved ethanol production by mixed immobilized cells of *Kluyveromyces marxianus* and *Saccharomyces cerevisiae* from cheese whey powder solution fermentation. Appl. Biochem. Biotechnol. 160, 532–538.

Hahn-Hägerdal, B., Galbe, M., Gorwa-Grauslund, M.F., Lindén, G., Zacchi, G., 2006. Bioethanol: the fuel of tomorrow from the residues of today. Trends Biotechnol. 24, 549–556.

Halleux, H., Lassaux, S., Renzoni, R., Germain, A., 2008. Comparative life cycle assessment of two biofuels ethanol from sugar beet and rapeseed methyl ester. Int. J. Life Cycle Assess. 13, 184–190.

Hamer, P.J.C., Carr, M.K.V., Wright, E., 1994. Crop production and water-use (2). The development and validation of a water-use model for sugar-beet. J. Agric. Sci. 123, 15–24.

He, M.X., Wu, B., Qin, H., Ruan, Z.Y., Tan, F.R., Wang, J.L., Shui, Z.X., Dai, L.C., Zhu, Q.L., Pan, K., Tang, X.Y., Wang, W.G., Hu, Q.C., 2014. *Zymomonas mobilis*: a novel platform for future biorefineries. Biotechnol. Biofuels 7 (1), 101.

Hickert, L.R., Da Cunha-Pereira, F., De Souza-Cruz, P.B., Rosa, C.A., Ayub, M.A.Z., 2013. Ethanogenic fermentation of co-cultures of *Candida shehatae* HM 52.2 and *Saccharomyces cerevisiae* ICVD254 in synthetic medium and rice hull hydrolysate. Bioresour. Technol. 131, 508–514.

Hofvendahl, K., Hahn-Hagerdal, B., 2000. Factors affecting the fermentative lactic acid production from renewable resources. Enz. Microb. Technol. 26, 87–107.

Huang, X., Li, D., Wang, L., 2017. Characterization of pectin extracted from sugar beet pulp under different drying conditions. J. Food Eng. 211, 1–6.

Ivetic´, D.Ž., Marina, B.S., Antov, M.G., 2012. Enzymatic hydrolysis of pretreated sugar beet shreds: statistical modelling of the experimental results. Biomass Bioenergy 7, 387–394.

Jarski, H.R., Heather, R., McGillivray, T.D., Terry, D., Groom, D.R., David, R., 2012. A comparison of anion concentrations across carbonation at the five valley factories of American Crystal Sugar Company. Int. Sugar J. 114 (1359), 156–159.

Jiang, Y., Liu, J., Jiang, W., Yang, Y., Yang, S., 2015. Current status and prospects of industrial bio-production of n-butanol in China. Biotechnol. Adv. 33 (7), 1493–1501.

Kamzon, M.A., Abderafi, S., Bounahmidi, T., 2016. Promising bioethanol processes for developing a biorefinery in the Moroccan sugar industry. Int. J. Hydrog. Energy 41, 20880–20896.

Kawa-Rygielska, J., Pietrzak, W., Regiec, P., Stencel, P., 2013. Utilization of concentrate after membrane filtration of sugar beet thin juice for ethanol production. Bioresour. Technol. 133, 134–141.

Kirdponpattara, S., Phisalaphong, M., 2013. Bacterial cellulose alginate composite sponge as a yeast cell carrier for ethanol production. Biochem. Eng. J. 77, 103–109.

Klement, T., Bachs, J., 2013. Itaconic acid: a biotechnological process in change. Bioresour. Technol. 135, 422–431.

Koch, T.J., Timo, J., Venus, J., 2014. Sugar beet syrups in lactic acid fermentation—Part II: saving nutrients by lactic acid fermentation with sugar beet thick juice and raw juice. Sugar Ind. 139 (11), 683–690.

Kosa, M., Ragauskas, A., 2011. Lipids from heterotrophic microbes: advances in metabolism research. Trends Biotechnol. 29, 53–61.

Kwiatkowski, J.R., MacAloon, A.J., Taylor, F., Johnston, D.B., 2006. Modelling the process and costs of fuel ethanol production by the dry-grind process. Ind. Crops Prod. 23, 288–296.

Leiper, K., Schlee, C., Tebble, I., Stewart, G.G., 2006. The fermentation of beet sugar syrup to produce bioethanol. J.I. Brewing J. 112, 122–133.

Li, M., Wang, L., Li, D., Cheng, Y., Adhikari, B., 2014. Preparation and characterization of cellulose nanofibers from de-pectinated sugar beet pulp. Carbohydr. Polym. 102, 136–143.

Lu, Y.S., Thomas, L., Schmidt, S., 2017. Differences in the thermal behaviour of beet and cane sucrose sources. J. Food Eng. 201, 57–70.

Magalhães-Guedes, K.T., Rodrigues, A.K., Gervasio, I.M., Gervasio, I., Nascimento, A.P., Schwan, R.F., 2013. Ethanol production from deproteinised cheese whey fermentations by co-cultures of *Kluyveromyces marxianus* and *Saccharomyces cerevisiae*. Afr. J. Microbiol. Res. 7, 1121–1127.

Maitah, M., ezbová, H., Smutka, L., Tomšík, K., 2016. European sugar production and its control in the world market. Sugar Technol. 18 (3), 236–241.

Mariam, I., Manzoor, K., Ali, S., Ul-Haq, I., 2009. Enhanced production of ethanol from free and immobilized *Saccharomyces cerevisiae* under stationary culture. Pak. J. Bot. 41, 821–833.

Märländer, B.B., Hoffmann, C., Koch, H.J., Ladewig, E., Merkes, R., Petersen, J., Stockfisch, N., 2003. Environmental situation and yield performance of the sugar beet crop in Germany: heading for sustainable development. J. Agron. Crop Sci. 189, 201–226.

Marx, S., Brandling, J., van der Gryp, P., 2012. Ethanol production from tropical sugar beet juice. Afr. J. Biotechnol. 11, 11709–11720.

McKinlay, J.B., Vieille, C., Zeikus, J., 2007. Prospects for a bio-based succinate industry. Appl. Microbiol. Biotechnol. 76, 727–740.

Merkes, R., Krohl, M., Mugele, H., Sauer, M., 2002. Sugar beet production technology in the year 2000: cost reduction, environmental protection, sustainability. Sugar Ind. 127 (7), 530–537.

Micard, V., Renard, C.M.G.C., Thibault, J., 1996. Enzymatic saccharification of sugar-beet pulp. Enz. Microb. Technol. 19, 162–170.

Minarovicova, L., Dandar, A., Jancurova, M., Schneiderova, R., 2007. Reduction of lime in the process of purification. Listy Cukrov. Repar. 123 (1), 27–30.

Mohd Azhar, S.H., Abdulla, R., Jambo, S.A., Marbawi, H., Azlan, J., Mohd Faik, A.A., Rodrigues, K.F., 2017. Yeasts in sustainable bioethanol production: a review. Biochem. Biophys. Rep. 10, 52–61.

Morocoima, J.A., Bertsch, A., Domínguez, G., Mazzani, C., Díaz, I., 2013. Optimization of *Aspergillus niger* and *Saccharomyces cerevisiae* co-culture for the production of ethanol from potato (*Solanum tuberosum*) agro-industrial processing wastes. Interciencia 38, 305–309.

Moukamnerd, C., Kawahara, H., Katakura, Y., 2013. Feasibility study of ethanol production from food wastes by consolidated continuous solid-state fermentation. J. Sust. Bioenergy Syst. 3, 143–148.

Munjal, N., Mattam, A.J., Pramanik, D., Srivastava, P.S., Yazdani, S.S., 2012. Modulation of endogenous pathways enhances bioethanol yield and productivity in *Escherichia coli*. Microb. Cell. Fact 11, 145–157.

Muñoz, I., Flury, K., Jungbluth, N., Rigarlsford, G., Milài Canals, L., King, H., 2014. Life cycle assessment of bio-based ethanol produced from different agricultural feedstocks. Int. J. Life Cycle Assess. 19, 109–119.

Ndaba, B., Chiyanzu, I., Marx, S., 2015. N-Butanol derived from biochemical and chemical routes: a review. Biotechnol. Rep. 8, 1–9.

Pacheco, A.M., Gondim, D.R., Gonçalves, L.R.B., 2010. Ethanol production by fermentation using immobilized cells of *Saccharomyces cerevisiae* in cashew apple bagasse. Appl. Biochem. Biotechnol. 161, 209–217.

Panella, L., 2010. Sugar beet as an energy crop. Sugar Technol. 12 (3–4), 288–293.

Park, E.Y., Naruse, K., Kato, T., 2012. One-pot bioethanol production from cellulose by co-culture of *Acremonium cellulolyticus* and *Saccharomyces cerevisiae*. Biotechnol. Biofuels 5 (1), 64.

Patrascu, E., Rapeanu, G., Bonciu, C., Hopulele, T., 2009. Bioethanol Production from Molasses by Different Strains of *Saccharomyces cerevisiae*. Paper presented at the International Symposium Euro-aliment 2009. The Annals of the University Dunarea de Jos of Galati, Galati-Romania, 9–10 October. pp. 49–56.

Pavlecic, M., Vrana, I., Vibovec, K., Ivancic Santek, M., Horvat, P., Santek, B., 2010. Ethanol production from different intermediates of sugar beet processing, ethanol from sugar beet intermediates. Food Technol. Biotechnol. 48, 362–367.

Pezzi, G., 2011. Considerations of the technological quality assessment of sugar beet. Sugar Ind. 136 (2), 85–89.

Pollach, G., Hein, W., Leitner, A., Zollner, P., 1978. Detection and control of strictly anaerobic, spore-forming bacteria in sugar beet tower extractors. Sugar Ind. 32 (2), 56–61.

Ranković, J., Dodić, J., Dodić, S., Popov, S., 2009. Bioethanol production from intermediate products of sugar beet processing with different types of *Saccharomyces cerevisiae*. Chem. Ind. Chem. Eng. Q 15, 13–16.

Razmovski, R., Vučurović, V., 2011. Ethanol production from sugar beet molasses by *S. cerevisiae* entrapped in an alginate maize stem ground tissue matrix. Enz. Microb. Technol. 48, 378–385.

Razmovski, R., Vučurović, V., 2012. 'Bioethanol production from sugar beet molasses and thick juice using *Saccharomyces cerevisiae* immobilized on maize stem ground tissue. Fuel 92, 1–8.

ezbová, H., Belová, A., Škubna, O., 2013. Sugar beet production in the European Union and its future trends. Agris On-line 5 (4), 165–178.

Rezic´, T., Oros, D., Markovi, I., Kracher, D., Ludwig, R., Šantek, B., 2013. Integrated hydrolyzation and fermentation of sugar beet pulp to bioethanol. J. Microbiol. Biotechnol. 23, 1244–1252.

Roa Engel, C.A., Straathof, A.J.J., Zijlmans, T.W., Van Gulik, W.M., van der Wielen, L., 2008. Fumaric acid production by fermentation. Appl. Microbiol. Biotechnol. 78, 379–389.

Rodriguez, L.A., Toro, M.E., Vazquez, F., Cerrea-Daneri, M.L., Gouriric, S.C., Vallejo, M.D., 2010. Bioethanol production from grape and sugar beet pomaces by solid state fermentation. Int. J. Hydrog. Energy 201, 5914–5917.

Rorick, R.E., Nabar, N., Pryor, S., 2009. Enzymatic hydrolysis and fermentation of sugar beet pulp. Paper presented at the American Society of Agricultural and Biological Engineers Annual International Meeting 2009. June 21-24, Reno (Nevada); United States; Code 79137.

Rozakis, S., Haque, I., Natsis, A., Borzecka-Walker, M., Mizak, K., 2013. Cost-effectiveness of bioethanol policies to reduce carbon dioxide emissions in Greece. Int. J. Life Cycle Assess. 18, 306–318.

Ruiz-Holst, M., Martin, F.D., Burba, M., Elias, P.D.Y., Diener, G., 2003. Assessment of the technical quality of sugar beet in Southern Spain. Sugar Ind. 128, 879–887.

Sanny, T., Arnaldos, M., Kunkel, S.A., Pagilla, K.R., Stark, B.C., 2010. Engineering of ethanolic *E. coli* with the *Vitreoscilla* haemoglobin gene enhances ethanol production from both glucose and xylose. Appl. Microbiol. Biotechnol. 88, 1103–1112.

Sarka, E., Bubnik, Z., Pour, V., 2015. Carbonatation precipitate: origin, separation and uses. Sugar Ind. 140 (7), 434–442.

Shaw, A.J., Podkaminer, K.K., Desai, S.G., Bardsley, J.S., Rogers, S.R., Thorne, P.G., Hogsett, D.A., Lynd, L.R., 2008. Metabolic engineering of a thermophilic bacterium to produce ethanol at high yield. Proc. Natl. Acad. Sci. USA 105, 13769–13774.

Shi, D.J., Wang, C.L., Wang, K.M., 2009. Genome shuffling to improve thermotolerance, ethanol tolerance and ethanol productivity of *Saccharomyces cerevisiae*. J. Ind. Microbiol. Biotechnol. 36, 139–147.

Shihadeh, J.K., Huang, H., Rausch, K.D., Tumbleson, M.E., Singh, V., 2014. Vacuum stripping of ethanol during high solids fermentation of corn. Appl. Biochem. Biotechnol. 173 (2), 486–500.

Singh, I., Langyan, S., Yadava, P., 2014. Sweet corn and corn-based sweeteners. Sugar Technol. 16 (2), 144–149.

Somda, M.K., Savadogo, A., Ouattara, C.A.T., Ouattara, A.S., Traore, A.S., 2011. Thermotolerant and alcohol-tolerant yeasts targeted to optimize hydrolyzation from mango peel for high bioethanol production. Asian J. Biotechnol. 3, 77–83.

Sun, Y., Cheng, J., 2002. Hydrolysis of lignocellulosic materials for ethanol production: a review. Bioresour. Technol. 83, 1–11.

Tan, L., Sun, Z., Okamoto, S., Takaki, M., Tang, Y., Morimura, S., Kida, K., 2015. Production of ethanol from raw juice and thick juice of sugar beet by continuous ethanol fermentation with flocculating yeast strain KF-7. Biomass Bioenergy 81, 265–272.

Taskin, M., Ortucu, S., Aydogan, M.N., Arslan, N.P., 2016. Lipid production from sugar beet molasses under non-aseptic culture conditions using the oleaginous yeast *Rhodotorula glutinis* TR29. Renew. Energy 99, 198–204.

Toğrul, H., Arslan, N., 2003. Production of carboxymethyl cellulose from sugar beet pulp cellulose and rheological behaviour of carboxymethyl cellulose. Carbohydr. Polym. 54 (1), 73–82.

Vaccari, G., Mantovani, G., Sgualdino, G., Dosi, E., 1996. Cooling crystallization of raw juice: laboratory investigations of sucrose crystal growth kinetics. Sugar Ind. 121 (2), 111–117.

Vaccari, G., Wawro, P., Tamburini, E., Sgualdino, G., Bernardi, T., 2002. Cooling crystallization of microfiltered raw juice and of traditional thick juice: a comparison. Sugar Ind. 127 (1), 22–28.

Vaccari, G., Tamburini, E., Sgualdino, G., Urbaniec, K., Klemes, J., 2005. Overview of the environmental problems in beet sugar processing: possible solutions. J. Clean. Prod. 13, 499–507.

Vargas-Ramirez, J.M., Haagenson, D.M., Pryor, S.W., Wiesenborn, D.P., 2013. Determination of suit-able storage conditions to preserve fermentable sugars in raw thick beet juice for ethanol production. Biomass Bioenergy 59, 362–369.

Vučurović, V., Razmovski, R., Miljić, U., Puškaš, V., 2013. Continuous ethanol production from sugar beet thick juice by *Saccharomyces cerevisiae* immobilized onto sugar beet pulp. Acta Period. Technol. 44, 313–321.

Vučurović, V., Razmovski, R., 2012. Sugar beet pulp as support for *Saccharomyces cerevisiae* immobilization in bioethanol production. Ind. Crops Prod. 39, 128–134.

Vukov, K., Monszpartne, Js., 1978. Production of sugar beet juice and apple juice with counter current extraction. Elelmezesi Ipar 32 (2), 56–61.

Wang, J., Liu, W., Ding, W., Zhang, G., Liu, J., 2013. Increasing ethanol titer and yield in a gpd1Δ gpd2Δ strain by simultaneous over expression of GLT1 and STL1 in *Saccharomyces cerevisiae*. Biotechnol. Lett. 5, 1859–1864.

Wawro, P., Gruska, R., Stanisz, M., 2012. Thickened raw juice: some aspects of thickening and use. Sugar Ind. 137 (2), 95–101.

Weinberg, J., Kaltschmitt, M., 2013. Greenhouse gas emissions from first generation ethanol derived from wheat and sugar beet in Germany: analysis and comparison of advanced by-product utilization pathways. Appl. Energy 102, 131–139.

Yarris, L., 1981. Sugar beet irrigation cut without sucrose loss. J. Agric. Res. 30 (6), 10–11.

Yeniel, N., Beyatli, Y., Aslim, B., Yuksekdag, Z.N., 2006. Designated production of exopolysaccharide (EPS) and lactic acid with sugar beet molasses by some lactic acid bacteria. Sugar Ind. 131 (12), 841–845.

Yomano, L.P., York, S.W., Zhou, S., Shanmugam, K.T., Ingram, L.O., 2008. Re-engineering *Escherichia coli* for ethanol production. Biotechnol. Lett. 30, 2097–2103.

Zabed, H., Sahu, J.N., Boyce, A.N., Faruq, G., 2016. Fuel ethanol production from lignocellulosic biomass: an overview on feedstocks and technological approaches. Renew. Sustain. Energy Rev. 66, 751–774.

Zheng, Y., Lee, C., Yu, C., Cheng, Y., Zhang, R., Jenkins, B.M., Vandergheynst, J.S., 2013. Dilute acid pre-treatment and fermentation of sugar beet pulp to ethanol. Appl. Energy 105, 1–7.

Zheng, J., Tashiro, Y., Wang, Q., Sonomoto, K., 2015. Recent advances to improve fermentative butanol production: genetic engineering and fermentation technology. J. Biosci. Bioeng. 119 (1), 1–9.

Zhong, C., Wang, C., Huang, F., Wang, F., Jia, H., Zhou, H., Wei, P., 2015. Selectivehydrolysis of hemicellulose from wheat straw by a nanoscale solid acid catalyst. Carbohydr. Polym. 131, 384–391.

Ziemiński, K., Kowalska-Wentel, M., 2017. Effect of different sugar beet pulp pre-treatments on biogas production efficiency. Appl. Biochem. Biotechnol. 181, 1211–1227.

Zigová, J., Šturdk, E., Vandák, D., Schlosser, A., 1999. Butyric acid production by Clostridium butyricum with integrated extraction and pertraction. Process Biochem. 34, 835–843.

CHAPTER

5

Sweet Sorghum for Bioethanol Production: Scope, Technology, and Economics

Ramesh C. Ray[*], Kiran Babu Uppuluri**, Chandrasekaran Trilokesh**, Claudia Lareo[†]*

*ICAR-Central Tuber Crops Research Institute (Regional Centre), Bhubaneswar, India

**SASTRA Deemed University, Thanjavur, India

[†]University of the Republic, Montevideo, Uruguay

5.1 INTRODUCTION

The plants that are well studied for bioethanol production include maize (*Zea mays* L.), sweet sorghum (*Sorghum bicolor* L.), sugarcane (*Saccharum officinarum* L.), cassava (*Manihot esculenta* Crantz) and sugar beet (*Beta vulgaris* L.). Among these plants, sweet sorghum has comparative advantages over others concerning yield of ethanol per ha of cultivation, better acclimatization in less favorable climatic conditions, resistance to drought, salinity, water logging, and pest and disease (Nadir et al., 2009; Reddy et al., 2007). The various parts of sweet sorghum cater to multiple purposes. For instance, the grains can be used as food for human consumption; the sugary rich stalks can be used to make a sugar syrup and also biofuels, such as bioalcohols and biodiesel, stovers, and bagasse for animal feed or even as natural fertilizers (Reddy et al., 2007). The yield per ha of sweet sorghum grains and stalks are 3–7 tonnes and 54–69 tonnes, respectively (Almodares and Hadi, 2009). The Brix content of sweet sorghum juice depends on the variety, and the general range is 16%–22% (Almodares and Sepahi, 1996;Ratnavathi, 2017;Reddy et al., 2007) and sugars rangeis 15%–21% (Guigou et al., 2011;Ratnavathi, 2017).Also, sweet sorghum produces a 4-carbon sugar compared to a 3-carbon sugar, that is, generated by most plants during photosynthesis, which in turn improves the photosynthetic efficiency (Reddy et al., 2007).

Bioethanol Production From Food Crops
http://dx.doi.org/10.1016/B978-0-12-813766-6.00005-9

5.2 SWEET SORGHUM AS A BIOFUEL CROP

The fuels arising from crude oil result in increased carbon emission. One of the significant alternatives to these nonrenewable fuels is biofuels, in particular, bioethanol. As discussed before, the sugary rich stalks of sweet sorghum can be used for bioethanol production. Though the Brix yield and feedstock yield for sweet sorghum are less than that of sugarcane, the water and fertilizer content required is one-third to that of sugarcane (Kim and Day, 2011). The existing sugarcane fields can be easily replaced by sweet sorghum and the cost associated with this process was estimated to be US$358/ha. This cost involves replacement of previous sugarcane stubble, seed cost for sweet sorghum, fertilizer, transportation, labor, and harvest (Viator et al., 2009).

The various parts of sweet sorghum contain sugars, which can be used to produce bioethanol. Sipos et al. (2009) reported that the ratio of stem, leaves, and panicles (grains) was found to be about 76%, 16%, and 8%, respectively. The juice extracted from the stem contains sucrose, glucose, and fructose and can therefore be readily fermented to alcohol (Guigou et al., 2011; Kim and Day, 2011; Sipos et al., 2009). The solid fraction left behind, called bagasse is a lignocellulosic residue, which can be processed to ethanol. The grains contain starch, which can be also fermented to ethanol. Table 5.1 shows the composition of sugars in stalk juice and fibers of sweet sorghum and Table 5.2, the ethanol production expressed in liters per ha.

TABLE 5.1 Composition of sugars in stalk juice and fibres of sweet sorghum

Component	Composition (%, dry weight)
Feedstock	
Juice	70.4-72.3 [a]
Fibres	12.0-14.0
Juice	
Sucrose	53.3-55.0
Glucose	16.5-18.9
Fructose	10.8-12.0
Total sugars	83.2-85.2 [b]
Fibres	
Cellulose	43.4-45.2
Hemicellulose	26.5-28.0
Lignin	18.5-21.4
Ash	0.3-0.5

[a] *Wet basis*

[b] *Total sugars were calculated in glucose equivalents as: glucose + fructose + 1.05 x sucrose*

Source: Laopaiboon, L., Nuanpeng, S., Srinophakun, P., Klanrit, P., Laopaiboon, P., 2009. Ethanol production from sweet sorghum juice using very high gravity technology: effects of carbon and nitrogen supplementations. Bioresour. Technol. 100, 4176–4182; Guigou, M., Lareo, C., Pérez, L.V., Lluberas, M.E., Vázquez, D., Ferrari, M.D., 2011. Bioethanol production from sweet sorghum: evaluation of post-harvest treatments on sugar extraction and fermentation. Biomass Bioenergy 35, 3058–3062; Kim, M., Day, D.F., 2011. Composition of sugar cane, energy cane, and sweet sorghum suitable for ethanol production at Louisiana sugar mills. J. Ind. Microbiol. Biotechnol. 38 (7), 803–807; Khalil, S.R.A., Abdelhafez, A.A., Amer, E.A.M., 2015. Evaluation of bioethanol production from juice and bagasse of some sweet sorghum varieties. Ann. Agric. Sci. 60 (2), 317–324; Barcelos, C.A., Maeda, R.N., Santa-Anna, L.M.M., Pereira, Jr., N., 2016. Sweet sorghum as a whole-crop feedstock for ethanol production. Biomass Bioenergy 94, 46–56; Ebrahimiaqda, E., Ogden, K.L., 2017. Simulation and cost analysis of distillation and purification step in production of anhydrous ethanol from sweet sorghum. ACS Sustain. Chem. Eng. 5 (8), 6854–6862.

An important aspect of the crop is the determination of its harvest time in the field. Currently, stem maturation, measured through the Brix (soluble solids content) of intermediary internodes and the identification of grain maturity is used as harvest indicators. However, it is believed that the determination of sweet sorghum ideal harvest time still imprecise and the methodologies used for decision making are inappropriate, because the yields at the mills are unsatisfactory. It is assumed that the combination between stem fresh biomass yield, juice extraction, and Brix of stem juice is more reliable and appropriate to determine sorghum harvest time (Teixeira et al., 2017).

5.3 PROCESSES FOR CONVERSION OF SWEET SORGHUM INTO BIOETHANOL

Different processes are adapted for bioprocessing stalk juice, grain, and bagasse for producing ethanol.

TABLE 5.2 Ethanol Production From Sweet Sorghum

Ethanol (L/ha) from					
Juice	Cellulose	Hemicellulose	Grains[e]	Total ethanol (L/ha)	References
2800[a]	2125[b]	1320[c]	—	6245	Kim and Day (2011)
2252–5414[a]	2154–6591[d]		430–1223	4867–13,032	Zhao et al. (2009)
3820	7410		2370	13,600	Barcelos et al. (2016)
5000–6300	3154–4163		—	8300–10,500	Castro et al. (2017)

—, not available.
[a] *Ethanol yield from sugar (L/ha) = total sugar content (%) in dry matter × dry biomass (tonne/ha) × 0.51 (conversion factor of ethanol from sugar) × 0.85 (process efficiency of ethanol from sugar) × 1000/0.79 (specific gravity of ethanol, g/mL).*
[b] *Ethanol yield from cellulose (L/ha) = cellulose content (%) in dry matter × dry biomass (tonne/ha) × 1.11 (conversion factor of glucose from cellulose) × 0.51 (conversion factor of ethanol from sugar) × 0.85 (process efficiency of ethanol from sugar) × 1000/0.79 (specific gravity of ethanol, g/mL).*
[c] *Ethanol yield from hemicellulose (L/ha) = xylan content (%) in dry matter × dry biomass (tonne/ha) × 1.14 (conversion factor of xylose from xylan) × 0.51 (conversion factor of ethanol from sugar) × 0.85 (process efficiency of ethanol from sugar) × 1000/0.79 (specific gravity of ethanol, g/mL).*
[d] *Ethanol yield from cellulose and hemicellulose (L/ha) = cellulose and hemicellulose content (%) in dry matter × dry biomass (tonne/ha) × 1.11 (conversion factor of sugar from cellulose and hemicellulose) × 0.85 (process efficiency of sugar from cellulose and hemicellulose) × 0.51 (conversion factor of ethanol from sugar) × 0.85 (process efficiency of ethanol from sugar) × 1000/0.79 (specific gravity of ethanol, g/mL).*
[e] *Ethanol yield from starch (L/ha) = starch content (%) in dry matter × grain dry yield (tonne/ha) × 1.11 (conversion factor of sugar from starch) × 0.51 (conversion factor of ethanol from sugar) × 0.85 (process efficiency of ethanol from sugar) × 1000/0.79 (specific gravity of ethanol, g/mL) sorghum and sugarcane molasses in India.*

5.3.1 Processes for Conversion of Sweet Sorghum Stalk Juice Into Bioethanol

Fig. 5.1 shows a simplified diagram of the different steps for ethanol production from sweet sorghum stalk juice.

5.3.1.1 Juice Extraction

The juice rich stalks are crushed using a series of mills, and later the juice is collected. The bagasse is separated from the juice. The juice is later screened to remove any particulate matter, followed by sterilization by heating at 100°C. This sterilized juice is then clarified (Quintero et al., 2008). The heavier bottom part of the clarifier section is sent for rotary vacuum filtration, and the uppermost part is further sent for evaporation to concentrate the clarified juice. When the juice is subjected to fermentation, the Brix levels need to be 16–18 Bx. When not in use and when kept for storage the Brix levels need to be 65–85 Bx (Almodares and Hadi, 2009).

5.3.1.2 Fermentation

Most studies on ethanol production from sweet sorghum juice have been carried out using free cells of *Saccharomyces cerevisiae* in batch processes (Bulawayo et al., 1996; Laopaiboon et al., 2007).

Ratnavathi et al. (2010) studied five genotypic variations of sweet sorghum, that is, Kelly, SSV 84, Wray, NSSH 104, and BJ 248 for ethanol production. Kelly showed maximum ethanol production of 9% w/v when a sugar concentration of 20% w/v in juice was used. But when unclean sweet sorghum juice (using the same variety) was used with a sugar concentration of 15% w/v an ethanol concentration of 6.47% w/v was obtained. Guigou et al. (2011) fermented sweet sorghum juices of M81, Topper and Theis varieties with a sugar concentration of 11.0% w/v, 19.1% w/v, and 13.8% w/v, respectively. Topper and Theis showed the best performance in terms of ethanol concentration (8.2% w/v and 7.1% w/v ethanol, respectively), and ethanol yield expressed as gram of ethanol produced per gram of total sugars consumed (0.42 and 0.48 g/g, respectively, which represented 82.2% and 93.9% of the theoretical value). Fernandes et al. (2014) studied four sorghum varieties (BRS 506, BRS 508, BRS 509, and BRS 511) and found similar results with fermentation times of 8 h.

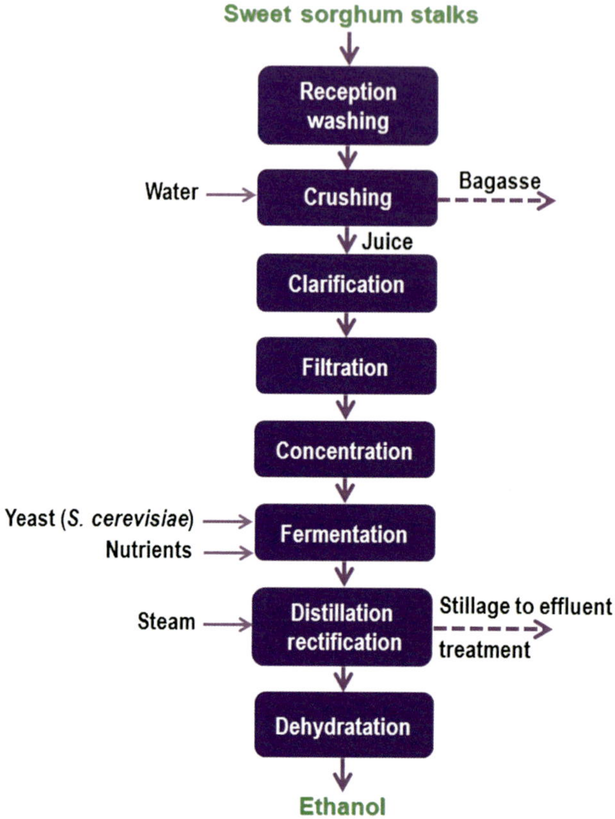

FIGURE 5.1 **Simplified diagram of ethanol production from sweet sorghum stalk juice.**

Very high gravity (VHG) technology, which is defined as the preparation and fermentation of media containing 270 g or more of dissolved solids per liter (Bayrock and Ingledew, 2001), presents several advantages for industrial applications, such as the increase in the ethanol concentration. This reduces capital and energy costs of the product recovery per liter of ethanol produced and also the risk of bacterial contamination is minimal (Bai et al., 2008; Thomas et al., 1996). However, this technology can produce slow and incomplete fermentation because the yeast cells are exposed to high dissolved solids concentration (high external osmotic pressure), and to high ethanol concentration, which can be toxic to the cells (Bai et al., 2008; Deesuth et al., 2015; Laopaiboon et al., 2009; Larnaudie et al., 2016). To use VHG technology with sweet sorghum juice, it should be concentrated, usually in a multiple effect evaporator. The energy saving in distillation using VHG conditions should offset the extra energy needed for juice concentration, the cost due to a slow fermentation and low substrate conversion into ethanol. In addition to the high content of sugars, concentrated crude materials also have a high level of naturally occurring compounds, such as organic acids and metallic ions that are potentially inhibitors of alcoholic fermentation. Few experimental data have been reported using concentrated natural materials from industrial facilities (Larnaudie et al., 2016). To obtain a high efficiency of ethanol production, many factors, such as nitrogen, trace elements, vitamins and/

or aeration on ethanol production, have been investigated under VHG conditions (Deesuth et al., 2015).

Bvochora et al. (2000) used a mixture of sweet sorghum juice and sorghum grain for ethanol production using VHG technology. Both malted and regular sorghum grains were used for fermentation. The strain used was *S. cerevisiae* N96. Sucrose was added to the mash for VHG fermentation to increase the sugar concentration to 34 g/100 mL. The total fermentation time was 96 h, and the obtained ethanol yield in VHG conditions was 16.8% v/v and 11% v/v for malted and regular sorghum grains.

An investigation of ethanol production from sweet sorghum juice was carried out by varying the concentration of dissolved solids, that is, one with 20%–24% suspended solids and another higher than 24% suspended solids (Khongsay et al., 2010). Though the ethanol efficiencies for both cases were similar, the ethanol productivity for 36 h for less dissolved solids, that is, 20%–24% was found to be higher than the juice with higher dissolved solids. For sweet sorghum juices containing up to 19% total sugars, almost total sugar consumption was observed by Larnaudie et al. (2016). The fermentation time increased with the initial sugar content, resulting in lag phases and uncompleted fermentations for high sugar content (46.5% sugar conversion for 32% w/v initial sugar concentration), and fermentation efficiencies in the range 84.2%–91.3% of the theoretical value. Similar results were described by Laopaiboon et al. (2009) and Deesuth et al. (2015) for VHG fermentations of sweet sorghum juice.

A comparison of batch and fed-batch process for ethanol production by *S. cerevisiae* TISTR 5048 was done (Laopaiboon et al., 2007). The fed-batch process yielded better results regarding ethanol concentration, ethanol yield, and productivity compared to batch process. The values were 100 g/L, 0.42 g/g, and 1.67 g/L/h for batch processing, whereas they were 120 g/L, 0.48 g/g, and 1.11 g/L/h for fed-batch processing, respectively. Immobilized *S. cerevisiae* cells with an optimum concentration of salts, such as KH_2PO_4, $(NH_4)_2SO_4$, and $MgSO_4$ increased the yield of ethanol to 91.6% (Liu et al., 2008). Fermentation parameters, such as temperature, agitation rate, pH, particle stuffing rate, were varied and studied for ethanol production from sweet sorghum from *S. cerevisiae* CICC 1308 (Liu and Shen, 2008).

Ethanol yields up to 6300 L/tonne from sweet sorghum juice have been reported (Castro et al., 2017).

5.3.1.3 Distillation and Dehydration

The ethanol produced in the fermented mash is purified by distillation to yield 95% ethanol, which is dehydrated to yield 99% purity (Larnaudie et al., 2016). The vinasse generated during fermentation can be used as fertilizer or can be combusted to produce steam (Christofoletti et al., 2013).

5.3.2 Sorghum Grains

Sorghum grain comprises three critical anatomical parts: pericarp, germ, and endosperm. The pericarp further includes epicarp, mesocarp, and endocarp. Among these components, the mesocarp and endosperm are rich in starch. The endosperm contains a kernel that includes starch. Overall the starch composition in sorghum would be around 60%–70%. The composition of maize and sorghum grain are almost similar with sorghum having 1% less fat and 1%–2% more protein than maize (Serna-Saldívar, 2010; Serna-Saldívar et al., 2012). Proteins are the second major component of sorghum grains. Lareo et al. (2012) reported 11.6% of protein, 1.45% of total starch free sugars (glucose, fructose, and sucrose), 68.7% of total fermentable sugars, and 26.4% no-fermentable solids for sweet sorghum grains in dry basis. The starch–protein interaction in sweet sorghum grains reduces the susceptibility of the starch to enzymatic hydrolysis, and therefore, it affects the conversion efficiency and ethanol yield. Besides the starch–protein relationship, some other factors, such as mash viscosity, presence of phenolic compounds, ratio of amylose to amylopectin, and formation of amylose-lipid complex in the mash, limit the rate of enzymatic hydrolysis and fermentation efficiency during bioethanol production (Ramírez et al., 2016; Serna-Saldívar et al., 2012).

5.3.2.1 Processes for Conversion of Sweet Sorghum Grains Into Bioethanol

Ethanol processing from sorghum grain is similar to that of maize and cassava, because the primary component present in all three crops is starch (Quintero et al., 2008; Serna-Saldívar et al., 2012). Starch is a homopolysaccharide made up of glucose. Starch cannot be broken down by yeast, such as *S. cerevisiae* to alcohol as it doesn't possess the required enzymes. Starch can be hydrolyzed by the combination of the following two enzymes, α-amylase and amyloglucosidase to yield glucose. Proteases can be added before or after liquefaction. This results in the breakdown of nitrogen to amino compounds, which act as a nitrogen source for the yeast, thereby enhancing their growth (Pérez-Carrillo and Serna-Saldívar, 2007). Fig. 5.2 shows a simplified scheme of the different steps for ethanol production from sweet sorghum grains. They include: milling, gelatinization, liquefaction (dextrinization)/saccharification, distillation, and dehydration. During the processing of grain to ethanol, a by-product called DDGS (Dried Distiller's Grains with Solubles) will be obtained, which is rich in nutrients and used as a valuable feedstock for livestock. Each of the individual steps is explained in the following section.

5.3.2.1.1 GRAIN CRUSHING AND MILLING

Milling can be done either as dry or wet manner. In dry grinding the grain is first size-reduced, followed by liquefaction, whereas in wet milling

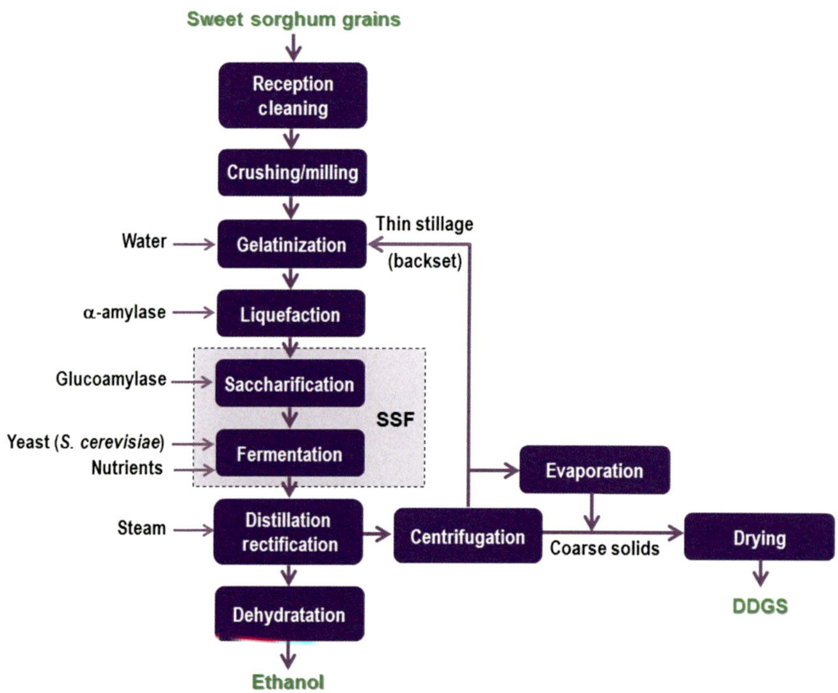

FIGURE 5.2 **Simplified diagram of ethanol production from sweet sorghum grains.**

the grains as a whole are soaked in water, to separate the germ from the kernel, followed by liquefaction. The advantages of dry milling are:

- Less cost concerning plant and equipment
- Simple processing
- Shorter time from construction to operation
- Minimal loss of starch

5.3.2.1.2 GELATINIZATION

In this stage, steam will be passed over the milled grains, and as a result, the starch would swell, decrystalize and solubilize in water. This paves the way for α-amylase to break down the starch to dextrins. The temperature should be maintained around 68–72°C (Ray and Ward, 2006).

5.3.2.1.3 LIQUEFACTION (DEXTRINIZATION)/ SACCHARIFICATION

In this step, the gelatinized starch is broken down to dextrins using thermostable α or β-amylases or glucoamylases. This, in turn, reduces the viscosity of the solution. This step has to be done immediately to prevent recrystallization of starch (Ray et al., 2008).

Lareo et al. (2012) reported 100% hydrolysis of sweet sorghum grains starch of a Topper variety, performed using commercial enzymes: α-amylase (Liquozyme SC, Novozymes: 86°C, 90 min, pH 5.8) for liquefaction and amyloglucosidase (Spirizyme Fuel, Novozymes: 60°C, 60 min, pH 4) for saccharification steps. A solid/water ratio of 1/5 was used, and a hydrolysate solution containing 157 g/L reducing sugars (mainly glucose) were obtained.

5.3.2.1.4 FERMENTATION AND DISTILLATION

The mash that is obtained in the previous step can be now sent for fermentation where yeast ferments glucose to alcohol. The ethanol produced is recovered by distillation and further dehydrated as indicated in a later section.

Lareo et al. (2012) performed fermentation studies of sweet sorghum grains of a Topper variety with a solid/water ratio of 1/5. An ethanol concentration of 60 g/L, an ethanol yield of 95% of the theoretical value and 420–440 L ethanol per ton of sweet sorghum grain in dry basis were obtained. Sugars were completely consumed after 24 h and 12 h for fermentations without and with addition of salts, respectively. Ethanol yields in the range 430–2370 L/ha have been reported for sweet sorghum grains (Barcelos et al., 2016; Castro et al., 2017), as shown in Table 5.2.

5.3.2.2 Techniques to Improve Ethanol Yield From Sorghum Grain

The following advanced techniques are followed to improve ethanol yield from sorghum grains.

5.3.2.2.1 BIOTECHNOLOGY TOOLS TO IMPROVE ETHANOL YIELD FROM SORGHUM GRAIN

The principal components that need to be studied for ethanol production from sorghum grain are starch, proteins, and tannin contents. Poor protein and starch digestibility affect the ethanol fermentation process. The use of transformation and genetic modification in sweet sorghum can produce sorghum varieties that can overcome the barriers to ethanol production. Waxy sorghums are genetically modified species that contains less amylose and more amylopectin. This has resulted in better starch digestibility, and thereby higher ethanol conversion (Wu et al., 2006). Differential scanning calorimetry (DSC) thermograms of starch from waxy sorghum and waxy maize showed similar peak patterns, but the natural sorghum starch showed addition peak indicating amylose–lipid complex. This limits the availability of starch. Thus waxy starch results in better conversion efficiencies to ethanol (Wu et al., 2010).

Sorghum varieties with high lysine, high protein digestibility have been developed (Wu et al., 2010). These varieties can swell and gelatinize at lower temperatures and possess high nutrient value and bioavailability that can be useful even for monogastrics. The major focus of sorghum breeding programs is to develop drought- and pest-tolerant hybrids with increased yields. Thus biotechnology tools can lead to increased yield of sorghum per ha and subsequently higher ethanol production (Rooney et al., 2007).

As explained before, protein digestibility is related to ethanol production and this digestibility, in turn, is related to the tendency of sorghum proteins to form web-like structures during mashing, which reduces the ability of enzymes to access starch. Protein solubility should decrease with the increase of protein crosslinking; thus this parameter can be used as a quality indicator in sorghum biorefineries (Zhao et al., 2009).

5.3.2.2.2 USE OF EXOGENOUS ENZYMES

Addition of proteases before starch gelatinization and starch liquefaction can improve starch hydrolysis and fermentation (Pérez-Carrillo and Serna-Saldívar, 2007). A study carried out by Pérez-Carrillo et al. (2008) with the addition of proteases on decorticated and whole sorghum meals revealed that the decorticated sorghum produced improved ethanol yields compared to the whole sorghum meals. The fermentation time also reduced to half of that required for conventional sorghum fermentation without the addition of proteases, thus leading to energy savings.

5.3.2.2.3 GERMINATION AND SPROUTING

Germinated or sprouted kernels from sorghum grain when used for ethanol fermentation resulted in higher yields and shorter fermentation time compared to the regular ones being used (Yan et al. 2009, 2010). This improved efficiency is because of the action of intrinsic enzymes in starch, proteins, and cell walls. Despite the advantages of purposely sprouting sorghum kernels, it should be remembered that large-scale sprouting requires a considerable amount of water, labor, and energy for drying.

5.3.2.2.4 PHYSICOMECHANICAL AND CHEMICAL TECHNOLOGIES TO IMPROVE ETHANOL YIELD

Apart from the aforementioned strategies, physicomechanical technologies can also be employed to eliminate the barriers for amylase enzymes to produce sugars, which in turn can be used for ethanol production. Some of these approaches include size reduction, steam flaking, and decortications.

Particle size reduction: Reduction in particle size of sorghum meals can pave the way for improved fermentation efficiencies. The gelatinization temperature is lesser for fine sorghum samples compared to larger particles (Wang and Liu, 2009). Also, size reduction has also paved the way for better access to amylase enzymes toward starch. A study by Mahasukhonthachat et al. (2010) found that diffusion is the mechanism for starch digestion and this is based on the average particle size.

Decortication: The outer cover of the grain that comprises of pericarp, aleurone, testa, and peripheral endosperm can hinder the digestibility of starch during liquefaction. Hence, it is important to remove them. That can be done using commercial machines equipped with abrasive disks. The resulting decorticated grain can be milled to a powder before ethanol fermentation. Decortication increases the amount of starch and reduces the fat, phenolics, and fiber (Rooney and Serna-Saldívar, 2000). This paves the way for better access to α-amylase enzymes to break down starch to simple sugars and thereby increased ethanol production.

Decorticated sorghum meals with starch were more susceptible to α-amylases and thereby resulted in high ethanol yields (Alvarez et al., 2010). During decortications, the bran will be separated, and it can be used as animal feed.

Ramírez et al. (2016) found that the removal of tannins using formaldehyde and an alkaline treatment presented a better performance of the simultaneous hydrolysis and fermentation, suggesting that those treatments may have eliminated inhibitors and also contributed to a partial depolymerization or loss of crystallinity of starch.

Steam-flaking: The steam-flaking method utilizes steam to break down the endosperm after which flaking is done through grooved mills. Steam is passed for 15–30 min. The moisture content of the sorghum should be around 21% before flaking. To avoid processing losses conditioning agent or surfactant, such as lecithin, is added to obtain flakes (Serna-Saldívar, 2010). After flakes are obtained, they can be powdered using mills. These steam-flaked sorghum grains are more easily accessible to amylase enzymes leading to the better digestibility of starch and thereby increased ethanol production. Steam-flaked sorghum yielded 40% more ethanol than conventional sorghum (Chuck-Hernandez et al., 2009). But the usage of steam will add to energy costs. The usage of steam for 1 ton of sorghum has been estimated to be US$7.5 (Chuck-Hernandez et al., 2009).

5.3.3 Processes for Conversion of Sweet Sorghum Bagasse Into Bioethanol

Ethanol production from sorghum bagasse follows the concept of lignocellulose-based ethanol production. It is a lignocellulose material that comprises mainly of cellulose, hemicelluloses, and lignin. The cellulose can be broken down by cellulase enzymes to yield glucose. But before that, the biomass needs to be pretreated by either acid, alkaline, steam explosion, or any other method to separate the different components and make the biomass susceptible to enzymatic hydrolysis. The separated lignin after pretreatment can be burned to generate heat. The glucose obtained can be utilized by *S. cerevisiae* to produce ethanol (Gnansounou et al., 2005). Fig. 5.3 shows simplified diagram of the different steps for ethanol production from sweet sorghum bagasse (SSB). Following are some of the commonly used pretreatments for SSB.

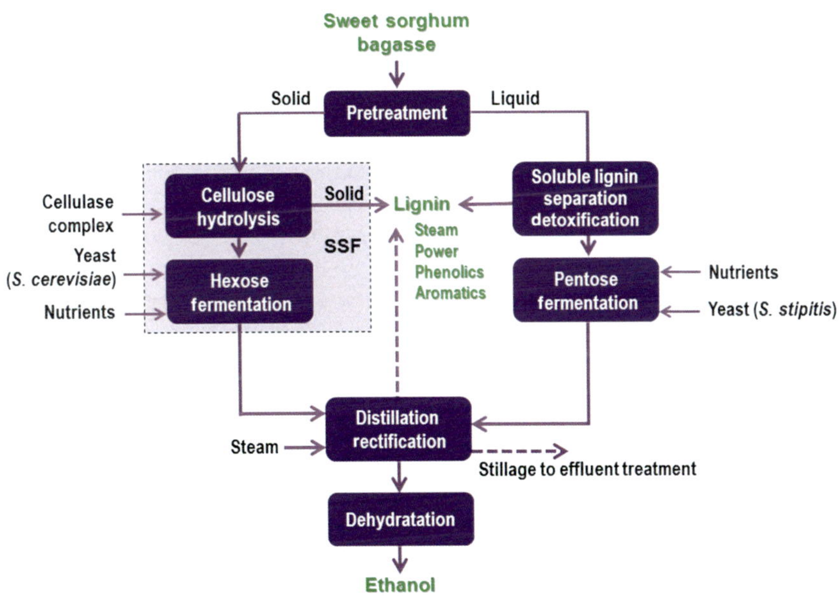

FIGURE 5.3 **Simplified diagram of ethanol production from sweet sorghum bagasse (cellulose and hemicellulose) using a biorefinery concept.**

5.3.3.1 Pretreatments for SSB

5.3.3.1.1 STEAM EXPLOSION

Steam-explosion pretreatment has shown to be effective for sorghum bagasse without catalyst addition (Tomás-Pejó et al., 2008). Castro et al. (2017) studied SSB pretreated by diluted phosphoric acid (0.5% w/w) steam explosion (5 min, 190°C). They found similar composition of the pretreated solids to the values reported by Pengilly et al. (2015) in a study of SSB pretreated with steam (5 min, 200°C). The recovery of xylose in the liquid fraction was in the range 35%–55%. Sipos et al. (2009) studied the ethanol production from SSB using steam explosion as pretreatment of the material at different temperatures and times. It was ground to powder and was impregnated with 3% w/w sulfur dioxide before the steam explosion. Two sets of conditions (190°C, 10 min and 200°C, 5 min) were found to be efficient to reach conversion of 85%–90% of cellulose. A study by Ballesteros et al. (2004) using steam pretreatment on SSB without sulfur dioxide impregnation resulted only in 20% cellulose solubilization. Hemicellulose was well solubilized. At the end of pretreatment, only 2% was left out in the fibros residue.

5.3.3.1.2 DILUTE ACID HYDROLYSIS

The most common pretreatment method is by using dilute acid. Besides, it is less expensive (Gnansounou et al., 2005). Sulfuric acid, hydrochloric, acetic and hydrofluoric acids are the most common acids used (Herrera et al., 2003). It involves the addition of dilute acid (0.1%–10% w/v) to the powdered raw material and introducing the sample to the autoclave. After acid pretreatment, the material predominantly comprises of cellulose and lignin, and the addition of enzymes at this stage will increase the number of fermentable sugars (Tellez-Luis et al., 2002). A study by Kurian et al. (2010) on ethanol from sorghum bagasse used sulfuric acid pretreatment at a concentration of 5 g/kg on sorghum bagasse. It was achieved 92 g/L of total sugars. The temperature was 140°C and time was 30 min.

5.3.3.1.3 ALKALI PRETREATMENT

Alkali pretreatment removes the lignin portion of the biomass and thereby enriching both cellulose and hemicelluloses. Alkali pretreatment prefers less harsh conditions that include time, temperature, and pressure. Commonly used alkali agents for pretreatment are sodium hydroxide, ammonia, and lime. The advantages of using these chemicals include low cost, high possibility of recycling water and chemicals (McIntosh and Vancov, 2010). The most common temperature used for alkali pretreatment is 60°C and 120°C for harsher conditions.

5.3.3.2 Enzyme Hydrolysis

Cellulose and hemicellulose can be converted to sugars using a mixture of pectinase, cellulases, and hemicellulases. The cellulases are an enzyme complex that comprises endoglucanase and exoglucanase enzymes (Lin et al., 2011) and hydrolyses cellulose to glucose. Despite being expensive, enzymes can lower the production cost (Gnansounou et al., 2005). Enzyme complex can be added before or after pretreatment. After chemical pretreatment, a solid and liquid fraction would be obtained. The liquid fraction is rich in xylose and can be fermented to ethanol using the pentose phosphate pathway or used as a medium for the production of cellulase or other valuable products. The solid fraction can be hydrolyzed to glucose and used for the fermentation of ethanol. An enzyme complex consisting of an endoglucanase, exoglucanase, xylanase, beta-glucosidase, and cellulase was added to an alkali pretreated sorghum bagasse and the enzyme hydrolysis yielded 540 g glucose/kg raw material. Similarly to alkali pretreated sorghum straw, 235 g xylose/kg was released (McIntosh and Vancov, 2010). Pengilly et al. (2015) optimized a combination of cellulase (Cellic CTec2) and endo-xylanase (Cellic HTec2) enzyme preparations adapted to steam treated sorghum bagasse. The optimum was defined as the minimum total enzyme dosage, that is, needed for a target glucan conversion of at least 80% for an incubation period of 72 h.

5.3.3.3 *Ethanol Fermentation From SSB*

Ethanol can be produced by fermentation of the hydrolyzed cellulosic and hemicellulose fractions of the bagasse. *S. cerevisiae* is the most common microorganism used to ferment glucose to ethanol. However, it cannot ferment xylose (major sugar from hemicellulose). *Scheffersomyces stipitis* is a fermenting xylose yeast, which could be used. On the other hand, during the pretreatments of the sorghum bagasse several inhibitors of the enzymatic hydrolysis and of the fermentation process can be produced. A detoxification process can be performed to remove these compounds to improve the ethanol yield and efficiency. Some researchers have used genetically modified microorganisms.

Ban et al. (2008) fermented sweet sorghum pretreated with phosphoric acid under mild conditions. The liquid hydrolyzate was fermented by *Pachysolen tannophylus* and *S. cerevisiae* with the addition of 60 FPU/g dry material of cellulase. The results showed that the optimal reaction conditions were 120°C, substrate concentration 80 g/L, 80 min and acid concentration 10%. Under these conditions, 0.30 g reducing sugar/g dry bagasse was obtained. Fermentation by *Pachysolen tannophyllus* resulted in an ethanol concentration of 14.5 g/L. A total ethanol yield of 0.147 g/g dry bagasse was obtained.

Sipos et al. (2009) studied steam pretreatment of SO_2-impregnated SSB at 190°C during 10 min and 200°C during 5 min. Cellulose conversion of 89% and 92% of the separated washed fiber fraction could be achieved, respectively. Applying separation step, cellulose fibers can be separated from inhibitory components of yeast ethanol fermentation, such as furfural, hydroxymethylfurfural, acetic acid, or formic acid. The advantageous effect of inhibitor removal in hydrolysis was observed only at harsh condition–pretreated substrates where the final conversions increased by about 7%. Fiber separation followed by enzymatic digestion resulted 80%–90% glucose to ethanol yields by *S. cerevisiae*. The pentoses can be separated before enzymatic hydrolysis and can be processed to several valuable products instead of fermented to ethanol. Thus the processing of SSB to ethanol finds applications in biorefinery.

The sweet sorghum varieties GK-Coba, Mn-1054, Ramada, Mn-4508, and SS-301, were analyzed for their ethanol productivity, sugar, and fiber contents. The striped stalks were squeezed to form bagasse. The bagasse and juice from stalks were used for ethanol production. All varieties significantly differed in yield of stripped stalk, juice, and bagasse. The microorganisms used were *S. cerevisiae* ATCC 7754 and *Zymomonas mobilis* ATCC 29191. Only the juices from the stalks of GK-coba, Mn-4508, and SS-301 were utilized for ethanol production as they had high quantities of sugars in them. The varieties Mn-1054, Ramada and SS-301, had high fiber content in their stalks, so they were squeezed to form bagasse, which was directly used for ethanol production. Bagasse was pretreated using 2% of sulfuric acid at 120°C for 60 min and later filtered. The sugar-rich filtrate was pH adjusted to 7.0 and nutrients required for ethanol production were added. Fermentation was carried out by both individuals and in mixed culture manner at a ratio of 1:1. The variety SS301 produced more ethanol from juice and bagasse than any other variety by mixed culture fermentation. The concentration of ethanol from juice and bagasse was 50.26 and 10.5 mL/L, respectively. It was estimated that 160 mL of bioethanol can be produced per kg of variety SS-301, when using both juice and bagasse (Khalil et al., 2015).

5.4 BIOTECHNOLOGY OF SORGHUM FERMENTATION

The following biotechnological approaches are adapted for improving sorghum fermentation.

5.4.1 Solid Substrate Fermentation

Ethanol is produced mostly by the process of submerged fermentation. It requires large amounts of water and energy when it comes to the large-scale manufacture of products. Solid state fermentation could be a good alternative to overcome the limitations of submerged fermentation. It is a process where microbial growth and product fermentation occurs on the surface of the solid substrate, in the absence of "free" water, where the moisture is absorbed to the solid matrix. Solid substrate fermentation has several potential advantages: (1) less requirement of water (especially attractive in summer months when water is scarce), (2) smaller volumes of the fermentation mash, (3) less physical energy requirement, (4) more modest capital investment, fewer operating costs, and lower space requirements, (5) reduced reactor volume, (6) more natural product recovery, and (7) less liquid water to be disposed of and hence fewer pollution problems (Ray et al., 2008). There are some reports of ethanol production from sorghum using solid substrate fermentation.

Chopped sweet sorghum particles were used as substrates for solid-state fermentation of ethanol by yeast strain (*S. cerevisiae*), which had ethanol tolerance in static flasks. The influence of various process parameters, such as temperature, yeast cell concentration, and moisture content, on the rate and extent of ethanol fermentation, was investigated. Optimal values of these parameters were found to be 35°C, 7×10^8 cells/g raw sorghum, and 70% moisture level, respectively (Kargi et al., 1985). In another study, chopped sorghum particles were used as a solid substrate for ethanol fermentation using ethanol-tolerant yeast strain in rotary drum fermentor. The effect of rotational speed on ethanol fermentation was studied, and it was found that increase in rotational speed reduced the ethanol fermentation (Kargi and Curme, 1985). Best results were obtained at 1 rpm with the rate of ethanol formation and yield to be ca. 3.1 g ethanol/L/h (based on expressed juice volume) and ca. 9.6 g ethanol/100 g mash.

Ethanol productivity by solid state fermentation depends on the following factors, that is, particle size, moisture content, size of the inoculum, temperature, and coculturing two or more microorganisms. In another study ethanol production from sweet sorghum by solid-state fermentation by a mixed culture of thermotolerant *S. cerevisiae* and an amylolytic *Bacillus* sp. (VB9) was studied. Results showed that mixed cultures yielded higher ethanol yield of 5 g ethanol/100 g of the substrate) than monoculture (3.5 g ethanol/100 g substrate) of the yeast alone. Other important parameters that were studied include temperature, 37°C, moisture, 70%, substrate particle size, 60–100 μm, and inoculum size, 10% (Sree et al., 1999).

The sugars in unsterilized sweet sorghum solids fermented readily during anaerobic solid-state fermentation. Products, based on 100 g hexose, were: 36.7 g volatiles (mass loss), 32.8 g ethanol, 2.8 g glycerol, 12.7 g lactic acid, 2.0 g acetic acid, 4.8 g dextrans, and 4.2 g mannitol, presumably from fermentation with naturally occurring heterolactic bacteria, such as *Leuconostoc dextranicum*, as well as wild yeasts. Inoculation of unsterilized sorghum solids with 0.007%–0.20% *S. cerevisiae* increased fermentation rate, gas evolution, and yields of ethanol, glycerol, and acetic acid, while other yields decreased. With 0.2% yeast, yields were 40.9 g ethanol, 4.8 g glycerol, and 45.5 g gases (Bryan, 1990).

Temperature, particle size, yeast inoculation rate, and water content were the parameters checked for yeast growth, CO_2, ethanol formation, and sugar use in a stable state fermentation using dry sweet sorghum (Shen and Liu, 2009). The effects of temperature and particle size on yeast growth and ethanol yield were significant. Despite yeast, inoculation rate and water content had less impact on ethanol formation on the two parameters influenced by yeast growth. For a good yeast growth, the temperature should be maintained at 25–30°C, and particle size should be around 1.6–2.5 mm. The obtained cell yield values were 0.27 and 0.35 mg/mg. Regarding ethanol production, the values of temperature and particle size were 35–40°C and 0.9–1.6 mm and higher ethanol yield values obtained were 0.24 and 0.27 mg ethanol/mg sugar. Higher yields values of 0.25 mg ethanol/mg sugar and 0.30 mg cell/mg sugar could be obtained when yeast inoculation rate was kept at 0.25. Around 76.47% of water content needs to be there for ethanol production and yeast growth.

5.4.2 Simultaneous Saccharification and Fermentation (SSF)

The advent of simultaneous SSF is a huge process development for enzymatic hydrolysis of biopolymers like starch, cellulose, and hemicellulose present in biomass (Ray and Naskar, 2008). This process reduces the number of reactors involved by employing thermotolerant yeast that eliminates the need for separate reactors for saccharification reaction. This also avoids product accumulation problems associated with enzymes, such as generation of amyloglucosidase and cellobiase that stops cellulose and starch hydrolysis (Chandel et al., 2007). In SSF, the enzymes for hydrolyzing polymers and the fermenting microorganisms are added simultaneously. The enzymes hydrolyze the polymers into simple sugars, and the fermenting microorganisms convert them into ethanol.

Some microorganisms can ferment cellulose into ethanol directly. A few fungal species, such as *Neurospora crassa* and *Fusarium oxysporium* were reported in the 1980s to have the ability to ferment cellulose directly to ethanol by SSF (Deshpande et al., 1986). A study on ethanol production from sweet sorghum was carried out using the strain *Fusarium oxysporum* F3 alone and in combination with *S. cerevisiae* 2541 or *Z. mobilis* Cp4 in fed-batch fermentation. Though SSF took place using the first strain, maximum yield was obtained with the mixed culture of fungus and yeast. The concentration of ethanol and its yield obtained under optimum conditions were 7.5% (w/v) and 29.7 g ethanol per 100 g sweet sorghum stalk. Further investigation needs to be done in this process for large-scale production of fuel ethanol from sweet sorghum (Christakopoulos et al., 1994; Lezinou et al., 1994). A mixed culture of *F. oxysporum* and *S. cerevisiae* was used for ethanol production from sweet sorghum by simultaneous SSF method. Initially, *F. oxysporum* was grown aerobically to produce enzymes necessary for the breakdown of cellulose and hemicellulose after which *S. cerevisiae* and *F. oxysporum* together can ferment the simple sugars obtained into ethanol. Three batches of sorghum were used in the study that was harvested at different periods of the year. The yield and concentration of ethanol obtained was 4 g ethanol/100 g of fresh sorghum and 9% (w/v), respectively. In all experiments, the practical ethanol yield exceeded the theoretical yield, by 1% due to bioconversion of polysaccharides to ethanol (Mamma et al., 1996).

5.4.3 Cell Immobilization

One important process development for alcoholic fermentation, which has gained scientific and technical interest is the immobilization of enzymes, cells, or both together (coimmobilization). The cell biomass or the enzyme can be concentrated inside the reactor by this method, and as a result, the bioconversion efficiency and reactor productivity increase whereas the size of the reactor decreases thus saving space and money (Nunez and Lema, 1987). Though immobilization could be carried out in a different number of ways, adsorption and entrapment in matrices are the most commonly used methods (Kim and Rhee, 1993; Yu et al., 1996).

Shen et al. (2009) found that the best conditions for maximum production of ethanol from sweet sorghum stalk juice using immobilized *S. cerevisiae* in shake flasks were temperature 34°C, agitation rate of 150–200 rpm, pH of 4.5, and particles stuffing rate of 25%. The ethanol yield, CO_2 weight loss rate and fermentation time were 95.15%, 0.508 g/h, and 6 h in shake flasks, respectively. But results in a 5 L nonagitated reactor were: ethanol yield 96.72% and fermentation time 10 h. Under the experimental conditions studied, ethanol fermentation by immobilized cells in shake flasks was found to be better than free cells in the conventional 5-L reactor.

Another study was done by Liu et al. (2008) on ethanol production from sweet sorghum stalk juice in shaking flasks and 10 L fluidized bed reactor using Yuantian No. 1 sweet sorghum cultivar with immobilized yeast cells. The order of influence of the nutrients for ethanol production is as follows: $(NH_4)_2SO_4$ > $MgSO_4$ > K_2HPO_4 in shake flasks. The optimum values of these nutrients for ethanol production were K_2HPO_4, 0%; $(NH_4)_2SO_4$, 0.2%; and $MgSO_4$, 0.05%. When the same quantity of nutrients was used for 10 L fluidized bed reactor, the ethanol content and yield became 6.2% v/v and 91.6% with the fermentation being run for 5 h. Moreover, the use of immobilized yeast compared to a conventional one resulted in fermentation times getting reduced 6–8 times in a fluidized bed reactor.

5.4.4 Use of Adjuncts

High ethanol concentrations are inhibitory for the microorganisms. Ethanol tolerance can be improved with appropriate supplementation of nutrients, such as various sources of nitrogen, vitamins, sterols, and unsaturated fatty acids. Also, the addition of small quantities of oxygen could improve the tolerance of yeast to high ethanol concentrations, which contribute to the stabilization of the plasma membrane by producing sterols (Deesuth et al., 2015; Yue et al., 2012). Many of the supplements used on a laboratory scale are too expensive for industrial use (Larnaudie et al., 2016).

Laopaiboon et al. (2009) studied ethanol production from sweet sorghum juice by *S. cerevisiae* NP01 under VHG fermentation with the addition of various carbon adjuncts and nitrogen sources. When sucrose was used as an adjunct, the fermentation yielded maximum ethanol concentration, productivity, and yield. The sweet sorghum juice contained a total sugar of 280 g/L, 3 g yeast extract/L, and 5 g peptone/L with ethanol concentration, productivity, and yield recorded as 120.7 g/L, 2.1 g/L/h, and 0.5 g/g, respectively. When sugarcane molasses was used as an adjunct, the recorded ethanol concentration, productivity, and yield were 109.3 g/L, 1.5 g/L/h, and 0.5 g/g, respectively. When ammonium sulfate was used as a nitrogen source in the fermentation medium, it resulted in a reduction of ethanol concentration and yielded for approximately 14% when compared to those of the unsupplemented juices.

5.5 TECHNOECONOMIC FEASIBILITY AND REAL-TIME APPLICATIONS

There is a rising demand for energy consumption and liquid fuels in rural areas of Brazil, and so an alternative to conventional fuels is required. Ethanol should be a good alternative to conventional fuels (Mayer et al., 2016). A study conducted by Rezende and Richardson (2017) evaluated the economic feasibility and the risks associated with ethanol production using sweet sorghum as raw material. The study was done in a representative sugar mill in Sao Paulo state, Brazil. The profit/loss for the sugar mill with and without using sweet sorghum was studied. During the sugarcane off season, 20% of the sugarcane land was cultivated with sweet sorghum. A Monte Carlo simulation was carried out to analyse the results, which showed that cultivation of sorghum on 20% of sugarcane land could increase the average net cash income and reduce the relative risk for net income and net present value. The analysis also showed that this kind of sorghum mix on sugarcane land would reduce the costs for the mill by spreading fixed costs across ethanol. The benefits of getting increased ethanol through this method can overshadow the costs associated with cultivating, harvesting, and producing ethanol from the crop. But despite these benefits, uncertainty about crop yields and labor scheduling has prevented the widespread adoption of sorghum has not occurred in southeastern Brazil. Yield improvements in sorghum can overcome this problem.

A technoeconomic study on producing ethanol from sweet sorghum juice during the time of lettuce off-season in Yuma, Arizona was studied. The off-season period was during November to February. The following sorghum varieties were studied for their ethanol yield, Dale, T-Sugar, M81E, and 350FS. At the end of fermentation, the broth has mixtures of water and ethanol. Thus the economic analysis mainly focused on the distillation and purification process to produce anhydrous ethanol. The two separation processes that were studied are distillation using extractive distillation using ethylene glycol and pressure swing adsorption (PSA). Design and operation parameters were studied using the software Aspen Plus (Version 11.1). To study thermodynamic properties, nonrandom two-liquid (NRTL) model was used. The size of the molecular sieves in PSA was 3 Å. Results from modeling yielded 99.8% wt. ethanol and also showed that extractive distillation results in 12% less cost compared to PSA. The cost associated with distillation and dehydration of ethanol from the fermentation broth of sweet sorghum is 40–50 cents/gal. Regarding extractive distillation, the solvent to feed molar ratio is the most important factor influencing the purity and total annualized cost for ethanol production (Ebrahimiaqda and Ogden, 2017).

5.5.1 Economics of Biofuel Production From Sweet Sorghum and Real-Time Applications in India

Pilot studies have been conducted by Indian Institute of Millets Research on sweet sorghum–based ethanol production in collaboration with distilleries in India. Some of them are M/S Renuka Sugars, Belgaum; Sagar Sugars, Chittoor, Andhra Pradesh; Praj Industries, Pune; National Sugar Institute, Kanpur; Somaiya Organo-chemicals, Sakarwadi; India Glycols Ltd, Khashipur; KCP Sugars, Laxmipuram, Andhra Pradesh; Nav Bharat Ventures, Samalkot, Andhra Pradesh in the period of 2001–06. The juice obtained from the crushed sweet sorghum stalks were used for ethanol production. The yield of ethanol was 25–40 L/tonne of crushed sweet sorghum stalks (Dayakar Rao et al., 2004). Another study conducted by a sugar factory in Karnataka, which used 356 tons of sweets sorghum stalk, obtained a yield of 38.89 L/tonne stalk (Hunsigi et al., 2010). Studies on technoeconomic feasibility of ethanol production from sweet sorghum have shown that it costs 0.03$ less than that from molasses (Umakanth, 2016) (Table 5.3).

The cultivation cost of sweet sorghum ranges US$108–122 with the cost of inputs, such as fertilizers accounts for 20%, and labor costs account for remaining the cost of cultivation. In India, agricultural irrigation has seen good improvements such that irrigation expenditure has come down roughly from 6.2% to 1.3%, which is approximately 800–1000 rupees/ha. Canal irrigation is an example of the improvized irrigation system. Table 5.4 shows the net returns involved in the cost of cultivation of sweet sorghum.

Based on the soil types used for sweet sorghum there was the difference in revenues obtained. The medium black soil generated revenue of US$63.5, while the revenues for shallow red soil and medium to shallow black soil was US$55.8 and US$37.3 (Umakanth, 2016).

Ethanol production from the juice of sweet sorghum was studied on a pilot scale by M/S Renuka sugars, Munoli, Belgaum. Two varieties SSV 74 and SSV 84 along with one hybrid Madhura were grown on more than 600 acres of dry land. Nearly 157 farmers from 66 villages were involved in this project of growing SSV 84 in 125 ha (Fig. 5.3). For one batch fermentation of sorghum 112 tonnes of cane was used and the Brix content in the juice was 18%. Fermentation

TABLE 5.3 Comparison of Particulars Per Liter Cost of Ethanol Production From Sweet Sorghum and Sugarcane Molasses in India

Particulars	Sweet sorghum (INR[a]/L)	Sugarcane molasses (INR/L)
Human power	0.5	0.25
Steam	1	1
Electricity	1	1
Yeast	0.1	0.1
Management/ administration	0.1	0.25
Pollution control	Nil	0.25
Raw material	10.41	12.13
Total	13.11	14.98

1 INR = US$0.015.

[a] *INR, Indian Rupees.*

Source: Sweet sorghum stalk @ INR 500 ton/ha; sugarcane molasses @ INR 2000 ton/ha (Dayakar Rao, B., Ratnavathi, C.V., Karthikeyan, K., Biswas, P.K., Rao, S.S., Vijay Kumar, B.S., Seetharama, N., 2004. Sweet sorghum cane for bio-fuel production: a SWOT analysis in Indian context. NRCS Technical Report No. 21/2004. National Research Centre for Sorghum, Rajendra Nagar, Hyderabad. 20 pp.).

was carried out for 48 h, and 9% of ethanol from juicewas obtained. The bagasse obtained was 46.44% with 2.58% sugars (Ratnavathi, 2017).

Compared to sugarcane, sweet sorghum yielded higher net returns to farmers in the area of a study done by Praj industries, Pune. It was US$256.5/ha for sweet sorghum compared to US$17.4/ha for sugarcane. But the gains for sweet sorghum were lower than sugarcane in catchment areas of Sagar sugars in Chittoor. Thus repeated field trials need to be done to ascertain the exact values of net gains (Ratnavathi, 2017).

The National Research Centre for Sorghum (NRCS), Rajendranagar, Hyderabad started producing SSV 84 breed of sorghum to sell them to various refineries that produce ethanol from sweet sorghum. Industries from Andhra Pradesh, Tamil Nadu, Orissa, Maharashtra, and Karnataka showed interest in buying these seeds. About 13.5 tonnes of seed were produced by NRCS and was sold at a rate of 40/kg.

TABLE 5.4 Net Returns Involved in Cost of Cultivation of Sweet Sorghum in India

Category	All categories	Marginal	Small	Medium
Grain yield (kg/ha)	346	255	449	299
Grain value (INR[a]/ha)	2214	1617	2865	1912
Stalk (q/ha)	141	150	136	145
Stalk value (INR/ha)	8504	9009	8170	8714
Gross income (INR/ha)	10,718	10,626	11,036	10,626
Total expenses excluding family labor (INR/ha)	7719	7641	7522	6729
Net income excluding family labor (INR/ha)	2999	2986	3514	3897

Note: Large farmers included in all category.

1 INR = US$0.015.

[a] *INR, Indian Rupees.*

Source: Basavaraj, G., Parthasarathy Rao, P., Basu, K., Reddy, C.R., Ashok Kumar, A., Srinivasa Rao, P., Reddy, B.V.S., 2013. Assessing viability of bio-ethanol production from sweet sorghum in India. Energy Policy. 56, 501–508. Open Access Repository, International Crops Research Institute for Semi-Arid Tropics, Patancheru.

5.6 CONCLUSIONS AND FUTURE PERSPECTIVES

Ethanol can be blended with gasoline in existing autos to reduce air pollution and to achieve fuel sufficiency. The high sugarcane/sugar beet cultivation cost, less yield due to climatic factors, instability in ethanol prices, and high molasses rates have opened other possible means to produce ethanol. An immense number of research and studies have taken place in ethanol production from sweet sorghum. The results have been favorable to obtain fuel ethanol from sweet sorghum. It has some advantages over other energy crops like rapid growth rate; limited fertilizer use, wide adaptability, early maturity, and higher water use efficiency. The government of India has set a target of 20% ethanol blending with gasoline by 2020. To achieve that ethanol has to be produced in large scale, it has to be economically viable. The government, research institutions, industries, and farmers can forge a partnership in developing a large-scale and economically feasible fuel ethanol from sweet sorghum. This can reduce our dependence on gasoline to some extent and can also minimize environmental pollution. Among farmers in India, more awareness needs to be created regarding ethanol production from sweet sorghum to cultivate it on a large scale.

References

Almodares, A., Hadi, M.R., 2009. Production of bioethanol from sweet sorghum: a review. Afr. J. Agr. Res. 4, 772–780.

Almodares, A., Sepahi, A., 1996. Comparison among sweet sorghum cultivars, lines and hybrids for sugar production. Ann. Plant Physiol. 10, 50–55.

Alvarez, M., Perez-Carrillo, E., Serna-Saldívar, S.O., 2010. Effect of decortication and protease treatment on the kinetics of liquefaction, saccharification, and ethanol production from sorghum. J. Chem. Technol. Biotechnol. 885 (8), 1122–1129.

Bai, F.W., Anderson, W.A., Moo-Young, M., 2008. Ethanol fermentation technologies from sugar and starch feedstocks. Biotechnol. Adv. 26, 89–105.

Ballesteros, M., Oliva, J.M., Negro, M.J., Manzanares, P., Ballesteros, I., 2004. Ethanol from lignocellulosic materials by a simultaneous saccharification and fermentation process (SFS) with *Kluyveromyces marxianus* CECT 10875. Process Biochem. 39 (12), 1843–1848.

Ban, J., Yu, J., Zhang, X., Tan, T., 2008. Ethanol production from sweet sorghum residual. Front. Chem. Eng. China 2, 452–455.

Barcelos, C.A., Maeda, R.N., Santa-Anna, L.M.M., Pereira, Jr., N., 2016. Sweet sorghum as a whole-crop feedstock for ethanol production. Biomass Bioenergy 94, 46–56.

Basavaraj, G., Parthasarathy Rao, P., Basu, K., Reddy, C.R., Ashok Kumar, A., Srinivasa Rao, P., Reddy, B.V.S., 2013. Assessing viability of bio-ethanol production from sweet sorghum in India. Energy Policy 56, 501–508, Open Access Repository, International Crops Research Institute for Semi-Arid Tropics, Patancheru.

Bayrock, D.P., Ingledew, W.M., 2001. Application of multistage continuous fermentation for production of fuel alcohol by very-high-gravity fermentation technology. J. Ind. Microbiol. Biotechnol. 27, 87–93.

Bryan, W.L., 1990. Solid state fermentation of sugars in sweet sorghum. Enzyme Microb. Technol. 12, 437–442.

Bulawayo, B., Bvochora, J.M., Muzondo, M.I., Zvauya, R., 1996. Ethanol production by fermentation of sweet-stem sorghum juice using various yeast strains. World J. Microbiol. Biotechnol. 12, 357–360.

Bvochora, J.M., Read, J.S., Zvauya, R., 2000. Application of very high gravity technology to the cofermentation of sweet stem sorghum juice and sorghum grain. Ind. Crops Prod. 11 (1), 11–17.

Castro, E., Nieves, I.U., Rondón, V., Sagues, W.J., Fernández-Sandoval, M.T., Yomano, L.P., York, S.W., Erickson, J., Vermerris, W., 2017. Potential for ethanol production from different sorghum cultivars. Ind. Crops Prod. 109, 367–373.

Chandel, A.K., Chan, E.S., Rudravaram, R., Narasu, M.L., Rao, L.V., Ravindra, P., 2007. Economics and environmental impact of bioethanol production technologies: an appraisal. Biotechnol. Mol. Biol. Rev. 2 (1), 14–32.

Christakopoulos, P., Kekos, D., Macris, B.J., 1994. Simultaneous saccharification and fermentation of sweet sorghum carbohydrates to ethanol in a fed-batch process. Biotechnol. Lett. 16, 983–988.

Christofoletti, C.A., Escher, J.P., Correia, J.E., Marinho, J.F.U., Fontanetti, C.S., 2013. Sugarcane vinasse: environmental implications of its use. Waste Manage. 33, 2752–2761.

Chuck-Hernandez, C., Perez-Carrillo, E., Serna-Saldivar, S.O., 2009. Production of bioethanol from steam-flaked sorghum and maize. J. Cereal Sci. 50 (1), 131–137.

Dayakar Rao, B., Ratnavathi, C.V., Karthikeyan, K., Biswas, P.K., Rao, S.S., Vijay Kumar, B.S., Seetharama, N., 2004. Sweet Sorghum Cane for Bio-Fuel Production: A SWOT Analysis in Indian Context, NRCS Technical Report No. 21/2004. National Research Centre for Sorghum, Rajendra Nagar, p. 20.

Deesuth, O., Laopaiboon, P., Klanrit, P., Laopaiboon, L., 2015. Improvement of ethanol production from sweet sorghum juice under high gravity and very high gravity conditions: effects of nutrient supplementation and aeration. Ind. Crops Prod. 74, 95–102.

Deshpande, V., Sulbha, K., Mishra, C., Rao, M., 1986. Direct conversion of cellulose/hemicellulose to ethanol by *Neurospora crassa*. Enzyme Microb. Technol. 8, 149–152.

Ebrahimiaqda, E., Ogden, K.L., 2017. Simulation and cost analysis of distillation and purification step in production of anhydrous ethanol from sweet sorghum. ACS Sustain. Chem. Eng. 5 (8), 6854–6862.

Fernandes, G., Braga, T.G., Fischer, J., Parrella, R.A.C., de Resende, M.M., Cardoso, V.L., 2014. Evaluation of potential ethanol production and nutrients for four varieties of sweet sorghum during maturation. Renew. Energy 71, 518–524.

Gnansounou, E., Dauriat, A., Wyman, C.E., 2005. Refining sweet sorghum to ethanol and sugar: economic trade-offs in the context of North China. Bioresour. Technol. 96, 985–1002.

Guigou, M., Lareo, C., Pérez, L.V., Lluberas, M.E., Vázquez, D., Ferrari, M.D., 2011. Bioethanol production from sweet sorghum: evaluation of post-harvest treatments on sugar extraction and fermentation. Biomass Bioenergy 35, 3058–3062.

Herrera, A., Téllez-Luis, S.J., Ramırez, J.A., Vázquez, M., 2003. Production of xylose from sorghum straw using hydrochloric acid. J. Cereal Sci. 37 (3), 267–274.

Hunsigi, G., Yekkeli, N.R., Kongawad, B.Y., 2010. Sweet stalk sorghum: an alternative sugar crop for ethanol production. Sugar Tech 12, 79–80.

Kargi, F., Curme, J.A., Sheehan, J., 1985. Solid state fermentation of sweet sorghum to ethanol. Biotechnol. Bioeng. 27, 34–40.

Kargi, F., Curme, J.A., 1985. Solid state fermentation of sweet sorghum to ethanol in a rotary-drum fermentor. Biotechnol. Bioeng. 27, 1122–1125.

Khalil, S.R.A., Abdelhafez, A.A., Amer, E.A.M., 2015. Evaluation of bioethanol production from juice and bagasse of some sweet sorghum varieties. Ann. Agric. Sci. 60 (2), 317–324.

Khongsay, N., Laopaiboon, L., Laopaiboon, P., 2010. Growth and batch ethanol fermentation of *Saccharomyces cerevisiae* on sweet sorghum stem juice under normal and very high gravity conditions. Biotechnology 9 (1), 9–16.

Kim, M., Day, D.F., 2011. Composition of sugar cane, energy cane, and sweet sorghum suitable for ethanol production at Louisiana sugar mills. J. Ind. Microbiol. Biotechnol. 38 (7), 803–807.

Kim, C.H., Rhee, S.K., 1993. Process development for simultaneous starch saccharification and ethanol fermentation by *Zymomonas mobilis*. Process Biochem. 28, 331–339.

Kurian, J.K., Ashok, M.K., Banerjee, A., Kishore, V.V.N., 2010. Bioconversion of hemicellulose hydrolysate of sweet sorghum bagasse to ethanol by using *Pichia stipitis* NCIM 3497 and *Debaryomyces hansenii* sp. Bioresources 5 (4), 2404–2416.

Lareo, C., Ferrari, M.D., Guigou, M., Larnaudie, V., Fajardo, L., Ramírez, M.B., 2012. Evaluación de materiales amiláceos experimentales desarrollados para bioenergía (boniato y grano de sorgo dulce) para la producción de bioetanol combustible. FPTA-INIA 39, ISSN 1688-924X.

Larnaudie, V., Rochón, E., Ferrari, M.D., Lareo, C., 2016. Energy evaluation of fuel bioethanol production from sweet sorghum using very high gravity (VHG) conditions. Renew. Energy 88, 280–287.

Laopaiboon, L., Thanonkeo, P., Jaisil, P., Laopaiboon, P., 2007. Ethanol production from sweet sorghum juice in batch and fed-batch fermentations by *Saccharomyces cerevisiae*. World J. Microbiol. Biotechnol. 23, 1497–1501.

Laopaiboon, L., Nuanpeng, S., Srinophakun, P., Klanrit, P., Laopaiboon, P., 2009. Ethanol production from sweet sorghum juice using very high gravity technology: effects of carbon and nitrogen supplementations. Bioresour. Technol. 100, 4176–4182.

Lezinou, V., Christakopoulos, P., Kekos, D., Macris, B.J., 1994. Simultaneous sacharification and fermentation of sweet sorghum carbohydrates to ethanol in a fed-batch process. Biotechnol. Lett. 16, 983–988.

Lin, Z.X., Zhang, H.M., Ji, X.J., Chen, J.W., Huang, H., 2011. Hydrolytic enzyme of cellulose for complex formulation applied research. Appl. Biochem. Biotechnol. 164 (1), 23–33.

Liu, R., Li, J., Shen, F., 2008. Refining bioethanol from stalk juice of sweet sorghum by immobilized yeast fermentation. Renew. Energy 33 (5), 1130–1135.

Liu, R., Shen, F., 2008. Impacts of main factors on bioethanol fermentation from stalk juice of sweet sorghum by immobilized *Saccharomyces cerevisiae* (CICC 1308). Bioresour. Technol. 99, 847–854.

Mahasukhonthachat, K., Sopade, P.A., Gidley, M.J., 2010. Kinetics of starch digestion in sorghum as affected by particle size. J. Food Eng. 96 (1), 18–28.

Mamma, D., Koullas, D., Fountoukidis, G., Kekos, D., Macris, B.J., Koukios, E., 1996. Bioethanol from sweet sorghum: simultaneous saccharification and fermentation of carbohydrates by a mixed microbial culture. Process Biochem. 31, 377–381.

Mayer, F.D., Brondani, M., Hoffmann, R., Feris, L.A., Marcilio, N.R., Baldo, V., 2016. Small-scale production of hydrous ethanol fuel: economic and environmental assessment. Biomass Bioenergy 93, 168–179.

McIntosh, S., Vancov, T., 2010. Enhanced enzyme saccharification of sorghum bicolor straw using dilute alkali pretreatment. Bioresour. Technol. 101 (17), 6718–6727.

Nadir, N., Mel, M., Karim, M.I.A., Yunus, R.M., 2009. Comparison of sweet sorghum and cassava for ethanol production by using *Saccharomyces cerevisiae*. J. Appl. Sci. 9, 3068–3073.

Nunez, M.J., Lema, J.M., 1987. Cell immobilization: application to alcoholic production. Enzyme Microb. Technol. 9, 642–650.

Pengilly, C., García-Aparicio, M.P., Diedericks, D., Brienzo, M., Görgens, J.F., 2015. Enzymatic hydrolysis of steam-pretreated sweet sorghum bagasse by combinations of cellulase and endo-xylanase. Fuel 154, 352–360.

Pérez-Carrillo, E., Serna-Saldívar, S.O., Álvarez, M.M., Cortes-Callejas, M.L., 2008. Effect of sorghum decortication and use of protease before liquefaction with thermo resistant alpha-amylase on efficiency of bioethanol production. Cereal Chem. 85 (6), 792–798.

Pérez-Carrillo, E., Serna-Saldívar, S.O., 2007. Effect of protease treatment before hydrolysis with alpha-amylase on the rate of starch and protein hydrolysis of maize, whole sorghum, and decorticated sorghum. Cereal Chem. 84 (6), 607–613.

Quintero, J.A., Montoya, M.I., Sanchez, O.J., Giraldo, O.H., Cardona, C.A., 2008. Fuel ethanol production from sugarcane and corn: comparative analysis for a Colombian case. Energy 33, 385–399.

Ramírez, M.B., Ferrari, M.D., Lareo, C., 2016. Fuel ethanol production from commercial grain sorghum cultivars with different tannin content. J. Cereal Sci. 69, 125–131.

Ratnavathi, C.V., 2017. Prospects of sweet sorghum as a raw material for ethanol production. Recent Adv. Petrochem. Sci. 3 (3), 555613.

Ratnavathi, C.V., Suresh, K., Vijay Kumar, B.S., Pallavi, M., Komala, V.V., Seetharama, N., 2010. Study on genotypic variation for ethanol production from sweet sorghum juice. Biomass Bioenergy 34, 947–952.

Ray, R.C., Mohapatra, S., Panda, S., Kar, S., 2008. Solid substrate fermentation of cassava fibrous residue for production of α-amylase, lactic acid and ethanol. J. Environ. Biol. 29 (1/2), 111–115.

Ray, R.C., Naskar, S.K., 2008. Bio-ethanol production from sweet potato (*Ipomoea batatas* L) by enzymatic liquefaction and simultaneous saccharification and fermentation. Dyn. Biotechnol. Proc. Biochem. Mol. Biol. 2, 47–49.

Ray, R.C., Ward, O.P., 2006. Post-harvest microbial biotechnology of topical root and tuber crops. Ray, Ramesh C., Ward, O.P. (Eds.), Microbial Biotechnology in Horticulture, vol. 1, Science Publishers, Enfield, New Hampshire, pp. 345–396.

Reddy, B.V.S., Ashok Kumar, A., Reddy, P.S., 2007. Sweet sorghum: a dryland-adapted, pro-poor bioethanol feedstock yielding both grain and fuel. Abstract of the Paper Presented at USDA Biofuel Conference. 20–22 August 2007, Minneapolis, Minnesota. .

Rezende, M.L., Richardson, J.W., 2017. Risk analysis of using sweet sorghum for ethanol production in southeastern Brazil. Biomass Bioenergy 97, 100–107.

Rooney, W., Blumenthal, J., Bean, B., Mullet, J.E., 2007. Designing sorghum as a dedicated bioenergy feedstock. Biofuels Bioprod. Bioref. 1 (2), 147–157.

Rooney, L., Serna-Saldívar, S., 2000. Sorghum. In: Kulp, K., Ponte, J. (Eds.), Handbook of Cereal Science and Technology. Marcel Dekker, New York, ISBN 0824782941.

Serna-Saldívar, S.O., 2010. Cereal Grains: Properties, Processing, and Nutritional Attributes. CRC Press, Boca Raton, FL.

Serna-Saldívar, S.O., Chuck-Hernández, C., Pérez-Carrillo, E., Heredia-Olea, E., 2012. In: Lima, M.A.P. (Ed.), Sorghum as a Multifunctional Crop for the Production of Fuel Ethanol: Current Status and Future Trends. InTech, ISBN: 978-953-51-0008-InTech, Available from: http://www.intechopen.com/books/bioethanol/sorghum-as-a-multifunctional-crop-for-the-production-of-fuelethanol-current-status-and-future-trend.

Shen, F., Liu, R., Wang, T., 2009. Effects of temperature, pH, agitation and particle stuffing on fermentation of sorghum stalk juice to ethanol. Energy Sources 31, 646–656.

Shen, F., Liu, R., 2009. Research on solid state ethanol fermentation using dry sweet sorghum stalk particles with active dry yeast. Energy Fuels 23, 519–525.

Sipos, B., Réczey, J., Somorai, Z., Kádár, Z., Dienes, D., Réczey, K., 2009. Sweet sorghum as feedstock for ethanol production: enzymatic hydrolysis of steam-pretreated bagasse. Appl. Biochem. Biotechnol. 153 (1–3), 151–162.

Sree, N.K., Sridhar, M., Rao, L.V., Pandey, A., 1999. Ethanol production in solid substrate fermentation using thermotolerant yeast. Process Biochem. 34, 115–119.

Teixeira, T.P.M., Pimentel, L.D., Dias, L.A.S., Parrella, R.A.C., da Paixão, M.Q., Biesdorf, E.M., 2017. Redefinition of sweet sorghum harvest time: new approach for sampling and decision-making in field. Ind. Crops Prod. 109, 579–586.

Tellez-Luis, S.J., Ramírez, J.A., Vázques, M., 2002. Mathematical modelling of hemicellulosic sugar production from sorghum straw. J. Food Eng. 52 (3), 285–291.

Thomas, K.C., Hynes, S.H., Ingledew, W.M., 1996. Practical and theoretical considerations in the production of high concentrations of alcohol by fermentation. Process Biochem. 31, 321–331.

Tomás-Pejó, E., Oliva, J.M., Ballesteros, M., 2008. Realistic approach for full-scale bioethanol production from lignocellulose: a review. J. Sci. Res. India 67, 874–884.

Umakanth, A.V., 2016. Sweet sorghum for bio-fuel production: industrial experiences and economics. Best Practices for Sorghum Cultivation and Importance of Value-Addition. ICAR-Indian Institute of Millets Research, Hyderabad, Training Manual 2016-17, ISBN: 81-89335-65-0.

Viator, H.P., Alison, M., Gravois, K., Han, K.J., Harrell, D., Hogan, A., Pittman, W., Salassi, M., Whatley, J., 2009. Sweet sorghum for biofuel production in Louisiana. LA Agric. 53, 16–17.

Wang, F., Liu, C.-Z., 2009. Development of an economic refining strategy of sweet sorghum in the Inner Mongolia region of China. Energy Fuel 23, 4137–4412.

Wu, X., Zhao, R., Wang, D., Bean, S.R., Seib, P.A., Tuinstra, M.R., Campbell, M., O'Brien, A., 2006. Effects of amylose, corn protein, and corn fiber contents on production of ethanol from starch-rich media. Cereal Chem. 83 (5), 569–575.

Wu, X., Jampala, B., Robbins, A., Hays, D., Yan, S., Xu, F., Rooney, W., Peterson, G., Shi, Y., Wang, D., 2010. Ethanol fermentation performance of grain sorghum (*Sorghum bicolor*) with modified endosperm matrices. J. Agric. Food Chem. 58 (17), 9556–9562.

Yan, S., Wu, X., MacRitchie, F., Wang, D., 2009. Germination-improved ethanol fermentation performance of high-tannin sorghum in a laboratory dry-grind process. Cereal Chem. 86 (6), 597–600.

Yan, S., Wu, X., Dahlberg, J., Bean, S.R., MacRitchie, F., Wilson, J.D., Wang, D., 2010. Properties of field-sprouted sorghum and its performance in ethanol production. J. Cereal Sci. 51 (3), 374–380.

Yu, B., Zhang, F., Zheng, Y., Wang, P., 1996. Alcohol fermentation from the mash of dried sweet potato with its dregs using immobilized yeast. Process Biochem. 31 (1), 1–6.

Yue, G., Yu, J., Zhang, X., Tan, T., 2012. The influence of nitrogen sources on ethanol production by yeast from concentrated sweet sorghum juice. Biomass Bioenergy 39, 48–52.

Zhao, Y.L., Dolat, A., Steinberger, Y., Wang, X., Osman, A., Xie, G.H., 2009. Biomass yield and changes in chemical composition of sweet sorghum cultivars grown for biofuel. Field Crops Res. 111, 55–64.

Further Reading

Jacques, K., Lyons, T.P., Kelsall, D.R. (Eds.), 1999. The Alcohol Textbook. third ed. Nottingham University Press, Nottingham, p. 388.

CHAPTER

6

Cassava as Feedstock for Ethanol Production: A Global Perspective

Sanette Marx

North-West University, Potchefstroom, South Africa

6.1 INTRODUCTION

The Paris Agreement limits annual global greenhouse gas (GHG) emissions to between 30 and 50 Gt CO_2eq by 2030 to prevent a global temperature rise of more than 2°C (Shepherd and Knox, 2016). It has been recognized that fossil-based transportation fuels are the largest contributors toward GHG emissions (United Nations, 2015) and therefore GHG emission targets can only be met if current liquid fuel vehicles are replaced by electric cars, or fossil-based fuels are replaced with biobased counterparts, such as bioethanol and biodiesel. Biofuels for transportation accounted for only 0.8% of global renewable energy consumption in 2015 (REN21, 2017). Many countries have opted for a switch to or increase in bioethanol production to meet their GHG targets as agreed upon by the COP21 Paris Agreement. The high oxygen content in bioethanol has the potential to lower harmful particulate emissions by 5%–10% (Thatoi et al., 2016) and GHG emissions by between 8% and 30% (Teodor et al., 2013). However, bioethanol consumption is reported to be 51% higher than conventional petroleum due to the lower energy value of ethanol (30 MJ/kg) (Sebayang et al., 2016).

The challenge for local sustainable bioethanol production is the choice of feedstock and appropriate processing technology to ensure GHG emissions compliance without compromising food and water security. The UN Committee on World Food Security (HLPE, 2013) recommended a combined food and energy policy for governments to ensure that resources necessary for food security, such as land, biodiversity, water, and labor are not at risk when biofuels production is increased to meet energy demands.

Lange (2011) showed that if land use is changed for the cultivation of energy crops, only crops grown on degraded grasslands will have a positive effect on GHG emission reduction. Countries wishing to expand their bioethanol production capacity thus need to consider crops, such as cassava (*Manihot esculenta* Crantz), that can be grown on degraded grasslands or marginal land (Nadir et al., 2009) and does not directly compete with food security. Baeyens

Bioethanol Production From Food Crops
http://dx.doi.org/10.1016/B978-0-12-813766-6.00006-0

et al. (2015) showed that cassava-based ethanol is much more economical to produce than ethanol from corn or sugarcane. However, when considering an energy crop for bioethanol production, factors, such as carbon footprint, water footprint, effect on food and feed chain, eutrophication potential as well as ease of processing, should be taken into account.

6.2 CASSAVA AS ENERGY CROP

Cassava is a shrub belonging to the family Euphoribiaceae that originated in South America (Li and Zhu, 2011; Marx and Nquma, 2013). The plants have tuberous roots with an average length of 5–10 cm and diameters of 15–35 cm (Okudoh et al., 2014). Cassava needs an annual rainfall of approximately 760–1015 mm and can grow in a wide range of soils (Thatoi et al., 2016) in tropical and subtropical regions. The plant's unique ability to utilize nutrients from both its stalks and storage roots when water becomes available makes it a hardy plant that can tolerate prolonged droughts (Nadir et al., 2009). Furthermore, it can be grown in marginal land of low fertility where other crops would fail and has a low input requirement for planting and harvesting (Jakrawatana et al., 2016; Nadir et al., 2009).

Cassava is the fifth most important source of plant-based starch in the world (Jansson et al., 2009). World cassava production reached approximately 250 million tons in 2014 with the main production areas being Africa (Nigeria, Cameroon), Asia (China, Thailand, Malaysia, and Indonesia), and Latin America (FAO, 2014). Africa produces approximately 50% of the world's cassava with Nigeria being the main producer with an annual production of more than 52 million tons in 2014 (Ibeto et al., 2011).

6.2.1 Carbon Footprint

Carbon footprints for different crops are very difficult to compare because the calculation is dependent on local yields and environmental conditions. Machado et al. (2017) compared the soil carbon loss for different crops using field trials in Brazil and found that soil carbon loss from cassava cultivation (14.5 ton/ha) was much higher than that of corn (4.7 ton/ha) and sugarcane (ton Mg/ha) but lower than that of sugar beets (14.5 ton/ha). It is estimated that GHG emissions from cassava-derived bioethanol produced in Malaysia could be between 58% and 72% lower compared to emissions from conventional gasoline use (Hanif et al., 2017; Numjuncharoen et al., 2015). Nguyen and Gheewala (2008) estimated that bioethanol produced from cassava in Thailand could lower GHG emissions by as much as 6%. Papong et al. (2017) compared the GHG emissions from cassava and sugarcane crops in Thailand and showed that when the entire value chain is taken into account both crops have approximately the same GHG emissions profile. The estimated GHG emissions associated with cassava-based ethanol from different countries is compared in Table 6.1.

When comparing the GHG emissions in Table 6.1 with the estimated GHG emissions of fossil-based petroleum (~1800 gCO_2eq/L fuel) most of the estimates of Cassava exceeds that of fossil fuel when GHG emissions associated with direct land use change (dLUC) and indirect land use change (iLUC) is taken into account in the life cycle assessment (LCA) (Machado et al., 2017). Nguyen et al. (2017) calculated that the CO_2 emissions debt due to cassava dLUC would take 25 years to pay back if the ethanol yield per ha can be maintained at 33 ton/ha. A number of studies showed that using water effluent from the cassava processing and ethanol production plants for biogas production could lower the GHG emissions to below that of the equivalent fossil fuel (Zhang et al., 2017). Furthermore, coprocessing of cassava wastes and peels to ethanol could significantly increase the ethanol yield per ha, which would also reduce the GHG emissions associated with a cassava crop.

TABLE 6.1 Comparison of Estimated GHG Emissions for Cassava-Based Bioethanol in Different Countries

Country	Cassava yield (ton/ha)	Ethanol yield (L/ha)	GHG emissions (gCO_2eq/L bioethanol)	References
Nigeria	10–40	1370–5480	n/a	Ohimain (2013)
	11.7	1603	n/a	Anyanwu et al. (2015)
Malaysia	92.2	3600	688	Hanif et al. (2017)
Colombia	10	1847[a]	488	Valencia and Cardona (2014)
Thailand	15.6–45.8	2558–7508[a]	2126	Kawasaki et al. (2015)
			1340	Lecksiwilai et al. (2017)
			1402–2863	Papong and Malakul (2010)
			790–1229	
			700–1400	Papong et al. 2017; Prapaspongsa and Gheewala (2016)
Vietnam	16–18	n/a	738	Le et al. (2013)
			763–2014	Nguyen et al. (2017)
China	40	150	n/a	Jansson et al. (2009)
			1166	Lauven et al. (2014)
			1662	Ren et al. (2015)
	16.4		1402–2863	Zhang et al. (2017)
	16.4		297–551	Zhang et al. (2017)
	12.1–19.3		n/a	Xie et al. (2017)

n/a, Not reported.
[a]*Calculated.*

6.2.2 Water Footprint

Feedstock growth is the most water-intense step in the cassava to ethanol value chain (Xie et al., 2017). Gerbens-Leenes et al. (2009) estimated that the global water footprint is approximately 2926 L/L of bioethanol produced. The blue water footprint was approximately 420 L of water per liter ethanol with the bulk of the water footprint obtained from rainfall. Cassava has the advantage over other energy crops that it can be grown on marginal land under rain-fed conditions and still produce a high yield (Papong et al., 2017). Some values for the water footprint of cassava-based ethanol as determined for different countries is given in Table 6.2.

Large variations in the reported water footprint values are mostly due to the assumptions made during the LCA, the cassava yield per ha as well as the conversion efficiency of the ethanol process. A few studies (Wang et al., 2017; Yang et al., 2017; Zhang et al., 2017) have combined biogas production with ethanol production to lower the water efficiency of the ethanol production process, but because the water footprint during ethanol production is negligible compared to the water footprint of cultivation and harvesting, this did not have a significant effect on the overall water footprint. However, Pingmuanglek et al. (2017) reported on an integrated ethanol from cassava system that made use of biogas production to supply the energy

TABLE 6.2 Comparison of Estimated Water Footprint for Cassava-Based Bioethanol in Different Countries

Country	Water footprint (L H_2O/L ethanol)	References
Nigeria	783	Adeoti (2010)
Thailand	2021–1471	Babel et al. (2011); Gheewala et al. (2013); Mangmeechai and Pavasantm (2013)
	2676	Papong et al. (2017)
	57.81[a]	Pingmuanglek et al. (2017)
China	484.62	Zhang et al. (2014)
	783–2926	Xie et al. (2017)
	2640	Hong et al. (2009)

[a] *Calculated from data.*

needs of the ethanol process. All wastewater and thin stillage from the cassava and ethanol processing plants were used for biogas production or used for irrigation in cassava cultivation. The latter, with optimized fertilizer and nutrient use, resulted in a significant reduction in freshwater requirements for the overall value chain as well as an increase in GHG reduction of 107 tCO_2eq per annum.

6.2.3 Eutrophication Potential

Eutrophication refers to the phenomenon of excessive enrichment of nutrients, such as phosphorous in rivers and bodies of water due to gray water run-off. Gray water that ends up in rivers and dams are mostly produced through agricultural practices (especially fertilizers) (Ruppenthal et al., 1997), poor sewage control and untreated chemical plant effluent. It is a very complex problem to evaluate and there are very few studies that investigated the gray water and pollution potential of cassava-based bioethanol. Xie et al. (2017) compared the different water footprints for sweet sorghum cassava and Jatropha and concluded that sweet sorghum has the highest blue water footprint and cassava has the largest gray water footprint. The authors defined a water stress index to compare the impact of biofuels production on water resources and showed that on average, cassava has a lower water stress index (0.02–0.05 L/MJ biofuel) than sweet sorghum (0.02–0.99 L/MJ biofuel) in most provinces of China (Xie et al., 2017). Gheewala et al. (2014) determined the water stress index for 25 watersheds in different provinces of Thailand and concluded that three watersheds have a high risk for cassava-based bioethanol production, that is, Mun, Chi, and Chao Phraya.

6.3 BIOETHANOL PRODUCTION FROM CASSAVA ROOTS

Cassava is processed into a wide variety of products for use in the food, feed, and energy industry, including chips, flakes, peel starch, and flour (Anyanwu et al., 2015; Ray and Swain, 2011). Most ethanol production plants using cassava as feedstock use chips because the dried chips can be stored under dry conditions until needed in the plant.

Composition of cassava roots will vary significantly, depending on the soil type, environmental conditions, amount of fertilizer used and variety cultivated. On average, cassava roots contain approximately 35% dry matter with a starch content of between 70 and 90 wt.% (Anyanwu et al., 2015; Benvenga et al., 2016;

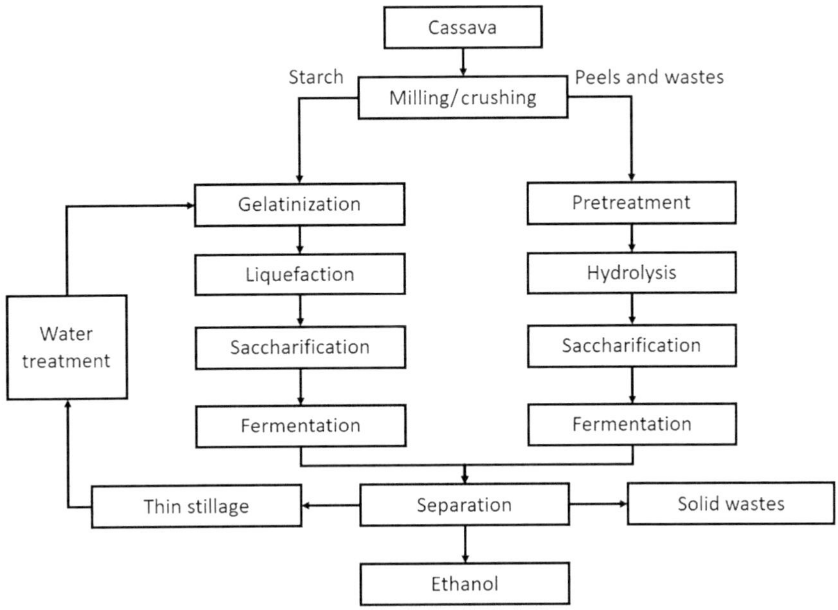

FIGURE 6.1 **Block diagram of processing steps in the cassava to ethanol production process.**

Hansupalak et al., 2015; Marx and Nquma, 2013; Poonsrisawat et al., 2016; Sebayang et al., 2016; Thatoi et al., 2016).

A generic process block diagram for the conversion of cassava to bioethanol is given in Fig. 6.1.

Research has been focused on optimizing the different process steps through optimization of energy usage and minimization of wastewater effluent. Starch is a polysaccharide that needs to be degraded into glucose, which can then be fermented to ethanol using microorganisms. This first step in the process is usually gelatinization, which involves swelling of the starch through cooking to sterilize the mash and break down the pectin. Gelatinization and liquefaction can be combined as liquefaction is usually done at temperatures above 50°C in the presence of α-amylase enzymes. Saccharification is done with glucoamylase enzymes to produce glucose. Hydrolysis, saccharification, and fermentation can be performed as three separate processing steps (SHF process) or the saccharification and fermentation steps can be combined as a single processing step (SSF) (Ray et al., 2008). Sebayang et al. (2016) and Thatoi et al. (2016) give an excellent summary of the details of cassava root fermentation.

6.3.1 New Developments

Lomthong et al. (2016) reported the use of a raw starch degrading enzyme derived from *Laceyella sacchari* LP175 bacteria supplemented with commercial glucoamylase to synergistically degrade starch directly into glucose. A maximum sugar yield of 157 g/L where obtained (66% hydrolysis). A modified SSF process was used in which the hydrolysate was first fermented using *Saccharomyces cerevisiae* M30 followed by a second fermentation using *Kluyveromyces marxianus* DMKU-KS07 to produce a final ethanol concentration of 90.9 g/L (88% theoretical yield). In a similar study, Khamkeaw and Phisalaphong (2016) immobilized cultures from Loog-Pang (Thai rice cake starter) on thin shell silk cocoon for the hydrolysis of cassava starch to glucose. *S. cerevisiae* M30 was used as fermenting organism and an ethanol yield of 71.2 g/L was obtained in a separate hydrolysis and fermentation process. Behera and Ray (2014) studied ethanol production from cassava flour using *S. cerevisiae* immobilized in calcium alginate beads. The growth of the yeast started in the log phage and maximum ethanol (189 ± 3.1 g ethanol/kg flour) production was achieved with 94.74% ± 2.187% sugar conversion during the stationary phase. Moshi et al. (2015) isolated α-amylase from *Bacillus* spp. WN11 and used it in a SSF process with *Caloramator boliviensis* as fermenting organism to directly ferment raw wild and domesticated cassava roots to ethanol. The system was able to ferment 76 g/L glucose to 33 g/L ethanol. In a follow-up study, Moshi et al. (2016) isolated an amylolytic strain of *Aspergillus* sp. from soil collected from the rhizosphere of wild cassava plants (*Manihot glaziovii*). The study assessed the effectiveness of in situ secretion of the amylolytic enzyme during coculture fermentation with *S. cerevisiae* M30 and found that a sugar yield equal to 66% of the theoretical sugar yield could be obtained. A cell wall degrading enzyme was isolated from *Aspergillus aculeatus* obtained from forest soil by Poonsrisawat et al. (2016) to modify the rheology of a cassava root mash. The enzyme was used in enzyme mixtures also containing *Trichoderma reesei* cellulose enzymes and *Aspergillus pectinase* in a ratio of 1:1:1 and SSF with *S. cerevisiae* resulted in a yield of 15.4% (w/w) ethanol (88.14% of theoretical yield). Seong et al. (2016) showed the over-expressions of the SPT15 gene in *S. cerevisiae* can increase the glucose consumption rate and ethanol productivity from cassava starch by 13.4% and 4.5%, respectively, even in the presence of high glucose and ethanol concentrations in the broth. Xu et al. (2016) used a purified raw starch-digesting glucoamylase from *Penicillium oxalicum* GXU20 expressed in *Pichia pastoris* for the direct SSF of raw starch to ethanol. The enzyme was able to hydrolyze both corn and cassava starch and was tolerant to a wide pH range (2–10.5). An ethanol yield of 57 g/L (93.5% fermentation efficiency) could be obtained directly from raw cassava starch. Wang et al. (2017) and Yang et al. (2017) used a combination of anaerobic digestion, aerobic digestion, and ion-exchange resins to recycle the thin stillage back to the gelatinization stage of the process. Yang et al. (2017) achieved an ethanol productivity of 0.9 g ethanol/L/h and could reduce the fermentation time by 28%. Wang et al. (2017) reported an ethanol yield of 11.43% (v/v) that could be maintained after treatment and recycling of the thin stillage with 322 mL of methane produced per gram of COD removed. Hansupalak et al. (2015) reported that the inclusion of anaerobic digestion in the ethanol production chain can lower the carbon footprint of cassava-based ethanol by 431–664 kg CO_2eq per ton starch.

6.4 BIOETHANOL FROM CASSAVA WASTES AND PEELS

The carbon footprint of cassava-based ethanol can be significantly reduced by increasing the ethanol yield per ha. This can be achieved by utilizing the cassava waste from starch-based ethanol plants and cassava processing plants for cellulosic ethanol production. Depending on the starch-to-ethanol process, cassava wastes can have a starch content as high as 60% (w/w) (Elemike et al., 2015). Waste from cassava processing plants include leaves, stems, and peels that can be used as second-generation ethanol feedstock. There are many different processing options for cassava wastes. One option is to first remove the starch through hydrolysis and saccharification followed by fermentation after removal by centrifugation (Ray et al., 2008). The residues can then be pretreated, hydrolyzed, and fermented. Elemike et al. (2015) and Martín et al. (2016) showed that it is better to first remove the starch component through starch hydrolysis and saccharification to prevent the glucose in the starch from degrading during acid or alkali pretreatment of the lignocellulose component. Dilute hydrochloric acid (0.5 M) was used for pretreatment and the hydrolysate after enzymatic hydrolysis using commercial enzymes was fermented using *S. cerevisiae* to obtain a total ethanol yield of 180 g/kg cassava wastes. Martín et al. (2016) used cassava wastes from an ethanol processing plant as feedstock. The starch was also removed through enzymatic hydrolysis and saccharification prior to pretreatment with 1% sulfuric acid solution. The resultant broth was enzymatically hydrolyzed with commercial cellulose enzymes (Behera and Ray, 2016). Sugar recovery was 85% with removal of starch compared to 52% without prior removal of starch. Zhang et al. (2014) introduced a novel process that incorporated a cellulolytic pretreatment step of cassava residues after the separation step and prior to anaerobic digestion. This additional step increased the ethanol yield to 317 g/kg ethanol and 68 g methane per kg of cassava.

A few studies reported on the use of hydrothermal pretreatment of cassava wastes for ethanol production. Akaracharanya et al. (2016) used thermal hydrolysis to remove starch from cassava waste pulp. The lignocellulose component was subsequently hydrolyzed with cellulose enzymes and fermented using a coculture of *S. cerevisiae* and *K. marxianus* to obtain an ethanol yield of 420 g/kg pulp. Nanssou et al. (2016) thermally treated cassava stems and peels obtained from a plantation. The authors used *Rhizopus* spp. for direct hydrolysis and fermentation of the hydrotreated broth and compared it to the ethanol yield obtained after enzymatic hydrolysis with commercial cellulose enzymes and fermentation with *S. cerevisiae*. Ethanol yields were generally very low (26 g/kg for the peels and 52.4 g/kg for the stems) with maximum yield obtained after enzymatic hydrolysis of the hydrotreated broth. Lu et al. (2017) optimized the hydrothermal pretreatment step for glucose recovery from cassava alcohol residues. The authors calculated an ethanol yield of 706 g/kg residue based on completed conversion of the maximum glucose and xylose yields (66.89% sugar recovery) obtained after enzymatic hydrolysis using a commercial cellulase enzyme mixture. Pooja and Padmaja (2015) compared hydrothermal treatment to microwave pretreatment and dilute acid pretreatment methods for liberation of reducing sugars from cassava leaves, stems, and peels. The authors showed that hydrothermal treatment resulted in the highest sugar yields of 576, 315, and 170 g/kg cassava waste for the peels, stems, and leaves, respectively. From these studies it can be concluded that hydrothermal treatment is the most effective pretreatment method to liberate lignocellulose sugars from cassava wastes and that higher ethanol yields can be obtained from wastes from cassava ethanol processing than fresh cassava wastes from plantations.

6.5 CASSAVA-BASED ETHANOL IN CHINA

Mainland China was the 14th largest global producer of cassava and the main importer of dried cassava chips from Thailand and Vietnam in 2013 (FAOSTAT, 2013). Twenty-nine percent of global GHG emissions originated in the People's Republic of China in 2013 (Hao et al., 2017). The Chinese government committed to a GHG emissions reduction of 60%–65% per unit GDP by 2013 compared to 2005 (United Nations, 2015). To meet this target, China is considering following in the UK's footsteps of banning the production and sale of fossil fuel–based cars in the near future. Additionally, China plans to increase its fuel ethanol production from 2.7 billion L in 2013 to 10 billion L in 2020. Because China has a large population with limited arable land and water resources, an increase in energy crop cultivation for increased biofuels production should be carefully considered. Soil degradation due to energy crop cultivation was identified by Ye et al. (2010) as the largest risk to food security in China. For this reason, the Chinese government, through their Development Plan for Energy Crops (Asian Development Bank, 2009a) has limited future energy crops production to noncereal feedstock cultivated on marginal land. Crops, such as cassava, which can be grown on marginal land, are thus being considered as feedstock for expanding China's biofuels industry (Cai et al., 2010; Jiang et al., 2014).

In 2007, China was the third-largest producer of biofuels, after Brazil and the United States, with an estimated production capacity of 1.33 million metric tons (Jansson et al., 2009), but ranked fifth in 2015 despite an increase in production capacity. China had four bioethanol plants in operation in 2007 and one cassava-based plant was scheduled to start production in 2008 (Asian Development Bank, 2009a). China's largest bioethanol plant is in Jilian and can produce 600,000 ton ethanol/year from maize. By 2015, China's bioethanol production capacity had increased to 2.5 million ton (Ye et al., 2017), but maize is still the main feedstock (Xie et al., 2017). Cassava is currently produced in Guangxi, Guangdong, Yunnan, Fujian, and Jiangxi provinces in China. A number of LCA studies have assessed the impact of increased cassava-based ethanol production on China's food–water–energy nexus (Xie et al., 2017; Zhang et al., 2017). It was concluded in these studies that cassava peels and wastes should be used for bioethanol production to lower the GHG emissions from dLUC and iLUC and that new developments for cassava-based ethanol should be limited to the Southern provinces where cassava is already cultivated to prevent stress on local water resources. Based on the LCA studies and the growth pattern in bioethanol production capacity, it would be difficult for China to meet its GHG emissions targets only from an increase in biofuels production.

6.6 CASSAVA-BASED ETHANOL IN THAILAND

Thailand is the second largest producer and largest exporter of cassava with an estimated production of 3 million tons in 2013 (FAOSTAT, 2013). The International Centre for Tropical Agriculture implemented a program between 1994 and 2003 to improve sustainability of cassava-based cropping systems in Thailand and Vietnam (CGIAR, 2006). The case study was aimed at improving soil management through improved farming practices. It was concluded from the study that cassava cultivation area and yields could be significantly improved through the adoption of improved cassava cultivars. The government initiated an ambitious 15-year alternative energy development plan (AEDP, 2009) that started in 2008. The policy was designed to be rolled out in three stages. During the first stage (2008–11) alternative energy technologies were to be promoted. During the second, midterm stage (2012–16), new alternative energy technologies were to be developed and utilization of the technologies is aimed for 2017–22. A midterm assessment of the success of the AEDP showed an increase in bioethanol production from 1.4 million L/day in 2012 to 3.21 million L/day in 2014 (AEDP, 2015). If this increase in biofuels production can be sustained, by linear projection, Thailand would be able to meet the projected 11.3 million L/day in 2036.

Thailand committed to a GHG reduction of 20%–25% by 2030. There are a number of incentives being implemented to meet this commitment, including expansion of the country's bioethanol industry. The government promotes the use of cassava and sugarcane as first-generation feedstock with a projected increase in cassava production of 29.5 million tons/year between 2015 and 2026. The projections assumed a constant cassava yield over this period. It is expected that only 5.71 million tons/year of cassava will be used for first-generation ethanol production in 2026 with 11.3 million L of ethanol to be produced from agricultural wastes in 2036. Thailand produced 3.21 million L/day of second-generation ethanol in 2014. It is expected that the AEDP policy would result in a total reduction of 140 million tons CO_2eq by 2036 (AEDP, 2015).

Usubharatana and Phungrassami (2015) reported that 40%–59% of carbon emissions from starch-based ethanol production plants can be attributed to the cassava cultivation and harvesting procedures resulting in a total estimated GHG emissions of 10 million tCO_2eq for starch production in Thailand in 2013. Silalertruksa and Gheewala (2011) assessed the GHG emissions implications of long-term ethanol production in Thailand and concluded the GHG emissions reduction benefit of cassava-based ethanol will decline over time as dLUC and iLUC emissions become more prominent due to planting area expansion to increase cassava production. The authors estimated that GHG emissions of 4600 tCO_2eq can be realized if fertilizer use during cultivation is minimized and biogas and energy from plants waste is used to replace coal usage in the production plants. Prapaspongsa and Gheewala (2016) used consequential LCA (CLCA) and a systematic iLUC model to assess the impact iLUC on GHG emissions in Thailand and concluded that iLUC emissions is equal to approximately 39%–76% of that of fossil-based petroleum. Pingmuanglek et al. (2017) assessed the impact of the AEDP on water resources in the Mun basin of Thailand. The authors concluded that additional irrigation would be required to supplement freshwater supply in this watershed if cassava cultivation were to be expanded to meet increased ethanol production demands. This confirmed the findings of an LCA study by Gheewala et al. (2014) that highlighted a threat to freshwater resources in the Mun, Chi, and Chao watersheds with an increase in energy crop cultivation. Lecksiwilai et al. (2017) used an ecological scarcity LCA approach and showed that the ecological impact of cassava and palm oil biofuels production is lower than that of fossil-based fuels currently in use. Kawasaki et al. (2015) used measured data at a research site in the Khon Kaen province of Thailand to estimate the environmental impact of cassava-based ethanol production compared to sugarcane as feedstock. The study showed that the overall footprint of cassava-based ethanol is smaller than that of sugarcane-based ethanol, primarily due to the high fertilizer requirements of sugarcane crops. It was reported by Jakrawatana et al. (2016) that ethanol yields from existing ethanol plants can be significantly increased by adopting very high gravity fermentation with SSF processes. It was suggested that cassava waste pulp from starch processing facilities can be used to supplement energy crops for ethanol production to meet the targets set by the AEDP without excessive expansion of cassava cultivation.

6.7 CASSAVA-BASED ETHANOL IN VIETNAM

Vietnam is the world's seventh-largest producer of cassava and the second-larger exporter of dried cassava chips (primarily to Thailand) (FAOSTAT, 2013). With a vibrant growing economy, Vietnam's annual energy demand is growing by 9%, which would result in Vietnam becoming a net importer of energy if additional sources of energy could not be secured (Asian Development Bank, 2009b). Furthermore, by ratifying the Paris Agreement (United Nations, 2015) the Vietnamese government had committed to a reduction of 8% in GHG emissions by 2030. In 2008, CO_2 emissions fossil fuels accounted for 25% of total GHG emissions associated with energy use (Le et al., 2013).

The Vietnamese government approved a National Program for Development of Biofuels up to 2015, with 2015 vision in 2007 (Decision no. 177/2007/QD-Tgg, 2007) with the aim of improving energy security by developing its biofuels industry. The policy aims to replace 5% of the transportation fuel supply with E5 ethanol by 2025 (1.8 million tons) (Nguyen, 2010). The bulk of the Vietnamese economy is based on agriculture and the country has vast amounts of organic matter available for ethanol production, including sources, such as rice straw and husks, sugarcane, and cassava. Approximately 560,000 ha of arable land was under cassava cultivation in 2008 with an average yield of 17.63 ton/ha. Although the CGIAR launched a project to increase cassava yields and improve agricultural practices to curb soil degradation (CGIAR, 2006), the project was not as successful as in Thailand. There are currently eight bioethanol plants in operation in Vietnam with a total capacity to produce 880 million L of ethanol per year (Trung et al., 2016). Four of these plants, with a capacity of producing 420 million L/year use cassava as primary feedstock. The largest of these is situated in Quang Nam province and has a capacity of 120 million L/year. Some studies have shown that as much as 3645 million L of second-generation ethanol could be produced from 28.67 million tons of rice straw that is annually generated within Vietnam. Rice straw–based ethanol does not require additional land use and is annually burned by farmers (Diep and Sakanishi, 2011) contributing to the country's GHG emissions inventory. Bioethanol from rice straw coupled to biogas production could significantly reduce GHG emissions and reduce eutrophication by using biogas digester effluent as fertilizer (Oritate et al., 2015). The technology for converting rice straw into ethanol is, however, not currently economically feasible (Kunimitsu and Ueda, 2013).

Le et al. (2013) evaluated the energy efficiency and potential GHG emissions savings from cassava-based ethanol, taking into account land use change because of switching to cassava as crop from rice or sugarcane. The authors concluded that E5 and E10 fuel using cassava-based ethanol will result in GHG emissions reductions only if fuel consumption does not increase more than 2.4% and 4.5% for E5 and E10, respectively, due to adoption of ethanol fuels. However, Nguyen et al. (2017) performing a more detailed LCA study showed that switching to cassava as bioethanol energy crop would results in a carbon debt for the country that would require a cassava yield of 33 ton/ha to repay in 25 years.

6.8 CONCLUDING REMARKS

Cassava is an important industrial crop in many countries. Africa produces more than 50% of the world's cassava, but uses it primarily as staple food. Cassava-based bioethanol is produced primarily in China, Thailand, and Vietnam. An increase in cassava-based ethanol production in South Eastern Asian countries has had a global market effect on the price of cassava as exports from mainly Thailand was redirected from Europe where it is used for feed to China for bioethanol production. China, Thailand, and Vietnam are investing in bioethanol to lower GHG emissions and meet the reduction targets as committed to by the Paris Agreement on Climate Change. Analyses have shown the risks posed by expanding land use for increased cassava production, especially in terms of the water–energy–food nexus. A number of LCA studies have also shown the importance of improved cassava varieties for higher crop yields and the use of an integrated biorefinery approach to utilize waste products to increase bioethanol yields per crop. Increased economic efficiency of second-generation ethanol processes could greatly increase the carbon efficiency of cassava ethanol and processing plant, thereby increasing the benefits that can be realized in terms of GHG emission reductions.

References

Adeoti, O., 2010. Water use impact of ethanol at a gasoline substitution ratio of 5% from cassava in Nigeria. Biomass Bioenerg. 34 (7), 985–992.

AEDP (Alternative Energy Development Plan). 2009. Department of Renewable Energy Development and Energy Efficiency, Minister of Energy, Thailand.

AEDP (Alternative Energy Development Plan). 2015. Department of Renewable Energy Development and Energy Efficiency, Minister of Energy, Thailand.

Akaracharanya, A., Krisomdee, K., Tolieng, V., 2016. Improved SSF-cellulosic ethanol production by the cellobiose fermenting yeast. Chiang Mai J. Sci. 43 (5), 985–996.

Anyanwu, C.N., Ibeto, C.N., Ezeoha, S.L., Ogbuagu, N.J., 2015. Sustainability of cassava (*Manihot esculenta* Crantz) as industrial feedstock, energy and food crop in Nigeria. Renew. Energ. 81, 745–752.

Asian Development Bank, 2009a. Status and Potential for the Development of Biofuels and Rural Renewable Energy: The People's Republic of China. Asian Development Bank, Mandaluyong City, Philippines.

Asian Development Bank, 2009b. Status and Potential for the Development of Biofuels and Rural Renewable Energy: Viet Nam. Asian Development Bank, Mandaluyong City, Philippines.

Babel, M.S., Shrestha, B., Perret, S.R., 2011. Hydrological impact of biofuel production: a case study of the Khlong Phlo Watershed in Thailand. Agr. Water Manag. 101 (1), 8–26.

Baeyens, J., Kang, Q., Appels, L., Dewil, R., Lv, Y., Tan, T., 2015. Challenges and opportunities in improving the production of bio-ethanol. Progress Energ. Combust. Sci. 47, 60–88.

Behera, S., Ray, R.C., 2014. Batch ethanol production from cassava (*Manihot esculenta* Crantz) flour by *Saccharomyces cerevisiae* cells immobilized in calcium alginate. Ann. Microbiol.doi: 10.1007/s13213-014-0918-8.

Behera, S.S., Ray, R.C., 2016. Solid state fermentation for production of microbial cellulases: recent advances and improvement strategies. Int. J. Biol. Macromol. 86, 656–669.

Benvenga, M.A.C., Librantz, A.F.H., Santana, J.C.C., Tambourgi, E.B., 2016. Genetic algorithm applied to study of the economic viability of alcohol production from Cassava root from 2002 to 2013. J. Clean Prod. 113, 483–494.

Cai, X., Zhang, X., Wang, D., 2010. Land availability for biofuels production. Environ. Sci. Technol. 45, 334–339.

CGIAR, 2006. Evidence from the CGIAR. Natural Resources Management Research Impact. Science Council Secretariat, Rome, Italy.

Decision no. 177/2007/QD-Tgg of Government of Vietnam, 2007. Approval of the scheme on development of biofuels up to 2015 and a vision to 2025. Government of Vietnam, November 20, 1–10.

Diep, N.Q., Sakanishi, K., 2011. Potential of bioethanol production from agricultural residues in the Mekong Delta, Vietnam. Int. Energy J. 12, 145–154.

Elemike, E.E., Oseghale, O.C., Okoye, A.C., 2015. Utilization of cellulosic cassava waste for bio-ethanol production. J. Environ. Chem. Eng. 3 (4), 2797–2800.

FAO, 2014. Food Outlook: Biannual Report on Global Food Markets. Food and Agriculture Organization of United Nations, Rome.

FAOSTAT, 2013. United Nations Food and Agriculture Organisation. Available from: http://www.fao.org/faostat/en/#rankings.

Gerbens-Leenes, P.W., Hoekstra, A.Y., Meer, T.V.D., 2009. The water footprint of energy from biomass: a quantitative assessment and consequences of an increasing share of bio-energy in energy supply. Ecol. Econ. 68, 1052–1060.

Gheewala, S.H., Silalertruksa, T., Nilsalab, P., 2013. Implications of the biofuels policy mandate in Thailand on water: the case of bioethanol. Bioresour. Technol. 150, 457–465.

Gheewala, S.H., Silalertruksa, T., Nilsalab, P., Mungkung, R., Perret, S.R., Chaiyawannakarn, N., 2014. Water footprint and impact of water consumption for food, feed, fuel crops production in Thailand. Water (Switzerland) 6 (6), 1698–1718.

Hanif, M., Mahlia, T.M.I., Aditiya, H.B., Abu Bakar, M.S., 2017. Energy and environmental assessments of bioethanol production from Sri Kanji 1 cassava in Malaysia. Biofuel. Res. J. 4 (1), 537–544.

Hansupalak, N., Piromkraipak, P., Tamthirat, P., Manitsorasak, A., Sriroth, K., Tran, T., 2015. Biogas reduces the carbon footprint of cassava starch: a comparative assessment with fuel oil. J. Clean. Prod. 134, 539–546.

Hao, M., Jiang, D., Wang, J., Fu, J., Huang, Y., 2017. Could biofuel development stress China's water resources? GCB Bioenergy 9 (9), 1447–1460.

HLPE, 2013. Biofuels and food security. A report by the High Level Panel of Experts on Food Security and Nutrition of the Committee on World Food Security, Rome.

Hong, Y., Yuan, Z., Liu, J., 2009. Land and water requirements of biofuel and implications for food supply and the environment in China. Energy Policy 37, 1876–1885.

Ibeto, C.N., Ofoefule, A.U., Agbo, K.E., 2011. A Global overview of biomass potentials for bioethanol production: a renewable alternative fuel. Trends App. Sci. Res. 6 (5), 410–425.

Jakrawatana, N., Pingmuangleka, P., Gheewala, S.H., 2016. Material flow management and cleaner production of cassava processing for future food, feed and fuel in Thailand. J. Clean. Prod. 134 (Part B), 633–641.

Jansson, C., Westerbergh, A., Zhang, J., Hu, X., Sun, C., 2009. Cassava, a potential biofuel crop in China. Appl. Energy 86, 595–599.

Jiang, D., Hao, M.M., Fu, J.Y., Huang, Y., 2014. Spatial-temporal variation of marginal land suitable for energy plants from 1990 to 2010 in China. China Sci. Rep. 4, 5816.

Kawasaki, J., Silalertruksa, T., Scheyvens, H., Yamanoshita, M., 2015. Environmental sustainability and climate benefits of green technology for bioethanol production in Thailand. J. Int. Soc. Southeast Asian Agr. Sci. 21 (1), 78–95.

Khamkeaw, A., Phisalaphong, M., 2016. Hydrolysis of cassava starch by co-immobilized multi-microorganisms of

Loog-Pang (Thai rice cake starter) for ethanol fermentation. Food Sci. Biotechnol. 25 (2), 509–516.

Kunimitsu, Y., Ueda, T., 2013. Economic and environmental effects of rice-straw bioethanol production in Vietnam. Paddy Water Environ. 11, 411–421.

Lange, M., 2011. The GHG balance of biofuels taking into account land use change. Energy Policy 39, 2373–2385.

Lauven, L.P., Liu, B., Geldermann, J., 2014. Determinants of economically optimal cassava-to-ethanol plant capacities with consideration of GHG emissions. Appl. Therm. Energy 70 (2), 1246–1252.

Le, L.T., Van Ierland, E.C., Zhu, X., Wesseler, J., 2013. Energy and greenhouse gas balances of cassava-based ethanol. Biomass Bioenergy 51, 125–135.

Lecksiwilai, N., Gheewala, S.H., Silalertruksa, T., Mungkalasiri, J., 2017. LCA of biofuels in Thailand using Thai Ecological Scarcity method. J. Clean. Prod. 142, 1183–1191.

Li, P., Zhu, M., 2011. A consolidated bio-processing of ethanol from cassava pulp accompanied by hydrogen production. Bioresour. Technol. 102, 10471–10479.

Lomthong, T., Lertwattanasakul, N., Kitpreechavanich, V., 2016. Production of raw starch degrading enzyme by the thermophilic filamentous bacterium *Laceyella sacchari* LP175 and its application for ethanol production from dried cassava chips. Starch/Staerke 68 (11–12), 1264–1274.

Lu, H., Lv, C., Zhang, M., Liu, S., Liu, J., Lian, F., 2017. Optimization of hydrothermal pretreatment for co-utilization C-5 and C-6 sugars of cassava alcohol residue. Energ. Conver. Manag. 132, 251–260.

Machado, K.S., Seleme, R., Maceno, M.M., Zattar, I.C., 2017. Carbon footprint in the ethanol feedstocks cultivation: agricultural CO_2 emission assessment. Agr. Syst. 157, 140–145.

Mangmeechai, A., Pavasantm, P., 2013. Water footprints of cassava- and molasses-based ethanol production in Thailand. Nat. Resour. Res. 22 (4), 1–10.

Martín, C., Wei, M., Xiong, S., Jönsson, L.J., 2016. Enhancing saccharification of cassava stems by starch hydrolysis prior to pretreatment. Ind. Crops Prod. 97, 21–31.

Marx, S., Nquma, Y.T., 2013. Cassava as feedstock for ethanol production in South Africa. Afr. J. Biotechnol. 12 (31), 4975–4983.

Moshi, A.P., Hosea, K.M.M., Elisante, E., Mamo, G., Mattiasson, B., 2015. High temperature simultaneous saccharification and fermentation of starch from inedible wild cassava (*Manihot glaziovii*) to bioethanol using *Caloramator boliviensis*. Bioresour. Technol. 180, 128–136.

Moshi, A.P., Hosea, K.M.M., Elisante, E., Mamo, G., Önnby, L., Nges, I.A., 2016. Production of raw starch-degrading enzyme by *Aspergillus* sp. and its use in conversion of inedible wild cassava flour to bioethanol. J. Biosci. Bioeng. 121 (4), 457–463.

Nadir, N., Mel, M., Karim, M.I.A., Yunus, R.M., 2009. Comparison of sweet sorghum and cassava for ethanol production by using *Saccharomyces cerevisiae*. J. Appl. Sci. 9, 3068–3073.

Nanssou, K.P.A., Nano, J.Y., Kapseu, C., 2016. Pretreatment of cassava stems and peelings by thermohydrolysis to enhance hydrolysis yield of cellulose in bioethanol production process. Renew. Energ. 97, 252–265.

Nguyen, A.P., 2010. Biofuels development from cassava and food security. In: Proceedings of the National Workshop on Policy for Biofuels Development in Vietnam. Institute of Policy and Strategy for Agricultural and Rural Development of Vietnam, Hanoi, Vietnam, 9 December 2010.

Nguyen, T.H., Williams, S., Paustian, K., 2017. Impact of ecosystem carbon stock change on greenhouse gas emissions and carbon payback periods of cassava-based ethanol in Vietnam. Biomass Bioenergy 100, 126–137.

Nguyen, T., Gheewala, S., 2008. Life cycle assessment of fuel ethanol from cassava in Thailand. Int. J. Life Cycle Assess. 13, 147–154.

Numjuncharoen, T., Papong, S., Malakul, P., Mungcharoen, T., 2015. Life-cycle GHG emissions of cassava-based bioethanol production. Energy Proced. 79, 265–271.

Ohimain, E., 2013. Environmental impacts of smallholder ethanol production from cassava feedstock for the replacement of kerosene household cooking fuel in Nigeria: energy sources, part A: recovery. Utiliz. Environ. Effects 35 (16), 1560–1565.

Okudoh, V., Trois, C., Workneh, T., Schmidt, S., 2014. The potential of cassava biomass and applicable technologies for sustainable biogas production in South Africa: a review. Renew. Sustain. Energ. Rev. 39, 1035–1052.

Oritate, F., Yuyama, Y., Nakamura, M., Yamaoka, M., Nguyen, P.D., Dang, V.B.H., et al., 2015. Regional diagnosis of biomass use in suburban village in Southern Vietnam. J. Jpn. Inst. Energy 94 (8), 805–829.

Papong, S., Malakul, P., 2010. Life-cycle energy and environmental analysis of bioethanol production from cassava in Thailand. Bioresour. Technol. 101, S112–S118.

Papong, S., Rewlay-ngoen, C., Itsubo, N., Malakul, P., 2017. Environmental life cycle assessment and social impacts of bioethanol production in Thailand. J. Clean. Prod. 157, 254–266.

Pingmuanglek, P., Jakrawatana, N., Gheewala, S.H., 2017. Freshwater use analysis of cassava for food feed fuel in the Mun River basin, Thailand. Int. J. Life Cycle Assess. 22, 1705–1717.

Pooja, N.S., Padmaja, G., 2015. Enhancing the enzymatic saccharification of agricultural and processing residues of cassava through pretreatment techniques. Waste Biomass Valoriz. 6 (3), 303–315.

Poonsrisawat, A., Wanlapatit, S., Wansuksri, R., Piyachomkwan, K., Paemanee, A., Gamonpilas, C., et al., 2016. Synergistic effects of cell wall degrading enzymes on rheology of cassava root mash. Process Biochem. 51 (12), 2104–2111.

Prapaspongsa, T., Gheewala, S.H., 2016. Risks of indirect land use impacts and greenhouse gas consequences: an assessment of Thailand's bioethanol policy. J. Clean. Prod. 134, 563–573.

Ray, R.C., Mohapatra, S., Panda, S., Kar, S., 2008. Solid substrate fermentation of cassava fibrous residue for production of α-amylase, lactic acid and ethanol. J. Environ. Biol. 29, 111–115.

Ray, R.C., Swain, M.R., 2011. Bio-ethanol, bioplastics and other fermented industrial products from cassava starch and flour. In: Colleen, M., Pace (Eds.), Cassava: Farming, Uses and Economic Impacts. Nova Science, New York, pp. 1–32.

Ren, J., Manzardo, A., Mazzi, A., Zuliani, F., Scipioni, A., 2015. Prioritization of bioethanol production pathways in China based on life cycle sustainability assessment and multicriteria decision-making. Int. J. Life Cycle Assess. 20 (6), 842–853.

REN21, 2017. Renewables 2017 Global Status Report. Renewable Energy Policy Network for the 21st century.

Ruppenthal, M., Leihner, D.E., Steinmüller, N., El Sharkawy, M.A., 1997. Losses of organic matter and nutrients by water erosion in cassava-based cropping systems. Expl. Agr. 33, 487–498.

Sebayang, A.H., Masjuki, H.H., Ong, H.C., Dharma, S., Silitonga, A.S., Mahlia, T.M.I., et al., 2016. A perspective on bioethanol production from biomass as alternative fuel for spark ignition engine. RSC Adv. 6 (18), 14964–14992.

Seong, Y.J., Park, H., Yang, J., Kim, S.J., Choi, W., Kim, K.H., et al., 2016. Expression of a mutated SPT15 gene in *Saccharomyces cerevisiae* enhances both cell growth and ethanol production in microaerobic batch, fed-batch, and simultaneous saccharification and fermentations. Appl. Microbiol. Biotechnol. 101, 3567–3575.

Shepherd, M., Knox, P., 2016. The Paris COP21 climate conference: what does it mean for the southeast? Southeast Geogr. 56, 147–151.

Silalertruksa, T., Gheewala, S.H., 2011. Long-term bioethanol system and its implications on GHG emissions: a case study of Thailand. Environ. Sci. Technol. 45 (11), 4920–4928.

Teodor, C., Gil, G., Camelia, V., 2013. Bioethanol production from residual lignocellulosic materials: a review part 2. Food Technol. 37, 25–38.

Thatoi, H., Dash, P.K., Mohapatra, S., Swain, M.R., 2016. Bioethanol production from tuber crops using fermentation technology: a review. Int. J. Sustain. Energ. 35 (5), 443–468.

Trung, K.H., Huong, C.T., Xuan, T.D., Trung, D.M., Khanh, T.D., 2016. Current status and future plan of development of bioenergy crops as renewable energy sources in Vietnam. J. Biol. Nature 5 (1), 1–8.

United Nations, 2015. Paris Agreement. United Nations, Paris.

Usubharatana, P., Phungrassami, H., 2015. Carbon footprint of cassava starch production in North-Eastern Thailand. Procedia CIRP 29, 462–467.

Valencia, M.J., Cardona, C.A., 2014. The Colombian biofuel supply chains: the assessment of current and promising scenarios based on environmental goals. Energy Policy 67, 232–242.

Wang, K., Yang, X., Ren, X., Zhang, J., Mao, Z., 2017. Development of a new cleaner production process for cassava ethanol. Chinese J. Chem. Eng. 25 (4), 493–498.

Xie, X., Zhang, T., Wang, L., Huang, Z., 2017. Biotechnology for Biofuels Regional water footprints of potential biofuel production in China. Biotechnol. Biofuel. 10 (95), 1–15.

Xu, Q.S., Yan, Y.S., Feng, J.X., 2016. Efficient hydrolysis of raw starch and ethanol fermentation: a novel raw starch-digesting glucoamylase from *Penicillium oxalicum*. Biotechnol. Biofuel. 9 (1), 216.

Yang, X., Wang, K., Wang, H., Zhang, J., Mao, Z., 2017. Novel process combining anaerobic-aerobic digestion and ion exchange resin for full recycling of cassava stillage in ethanol fermentation. Waste Manag. 62, 241–246.

Ye, L., Yang, J., Verdoodt, A., Marssadek, R., Van Ranst, E., 2010. China's food security threatened by soil degradation and biofuels production. In: Proceedings of the 19th World Congress of Soil Science, Soil Solutions for a Changing World, Symposium 4.2.1: Soil, Energy and Food Security, pp. 5–8.

Ye, F., Li, Y., Lin, Q., Zhan, Y., 2017. Modeling of China's cassava-based bioethanol supply chain operation and coordination. Energy 120, 217–228.

Zhang, T., Xie, X., Huang, Z., 2014. Life cycle water footprints of non-food biomass fuels in China. Environ. Sci. Technol. 48 (7), 4137–4144.

Zhang, T., Xie, X., Huang, Z., 2017. The policy recommendations on cassava ethanol in China: analyzed from the perspective of life cycle "2E&W". Resour. Conserv. Recy. 126, 12–24.

CHAPTER

7

Sweet Potato as a Bioenergy Crop for Fuel Ethanol Production: Perspectives and Challenges

Claudia Lareo, Mario D. Ferrari

Instituto de Ingeniería Química, Universidad de la República, Montevideo, Uruguay

7.1 INTRODUCTION

Sweet potato (*Ipomoea batatas* L.) has been considered a promising raw material for ethanol fuel production. It has a higher starch yield per unit land cultivated than grains (Comis, 2008; Duvernay et al., 2013; Lee et al., 2012; Srichuwong et al., 2009; Ziska et al., 2009). It is an herbaceous plant and an important food crop in the world, which belongs to the Convolvulaceous family (International Potato Center, 2017; Torquato-Tavares et al., 2017).

Sweet potato is a root crop cultivated in tropical and subtropical regions as a perennial crop, whereas it is an annual crop in temperate regions (El Sheikha and Ray, 2017; Jin et al., 2012; Kays and Wang, 2001; Ray and Tomlins, 2010; Vilaró et al., 2009; Zabed et al., 2017). In these regions, the harvest period is approximately 4–6 months. Adequate postharvest conservation practices and highly productive genetic material are used to achieve supply throughout the year (Jin et al., 2012; Vilaró et al., 2009). In Southern Uruguay, the harvest period of sweet potato is from February to May inclusive. In the north, with early cultivars it is possible to obtain two harvests per season (Lareo et al., 2013; Vilaró et al., 2009).

7.2 GENERAL CHARACTERISTICS OF THE SWEET POTATO CROP

Sweet potato crops have several agronomic characteristics that determine their wide adaptation to marginal lands, such as drought resistant, saline-base tolerance, high multiplication rate, and low degeneration of the propagation material, short grow cycle, low illness incidence, and plagues (Mukherjee et al., 2012, 2015). It covers rapidly the soil and therefore protects it from the erosive rains and controlling the weed problem (Cao et al., 2011; Duvernay et al., 2013; Vilaró et al., 2009).

Bioethanol Production From Food Crops
http://dx.doi.org/10.1016/B978-0-12-813766-6.00007-2

It can grow at altitudes from sea level to 2500 m (International Potato Center, 2017).

There is a large genetic variability in characteristics of color (white to yellow, orange, and deep purple) and pulp, texture, and dry matter content of sweet potato, associated to its uses. There are many varieties of sweet potatoes, which differ in taste, texture, composition, and yields. Different countries and regions have preferences in varietal types, associated with local uses and customs. In most countries, the cultivation of local populations or introduced old varieties predominates. These, in general, present a long productive cycle, low productivity, excessive foliage, poor conservation, and poor commercial quality (Ray and Tomlins, 2010). The genetic improvement in the crop has allowed to improve the commercial quality, precocity of harvest, yield, conservation, and availability of the product throughout the year. In general, these cultivars also exhibit a significant improvement in agronomic aspects, facilitating their cultivation and harvesting in mechanized form (Vilaró et al., 2009). The type of sweet potato with purple skin, yellow pulp, and semimoist texture is traditionally preferred for boiled or baked consumption. There is a growing demand for sweet potatoes of orange pulp and moist texture. This type of cultivar also allows a very significant source of beta carotene to be included in a healthy diet (El Sheikha and Ray, 2017; Vilaró et al., 2009).

7.3 WORLD PRODUCTION AND CURRENT USES

Sweet potato is typically a small-farmer crop and often grown on marginal soils with limited outputs. It is mainly used as food, being basically commercialized as fresh and in minor quantities as dried tubers, whether or not sliced or in the form of pellets (Ray and Ravi, 2005). Although the crop is widely cultivated in the world, production tends to be concentrated in those countries with lower per capita income. Sweet potato ranks as the world's seventh most important food crop, and has major contribution to energy and phytochemical source of nutrition (Shekhar et al., 2015). Sweet potato production and the harvested area has declined gradually between 1974 and 2014, reaching values in the order of 105 million tonnes and 8.2 million ha in the period 2005–14, respectively (Fig. 7.1). More than 90% of the world's current sweet potato production takes place in Asia and Africa, with China being the largest producing country. In 2014, China produced 67% of world production. Other relevant producers are Indonesia, Vietnam, and India in Asia; Nigeria, Uganda, Tanzania, Ethiopia, Tanzania, Madagascar, Angola, and Mozambique in Africa; and Brazil and USA in the Americas (Table 7.1). Although the world production of sweet potatoes and their consumption per capita shows a tendency to decline, the behavior depends on the region or country considered. While in Asia, basically due to the predominance of China, there is a decrease in its production, in Africa it shows an increase, being for some countries a basic source of food (Table 7.2).

Although in some countries the production of sweet potato is low on a global scale, it is very important locally as food. Sweet potato has a relevant contribution in the daily calorie intake in Tanzania, Angola, Uganda, Rwanda, Mozambique, Haiti, and Solomon Islands (Table 7.2). The sweet potato contains important components, which meet human nutritional needs (El Sheikha and Ray, 2017). The sweet and attractive taste, longer storability with very simple preparation are also reasons why sweet potato is a suitable supplemental food for poor people in urban and suburb areas (Widodo et al., 2015). It plays an important role in food security (Mukherjee et al., 2012; Wang et al., 2016).

Approximately 44% of the sweet potato produced in 2014 in Asia was destined for animal feed, with China being the country with the greatest use for that purpose. Both roots and

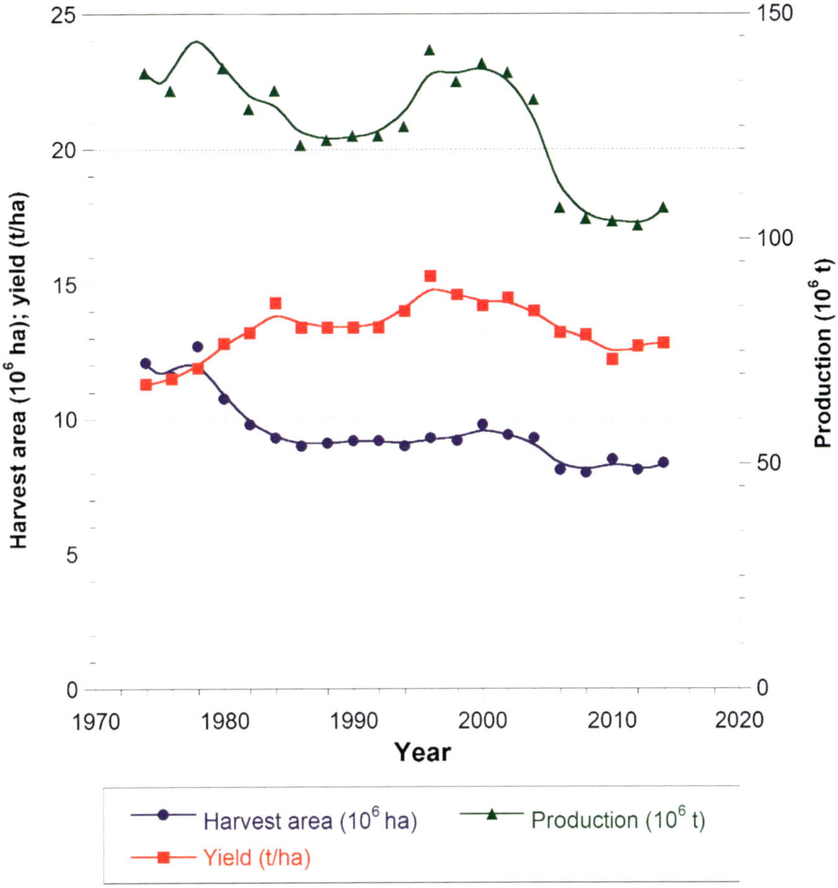

FIGURE 7.1 **World production, harvested area, and agronomical yield of sweet potato from 1974 to 2014.** *Source: Data taken from FAOSTAT, 2014. Statistics division. Food and Agriculture Organization of the United Nations. Crops. Available from: http://www.fao.org/faostat/en/#data/QC.*

vines are used. The roots can be given to all ruminants as fresh, chopped tubers, dried chips, and silage along with locally available grasses during the dry season. Sweet potato roots and vines are common feeds for pigs and other livestock in many countries, including China, India, Indonesia, Uganda, and Vietnam. The use of sweet potato as an animal feed is conditioned by the availability of other foods, such as cereals and the digestibility of planted varieties (Murugan et al., 2012).

Most of the sweet potato produced is consumed locally and with little processing. The sweet potato trade is very small compared to the total amount produced. In 2014, only 0.4% of sweet potatoes produced was exported (Table 7.3).

A reaction to a declining production or demand for sweet potatoes in some countries is its use for fuel ethanol within a policy of renewable energy sources. Another strategy is to increase the demand for sweet potatoes as food by developing varieties with nutritional features ("functional food"), taste, color, and aspects that are more attractive to consumers. New products are also reported, such as anthocyanin pigments in the purple varieties for food colorings and use in the

TABLE 7.1 Main Sweet Potato Producers (2014)

Region/country	Production (10^6 t)	Yield (t/ha)
Asia	79.2	19.7
China	71.3	21.1
Indonesia	2.4 (1.9)[a]	15.2 (10.7)[a]
Vietnam	1.4	10.7
India	1.1	10.3
Africa	22.6	5.8
Nigeria	3.8	2.6
Tanzania	3.5	4.8
Ethiopia	2.7	5.2
Mozambique	2.4	33.6
Angola	1.9	11.4
Uganda	1.9	4.1
Madagascar	1.1	7.9
The Americas	3.8	13.3
United State of America	1.3	13.3
Cuba	0.5	12.6
Brazil	0.5	13.2
Haiti	0.5	6.7
Oceania	0.9	5.9
Europe	0.05	11.6
World	106.6	12.8

[a] *Jusuf and Ginting (2014).*
Source: FAOSTAT, 2014. Statistics Division. Food and Agriculture Organization of the United Nations. Crops. Available from: http://www.fao.org/faostat/en/#data/QC.

TABLE 7.2 Energy Supply of Sweet Potato as Food (kcal/Capita/Day)

	Year		
Region/country	1993	2003	2013
Asia	56	41	25
China	137	106	64
Indonesia	24	20	21
Vietnam	73	16	13
India	3	3	2
Africa	20	30	38
Nigeria	6	39	37
Tanzania	23	14	123
Ethiopia	7	17	36
Mozambique	9	104	82
Angola	39	82	129
Uganda	225	217	123
Rwanda	363	262	227
Madagascar	39	27	57
The Americas	5	6	7
United State of America	3	4	6
Cuba	32	89	73
Brazil	3	3	2
Haiti	51	40	118
Oceania	10	13	13
Solomon Islands	481	507	469
Europe	0	0	0
World	37	30	22

Source: FAOSTAT, 2013. Statistics Division. Food and Agriculture Organization of the United Nations. Food Balance Sheets. Available from: http://www.fao.org/faostat/en/#data/FBS.

cosmetics industry (CTA, 2017; El Sheikha and Ray, 2017).

The world agronomical yield of the sweet potato has remained in the range of 12–15 t/ha in the last 30 years, without substantial improvements (Fig. 7.1). Some countries have yields well below the world average of 13 t/ha. China is one of countries showing high yields, over 20 t/ha (Table 7.1). The highest yields obtained are generally due more to the agronomic practices employed than to the selection of planted varieties.

7.4 INDUSTRIAL SWEET POTATO

Based on its low requirements for cultivation and high starch content, the sweet potato is a promising raw material for the production of fuel ethanol. Industrial sweet potatoes developed for the production of bioenergy are not intended for use as a food crop, without

TABLE 7.3 Distribution of Uses of Sweet Potato (2013)

	Asia		Africa		America		World	
	10^3 t	Percentage	10^3 t	Percentage	10^3 t	Percentage	10^3 t	Percentage
Production	78,549	–	18,186	–	3,615	–	100,472	–
Export	140	0.2	18	0.1	156	4.3	360	0.4
Food	39,578	50.0	14,506	79.8	2,767	76.5	57,061	56.8
Feed	34,472	43.9	368	2.0	318	8.8	35,259	35.1
Seed	261	0.3	212	1.2	119	3.3	594	0.6
Processing/ other uses	77	0.1	0	0	29	0.8	107	0.1
Losses	4,031	5.1	2,088	11.5	303	8.4	6,432	6.4

It may include official, semi-official, estimated, or calculated data.
Source: FAOSTAT, 2013. Statistics Division. Food and Agriculture Organization of the United Nations. Food Balance Sheets. Available from: http://www.fao.org/faostat/en/#data/FBS.

interest in its attractiveness in visual aspect, color, taste, and are selected for their higher content of starch and better agricultural yields (Viláró et al., 2009). For a sustainable production of fuel ethanol, it is necessary to obtain sweet potatoes with high starch content, high agronomical yield, low moisture level, small size to economize transport and storage costs, low temperature gelatinization starch, and low fiber content. Generally, white fleshed types contain higher amounts of starch than other varieties, they are edible but are not palatable (Duvernay et al., 2013; Thatoi et al., 2014). Fig. 7.2 shows an experimental industrial sweet potato variety (K9807.1) grown for alternative uses, such as ethanol production. Therefore, they offer potentially greater fermentable sugar yields than a common sweet potato crop for industrial conversion processes and the opportunity to increase planted acreage on marginal lands beyond what is in place for food (Lareo et al., 2013; Torquato-Tavares et al., 2017).

There are several recent experimental studies of selection of sweet potato varieties as a bioenergy crop for fuel ethanol production. In some countries, there is a national program of sweet potato genetic improvement for bioenergy production that seeks to obtain varieties with high yields of dry biomass and starch (Viláró et al., 2009). It has been reported that some industrial sweet potato breeding lines developed could produce ethanol yields of 4500–6500 L/ha compared to 2800–3800 L/ha for corn (Duvernay et al., 2013; Ziska et al., 2009). The biomass yields reached are higher than cultivated varieties for human consumption; that is, in Uruguay, 10 t/ha and up to 4.7 t/ha (dry basis), respectively (http://www.mgap.gub.uy/portal/hgxpp001.aspx?7,5,659,O,S,0,MNU;E;27;8;MNU). Some researchers have found promising varieties with agronomic yields higher than 30 t/ha and high starch content (Table 7.4). Maluf (2009) describes an experimental study carried out at the University of Lavras (Minas Gerais, Brazil) comprising 15 varieties of sweet potato which were classified according to their suitability for human consumption, animal feed or as biofuel crop for ethanol production based on agricultural yield of marketable fresh roots (t/ha), agricultural yield of protein in vines and leaves (kg/ha) and agroindustrial ethanol yield (L/ha), respectively. Four varieties were potentially suitable for specific use (market or animal feed), the remaining eleven could be used for more than one

FIGURE 7.2 **Industrial sweet potato variety (K9807.1) selected as bioenergy crop.**

TABLE 7.4 Sweet Potato Fresh Root Yield of Experimental Crops Evaluated for Fuel Bioethanol Production

Variety	Origen	Yield (t/ha)	Observations	References
–	Palmas, Tocantins, Brazil	32.2 – 65.5	Starch: 18.8% – 30.2% (fresh root basis)	Maluf (2009)
NS007	Sichuan, China[a]	26.16	Average fermentable sugars: 34.66% (fresh root basis)	Jin et al. (2012)
SS19	Sichuan, China[a]	33.75	Average fermentable sugars: 28.51 (fresh root basis)	Jin et al. (2012)
Shiroyutaka	Indonesia	> 30	Starch: 22% (fresh root basis)	Jusuf & Ginting (2014)
Sukuh	Indonesia	> 30	Starch: 19% (fresh root basis)	Jusuf & Ginting (2014)
Jago	Indonesia	> 30	Starch: 18% (fresh root basis)	Jusuf & Ginting (2014)
Darma MPand1 CCRI-01 CTRA-01	Jatinangor, Indonesia[b]	41.4 – 49.0	Starch: 13.5% – 17.9% (fresh root basis)	Waluyo et al. (2015)
K 9807.1	Uruguay	30	Total sugars: 22.5% (fresh root basis, moisture 70%)[c]	Lareo et al. (2013)
Beauregard	Alabama, USA	43	Yield calculated assuming moisture of 30%. Total sugars: 9.38% (dry root basis).	Ziska et al. (2009)
Beauregard	Maryland, USA	47	Yield calculated assuming moisture of 30%. Total sugars: 12.7% (dry root basis).	Ziska et al. (2009)
CX1	South Carolina, USA	46.9	Starch: 8.9% (fresh root basis)	Mussoline et al. (2017)

[a] *The best two varieties for fuel ethanol production among 10 varieties tested.*
[b] *The best four varieties in terms of liter of ethanol produced per hectare among 50 clones tested.*
[c] *More detailed information on the composition is given in Table 7.5.*

purpose. The varieties with greater aptitude for the production of ethanol showed yields in the range of 6200–13,600 L/ha. Although the experimental tests are generally successful, scaling up of sweet potato production into large plantation is difficult (Widodo et al., 2015).

7.5 CHEMICAL COMPOSITION OF SWEET POTATO

The chemical and nutritional composition of sweet potato has been recently reviewed (El Sheikha and Ray, 2017; Wang et al., 2016; Zhu and Wang, 2014). Fresh sweet potato contains high water content. Most of its dry matter is starch, which accounts for up to approximately 80% of the dry matter. It also contains simple soluble sugars, such as glucose, fructose, and sucrose, which result in the sweet taste of sweet potato root, lower amounts of fiber, proteins, and minor quantities of carotenes, thiamine, riboflavin, niacin, potassium, zinc, calcium, iron, and vitamins A and C (El Sheikha and Ray, 2017; Wang et al., 2016). Table 7.5 summarizes the content of simple sugars, starch, fiber, proteins, lipids, and ash of sweet potato found in the literature. These components should be analyzed taking into account that those were obtained using different methodologies, with little standardization among the different authors. In some cases, it is not clearly specified whether the values are expressed on a dry or wet basis. The total simple sugar and starch content varied in a wide range (3.6%–29.6% and 24.6%–83.7%, respectively). However, the total sugar contents (glucose equivalents) were in a narrower range 64.6%–88.6%. Table 7.6 presents the average composition of sweet potato as bioenergy crop.

The fermentable components in sweet potatoes are starch and simple sugars. The starch content and composition varies widely among different genotypes of sweet potato, and with the harvest time. Starch is a glucose polymer, which consists of two major components: amylose and amylopectin (Panda et al., 2012). Amylose is fundamentally a linear molecule of α-1,4-linkedglucan and represents up to 34%–38% of starch in sweet potato (Moorthy, 2002; Zhu and Wang, 2014), while amylopectin, the major component (generally 70%–85%), is a larger molecule with highly α-1,6 branched chains. The water content, amylose–amylopectin ratio, structures of amylose and amylopectin, the packing of amylose and amylopectin chains in the granules, the presence of minor components (e.g., phosphorous and lipids), the environment (e.g., pressure, heating, and cooling rates) have great influence on the physical changes of starch during heating and cooling (Hoover, 2001; Srichuwong and Jane, 2007). Nevertheless, it is not clear what structural features affect the degree of digestion (Srichuwong et al., 2005).

The bioavailability of the starch may differ with the cultivars and this may affect the rate of conversion and final yield of ethanol (Moorthy, 2002). The susceptibility of starch to hydrolysis by α-amylase has been shown to vary with botanical origin. Most studies of starch hydrolysis use enzymes, temperature conditions, and reaction times, which have been developed for corn. The starch of sweet potatoes is considered more complex than cereal starches, making it more challenging to hydrolyze into fermentable sugars. It can present high amylose content, and contains longer and more complex amylopectin chains, making it more challenging to hydrolyze into fermentable sugars (Duvernay et al., 2013; Hoover, 2001; Moorthy, 2002; Robertson et al., 2006; Srichuwong et al., 2005). However, among other tubers, sweet potato starch was found to be more susceptible than cassava starch to degradation by α-amylase and glycoamylase (Moorthy, 2002). Noda et al. (1992) found no significant correlation between digestibility and amylose content. Srichuwong et al. (2005) found that the relative abundances of amylopectin unitchains were significantly correlated with gelatinization, retrogradation properties, and enzyme digestibility.

TABLE 7.5 Sweet Potato Composition (Fresh Root and Flour)

Moisture content (%)	Free sugars (% w/w db)			Total free sugars (% w/w db)	Starch (% w/w db)	Total sugars (% w/w db)	Fiber (% w/w db)	Proteins (% w/w db)	Lipids (% w/w db)	Ash (% w/w db)	References
	Glucose	Fructose	Sucrose								
Fresh root											
66.2–68.9	—	—	—	—	63.9–64.9	—	2.3–2.7	4.3–5.1	—	2.1–2.5	Dincer et al. (2011)
67.2–82.0	—	—	—	—	35.9–83.7[a]	—	8.9–18.4[a]	4.5–10.8[a]	0.2–1.0[a]	3.0–6.2[a]	Hernández-Suárez et al. (2016)
69.9–70.9	—	—	—	—	—	75.4–83.5[b]	7.6–11.7[a]	4.2–7.3[a]	0.7–1.1[a]	3.6–4.9[a]	Ishida et al. (2000)
65.8–73.1	2.4-4.4	2.6-2.9	5.6–8.0	10.9–15.7	55.5–61.4	75.0–81.3	1.0–1.3	0.2–3.5	0.2–0.4	2.2–4.1	Lareo et al. (2012, 2013)
59.9	—	—	—	7.8[a]	63.7[a]	78.5[a]	1.6[a]	2.6[a]	0.2[a]	3.1[a]	Masiero et al. (2014)
62.8–69.4	—	—	—	3.7–9.9	63.1–77.3	—	1.9–3.5	3.0–7.2	1.1–2.1	2.3–4.2	Ravindran et al. (1995)
59.4–59.5	0.6-0.7	0.4	2.4–3.2	3.6-4.3	68.7–70.1	79.9–82.1	—	—	—	—	Srichuwong et al. (2012)
77.3	—	—	—	—	—	88.6[a]	13.2[a]	6.9[a]	0.2[a]	4.4[a]	USDA (2014)
58.6–82.0	—	—	—	—	13.8–79.1[a,c]	—	—	—	—	—	Waluyo et al. (2015)
65–80	0-6.8[k]	0–5.5[k]	8.1–22.4[k]	—	34.5–68.3	—	2.9–5.6	5.2–10.1	0.7–0.9	3.3-4.6	Wu and Bagby (1987)
66.5–75.7	4.4	4.8	3.4	4.8–12.7	46.8–73.6[d]	64.6–86.5	—	—	—	—	Zhang et al. (2002)
61.5	20.1[i]	1.2	2.8	19.7[j]	62.1	88.6	2.2[h]	0.7[h]	0.2[h]	1.6[h]	Zhang et al. (2011)
64.9	—	—	—	8.8	67.5	—	—	12.5	—	—	Zhang et al. (2013)
Flour											
8.1–12[f]	—	—	—	—	—	74.6–90.9	0.1–5.5	0.6–5.9	0.04–1.5	0.2–2.1	Olatunde et al. (2016)
7.7	2.1	1.6	15.8	20.3	51.1	77.0	3.0	6.6	1.8	2.7	Lareo et al. (2012, 2013)

—	1.2–8.7	1.2-8.8	6.7–15.2	9.3–29.6	24.6–79.2	—	—	—	—	—	Montefusco et al. (2014)
—	—	—	—	6.6	72.1	—	—	4.5	—	—	Mussoline et al. (2017)
—	—	—	—	—	—	73.0[g]	17.5[g]	6.6[g]	1.0[g]	1.0[g]	RamesYadav et al. (2006)
—	1.6/1.9[e]	—	28.8/27.2[e]	—	41.8/61.7[e]	—	—	—	—	—	Ziska et al. (2009)

[a] *Calculated in dry weight from the reference*
[b] *Calculated by difference with the other components: 100 – (protein + lipid + fibre + ash)*
[c] *Calculated using the mean moisture content (65.2%)*
[d] *Composition at harvest time*
[e] *Values for varieties from Alabama/Maryland*
[f] *Flour, prepared from sweet potato peeled dried under sun for 3-4 days and oven at 50 °C for 5 h*
[g] *Peeled*
[h] *Values given in % (v/v)*
[i] *This value may include the glucose from cellulose*
[j] *Calculated as the difference between the total sugar and starch composition x 1.11*
[k] *Determined from freeze-dried samples.*

Total sugars and total free sugars were calculated in equivalents of glucose as 1.05 x sucrose +glucose + fructose, and 1.11 x starch + 1.05 x sucrose + glucose + fructose, respectively.

db, dry base; —, not available.

TABLE 7.6 Average Composition of Sweet Potato as Bioenergy Crop

Moisture content (%)	Starch (% w/w db)	Total sugars (% w/w db)	Fiber (% w/w db)	Proteins (% w/w db)	Lipids (% w/w db)	Ash (% w/w db)
70	64	80	3.5	4.5	0.7	3.1

Total sugars were calculated in equivalents of glucose as 1.11 x starch + 1.05 x sucrose +glucose + fructose.
db, dry base.

TABLE 7.7 Composition of Sweet Potatoes From Two-Year Field Trials

Year	Cultivar	Origin	Moisture content (%)	Total free sugars (% w/w db)	Starch (% w/w db)	Total sugars (% w/w db)	Fiber (% w/w db)	Proteins (% w/w db)	Lipids (% w/w db)	References
2009	K 9807.1	Uruguay	73.0	13.4	56.3	75.3	1.0	3.5	0.3	Lareo et al. (2012)
2010			68.6	13.2	61.4	81.3	1.3	0.2	0.2	
2014	CX1	South Carolina, USA	72.2	6.4	73.5	88.0	—	6.5	—	Mussoline et al. (2017)
2015			73.8	6.6	72.1	86.6	—	4.5	—	

Total sugars and total free sugars were calculated in equivalents of glucose as 1.05 x sucrose +glucose + fructose, and 1.11 x starch + 1.05 x sucrose + glucose + fructose, respectively.
db, dry base; —, not available.

The chemical composition of sweet potato can vary widely depending on the planting strategy, climatic conditions, cultivar, geographic region, soil quality, fertilization rate, irrigation patterns, drainage conditions, harvest time, and postharvest storability time and environmental conditions (Mussoline et al., 2017). An example of variation of the composition of two industrial sweet potato cultivars selected as alternative crops for bioethanol production and harvested in different years can be seen in Table 7.7.

Starch is generally easier to store than fermentable not polymeric sugars. The postharvest storability is one advantageous characteristic that supports the use of industrial sweet potatoes for ethanol production. The starch stability has been related to certain differences in its structure, particularly higher amylose content has been correlated with higher resistant starch concentrations (Berry, 1986; Mussoline et al., 2017), which could reduce ethanol productivity (Sharma et al., 2010). The roots can potentially be stored up to 6 months and used as a continual feedstock supply for ethanol for ethanol production (Mussoline et al., 2017; Vilaró et al., 2009). Zhang et al. (2002) studied the variation of sugar content (starch and free sugars, such as glucose, fructose, and sucrose) during storage up to 120 days from harvest. During storage, five of the six genotypes studied showed slight decreases in starch content (from 47%–58% to 38%–50%), while the genotype, which presented the highest starch content at the time of harvest, decreased from 73.6% to 50% (dry basis) after 6 months. The hydrolysis of starch to maltose and oligosaccharides in stored sweet potato is due to the activity of amylases in root (Morrison et al., 1993). Zhang et al. (2002) observed that the α-amylase activity increased from harvest to 60 days and gradually declined to harvest levels after 180 days of storage. Also, Reddy and Sistrunk (1980) and Sistrunk (1971) found that the average starch content diminished 44%–50% after 6–7 months of storage. However, other researchers have found that the starch content in sweet potato roots was stable after a 6-month storage period (Lareo et al., 2013; Mussoline et al., 2017). Jin et al. (2012) assessed the fermentable sugar, fiber, and pectin content of fresh sweet potato of 10 different sweet potato varieties for three harvest times (100, 130, and 160 days). In most varieties, the ethanol production corresponding to day 130 was the highest. Therefore, the screening of genotypes with high dry matter and starch content should take into account the stability of these compounds over time (Zhang et al., 2002).

7.6 PROCESSES FOR CONVERSION OF SWEET POTATO INTO BIOETHANOL

The conventional process for bioethanol production from starch-based materials, such as sweet potato, involves the conversion of starch into fermentable sugars, which generally takes place in two enzymatic steps: liquefaction using α-amylase and saccharification by addition of amyloglucosidase (AMG). The glucose from the starch hydrolysis is fermented together with the free soluble sugars present in the sweet potato. Fig. 7.3 shows a simplified diagram for the fuel ethanol production from the sweet potato. The sweet potato can be processed fresh or dried as flour.

7.6.1 Pretreatment

Previous transformation of the raw material into chips or flour (dried and milled into powder) can be done in order to facilitate its transport, storing, and/or processing. The use of flour of sweet potatoes would allow working with higher sugar concentration during the fermentation [conditions very high gravity (VHG)]

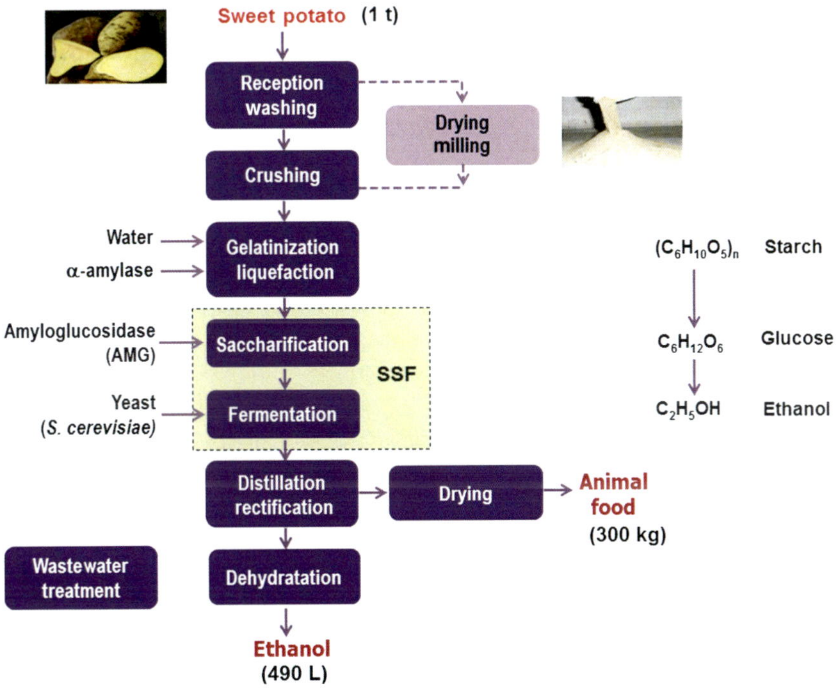

FIGURE 7.3 **Diagram for the fuel ethanol production from fresh sweet potato.** The *dashed blue line* indicates the drying and milling steps needed to obtain flour of sweet potato. *Source: Data in parentheses taken from Lareo, C., Ferrari, M.D., Guigou, M., Larnaudie, V., Fajardo, L., Ramírez, M.B., 2012. Evaluación de materiales amiláceos experimentales desarrollados para bioenergía (boniato y grano de sorgo dulce) para la producción de bioetanol combustible. FPTA-INIA 39, ISSN:1688-924X.*

than fresh sweet potatoes without the addition of water. In this case, it should be assessed the energy saving of manipulating lesser amount of material, the handling of high viscous material, the extra cost of drying, and the effect of drying on the performance of the process (conversion of starch to fermentable sugars) (Ferrari et al., 2013; Moorthy, 2002). Ferrari et al. (2013) estimated the energy consumption of the different stages of the industrial bioethanol process using a process simulation software and experimental data. High energy consumption was found in the drying and milling steps to obtain the sweet potato flour from fresh material, in the range 60.2%–80.9% of the global process (23.0–30.8 GJ/m^3 ethanol) for all studied conditions. Considering that the lower heating value of ethanol is 21.1 GJ/m^3 ethanol, it was observed that the energy consumption of drying was greater than the energy that could be obtained from the ethanol produced. Thus, if a drying step is needed, the use of renewable energy sources is required to have a favorable overall energy balance (see other sections for further discussion).

In the process of converting sweet potatoes to ethanol, the raw material is weighted, cleaned to remove foreign objects, and washed. The cleaned sweet potato is crushed and can have two destinations: it can be discharged in the liquefaction tank to continue in the process as fresh sweet potatoes, or it can be sent to a dryer and continues in the process as flour of sweet potato (final humidity in the range 8%–12%). The dried sweet potato continues in the processing line, passing through a milling step. The flour is storage, before it is brought to the liquefaction step (Ferrari et al., 2013; Olatunde et al., 2016).

The effect of the drying of sweet potatoes on the sugar composition and ethanol yield has been studied by some researchers. Lareo et al. (2013) found that drying sweet potatoes at 55 and 95°C in a forced convection tray dryer (28 and 18 h, respectively), neither affected the sugar content nor the starch enzymatic hydrolysis efficiency, which reached values close to 100%. Similar results were found by Ramesh Yadav et al. (2006), who did not find differences in the composition (carbohydrates, protein, fat, and total dietary fiber) of raw and processed sweet potato flours prepared using a drum drier and a batch type cross flow hot-air drier. They also found that the degradation of starch by native amylases during hot-air drying (65°C, 7–8 h) increased the digestibility.

7.6.2 Gelatinization and Liquefaction

The liquefaction step generally involves the partial starch hydrolysis by the addition of the α-amylase at high temperature. The appropriate quantity of sweet potato (fresh or flour) is sent to liquefaction tanks along with water, and mixed. The total water quantity added depends on the ratio dry material to water required. The pH is adjusted to the optimum of the α-amylase activity by adding lime or sulfuric acid accordingly. The suspension is heated for gelatinization and α-amylase is added and mixed. During the gelatinization step, there is a sharp rise in the slurry viscosity that is rapidly decreased as the α-amylase hydrolyzes the starch. A better homogenization of the mash has been observed with increasing temperature (Lareo et al., 2013). Gelatinization would allow enzymes to penetrate easily into starch structures contributing to a more efficient reaction (Delgado et al., 2009; Hansen et al., 2008). It has been found that the heat treatment used during the gelatinization step, without enzyme addition, produces a partial starch hydrolysis. Lareo et al. (2013) found a starch hydrolysis percentage in the range of 47%–61% and 34%–45 % for fresh material and flour respectively, during the gelatinization process at 90–121°C, without enzyme addition.

Lareo et al. (2013) studied the liquefaction of sweet potato without a pregelatinization step (the enzyme was added before heating the sample to the liquefaction temperature, 86°C). No significant difference was found for the starch hydrolysis percentages and reducing sugar concentration values with or without gelatinization step. In these studies, it was concluded that the gelatinization step before the addition of the amylase would not be necessary. It also allows working with sweet potato mashes with lower viscosity, which improves the manipulation of the material especially for VHG conditions, in particular its homogeneity and transport, allowing the acquisition of more consistent results and reducing the energy consumption of the process.

7.6.3 Saccharification and Fermentation

The saccharification (addition of AMG) and fermentation steps can be performed separately [separated hydrolysis and fermentation (SHF)] or simultaneously [simultaneous saccharification and fermentation (SSF)]. The separation of hydrolysis and fermentation offers processing advantages. It enables enzymes and microorganism to operate at their optimum temperature for increased performance, 85–90°C and ~30°C, respectively, optimizing the utilization of sugars. During saccharification, hydrolysis percentages close to 100% for both fresh sweet potato and flour have been found (Lareo et al., 2013). Some researchers added the AMG while the α-amylase was still acting [presaccharification and simultaneous fermentation (PSSF)] (Mojovic´ et al., 2006; Montesinos and Navarro, 2000; Zhang et al., 2011) for other starchy materials. This reduces the inhibition of the AMG activity by the presence of high glucose concentrations.

SSF has been considered a good choice to reduce: (1) the osmotic pressure caused by high initial concentration of dissolved sugars in batch ethanol fermentation under VHG conditions (Cao et al., 2011; Ray and Naskar, 2008; Shen et al., 2011; Srichuwong et al., 2009; Zhang et al., 2011), (2) the feedback inhibition of the AMG by the presence of high concentrations of glucose (Cao et al., 2011; Mojovic´ et al., 2006), and (3) the time and energy of the complete process since the saccharification step separated from fermentation at temperatures above 50°C is eliminated (Zhang et al., 2010). However, in an SSF process, the temperature and pH are more favorable for the yeast growth rather than for the AMG activity. Shen et al. (2012a) have suggested the control of the glucose-releasing rates by incremental addition of glucoamylase to reduce cell inhibition.

Under VHG conditions, it has been found that the behavior of the SSF was better than SHF for fresh sweet potato, in terms of sugars, ethanol, and ethanol yield (Cao et al., 2011; Lareo et al., 2013; Zhang et al., 2011). Zhang et al. (2011) found that the process of SSF was completed in 27 h as compared with SHF and PSSF in 33 h (30 h for fermentation and 3 h for saccharification). The maximum ethanol concentration (128.5 and 128.0 g/L), average ethanol productivity (4.8 and 3.9 g/L/h), and the ethanol fermentation efficiency (90%) were almost the same in SSF and PSSF, but better than that in SHF (93.6 g/L, 2.8 g/L/h, 67%, respectively) from 28% (w/v) of initial carbohydrates from raw sweet potato.

An ideal industrial fermenting microorganism should achieve high ethanol yields and productivity, and be tolerant to high ethanol concentration. Yeast strains, such as *Saccharomyces cerevisiae*, are currently used in the industrial processes due their high conversion efficiency and adaptability under fermentation conditions. However, it does not ferment starch directly. Also, the bacterium, *Zymomonas mobilis*, has been used, which presents some physiological advantages, such as higher ethanol yield and tolerance, faster sugar uptake rate, and lower biomass production (He et al., 2009a; Thatoi et al., 2014; Zhang and Feng, 2010). The direct production of ethanol from sweet potato starch using a genetically engineered *Z. mobilis* expressing glucoamylase genes from *Aspergillus awamori* has been reported (He et al., 2009a), although at a low level and volumetric productivity (fermentation time was 6 days).

In general, the fermentation time varies in the range 14–72 h depending on the raw material and the dry matter ratio of sweet potato to water used, the fermentation efficiency in the range 70%–92% and ethanol concentration up to 134 g/L can be found (Table 7.8). Inhibition and low sugar conversion were found for initial sugar concentrations higher than 250 g/L (VHG conditions) (Lareo et al., 2013; Montefusco et al., 2014). Industrial and agroindustrial yields were the ranges 300–520 L ethanol per tonne sweet potato dry base and 3020–4900 L ethanol/ha, respectively. Table 7.8 shows a summary of fermentation results found in the literature.

Mixed cultures of ethanol producing yeast or bacteria with saccharolytic microorganisms have also been studied to carry out the starch hydrolysis and fermentation processes in one step. Because the strains used in mixed-cultures do not always have similar culture requirements, such as pH, temperature, nutrient, and oxygen demand, it is very difficult to optimize the conditions for one strain without affecting the other strains. Immobilization of cells could mitigate inhibition caused by high substrate and ethanol concentration. The use of the mixed cultures coimmobilized would allow to implement two stage fermentation process in a single step, which could reduce the energy input of the process and resolve the aforementioned problem (Lee et al., 2012). Generally, this type of culture presents longer fermentation time and low ethanol concentration than pure cultures (see Table 7.8).

7.6.4 Ethanol Recovery

After fermentation, the fermented medium is sent through a degasser to flash off the vapors. The vapor stream is primarily ethanol and water with some residual carbon dioxide. The ethanol and water vapors are then condensed and recombined with the liquid stream prior to distillation. Any uncondensed vapor is combined with the carbon dioxide produced during fermentation and sent through a carbon dioxide scrubber prior to venting or recovery. The outlet from the bottom of the column is sent to the process condensate tank (Ferrari et al., 2013).

The medium fermented is heated to ~80°C in the degasser before entering to the beer column, which captures nearly all of the ethanol produced during fermentation. An almost equal amount of water is also distilled that must be separated from the ethanol in the next stage of rectification/stripping. The outlet from the bottom of the beer column contains a considerable amount of water and nonfermentable material, such as protein, yeast, oil, fibers, and residual chemicals unconsumed during fermentation, and it is sent to separation steps to recover as coproduct. The subsequent steps for the final ethanol recovery include the combined action of a rectifier, stripper, and molecular sieves. Most of the ethanol goes out of the top of the rectifier as distillate (Ferrari et al., 2013; Kwiatkowski et al., 2006).

The outlet vapor stream of the beer column is mixed with the streams that comes from the top of the stripping and from the molecular sieves, and enters to the rectifier column. The outlet from the bottom of the rectifier column is fed to the stripping column to remove additional water. The ethanol distillate from stripping is recombined with the feed to the rectifier. The distillate of the rectifier containing primarily ethanol feeds the molecular sieves, which captures the last bit of water creating 99.2% (v/v) pure ethanol. The water molecules trapped and adsorbed inside the microporous beads of the molecular sieves are regenerated by heating in an offline operation, and this stream is combined with the process condensate stream (Ferrari et al., 2013).

7.6.5 Coproducts From Sweet Potato Stillage Recovery

After ethanol is distilled, residual stillage from the bottom of the beer column is separated by centrifugation producing wet solids (Fig. 7.3). The solids recovered represent valuable protein-rich by-products, which may have commercial

TABLE 7.8 Fermentation Results for Fresh Sweet Potato And Flour at Different Dry Matter to Water Ratios

Sweet potato	Microorganism/ experimental conditions	Dry matter of sweet potato to water ratio (%, w/v)	Dry matter of sweet potato to water ratio (w/v)	Ethanol (g/L)	Sugar conversion[a] (%)	Fermentation efficiency[b] (%)	Volumetric productivity (g/L/h)	Industrial yield[c] (L ethanol/t sweet potato dry base)	Agroindustrial yield[c,d] (L ethanol/ ha)	References
Single cultures										
Flour	*Saccharomyces cerevisiae*/ cellulase[h]: 50°C; α-amylase: 95°C, 0.5 h; glucoamylase 60°C; SHF: 30°C	50	1:2	113[c]	78[r]	—	—	286	—	Cao et al. (2011)
		43	1:2.3	114[c]	93[r]	—	—	336	—	
	Saccharomyces cerevisiae/ cellulase[h]: 50°C; α-amylase: 95°C, 0.5 h; SSF: 30°C	50	1:2	120 [c]	88	86	2.3	304 [l]	—	
		43	1:2.3	117[c]	92	87	2.6	344[l]	—	
		38	1:2.6	106[c]	95	87	3.3	353[l]	—	
		34	1:2.9	93[c]	93	87	3.3	347[l]	—	
Fresh	*Saccharomyces cerevisiae*/ α-amylase: 85°C, 3 h; glucoamylase: 65°C, 48 h; SHF: 35°C	12.5	1:8	35[k]	100	100	0.7[k]	355[k,l]	—	Duvernay et al. (2013)
Flour		25	1:4	68[k]	100	95	1.4[k]	344[k,l]	—	
Starch	Genetically engineered *Zymomonasmobilis* with-glucoamylase genes from *Aspergillus awamori* (without addition of enzymes), 6 days	(50 g/L sweet potato starch)	—	14	—	49	0.1	—	—	He et al. (2009a)
Fresh	*Saccharomyces cerevisiae*/ α-amylase 85°C, 10 min; SSF: 30°C	—	—	—	—	84-92	—	305-701[j]	3020-4370	Jin et al. (2012)
Fresh	*Saccharomyces cerevisiae*/ α-amylase: 86°C, 90 min; SSF: 30°C	45.5[e]	1:2.2[e]	100	67	92	2.1	320	3170	Lareo et al. (2013)
		20	1:5	63	88	82	2.6	380	3730	
		12.5	1:8	45	100	84	3.2	490	4790	

(Continued)

TABLE 7.8 Fermentation Results for Fresh Sweet Potato and Flour at Different Dry Matter to Water Ratios (*cont.*)

Sweet potato	Microorganism/ experimental conditions	Dry matter of sweet potato to water ratio (%, w/v)	Dry matter of sweet potato to water ratio (w/v)	Ethanol (g/L)	Sugar conversion[a] (%)	Fermentation efficiency[b] (%)	Volumetric productivity (g/L/h)	Industrial yield[c] (L ethanol/t sweet potato dry base)	Agroindustrial yield[c,d] (L ethanol/ ha)	References
Flour	*Saccharomyces cerevisiae*/ α-amylase: 86°C, 90 min; SSF: 30°C	50	1:2	99	77	79	2.1	305	2990	Lareo et al. (2013)
		33	1:3	97	100	92	2.7	460	4490	
		20	1:5	58	99	90	3.6	425	4170	
		12.5	1:8	38	99	84	2.5	410	4020	
Fresh[f]	*Saccharomyces cerevisiae* CAT-1/ mixture of α-amylase and amyloglucosidase 52°C, 1 h, SHF 32°C, 72 h	20	—	14–22	—	53–78	0.3–0.5	306	4900	Masiero et al. (2014)
Flour[m]	*Saccharomyces cerevisiae immobilized*/ α-amylase: 80°C, 45 min; amyloglucosidase: 50°C, 2 h; SHF	—	—	97	95	100	1.2[n]	381	2032	Montefusco et al. (2014)
		—	—	87	100	97	1.2	—	—	
		—	—	53	100	84	1.1	—	—	
		—	—	39	100	75	1.2	—	—	
		—	—	24	100	70	0.7	—	—	
Flour	*Saccharomyces cerevisiae*/ α-amylase: 95°C, 3 h; glucosidase 65°C, 3 h; SHF and SSF 32-35°C, 24 h	25	—	26–38	—	76–97	1.1–1.6	—	—	Purohit and Mishra (2012)
Fresh	*Saccharomyces cerevisiae*/ α-amylase: 90°C, 1 h; SSF 40°C, 96 h	—	1:6[p]	16	95	94	0.2	400[j]	—	Ray and Naskar (2008)
Flour		—	1:6[p]	40	91	84	0.4	359	—	
Flour	*Saccharomyces cerevisiae*/ α-amylase: 85°C, 2 h; SSF 30°C, 64-80 h	(260 g/kg glucose)	—	120	—	90	1.48[t]	—	—	Shen et al. (2011b)

Flour	*Saccharomyces cerevisiae*/ α-amylase: 85°C, 2 h; glucoamylase 60°C, 4 h; SHF 30°C.	(120 g/kg glucose)	—	56	100	91	1.4	—	—	Shen et al. (2012a)
		(250 g/kg glucose)	—	87	75	91	1.1	—	—	
Flour	*Saccharomyces cerevisiae*/ α-amylase: 85°C, 2 h; glucoamylase 55°C, 4 h;SHF 30°C, 72 h	(210 g/kg glucose)	—	105	100	98	1.45	—	—	Shen et al. (2012b)
Fresh	*Saccharomyces cerevisiae*/ pectinase, cellulase, glucanase, hemicellulase: 50°C, 3 h; α-amylase: 60°C, 2 h; SSF 30°C, 48-72 h[i]	33-37 (260-270 g/L glucose)	—	119–122	—	89–91	1.7–2.5	440/450	—	Srichuwong et al. (2012)
Flour[f,q]	*Saccharomyces cerevisiae*/ pectinase, α-amylase, glucoamylase; SHF	(181 g/kg glucose)	—	83	94	95	—	—	—	Vijaya et al. (1997)
Fresh	*Saccharomyces cerevisiae*/ pectinase: 50°C, 2 h; α-amylase: 90°C, 1h; glucoamylase: 60°C, 2 h; SHF: 30°C, 66 h	20	—	71	99	95	1.1	446[l]	—	Wu and Bagby (1987)
		26.6	—	79	99	84	1.2	376[l]	—	
		35	—	73	99	83	1.1	264[l]	—	
Flour	*Saccharomyces cerevisiae* immobilized (packed bed reactor)/ saccharification: 60°C 15 min; SHF: 35°C, reactors of 1 L and 1000 L	(11.8-12.5% w/v total sugars)	1:4	52–60	87–90	94-100	2.4–2.8[s]	—	—	Yu et al. (1996)
Fresh	*Saccharomyces cerevisiae*/ xylanase 44°C, 88 min; α-amylase: 80°C; SSF 30°C, 36 h[g]	30	—	134[c]	94	91	3.7	496	—	Zhang et al. (2010)

(*Continued*)

TABLE 7.8 Fermentation Results for Fresh Sweet Potato and Flour at Different Dry Matter to Water Ratios (*cont.*)

Sweet potato	Microorganism/ experimental conditions	Dry matter of sweet potato to water ratio (%, w/v)	Dry matter of sweet potato to water ratio (w/v)	Ethanol (g/L)	Sugar conversion[a] (%)	Fermentation efficiency[b] (%)	Volumetric productivity (g/L/h)	Industrial yield[c] (L ethanol/t sweet potato dry base)	Agroindustrial yield[c,d] (L ethanol/ ha)	References
Fresh	*Saccharomyces cerevisiae*/ α-amylase: 85°C, 20 min; SSF with xylanase, 30°C[g]	60-70 (20-28% carbohydrates)	—	128 (lab scale)	99	91	4.8	—	—	Zhang et al. (2011)
		(20-24% carbohydrates)	—	90–109 (pilot scale)	98-99	90-91	4.2-5.0	—	—	
		(15-21.5% carbohydrates)	—	70–98 (commercial scale)	91-92	90-91	3.4-4.2	—	—	
Flour	*Zymomomasmobilis*/ α-amylase 100°C, 3 h; glucoamylase 60°C, 2 h; SHF	20 (150 g/L glucose)	—	66	94	94	2.8	418[l]	—	Zhang and Feng (2010)
		30 (175 g/L glucose)	—	79	94	96	2.2	334[l]	—	
		40 (200 g/L glucose)	—	99	92	92	2.1	313[l]	—	
Co- and mixed-cultures										
Flour	*Paenibacillus*sp. and*Zymomonasmobilis*, 120 h	(50 g/L sweet potato starch)	—	7	—	23				He et al. (2009b)
Flour	Immobilization of *A. oryzae, Monascuspurpureus* and *S. cerevisiae*/ 30°C, 9 days	10	—	32[c]	99	89	0.1	—	—	Lee et al. (2012)

Flour[t]	*Trichoderma* sp. and *S. cerevisiae*/ 30°C, 72 h	20	—	(172 g/kg substrate)	—	72	(2.8 g/kg substrate/h)	218	—	Swain et al. (2013)
Fresh	*Aspergillus niger*and*Zymomomasmobilis*/ 30°C, 60 h	—	—	104[c]	—	87	1.7	520	—	Zhang et al. (2013)

[a] *Sugar conversion based on the total sugar present in the raw material (fresh or flour)*
[b] *Efficiency based on 0.511 g ethanol/g sugars as glucose*
[c] *Calculated using the ethanol density at 20 °C (0.7894 kg/L)*
[d] *Calculated based on an agriculture yield of 10 t/ha (dry matter) (Vilaró et al., 2009) and a distillation efficiency of 98%*
[e] *Fresh sweet potato without addition of water*
[f] *Sweet potato peeled*
[g] *Xylanase was used to reduce viscosity*
[h] *Cellulases were added to reduce viscosity*
[i] *Pectinase, cellulase, glucanase and hemicellulase were added to reduce viscosity*
[j] *Calculated assuming 70% moisture content*
[k] *Results for media with addition of salts*
[l] *Calculated using the ethanol density at 20 °C (0.7894 kg/L), the dry matter of sweet potato to water ratio (%, w/v) and assuming that the volume of water added was equal to the volume of the mash*
[m] *Sweet potato from local farmers, province Lecce, Italy*
[n] *Calculated at 72 h fermentation*
[p] *Wet basis (flour 8-10% moisture content)*
[q] *Red variety purchased from local market*
[r] *Calculated using the residual reducing sugars informed*
[s] *Values based on the total volume of the reactor*
[t] *Calculated assuming that the density of the suspension is 1000 kg/m^3.*

SHF, separated hydrolysis (saccharification) and fermentation; SSF, simultaneous saccharification and fermentation; PSSF, addition of AMG while the α-amylase was still acting ; —, not available..

potential (Wu and Bagby, 1987). Following similar stages than that for producing distiller's dried grains with solubles from corn (DDGS) (Kwiatkowski et al., 2006), the water of the liquid product from centrifugation, known as thin stillage, can be removed in an evaporator and recovered, and the concentrate dried further. The concentrate from the evaporator is mixed with the wet distiller's solids coming from the centrifuge and sent to a dryer that reduces the moisture content to ~10%. This becomes the coproduct protein-rich that can be sold as an animal feed.

7.6.6 Wastewater Treatment

All process wastewaters considered as contaminated wastes, such as washing waters, should be treated by aerobic or anaerobic digestion.

7.7 EFFECT OF THE MAIN VARIABLES OF THE CONVERSION PROCESS: SOLID TO LIQUID RATIO, ENZYMES, AND TEMPERATURE

An important aspect in the production of bioethanol is that the process must have a minimum consumption of energy. Distillation is one of the stages that consume more energy. Energy saving in distillation can be achieved with high ethanol concentration after fermentation by adding a minimal amount of water to the raw material or sweet potato mashes (Cao et al., 2011; Ferrari et al., 2013; Shen et al., 2011). VHG fermentation technology should then be used (Bai et al., 2008). In the case of ethanol production from root and tuber crops, the use of a VHG medium implies manipulation of a material with high solid content and high viscosity. The high viscous nature causes several handling difficulties during process (Zhang et al., 2011). The addition of α-amylase (liquefaction) reduces the starch-paste viscosity (Lareo et al., 2013). However, some researchers have studied the addition of enzyme preparations to reduce further the viscosity from mashes, such as pectinase, cellulase, and hemicellulase (Srichuwong et al., 2009, 2012), and xylanase (Zhang et al., 2010, 2011), to disrupt the cell-wall matrix (Table 7.8). Zhang et al. (2011) used xylanase simultaneously with fermentation and saccharification of fresh sweet potato, at labs, pilot and industrial scale (20 m^3 fermenters). They found that the use of xylanase at pilot scale not only decreased the viscosity sharply from approximately 5000–800 cp, but also the fermentation time was reduced.

Besides, VHG technology can show incomplete hydrolysis of starch to fermentable sugars (Watanabe et al., 2010; Zhang et al., 2011) and fermentation, because the yeast cells are exposed to several stresses (high concentration of dissolved solids, which increases external osmotic pressure, high ethanol concentration can be toxic to the cells) (Lareo et al., 2013; Pradeep and Reddy, 2010; Reddy and Reddy, 2006). In Table 7.8, it can be seen that fermentations with high dry matter of sweet potato to water ratio presented low sugar conversion (Cao et al., 2011; Lareo et al., 2013; Shen et al., 2012a; Zhang and Feng, 2010; Zhang et al., 2010). Montefusco et al. (2014) found that initial sugar concentration of 267 g/L resulted in a marked inhibition of the fermentation process. The initial substrate concentration influences the time required for complete sugar fermentation and the ethanol yield. Cao et al. (2011) found that when the solid: liquid ratio was increased to 1:1.7, fermentation did not occur because the broth became too thick with a high level of initial dissolved solids. This could be overcome by using yeast with a high tolerance to these stress conditions. It has been suggested that ethanol tolerance of yeast can be improved with appropriate supplementation of nutrients, such as various sources of nitrogen, vitamins, sterols, and unsaturated fatty acids. It has also been reported that supplying small quantities of oxygen could improve the tolerance of yeast to high ethanol concentrations, which contribute to the stabilization of the plasma membrane by producing sterols (Breisha, 2010; Deesuth et al., 2015; Pereira et al., 2010; Srichuwong et al., 2009). Unfortunately, special yeasts tolerant to stress conditions are not commercially available and many of the supplements used on a laboratory scale, are too expensive for industrial use (Larnaudie et al., 2016).

Starch granules are not soluble in water at ambient temperature because of its semicrystalline structure. When they are placed in sufficient water, starch granules absorb a small amount of water and swell. If heated above the gelatinization temperature, starch losses the native crystalline structure and solubilization occurs. Gelatinization of starch is an important physical transformation, which converts semicrystalline starch granule to the amorphous conformation, which is an enzyme susceptible form (Srichuwong and Jane, 2007). The gelatinization of sweet potato starch occurs in the range 60–88°C (Kitahara et al., 2005; Moorthy, 2002; Srichuwong et al., 2005; Zhu and Wang, 2014). Temperatures higher than the starch gelatinization range are usually used in liquefaction process to ensure complete gelatinization of starch (Srichuwong et al., 2012). One promising technology, which would reduce the manipulation of viscous material and the costs due to the high energy demand of starch-based ethanol production, is known as granular starch hydrolysis. In that process, the starch hydrolysis is performed with a mixture of α-amylase and glucoamylase, which is capable of hydrolyzing raw starch at temperatures below its gelatinization temperature, ~60°C (Cinelli et al., 2015; Robertson et al., 2006; Vidal et al., 2009). This is an emerging technology, which is enabled by several enzymes that act on starch in its granular form and do not need the liquefaction and cooking steps of the conventional process, allowing a reduction in energy consumption. It comprises amylolytic enzymes that act synergistically to breakdown starch to glucose: endoamylases, exoamylases, and debranching enzymes (Castro et al., 2011; Cinelli et al., 2015). However, its low hydrolysis rate and/or incomplete hydrolysis due to the structural heterogeneity and crystalline nature of starch molecules are challenges of this technology (Robertson et al., 2006).

There are still challenges to be overcome to make cold hydrolysis economically competitive when compared to the conventional process. The low temperature employed usually generates microbial contamination problems. Besides, granular starch remains solid and therefore, the degradation reaction takes place in a solid–liquid system, where mass transfer can be a limitation. Therefore, both greater amount of enzymes, which increases the cost, and longer times are needed for complete glucose conversion (Masiero et al., 2014; Vidal et al., 2009). A preheating step at subgelatinization temperature has been studied to minimize these problems (Masiero et al., 2014). Some researchers have also used auxiliary enzymes, such as cellulases and pectinases to reduce viscosity and to improve the hydrolysis performance (Srichuwong et al., 2012).

Masiero et al. (2014) studied simultaneous cold hydrolysis (52°C, 1 h) and fermentation of fresh sweet potato. They reached low substrate concentration, low ethanol concentrations (in the range of 14–22 g/L) and ethanol fermentation efficiencies of 53%–79% in 96 h fermentation from a mash containing 200 g/L sweet potato (Table 7.8). In order to increase the substrate concentration in the medium and also to decrease the enzyme consumption, higher processing degree of the raw material (drying and grinding) and/or utilization of auxiliary enzymes (cellulases and pectinases) should be evaluated. Besides, the fermentation efficiency has also to be enhanced.

New sweet potatoes varieties containing starch that can gelatinize at low temperature have been found and developed in Japan. The gelatinization temperature of the starch was 10–20°C lower than that of ordinary sweet potato starch (Kitahara et al., 2005; Srichuwong et al., 2012). Srichuwong et al. (2012) studied the newly developed sweet potato K159, which presented a gelatinization temperature in the range 49.5–66.2°C. They found that liquefaction at 60°C was sufficient to reach an ethanol concentration of 119–122 g/L and 88.8%–90.6% theoretical fermentation efficiency were obtained during 48–72 h (Table 7.8). An ordinary sweet potato type requires a higher liquefaction temperature (~ 90°C) to achieve a similar ethanol yield.

7.8 BY-PRODUCTS: USE OF RESIDUAL SOLIDS FOR ANIMAL FEED

After ethanol is distilled, residual stillage can be separated into filter cake, centrifuged solids, and stillage solubles, which play an important role in the commercial success of the ethanol industry. The residues contain nonfermentable sweet potato material, yeast, enzymes, and salts. Thus, filter cakes had much higher protein, fat, fiber, and ash contents than the original sweet potatoes. Sweet potato filter cake may have food potential, since it has high protein content and an amino acid balance superior to that of cereal grains. Wu and Bagby (1987) reported that nitrogen from yeast, amylase, and glucoamylase accounted for 4.1%–9.8% of total sweet potato nitrogen. They found that filter cake and centrifuged solids had protein contents in the range 22%–32% and 42%–57%, respectively, which account for 44%–85% and 0%–17% for total sweet potato nitrogen. Sweet potatoes and their fermented products had 4.3–7.6 g of lysine/16 g of N and are expected to have good nutritional value. The relatively low fat content of sweet potato filter cake, compared with that of corn distillers' grains, may improve storage stability (Wu and Bagby, 1987). Sweet potato stillage solubles contain 4%–6% dry matter; evaporation of water to recover solids for potential food or feed uses is expensive. Solids can also be recovered by ultrafiltration or reverse osmosis, which do not evaporate water, and the energy consumption and cost could be reduced (Wu and Bagby, 1987).

Lareo et al. (2012) also studied the residue of ethanol fermentation of sweet potato flour with a ratio of dry matter to water of 1:5 (20% w/v). For each 100 g of sweet potato (dry basis), 20 g of suspended solids separated by filtration were obtained after fermentation. The composition was: 3.7% (w/w) protein, 6.1% (w/w) fat, and 15% (w/w) fibers. Also, 10 g of soluble solids were obtained. These results corresponded to residues of the medium unfermented and included the yeast developed during fermentation.

Table 7.9 shows the composition of materials commonly used for livestock ration in the literature and composition of fermentation residues. Based on these contents the subsequent formulation of the ration should be made to suit the nutritional requirements of the animals. DDGS from corn normally has a protein content of 27.8% (Kwiatkowski et al., 2006) and corn distillers 29.4% (Harris, 2003). Similar protein content was found by Wu and Bagby (1987) in the residues after fermentation (total protein content in the range 25.6%–27.7% dry base; Table 7.9). Lower values were found by Lareo et al. (2013). Changes in the composition greatly impact the cost of ethanol production and the DDGS (Kwiatkowski et al., 2006), which should be taken into account in the selection of the raw material.

7.9 TECHNOECONOMIC FEASIBILITY

The sweet potato cost is a key factor for ethanol production (Shumaker et al., 2007). Its contribution depends on the local conditions of the crop (inputs, agricultural practices, and yields) and the size of the industrial facility considered. Labandera et al. (2010) carried out an economic feasibility study of the production of anhydrous ethanol for a facility size adequate for the market and agroindustrial conditions existing in Uruguay. The main cost factor was

TABLE 7.9 Composition of Materials Used as Cattle Ration and Residues After Fermentation of Sweet Potato

Material	Dry matter of wet recovered solid (%)	Fermentation residues based on initial dry matter (%w/w)	Type of solids	Crude protein (% db)	Fat (% db)	Fiber (% db)	Ash (% db)	References
Sweet potato (variety NS 007 and SS19) fermentation residues	—	24.1-30.1[b]	—	—	--	—	—	Jin et al. (2012)
Sweet potato (30% w/w db) fermentation residues[a]	14	20	Suspended	4.0	6.6	16.3	—	Lareo et al. (2012)
	—	10	Soluble	—	—	—	—	
Sweet potato (18.3% (w/w db)) fermentation residues[a]	8.9	16.2	Filter cake	12.1	1.8	13.9	3.5	Wu and Bagby (1987)
	31.2	3.6	Centrifuged	4.8	0.3	1.1	0.3	
	—	16.2	Soluble	10.8	—	—	13.9	
Sweet potato (20% (w/w db)) fermentation residues *treated with pectinase*[a]	14.9	22.0	Filter cake	19.8	2.0	11.8	11.8	
	30.7	0.1	Centrifuged	0.2	—	—	—	
	—	14.0	Soluble	5.6	—	14.8	14.8	
Sweet potato (27.4% (w/w db)) fermentation residue[a]	8.7	19.5	Filter cake	17.6	1.6	14.2	7.5	
	32.3	2.2	Centrifuged	3.8	0.2	0.7	0.3	
	—	10.2	Soluble	5.6	—	—	10.6	
Sweet potato (26.6% (w/w db)) fermentation residues *treated with pectinase*[a]	15.9	21.1	Filter cake	19.5	1.8	13.5	8.3	
	26.3	0.06	Centrifuged	0.1	—	—	—	
	—	10.9	Soluble	4.7	—	—	10.8	
Sweet potato (35.6% (w/w db)) fermentation residue[a]	12.5	18.8	Filter cake	16.7	2.2	11.9	9.3	
	38.8	0.5	Centrifuged	0.9	—	—	0.1	
	—	6.0	Soluble	2.4	—	—	6.6	
Sweet potato (35.0% (w/w db)) fermentation residues *treated with pectinase*[a]	18.0	13.2	Filter cake	16.7	1.9	10.8	5.9	
	31.6	0.2	Centrifuged	0.4	—	—	0.1	
	—	11.5	Soluble	3.6	—	—	16.1	
Alfalfa silage	35	—	—	17.1	5.1	—	—	Harris (2003)
Carrot roots	12	—	—	10.0	1.7	—	—	
Corn distillers	92	—	—	29.4	9.8	—	—	
Sorghum, grain, silage	30	—	—	8.0	—	—	—	
Sorghum, forage silage	30	—	—	7.0	—	—	—	
Wheat silage	26	—	—	11.5	3.8	—	—	
Yeast, brewers	93	—	—	47.3	1.1	—	—	

[a] *Fermentation with a 20% (w/v dry base) sweet potato.*
[b] *Calculated assuming 70% moisture content of sweet potato.*
db, dry base; —, not available/do not correspond. Filter cake corresponds to the solids obtained after filtration of the stillage through a cheesecloth. Centrifuged solids correspond to the solids obtained after centrifugation of the filterer liquid through the cheesecloth.

TABLE 7.10 Share of Different Expenses to the Final Cost of Fuel Anhydrous Ethanol From Sweat Potato for a Projected Facility of 10,000 m^3/year (Uruguayan Conditions)

Item	Share (%)
Variable costs	
Raw material[a,b]	66.2
Operating labor	6.8
Wood (for steam generation)	5.0
Electricity	2.0
Yeast[c]	2.7
Enzymes	0.7
Plant overhead, maintenance, etc.	2.9
Total variable costs	86.3
Fixed costs	
Depreciation	10.3
Interest on capital	3.4
Total fixed costs	13.7

[a] *Cost based on the price paid at the plant (price includes transportation cost).*
[b] *Variety of sweet potato selected as an bioenergy crop.*
[c] *Assuming yeast recycled seven times.*
Source: Labandera et al. (2010).

the sweet potato accounting for 66% of the total cost (Table 7.10). Shumaker et al. (2007) investigated the cost of producing ethanol in four different sized plants (38,000–140,000 m^3 per year) and concluded that the breakeven sweet potato price should be 18%–20% of the production cost of bulk harvested sweet potato. Another aspect to consider in the price of the sweet potato to be paid at the plant is the cost of transportation from the field. Sweet potato has high moisture content and low bulk density, which increase the transport cost. This aspect limits the area to be cultivated to supply an industrial plant. For a large-scale production, it would be convenient to develop collection centers where the sweet potato is dried and then, transported as chips or flour to the processing plants. The production of flour allows a reduction of volume for its transport and storage and facilitates its conservation avoiding the postharvest spoilage with decrease in starch and total sugars. An economic balance between the extra cost of drying and the economy in volume reduction during transport and storage is necessary. On the other hand, if it is not possible to meet the demand of sweet potato for a large-scale plant, a strategy may be coprocessing with other starchy raw matter, such as corn or grain sorghum (Shumaker et al., 2007).

Energy cost is generally the second largest component of total ethanol production cost after that of the raw material (Bai et al., 2008; Kwiatkowski et al., 2006; Shumaker et al., 2007; Srichuwong et al., 2012). This cost depends heavily on the price and local availability of energy sources. The most common sources used are natural gas, coal, firewood, waste (bagasse, straw, sawdust, etc.) and electricity. The stages of higher energy consumption in the conversion of fresh sweet potato to ethanol are distillation, treatment of wastewater (stillage concentration), gelatinization, and liquefaction. The energy consumption can be increased if a previous stage of drying of the sweet potato is included as previously mentioned in an earlier section (Ferrari et al., 2013; Schweinberger et al., 2016). Labandera et al. (2010) found that the cost of energy (firewood and electricity) accounted for 7% of the total cost (Table 7.10). Schweinberger et al. (2016), using conventional gelatinization and low temperature saccharification, estimated that the energy cost (electricity and firewood) was 7.6% of variable costs for the hydrous ethanol production.

Fixed costs (capital cost and depreciation) generally represent an important component of the total cost of production (Labandera et al., 2010; Shumaker et al., 2007). These costs depend on the scale of production, the efficiency and rate (productivity) of the industrial conversion processes. Capital costs per unit of capacity decline significantly as capacity increases. Shumaker et al. (2007) reported that the fixed cost of ethanol decreases from US\$52 to US\$32/m^3 when the capacity of the industrial plant increases from 38,000 to 140,000 m^3/year. Sweet potatoes are traditionally cultivated by small producers. The excess of production against the development of other foods is causing in some regions discouragement for its cultivation and an increase of material discarded for human consumption. These conditions could make the small-scale ethanol production attractive. A "microdistillery" could locally supply liquid fuel that has no transport costs from remote and traditional regions, where ethanol is actually produced. In addition, the byproducts (vines, leaves, stillage, etc.) constitute a valuable source of raw material for other products, such as livestock, biogas, or liquid fertilizer. However, the process inefficiencies and the low quality of the ethanol produced at a small scale significantly impact costs and the use as motor fuel. Virgínio e Silva et al. (2017) conducted tests of production of ethanol from sweet potatoes at a small-scale plant in Brazil (3 m^3/day), and obtained results that were considered satisfactory in terms of yields. However, they observed difficulties in achieving high alcohol content in the distillates (average grade of 90 GL). Dias Mayer et al. (2016) investigated the small-scale ethanol production using sugarcane and sweet sorghum as feedstocks under the existing conditions in Brazil. They concluded that the sale of products coordinated by cooperatives can be economically feasible. This involves the direct trade between the producer and the final consumer, without the incidence of taxes. However, even this scenario can present problems if the feedstock price increases even slightly. This concept could be applied to small-scale ethanol production from sweet potato.

The cost of yeast is also an important component of the total production cost. The high solids concentration of the fermented media makes yeast recovery and reuse difficult. Generally, yeast is not separated, being incorporated in the byproducts increasing its protein content as previously described in an earlier section. Labandera et al. (2010) found that the cost of yeast accounted for 2.6% of the total production cost despite considering that it is recycled seven times (Table 7.10). Schweinberger et al. (2016), in their work referred previously, reported that the cost of yeast is 36.5% of the variable production cost, being the main cost. If the yeast is recycled 11 times, its contribution to the variable cost decreases drastically to 5%. The use of fermentation media with high solid to liquid ratio (VHG condition) makes it even more difficult to recycle yeast or to use immobilized yeast.

Table 7.11 summarizes the main technological challenges to reduce the cost of producing ethanol from sweet potato. Because the cost of the raw material is a critical factor, it is necessary to select varieties of sweet potato grown with low levels of inputs, having high agricultural yield, high starch content, and low moisture. Its size and shape is also important in order to occupy less space during transportation and storage. Small, elongated, and smooth sweet potatoes are preferable. Energy consumption for the gelatinization and liquefaction steps can be reduced by using sweet potato cultivars containing a starch with low gelatinization temperature and cold starch hydrolysis. Energy savings on distillation and wastewater treatment can be also achieved with high ethanol concentration using a VHG fermentation technology (see earlier sections). The high viscosity would cause several handling difficulties in ethanol production. Mash viscosity can be reduced by the pretreatment with mixed enzyme preparations of pectinase, cellulase, and xylanase. The development of enzymes, yeast showing both high rates and high efficiencies of conversion in concentrated media (low inhibition effects) are important in order to reduce feedstock and capital costs.

Ferrari et al. (2013) evaluated the effect of solid-to-liquid ratio on energy consumption (steam and electricity) in the production of ethanol from fresh sweet potato, using experimental data and process simulation. Fig. 7.4 shows that the energy consumption decreased working with mashes containing high solids concentrations; in particular the effect was more significant in the range 18%–25% w/v, assuming a constant

TABLE 7.11 Main Strategies to Overcome the Key Technological Challenges to Reduce the Cost of Producing Ethanol From Sweet Potato

Factor	Description	References
Raw matter	Varieties of sweet potato developed locally with specific characteristics (agronomical yield, culture practices, starch, moisture, size, sharp, etc.) to be used for fuel ethanol (bioethanol crop)	Jin et al. (2012) Maluf (2009) Mussoline et al. (2017) Jusuf and Ginting (2014) Waluyo et al. (2015) Widodo et al. (2015) Ziska et al. (2009)
Energy saving in gelatinization and liquefaction	Sweet potato developed having low temperature-gelatinizing starch in order to perform a low temperature-liquefying process and the subsequent VHG fermentation.	Srichuwong et al. (2012)
	Simultaneous cold hydrolysis and fermentation of fresh sweet potato.	Masiero et al. (2014)
	Co-culture of an *Aspergillus niger* strain showing high levels of glucoamylase activity and *Zymomonas. mobilis* for ethanol fermentation, avoiding gelatinization/liquefaction step.	Zhang et al. (2013)
	Use of microwave as heating method.	Schweinberger et al. (2016)
Energy saving in distillation/stillage concentration	VHG strategy applying cellulase to reduce high viscosity of sweet potato mash and optimal amylase and glucoamylase dosage.	Cao et al. (2011)
	VHG strategy applying xylanase to reduce high viscosity of sweet potato mash under optimized conditions of enzyme dose, pretreatment time and temperature.	Zhang et al. (2010) Zhang et al. (2011)
	Optimization of amylase and glucoamylase loading for VHG conditions.	Lareo et al. (2012) Lareo et al. (2013)
	Use of sweet potato flour (powder) to increase solid to liquid ratio in mashes.	Lareo et al. (2012) Lareo et al. (2013)
Industrial yield improvement	Fermentation mode: simultaneous sacarification and fermentation (SSF), separate hydrolysis and fermentation (SHF), and partial-SSF (PSSF).	Lareo et al. (2012) Lareo et al. (2013) Zhang et al. (2011)
	Algae biomass as a low-cost nitrogen source fed in mash to improve ethanol fermentation yield.	Shen et al. (2012b)
	Simultaneous liquefaction, saccharification and fermentation with co-immobilized fungi (*Aspergillus oryzae or Monascuspurpureus*) and yeast (*Saccharomyces cerevisiae*).	Lee et al. (2012)

VGH, very high gravity.

fermentation efficiency of 0.85 and a feedstock containing 75% of total sugars (dry basis). The maximum solid to liquid ratio evaluated was 37% w/v that corresponds to the fresh sweet potato without addition of water.

7.10 LIFE CYCLE ANALYSIS

The incorporation of renewable fuels into the energy matrix of the transport sector has several driving forces: less use of fossil fuels, reduction

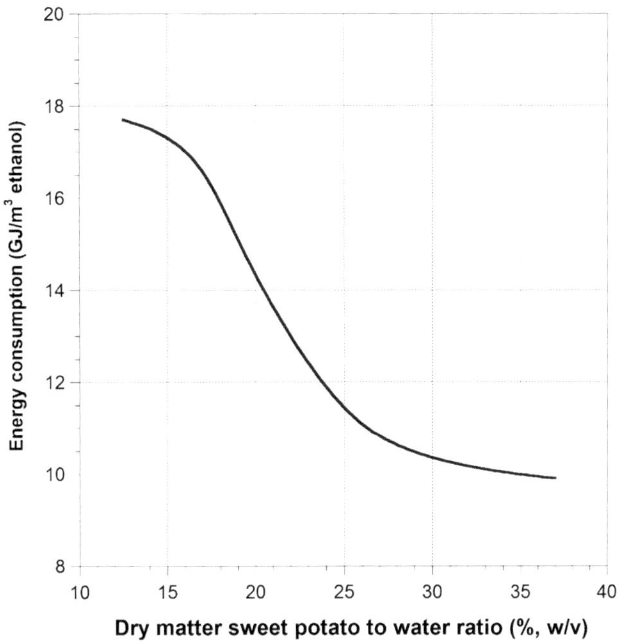

FIGURE 7.4 **Effect of dry matter to water on energy consumption (steam and electricity) for the ethanol production from fresh sweet potato, assuming conversion efficiency of 0.85.** *Source: Data taken from Ferrari, M.D., Guigou, M., Lareo, C., 2013. Energy consumption evaluation of fuel bioethanol production from sweet potato. Bioresour. Technol. 136, 377–384.*

of pollution particularly greenhouse gas (GHG) emissions, energy security, and development of rural areas. For a renewable fuel to be sustainable, economic, social, and environmental aspects must be addressed. However, the renewable nature of the fuel does not necessarily imply that its production is sustainable or that it contributes to the aforementioned objectives. It should be assessed whether the use of a renewable fuel, for example, effectively reduces fossil fuel use, generates a net reduction of GHG emissions, and does not lead to increases price and shortages of foods. While there are many articles published on the environmental assessment of the production of fuel ethanol from sugary or starchy materials, mainly sugarcane and corn, studies on sweet potato ethanol are very scarce.

Life cycle assessment (LCA) is a common and useful methodology for the evaluation of the environmental impacts of a product, service or process. The LCA is strongly dependent on the agroclimatic conditions and the productive model in each region or country, the reliability of the information available, the assumptions made and the definition and limits of the system. Important aspects are the energy source used in the agricultural and industrial stages and the allocation of the energy use or the emissions to coproducts (electricity, biogas, animal feed, etc.). Fossil fuel savings and emission reductions are greater when renewable energy sources, such as wood, agricultural waste, biogas, or electricity from renewable sources (hydro, wind, solar, etc.), are used for the most energy intensive stages.

Wang et al. (2013) carried out a LCA to evaluate the energy efficiency and environmental impacts of sweet potato-based bioethanol production. The study was carried out considering a variety of sweet potatoes specifically selected for the production of ethanol (45 t/ha, western China), an average transport distance of 150 km and coal, electricity, and cogenerated biogas as energy sources for the bioethanol conversion step. They concluded that eutrophication was the most significant environmental impact category, followed by acidification, global warming, human toxicity (mainly due to emissions of NO_x, SO_2, PM_{10}, and pesticides), and photochemical oxidation (associated with emissions of NO_x, VOC, CH_4, and CO). Emissions from coal combustion are the main contribution to acidification, human toxicity, and photochemical oxidation. Sensitivity analysis revealed that steam consumption during bioethanol conversion exerts the most effect on the results, followed by sweet potato yields and fertilizers input. The highest level of both fossil energy consumption and GHG emissions takes place in the agricultural phase, mainly due to the use and application of fertilizers. The fertilizer industry is energy intensive. This behavior was also found for other crops (Belboom et al., 2015; de Vries et al., 2010; Macedo et al., 2008; Olave, 2015). The values obtained of energy ratio (renewable energy output/fossil fuel energy throughout the LCA) and GHG emissions were 1.48 MJ/MJ and 69 g CO_2 equivalents/MJ, respectively. Carrasco-Letelier et al. (2013) analyzed the renewable energy yield in a LCA for the production of ethanol from a sweet potato specially selected as bioenergy crop (30 t/ha) for Uruguayan conditions and considering an average transport distance of 50 km. They obtained an energy ratio of 1:16. In this case the highest energy consumption was due to fertilization.

Su et al. (2015) carried out a water footprint analysis of bioethanol energy crops in Taiwan. They reported that sweet potatoes have the smallest total water consumption (572–687 $L_{water}/L_{ethanol}$), followed by corn (1904–1808 $L_{water}/L_{ethanol}$), sugar cane (2670–3592 $L_{water}/L_{ethanol}$), and sweet sorghum (4394 $L_{water}/L_{ethanol}$). A LCA conducted by Fingerman et al. (2010) in California (USA), found that bioethanol from sugarcane and corn had lower values of total water consumption, 1271 L_{water} and 1533 L_{water}/L of ethanol, respectively.

Table 7.12 shows energy and GHG emission values for the production of fuel ethanol

TABLE 7.12 Energy Ratio and GHG Emissions for the Production of Fuel Bioethanol From Different Food Crops

Crop	Country/region	Energy ratio (MJ/MJ)	GHG emissions (g CO_2 eq./MJ)	References
Sweet potato	China	1.48	69[a]	Wang et al. (2013)
Sweet potato	Uruguay	1.16	–	Carrasco-Letelier et al. (2013)
Sugarcane	Brazil	8 – 9	20 – 21	de Vries et al. (2010) Macedo et al. (2008)
Sugarbeet	NorthwestEurope	1.5	48	de Vries et al. (2010)
Sweet sorghum	China	6.8	22	de Vries et al. (2010)
Corn	USA	1.1 – 1.6[b]	87	de Vries et al. (2010) Gnansounou and Dauriat (2004) Kim and Dale (2005)
Wheat	NorthwestEurope	1.4 – 2.2[b]	40 – 54	Belboom et al. (2015) de Vries et al. (2010) Gnansounou et al. (2009)
Sorghum (grain)	Uruguay	2.5	36	Olave (2015)
Cassava	Thailand	2.0	40	de Vries et al. (2010)

[a] *Value calculated from the data reported by the authors using a lower calorific value of ethanol of 21.2 MJ/L.*

[b] *Range of typical values, some authors report values lower than 1, which is equivalent to a negative energy yield (see Gnansounou and Dauriat, 2004).*

from different crops used as food. These values should be analyzed with caution because they were obtained using different methodologies, boundaries, and system definitions. However, it can be observed that the values reported for the sweet potato are similar to those obtained for other starchy materials. While positive energy yields and GHG gas emissions lower than the emissions generated by the combustion of the equivalent quantity gasoline replaced (there is a net reduction of GHG emissions) are reported, such values show that it is necessary to make improvements in several issues throughout the life cycle. The critical aspects to improve are: agricultural yields using low fertilizer consumption, energy efficiency, and substitution of fossil fuels (coal, residual oils) by renewable energy sources.

7.11 CONCLUSIONS AND FUTURE PERSPECTIVES

The sweet potato is one of the most important food crops. It has applications as food and animal feed and can be processed into other products of higher value-added. However, only in the last 20 years has there been an intensification of studies aimed at improving their yields and obtaining different varieties that allow the diversification of their use, including the whole plant (roots, leaves, and vines). Many researchers have developed varieties of sweet potato with a higher agronomic yield than the typical national yields reported (see Tables 7.2 and 7.4). However, these experimental crops are limited both regionally and in their short period of evaluation. In many cases, very high yields have been reported but the associated uncertainty was not informed. Therefore, larger scale and longer studies are required to confirm the auspicious values of yield and starch content. There is also little standardization among the different authors for the chemical characterization of sweet potatoes for the production of ethanol. In some cases, it is not clearly specified whether the values are expressed on a dry or wet basis. The reported values of starch, free simple sugars, total sugars or moisture content in Tables 7.4 and 7.5 were obtained using different analytical methods. This makes it difficult to compare the values reported by the different authors and to make the ethanol yield calculations based on mass balances (Table 7.8). Further standardization of the analytical methods used for sweet potato characterization is required, comprising also the conservation and preparation of samples. In particular, the selection of analytical methods is a critical aspect, which should be effective in evaluating the performance of sweet potato in the industrial process to be used.

To promote the use of sweet potatoes as a bioenergy crop for fuel ethanol, it is necessary to attend to these critical aspects:

- Selection of varieties of sweet potato nonattractive for food, grown with low input level of fertilization and irrigation, having high agricultural yield, high starch content, and low moisture.
- Development of coproducts related to fermentation residues, leaves, and vines within to the biorefinery concept.
- Energy saving, particularly in the stages of gelatinization, liquefaction, distillation, and stillage concentration.
- Use of renewable energy source in substitution of fossil fuel (coal, residual oil, etc.) during the bioethanol life cycle, particularly in fertilizer production and industrial processing.
- Saccharification and fermentation process having a high rate in order to use industrial plants more compact and reduce fixed production costs.
- Definition of the productive model to be applied attending the existing local conditions, for example: microdistillery or large-scale production, fuel ethanol grade (hydrous or anhydrous), coprocessing with other starchy materials, and optimal transport distance from field to plant.

References

Bai, F.W., Anderson, W.A., Moo-Young, M., 2008. Ethanol fermentation technologies from sugar and starch feedstocks. Biotechnol. Adv. 26, 89–105.

Belboom, S., Bodso, B., Leonard, L., 2015. Does the production of Belgian bioethanol fit with European requirements on GHG emissions? Case of wheat. Biomass Bioenergy 74 (58), 65.

Berry, C.S., 1986. Resistant starch: formation and measurement of starch that survives exhaustive digestion and amylolytic enzymes during the determination of dietary fiber. J. Cereal Sci. 4, 301–314.

Breisha, G.Z., 2010. Production of 16% ethanol from 35% sucrose. Biomass Bioenergy 34, 1243–1249.

Cao, Y., Tian, H., Yao, K., Yuan, Y., 2011. Simultaneous saccharification and fermentation of sweet potato powder for the production of ethanol under conditions of very high gravity. Front. Chem. Sci. Eng. 5 (3), 318–324.

Carrasco-Letelier, L., Vázquez, D., D'Ottone, F., Resquín, F., Scoz, R., Vilaró, F., Rodríguez, G., Vicente, E., Terra, J., 2013. Balance energético de cadenas agro-industriales de interéspara la producción de bioenergías. Revista INIA 32, 46–50.

Castro, A.M., Castilho, L.R., Freire, D.M.G., 2011. An overview on advances of amylases production and their use in the production of bioethanol by conventional and non-conventional processes. Biomass Convers. Bioref. 1, 245–255.

Cinelli, B.A., Castilho, L.R., Freire, D.M.G., Castro, A.M., 2015. A brief review on the emerging technology of ethanol production by cold hydrolysis of raw starch. Fuel 150, 721–729.

Comis, D., 2008. Sweet Potato Out-Yields Corn in Ethanol Production Study. Agriculture Research Service, US Department of Agriculture, Washington, DC. Available from: https://www.ars.usda.gov/news-events/news/research-news/2008/sweet-potato-out-yields-corn-in-ethanol-production-study/.

CTA, 2017. Sweet-Potato Value Chain Management. Technical Centre for Agricultural and Rural Cooperation. Available from: http://www.cta.int/en/article/2017-07-05/value-chain-management.html.

de Vries, S.C., van de Ven, G.W.J., van Ittersum, M.K., Giller, K.E., 2010. Resource use efficiency and environmental performance of nine major biofuel crops, processed by first-generation conversion techniques. Biomass Bioenergy 34, 588–601.

Deesuth, O., Laopaiboon, P., Klanrit, P., Laopaiboon, L., 2015. Improvement of ethanol production from sweet sorghum juice under high gravity and very high gravity conditions: effects of nutrient supplementation and aeration. Ind. Crops Prod. 74, 95–102.

Delgado, R., Castro, A.J., Vázquez, M., 2009. A kinetic assessment of the enzymatic hydrolysis of potato (*Solanum tuberosum*). LWT Food Sci. Technol. 42, 797–804.

Dias Mayer, F., Brondani, M., Hoffmann, R., AmaralFeris, L., RomeuMarcilio, N., Baldo, V., 2016. Small-scale production of hydrous ethanol fuel: economic and environmental assessment. Biomass Bioenergy 93, 168–179.

Dincer, C., Karaoglan, M., Erden, F., Tetik, N., Topuz, A., Ozdemir, F., 2011. Effects of baking and boiling on the nutritional and antioxidant properties of sweet potato [Ipomoea batatas (L.) Lam.] cultivars. Plant Foods Hum. Nutr. 66, 341–347.

Duvernay, W.H., Chinn, M.S., Yencho, G.C., 2013. Hydrolysis and fermentation of sweet potatoes for production of fermentable sugars and ethanol. Ind. Crop Prod. 42, 527–537.

El Sheikha, A.F., Ray, R.C., 2017. Potential impacts of bioprocessing of sweet potato: review. Crit. Rev. Food Sci. Nutr. 57 (3), 455–471.

FAOSTAT, 2013. Statistics Division. Food and Agriculture Organization of the United Nations. Food Balance Sheets. Available from: http://www.fao.org/faostat/en/#data/FBS.

FAOSTAT, 2014. Statistics Division. Food and Agriculture Organization of the United Nations. Crops. Available from: http://www.fao.org/faostat/en/#data/QC.

Ferrari, M.D., Guigou, M., Lareo, C., 2013. Energy consumption evaluation of fuel bioethanol production from sweet potato. Bioresour. Technol. 136, 377–384.

Fingerman, K.R., Torn, M.S., O'Hare, M.H., Kammen, D.M., 2010. Accounting for the water impacts of ethanol production. Environ. Res. Lett. 5, 014020.

Gnansounou E. and Dauriat A. Energy Balance of Bioethanol: A Synthesis, 2004. Lasen, Ecole Polytechnique Fédérale de Lausanne, Lausanne, Switzerland.

Gnansounou, E., Dauriat, A., Villegas, J., Panichelli, L., 2009. Life cycle assessment of biofuels: energy and greenhouse gas balances. Bioresour. Technol. 100, 4919–4930.

Hansen, M.R., Blennow, A., Pedersen, S., Nørgaard, L., Engelsen, S.B., 2008. Gel texture and chain structure of amylomaltase-modified starches compared to gelatin. Food Hydrocoll. 22, 1551–1566.

Harris B., 2003. Nutrient requirements of dairy cattle. Available from: http://ufdcimages.uflib.ufl.edu/IR/00/00/47/28/00001/DS08700.pdf.

He, M.-X., Feng, H., Bai, F., Li, Y., Liu, X., Zhang, Y.-Z., 2009a. Direct production of ethanol from raw sweet potato starch using genetically engineered *Zymomonas mobilis*. Afr. J. Microbiol. Res. 3 (11), 721–726.

He, M.-X., Li, Y., Liu, X., Bai, F., Feng, H., Zhang, Y.-Z., 2009b. Ethanol production by mixed-cultures of Paenibacillussp. and Zymomonasmobilis using the raw starchy material from sweet potato. Ann. Microbiol. 59 (4), 749–754.

Hernández-Suárez, M., Montes-Hernández, A.I., Rodríguez-Galdón, B., Hernández-Rodríguez, L., Medina-Cabrera, C.E., Ríos-Mesa, D., Rodríguez-Rodríguez, E.M., Díaz-Romero, C., 2016. Application of multidimensional scaling technique to differentiate sweet potato (Ipomoea batatas (L.) Lam) cultivars according to their chemical composition. J. Food Composit. Anal. 46, 43–49.

Hoover, R., 2001. Composition, molecular structure, and physicochemical properties of tuber and root starches: a review. Carbohydr. Polym. 45, 253–267.

International Potato Center, 2017. Facts and Figures about Sweet Potato. Available from: http://cipotato.org/sweetpotato/facts.

Ishida, H., Suzuno, H., Sugiyama, N., Innami, S., Tadokoro, T., Maekawa, A., 2000. Nutritive evaluation on chemical components of leaves, stalks and stems of sweet potatoes (Ipomoea batataspoir). Food Chem. 68, 359–367.

Jin, Y., Fang, Y., Zhang, G., Zhou, L., Zhao, H., 2012. Comparison of ethanol production performance in 10 varieties of sweet potato at different growth stages. Acta Oecol. 44, 33–37.

Jusuf, M., Ginting, E., 2014. The prospects and challenges of sweet potato as bioethanol source in Indonesia. Energy Procedia 47, 173–179.

Kays, S.J., Wang, Y., 2001. Sweetpotato quality: its importance, assessment and selection inbreeding programs. I. Int. Conf. Sweet Pot. Food Health Future 583, 187–193.

Kim, S., Dale, B.E., 2005. Environmental aspects of ethanol derived from no-tilled corn grain: nonrenewable energy consumption and greenhouse gas emissions. Biomass Bioenergy 28, 475–489.

Kitahara, K., Fukunaga, S., Katayama, K., Takahata, Y., Nakazawa, Y., Yoshinaga, M., Suganuma, T., 2005. Physicochemical properties of sweet potato starches with different gelatinization temperatures. Starch 57, 473–479.

Kwiatkowski, J.R., MacAloon, A.J., Taylor, F., Johnston, D.B., 2006. Modeling the process and costs of fuel ethanol production by the corn dry-grind process. Ind. Crop Prod. 23, 288–296.

Labandera, A., Mathisson, J., Damonte, D., Pereira, C., Bonner, R., 2010. ProyectoBioet: Final Report of Chemical Engineering Degree. Instituto de Ingeniería Química, Montevideo, Uruguay.

Lareo, C., Ferrari, M.D., Guigou, M., Larnaudie, V., Fajardo, L., Ramírez, M.B., 2012. Evaluación de materiales amiláceos experimentales desarrollados para bioenergía (boniato y grano de sorgo dulce) para la producción de bioetanol combustible. FPTA-INIA 39, ISSN:1688-924X.

Lareo, C., Ferrari, M.D., Guigou, M., Fajardo, L., Larnaudie, V., Ramírez, M.B., Martínez-Garreiro, J., 2013. Evaluation of sweet potato for fuel bioethanol production: hydrolysis and fermentation. Springer Plus 2, 493.

Larnaudie, V., Rochón, E., Ferrari, M.D., Lareo, C., 2016. Energy evaluation of fuel bioethanol production from sweet sorghum using very high gravity (VHG) conditions. Renew. Energy 88, 280–287.

Lee, W.-S., Chen, I.-C., Chang, C.-H., Yang, S.-S., 2012. Bioethanol production from sweet potato by co-immobilization of saccharolytic molds and *Saccharomyces cerevisiae*. Renew. Energy 39, 216–222.

Macedo, I.C., Seabra, J.E.A., Silva, J.E.A.R., 2008. Greenhouse gases emissions in the production and use of ethanol from sugarcane in Brazil: the 2005/2006 averages and a prediction for 2020. Biomass Bioenergy 32, 582–595.

Maluf, W.R., 2009. Proving Interdisciplinary Education in Biofuels Technology: 1st Brazil-US Biofuels Short Course. Universidade de São Paulo, Brazil. Available from: http://www.iea.usp.br/midiateca/apresentacao/malufbiofuels.pdf.

Masiero, S.S., Peretti, A., Trierweiler, L.F., Trierweiler, J.O., 2014. Simultaneous cold hydrolysis and fermentation of fresh sweet potato. Biomass Bioenergy 70, 174–183.

Mojovic´, L., Nikolic´, S., Rakin, M., Vukasinovic´, M., 2006. Production of bioethanol from corn meal hydrolyzates. Fuel 85, 1750–1755.

Montefusco, A., Durante, M., Grassi, S., Piro, G., Dalessandro, G., Lenucci, M.S., 2014. Assessment of sweet potato [*Ipomoea batatas* (L.) Lam] for bioethanol production in southern Italy. Plant Biosyst. 148 (6), 1117–1126.

Montesinos, T., Navarro, J.-M., 2000. Production of alcohol from raw wheat flour by amyloglucosidase and *Saccharomyces cerevisiae*. Enzyme Microb. Technol. 27, 362–370.

Moorthy, S.N., 2002. Physicochemical and functional properties of tropical tuber starches: a review. Starch 54, 559–592.

Morrison, T.A., Pressey, R., Kays, S.J., 1993. Changes in α- and β-amylase during storage of sweet potato lines with varying starch hydrolysis potential. J. Am. Soc. Hortic. Sci. 118, 236–242.

Mukherjee, A., Naskar, S.K., Rao, K.R., Ray, R.C., 2012. Sweet potato gains through biotechnology. Fruit. Veg. Cereal Sci. Biotechnol. 6 (Suppl. 1), 30–42.

Mukherjee, A., Naskar, S.K., Ray, R.C., Pati, K., Mukherjee, A., 2015. Sweet potato and taro resilient to stresses: sustainable livelihood in fragile zones vulnerable to climate changes. J. Environ. Sociobiol. 12, 53–64.

Murugan, S., Paramasivam, S.K., Nedunchezhiyan, M., 2012. Sweet potato as animal feed and fodder. Fruit. Veg. Cereal Sci. Biotechnol. 6, 106–114.

Mussoline, W.A., Bohac, J.R., Boman, B.J., Trupia, S., Wilkie, A.C., 2017. Agronomic productivity, bioethanol potential and postharvest storability of an industrial sweet potato cultivar. Ind. Crop Prod. 95, 96–103.

Noda, T., Takahata, Y., Nagata, T., 1992. Developmental changes in properties of sweet potato starch. Starch/Stärke 44, 405–409.

Olatunde, G.O., Henshaw, F.O., Idowu, M.A., Tomlins, K., 2016. Quality attributes of sweet potato flour as influenced by variety, pretreatment and drying method. Food Sci. Nutr. 4 (4), 623–635.

Olave, M.P., 2015. Análisis de ciclo de vida del bioetanol combustible producido a partir de sorgo grano: balance de energía y emisión de gases de efecto invernadero. Thesis. Master in Energy Engineering. Facultad de Ingeniería, UdelaR, Montevideo, Uruguay.

Panda, S.K., Swain, M.R., Singh, S., Ray, R.C., 2012. Proximate compositions of a purple sweet potato (*Ipomoea batatas* L.) herbal wine. J Food Proc. Preserv. doi: 10.1111/j.1745-4549.2012.00681.x.

Pereira, F.B., Guimarães, P.M.R., Teixeira, J.A., Domingues, L., 2010. Optimization of low-cost medium for very high gravity ethanol fermentations by *Saccharomyces cerevisiae* using statistical experimental designs. Bioresour. Technol. 101, 7856–7863.

Pradeep, P., Reddy, O.V.S., 2010. High gravity fermentation of sugarcane molasses to produce ethanol: effect of nutrients. Indian J. Microbiol. 50 (Suppl. 1), S82–S87.

Ramesh Yadav, A., Guha, M., Tharanathan, R.N., Ramteke, R.S., 2006. Changes in characteristics of sweet potato flour prepared by different drying techniques. LWT Food Sci. Technol. 39, 20–26.

Ravindran, V., Ravindran, G., Sivakanesan, R., Rajaguru, S.B., 1995. Biochemical and nutritional assessment of tubers from 16 cultivars of sweet potato (Ipomoea batataasL.) J. Agric. Food Chem. 43, 2646–2651.

Ray, R.C., Naskar, S.K., 2008. Bio-ethanol production from sweet potato (*Ipomoea batatas* L.) by enzymatic liquefaction and simultaneous saccharification and fermentation. Dyn. Biotechnol. Proc. Biochem. Mol. Biol. 2, 47–49.

Ray, R.C., Ravi, V., 2005. Spoilage of sweet potato in tropics and control measures: review. Crit. Rev. Food Sci. Nutr. 45, 623–644.

Ray, R.C., Tomlins, K.I., 2010. Sweet Potato: Post-Harvest Aspects in Food, Feed and Industry. Nova Science Publishers, New York, 316 pp..

Reddy, L.V.A., Reddy, O.V.S., 2006. Rapid and enhanced production of ethanol in very high gravity (VHG) sugar fermentation by *Saccharomyces cerevisiae*: Role of finger millet (*Eleusinecoracana* L.) flour. Process Biochem. 41, 726–729.

Reddy, N.N., Sistrunk, W.A., 1980. Effect of cultivar, size, storage, and cooking method on carbohydrates and some nutrients of sweet potatoes. J. Food Sci. 45, 682–684.

Robertson, G.H., Wong, D.W.S., Lee, C.C., Wagschal, K., Smith, M.R., Orts, W.J., 2006. Native or raw starch digestion: a key step in energy efficient biorefining of grain. J. Agric. Food Chem. 54, 353–365.

Schweinberger, C.M., Putti, T.R., Susin, G.B., Trierweiler, J.O., Trierweiler, L., 2016. Ethanol production from sweet potato: the effect of ripening, comparison of two heating methods, and cost analysis. Can. J. Chem. Eng. 94, 716–724.

Sharma, V., Rausch, K.D., Graeber, J.V., Schmidt, S.J., Buriak, P., Tumbleson, M.E., Singh, V., 2010. Effect of resistant starch on hydrolysis and fermentation of corn starch for ethanol. Appl. Biochem. Biotechnol. 160, 800–811.

Shekhar, S., Mishra, D., Buragohain, A.K., Chakraborty, S., Chakraborty, N., 2015. Comparative analysis of phytochemicals and nutrient availability in two contrasting cultivars of sweet potato (*Ipomoea batatas* L.). Food Chem. 173, 957–965.

Shen, Y., Zhang, H., Zheng, X., Zhang, X., Guo, J., Chen, Y., 2011. Very high gravity fermentation using sweet potato for fuel ethanol production. Adv. Mater. Res., 236–238, 59–62.

Shen, Y., Guo, J.S., Chen, Y.P., Zhang, H.D., Zheng, X.X., Jiang, Y., Zhang, X.M., Li, X.H., 2012a. Effects of glucose releasing rate on cell growth and performance of simultaneous saccharification and fermentation (SSF) in sweet potato medium for fuel ethanol production using *Saccharomyces cerevisiae*. Int. J. Chem. Reactor Eng. 10. doi: 10.1515/1542-6580.2976.

Shen, Y., Guo, J.S., Chen, Y.P., Zhang, H.D., Zheng, X.X., Zhang, X.M., Bai, F.W., 2012b. Application of low-cost algal nitrogen source feeding in fuel ethanol production using high gravity sweet potato medium. J. Biotechnol. 160, 229–235.

Shumaker, G.A., Luke-Morgan, A., McKissick, J.C., 2007. The Economic Potential of Ethanol Production from Sweet and Irish Potatoes in Georgia. Feasibility Report: FR-07-08. Center for Agribusiness and Economic Development, University of Georgia, Athens, Georgia, USA.

Sistrunk, W.A., 1971. Carbohydrate transformations, color and firmness of canned sweet potatoes as influenced by variety, storage, pH and treatment. J. Food Sci. 36, 39–42.

Srichuwong, S., Jane, J., 2007. Physicochemical properties of starch affected by molecular composition and structures: a review. Food Sci. Biotechnol. 16 (5), 663–674.

Srichuwong, S., Sunarti, T.C., Mishima, T., Isono, N., Hisamatsu, M., 2005. Starches from different botanical sources I: contribution of amylopectin fine structure to thermal properties and enzyme digestibility. Carbohydr. Polym. 60, 529–538.

Srichuwong, S., Fujiwara, M., Wang, X., Seyama, T., Shiroma, R., Arakane, M., Mukojima, N., Tokuyasu, K., 2009. Simultaneous saccharification and fermentation (SSF) of very high gravity (VHG) potato mash for the production of ethanol. Biomass Bioenergy 33, 890–898.

Srichuwong, S., Orikasa, T., Matsuki, J., Shiina, T., Kobayashi, T., Tokuyasu, K., 2012. Sweet potato having a low temperature-gelatinizing as a promising feedstock for bioethanol production. Biomass Bioenergy 39, 120–127.

Su, M.-H., Huang, C.-H., Li, W.-Y., Tso, C.-T., Lur, H.-S., 2015. Water footprint analysis of bioethanol energy crops in Taiwan. J. Clean. Prod. 88, 132–138.

Swain, M.R., Mishra, J., Thatoi, H., 2013. Bioethanol production from sweet potato (Ipomoea batatas L.) flour using co-culture of Trichoderma sp. and Saccharomyces cerevisiae in solid-state fermentation. Braz. Arch. Biol. Technol. 56 (2), 171–179.

Thatoi, H., Dash, P.K., Mohapatra, S., Swain, M.R., 2014. Bioethanol production from tuber crops using fermentation technology: a review. Int. J. Sustain. Energy 35 (5), 443–468.

Torquato-Tavares, A., Rodrigues-do Nascimento, I., Pascual-Reyes, I.D., Rosa-de Santana, W., da Silveira, M.A., 2017. Potential for sweet potato (*Ipomoea batatas*(L.) Lam.) single crosses to improve ethanol production. Revist. Chap. Serie Hortic. 23 (1), 59–74.

USDA, National Nutrient Database for Standard Reference Release, 2017, United States Department of Agriculture, Agricultural Research Service, 28, Available from: https://ndb.nal.usda.gov/ndb/foods/show/3207?manu=&fgcd=&ds=.

Vidal, Jr., B.C., Rausch, K.D., Tumbleson, M.E., Singh, V., 2009. Kinetics of granular starch hydrolysis in corn dry grind process. Starch/Stärke 61, 448–456.

Vilaró, F., Vicente, E., Rodríguez, G., Pereira, G., Spina, W., Reggio, A., Ibañez, F., 2009. Nuevas variedades de boniato para usos diversos. Revista INIA 20, 44–47.

Virgínio e Silva, J.O., Almeida, M.F., da ConceiçãoAlvim-Ferraz, M., Dias, J.M., 2017. Integrated production of biodiesel and bioethanol from sweet potato. Renew. Energy. doi: 10.1016/j.renene.2017.07.052.

Waluyo, B., Roosda, A.A., Istifadah, N., Ruswandi, D., Karuniawan, A., 2015. Identification of fifty sweet potato (Ipomoea batatas (L.) Lam.) promising clones for bioethanol raw materials. Energy Procedia 65, 22–28.

Wang, M., Shi, Y., Xia, X., Li, D., Chen, Q., 2013. Life-cycle energy efficiency and environmental impacts of bioethanol production from sweet potato. Bioresour. Technol. 133, 285–292.

Wang, S., Nie, S., Zhu, F., 2016. Chemical constituents and health effects of sweet potato. Food Res. Int. 89, 90–116.

Watanabe, T., Srichuwong, S., Arakane, M., Tamiya, S., Yoshinaga, M., Watanabe, I., Yamamoto, M., Ando, A., Tokuyasu, K., Nakamura, T., 2010. Selection of stress-tolerant yeasts for simultaneous saccharification and fermentation (SSF) of very high gravity (VHG) potato mash to ethanol. Bioresour. Technol. 101, 9710–9714.

Widodo, Y., Wahyuningsih, S., Ueda, A., 2015. Sweet potato production for bio-ethanol and food related industry in Indonesia: challenges for sustainability. Proced. Chem. 14, 493–500.

Wu, Y.V., Bagby, M.O., 1987. Recovery of protein-rich by-products from sweet potato stillage following alcohol distillation. J. Agric. Food Chem. 35 (3), 321–325.

Zabed, H., Sahu, J.N., Suely, A., Boyce, A.N., Faruq, G., 2017. Bioethanol production from renewable sources: current perspectives and technological progress. Renew. Sustain. Energy Rev. 71, 475–501.

Zhang, K., Feng, H., 2010. Fermentation potentials of *Zymomonas mobilis* and its application in ethanol production from low-cost raw sweet potato. Afr. J. Biotechnol. 9 (37), 6122–6128.

Zhang, P., Chen, C., Shen, Y., Ding, T., Hua, Z., Sun, 2013. Starch saccharification and fermentation of uncooked sweet potato roots for fuel ethanol production. Bioresour. Technol. 128, 835–838.

Zhang, Z., Wheatley, C.C., Corke, H., 2002. Biochemical changes during storage of sweet potato roots differing in dry matter content. Postharv. Biol. Technol. 24, 317–325.

Zhang, L., Chen, Q., Jin, Y., Xue, H., Guan, J., Wang, Z., Zhao, H., 2010. Energy-saving direct ethanol production from viscosity reduction mash of sweet potato at very high gravity (VHG). Fuel Process. Technol. 91, 1845–1850.

Zhang, L., Zhao, H., Gan, M., Jin, Y., Gao, X., Chen, Q., Guan, J., Wang, Z., 2011. Application of simultaneous saccharification and fermentation (SSF) from viscosity reducing of raw sweet potato for bioethanol production at laboratory, pilot and industrial scales. Bioresour. Technol. 102, 4573–4579.

Zhu, F., Wang, S., 2014. Physicochemical properties, molecular structure, and uses of sweet potato starch. Trends Food Sci. Technol. 36, 68–78.

Ziska, L.H., Runion, G.B., Tomecek, M., Prior, S.A., Torbet, H.A., Sicher, R., 2009. An evaluation of cassava, sweet potato and field corn as potential carbohydrate sources for bioethanol production in Alabama and Maryland. Biomass Bioenergy 33 (11), 1503–1508.

CHAPTER

8

Jerusalem Artichoke: An Emerging Feedstock for Bioethanol Production

*Konstantinos V. Kotsanopoulos**, Ramesh C. Ray***, Sudhanshu S. Behera*[†]

*University of Thessaly, Volos, Hellas, Greece

**ICAR-Central Tuber Crops Research Institute (Regional Centre), Bhubaneswar, India

[†]Department of Fisheries and Animal Resource Development, Government of Odisha, Cuttack, India

ABBREVIATIONS

ABE	Acetone–butanol–ethanol
FOS	Fructooligosaccharides
HMF	Hydroxymethyfurfural
HPLC	High performance liquid chromatography
JA	Jerusalem artichoke
MS	Mass spectrometry

8.1 INTRODUCTION

Jerusalem artichoke (*Helianthus tuberosus* L.) (JA) is a plant native to temperate North America, and has been advocated for many years to be a possible feedstock for bioethanol production (Bajpai and Bajpai, 1991; Bajpai and Margaritis, 1982; Guo et al., 2013; Margaritis et al., 1981; Nakamura et al., 1996). It is characterized by high tolerance to drought, frost, and salinity as well as plant diseases and insect pests. JA has numerous other advantageous characteristics over traditionally grown crops, such as elevated growth rate, high productivity, with minimal to zero fertilizer requirements, climate resilience, and ease harvesting. Conventionally, JA has been used for food and animal feeds and for the past few decades for the production of bioactive food ingredients, such as inulin, oligofructans, and most importantly as raw material for bioethanol production (Yang et al., 2015).

Bioethanol Production From Food Crops
http://dx.doi.org/10.1016/B978-0-12-813766-6.00008-4

The biomass of JA is mainly sourced from its tubers (Fig. 8.1), which contain inulin as a major component that can be used through hydrolysis for the production of fermentable sugars, while no energy-intensive pretreatment is required (Long et al., 2016). During and after World War I, extensive tests conducted by the British Department of Scientific and Industrial Research demonstrated that JA could lead to the production of approximately the same quantities of fermentable carbohydrate per acre as sugar beet and higher quantities than Irish potato (Sachs et al., 1981). The production of bioethanol from JA inulin is a complex process, since unlike many sugar and starch-based feedstocks it is to be broken down into hexose sugars in the initial step, and then to be fermented by ethanologenic microorganisms, such as yeasts, *Saccharomyces cerevisiae*, *Kluveromyces marxianus*, or bacterium *Zymomonas mobilis*. The whole process of fermentation can be carried out either as separate hydrolysis and fermentation (SHF), simultaneous saccharification and fermentation (SSF), or as consolidated bioprocessing (CBP). In this chapter we discuss the bioprocess characteristics, microorganisms, and technological aspects of bioethanol and biobutanol production from JA.

8.2 CHARACTERISTICS OF JERUSALEM ARTICHOKE

JA contains more than 50% (dry basis) inulin in its tubers and seems to be an excellent candidate for producing ethanol (Wang et al., 2014). The growing cycle of the JA is characterized by the accumulation of fructose polymers in the stalks, which are then transferred and remain in the tubers. The JA could also be cultivated to produce sugar but harvesting of the stalks around the flowering period would be required (D'Egidio et al., 1998). Its crop growth and yield were examined in field conditions harvested periodically during a 2-year period by Schorr-Galindo and Guiraud (1997). The investigations were focused on: (1) cultivating JA and (2) optimizing the analytical techniques for determining sugars, aiming at finally determining cultivars and harvest periods for optimum sugar production. It was demonstrated that this crop offers several benefits when biomass yield is considered, reaching 100–130 t/ha for the optimum cultivars. Analytical techniques, such as high performance liquid chromatography (HPLC), have been implemented for the analysis of inulin hydrolysis throughout the growing season of

FIGURE 8.1 **Jerusalem artichoke tubers.**

JA tubers. The depolymerization of the polyfructosans led to the production of juices with various polymers contents depending on the harvest date. That decreased considerably the fructose/glucose ratio from 11 to 3 during the harvesting season and this effect would therefore need to be considered when the use of tubers is considered for the production of sugars or alcohol.

It is essential to know both the total sugar levels of the crops and their composition, in order to optimize the production of bioethanol from JA tubers. Many methods have been employed to analyze the carbohydrates present in JA tubers. However, these analyses can be developed further, through the use of faster and more reliable identification techniques and peak confirmations, for example, via the use of mass spectrometry (MS) (Matías et al., 2011). Kocsis et al. (2007) examined how seasonal variation in the concentration and profile of soluble carbohydrates are represented in JA tubers. A high percentage (60%–65% dry mass) of water-soluble carbohydrates was detected both in early harvested varieties (Bella and Bianka) and product harvested 22–25 weeks following plantation (Topstar and Gigant). Similar levels were detected in late varieties (Waldspindel, Violet de Rennes, and Rote Zonenkugel) (55%–60% of dry mass) harvested 29–33 weeks following plantation (Kocsis et al., 2007). There were, however, significant effects on the maturing process as well as frost period alternations that led to the conversion of highly polymerized inulin into low polymer inulin and sucrose. Early varieties and those harvested 22–25 weeks after plantation presented a correlation between sucrose and inulin content ($r = -0.952^{**}$) with a linear regression of $y = -1.35x + 62.32$. The degree of polymerization$_n$ (DP) of inulin fell from 12–14 to 6–8. In late cultivars this correlation was not as exact ($r = -0.502^{**}$), with DP_n of inulin decreasing from 12–16 down to 7–10. These data that relate to the carbohydrate profiles of JA can facilitate the selection of certain cultivars and can allow decisions to be taken with regards to the most suitable harvest time, to optimize the production of bioethanol.

A study was conducted by Modler et al. (1993) for determining how factors, such as harvest period, storage temperature, and overwintering of plants those affect the fructoologosaccharides (FOS) of JA tubers. The researchers also conducted experiments aiming at determining whether alteration in the FOS content are associated with problems in relation to the spray-drying process of the tubers that is undertaken for the production of JA tubers flour. The harvesting of tubers from four JA cultivars (Columbia, Challengers, Sunroot, and Fusil) was followed by trimming, washing, and placement of the materials in polyethylene bags. The storage stability and any alterations of the FOS profile were then evaluated. Out of the five storage conditions examined (5°C, 2°C, −10°C, program cooled to −10°C, and ambient), the treatment at 2°C led to tubers of optimum quality after 12 months of storage. Tubers stored at 5°C presented a degree of sprouting following half-year of storage and even spoilage following storage for 1 year. Unsatisfactory results were obtained from the other treatments. After storage for a short period of time (18 weeks) the inulin of the JA tubers was consisted mainly of shorter chain FOS. Storage of tubers for 1 year and 4 months at 5°C led to the elimination of FOS with a DP≥10, but significant levels of FOS with DP 1–4 were accumulated.

8.3 CHARACTERISTICS OF INULIN AND ITS YIELDS FROM JERUSALEM ARTICHOKE

Inulins are β (2→1) fructans that often terminate in a glucose linked by an α (1↔2) bond. The fructose units in their furanose ring structure are linked together like a polyethylene oxide linear chain as shown in Fig. 8.2.

The solubility of inulin in water decreases with increase in DP. When their DP is 2–9, they can be classified as FOS or oligofructose (OF) (Biedrzycka and Bielecka, 2004). Inulins become insoluble in water near 23 DP. They form a gel

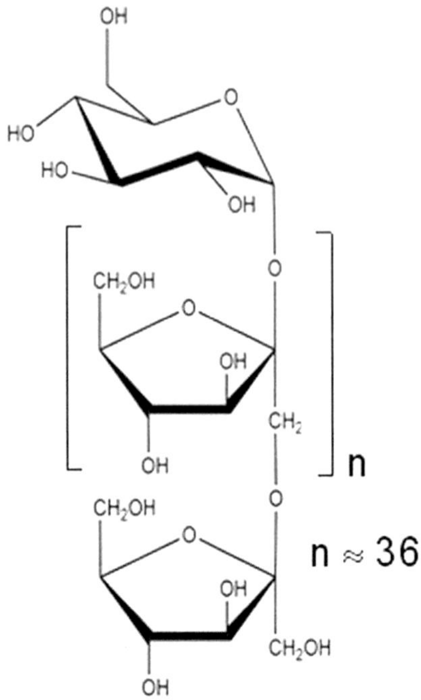

FIGURE 8.2 **Structure of inulin.**

in water when their concentration at room temperature is higher than 10%–15% (Glibowski and Wasko, 2008). Inulin production is initiated close to the flowering period. Li et al. (2015) found that inulin content was about 3.5% in tubers, 10 days before flowering and reached a maximum of 12.21%, 40 days after flowering, while decreased at 7.3%, 80 days after flowering, with all percentages based on the weight of wet tubers and having a wide-range of DP (Gunnarsson et al., 2014a,b).

8.4 BIOETHANOL PRODUCTION FROM JERUSALEM ARTICHOKE TUBERS

Yeasts, such as *S. cerevisiae* and *K. marxianus* and the bacterium *Z. mobilis*, have metabolic pathways for the efficient conversion of hexoses into pyruvates. *S. cerevisiae* carries out this function through the Embden-Meyerhof-Parnas (EMP) glycolytic pathway, whereas *Zymomonas* does it through the Entner-Doudoroff (ED) glycolytic pathway. The latter generates only one adenosine triphosphate (ATP) compared to two ATPs in the case of the EMP pathway, which results in less cell mass and high ethanol productivity in fermentation (Clomburg and Gonzalez, 2010). Under anaerobic conditions, metabolic flux is driven to produce acetaldehyde through the action of pyruvate decarboxylase, which is then converted by alcohol dehyogenase to produce ethanol and carbon dioxide (Ray and Joshi, 2014). These organisms do not naturally metabolize pentoses like xylose and arabinose, but this is not a problem in the fermentation of JA tuber inulins as they are composed solely from fructose and glucose. Moreover, the JA flour or juice may contain 6%–7% protein, and therefore less nitrogen needs to be added to allow optimum cell growth (Cieślik et al., 2011). However, one issue with fermentation of JA sugars is that they are richer in fructose than glucose. A portion of sugars can remain unhydrolyzed in the fermenter hexose transporters of *K. marxianus*. The activation of these transporters needs to be investigated to allow a more efficient ethanol fermentation to be achieved (Yuan et al., 2012). Table 8.1 provides a comprehensive list of organisms, substrates, ethanol yields, and volumetric ethanol productivity detailed in several studies.

8.4.1 Separate Hydrolysis and Fermentation (SHF)

The acid hydrolysis of the SHF approach offers certain benefits, such as shorter reaction times and low cost. Kim and Hamdy (1986) suggested that acid hydrolysis on inulin from JA tubers is carried out in 0.1 *N* hydrochloric acid for 15 min at 97°C for maximum recovery of reducing sugar. Sarchami and Rehmann (2015) maximized fructose recovery (98.5%) from inulin derived from JA at pH 2 and 97°C in 35 min using mineral acids. However, the authors did observe byproduct formation that worsened

TABLE 8.1 Reprentative List of Microorganisms, Substrates, Fermentation Conditions, Ethanol Yield, and Volumetric Ethanol Productivity Reported in Published Articles

Microorganisms	Substrate	Fermentation process and condition	Maximum ethanol productivity and/or concentration and yield	References
Kluyveromyces fragilis LM92	Glucose and inulin	2-L fermenter; 28°C	4.3 g/L/h; 94%[a]	Guiraud et al. (1987)
Zymomonas mobilis with inulinase	JA juice; 154.2 g/L	SSF; continuous; 35°C	55.1 g/L/h; 95%[a]	Kim and Rhee (1990)
K. fragilis with *Z. mobilis* with inulinase	0.02 g enzyme/kg tubers	Batch; 30°C, 72 h	0.48 g/g; 94%[a]	Szambelan et al. (2004)
Aspergillus niger SL-09 with *Saccharomyces cerevisiae* Z-06	JA	SSF; fed-batch; 30°C, 48 h	19.6 % (v/v); 90%[a]	Ge and Zhang (2005)
Saccharomyces sp. W0 with *recombinant* inulinase	50 g/100 mL tuber meal of JA	28°C,144 h	0.319 ± 0.9 g/g; 62.5%[a]	Zhang et al. (2010)
K. marxianus PT-1	200 g/L tuber flour	40°C, 84 h	73.6 g/L; 90%[a]	Hu et al. (2012)
S. cerevisiae JZ1C	200 g/L tuber flour	40°C, 84 h	65.2 g/L; 79.7%[a]	Hu et al. (2012)
S. cerevisiae and *K. marxianus* strain Y00959	JA extract 24% (w/w)	148 h	10.67% (v/v); 76%	Dénes et al. (2013)
S. cerevisiae DQ1	35% (w/w) inulin-type sugars	Batch; 30°C, 28 h	128.7 g/L; 73.5%[a]	Guo et al. (2013)
Saccharomyces sp. W0	30.0% (w/w) tuber meal of JA	30°C, 120 h	0.331 g/g; 65%[a]	Li et al. (2013)
K. marxianus	220 g/L dry meal of JA	35°C; 48 h	62–69 g/L; 1.29–1.44 g/L/h; 87%[a]	Yuan et al. (2013a, b)
K. marxianus CBS1555 (KCTC7001)	8% (w/v) tuber	30°C; 24 h; 200 rpm[b]; 5 L fermenter	1.51 g/L/h; 45.3 g/L	Kim and Kim (2014)
K. marxianus Y179	Inulin-type sugars (235 g/L)	30°C; 48 h; 3 L fermenter	1.63 g/L/h; 98 g/L	Gao et al. (2015)
S. cerevisisiae MK 01	JAS[c] (20 w/v)	Fed-batch; 30°C, 48 h	38.3 g/L; 71%[a]	Khatun et al. (2015)
S. cerevisisiae 6525	JA dry powder (240 g/L)	Fed-batch; 72 h; pH 5.5; 30°C; 80 rpm[b]	84.3 g/L; 88.6%[a]	Wang et al. (2015)
S. cerevisisiae	20% (w/v) of inulin	Batch; 30°C, 60 h; 250 rpm[b]	80.2 g/L	Hong et al. (2015)
S. cerevisisiae KCTC 7606	JA stalks 1% (w/v)	32°C; 48 h; 190 rpm[b]	0.331 g/g; 84%[a]	Song et al. (2016)

(Continued)

TABLE 8.1 Reprentative List of Microorganisms, Substrates, Fermentation Conditions, Ethanol Yield, and Volumetric Ethanol Productivity Reported in Published Articles (*cont.*)

Microorganisms	Substrate	Fermentation process and condition	Maximum ethanol productivity and/or concentration and yield	References
Recombinant *S. cerevisisiae* 6525	JA dry powder	25°C; 48 h; 50 L fermenter	85.67 g/L; 90.1%[a]	Zou et al. (2016)
K. marxianus Y179	Inulin sugar (230 g/L)	30°C; 16–18 h; 150 rpm[b]	2.04 g/L/h; 97.7 g/L	Gao et al. (2016)
S. cerevisisiae KCTC 7906	JA dry powder (0.35–0.45 mm)	30°C; 24 h	24.5 mg/mL; 85.3%[a]	Song et al. (2017)

Concentration in g/L or g/g or v/v; ethanol productivity in g/L/h.
JA, Jerusalem artichoke; *L*, liter; *SSF*, simultaneous saccharification and fermentation.
[a] *Theoretical yield.*
[b] *Agitation.*
[c] *Jerusalem artichoke stalks.*

with the reaction duration. Eskandari Nasab et al. (2009) used response surface methodology and found that maximum inulin hydrolysis using hydrochloric acid needed pH < 2 for 60 min at temperatures higher than 90°C. Additionally, catalysts are often used in chemical hydrolysis to improve product selectivity. Abasaeed and Lee (1995) reported a moderate fructose recovery (75%) from JA, when hydrolysis was carried out with an acidic Zeolite LZ-M-8 catalyst over 150 min. Interestingly, the by-products that were identified as problematic in the Sarchami and Rehmann (2015)'s study were not detected even though the reaction was four times as long (Abasaeed and Lee, 1995). Razmovski et al. (2011) found that a temperature of 126°C for 60 min with 1:1 ratio of JA: water at pH 2.0 with HCl was the best for optimum inulin hydrolysis and it also kept 5-(Hydroxymethylfurfural) HMF concentrations at levels lower than 0.2 g/L.

For enzymatic SHF, Parekh and Margaritis (1986) carried out enzymatic hydrolysis of JA inulin using immobilized dead cells of *K. marxianus* in alginate beads. In a packed bed reactor, a volumetric productivity of 136 g/L/h of total reducing sugars were found with 98% conversion of inulins. The enzyme activity half-life was 28 days. The same biocatalyst gave a half-life of 14 days when recycled in a batch process with 20% JA fructans and 98% sugar conversion. Kim and Rhee (1990) carried out fermentation with free or immobilized *Z. mobilis* ZM4. In two of the methods, inulin was prehydrolyzed either with sulfuric acid pH 1.5 or enzymatically with inulinase from *Aspergillus ficuum* at 60°C for 48 h, and autoclaving followed at 121°C for 15 min in both methods. Enzymatic hydrolysis was superior to acid hydrolysis due to formation of by-products in the latter that reduced ethanol yields from 88% to 78%. In a recent study, hydrogen peroxide–acetic acid pretreatment was used to enhance the enzymatic hydrolysis and to effectively remove the lignin of JA. With optimized enzyme doses, synergy was observed from the combination of three different enzymes (RUT-C30, pectinase, and xylanase), which provided a conversion rate was approximately 30% higher than the rate with from treatment with RUT-C30 alone. Fermentation of the JA hydrolysates by *S. cerevisiae* produced a fermentation yield of approximately 84% (Song et al., 2016).

Saccharomyces sp. W0 was used for the production of high concentrations of ethanol (Zhang et al., 2010). Nevertheless, this yeast strain was unable to secrete inulinase. As a result, hydrolysis of inulin into reducing sugars was carried out by the recombinant inulinase obtained from *Pichia pastoris* X-33/pPICZaA-*INU1*. It was shown that 38.2 U of the recombinant inulinase/gram of inulin could be successfully used to hydrolyze inulin and produce ethanol using *Saccharomyces* sp. W0 (Zhang et al., 2010). The fermentation lasted for 120 h and led to the production of more than 14.6 ml of ethanol/100 mL of the fermentation medium from 30% inulin, while the ethanol productivity was over 0.384 g of ethanol/g of inulin. In a later study, after the endo-inulinase gene from *Arthrobacter* sp. was ligated the expression vectors pMIDSC31 and pMIRSC31, the endo-inulinase gene was inserted into the chromosomal DNA of *Saccharomyces* sp. W0. It was found that the inulinase activity of the recombinant yeast D5 in which the endo-inulinase gene was inserted into the delta sequence was higher than that of the recombinant yeast R1 in which the endo-inulinase gene was inserted into 18S rDNA sequence. More ethanol from inulin was produced by the recombinant yeast D5 than by the recombinant yeast R1. But *Saccharomyces* sp. W0 produced the lowest inulinase activity and concentration of ethanol. The recombinant yeast D5 could produce 13.6 mL of ethanol per 100 mL of the fermented medium from 30% inulin (Li et al., 2013).

More than 98.8% of total sugar was used. Szambelan et al. (2004) carried out SHF with cocultures of *Kluyveromyces fragilis* with either *S. cerevisiae*, or *Z. mobilis* and found that mixed cultures had 2%–12% higher ethanol yield than single cultures, tested on tubers from *Albik* and *Rubik* cultivars that were hydrolyzed by inulinase from *Aspergillus niger* (20 mg enzyme/kg tubers) prior to fermentation. Moreover, *Z. mobilis* combination yielded 4%–8% more ethanol than *S. cerevisiae* combination.

8.4.2 Simultaneous Saccharification and Fermentation (SSF)

In the 1980s, fermentations were carried out on juice recovered after pressing the tubers in a hydraulic or screw press (Duvnjak et al., 1982; Guiraud et al., 1982; Rosa et al., 1987). In several publications from Margaritis and coworkers 1:1 ratio of ½ sliced tubers was incubated in water at 75°C for 1 h that leading to 94% inulin recovery followed by filtration and sterilization at 120°C for 20 min (Bajpai and Bajpai, 1991; Bajpai and Margaritis, 1982; Margaritis et al., 1981; Margaritis and Bajpai, 1982c). Also, Chabbert and coworkers extracted inulins by a continuous diffusion process using boiling water (Chabbert et al., 1983, 1985a,b).

In a SSF approach, Nakamura et al. (1996) used inulinase producing *A. niger* 817 that had four-fold higher levels of inulinase activity in comparison to the wild strain, along with *S. cerevisiae* 1200, in order to produce ethanol. In their study, inulinase powder from *A. niger* and mashed tuber yielded 92% ethanol yield but only 52% yield was obtained using inulinase powder and juice concentrate with *S. cerevisiae*, due to inhibition from reducing sugars. When *A. niger* and *S. cerevisiae* were used together, 80% ethanol yield was obtained from ball-milled JA tuber flour.

A strain of *S. cerevisiae*, KCCM50549, was found to efficiently ferment the inulin-containing carbohydrates in JA without acidic or enzymatic pretreatment prior to fermentation. *S. cerevisiae* KCCM50549 could utilize almost completely the fructooligosaccharides (FOS) present in JA (up to DP of 15), in contrast to the other *S. cerevisiae* strain, such as NCYC625 that fermented the FOS with DP of up to around six. Inulin-fermenting *S. cerevisiae* KCCM50549 produced ca.1.6 times more ethanol from JA compared with *S. cerevisiae* NCYC625. Direct ethanol fermentation of JA flour at 180 g/L without any supplements or pretreatments by *S. cerevisiae* KCCM50549 in a 5-L jar fermentor yielded 36.2 g/L of ethanol within 36 h. The conversion efficiency of inulin-type sugars to ethanol was 70% of the theoretical ethanol yield (Lim et al., 2011).

8.4.3 Consolidated Bioprocessing (CBP)

Some relatively new studies have employed the CBP approach for ethanol production using JA tuber flour. In these studies, tubers were washed, cut or chopped in a grinder, dried, and then milled to obtain tuber flour (Guo et al., 2013; Sarchami and Rehmann, 2014; Yuan et al., 2008, 2013a). Dried slices (Yuan et al., 2013b) or tuber mash after grinding (Charoensopharat et al., 2015) have also been the source of inulins.

Barthomeuf et al. (1991) reported ethanol titer was 91 g/L using *S. cerevisiae* Y-481 from JA mash. Also, Zhang et al. (2010) managed to engineer a *S. cerevisiae* strain through the expression of an inulinase gene to create a recombinant strain *S. cerevisiae* W0. The use of this strain led to the production of 95.5 g/L of ethanol after 120 h. Lim et al. (2011) reported the use of *S. cerevisiae* KCCM50549 as the CBP strain for producing ethanol from JA, receiving 36.2 g/L ethanol. Du et al. (2014) on the other hand engineered a recombinant *S. cerevisiae* strain producing exo-inulinase that could produce 91.8 g/L ethanol.

Kim and Kim (2014) used both aboveground biomass and tuber for ethanol production. They pretreated the biomass using 0.5% H_2SO_4 at 121°C for 60 min followed by 1M NaOH for another 60 min at 121°C, then mixed the pretreated solids with tubers in a 10:1 ratio and performed batch SSF and fed-batch SSF with Cellic CTec2 cellulase (80 FPU per g total biomass) and *K. marxianus* CBS1555 at 37°C. They obtained 0.497 g and 0.361 g ethanol/g glucose that corresponded to 83.6% and 70.8% sugar conversion in batch and fed-batch SSF, respectively. Species of the yeast *Kluyveromyces*, that is, *K. marxianus*, *K. cicerisporus*, and *K. fragilis* naturally produce inulinase that allows CBP of JA inulins. *K. marxianus* ATCC 12708 produced the highest ethanol yield of 87.4% of the theoretical limit with a productivity of 1.68 g ethanol/g feedstock/h (Duvnjak et al., 1981). In another study, *K. marxianus* UCD (FST) 55–82 and *Saccharomyces rosei* UWO (PS) 80–38 both gave ethanol yield of 88%, but the former presented higher growth rates and led to higher ethanol concentration (Margaritis and Bajpai, 1982c).

Yuan et al. (2008, 2012) designed *Kluyveromyces* strains that were able to produce inulinase and ethanol levels of 96.3 g/L medium using JA as a substrate. Yu et al. (2010) reported that another *Kluyveromyces* strain can produce ethanol at levels of up to 97.3 g/L medium using again JA. *Kluyveromyces* strains commonly have an inulinase activity, even though their low ethanol tolerance can lead to lower ethanol fermentation titer comparing to *S. cerevisiae* strains (Lane and Morrissey, 2010).

Zhang et al. (2010) took both the SHF and CBP approaches. In the case of the SHF, *P. pastoris* X-33 with the cloned INU1 gene was used for inulinase production and *Saccharomyces* sp. W0 for ethanol production. In CBP, *Saccharomyces* sp. W0/YCPlac33 PGK/CYC1-INU1 was used in both the SHF and CBP approaches. Yields of 0.384 g ethanol/g inulin and 0.319 g/g sugar were obtained using the SHF and CBP methods, respectively. Wang et al. (2016) introduced an inulinase gene from *K. marxianus* into *S. cerevisiae* diploid strain JCD, and repressed the vacuolar proteinase gene PEP4 to increase heterologous protein production. The resulting strain (JZD-InuMKCP) had an ethanol productivity of 3.2 g/L/h in 24 h and 2.44 g/L/h in 36 h with 95% theoretical ethanol yield solely fermented on JA tuber flour. Guo et al. (2013) used engineered *S. cerevisiae* DQ1 for CBP at 30°C and pH 5.5. The optimum activity of the inulinase from this strain was presented at 50–55°C and pH 5.0. This study is noteworthy as ethanol yield of 73.5% was achieved at the highest treated tuber loading of 35% w/w on a dry basis in helical ribbon bioreactor that allowed better mixing at high solids loading. Yuan et al. (2013a) performed chromosome integration of the one-liners gene in *K. marxianus* ATCC 8554 to produce K/INU2 strain that increased inulinase secretion from 2.4 to 3.7 U/mL during fermentation using JA inulin and 3.1–6.8 U/mL on JA tuber mash. The ethanol productivity improved from 1.12 to 1.34 g/L/h when fed on JA inulin and from 1.29 to 1.44 g/L/h when tuber mash was used. They mentioned that inulinase production improved using a similar approach as in the case of *S. cerevisiae* but ethanol production was not affected. CBP using *K. marxianus* immobilized in alginate beads has also been carried out (Margaritis and Bajpai, 1982b). The result of immobilization was the high ethanol productivity of 118 g/L/h with JA tuber extract in a continuous packed bed reactor. Only 15% loss of ethanol productivity occurred after 30 days. In another study, a half-life of 72 days for the same immobilized yeast cells was reported with a volumetric ethanol productivity of 104 g/L/h (Margaritis and Bajpai, 1982b). Also, in the study of Margaritis et al. (1981), immobilized *K. fragilis* gave 96% of the theoretical ethanol yield with an ethanol productivity of 13.5 g/L/h. More recently, Khatun et al. (2017) and Li et al. (2016) explored lignocellulosic ethanol production solely from JA stalk by an engineered inulinase-producing yeast *S. cerevisiae*.

A schematic representation of bioethanol production from JA tuber is given in Fig. 8.3.

8.5 BUTANOL

Butanol is perceived to be superior to ethanol mainly as a result of its higher energy content and lower hydroscopic behavior. This has resulted in an increased research interest in butanol production from renewable biomass (Jang et al., 2012). Although butanol is a petroleum-derived product, it can be obtained through microbial fermentation, albeit, the cost of production is high. JA can be used as feedstock for biobutanol production in industrial-scale fermentation of butanol by solventogenic clostridia (Gu et al., 2011). An acetone–butanol–ethanol(ABE) yield of 0.31 g/g and an overall fermentation productivity of 0.25 g/L/h were obtained from acid hydrolysis of inulin derived from JA flour, also demonstrating that the feedstock is suitable for the production of ABE through fermentation (Sarchami and Rehmann, 2014, 2015).

In a different study, *Clostridum acetobutylicum* L7 was found to produce low concentrations of butanol from JA biomass feedstock. However, when 0.001B g/L $ZnSO_4B{\cdot}7H_2O$ was added into medium that contained fructose as the sole carbon source, butanol levels of 12.8 g/L were obtained, while the productivity was 0.089 g/L/h in comparison to only 4.5 g/L of butanol obtained and butanol productivity of 0.028B g/L/h in the control sample that was not supplemented with zinc. The zinc micronutrient also enhanced the butanol production to a maximum of 8.3 g/L (45.2 g/L xylose). Micronutrient zinc supplementation could therefore be a solution in rendering the production of butanol from JA feedstock financially sustainable (Wu et al., 2016).

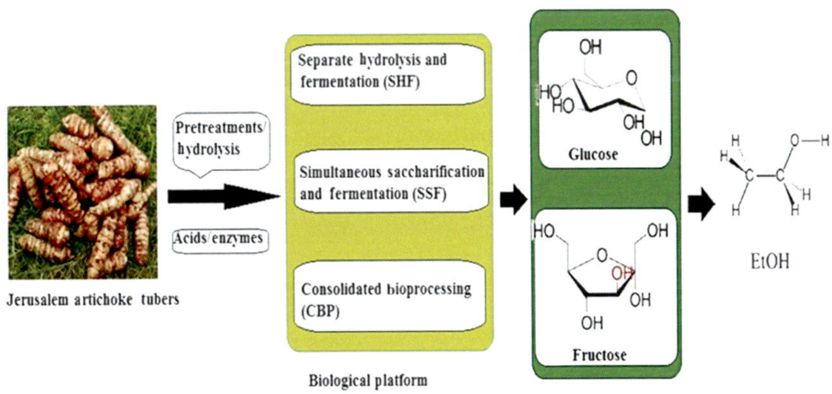

FIGURE 8.3 **Simplified process of producing ethanol from JA tubers.**

8.6 TECHNOECONOMIC FEASIBILITY

The feasibility of ethanol production through fermentation of JA tubers and the ethanol yield obtained do not differ significantly when glucose or fructose are used as substrate (Zhou et al., 2008). In addition, the prices of the biofuels and the need for increased profits from the valorization of byproducts can be of high importance. The level of JA raw material currently required is approximately 7.3 ton/ton of bioethanol, while its cost is estimated at 83 €/ton (Song et al., 2017). JA is available in abundance, making it probably the cheapest bioethanol-producing feedstock with the production cost of bioethanol being at 20 €/ton, when the cost of producing bioethanol from other feedstocks like cassava can be as high as 820 €/ton (Song et al., 2017). Moreover, JA contains a number of soluble proteins, amino acids, and minerals; these might be transformed into ethanol by bacteria in CBP (Zhou et al., 2008).

8.7 CONCLUSION AND FUTURE PERSPECTIVES

JA tubers can give ethanol yields of 2500–6500 L/ha. In comparison, corn and sugar cane can be used to produce 4182 L/ha and 6471 L/ha of ethanol in USA and Brazil, respectively (Goldemberg and Guardabassi, 2010). The inulinase-expressing microorganisms that have been reported to date include only few strains of *Kluyveromyces*, *Cryptococcus*, and *Pichia*, but these strains cannot produce as much ethanol as *S. cerevisiae* due to their low ethanol tolerance (Chi et al., 2011). Strains, such as *S. cerevisiae* Y-481 and *S. cerevisiae* KCCM50549 fermented JA inulins and produced ethanol in quantities significantly lower than those produced by the commercial ethanogenic strains. Moreover, the cost of commercial inulinase is one of the most important limiting factors for its use in the production of ethanol from JA. Therefore, further research is required on the fermentation process of this plant and particular emphasis should be given in achieving fast and efficient conversion of inulin into ethanol, while maintaining the cost of the industrial production at acceptable levels. Further studies are also needed to maximize the sugar quantities obtained from this plant.

References

Abasaeed, A.E., Lee, Y.Y., 1995. Inulin hydrolysis to fructose by a novel catalyst. Chem. Eng. Technol. 18, 440–444.

Bajpai, P.K., Bajpai, P., 1991. Cultivation and utilization of Jerusalem artichoke for ethanol, single cell protein, and high-fructose syrup production. Enzyme Microb. Technol. 13, 359–362.

Bajpai, P., Margaritis, A., 1982. Ethanol inhibition kinetics of *Kluyveromyces marxianus* grown on Jerusalem artichoke juice. Appl. Environ. Microbiol. 44, 1325–1329.

Barthomeuf, C., Regerat, F., Pourrat, H., 1991. High-yield ethanol production from Jerusalem artichoke tubers. World J. Microbiol. Biotechnol. 7, 490–493.

Biedrzycka, E., Bielecka, M., 2004. Prebiotic effectiveness of fructans of different degrees of polymerization. Trends Food Sci. Technol. 15, 170–175.

Chabbert, N., Braun, P., Guiraud, J.P., Arnoux, M., Galzy, P., 1983. Productivity and fermentability of Jerusalem artichoke according to harvesting date. Biomass 3, 209–224.

Chabbert, N., Guiraud, J.P., Arnoux, M., Galzy, P., 1985a. Productivity and fermentability of different Jerusalem artichoke (*Helianthus tuberosus*) cultivars. Biomass 6, 271–284.

Chabbert, N., Guiraud, J.P., Arnoux, M., Galzy, P., 1985b. The advantageous use of an early Jerusalem artichoke cultivar for the production of ethanol. Biomass 8, 233–240.

Charoensopharat, K., Thanonkeo, P., Thanonkeo, S., Yamada, M., 2015. Ethanol production from Jerusalem artichoke tubers at high temperature by newly isolated thermotolerant inulin-utilizing yeast *Kluyveromyces marxianus* using consolidated bioprocessing. Anton. Leeuw. 108, 173–190.

Chi, Z.M., Zhang, T., Cao, T.S., Liu, X.Y., Cui, W., Zhao, C.H., 2011. Biotechnological potential of inulin for bioprocesses. Bioresour. Technol. 102, 4295–4303.

Cieślik, E., Gębusia, A., Florkiewicz, A., Mickowska, B., 2011. The content of protein and of amino acids in Jerusalem artichoke tubers (*Helianthus tuberosus* L) of red variety Rote Zonenkugel. Acta Sci. Polon. Technol. Aliment. 10, 433–441.

Clomburg, J.M., Gonzalez, R., 2010. Biofuel production in *Escherichia coli*: the role of metabolic engineering and synthetic biology. Appl. Microbiol. Biotechnol. 86, 419–434.

D'Egidio, M.G., Cecchini, C., Cervigni, T., Donini, B., Pignatelli, V., 1998. Production of fructose from cereal stems and polyannual cultures of Jerusalem artichoke. Ind. Crop Prod. 7, 113–119.

Dénes, K., Farkas, C., Hoschke, Á., Rezessy-Szabó, J., Nguyen, Q., 2013. Bioethanol fermentation of Jerusalem artichoke using mixed culture of *Saccharomyces cerevisiae* and *Kluyveromyces marxianus*. Acta Aliment. Hung. 42, 10–18.

Du, Y.C., Long, X.H., Liu, Z.P., Shao, H.B., 2014. Optimizing medium for producing ethanol from industrial crop Jerusalem artichoke by one-step fermentation and recombinant *Saccharomyces cerevisiae*. Plant Biosyst. 148, 118–126.

Duvnjak, Z., Kosaric, N., Hayes, R.D., 1981. Kinetics of ethanol production from Jerusalem artichoke juice with some *Kluyveromyces* species. Biotechnol. Lett. 3, 589–594.

Duvnjak, Z., Kosaric, N., Kliza, S., Hayes, D., 1982. Production of alcohol from Jerusalem artichokes by yeasts. Biotechnol. Bioeng. 24, 2297–2308.

Eskandari Nasab, E., Habibi-Rezaei, M., Khaki, A., Balvardi, M., 2009. Investigation on acid hydrolysis of inulin: a response surface methodology approach. Int. J. Food Eng. 5 doi: 10.2202/1556-3758.1676.

Gao, J., Yuan, W., Kong, L., Xiang, R., Zhong, S., 2015. Efficient ethanol production from inulin by two-stage aerate strategy. Biomass Bioenergy 80, 10–16.

Gao, J., Yuan, W., Li, Y., Bai, F., Zhong, S., Jiang, Y., 2016. Application of redox potential control to improve ethanol productivity from inulin by consolidated bioprocessing. Process Biochem. 51, 1544–1551.

Ge, X., Zhang, W., 2005. A shortcut to the production of high ethanol concentration from Jerusalem artichoke tubers. Food Technol. Biotechnol. 43, 241–245.

Glibowski, P., Wasko, A., 2008. Effect of thermochemical treatment on the structure of inulin and its gelling properties. Int. J. Food Sci. Technol. 43, 2075–2082.

Goldemberg, J., Guardabassi, P., 2010. The potential for first-generation ethanol production from sugarcane. Biofuels Bioprod. Biorefin. 4, 17–24.

Gu, Y., Jiang, Y., Wu, H., Liu, X., Li, Z., Li, J., Li, Y., 2011. Economical challenges to microbial producers of butanol: feedstock, butanol ratio and titer. Biotechnol. J. 6, 1348–1357.

Guiraud, J.P., Bourgi, J., Stervinou, M., Claisse, M., Galzy, P., 1987. Isolation of a respiratory-deficient *Kluyveromyces fragilis* mutant for the production of ethanol from Jerusalem artichoke. Biotechnol. Bioenergy 29, 850–858.

Guiraud, J.P., Caillaud, J.M., Galzy, P., 1982. Optimization of alcohol production from Jerusalem artichokes. Appl. Microbiol. Biotechnol. 14, 81–85.

Gunnarsson, I.B., Karakashev, D., Angelidaki, I., 2014a. Succinic acid production by fermentation of Jerusalem artichoke tuber hydrolysate with *Actinobacillus succinogenes* 130Z. Ind. Crop Prod. 62, 125–129.

Gunnarsson, I.B., Svensson, S.E., Johansson, E., Karakashev, D., Angelidaki, I., 2014b. Potential of Jerusalem artichoke (*Helianthus tuberosus* L.) as a biorefinery crop. Ind. Crop Prod. 56, 231–240.

Guo, L., Zhang, J., Hu, F., Dy Ryu, D., Bao, J., 2013. Consolidated bioprocessing of highly concentrated Jerusalem artichoke tubers for simultaneous saccharification and ethanol fermentation. Biotechnol. Bioenergy 110, 2606–2615.

Hong, S.J., Kim, H.J., Kim, J.W., Lee, D.H., Seo, J.H., 2015. Optimizing promoters and secretory signal sequences for producing ethanol from inulin by recombinant *Saccharomyces cerevisiae* carrying *Kluyveromyces marxianus* inulinase. Bioprocess. Biosyst. Eng. 38, 263–272.

Hu, N., Yuan, B., Sun, J., Wang, S.A., Li, F.L., 2012. Thermotolerant *Kluyveromyces marxianus* and *Saccharomyces cerevisiae* strains representing potentials for bioethanol production from Jerusalem artichoke by consolidated bioprocessing. Appl. Microbiol. Biotechnol. 95, 1359–1368.

Jang, Y.S., Malaviya, A., Cho, C., Lee, J., Lee, S.Y., 2012. Butanol production from renewable biomass by clostridia. Bioresour. Technol. 123, 653–663.

Khatun, M.M., Li, Y.H., Liu, C.G., Zhao, X.Q., Bai, F.W., 2015. Fed-batch saccharification and ethanol fermentation of Jerusalem artichoke stalks by an inulinase producing *Saccharomyces cerevisiae* MK01. RSC Adv. 5, 107112–107118.

Khatun, M.M., Liu, C.G., Zhao, X.Q., Yuan, W.J., Bai, F.W., 2017. Consolidated ethanol production from Jerusalem artichoke tubers at elevated temperature by *Saccharomyces cerevisiae* engineered with inulinase expression through cell surface display. J. Ind. Microbiol. Biotechnol. 44, 295–301.

Kim, C.H., Rhee, S.K., 1990. Ethanol production from Jerusalem artichoke by inulinase and *Zymomonas mobilis*. Appl. Biochem. Biotechnol. 23, 171–180.

Kim, K., Hamdy, M.K., 1986. Acid hydrolysis of Jerusalem artichoke for ethanol fermentation. Biotechnol. Bioeng. 28, 138–141.

Kim, S., Kim, C.H., 2014. Evaluation of whole Jerusalem artichoke (*Helianthus tuberosus* L.) for consolidated bioprocessing ethanol production. Renew. Energy 65, 83–91.

Kocsis, L., Liebhard, P., Praznik, W., 2007. Effect of seasonal changes on content and profile of soluble carbohydrates in tubers of different varieties of Jerusalem artichoke (*Helianthus tuberosus* L.). J. Agric. Food Chem. 55, 9401–9408.

Lane, M.M., Morrissey, J.P., 2010. *Kluyveromyces marxianus*: a yeast emerging from its sister's shadow. Fungal Biol. Rev. 24, 17–26.

Li, Y., Liu, G.L., Chi, Z.M., 2013. Ethanol production from inulin and unsterilized meal of Jerusalem artichoke tubers by *Saccharomyces* sp. W0 expressing the endo-inulinase gene from *Arthrobacter* sp. Bioresour. Technol. 147, 254–259.

Li, W., Zhang, J., Yu, C., Li, Q., Dong, F., Wang, G., Guo, Z., 2015. Extraction, degree of polymerization determination and prebiotic effect evaluation of inulin from Jerusalem artichoke. Carbohyd. Polym. 121, 315–319.

Li, K., Qin, J.C., Liu, C.G., Bai, F.W., 2016. Optimization of pretreatment, enzymatic hydrolysis and fermentation for more efficient ethanol production by Jerusalem artichoke stalk. Bioresour. Technol. 221, 188–194.

Lim, S.H., Ryu, J.M., Lee, H., Jeon, J.H., Sok, D.E., Choi, E.S., 2011. Ethanol fermentation from Jerusalem artichoke powder using *Saccharomyces cerevisiae* KCCM50549 without pretreatment for inulin hydrolysis. Bioresour. Technol. 102, 2109–2111.

Long, X.H., Shao, H.B., Liu, L., Liu, L.P., Liu, Z.P., 2016. Jerusalem artichoke: a sustainable biomass feedstock for biorefinery. Renew. Sust. Energy Rev. 54, 1382–1388.

Margaritis, A., Bajpai, P., 1982b. Continuous ethanol production from Jerusalem artichoke tubers. I: use of free cells of *Kluyveromyces marxianus*. Biotechnol. Bioeng. 24, 1473–1482.

Margaritis, A., Bajpai, P., 1982c. Ethanol production from Jerusalem artichoke tubers (*Helianthus tuberosus*) using *Kluyveromyces marxianus* and *Saccharomyces rosei*. Biotechnol. Bioeng. 24, 941–953.

Margaritis, A., Bajpai, P., Cannell, E., 1981. Optimization studies for the bioconversion of Jerusalem artichoke tubers to ethanol and microbial biomass. Biotechnol. Lett. 3, 595–599.

Matías, J., González, J., Royano, L., Barrena, R.A., 2011. Analysis of sugars by liquid chromatography-mass spectrometry in Jerusalem artichoke tubers for bioethanol production optimization. Biomass Bioenergy 35, 2006–2012.

Modler, H.W., Jones, J.D., Mazza, G., 1993. Observations on long-term storage and processing of Jerusalem artichoke tubers (*Helianthus tuberosus*). Food Chem. 48, 279–284.

Nakamura, T., Ogata, Y., Hamada, S., Ohta, K., 1996. Ethanol production from Jerusalem artichoke tubers by *Aspergillus niger* and *Saccharomyces cerevisiae*. J. Ferment. Bioeng. 81, 564–566.

Parekh, S.R., Margaritis, A., 1986. Application of immobilized cells of *Kluyveromyces marxianus* for continuous hydrolysis to fructose of fructans in Jerusalem artichoke extracts. Int. J. Food Sci. Technol. 21 (4), 509–515.

Ray, R.C., Joshi, V.K., 2014. Fermented foods: past, present and future scenario. In: Ray, R.C., Montet, D. (Eds.), Microorganisms and Fermentation of Traditional Foods. CRC Press, Boca Raton, FL, United States, pp. 1–36.

Razmovski, R.N., Šćiban, M.B., Vučurović, V.M., 2011. Bioethanol production from Jerusalem artichoke by acid hydrolysis. Roman. Biotechnol. Lett. 16, 6497–6503.

Rincón, L.E., Jaramillo, J.J., Cardona, C.A., 2014. Comparison of feedstocks and technologies for biodiesel production: an environmental and techno-economic evaluation. Renew. Energy 69, 479–487.

Rosa, M.F., Correia, I.S., Novais, J.M., 1987. Production of ethanol at high temperatures in the fermentation of Jerusalem artichoke juice and a simple medium by *Kluyveromyces marxianus*. Biotechnol. Lett. 9, 441–444.

Sachs, R., Low, C., Vasavada, A., Sully, M., Williams, L., Ziobro, G., 1981. Fuel alcohol from Jerusalem artichoke. Calif. Agr. 35, 4–6.

Sarchami, T., Rehmann, L., 2014. Optimizing enzymatic hydrolysis of inulin from Jerusalem artichoke tubers for fermentative butanol production. Biomass Bioenergy 69, 175–182.

Sarchami, T., Rehmann, L., 2015. Optimizing acid hydrolysis of Jerusalem artichoke-derived inulin for fermentative butanol production. BioEnergy Res. 8, 1148–1157.

Schorr-Galindo, S., Guiraud, J.P., 1997. Sugar potential of different Jerusalem artichoke cultivars according to harvest. Bioresour. Technol. 60, 15–20.

Schorr-Galindo, S., Guiraud, J.P., 1997. Sugar potential of different Jerusalem artichoke cultivars according to harvest. Bioresour. Technol. 60, 15–20.

Song, Y., Wi, S.G., Kim, H.M., Bae, H.J., 2016. Cellulosic bioethanol production from Jerusalem artichoke (*Helianthus tuberosus* L) using hydrogen peroxide-acetic acid (HPAC) pretreatment. Bioresour. Technol. 214, 30–36.

Song, Y., Oh, C., Bae, H.J., 2017. Simultaneous production of bioethanol and value-added d-psicose from Jerusalem artichoke (*Helianthus tuberosus* L) tubers. Bioresour. Technol. 244, 1068–1072.

Szambelan, K., Nowak, J., Czarnecki, Z., 2004. Use of *Zymomonas mobilis* and *Saccharomyces cerevisiae* mixed with *Kluyveromyces fragilis* for improved ethanol production from Jerusalem artichoke tubers. Biotechnol. Lett. 26, 845–848.

Wang, L., Du, Y.C., Meng, X.F., Long, X.H., Liu, Z.P., Shao, H., 2014. Direct production of bioethanol from Jerusalem artichoke inulin by gene-engineering *Saccharomyces cerevisiae* 6525 with exoinulinase gene. Plant Biosyst. 148, 133–139.

Wang, D., Li, F.L., Wang, S.A., 2016. Engineering a natural *Saccharomyces cerevisiae* strain for ethanol production from inulin by consolidated bioprocessing. Biotechnol. Biofuels 9, 96.

Wang, Y.Z., Zou, S.M., He, M.L., Wang, C.H., 2015. Bioethanol production from the dry powder of Jerusalem artichoke tubers by recombinant *Saccharomyces cerevisiae* in simultaneous saccharification and fermentation. J. Ind. Microbiol. Biotechnol. 42, 543–551.

Wu, Y.D., Xue, C., Chen, L.J., Bai, F.W., 2016. Impact of zinc supplementation on the improved fructose/xylose utilization and butanol production during acetone–butanol–ethanol fermentation. J. Biosci. Bioeng. 121, 66–72.

Yang, L., He, Q.S., Corscadden, K., Udenigwe, C.C., 2015. The prospects of Jerusalem artichoke in functional food ingredients and bioenergy production. Biotechnol. Rep. 5, 77–88.

Yu, J., Jiang, J., Fang, Z., Li, Y., Lv, H., Liu, J., 2010. Enhanced expression of heterologous inulinase in *Kluyveromyces lactis* by disruption of HAP1 gene. Biotechnol. Lett. 32, 507–512.

Yuan, W.J., Zhao, X.Q., Ge, X.M., Bai, F.W., 2008. Ethanol fermentation with *Kluyveromyces marxianus* from Jerusalem artichoke grown in salina and irrigated with a mixture of seawater and freshwater. J. Appl. Microbiol. 105, 2076–2083.

Yuan, W.J., Chang, B.L., Ren, J.G., Liu, J.P., Bai, F.W., Li, Y.Y., 2012. Consolidated bioprocessing strategy for ethanol production from Jerusalem artichoke tubers by *Kluyveromyces marxianus* under high gravity conditions. J. Appl. Microbiol. 112, 38–44.

Yuan, B., Wang, S.A., Li, F.L., 2013. Improved ethanol fermentation by heterologous endoinulinase and inherent invertase from inulin by *Saccharomyces cerevisiae*. Bioresour. Technol. 139, 402–405.

Yuan, W., Zhao, X., Chen, L., Bai, F., 2013a. Improved ethanol production in Jerusalem artichoke tubers by overexpression of inulinase gene in *Kluyveromyces marxianus*. Biotechnol. Bioprocess. Eng. 18, 721–727.

Yuan, X., Cheng, M., Gao, M., Zhuo, R., Zhang, L., Xiao, H., 2013b. Cytotoxic constituents from the leaves of Jerusalem artichoke (*Helianthus tuberosus* L) and their structure–activity relationships. Phytochem. Lett. 6, 21–25.

Zhang, T., Chi, Z., Zhao, C.H., Chi, Z.M., Gong, F., 2010. Bioethanol production from hydrolysates of inulin and the tuber meal of Jerusalem artichoke by *Saccharomyces* sp. W0. Bioresour. Technol. 101, 8166–8170.

Zhang, X., Xu, H., Fu, R., Jin, L., Zhao, H., Wang, L., Liang, M., 2016. Bioethanol production by heterologous expression of two individual 1-FEH genes from *Helianthus tuberosus* in *Saccharomyces cerevisiae* 6525. BioEnergy Res. 9, 884–893.

Zhou, Z., Cao, H.L., Zhu, Y., Li, S.G., Bai, X.F., Zhao, X.M., Du, Y.G., 2008. The primary study of ethanol production by fermenting of Jerusalem artichoke instead of corn. Acta Agric. Bor. Sin. 4, 67–70.

Zou, S., Wang, Y., He, M., Deng, X., Wang, C., 2016. Scale-up batch fermentation of bioethanol production from the dry powder of Jerusalem artichoke (*Helianthus tuberosus* L) tubers by recombinant *Saccharomyces cerevisiae*. J. Ind. Brew. 122, 261–267.

Further Readings

Liu, Z.X., Steinberger, Y., Chen, X., Wang, J.S., Xie, G.H., 2015. Chemical composition and potential ethanol yield of Jerusalem artichoke in a semi-arid region of China. Ital. J. Agron. 10, 34–43.

Margaritis, A., Bajpai, P., 1982a. Continuous ethanol production from Jerusalem artichoke tubers. II. Use of immobilized cells of *Kluyveromyces marxianus*. Biotechnol. Bioeng. 24, 1483–1493.

SECTION III

BIOETHANOL FROM LIGNOCELLULOSES

9 *Lignocellulosic Ethanol: Feedstocks and Bioprocessing* 165

10 *Bioethanol From Sugarcane Bagasse: Status and Perspectives* 187

11 *Bioethanol Production From Rice- and Wheat Straw: An Overview* 213

12 *Forest Bioresources for Bioethanol and Biodiesel Production With Emphasis on Mohua* (Madhuca latifolia *L.) Flowers and Seeds* 233

13 *Microbial Enzyme Applications in Bioethanol Producing Biorefineries: Overview* 249

14 *Application of Fungal Pretreatment in the Production of Ethanol From Crop Residues* 267

15 *Currently Used Microbes and Advantages of Using Genetically Modified Microbes for Ethanol Production* 293

CHAPTER

9

Lignocellulosic Ethanol: Feedstocks and Bioprocessing

Richa Arora[*,†], *Nilesh K. Sharma*[*], *Sachin Kumar*[*,**], *Rajesh K. Sani*[**]

[*]Sardar Swaran Singh National Institute of Bio-Energy, Kapurthala, India

[**]South Dakota School of Mines and Technology, Rapid City, SD, United States

[†]School of Bioengineering and Biosciences, Lovely Professional University, Phagwara, India

ABBREVIATIONS

ADH Alcohol dehydrogenase
AFEX Ammonia fiber explosion
ARP Ammonia recycle percolation
CBP Consolidated bioprocessing
F6P Fructose-6-phosphate
G3P Glyceraldehyde-3-phosphate
HMF Hydroxymethyl furfural
LCB Lignocellulosic biomass
PPP Pentose phosphate pathway
SHF Separate hydrolysis and fermentation
SSCF Simultaneous saccharification and cofermentation
SSF Simultaneous saccharification and fermentation

9.1 INTRODUCTION

The comprehensive rise in energy utilization, envisaged boost in energy demands, depletion of fossil fuels, and global warming are the driving forces to search for renewable and green energy resources (Agbor et al., 2011). Several governmental policies have been designed to promote the fabrication and utilization of renewable energy resources, considering their thump on the environment (Glithero et al., 2013). Approximately 30% of the global energy prerequisite will be anticipated by biofuel, including bioethanol, biogas, biodiesel, biobutanol, and biohydrogen, which will be economical, safe, clean, competent, and abundantly available (Guo et al., 2015; Limayem and Ricke, 2012). However, liquid biofuels have accomplished the requirement of transportation in marine and aviation sectors and on roads by approximately 3% (IEA, 2011).

In the midst of liquid biofuels, bioethanol possesses immense potential as a substitute to transportation fuel, thereby, replacing the fossil fuels that release considerable amount of greenhouse gases into the environment (Dias et al., 2013; Nigam and Singh, 2011). The fuel properties, namely, octane number and efficient combustion can be improved significantly when bioethanol is blended with gasoline (Pejin et al., 2009). Also,

Bioethanol Production From Food Crops
http://dx.doi.org/10.1016/B978-0-12-813766-6.00009-6

blending of gasoline below 10% of bioethanol does not require any engine modification. On the other hand, higher blending levels (more than 10%) require several modifications in the engines (Balat, 2009). Blending of bioethanol with gasoline E2–E20 has been made obligatory by many countries under biofuel program (REN21, 2014).

Current research on bioethanol mainly focuses on lignocellulosic biomass (LCB) owing to their profusion and renewability. Efforts are being made for mounting and combining the critical steps of pretreatment, saccharification, and fermentation (Behera et al., 2014; Mathew et al., 2013; Matsushita et al., 2013; Saratale and Oh, 2012). The foremost benefit of utilizing LCB is to overcome the competition between the food and fuel (Nigam and Singh, 2011; Singh et al., 2010).

Another important shift in lignocellulosic bioethanol production is the utilization of thermophilic/thermotolerant microorganisms, which are reported to have broad substrate range and can tolerate wide range of acids, bases, salts, and solvents (Kuhad et al., 2011). Both saccharification and fermentation at high temperature have several advantages (Klein-Marcuschamer et al., 2012). The enzymes required for saccharification, that is, cellulases and hemicellulases are very active and stable at high temperatures. High specific activity, reduced viscosity, rise in diffusion coefficient are favored at elevated temperature for efficient saccharification (Behera et al., 2016; Kumar et al., 2009b; Limayem and Ricke, 2012; Menon and Rao, 2012; Talebnia et al., 2010). Apart from this, downstream processing can be made easily at high temperature, thereby, reducing the toxic level of ethanol and improving the overall process economics (Arora et al., 2015a). Also, the growth rate and productivity of the culture also increase at elevated temperatures owing to swift diffusion of the nutrients (Balat, 2011). Several thermophilic/thermotolerant microorganisms have been reported to coferment both hexose and pentose sugars, which further aid in the overall process economics (Frock and Kelly, 2012; Hasunuma and Kondo, 2012).

The present chapter deals with various pretreatment processes, critical factors affecting pretreatment type and feasibility of the processes at industrial scale. Along with pretreatment, stress has been laid on the ethanologens, which could facilitate saccharification and ethanol fermentation at high temperature for the development of economical bioprocess.

9.2 LIGNOCELLULOSIC BIOETHANOL PRODUCTION: AN OVERVIEW

The key steps for biochemical conversion of LCB to bioethanol include: (1) pretreatment to release holocellulose (cellulose and hemicellulose) from the recalcitrant matrix; (2) saccharification to hydrolyze the complex polymers to simple sugars; and (3) cofermentation of pentose and hexose sugars (Fig. 9.1). Table 9.1 shows the ethanol fermentation at high temperature with a variety of LCBs.

The main aim of pretreatment is to increase the biomass porosity and to solubilize the lignin, thereby, increasing the accessibility of the enzymes to cellulose and hemicelluloses (Behera et al., 2014). Among the various pretreatment technologies, chemical pretreatment offers the promising potential in terms of biomass porosity and solid separation (Yang and Wyman, 2008). However, its major setback lies in the recycling of the chemicals and it is unfriendly to the environment in terms of disposal (Brodeur et al., 2011; Harmsen et al., 2010).

The second step in lignocellulosic bioethanol production is the hydrolysis of carbohydrates to simple sugars with the help of synergistic action of lignocellulolytic enzymes, that is, cellulases and hemicellulases (Fisher and Tullman-Ercek, 2013; Wongwilaiwalin et al., 2010). However, in spite of the far-reaching research on saccharification of cellulosic and hemicellulosic

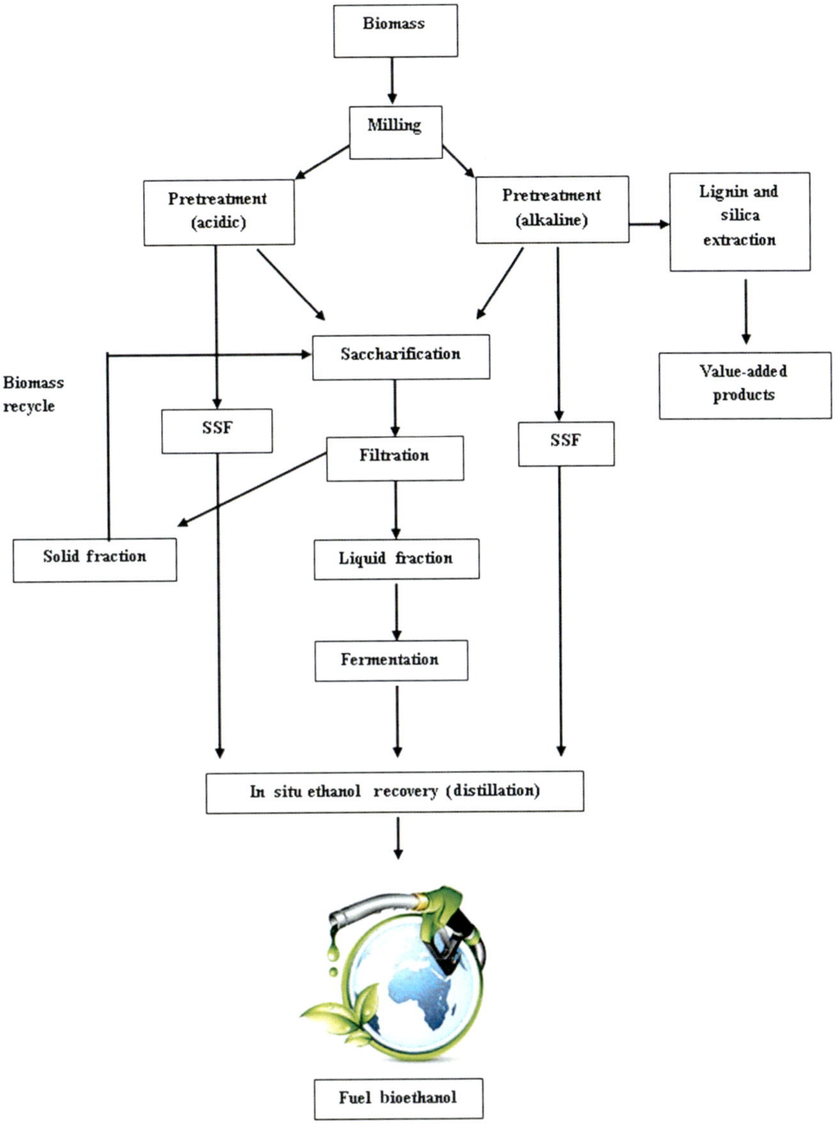

FIGURE 9.1 The schematic representation of bioethanol production from lignocellulosic biomass.

fractions, the technology still faces certain bottlenecks. The high cost of the enzymes for saccharification at commercial scale makes the process less doable. Various parameters elevate the enzyme cost, including regulation of enzyme induction, enzyme repression, prevention of feedback inhibition, pricey monitoring systems, unusual culturing conditions, elimination of unwanted products, expensive inducers, namely, lactose, cellulose.

Moreover, hydrolysis with conventional mesophilic enzyme producers is exigent owing to the requirement of high enzyme loading, long period of incubation, and less mass transfer rate

TABLE 9.1 Ethanol Production From Various Sugars and Biomass Sources Under Different Fermentation Conditions

Substrate	Fermentation temperature (°C)/pH	Microorganisms	Final ethanol concentration (g/L)	Ethanol yield (% of theoretical yield)	By-products produced	References
Cellulose	72/6.75	Cocultures of *Caldicellulosiruptor* DIB 087C and *Thermoanaerobacter* DIB 097X	n.d.[a]	n.d.	Acetate and lactate	Svetlitchny et al. (2013)
Bagasse hydrolysate	45/5.0	*Kluyveromyces* sp. IIPE453	14.8 (batch process)	84.15	Xylitol	Kumar et al. (2015a)
Bagasse hydrolysate	45/5.0	*Kluyveromyces* sp. IIPE453	56.3 (continuous with gas stripping)	96.0	Xylitol	Kumar et al. (2015b)
Bagasse pith hydrolysate	45/4.5	*Kluyveromyces* sp. IIPE453	17.40	88.00	n.d.	Dasgupta et al. (2013)
Cellulose hydrolysates	45/5.0	*Clostridium* strain AK1	n.d.	66.75	Acetate and hydrogen	Orlygsson (2012)
Cellulose	65/7.0	*Thermoanaerobacter* strainJ1	0.35	n.d.	Acetate, hydrogen, and methane	Jessen and Orlygsson (2012)
Cellulose hydrolysates	65/5.0	*Thermoanaerobacteriumaciditolerans*	1.11	n.d.	Hydrogen and acetate	Sigurbjornsdottir and Orlygsson (2012)
Cellulose hydrolysates	65/6.0	*T. xylanolyticum, T. islandicum*	1.43	75.00	Acetate, hydrogen	Almarsdottir et al. (2012)
Napier grass	60/n.d.	*Clostridium thermocellum*	n.d.	35.20	n.d.	Lin et al. (2011)

[a] *n.d.—not defined.*

(Ju and Afolabi, 1999; Lambertz et al., 2014; Rollin et al., 2011; Shahriarinour et al., 2011; Sukumaran et al., 2005). To overcome all these technological setbacks, current research on saccharification is based on the employment of thermozymes owing to their effective catalysis, better specific activity and diffusion rate, thereby, aid in overall process economics (Anbar et al., 2012; Arora et al., 2014a,b; Wu and Arnold, 2013).

Finally, the reducing sugars so obtained are fermented by ethanol fermenting microorganisms. Conventional microorganisms for bioethanol production are *Zymomonas mobilis*, *Saccharomyces cerevisiae*, and some genetically engineered microorganisms that have a narrow range of substrate utilization at ambient temperature range (28–37°C). Moreover, these microbes have lesser growth rates, cannot ferment both pentose and hexose sugars, and cannot tolerate high sugar concentration. On the contrary, thermotolerant/thermophilic microorganisms have a wide range of substrate, better yield, and productivity, and have lesser cost in terms of bioprocessing of LCB to bioethanol (Arora et al., 2015a).

9.3 FEEDSTOCKS

Generally, LCBs are composed of three major fractions, namely, cellulose 38%–50%, hemicelluloses 23%–32% and lignin 15%–30% (Arora et al., 2015a). The fermentable sugars are obtained from cellulose and hemicelluloses. The breakdown of lignin releases large numbers of polyphenols. LCB is abundantly available as forestry residues, agricultural waste, urban and industrial waste, and several energy crops. Table 9.2 shows the composition of different LCBs along with their theoretical ethanol production potential. The residues obtained from the forests, include prunings, dead trees, hardwood, softwood, wood chips, and slashes (Prasad et al., 2007).

Wastes from agriculture include wheat straw, paddy straw, groundnut shell, alfalfa fiber, corn stover, sunflower hulls, cotton stalks, sunflower stalks, sugarcane bagasse, palm bagasse, barley hulls, wheat bran, corn-cobs, rice husks and so on (Carvalho et al., 2009; Kim et al., 2008; Kumar et al., 2008; Wang et al., 2009). Industrial and urban waste generally contain vegetable processing waste, fruit processing waste, food processing waste, cotton linters, processing papers, pulps, cotton linters, and household waste (Dominguez-Bocanegra et al., 2015; Gupta and Verma, 2015). In addition, energy crops for bioethanol production include sweet sorghum, cassava, and switchgrass (Basavaraj et al., 2013; Zhang et al., 2013).

9.4 PRETREATMENT

The foremost objective of pretreatment is to break the recalcitrance of LCB by releasing lignin from cellulose and hemicellulose, tumbling the crystalline structure of cellulose so that hydrolytic enzymes can easily attack the cellulosic and hemicellulosic fibers (Behera et al., 2014; Brodeur et al., 2011). Moreover, pretreatment increases the porosity and surface area of the biomass, thereby, altering the whole structure (Zhang et al., 2009). Different types of pretreatment processes may be used in combination to speed up the degradation of recalcitrance of LCB before the process of saccharification (Mirahmadi et al., 2010).

The bioprocessing of LCB to bioethanol can be made efficient and economical by exploiting the diversifying the pretreatment techniques as there is a direct correlation between the elimination of hemicelluloses, lignin, and digestibility of cellulose (Alvira et al., 2010; Carvalheiro and Duarte, 2008; Narayanaswamy et al., 2013). Notionally, the cellulose remains intact after pretreatment while most of the lignin and hemicelluloses are solubilized (Amidon et al., 2011).

9.4.1 Techniques

Techniques for pretreatment can be classified as physical (grinding, milling, etc.), chemical (dilute alkali, acid, oxidizing agents, organic solvents, etc.), physicochemical (steam explosion, wet oxidation, hydrothermolysis, etc.), biological, or a permutation of these techniques (Alvira et al., 2010; Chiaramonti et al., 2012). Various pretreatment techniques are described in the next sections.

9.4.1.1 Physical Pretreatment

This pretreatment technique increases the pore size and surface area of biomass, limits the degree of polymerization and reduces the crystallinity of cellulose (Harmsen et al., 2010). Further, hydrolysis of biomass can be enhanced by various other processes like grinding, irradiating, milling, and chipping. Even though, size reduction of biomass during pretreatment is a costly and energy intensive step but overall it is considered to be economical (Hendriks and Zeeman, 2009; Zhu et al., 2010).

TABLE 9.2 Composition of Potential Lignocellulosic Raw Materials for Ethanol Production

LCB	% Dry weight basis						Theoretical ethanol yield (gallons per ton of dry feedstock)[a]	References
	Glucose	Mannose	Galactose	Xylose	Arabinose	Lignin		
Wheat straw	36.6	0.8	2.4	19.2	2.4	14.5	107.0	Lee (1997)
Corn fiber	37.19	NA[b]	3.59	17.58	11.25	7.78	124.2	Saha et al. (1998)
Poplar	39.8	2.4	0.0	14.8	1.2	29.1	101.2	Chung et al. (2005)
Sugarcane bagasse	38.1	NA	1.1	23.3	2.5	18.4	113.3	Lee (1997)
Corn stover	36.1	1.8	2.5	21.4	3.5	17.2	113.8	Laureano-Perez et al. (2005)
Switchgrass	32.2	0.4	0.0	20.3	3.7	23.2	98.7	Chung et al. (2005)

[a] *Theoretical ethanol yield calculated by theoretical ethanol yield calculator.*
[b] *Not available.*

Besides size reduction, an extruder system could be utilized for extrusion, which is quite beneficial owing to its potential of heat and mass transfer with high shear (Karunanithy and Muthukumarappan, 2010).

9.4.1.2 Chemical Pretreatment

One of the promising substitutes for increasing the degradation of biomass is pretreatment with the help of chemicals (Agbor et al., 2011). Chemical pretreatment is extensively employed in the pulp and paper industry for delignification process (Zhang et al., 2009). This pretreatment technique employs various alkalis, acids, salts, oxidizing agents, and so on. Besides, several organic acids, that is, oxalic acid, salicylic acid, and acetylsalicylic acid have also been employed as catalysts (Singh et al., 2011). On the other hand, concentrated acids are usually evaded to avoid corrosion of the reactor and problems during downstream processing (Tabil et al., 2011).

9.4.1.3 Physicochemical Pretreatment

Physicochemical pretreatment is a combination of physical and chemical processes for degradation of recalcitrant structure of LCB, and thereby aids in the hydrolysis of LCB (Hendriks and Zeeman, 2009). Various techniques involved in physicochemical pretreatment include organosolv process, steam explosion, ammonia fiber explosion (AFEX), wet oxidation, carbon-dioxide explosion, and ammonia recycle percolation (ARP) (Brodeur et al., 2011). Recently, pretreatment has been focused on the use of ionic liquids and cellulose solvent-based fractionation techniques (Sathitsuksanoh et al., 2012).

9.4.1.4 Biological Pretreatment

Biological pretreatment involves the action of various enzymes obtained from different microorganisms, particularly fungi, for degradation of hemicelluloses and lignin fractions of LCB. Biological pretreatment has a better score over physical and chemical pretreatments in terms of higher specificity, higher product yield, lesser energy requirement, and generation of nontoxic components (Narayanaswamy et al., 2013; Saritha and Arora, 2012). On the other hand, its major setback is the establishment of aseptic conditions for microorganisms and it is a slow process. Besides, several lignin consuming microbes also guzzle hemicellulosic and cellulosic fractions present in LCB (Kuhar et al., 2008). Considering the technoeconomic challenges of biological pretreatment, it is not exploited at industrial level. However, research efforts are being made in this field for the development of safe, sustainable, and eco-friendly processes (Ma and Liu, 2010). Various delignifying microorganisms include *Phanerochaete chrysosporium*, *Ceriporiopsis subvermispora*, *Cyathus stercolerus*, *Ceriporia lacerate*, *Pleurotus ostreaus*, and *Pycnoporus cinnarrbarinus* (Saritha and Arora, 2012; Vijaya and Reddy, 2012).

9.4.2 Parameters for Pretreatment

Several parameters that influence the complexity of the pretreatment include biomass crystallinity, degree of polymerization (DP), surface area accessibility, hemicellulose and lignin content. These are described in the next sections.

9.4.2.1 Biomass Crystallinity

Lignocellulosic components, namely, cellulose, hemicellulose, and lignin are allied to each other by varying degrees and subjective relative composition depending on the species, source, and type of biomass (Carere et al., 2008; Chandra et al., 2007; Mckendry, 2002). The crystallinity of cellulose makes the biomass highly resistant to enzymatic and chemical attack and thus, reduces the biomass hydrolysis. The dense network of hydrogen bonds in the crystalline cellulose makes it highly resistant to enzymatic attack. On the other hand, the hydrophobic side of the crystalline cellulose fibrils makes it less susceptible to chemical hydrolysis (Fan et al., 1980; Himmel et al., 2007; Nishiyama et al., 2002).

9.4.2.2 Degree of Polymerization (DP)

Degree of polymerization (DP) refers to the number of monomeric units present in the polymer and it strongly determines the physical properties of biomass (Chang and Holtzapple, 2000). During enzymatic attack, the endoglucanases are principally responsible for reducing the DP (Mansfield et al., 1999). Moreover, the removal of xylan has a significant role in reducing the DP of crystalline cellulose as compared to lignin removal during pretreatment of biomass (Kumar et al., 2009a).

9.4.2.3 Accessible Surface Area

Efficient enzymatic hydrolysis takes place by physical contact between the cellulosic fibrils and the enzymatic subunits (Fan et al., 1980; Sun and Cheng, 2002; Wyman, 1996). LCB have two accessible surfaces, namely, internal and external, where the internal is related to the capillary pores of the cellulosic fibrils and external depends on the shape and size of fibrils (Taherzadeh and Karimi, 2008). For efficient enzymatic hydrolysis, the external area should be more than internal area for the prevention of enzyme entrapment in the internal area (Zhang and Lynd, 2004). Moreover, the accessibility of the enzymes to the surface area can be improved by increasing the pore size of the cellulose via hemicelluloses removal (Gregg and Saddler, 1996; Grethlein, 1985; Palonen et al., 2004). Furthermore, drying of pretreated LCB reduces the accessibility of the enzymes as the pore structure gets collapsed (Grous et al., 1986).

9.4.2.4 Lignin and Hemicelluloses Contents

The LCB is recalcitrant owing to the content and distribution of lignin on the cellulosic and hemicellulosic fibers, thereby, limiting the accessibility of the enzymes and thus, reduces the enzymatic saccharification (Chang and Holtzapple, 2000; Lynd et al., 2002; Perez et al., 2002; Taherzadeh and Karimi, 2007). Thus, delignification of LCB aids the rate of enzymatic saccharification (Esteghlalian et al., 2001; Jeoh et al., 2007).

Hemicelluloses provide more rigidity to the complex network of LCB and are sensitive to temperature, moisture content, pH, and chemicals (Laureano-Perez et al., 2005). The type of pretreatment should be warily designed to prevent the degradation of hemicelluloses to fermentation inhibitors like furfural and hydroxymethyl furfural (HMF) (Palmqvist and Hahn-Hägerdal, 2000a,b).

9.4.3 Feasibility at Industrial Scale

Pretreatment of LCB seems to have the highest capital and operational cost among the three critical steps of bioethanol production (Goh et al., 2010; Mosier et al., 2005). Different pretreatment processes, namely, physical, chemical, physicochemical, and biological, may be selected depending on the composition and location of a particular feedstock (Mussatto et al., 2010). However, apart from the type of feedstock, the selective pretreatment technique also depends on the conversion process. For a particular LCB, best pretreatment strategy can be chosen on the basis of its effect on overall economy (Aden et al., 2002). Costly equipment are required for some pretreatment techniques, such as AFEX and steam explosion, whereas milling requires relatively cheaper equipment (Cardona and Sanchez, 2007). Besides, another major cost involving part of pretreatment is the use of chemicals. Acquisition of acids and bases for various pretreatments and subsequent neutralization further aids in the overall cost. Moreover, the expenditure on oxygen and power utilization for oxygen compressor leads to costly wet oxidation process (Littlewood et al., 2013). Furthermore, the retention time for pretreatment also aids in the overall cost of the process. Table 9.3 shows the effect of different pretreatment processes on LCB along with the required accessories.

TABLE 9.3 Responses of Various Pretreatment Techniques Along With the Equipments Required

Pretreatment type	Response of pretreatment	Equipment/others
Milling	Increases surface area; reduces power requirements	Hammermill; ball mill; two-roll mill; colloid mill; vibratory ball mill
Mechanical	Destroys cell structure; decreases crystallinity of LCB	Feeders; conveyors; separators; hoppers; washer; chipper
Dilute alkali	Solublizes and extracts lignin; separates linkages between lignin and carbohydrates; removes acetyl groups of hemicellulose	Pretreatment reactor; Pneumapress filter; shredder; screw conveyors; tanks; dilute alkali
Dilute acid	Hydrolysis of hemicelluloses to soluble sugars; alters lignin structure	Pretreatment reactor; blowdown and conditioning tanks; pneumapress filter; shredder; screw conveyors; percolation, plug flow, shrinking-bed, batch and countercurrent reactors; dilute acid
Ion exchange	Removal of acetic acid, H_2SO_4, phenolic compounds	ISEP continuous ion exchange Reactor; weak base amberlist A20, regenerated with ammonia
Lime	No inhibitor formation; aids in detoxification; removes amorphous substances, such as lignin; removes acetyl groups from hemicelluloses	Overliming tank; hydrocyclone, rotary filter
Liquid hot water	88%–98% recovery of xylose; environmentally attractive; water itself acts as dilute acid at high temperatures; hydrolysis of hemicellulose to oligomers, acetic acid and other acids; avoid the formation of inhibitors; enhances enzymatic digestibility	Plug-flow pretreatment reactor; recycled hot water from the distillation column bottoms
Steam explosion with H_2SO_4 catalyst (SECA)	Effective on feedstock with large particle sizes with lesser energy requirement; improve the enzymatic hydrolysis; decrease the production of inhibitory compounds; complete removal of hemicelluloses	Pretreatment reactor; stream of dilute sulfuric acid; blow down tank, conditioning tank, ammonia for neutralization
Steam explosion without catalyst	Higher solubility of hemicellulose; water itself acts as dilute acid at high temperatures; rapid pressure release breakdown the biomass structure; lower environmental impact; lower capital investment; more potential for energy efficiency; complete sugar recovery	Steam explosion reactor; high pressure steam; blow down tank; flash tank; vapor compressor
Steam explosion and solvent pretreatment	Xylose production; delignification of biomass	SSF reactor, vacuum cycling
Wet oxidation	Solubilization of hemicelluloses and lignin; increased digestibility of cellulose; lesser inhibitor formation; minimizes the energy demand	Pretreatment reactor, oxygen or air under pressure
Ammonia fiber explosion (AFEX)	Disruption of lignin carbohydrate linkages and reduction in lignin content; deacetylation and removal of hemicellulose; decrystallization of cellulose; increased digestibility of biomass; low formation of inhibitors	AFEX reactor; liquid ammonia; blow down tank; flash drum
Ammonia recycle percolation (ARP)	Solubilization of hemicellulose; formation of short-chained cellulosic material with high glucan content	Flow-through column reactor; crystallizer

9.5 SACCHARIFICATION

Saccharification of LCB is dependent on a number of parameters, including cellulose crystallinity, accessibility of active sites on cellulose fibers, DP, and total amount of lignin present. Hydrolysis of LCB is carried out by two enzymes, namely, cellulases and hemicellulases, which are discussed in the next sections.

9.5.1 Cellulose Hydrolysis

Cellulases can be categorized into three main types, namely, endoglucanases, which cleaves the internal bonds of the molecule; exoglucanase, which attacks at the ends of the chain and β-glucosidase, which catalyses the cleavage of cellobiose to release glucose moieties (Arora et al., 2015b; Kumar et al., 2008; Prasetyo et al., 2011; Ray and Behera, 2016). Cellulase-producing microorganisms are either aerobic or anaerobic.

9.5.1.1 Aerobic Microorganisms

Among the aerobic microorganisms, cellulolytic activity is mostly found in *Actinomycetales. Actinomycetales* are aerobic microorganisms, which release extracellular enzymes and commence hydrolysis to release cellodextrins from the cellulose fiber and further oxidized to carbon dioxide and water inside the cell (Horn et al., 2012; Lynd et al., 2002).

9.5.1.2 Anaerobic Microorganisms

Among the anaerobic microorganisms, cellulolytic activity is mostly found in *Clostridiales*. Anaerobic microorganisms form cellulosomes for adherence with the cellulose fibers, thereby, releasing cellulolytic enzymes. The hydrolyzed molecules released from the cellulose fiber undergo fermentation. The products of fermentation may also be used for cross-feeding other noncellulolytic microorganisms. Apart from the cellulosomes, adhesion to the cellulose fibers can also be accomplished by carbohydrate epitope of glycocalyx, pili, or fimbriae (Sadhu and Maiti, 2013).

9.5.2 Hemicellulose Hydrolysis

The second abundant polymer in LCB is hemicelluloses, which is mainly composed of xylan. Hemicelluloses should also be exploited either for bioethanol or for other value-added products for economical bioprocessing. Hydrolysis of xylan can be done by several enzymes mainly, endoxylanase, which cleaves the xylan backbone to release xylooligosaccharides; β-xylosidase, which degrades the xylooligosaccharides and various glycosidases including α-4-O-methyl glucuronosidases, *P*-coumaric acid and ferulic, acetyl xylan esterase, α-1-arabinofuranosidase, which degrade the side groups in the xylan polymer (Arora et al., 2015b; Kumar et al., 2008).

Generally, hemicellulases are found in the order *Thermoanaerobacteriales* (Lynd et al., 2008). Although many studies have been reported for the isolation and characterization of hemicellulases, yet, hemicellulolytic systems are lesser exploited and understood. Therefore, more studies need to be done to investigate the regulation mechanism of hemicellulases.

9.6 FERMENTATION

Conventionally fermentation is carried out by *S. cerevisiae*, *Z. mobilis*, and some genetically engineered microbes. However, the problems with such microbes are the high cost of converting biomass to sugars, inability to ferment pentoses, lesser tolerance to high sugar concentration. Moreover, many genetically engineered microorganisms are very unstable by nature. To overpower these technological barriers, recent fermentation studies have been carried out using extremophilic microorganisms (Arora et al., 2015c).

9.6.1 Thermotolerant/Thermophilic Ethanologens

On the basis of optimal growth temperature, thermophiles can be classified into three categories: moderate (50–64°C), extreme (65–79°C), and hyper (above 80°C) (Eram and Ma, 2013). Longer half-life and high transition temperature allows the microorganisms to remain stable at high temperatures (Satyanarayana et al., 2013).

Thermotolerant/thermophilic ethanologens have the potential to ferment both cellulosic and hemicellulosic fractions of LCB. A diverse range of thermophilic/thermotolerant ethanologens have been reported in the literature, which include *Clostridium acetobutylicum*, *C. thermosulfurogenes*, *C. thermosaccharolyticum*, *C. thermohydrosulfurium*, *C. tetani*, *Kluyveromyces marxianus*, *Thermoanaerobacterium saccharolyticum*, *Aeromonas hydrophila*, *Thermoanaerobacter ethanolicus*, *Geobacillus* sp., *Erwinia* sp. (Chandel et al., 2011; Shaw et al., 2008). Some of the thermoethanologenic microorganisms are discussed in the next sections.

9.6.1.1 *Clostridium* Species

Clostridium belonging to family *Clostridiaceae* and order *Clostridiale* is the oldest known thermophilic ethanologen. Special features of these species include cofermentation of both pentose and hexose sugars, high ethanol yield, cellulolytic activity, and acquiescent to genetic modification. However, the culturing of *Clostridium* species requires anaerobic environment (Argyros et al., 2011; Orlygsson, 2012).

9.6.1.2 *Thermoanaerobacterium* Species

Thermoanaerobacterium species can ferment both hexoses and pentoses to ethanol with relatively high substrate affinity and high ethanol yield. Apart from this, it also possesses vitamin B12 biosynthesis pathway. However, major limitations with these species include accumulation of by-products, rapid cell lysis, and lower ethanol yield. Moreover, the physiology of this culture is less explored (Chang and Yao, 2011; Orlygsson et al., 2010).

9.6.1.3 *Thermoanaerobacter* Species

These species are reported to be the promising species due to its several characteristics including alcohol dehydrogenases (ADH), complete vitamin B12 biosynthesis pathway, modifications to pentose metabolism, xylose transporters, and low ethanol tolerance (Orlygsson et al., 2010; Tomas et al., 2013).

9.6.1.4 *Kluyveromyces* Species

Kluyveromyces species have unique characteristics including fermentation of hexose and pentose sugars, reduced cost of cooling and contamination, insensitivity to catabolite repression, submissive to genetic modification. However, a major lacuna with these species is that they form xylitol along with ethanol during xylose fermentation and it has low tolerance to ethanol (Oda et al., 2010; Rodrussamee et al., 2011).

9.6.1.5 *Pichia* Species

These species are widely known for the fermentation of pentose sugars. Moreover, they possess cellulase enzyme system. However, they cannot tolerate more than 40 g/L ethanol concentration and are highly sensitive to chemical inhibitors. Besides this, the ethanol formation rate is very low and it reassimilates the formed ethanol (Buaban et al., 2010; Dhaliwal et al., 2011; Dubey et al., 2012; Kitagawa et al., 2010; Kwon et al., 2011; Lin et al., 2011; Shields and Boopathy, 2011; Shupe and Liu, 2012).

9.6.2 Metabolic Pathway

Hexose sugars, namely, glucose, galactose, mannose are metabolized by glycolytic pathway, while, pentose sugars, namely, xylose, arabinose are undertaken pentose phosphate pathway (PPP) (Chang and Yao, 2011). However, bacteria utilize isomerase pathway, for converting xylose to xylitol whereas yeast and fungi undergo reductase pathway where xylose is reduced to xylitol and subsequently oxidized to xylulose (Jeffries, 2006; Margeot et al., 2009). The intermediates produced in PPP [glyceraldehyde-3-P (G3P) and fructose-6-P (F6P)] enter into the glycolytic pathway for the production of ethanol (Dellomonaco et al., 2010; Kuhad et al., 2011) (Fig. 9.2).

Pyruvate is a key intermediate in the metabolism of sugars to ethanol. However, in case of thermophilic/thermotolerant microorganisms, there is formation of acetyl CoA by pyruvate ferridoxin oxidoreductase or pyruvate formate lyase, which is further reduced to acetaldehyde (Eram and Ma, 2013). On the other hand, mesophilic microorganisms possess pyruvate decarboxylase enzyme, which involves direct conversion of pyruvate to acetaldehyde. Further, acetaldehyde in both mesophilic and thermophilic microorganisms is reduced to ethanol by ADH.

9.6.3 Process Configurations

Different steps for bioethanol production, namely, pretreatment, saccharification, hexose fermentation, pentose fermentation may be integrated for economical production of bioethanol (Dien et al., 2006; Sun and Cheng, 2002). Most commonly used integrated processes are

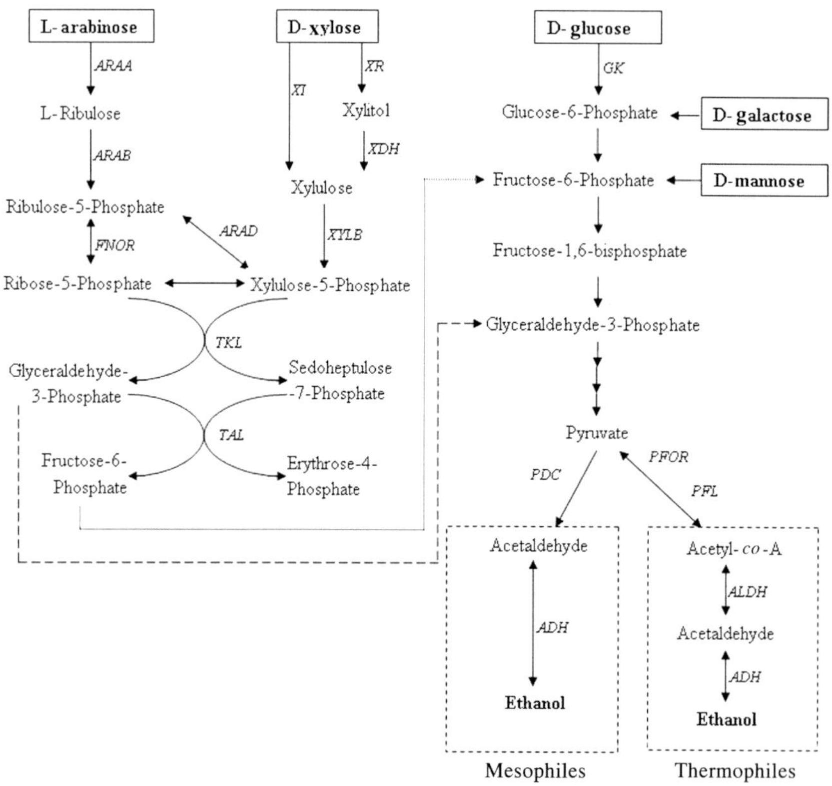

FIGURE 9.2 **Carbohydrate metabolism in microorganisms for ethanol production.** *Source: Arora, R., Behera, S., Kumar, S., 2015a. Bioprospecting thermophilic/thermotolerant microbes for production of lignocellulosic ethanol: a future perspective. Renew. Sust. Energy Rev. 51, 699–717; Chang, T., Yao, S., 2011. Thermophilic, lignocellulolytic bacteria for ethanol production: current state and perspectives. Appl. Microbiol. Biotechnol. 92, 13–27.*

separate hydrolysis and fermentation (SHF), simultaneous saccharification and fermentation (SSF), simultaneous saccharification and cofermentation (SSCF), and consolidated bioprocessing (CBP).

Viola et al. (2013) employed a two-chamber bioreactor for bioethanol production, which resulted in a 20% increase in the ethanol yield as compared to SSF, which was carried out in a single reactor. Moreover, lignin got confined in one chamber making the product recovery easier. Besides, SHF, SSF configuration has been browbeaten in the recent years. Lin et al. (2013) evaluated ethanol production by employing SSF using *K. marxianus* and commercial Accellerase 1000 at 37, 42, and 45°C, which resulted in 85.1%, 92.2%, and 76.2% of the theoretical ethanol yield, respectively. Cha et al. (2015) reported ethanol production from *Miscanthus* by using *S. cerevisiae* mbc 2, which resulted in 15.3 g/L of ethanol at 42°C after 48 h.

CBP is the finest configuration, which integrates all the steps of bioethanol production, involving an engineered microorganism capable of saccharification and cofermentation with high ethanol yield (Hasunuma and Kondo, 2012). Nevertheless, this configuration is not completely developed at commercial level owing to long fermentation period, poor ethanol yield, low tolerance to ethanol, and formation of fermentation inhibitors (Zaldivar et al., 2001). Sakamoto et al. (2012) employed CBP configuration by using genetically modified *S. cerevisiae* for the conversion of hemicellulosic fraction of paddy straw to ethanol, which resulted in 82% of the theoretical ethanol yield based on the polysaccharides present.

9.7 CHALLENGES

The lignocellulosic bioethanol production suffers from several confronts. An efficient system for biomass collection and its storage is very critical for the establishment of the commercial plant. Moreover, a strong policy is needed for the collection and segregation of organic waste as well as blending of bioethanol. Besides, the technological gaps need to be filled up for the production of ethanol with high yield and valorization of by-products for process economy.

9.7.1 Cofermentation

Cofermentation of both pentose and hexose sugars is the major challenge for bioethanol production at commercial scale (Kazi et al., 2010). Many approaches are being used for the development of efficient pentose fermenting microorganism. One such approach is the evolutionary adaptation where the microorganism is adapted to grow in a xylose containing medium (Sharma et al., 2016a,b; Wisselink et al., 2009). Besides this, microorganisms may be subjected to mutation and genetic engineering before adaptation for increasing the growth rate (Liu and Hu, 2010). Many researchers have introduced the genes for xylose fermentation, namely, xylose reductase, xylose isomerase, xylitol dehydrogenase from native xylose fermenting microorganisms into conventional ethanologens (Martin et al., 2007; Tomas-Pejo et al., 2008). Even though momentous work has been done for biochemical conversion of pentose sugars to bioethanol, yet, commercialization is still a challenge.

9.7.2 Sensitivity to Products and Inhibitors

Another setback in the commercialization of bioethanol is the low tolerance of the ethanologens to the product and inhibitors formed during process. Product inhibition begins at low concentration at 5%, which decreases the rate of fermentation. The cell viability of the culture is also reduced at higher concentrations of product and inhibitors produced.

However, the microorganisms may be adapted to a wide range of fermentation conditions and inhibitors like furfurals, hydroxymethyl furfurals, and organic acids are produced (Parawira and Tekere, 2011). Ibraheem and Ndimba (2013) described the adaptation mechanism at molecular level, which includes maintenance of integrity of the cell membrane, activation, and regulation in response to stress, substrate-and-inhibitor-specific degradation pathways, and maintenance of intracellular pH homeostasis.

Zhu et al. (2009) reported an escalation in ethanol yield and productivity by using adapted cells of *Pichia stipitis* and *Candida shehatae* in the presence of inhibitors. Furthermore, Agbogbo et al. (2008) evaluated higher ethanol production (19.4 g/L) by using adapted cells of *P. stipitis* CBS 6054 on solid agar than liquid medium (18.4 g/L) and unadapted cells (16.3 g/L).

9.7.3 Genetic Engineering

To overcome the challenges posed in the commercialization of bioethanol production, namely low ethanol yield and productivity, narrow substrate range, less tolerance to high temperature, inhibitors and products, several approaches are being made to modify the microorganisms at molecular level. Three pyruvate assimilating enzymes, namely, pyruvate ferridoxin oxidoreductase, pyruvate dehydrogenase, and pyruvate decarboxylase have been modified in thermophilic microorganisms for >90% of the theoretical yield (Olson et al., 2015). In another study, genetically engineered strains of *T. saccharolyticum* and *C. thermocellum* resulted in 38 g/L cellulosic ethanol in 146 h (Argyros et al., 2011).

Further, molecular approaches have been made in shrinking the carbohydrate metabolizing pathway to lesser reactions by using gene knockout mutations in genetically engineered *Escherichia coli* for both pentose and hexose fermentations to ethanol with high yield (Clomburg and Gonzalez, 2010; Trinh and Srienc, 2009; Trinh et al., 2008). Watanabe et al. (2010) introduced a mutant strain of *Candida glabrata* (Cgrd1) to SSF at high temperature, where 17.0 g/L of ethanol was produced in 48 h, with a theoretical yield of 66.6% and ethanol productivity of 0.35 g/L/h.

Studies have also been carried out to display the cellulolytic enzymes (EGII from *Tricoderma reesei* and BGL1 from *Aspergilus aculeatus*) on the surface of cells of genetically engineered *K. marxianus*, which resulted in an ethanol concentration of 20.4 g/L at 45°C in CBP configuration (Yanase et al., 2010). For the improvement of pentose metabolism, enzymes of PPP have been inserted in ethanol red yeast GS1.11–26, which exhibits complete utilization of 35 g/L of xylose in 17 h with xylose utilization rate of 1.1 g/L/h (Demeke et al., 2013).

9.8 ECONOMIC ASSESSMENT

The process economics and environment friendly process are the critical factors for the establishment of a technology at commercial scale (Margeot et al., 2009). However, bioethanol production from lignocelluloses is quite exigent due to expensive feedstocks and technological barriers described above (Balat, 2011; Goh et al., 2010). Cost estimation of lignocellulosic ethanol process is quite cumbersome as it involves several pretreatment techniques and conversion processes.

According to Hamelinck et al. (2005) the process economics for lignocellulosic bioethanol can be estimated by the total annual costing of each system per quantity of bioethanol produced. However, among several steps of bioethanol production, the pretreatment process has the highest capital and operational cost (Goh et al., 2010; Mosier et al., 2005). Finicky economic analysis is indispensable to choose the pretreatment strategy for a specific feedstock, which is entirely dependent on the site as well as accessibility (Aden et al., 2002; Mussatto et al., 2010). Further, by employing the integrated processes and utilization of effective saccharifying enzymes and high yielding ethanologens, the overall process can be made economical (Dien et al., 2006; Sun and Cheng, 2002).

Several thermotolerant/thermophilic microorganisms, including *C. thermohydrosulfuricum*, *K. marxianus*, *T. ethanolicus*, have been studied for their potential to produce biethanol (Balat, 2011). However, these ethanologens release heat during growth and fermentation, which may be utilized to uphold the completely insulated reactor temperature, and therefore, aids in overall process economics (Kumar et al., 2013).

9.9 CONCLUDING REMARKS

Lignocellulosic feedstocks proffer mammoth prospective for biofuel production, especially liquid biofuels, which allow having safe, clean, and convenient bioprocess. However, certain technological fissures need to be filled through several approaches at the molecular level. Improved expression systems for enzymatic saccharification and better strains to endure cofermentation are in progress. Further, metabolic engineering techniques allow the development of stout strains. The approaches of metabolic flux analysis help in better understanding of the physiological systems for boosting of ethanol yields and productivities from different feedstocks.

Acknowledgments

Richa Arora is very thankful to Sardar Swaran Singh National Institute of Bio-Energy, Kapurthala for providing Senior Research Fellowship. Authors are also gratefully acknowledged the Ministry of New and Renewable Energy, Govt. of India for providing financial supports to carry out the research activities.

References

Aden, A., Ruth, M., Ibsen, K., Jechura, J., Neeves, K., Sheehan, J., Wallace, B., Montague, L., Slayton, A., Lukas, J., 2002. Lignocellulosic Biomass to Ethanol Process Design and Economics Utilizing Co-Current Dilute Acid Prehydrolysis and Enzymatic Hydrolysis of Corn Stover. Biotechnology Center for Fuels and Chemicals, Golden, CO, United States, NREL/TP-510-32438.

Agbogbo, F.K., Haagensen, F.D., Milam, D., Wenger, K.S., 2008. Fermentation of acid pretreated corn stover to ethanol without detoxification using *Pichia stipitis*. Appl. Biochem. Biotechnol. 145, 53–58.

Agbor, V.B., Cicek, N., Sparling, R., Berlin, A., Levin, D.B., 2011. Biomass pretreatment: fundamentals toward application. Biotechnol. Adv. 29, 675–685.

Almarsdottir, A.R., Sigurbjornsdottir, M.A., Orlygsson, J., 2012. Effects of various factors on ethanol yields from lignocellulosic biomass by *Thermoanaerobacterium* AK17. Biotechnol. Bioeng. 109, 6686–6694.

Alvira, P., Tomas-Pejo, E., Ballesteros, M., Negro, M.J., 2010. Pretreatment technologies for an efficient bioethanol production process based on enzymatic hydrolysis: a review. Bioresour. Technol. 101, 4851–4861.

Amidon, T.E., Bujanovic, B., Liu, S., Howard, J.R., 2011. Commercializing biorefinery technology: a case for the multi-product pathway to available biorefinery. Forests 2, 929–947.

Anbar, M., Gul, O., Lamed, R., Sezerman, U.O., Bayer, E.A., 2012. Improved thermostability of *Clostridium thermocellum* endoglucanase Cel8A using consensus-guided mutagenesis. Appl. Environ. Microbiol. 287, 9213–9221.

Argyros, D.A., Tripathi, S.A., Barrett, T.F., Rogers, S.R., Feinberg, L.F., Olson, D.G., Foden, J.M., Miller, B.B., Lynd, L.R., Hogsett, D.A., Caiazza, N.C., 2011. High ethanol titers from cellulose by using metabolically engineered thermophilic, anaerobic microbes. Appl. Environ. Microbiol. 77, 8288–8294.

Arora, R., Behera, S., Kumar, S., 2014a. Comparative study of fermentation efficiency for bio-ethanol production by isolates. In: Kumar, S., Sarma, A.K., Tyagi, S.K., Yadav, Y.K. (Eds.), Recent Advances in Bio-energy Research. SSS-NIRE, Kapurthala, India, pp. 149–155.

Arora, R., Behera, S., Kumar, S., 2015a. Bioprospecting thermophilic/thermotolerant microbes for production of lignocellulosic ethanol: a future perspective. Renew. Sust. Energy Rev. 51, 699–717.

Arora, R., Behera, S., Kumar, S., 2015b. Bioprospecting thermostable cellulosomes for efficient biofuel production from lignocellulosic biomass. Bioresour. Bioprocess 2, 12.

Arora, R., Behera, S., Sharma, N.K., Kumar, S., 2015c. A new search for thermotolerant yeasts, its characterization and optimization using response surface methodology for ethanol production. Frontiers Microbiol. 6, 889.

Arora, R., Behera, S., Sharma, N.K., Singh, R., Yadav, Y.K., Kumar, S., 2014b. Biochemical conversion of rice straw (*Oryza sativa* L) to bioethanol using thermotolerant isolate K. marxianus NIRE-K3. In: Sharma, N.R., Thakur, R.C., Sharma, M., Parihar, L., Kumar, G. (Eds.), In: Proceedings of Exploring Basic and Applied Sciences for Next Generation Frontiers. Elsevier, New Delhi, pp. 143–146.

Balat, M., 2009. Bioethanol as a vehicular fuel: a critical review. Energy Source Part-A 31, 1242–1255.

Balat, M., 2011. Production of bioethanol from lignocellulosic materials via the biochemical pathway: a review. Energy Convers. Manage. 52, 858–875.

Basavaraj, G., Rao, P.P., Basu, K., Reddy, Ch.R., Kumar, A.A., Rao, P.S., Reddy, B.V.S., 2013. Assessing viability of bioethanol production from sweet sorghum in India. Energy Policy 56, 501–508.

Behera, S., Arora, R., Nandhagopal, N., Kumar, S., 2014. Importance of chemical pretreatment for bioconversion of lignocellulosic biomass. Renew. Sust. Energy Rev. 36, 91–106.

Behera, S., Sharma, N.K., Arora, R., Kumar, S., 2016. Effect of evolutionary adaption on xylosidase activity in thermotolerant yeast isolates *Kluyveromyces marxianus* NIRE-K1 and NIRE-K3. Appl. Biochem. Biotechnol. 179, 1143–1154.

Brodeur, G., Yau, E., Badal, K., Collier, J., Ramachandran, K.B., Ramakrishnan, R., 2011. Chemical and physiocochemical pretreatment of lignocellulosic biomass: a review. Enz. Res. 11, 1–17.

Buaban, B., Inoue, H., Yano, S., Tanapongpipat, S., Ruanglek, V., Champreda, V., Pichyangkura, R., Rengpipat, S., Eurwilaichitr, L., 2010. Bioethanol production from ball milled bagasse using an on-site produced fungal enzyme cocktail and xylose-fermenting *Pichia stipitis*. J. Biosci. Bioeng. 110, 18–25.

Cardona, C.A., Sanchez, O.J., 2007. Fuel ethanol production: process design trends and integration opportunities. Bioresour. Technol. 98, 2415–2457.

Carere, C.R., Sparling, R., Cicek, N., Levin, D.B., 2008. Third generation biofuels via direct cellulose fermentation. Int. J. Mol. Sci. 9, 1342–1360.

Carvalheiro, F., Duarte, L.C., 2008. Hemicellulose biorefineries: a review on biomass pretreatments. J. Sci. Ind. Res. 67, 849–864.

Carvalho, L.G., Gomes, A.M., Aranda, D.A.G., Pereira, N., 2009. Ethanol from lignocellulosic residues of palm oil industry. 11th International Conference on Advanced Materials, Rio de Janeiro, Brazil, September 20–25, 2009.

Cha, Y.L., An, G.H., Yang, J., Moon, Y.-H., Yu, G.-D., Ahn, J.-W., 2015. Bioethanol production from *Miscanthus* using thermotolerant *Saccharomyces cerevisiae* mbc 2 isolated from the respiration-deficient mutants. Renew. Energy 80, 259–265.

Chandel, A.K., Chandrasekhar, G., Radhika, K., Ravinder, R., Ravindra, P., 2011. Bioconversion of pentose sugars into ethanol: a review and future directions. Biotechnol. Mol. Biol. Rev. 6, 8–20.

Chandra, R., Bura, R., Mabee, W., Berlin, A., Pan, X., Saddler, J., 2007. Substrate pretreatment: the key to effective enzymatic hydrolysis of lignocellulosics. Adv. Biochem. Eng. Biotechnol. 108, 67–93.

Chang, T., Yao, S., 2011. Thermophilic, lignocellulolytic bacteria for ethanol production: current state and perspectives. Appl. Microbiol. Biotechnol. 92, 13–27.

Chang, V., Holtzapple, M., 2000. Fundamental factors affecting biomass enzymatic reactivity. Appl. Biochem. Biotechnol. 84/86, 5–37.

Chiaramonti, D., Prussi, M., Ferrero, S., Oriani, L., Ottonello, P., Torre, P., Cherchi, F., 2012. Review of pretreatment processes for lignocellulosic ethanol production, and development of an innovative method. Biomass Bioenergy 46, 25–35.

Chung, Y.C., Bakalinsky, A., Penner, M.H., 2005. Enzymatic saccharification and fermentation of xylose-optimized dilute acid-treated lignocellulosics. Appl. Biochem. Biotechnol. 121-124, 947–961.

Clomburg, J.M., Gonzalez, R., 2010. Biofuel production in *Escherichia coli*: the role of metabolic engineering and synthetic biology. Appl. Microbiol. Biotechnol. 86, 419–434.

Dasgupta, D., Suman, S.K., Pandey, D., Ghosh, D., Khan, R., Agrawal, D., Jain, R.K., Vadde, V.T., Adhikari, D.K., 2013. Design and optimization of ethanol production from bagasse pith hydrolysate by a thermotolerant yeast *Kluyveromyces* sp. IIPE453 Using Response Surface Methodology, 2, Springerplus, New York, NY, United States, p. 159.

Dellomonaco, C., Fava, F., Gonzalez, R., 2010. The path to next generation biofuels: successes and challenges in the era of synthetic biology. Microb. Cell Fact. 9, 3.

Demeke, M.M., Dietz, H., Li, Y., Foulquie-Moreno, M.R., Mutturi, S., Deprez, S., Abt, T.D., Bonini, B.M., Liden, G., Dumortier, F., Verplaetse, A., Boles, E., Thevelein, J.M., 2013. Development of a D-xylose fermenting and inhibitor tolerant industrial *Saccharomyces cerevisiae* strain with high performance in lignocellulose hydrolysates using metabolic and evolutionary engineering. Biotechnol. Biofuels 6, 89.

Dhaliwal, S.S., Oberoi, H.S., Sandhu, S.K., Nanda, D., Kumar, D., Uppal, S.K., 2011. Enhanced ethanol production from sugarcane juice by galactose adaptation of a newly isolated thermotolerant strain of *Pichia kudriavzevii*. Bioresour. Technol. 102, 5675–5968.

Dias, M.O.S., Junqueira, T.L., Rossell, C.E.V., Filho, R.M., Bonomi, A., 2013. Evaluation of process configurations for second generation integrated with first generation bioethanol production from sugarcane. Fuel Process. Technol. 109, 84–89.

Dien, B.S., Jung, H.J.G., Vogel, K.P., Casler, M.D., Lamb, J.A.F.S., Iten, L., Mitchell, R.B., Sarath, G., 2006. Chemical composition and response to dilute acid pretreatment and enzymatic saccharification of alfalfa, reed canarygrass and switchgrass. Biomass Bioenergy 30, 880–891.

Dominguez-Bocanegra, A.R., Torres-Munoz, J.A., Lopez, R.A., 2015. Production of bioethanol from agro-industrial wastes. Fuel 149, 85–89.

Dubey, A.K., Gupta, P.K., Garg, N., Naithani, S., 2012. Bioethanol production from waste paper acid pretreated hydrolyzate with xylose fermenting *Pichia stipitis*. Carbohydr. Polym. 88, 825–829.

Eram, M.S., Ma, K., 2013. Decarboxylation of pyruvate to acetaldehyde for ethanol production by hyperthermophiles. Biomolecules 3, 578–596.

Esteghlalian, A.R., Srivastava, V., Gilkes, N., Gregg, D.J., Saddler, J.N., 2001. An overview of factors influencing the enzymatic hydrolysis of lignocellulosic feedstocks. In: Himmel, M.E., Baker, W., Saddler, J.N. (Eds.), Glycosyl Hydrolases for Biomass Conversion. ACS, Washington, DC, pp. 100–111.

Fan, L.T., Lee, Y., Beardmor, D.H., 1980. Mechanism of the enzymatic hydrolysis of cellulose: effects of major structural features of cellulose on enzymatic hydrolysis. Biotechnol. Bioeng. 22, 177–199.

Fisher, M.A., Tullman-Ercek, D., 2013. Change, exchange, and rearrange: protein engineering for the biotechnological production of fuels, pharmaceuticals and other chemicals. Curr. Opin. Biotechnol. 24, 1010–1016.

Frock, A.D., Kelly, R.M., 2012. Extreme thermophiles: moving beyond single-enzyme biocatalysis. Curr. Opin. Chem. Eng. 1, 1–10.

Glithero, N.J., Wilson, P., Ramsden, S.J., 2013. Straw use and availability for second generation biofuels in England. Biomass Bioenergy 55, 311–321.

Goh, C.S., Tan, K.T., Lee, K.T., Bhatia, S., 2010. Bio-ethanol from lignocellulose: status, perspectives and challenges in Malaysia. Bioresour. Technol. 101, 4834–4841.

Gregg, D., Saddler, J.N., 1996. A techno-economic assessment of the pretreatment and fractionation steps of a biomass-to-ethanol process. Appl. Biochem. Biotechnol. 57, 711–727, 58.

Grethlein, H.E., 1985. The effect of pore size distribution on the rate of enzymatic hydrolysis of cellulose substrates. Nat. Biotechnol. 3, 155–160.

Grous, W.R., Converse, A.O., Grethlein, H.E., 1986. Effect of steam explosion pretreatment on pore size and enzymatic hydrolysis of poplar. Enzyme Microb. Technol. 8, 274–280.

Guo, M., Song, W., Buhain, J., 2015. Bioenergy and biofuels: history, status and perspective. Renew. Sust. Energy Rev. 42, 712–725.

Gupta, A., Verma, J.P., 2015. Sustainable bio-ethanol production from agro-residues: a review. Renew. Sust. Energy Rev. 41, 550–567.

Hamelinck, C.N., Van Hooijdonk, G., Faaij, A.P.C., 2005. Ethanol from lignocellulosic biomass: techno-economic performance in short-, middle- and long-term. Biomass Bioenergy 28, 384–410.

Harmsen, P., Huijgen, W., Bermudez-Lopez, L., Bakker, R., 2010. Literature Review of Physical and Chemical Pretreatment Processes for Lignocellulosic Biomass, first ed. Wageningen: Food and Biobased Research, The Netherlands, 857-570.

Hasunuma, T., Kondo, A., 2012. Consolidated bioprocessing and simultaneous saccharification and fermentation of lignocellulose to ethanol with thermotolerant yeast strains. Process Biochem. 47, 1287–1294.

Hendriks, A., Zeeman, G., 2009. Pretreatments to enhance the digestibility of lignocellulosic biomass. Bioresour. Technol. 100, 10–18.

Himmel, M.E., Ding, S.Y., Johnson, D.K., Adney, W.S., Nimlos, M.R., Brady, J.W., Foust, T.D., 2007. Biomass recalcitrance: engineering plants and enzymes for biofuels production. Science 315, 804–807.

Horn, J.H., Vaaje-Kolstad, G., Westereng, B., Eijsink, V.G.H., 2012. Novel enzymes for the degradation of cellulose. Biotechnol. Biofuels 5, 1–12.

Ibraheem, O., Ndimba, B.K., 2013. Molecular adaptation mechanisms employed by ethanologenic bacteria in response to lignocellulose-derived inhibitory compounds. Int. J. Biol. Sci. 9, 598–612.

IEA, 2011. Biofuels for Transport. International Energy Agency, Paris, France.

Jeffries, T.W., 2006. Engineering yeasts for xylose metabolism. Curr. Opin. Biotechnol. 17, 320–326.

Jeoh, T., Ishizawa, C.I., Davis, M.F., Himmel, M.E., Adney, W.S., Johnson, D.K., 2007. Cellulase digestibility of pretreated biomass is limited by cellulose accessibility. Biotechnol. Bioeng. 98, 112–122.

Jessen, J.E., Orlygsson, J., 2012. Production of ethanol from sugars and lignocellulosic biomass by *Thermoanaerobacter* J1 isolated from a hot spring in Iceland. J. Biomed. Biotechnol. 2012, 1–7.

Ju, L.K., Afolabi, O.A., 1999. Wastepaper hydrolysate as soluble inducing substrate for cellulase production in continuous culture of *Trichoderma reesei*. Biotechnol. Prog. 15, 91–97.

Karunanithy, C., Muthukumarappan, K., 2010. Influence of extruder temperature and screw speed on pretreatment of corn stover while varying enzymes and their ratios. Appl. Biochem. Biotechnol. 162, 264–279.

Kazi, F.K., Fortman, J.A., Anex, R.P., Hsu, D.D., Aden, A., Dutta, A., Kothandaraman, G., 2010. Techno-economic comparison of process technologies for biochemical ethanol production from corn stover. Fuel 89, S20–S28.

Kim, T.H., Taylor, F., Hicks, K.B., 2008. Bioethanol production from barley hull using SAA (soaking in aqueous ammonia) pretreatment. Bioresour. Technol. 99, 5694–5702.

Kitagawa, T., Tokuhiro, K., Sugiyama, H., Kohda, K., Isono, N., Hisamitsu, M., Takahashi, H., Imaeda, T., 2010. Construction of β-glucosidase expression system using the multistress-tolerant yeast *Issatchenkia orientalis*. Appl. Microbiol. Biotechnol. 87, 1841–1853.

Klein-Marcuschamer, D., Oleskowicz-Popiel, P., Simmons, B.A., Blanch, H.W., 2012. The challenge of enzyme cost in the production of lignocellulosic biofuels. Biotechnol. Bioeng. 109, 1083–1087.

Kuhad, R.C., Gupta, R., Khasa, Y.P., Singh, A., Percival Zhang, Y.H., 2011. Bioethanol production from pentose sugars: current status and future prospects. Renew. Sust. Energy Rev. 15, 4950–4962.

Kuhar, S., Nair, L.M., Kuhad, R.C., 2008. Pretreatment of lignocellulosic material with fungi capable of higher lignin degradation and lower carbohydrate degradation improves substrate acid hydrolysis and the eventual conversion to ethanol. Can. J. Microbiol. 54, 305–313.

Kumar, R., Mago, G., Balan, V., Wyman, C.E., 2009a. Physical and chemical characterizations of corn stover and poplar solids resulting from leading pretreatment technologies. Bioresour. Technol. 100, 3948–3962.

Kumar, R., Singh, S., Singh, O.V., 2008. Bioconversion of lignocellulosic biomass: biochemical and molecular perspectives. J. Ind. Microbiol. Biotechnol. 35, 377–391.

Kumar, S., Singh, S.P., Mishra, I.M., Adhikari, D.K., 2009b. Recent advances in production of bioethanol from lignocellulosic biomass. Chem. Eng. Technol. 32, 517–526.

Kumar, S., Dheeran, P., Singh, S.P., Mishra, I.M., Adhikari, D.K., 2013. Cooling system economy in ethanol production using thermotolerant yeast *Kluyveromyces* sp IIPE453. Am. J. Microbiol. Res. 1, 39–44.

Kumar, S., Dheeran, P., Singh, S.P., Mishra, I.M., Adhikari, D.K., 2015a. Bioprocessing of bagasse hydrolysate for ethanol and xylitol production using thermotolerant yeast. Bioprocess Biosyst. Eng. 38, 39–47.

Kumar, S., Dheeran, P., Singh, S.P., Mishra, I.M., Adhikari, D.K., 2015b. Continuous ethanol production from sugarcane bagasse hydrolysate at high temperature with cell recycle and in-situ recovery of ethanol. Chem. Eng. Sci. 138, 524–530.

Kwon, Y.J., Ma, A.Z., Li, Q., Wang, F., Zhuang, G.Q., Liu, C.Z., 2011. Effect of lignocellulosic inhibitory compounds on growth and ethanol fermentation of newly-isolated thermotolerant *Issatchenkia orientalis*. Bioresour. Technol. 102, 8099–8104.

Lambertz, C., Garvey, M., Klinger, J., Hessel, D., Klose, H., Fischer, R., Commandeur, U., 2014. Challenges and advances in the heterologous expression of cellulolytic enzymes: a review. Biotechnol. Biofuels 7, 135.

Laureano-Perez, L., Teymouri, F., Alizadeh, H., Dale, B.E., 2005. Understanding factors that limit enzymatic hydrolysis of biomass. Appl. Biochem. Biotechnol. 121, 1081–1099.

Lee, J., 1997. Biological conversion of lignocellulosic biomass to ethanol. J. Biotechnol. 56, 1–24.

Limayem, A., Ricke, S.C., 2012. Lignocellulosic biomass for bioethanol production: current perspectives, potential issues and future prospects. Prog. Energy Combust. 38, 449–467.

Lin, C.W., Wu, C.H., Tran, D.T., Shih, M.C., Li, W.H., Wu, C.F., 2011. Mixed culture fermentation from lignocellulosic materials using thermophilic lignocelluloses-degrading anaerobes. Process Biochem. 46, 489–493.

Lin, Y.S., Lee, W.C., Duan, K.J., Lin, Y.H., 2013. Ethanol production by simultaneous saccharification and fermentation in rotary drum reactor using thermotolerant *Kluveromyces marxianus*. Appl. Energy 105, 389–394.

Littlewood, J., Murphy, R.J., Wang, L., 2013. Importance of policy support and feedstock prices on economic feasibility of bioethanol production from wheat straw in the UK. Renew. Sust. Energy Rev. 17, 291–300.

Liu, E., Hu, Y., 2010. Construction of a xylose-fermenting *Saccharomyces cerevisiae* strain by combined approaches of genetic engineering, chemical mutagenesis and evolutionary adaptation. Biochem. Eng. J. 48, 204–210.

Lynd, L.R., Weimer, P.J., van Zyl, W.H., Pretorius, I.S., 2002. Microbial cellulose utilization: fundamentals and biotechnology. Microbiol. Mol. Biol. Rev. 66, 506–577.

Lynd, R., Currie, D., Ciazza, N., Herring, C., Orem, N., 2008. Consolidated bioprocessing of cellulosic biomass to ethanol using thermophilic bacteria. In: Wall, J.D., Harwood, C.S., Demain, A. (Eds.), Bioenergy. American Society for Microbiology, Washington, DC, pp. 55–73.

Ma, M., Liu, Z.L., 2010. Mechanisms of ethanol tolerance in *Saccharomyces cerevisiae*. Appl. Microbiol. Biotechnol. 87, 829–845.

Mansfield, S.D., Mooney, C., Saddler, J.N., 1999. Substrate and enzyme characteristics that limit cellulose hydrolysis. Biotechnol. Prog. 15, 804–816.

Margeot, A., Hahn-Hagerdal, B., Edlund, M., Slade, R., Monot, F., 2009. New improvements for lignocellulosic ethanol. Curr. Opin. Biotechnol. 20, 372–380.

Martin, C., Marcet, M., Almazan, O., Jonsson, L.J., 2007. Adaptation of a recombinant xylose-utilizing *Saccharomyces cerevisiae* strain to a sugarcane bagasse hydrolysate with high content of fermentation inhibitors. Bioresour. Technol. 98, 1767–1773.

Mathew, A.K., Crook, M., Chaney, K., Humphries, A.C., 2013. Comparison of entrapment and biofilm mode of immobilization for bioethanol production from oilseed rape straw using *Saccharomyces cerevisiae* cells. Biomass Bioenergy 52, 1–7.

Matsushita, Y., Jo, E.K., Inakoshi, R., Yagami, S., Takamoto, N., Fukushima, K., Lee, S.C., 2013. Hydrothermal reaction of sulfuric acid lignin generated as a by-product during bioethanol production using lignocellulosic materials to convert bioactive agents. Ind. Crop Prod. 42, 181–188.

Mckendry, P., 2002. Energy production from biomass (part 1) overview of biomass. Bioresour. Technol. 83, 37–46.

Menon, V., Rao, M., 2012. Trends in bioconversion of lignocellulose: biofuels, platform chemicals & biorefinery concept. Prog. Energy Combust. 38, 522–550.

Mirahmadi, K., Kabir, M.M., Jeihanipour, A., Karimi, K., Taherzadeh, M.J., 2010. Alkaline pretreatment of spruce and birch to improve bioethanol and biogas production. Bioresources 5, 928–938.

Mosier, N., Wyman, C., Dale, B., Elander, R., Lee, Y.Y., Holtzapple, M., Ladisch, M., 2005. Features of promising technologies for pretreatment of lignocellulosic biomass. Bioresour. Technol. 96, 673–686.

Mussatto, S.I., Dragone, G., Guimarães, P.M.R., Silva, J.P.A., Carneiro, L.M., Roberto, I.C., Domingues, T., Teixeira, J.A.,

2010. Technological trends, global market, and challenges of bio-ethanol production. Biotechnol. Adv. 28, 817–830.

Narayanaswamy, N., Dheeran, P., Verma, S., Kumar, S., 2013. Biological pretreatment of lignocellulosic biomass for enzymatic saccharification. In: Fang, Z. (Ed.), Pretreatment Techniques for Biofuels and Biorefineries. Springer Verlag, Berlin, Heidelberg, pp. 3–34.

Nigam, P.S., Singh, A., 2011. Production of liquid biofuels from renewable resources. Prog. Energy Combust. 37, 52–68.

Nishiyama, Y., Langan, P., Chanzy, H., 2002. Crystal structure and hydrogen bonding system in cellulose Iβ from synchrotron X-ray and neutron fiber diffraction. J. Am. Chem. Soc. 124, 9074–9082.

Oda, Y., Nakamura, K., Shinomiya, N., Ohba, K., 2010. Ethanol fermentation of sugar beet thick juice diluted with crude cheese whey by the flex yeast *Kluyveromyces marxianus* KD-15. Biomass Bioenergy 34, 1263–1266.

Olson, D.G., Sparling, R., Lynd, L.R., 2015. Ethanol production by engineered thermophiles. Curr. Opin. Biotechnol. 33, 130–141.

Orlygsson, J., Sigurbjornsdottir, M.A., Bakken, H.E., 2010. Bioprospecting thermophilic ethanol and hydrogen producing bacteria from hot springs in Iceland. Icel. Agric. Sci. 23, 73–85.

Orlygsson, J., 2012. Ethanol production from biomass by a moderate thermophile *Clostridium* AK_1. Icel. Agr. Sci. 25, 25–35.

Palmqvist, E., Hahn-Hagerdal, B., 2000a. Fermentation of lignocellulosic hydrolysates I: inhibition and detoxification. Bioresour. Technol. 74, 17–24.

Palmqvist, E., Hahn-Hägerdal, B., 2000b. Fermentation of lignocellulosic hydrolysates II: inhibitors and mechanisms of inhibition. Bioresour. Technol. 74, 25–33.

Palonen, H., Thomsen, A.B., Tenkanen, M., Schmidt, A.S., Viikari, L., 2004. Evaluation of wet oxidation pretreatment for enzymatic hydrolysis of softwood. Appl. Biochem. Biotechnol. 117, 1–17.

Parawira, W., Tekere, M., 2011. Biotechnological strategies to overcome inhibitors in lignocelluloses hydrolysates for ethanol production: review. Crit. Rev. Biotechnol. 31, 20–31.

Pejin, D., Mijovic, L.C., Vucurovic, V., Pejin, J., Dencic, S., Rakin, M., 2009. Fermentation of wheat and *triticale* hydrolysates: a comparative study. Fuel 8, 625–628.

Perez, J., Dorado, J.M., Rubia, T.D., Martinez, J., 2002. Biodegradation and biological treatment of cellulose, hemicellulose and lignin: an overview. Microbiol. Int. 5, 53–63.

Prasad, S., Singh, A., Joshi, H.C., 2007. Ethanol as an alternative fuel from agricultural, industrial and urban residues. Resour. Conserv. Recyc. 50, 1–39.

Prasetyo, J., Naruse, K., Kato, T., Boonchird, C., Harashima, S., Park, E.Y., 2011. Bioconversion of paper sludge to biofuel by simultaneous saccharification and fermentation using a cellulase of paper sludge origin and thermotolerant *Saccharomyces cerevisiae* TJ14. Biotechnol. Biofuels 4, 35.

Ray, R.C., Behera, S., 2016. Solid state fermentation for production of microbial cellulases. In: Brahmachari, G., Demain, A. (Eds.), Biotechnology of Microbial Enzymes: Production, Biocatalysis and Industrial Applications. Academic Press, Cambridge, MA, pp. 43–79.

REN21, 2014, Renewables 2014 Global Status Report. Renewable Energy Policy Network for the 21st Century. Available from: http://www.ren21.net/portals/0/documents/resources/gsr/2014/gsr2014_full%20report_low%20res.pdf

Rodrussamee, N., Lertwattanasakul, N., Hirata, K., Suprayogi, Limtong, S., Kosaka, T., Yamada, M., 2011. Growth and ethanol fermentation ability on hexose and pentose sugars and glucose effect under various conditions in thermotolerant yeast *Kluyveromyces marxianus*. Appl. Microbiol. Biotechnol. 90, 1573–1586.

Rollin, J.A., Zhu, Z., Sathisuksanoh, N., Zhang, H.P., 2011. Increasing cellulose accessibility is more important than removing lignin: a comparison of cellulose solvent-based lignocellulose fractionation and soaking in aqueous ammonia. Biotechnol. Bioeng. 108, 22–30.

Sadhu, S., Maiti, T.K., 2013. Cellulase production by bacteria: a review. British Microbiol. Res. J. 3, 235–258.

Saha, B.C., Dien, B.S., Bothast, R.J., 1998. Fuel ethanol production from corn fiber: current status and technical prospects. Appl. Biochem. Biotechnol., 70–72, 115–125.

Sakamoto, T., Hasunuma, T., Hori, Y., Yamada, R., Kondo, A., 2012. Direct ethanol. production from hemicellulosic materials of rice straw by use of an engineered yeast strain codisplaying three types of hemicellulolytic enzymes on the surface of xylose-utilizing *Saccharomyces cerevisiae* cells. J. Biotechnol. 158, 203–210.

Saratale, G.D., Oh, S.E., 2012. Lignocellulosics to ethanol: the future of the chemical and energy industry. Afr. J. Biotechnol. 11, 1002–1013.

Saritha, M., Arora, A.L., 2012. Biological pretreatment of lignocellulosic substrate for enhanced delignification and enzymatic digestibility. Ind. J. Microbiol. 52, 122–130.

Sathitsuksanoh, N., Zhu, Z., Zhang, Y.H., 2012. Cellulose solvent and organic solvent-based lignocelluloses fractionation enabled efficient sugar release from a variety of lignocellulosic feedstocks. Bioresour. Technol. 117, 228–233.

Satyanarayana, T., Littlechild, J., Kawarabayasi, Y., 2013. Thermophilic microbes in environmental and industrial biotechnology, 2 Biotechnology of Thermophiles. Springer, New York, London.

Shahriarinour, M., Wahab, M.N.A., Mohamad, R., Mustafa, S., Ariff, A.B., 2011. Effect of medium composition and cultural condition on cellulase production by *Aspergillus terreus*. Afr. J. Biotechnol. 10, 7459–7467.

Sharma, N.K., Behera, S., Arora, R., Kumar, S., 2016a. Enhancement in xylose utilization using *Kluyveromyces marxianus* NIRE-K1 through evolutionary adaptation approach. Bioprocess Biosyst. Eng. 39, 835–843.

Sharma, N.K., Behera, S., Arora, R., Kumar, S., 2016b. Potential role of xylose transporters in industrial yeast for bioethanol production: a perspective review. In: Kumar, S., Khanal, S., Yadav, Y. (Eds.), In: Proceedings of the First International Conference on Recent Advances in Bioenergy Research. Springer Proceedings in Energy, Springer, New Delhi, pp. 81–93.

Shaw, A.J., Jenney, F.E., Adams, M.W.W., Lynd, L.R., 2008. End-product pathways in the xylose fermenting bacterium, *Thermoanaerobacterium saccharolyticum*. Enz. Microb. Technol. 42, 453–458.

Shields, S., Boopathy, R., 2011. Ethanol production from lignocellulosic biomass of energy cane. Int. Biodeter. Biodegr. 65, 142–146.

Shupe, A.M., Liu, S., 2012. Effect of agitation rate on ethanol production from sugar maple hemicellulosic hydrolysate by *Pichia stipitis*. Appl. Biochem. Biotechnol. 168, 29–36.

Sigurbjornsdottir, M.A., Orlygsson, J., 2012. Combined hydrogen and ethanol production from sugars and lignocellulosic biomass by *Thermoanaerobacterium* AK_{54}, isolated from hot spring. Appl. Energy 97, 785–791.

Singh, A., Pant, D., Korres, N.E., Nizami, A.S., Prasad, S., Murphy, J., 2010. Key issues in life cycle assessment of ethanol production from lignocellulosic biomass: challenges and perspectives. Bioresour. Technol. 101, 5003–5012.

Singh, A., Tuteja, S., Singh, N., Bishnoi, N.R., 2011. Enhanced saccharification of rice straw and hulls by microwave-alkali pretreatment and lignocellulolytic enzyme production. Bioresour. Technol. 102, 1773–1782.

Sukumaran, R.K., Singhania, R.R., Pandey, A., 2005. Microbial cellulases: production, application and challenges. J. Sci. Ind. Res. 64, 832–844.

Sun, Y., Cheng, J., 2002. Hydrolysis of lignocellulosic materials for ethanol production: a review. Bioresour. Technol. 83, 1–11.

Svetlitchny, V.A., Kensch, O., Falkenhan, D.A., Korseska, S.G., Lippert, N., Prinz, M., Sassi, J., Schickor, A., Curvers, S., 2013. Single-step ethanol production from lignocelluloses using novel extremely thermophilic bacteria. Biotechnol. Biofuels 6, 31.

Tabil, L., Adapa, P., Kashaninejad, M., 2011. Biomass feedstock pre-processing—part 1: pretreatment. In: Bernaedes, M.A.D.S. (Ed.), Biofuel's Engineering Process Technology. PN InTech, pp. 411–438.

Taherzadeh, M.J., Karimi, K., 2007. Acid based hydrolysis processes for ethanol from lignocellulosic materials: a review. Bioresources 2, 472–499.

Taherzadeh, M.J., Karimi, K., 2008. Pretreatment of lignocellulosic wastes to improve ethanol and biogas production: a review. Int. J. Mol. Sci. 9, 1621–1651.

Talebnia, F., Karakshev, D., Angelidaki, I., 2010. Production of bioethanol from wheat straw: an overview on pretreatment, hydrolysis and fermentation. Bioresour. Technol. 101, 4744–4753.

Tomas, A.F., Karagoz, P., Karakashev, D., Angelidaki, I., 2013. Extreme thermophilic ethanol production from rapeseed straw: using the newly isolated *Thermoanaerobacter pentosaceus* and combining it with *Saccharomyces cerevisiae* in a two-step process. Biotechnol. Bioeng. 110, 1574–1582.

Tomas-Pejo, E., Oliva, J.M., Ballesteros, M., Olsson, L., 2008. Comparison of SHF and SSF processes from steam-exploded wheat straw for ethanol production by xylose-fermenting and robust glucose-fermenting *Saccharomyces cerevisiae* strains. Biotechnol. Bioeng. 100, 1122–1131.

Trinh, C.T., Srienc, F., 2009. Metabolic engineering of *Escherichia coli* for efficient conversion of glycerol to ethanol. Appl. Environ. Microbiol. 75, 6696–6705.

Trinh, C.T., Unrean, P., Srienc, F., 2008. Minimal *Escherichia coli* cell for the most efficient production of ethanol from hexoses and pentoses. Appl. Environ. Microbiol. 74, 3634–3643.

Vijaya, C., Reddy, M., 2012. Biodelignification ability of locally available edible mushrooms for the biological treatment of crop residues. Ind. J. Biotechnol. 11, 191–196.

Viola, E., Zimbardi, F., Valerio, V., Nanna, F., Battafarano, A., 2013. Use of a two-chamber reactor to improve enzymatic hydrolysis and fermentation of lignocellulosic materials. Appl. Energy 102, 198–203.

Wang, L., Hanna, M.A., Weller, C.L., Jones, D.D., 2009. Technical and economical analyses of combined heat and power generation from distillers grains and corn stover in ethanol plants. Energy Convers. Manage. 50, 1704–1713.

Watanabe, I., Nakamura, T., Shima, J., 2010. Strategy for simultaneous saccharification and fermentation using a respiratory-deficient mutant of *Candida glabrata* for bioethanol production. J. Biosci. Bioeng. 110, 176–179.

Wisselink, H.W., Toirkens, M.J., Wu, Q., Pronk, J.T., van Maris, A.J., 2009. A novel evolutionary engineering approach for accelerated utilization of glucose, xylose and arabinose mixtures by engineered *Saccharomyces cerevisiae*. Appl. Environ. Microbiol. 75, 907–914.

Wongwilaiwalin, S., Rattanachomsri, U., Laothanachareon, T., Eurwilaichitr, L., Igarashi, Y., Champreda, V., 2010. Analysis of a thermophilic lignocelluloses degrading microbial consortium and multi-species lignocellulolytic enzyme system. Enz. Microb. Technol. 47, 283–290.

Wu, I., Arnold, F.H., 2013. Engineered thermostable fungal Cel6A and Cel7A cellobiohydrolases hydrolyze cellulose efficiently at elevated temperatures. Biotechnol. Bioeng. 110, 1874–1883.

Wyman, C.E., 1996. Handbook on Bioethanol: Production and Utilization. Taylor & Francis, Washington, DC.

Yanase, S., Hasunuma, T., Yamada, R., Tanaka, T., Ogino, C., Fukuda, H., Kondo, A., 2010. Direct ethanol production from cellulosic materials at high temperature using the thermotolerant yeast *Kluyveromyces marxianus* displaying cellulolytic enzymes. Appl. Microbiol. Biotechnol. 88, 381–388.

Yang, B., Wyman, C.E., 2008. Pretreatment: the key to unlocking low-cost cellulosic ethanol. Biofuels Biopr. Bioref. 2, 26–40.

Zaldivar, J., Nielsen, J., Olsson, L., 2001. Fuel ethanol production from lignocelluloses: a challenge for metabolic engineering and process integration. Appl. Microbiol. Biotechnol. 56, 17–34.

Zhang, J., Osmani, A., Awudu, I., Gonela, V., 2013. An integrated optimization model for switchgrass-based bioethanol supply chain. Appl. Energy 102, 1205–1217.

Zhang, Y.H., Berson, E., Sarkanen, S., Dale, B.E., 2009. Pretreatment and biomass recalcitrance: fundamentals and progress. Appl. Biochem. Biotechnol. 153, 80–83.

Zhang, Y.H., Lynd, L.R., 2004. Toward an aggregated understanding of enzymatic hydrolysis of cellulose: noncomplexed cellulase systems. Biotechnol. Bioeng. 88, 797–824.

Zhu, J.J., Yong, Q., Xu, Y., Chen, S.X., Yu, S.Y., 2009. Adaptation fermentation of *Pichia stipitis* and combination detoxification on steam exploded lignocellulosic prehydrolyzate. Nat. Sci. 1, 47–54.

Zhu, W., Zhu, J.Y., Gleisner, R., Pan, X.J., 2010. On energy consumption for size-reduction and yields from subsequent enzymatic saccharification of pretreated lodgepole pine. Bioresour. Technol. 101, 2782–2792.

CHAPTER

10

Bioethanol From Sugarcane Bagasse: Status and Perspectives

*Sadat M.R. Khattab**,**, *Takashi Watanabe***

*Al-Azhar University, Assiut Campus, Egypt

**Research Institute for Sustainable Humanosphere (RISH), Kyoto University, Uji Campus, Kyoto, Japan

10.1 INTRODUCTION

With industrial modernization, serious concerns have developed in many countries over the worldwide growing demand and uncertainty of petroleum supplies required for transportation, heating, and industrial processes. Furthermore, excessive use of fossil fuels has resulted in global warming and greenhouse gas (GHG) emissions (Saini et al., 2015; Watanabe, 2013). Fossil fuels (i.e., oil, coal, and natural gas) represent about 80% of the global energy supply and can be defined as depletable, or nonrenewable resources. According to the 2011 consumption rate, fossil fuels are predicted to sustain current oil usage for 54 years, current natural gas usage for 64 years, and current coal usage for 112 years. These estimates are further exacerbated by several other factors, including the use of local technologies, low-energy conversion efficiencies, and excessive losses of indigenous resources. Fossil fuels are derived from biological materials that have absorbed carbon dioxide (CO_2) from the atmosphere, but the time scale for conversion to fossil fuels occurs over many millions of years (Singh and Singh, 2012; Sugathapala, 2013).

Of the 3,850,000 exajoules (EJ) of total solar energy absorbed by Earth each year, photosynthesis captures approximately 3000 EJ per year in biomass, and produces more than 100 billion tons of dry biomass annually (Barber, 2009; Najafpour and Pashaei, 2012). Therefore, it seems inevitable that the aforementioned problems may be solved by developing clean and renewable energy sources using improved energy efficiencies from emerging biomass-to-energy conversion technologies, as lignocellulosic biomass is considered one of the most important renewable natural resources (Watanabe, 2013).

Sugarcane (*Saccharum officinarum* L.) is an indispensable crop, representing a significant proportion of cultivated crops worldwide. Its annual global production of ~1.6 billion tons

Bioethanol Production From Food Crops
http://dx.doi.org/10.1016/B978-0-12-813766-6.00010-2

generates several hundred million metric tons of bagasse, or remaining fiber after the extraction of the sugar-bearing juice (Hatfield and Fukushima, 2005). Bagasse is composed of up to 70% holocellulose, which is represented by high tensile strength crystalline cellulose with intervening amorphous regions gathered into bundles of microfibrils embedded in an amorphous matrix of hemicellulose and further crosslinked lignin. Small amounts of pectin, extractives, and ashes also are included in the biomass composition. Lignin acts as a physical barrier and lowers the biodegradability of both cellulose and hemicelluloses (Bidlack et al., 1992; Fox et al., 1987; Jørgensen et al., 2007; Pandey et al., 2000).

Monumental efforts using various strategies have been envisioned to discover the best pretreatment method for converting biomass into fermentable sugars. Pretreatment technologies are under development using physical, chemical, or biological methodologies, or even combinations of these methods. Multiple pretreatment steps are needed currently to overcome the strength of the bagasse. For example, the hydrolysis of hemicelluloses and/or removal of lignin are used to increase surface area and decrease the crystallinity of the cellulose without forming inhibitor byproducts in an effort to improve enzyme saccharification and the fermentation process. Pretreatments also have great potential for improving efficiency and lowering saccharfication doses and total costs of the fermentation process (Khattab, 2015; Lee et al., 1994; Lynd, 1996). Researchers are striving to reduce pretreatment costs via different strategies: mechanical pulverization; hydrothermal pretreatments, including steam explosion; chemical treatments with acid, alkali, hydrogen peroxide, ammonia fiber explosion, ammonia recycle percolation, wet-oxidation, lime, CO_2 explosion, or organic solvents, and biological treatments accompanied by conventional heating, microwaves, or ultrasonic waves. Each of these pretreatment methods has its distinct advantages and disadvantages (Martín et al., 2002; Ramos and Fontana, 1996; Sasaki et al., 2011).

On the other hand, finding robust microbial strains that have the abilities to ferment hydrolyzed sugars efficiently has been another obstacle for the second generation of bioethanol production. Intensive studies were implemented to select native microorganisms or to genetically engineer microorganisms that could ferment efficiently the hydrolyzed sugars in biomass to bioethanol. *Saccharomyces cerevisiae* is known as the top fermenting yeast capable of fermenting C_6 sugars (i.e., glucose, fructose) and tolerates the action of inhibitors present in the media or hydrolysates. In addition, it is eco-friendly and safe, but it cannot ferment xylose, the most abundant sugar in biomass after glucose. Extensive trials have been carried out to engineer robust recombinant xylose-fermenting strains to maximize bioethanol production from the released sugars of hydrolyzed biomass. This chapter glance explore the bioethanol production from sugarcane bagasse by highlighting the most common pretreatment methods with tracking the evolution of recombinant *S. cerevisiae* toward the second generation of bioethanol production.

10.2 WHY BIOETHANOL?

10.2.1 Bioethanol for Sustainable Fuel and Society

With the oil crises of the 1970s, ethanol was established as an alternative fuel. Many countries began programs to study and develop more economic fuel production from available raw materials. Bioethanol was declared a promising renewable energy source, with better antiknock characteristics than traditional oil-based fuels (i.e., high octane number and high heat of vaporization), thus the world's nations were prompted to expand bioethanol use and production. Bioethanol is unrivaled when considering negative environmental impacts. With

35% oxygen, there is a complete combustion of the fuel, thus efficiently providing energy while reducing particulate emissions that pose threats to human health. Subsequently, the toxicity of exhaust emissions is lower compared to those from petroleum sources. The use of bioethanol began with the production of motor fuel for use in the internal combustion engine in 1897 (Fulton et al., 2004; Taherzadeh and Karimi, 2008).

From an environmental aspect, using bioethanol for transportation fuel reduces GHG emissions by approximately 30%–85% compared to gasoline (Wyman and Hinman, 1990). Even 10% ethanol–gasoline blends reduce GHG emissions by 12%–19% compared to conventional fossil fuels. Using E 85 (85% ethanol) further reduces the nitrogen, oxide, and sulfate emissions by 10%, 20%, and 80%, respectively, compared to conventional gasoline (Saini et al., 2015). The global production of bioethanol increased rapidly from 15.5 billion L in 2000 to 51.5 billion L in 2007 (Fulton et al., 2004), and continued to rise showing an increase in global bioethanol production to 85 billion L by 2011 (Avci et al., 2013; Singh and Bishnoi, 2012).

Currently, more than 100 billion L of biofuels are produced annually, yet this accounts for only a small fraction (2.7%) of the total energy used for transportation. Together, the United States and Brazil, the world's top producers of bioethanol, account for 90% of its global production (Buijs et al., 2013). Once disclosed, the dilemma of conversion cost of lignocellulosic biomass to fermentable sugars, the global bioethanol production from total agricultural residue will increment dramatically to approximately 491 billion L annually (Kim and Dale, 2004; Sarkar et al., 2012).

10.2.2 Bioethanol as Feedstock for the Chemical Industry

Many industries have converted to biorefineries of lignocellulosic biomass. Biobased chemicals from lignocellulose are one of the platforms that are being initiated actively (Watanabe, 2013). The following chemicals are produced using bioethanol used as feedstock:

- *Triethylamine* was industrialized by Daicel Co. Ltd. in Japan in 2007 as a feedstock to produce quaternary ammonium compounds for textile auxiliaries and quaternary ammonium salts of dyes. Moreover, it can be used as a catalyst and acid neutralizer for condensation reactions, as well as an intermediate for manufacturing medicines, pesticides, and other chemicals.
- *Ethyl acetate* was industrialized by Shandong Haihua Company in China and Daicel Co. Ltd.
- *Polyvinylchloride* can be produced from brine-derived chlorine and bioethanol through ethylene production, and was planned by Solvay in Belgium.
- *Acetaldehyde, acetic acid, epoxyethane*, and *ethylene glycol* are potential chemicals products of bioethanol processing. *Ethylene glycol* can be incorporated into polymers, such as *poly (ethylene terephthalate)*.

Thus, in recent years, much attention regarding the production of chemicals from bioethanol have been launched worldwide (Watanabe, 2013).

10.3 WHY SUGARCANE BAGASSE?

UN data estimates ~493 million metric tons of bagasse generates globally from the sugar industry (UN Energy Statistics Database, 2013). However, there are several variations in the chemical composition of bagasse, as shown in Table 10.1 that are attributed to the particular sugarcane species, the occurrence of hybrids, environmental conditions, harvesting systems, and crop ages (Hatfield and Fukushima, 2005). In general, bagasse is composed primarily of cellulose, hemicelluloses, and lignin in an

TABLE 10.1 Chemical Composition (% w/w, Dry Basis) of Sugarcane Bagasse

	Chemical composition (%, dry mass)				
References	**Cellulose**	**Hemicellulose**	**Lignin**	**Extractives**	**Ash**
Saad et al. (2008)	36.1	26.9	26.2	5.3	2.1
Brienzo et al. (2009)	42.4	25.2	19.6	–	1.6
Zhao et al. (2009)	45.0	31.8	20.3	–	2.9
da Silva et al. (2010)	38.8	26.0	32.4	–	2.8
Canilha et al. (2011)	45.0	25.8	19.1	9.1	1.0
Rabelo et al. (2011)	38.4	23.2	25.1	–	1.5
Rocha et al. (2011)	45.5	27.0	21.1	4.6	2.2
Santos et al. (2011)	46.2	27.8	22.1	–	1.0
Rocha et al. (2012)	45.0	26.4	23.4	2.7	2.5
Benjamin et al. (2013)	39.6	24.0	22.4	5.0	1.3
Maryana et al. (2014)	44.4	22.9	17.5	14.1	–
Brienzo et al. (2015)	47.3	26.4	14.1	11.4	–
Guilherme et al. (2015)	38.5	27.8	17.7	2.7	8.8
Khattab (2015)	37.3	30.5	30.2	–	1.98
Rocha et al. (2015)	36.8	28.3	24.7	6.8	3.1
Minimum	36.1	22.9	14.1	2.7	0.8
Maximum	47.3	31.8	32.4	14.1	8.8
Average	*41.75*	*26.66*	*22.39*	*6.8*	*2.52*

approximate 4:3:3 ratios, and includes some other minor miscellaneous compounds, such as silica and extractives.

Sugarcane belongs to the family *Poaceae*, which are grasses. The tissues within these types of plants are composed of cells with Type II cell walls, meaning the cellulose fibers are within a glucurono-arabino-xylan matrix with high levels of hydroxycinnamates. As the sugarcane matures, secondary cell walls are created between the primary cell wall and the plasma membrane, which consist of cellulose, hemicellulose (mainly glucuronoarabinoxylan), and lignin (Zhang et al., 2016).

10.3.1 Cellulose

Cellulose is a linear homopolymer composed of D-glucopyranose units linked by α-1, 4-glycosidic bonds to form the subunit dimer cellobiose, which links together to form cellulose (Fig. 10.2A). The chemical formula of cellulose is $(C_6H_{10}O_5)_n$, where n is called the degree of polymerization (DP), represents the number of glucose units, which can range from hundreds to thousands, or even tens of thousands of dehydrated glucose units. Cellulose chains are highly ordered and very cohesive with rigid structure (crystalline) or less ordered (amorphous) regions. The cellulose (glucose) chains are precisely aggregated using a strong and complex intra- and intermolecular hydrogen bond network and van der Waals forces to form a crystalline core of elementary fibril (Abdul Khalil et al., 2015; Ding and Himmel, 2006; Himmel et al., 2007; Zhang et al., 2016). In addition, hydroxyl groups of cellulose chains forming covalent bonds with hemicellulose sheath the elementary fibrils. Lignin matrix entwined hemicellulose and encapsulated elementary fibrils to form a microfibril as shown in Fig. 10.1 (Abdul Khalil et al., 2015; Fengel and Wegener, 1984; Ohad and Danon, 1964). Strong intermolecular hydrogen bonds and weaker hydrophobic interactions between cellulose sheets provide a resistance to acid hydrolysis (Wang and Sun, 2010). Microfibrils are further ordered by hydrogen bonding to a hemicellulose matrix and enclosed by either a lignin or pectin matrix into larger aggregates (macrofibrils) of 50–250 nm in diameter (Ding and Himmel, 2006).

10.3.2 Hemicellulose

It is short (50–200 units), highly branched heteropolymer units of sugars (mainly D-xylose,

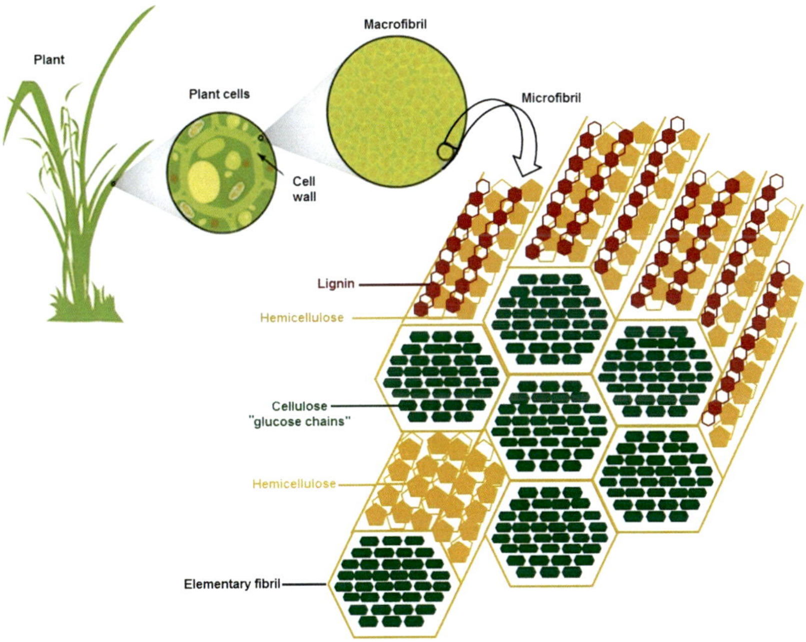

FIGURE 10.1 **The schematic design of a lignocellulosic complex showing magnification of the shape and arrangement of main constituents: macrofibril, microfibril, and elementary fibrils.** *Source:* A part of this figure quoted from Rubin, EM., 2008. Genomics of cellulosic biofuels. Nature 454(7206), 841–845. doi:10.1038/nature07190.

followed by D-glucose, and then L-arabinose) and lesser D-mannose, D-galactose units. Hemicelluloses are the second most abundant polysaccharides in plant cell walls. They are classified usually according to the main residues of the sugars present in the backbone of the structural polymer. In bagasse: arabinoxylan (AX) is the most abundant hemicellulose with a β-(1 → 4)-linked backbone and a 2-*O*- or 3-*O*-α-L-arabinose side chain (Fig. 10.2B). Mixed-linkage-β-glucan (BG) and xyloglucan (XG) are also the major hemicellulose components associated with leaf and culm cellulose in sugarcane. In cell walls of graminaceous plants, including sugarcane, ferulic acid and *P*-coumaric acid are covalently linked to arabinofuranose residues in AX by ester bonds, and to components of lignin by ester or ether bonds (Koshijima and Watanabe, 2003; Sun et al., 2004). *P*-coumaric acid is mostly esterified to lignin, while ferulic acid is linked to AX and partially to lignin. During cell wall biosynthesis, ferulated AXs

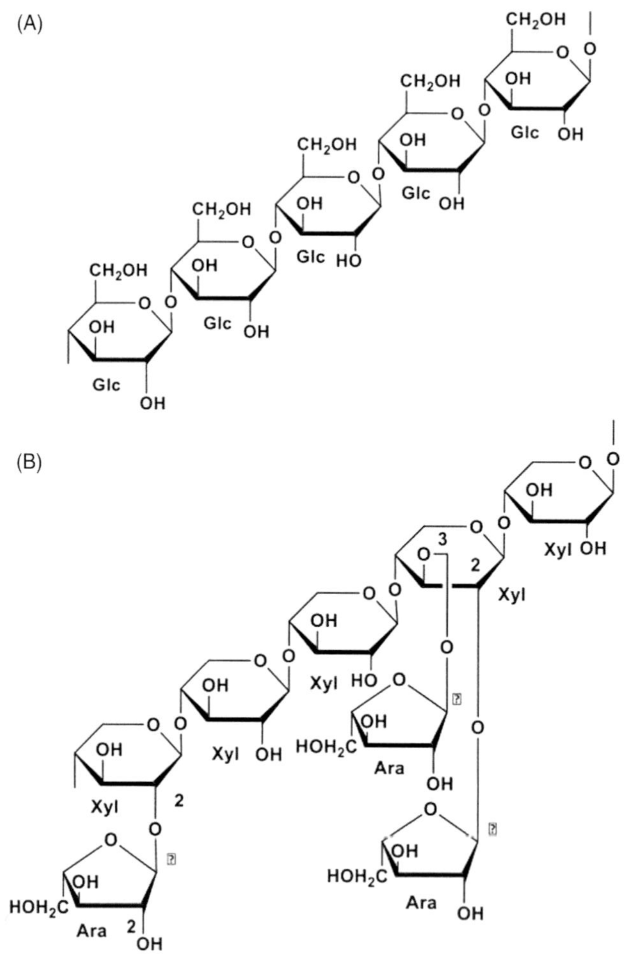

FIGURE 10.2 Partial chemical structures of (A) cellulose and (B) arabinoxylan in sugarcane bagasse.

are copolymerized with monolignols by radical coupling or nucleophilic addition to quinone-methide intermediates to form lignin-carbohydrate complexes, giving physical and biological resistance to the plant cell walls.

10.3.3 Lignin

Lignin is the third most abundant organic (aromatic) compound in bagasse, after cellulose and hemicellulose. Lignin reinforces the cellulose–hemicellulose networks and encloses both microfibrils and macrofibrils to provide rigidity to the biomass. These physical distributions protect the polysaccharide chains from chemical depolymerization and microbial attack, and confer recalcitrance to enzymatic hydrolysis for the bagasse. In the cell walls of sugarcane, three monolignols, *P*-coumaryl alcohol, coniferyl alcohol, and synapyl alcohol, polymerize by radical coupling reactions to form the heterogeneous aromatic polymer, lignin (Figs. 10.3 and 10.4). Thus the lignin in sugarcane is composed of *P*-hydroxy phenyl (H), guaiacyl (G), and syringyl (S) nucleus (Brienzo et al., 2015; Chiang and Funaoka, 1990; Koshijima and Watanabe, 2003; Oshiro et al., 2017).

10.4 CONVERSION TECHNOLOGIES OF SUGARCANE BAGASSE

Sugarcane bagasse, composed of lignocellulosic materials, has a higher degree of complex inherent structure and does not contain monosaccharides that are accessible to microbes during the fermentation process. Because of this, a cost-effective pretreatment is necessary to break down the material using physicochemical processes prior to introducing hydrolytic enzymes. Pretreatment is the central technology in a conversion scenario prior to enzymatic saccharification of lignocellulose. Pretreatment technologies are classified as physical, chemical, or a combination of the two. Chemical pretreatments play a primary role and usually are conducted above the boiling point of the solvent. Pressure is often added as an accompanying factor, with or without the assistance of catalysts. Furthermore, integrating previously known strategies with newly emerging technologies, such as microwaves and ultrasonic waves, increases the number of trial possibilities. Therefore, numerous pretreatment trials have been proposed and researched to develop an ideal scenario for hydrolyzing bagasse to fermentable sugars prior

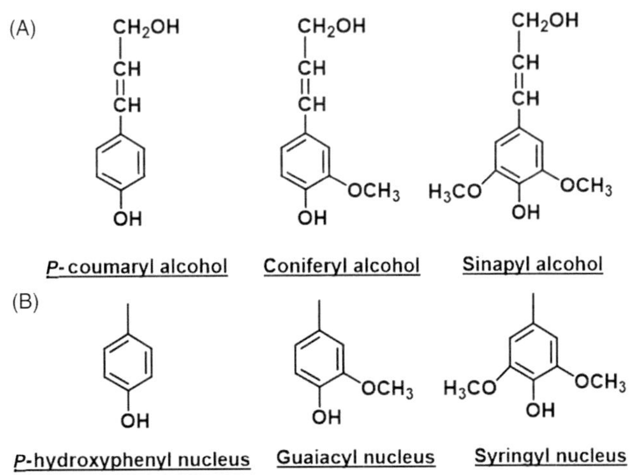

FIGURE 10.3 **Structural units of lignin.** (A) precursor of lignin biosynthesis (monolignols) and (B) aromatic nuclei.

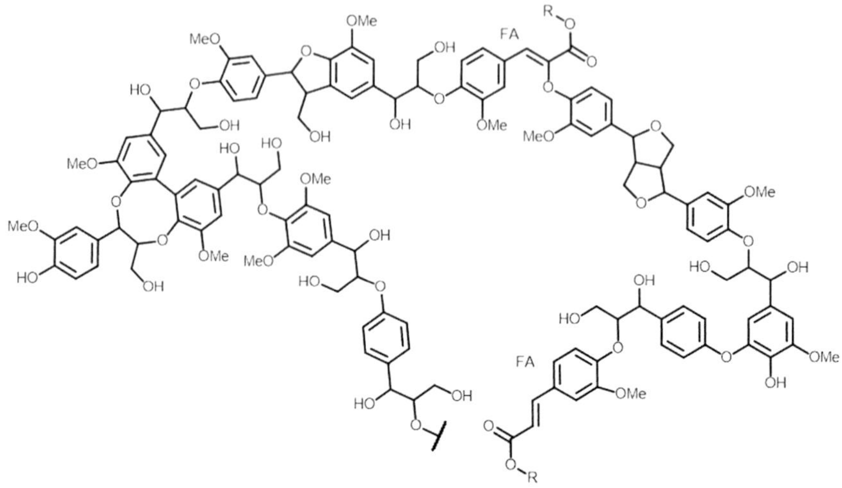

FIGURE 10.4 **Partial chemical structure of lignin in graminaceous plants.**

to fermentation to bioethanol. In this chapter, we explore common pretreatment methods and techniques that are used during the conversion of bagasse to fermentable sugars, highlighting the most promising approaches.

10.4.1 Physical Pretreatment

10.4.1.1 Mechanical Pulverization

Comminution of lignocellulosic biomass is one of the considered strategies using a combination of chipping, grinding, and milling to reduce cellulose crystallinity. The effectiveness of ball milling (BM) and wet disk milling (WDM) during pretreatments of sugarcane bagasse has been confirmed enzymatically using 10 filter paper unit (FPU) *Acremonium* cellulase per gram of biomass, with 5% enzyme loading assisted by 0.4% (v/v) Optimash BG of β-xylosidase. It is clear that BM and WDM pretreatments increase the enzymatic digestibility of bagasse by increasing the surface area available to cellulase enzymes. The modes of action for BM and WDM include reduced crystallinity and defibrillation, respectively. Although BM treatments completely disrupt the native structure of bagasse, glucose, and xylose hydrolysis yields at optimum conditions were 78.7% and 72.1%, respectively. BM-treated bagasse requires intensive energy, resulting in an expensive total pretreatment method that is not feasible for use in the industry. The WDM treatment has shown potential as a pretreatment for other biomasses, but it did not perform well for bagasse, where the maximum glucose and xylose yields were 49.3% and 36.7%, respectively, consuming 48 MJ/kg of bagasse. Therefore, further modifications to mechanical comminution methods and further exploration of other, more economical pulverization processes are necessary for the pretreatment of bagasse (da Silva et al., 2010).

10.4.2 Chemical Pretreatment

10.4.2.1 Acid Hydrolysis

One of the widespread pretreatment methods for bagasse is acid hydrolysis. While phosphoric acid, hydrochloric acid, and nitric acid all have been reported as chemical agents for the pretreatment of sugarcane bagasse (Gámez et al., 2006; Rodríguez-Chong et al., 2004), sulfuric acid is the most commonly used acid for this purpose (Lavarack and Griffin, 2002). Also, acetic acid combined with hydrogen peroxide was reported as a delignifying agent, but the delignification rate was low (Tan et al., 2010).

10.4.2.1.1 PHOSPHORIC ACID

Phosphoric acid also has been analyzed for its use in the pretreatment of bagasse. The optimal conditions selected in that study were 4% H_3PO_4 at 122°C for 300 min (Gámez et al., 2006). The efficiency reported under these conditions was 4.46 g sugars/g inhibitors, greater than 55% of the solid biomass was hydrolyzed to the liquid phase, represented by 17.6 g of xylose/L, 2.6 g of arabinose/L, 3.0 g of glucose/L, 1.2 g furfural/L, and 4.0 g acetic acid/L. Conversion of whole bagasse to fermentable sugars was later studied where 0.20% phosphoric acid depolymerized 96% and 98% of the bagasse hemicellulose to its monomers at 186°C after 8 and 24 min, respectively. In addition, cellulose and lignin were solubilized by 20% and 14%, respectively. However, the efficacy of this method to adequately hydrolyze hemicellulose to supply the necessary phosphorous requirement for the subsequent fermentation processes is limited, and only 56.38% of the residual cellulose was converted to glucose using 20 FPU/g bagasse of Celluclast in the enzymatic hydrolysis process (de Vasconcelos et al., 2013).

10.4.2.1.2 HYDROCHLORIC ACID

Hydrochloric acid also has been used for pretreating bagasse. After screening different concentrations of hydrochloric acid (2%–6%), reaction times (0–300 min), and temperatures (100–128°C), the optimum pretreatment conditions were found at 2% HCl, 51.1 min, and 128°C, resulting in 22.6 g xylose/L, 3.31 g arabinose/L, 3.77 g glucose/L, 3.59 g acetic acid/L, and 1.54 g furfural/L in the soluble fraction (Bustos et al., 2003). Also, 90 mg HCl/g of dry bagasse in an autoclave (121°C and 1.1 kg/cm^2) for 4 h yielded a higher level of reducing sugars, with 37.21% for depithed bagasse and 35.37% for bagasse with pith intact (Hernández-Salas et al., 2009).

10.4.2.1.3 NITRIC ACID

Nitric acid was used similarly to the aforementioned hydrochloric acid pretreatment conditions and the optimal conditions selected were: 6% HNO_3 at 122°C for 9.3 min. Under these conditions, 18.6 g xylose/L, 2.04 g arabinose/L, 2.87 g glucose/L, 0.9 g acetic acid/L, and 1.32 g furfural/L were obtained. Greater catalytic efficiency of nitric acid over sulfuric and hydrochloric acids was highlighted in the context (Rodríguez-Chong et al., 2004).

10.4.2.1.4 SULFURIC ACID

Sulfuric acid (H_2SO_4) is generally the most popular acid for pretreatments of lignocellulosic biomass, particularly bagasse (Aguilar et al., 2002; Alves et al., 1998, 2002; Gong et al., 1993; Khattab, 2015; Neureiter et al., 2002), because this is the most cost-effective method among all others described (Chen et al., 2011b). Sulfuric acid prehydrolysis conditions generally occur with a 1%~2% solution, with fluctuations in temperature 100~200°C, and with varying treatment times. Under optimum conditions, hydronium ions, with the assistance of heat, microwaves, or ultrasonic waves, depolymerize the hemicellulose to its monomers. Pattra et al. (2008) screened the optimal pretreatment conditions for bagasse using H_2SO_4 and reported that 0.5% H_2SO_4 for 60 min at 121°C in an autoclave worked best among the various concentrations (0.25%–7.0% volume) and reaction times (15–240 min) tested. These conditions yielded 24.5 g/L of total sugar that fractionated to 11 g/L glucose; 11.29 g/L xylose; 2.22 g/L arabinose; 2.48 g/L acetic acid, and 0.12 g/L furfural. It is important to note here that xylose is a major fraction recovered from the hemicellulose hydrolysate of bagasse. Consistent with these results, Khattab (2015) reported that xylose was the main constituent of hemicellulose-hydrolyzed sugars (18.55 g/L), matching 15.16% of total weight of the bagasse after treatments with 1% sulfuric acid for 25 min at 135°C in an autoclave. Glucose was the next most abundant recovered sugar at 8.11 g/L, which represented 6.627% of the total bagasse. Another important point to mention is the diminution of glucose and xylose ratios in the treatments with excess acid concentration and/or rise in temperature certainly due to decomposition of xylose and glucose to furfural and hydroxy methyl furfural (Cheng et al., 2008; Rodríguez-Chong et al., 2004).

10.4.2.2 Alkaline Pretreatment

Alkaline pretreatments are reasonable and ubiquitous at the industrial level, for example, pulping and paper industries, nevertheless, their application in regards to industrial bioethanol production from lignocellulosic biomass is still under debate and requires further investigation. The mode of alkaline action on bagasse is the hydrolysis of uronic and acetic esters bonds. Breaking apart these bonds dissolves hemicelluloses, lignin, and their cellulose-based networks that eventually swell, and then subsequently cause cell walls to burst (Fan et al., 1987; Pandey et al., 2000; Quintero et al., 2008; Soccol et al., 2011). Evaluation of alkaline pretreatments of sugarcane bagasse showed that cooking the bagasse in 1N NaOH for 30 min decreased the lignin content optimally to its lowest level at, 7.16% of its original weight (Maryana et al., 2014).

10.4.3 Oxidation Bleaching

Enhancement of cellulose hydrolysis in sugarcane bagasse has been confirmed by the selective removal of lignin with sodium chlorite and acetic acid. After 72 h of hydrolysis using cellulose, either with or without β-glucosidase, maximal conversions of 90% were obtained after 70% of the lignin was removed with only slight hemicellulose solubilization. By bleaching lignin, 58.8% cellulose hydrolysis occurred within the first 2 h of treatment, and then increased to 71% after 4 h (Siqueira et al., 2013). Also, delignification was accomplished by three oxidation-bleaching cycle treatments using sodium chlorite and acetic acid at 70°C for 1 h per cycle. The swelling action of the sodium bicarbonate liberated the integrated lignin in the final oxidation cycle, which enhanced enzymatic saccharification by 37% and, consequently, increased bioethanol production by 28% (Khattab, 2015).

10.4.4 Organosolv Pretreatment

Organic solvents, commonly known as organosolv pretreatments, have gained much attention recently for their enhancement of enzymatic saccharification of lignocellulose due to a higher selectivity toward lignin removal with lower sulfur contents in products (Brosse, 2012; Calvo-Flores and Dobado, 2010). These results are promising for the pulp and paper industries, and increase its prospects as a pretreatment strategy for the production of bioethanol from lignocelluloses. Organic solvents, such as methanol, ethanol, acetone, acetic acid, and glycerol, are used commonly in this regard; however, glycerol has shown the most promising results among the organic solvents, and discussion of this follows.

Potential use of glycerol as a pretreatment was confirmed after delignifying 81.4% of depithed bagasse using 80% glycerol at 198.3°C for 150 min with no catalyst and the pulp yield was 54.4%. In addition, 13.7% of the polyoses from the biomass emerged during this pretreatment (Novo et al., 2011). With increasing glycerol concentrations, there is increasing lignin solubilization, but the opposite is true for xylan, indicating that high glycerol concentrations (80%) act more selectively on lignin than on xylan. Furthermore, when glycerol was combined with H_2SO_4 or NaOH during bagasse pretreatments, it enhanced the delignification action. Xylan solubilization decreased when glycerol was combined with NaOH, and this pattern continued with increasing glycerol concentrations.

Ultimately, enzymatic convertibility of residual pretreated bagasse using pure glycerol reached 85%–94%. This convertibility, as measured by pulping ratios and released compounds in the soluble fraction, was higher than in analogous pretreatments combining glycerol with NaOH or H_2SO_4 (Martín et al., 2011).

Acidified aqueous glycerol solution, containing 1.2 wt.% HCl, was evaluated on both laboratory and pilot scales, 4 g and 10 kg, respectively, for the pretreatment of sugarcane bagasse at 130°C for 60 min. Glucan digestibilities were 88% and 90%, respectively, with laboratory and pilot scales. Pretreatment of sugarcane bagasse with acid-catalyzed glycerol solutions likely produces glycerol-glycosides, which together with hydrolyzed lignin, are potential substrates to produce biopolymers (Zhang et al., 2013). The same research group conducted steam explosion pretreatments with acidified aqueous glycerol to compare the effect of pretreatments on various doses of saccharification enzyme mixtures. The glycerol/HCl pretreatment revealed the best and maximum glucan conversion at 100% with Celluclast 1.5 L (15–25 FPU/g glucan). Conversion rates were in the following order: glycerol/HCl > HCl > H_2SO_4 > NaOH; therefore, NaOH-treated substrate achieved optimum saccharification with only 8 FPU/g glucan. This study concluded that acidified glycerol pretreatments offer an opportunity to simplify the complexities of enzyme mixtures required for the saccharification of lignocellulosics (Harrison et al., 2013).

10.4.5 Ultrasonic Assisting Chemical Pretreatments

Radiation emitted at ultrasound frequencies generates acoustic cavitation phenomenon, which induces the formation of millions of microbubbles (100 microns in diameter). The bubbles implode violently in less than a microsecond causing the gases and vapors inside the cavity to be compressed, generating intense heat that raises the temperature of the liquid immediately surrounding the cavity. This creates a local hot spot that heats the contents to 5500°C (about the temperature of the sun's surface) with negative pressures as high as 100 MPa. Nevertheless, the region itself is so small that the heat dissipates quickly and the bulk of the liquid remains at the ambient temperature. As a result, the generation of high-speed microjets (physical effect) and hydroxyl radicals occurs due to the pyrolysis in water (Ramadoss and Muthukumar, 2016; Suslick, 1989).

Ultrasonic radiation also has been applied recently to improve the efficiency of bagasse pretreatments. In a comparative study with steam and alkali-steam explosion, sonoassisted alkaline pretreatments yielded higher glucose recovery and fermentation efficiencies, with 95.8 wt.% total glucose recovery from the total glucan content of the original bagasse, and with 91.4% fermentation efficiency. Sonoassisted alkali pretreatment of cane bagasse emerges as a simpler, more efficient, and more sustainable method for bioethanol production, particularly considering the capital cost involved with steam explosion. Furthermore, lower enzyme loading (33 mg) for hydrolysis corresponded to a total cellulase activity of 7.1 FPU/g of Cellic CTec3 (Novozymes) and Thermossac Dry (Lallemand) (da Silvaa et al., 2016). Also, ultrasound and titanium dioxide (TiO_2) were used in conjunction with hydrogen peroxide to form hydroxyl, methoxy, and carbonyl radicals to delignify 79% of the bagasse and recover 95% of the original holocellulose (Ramadoss and Muthukumar, 2016).

10.4.6 Microwave Assisting Chemical Pretreatments

In recent years, there is increasing interest in using electromagnetic waves, with frequencies between 300 and 300 GHz, instead of conventional thermal heating methods for both synthetic chemistry and biomass pretreatment methods (Fan et al., 2013; Macquarrie et al., 2012). The dipole moment and the mobility of a substance support the electromagnetic field that induces friction between molecules, subsequently increasing temperatures more directly, more rapidly, and more uniformly compared to conventional heating (Lancaster, 2002). Continuous-flow types and large-scale batch types (50 L) of microwave irradiators have been developed for biomass pretreatments and have been applied to procedures for bioethanol production from lignocellulosics (Hasegawa et al., 2014; Watanabe and Mitani, 2012).

Microwave-assisted H_2SO_4 (0.2 M) treatments for 5 min at 190°C have been used to disrupt the β-glycosidic bonds and crystalline structure of cellulose to solubilize nearly all hemicellulose in bagasse; however, the lignin content clearly remained (Chen et al., 2011a). In a comparative study, a fourfold increase in released sugars within one sixth of times was obtained from pretreated sugarcane bagasse using microwave-assisted H_2SO_4 or NaOH compared with that of conventional heating methods (Zhu et al., 2016). Sugarcane bagasse was pretreated with a selective white-rot fungus, *Ceriporiopsis subvermispora*, and microwave hydrothermolysis. The combined pretreatment systems caused more defibration and thinning of the plant tissues than without the fungal treatment, and increased the bioethanol yield by 32.6% in simultaneous saccharification and fermentation using *S. cerevisiae* (Sasaki et al., 2011).

Hitherto, the higher costs of converting bagasse to bioethanol hinder its production at industrial scale. Nonetheless, the development of promising strategies for integrating the production of bioethanol with other biobased chemicals has been initiated to make the process cost competitive and economically viable. For instance, use of an efficient pretreatment method, like organosolv assisted by microwave or ultrasound, plus the addition of catalysts to release sufficient fermentable sugars without inhibitors and provide an adequate quantity of value-added chemicals from lignin depolymerization, all at the same time, is a more cost-efficient strategy. Furthermore, the development of cellulase enzymes to structurally and functionally improve their catalytic activities by reducing their adhesive tendencies toward lignin would allow a reduction in enzyme doses, subsequently reducing saccharification costs. In addition, the engineered evolution of robust recombinant fermenting yeasts, especially those using xylose, can achieve efficient fermentation undersuch harsh conditions.

10.5 EVOLUTION OF *SACCHAROMYCES CEREVISIAE* TOWARD SECOND-GENERATION BIOETHANOL PRODUCTION FROM LIGNOCELLULOSIC BIOMASS

S. cerevisiae has very long history as being used for fermentation (Frazer, 2013). It also has been the most thoroughly investigated eukaryotic microorganism for several centuries due to a number of reasons:

- A long history with the production of food and alcoholic beverages.
- Generally regarded as safe (GRAS organism).
- Well-established fermentation process technology for large-scale production.
- Used in several different processes for the pharmaceutical industry.
- Susceptible to genetic modifications by recombinant DNA technology, which has been even further facilitated by the availability of the complete genome sequence of *S. cerevisiae* (Goffeau et al., 1996; Ostergaard et al., 2000).

With the second generation of bioethanol production, *S. cerevisiae* is considered a workhorse of the incipient lignocellulosic biofuel industry (Klinke et al., 2004) due to its ability to voraciously ferment six-carbon sugars, its robustness, and tolerance to stress compared to other fermenting microbes. However, *S. cerevisiae* is particular regarding it carbon sources, ignoring xylose and other five-carbon sugars that can make up nearly half of all available plant sugars. Furthermore, native *S. cerevisiae* is powerless for hydrolyzing cellulose, hemicellulose, or lignin. For cellulosic biofuels to become economically feasible, microbes must be able to hydrolyze all of a plant's components to sugars and ferment those released sugars, including xylose and arabinose, into fuel. So far, finding or engineering a robust microorganism with the "super power" to ferment all released biomass sugars synchronously with the hydrolysis of lignocellulosic components has been a challenge for interested scientists. Laborious efforts have been exerted in this field, many of which have overcome thresholds and paved paths for further improvements in the field.

Many avenues have been pursued in the efforts to engineer a version of *S. cerevesiae* for the second generation of bioethanol production from lignocellulose, including genetic recombination, mutations, hybridization, genome shuffling, or integrating some or even all of these options, as:

- Emergence and expansion of recombinant pentose-fermenting strains, specifically xylose.
- Improvements in fermentative abilities through overexpression or amendment of some metabolic pathways, or deletions of genes with undesired products.
- Insertions and/or augmentation of sugar transporter genes to further increase the rates of fermentation.
- Strengthen inherited stress tolerance toward (1) inhibitors that emerge from biomass hydrolysis and (2) osmotolerance toward either substrates or ethanol.

10.5.1 Endeavors of Evolution in Xylose-Fermenting *S. cerevisiae*

Wild-type *S. cerevisiae* voraciously ferments glucose, mannose, and fructose, as well as the disaccharides sucrose and maltose via the Emden-Meyerhof glycolytic pathway. D-galactose is fermented via combined action of the Leloir pathway and glycolysis, but *S. cerevisiae* lacks the ability to ferment D-xylose, L-arabinose, galacturonic acid, and L-rhamnose (van Maris et al., 2006). The link between xylose and xylulose, a bacterial metabolic pathway of xylose, was described first by Gunsalus et al. (1955), when the enzyme xylose (glucose) isomerase was added to xylose in vitro in the presence of yeast, producing ethanol. Furthermore, increasing xylulokinase activity with overexpression of a strong promoter increases the fermentation rate of xylose and xylulose to more than twice as fast as the parent yeast (Deng and Ho, 1990). Disappointing results have been observed from the introduction of a xylose isomerase (E.C. 5.3.1.5) in *S. cerevisiae* (Gárdonyi and Hahn-Hägerdal, 2003; Harhangi et al., 2003; Hector et al., 2013; Jeffries and Shi, 1999; Kuyper et al., 2003, 2005; Lönn et al., 2002; Moes et al., 1996; Peng et al., 2015; Sarthy et al., 1987; Walfridsson et al., 1996) due to an inappropriate heterologous expression system that has been reported to cause erroneous protein folding, posttranslational modifications, disulfide-bridge formation (Sarthy et al., 1987; van Maris et al., 2006), or other driving forces necessary for promoting this reaction toward xylulose, where at equilibration, the isomerization energetics between xylose and xylulose favors xylose formation by 83:17 (Jeffries and Jin, 2004). It deserves to mention that a little progress in the expression of xylose isomerase in *S. cerevisiae* has been reported, although, not enough to be auspicious since all enzymes involved in the conversion of xylulose to glycolysis intermediates were overexpressed (Walfridsson et al., 1996). However, successful outcomes were reported recently when this technique was integrated with evolutionary engineering, which we explore later (Demeke et al., 2013; Wisselink et al., 2009).

In Fungi, two oxidoreductases, xylose reductase (XR) and xylitol dehydrogenase (XDH), are involved in the fermentation of xylose to xylulose. Xylose is first reduced to xylitol by the action of XR, where NADPH is the preferable cofactor for most XR enzymes (Rizzi et al., 1988). Xylitol then is oxidized to xylulose by the action of xylitol dehydrogenase and the NAD^+cofactor (Rizzi et al., 1989). Xylulose, unlike xylose, can then be naturally utilized by *S. cerevisiae* and several other yeast species (Bruinenberg et al., 1983a). When oxidoreduction is carried out with different cofactor specificities, a conversion yields one $NADP^+$ and one NADH. In the second cycle of conversion under anaerobic conditions, NADPH and NAD^+ should be regenerated; otherwise, xylose fermentation will halt if there is no surplus of NADPH or NAD^+ and an absence of electron acceptors to recycle them in the cell. NADPH is generated typically through the oxidative part of the pentose phosphate pathway (PPP) by the action of glucose-6-phosphate dehydrogenase (ZWF1) (Verho et al., 2003). The cofactors can be regenerated by externally added electron acceptors, or the introduction of acetoin, furfural, or acetaldehyde to NADH, as well as, 5-hydroxymethylfurfural (HMF) to NADPH (Bruinenberg et al., 1983a; Wahlbom and Hahn-Hägerdal, 2002). Whereas, under aerobic condition, NADH can be reoxidized internally via an oxygen acceptor of the respiratory chain (van Maris et al., 2006). Conversion of NADH to NADPH with a transhydrogenase is another uncommon strategy, however, *S. cerevisiae* does not inherently possess this gene (Blank et al., 2005; Bruinenberg et al., 1983b), but with the introduction of a heterologous gene, a recombinant *S. cerevisiae* favors NADPH production (Nissen et al., 2001).

Efficient fermentation of xylose anaerobically is not only due to the absence of key enzymes of the fermentation pathway, but also to the overall redox unbalance, which is another important key factor for dictating conversion efficiencies. Considerable efforts to avoid discrepancies between XR and XDH cofactors have depended on the selection and substitution of different XRs and XDHs from different yeasts, along with the mutations of their proteins and/or different promoter expression levels. These efforts have been the most evaluated, as we discuss subsequently.

The initial development of a recombinant xylose-fermenting *S. cerevisiae* was started in 1990 by transforming two oxidoreductase genes, xylose reductase (*Ps*XR) xylitol dehydrogenase (*Ps*XDH), from *Pichia stipitis* (Amore et al., 1991; Kötter et al., 1990). Lower ratios of XR:XDH (indicating greater XDH activity) not only prevented xylitol and glycerol excretion, but also produced more ethanol than other strains (Walfridsson et al., 1997). With stronger expression of XDH over xylulokinase (XK), an accumulation of xylulose was observed, indicating that XK is another limiting factor for xylose fermentation accompanied by higher XDH activity (Jin and Jeffries, 2003). *S. cerevisiae* is capable of metabolizing xylulose, but at very low rates (Hsiao et al., 1982; Wang and Schneider, 1980). Xylulose enters the PPP after phosphorylation by xylulokinase (XK). As a result, better xylose fermentation performance was obtained from a recombinant *Saccharomyces* sp. strain 1400 (Ho et al., 1998), where overexpressed XK was integrated with both *Ps*XR and *Ps*XDH after it was fused to signals controlling gene expression from *S. cerevisiae* glycolytic genes. Furthermore, it effectively utilized xylose for aerobic growth, synthesis of the xylose-metabolizing enzymes directed by the cloned genes, and did not require the presence of xylose for induction, nor was the synthesis repressed by the presence of glucose in the medium. These properties make the recombinant yeasts able to efficiently ferment xylose to ethanol, and able to efficiently coferment glucose and xylose to ethanol simultaneously in the same medium (Ho et al., 1998). Likewise, conversions of xylose to ethanol by recombinant *S. cerevisiae* harboring *Ps*XR and *Ps*XDH have been reported with an emphasis on the importance of XK to enhance bioethanol production from xylose (Eliasson et al., 2001; Toivari et al., 2001). High activities of both XR and XDH, and moderate XK activity, provide the most efficient expression levels to enhance ethanol yields and decrease xylitol production (Matsushika and Sawayama, 2008). Contradictorily, overexpression of the *Sc*XK gene inhibits growth on xylulose and considerably reduces xylose consumption by recombinant *S. cerevisiae* (Jin et al., 2003; Johansson et al., 2001; Rodriguez-Pena et al., 1998), possibly due to abnormal accumulation of X5P and depletion of ATP (Jin et al., 2003; Johansson et al., 2001), therefore, optimum expression of this gene is strongly required for enhanced bioethanol production from xylose.

Xylitol is one of the undesired products resulting from the conversion of xylose to bioethanol. Overwhelmingly, xylitol is produced due to:

1. Redox imbalance between two oxidoreductase genes, XR and XDH (Fig. 10.5).
2. Higher expression level of XR than XDH.
3. Shortage of cofactors

Therefore, several metabolic engineering strategies have been proposed to solve the aforementioned issues in an effort to reduce xylitol excretion during xylose fermentation:

10.5.1.1 Overhaul of the Coenzyme Specificity of Both XR and XDH

A series of trials have been carried out via site-directed mutagenesis to alter or reverse XR and/or XDH to optimize the recycling of cofactors. Most of these trials have targeted NADH/$NAD^+$by diminishing or remove NADP-binding sites of XR (Fig. 10.6). In this regard, the expression of a *Ps*XR mutant with a higher *km* for NADPH increases ethanol production from xylose in recombinant *S. cerevisiae* (Jeppsson et al., 2006). NAD(P)H-dependent *Ps*XR was altered using a semirational mutagenesis approach (CASTing) and a best mutant was obtained with four mutations: K270S, N272P, S271G, and R276F. This mutant showed a preference toward NADH over NADPH by a factor of about 13-fold, but the effect of that preference on improved xylose fermentation has not been studied at this point (Liang et al., 2007).

The positive effects of decreased NADPH-dependency by mutated *Ps*XR (K270R) were reflected by a 31% reduction in xylitol excretion and a 5.1% improvement in bioethanol production (Watanabe et al., 2007a). Likewise, the catalytic efficiency of the R276H mutant was improved 52-fold regarding the ratio of NADH/NADPH compared with the wild type, which was due to a decrease of *kcat* with NADPH. The most positive effect on xylose fermentation was a 20% increase in ethanol production and a 52% decrease in xylitol excretion, compared with the Y-WT strain expressing *Ps*XR WT and *Ps*XDH WT (Watanabe et al., 2007b). Zeng et al. (2009) completely reversed the coenzyme specificity of *Ps*XR toward NADH and improved catalytic efficiency. Initial experiments showed their catalytic efficiencies for NADPH were nearly zero. The catalytic efficiency of the double mutant (K21A/N272D) for NADH was almost ninefold compared with that of the wild-type enzyme; however, the effect of that mutant on xylose fermentation was not reported.

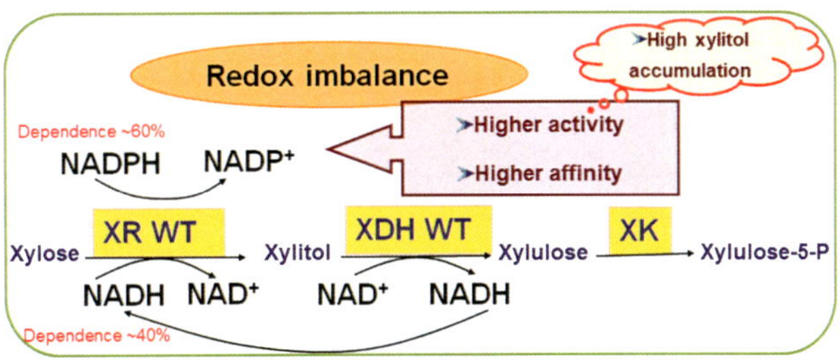

FIGURE 10.5 **Schematic diagram for xylitol excretion resulting from a redox imbalance between wild types *Ps*XR and *Ps*XDH during xylose fermentation in recombinant *S. cerevisiae*.**

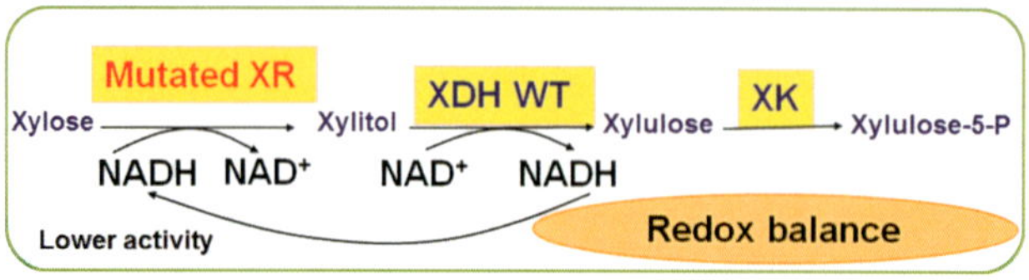

FIGURE 10.6 Schematic diagram for overcoming xylitol accumulation with a mutated*Ps*XR toward exclusively utilizing NADH, to be recycled with wild type *Ps*XDH during xylose fermentation in recombinant *S. cerevisiae*.

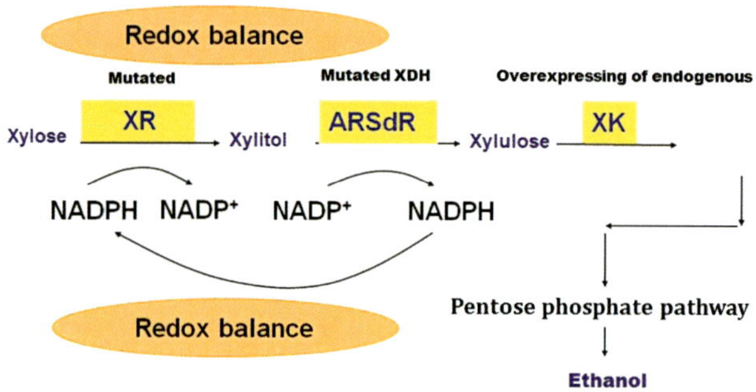

FIGURE 10.7 Schematic diagram for overcoming xylitol accumulation with a mutated *Ps*XDH toward the dual utilization of NAD^+ and $NADP^+$, to be recycled with wild type *Ps*XR during xylose fermentation in recombinant *S. cerevisiae*.

Another method utilizing matched dual cofactors, NAD(H) and NADP(H), also has been checked (Fig. 10.7). This method increases the affinity of XDH for the cofactor NADP. An NADP recognition sequence from *Thermoanaerobium brockii* was introduced, resulting in equal apparent *Km* values for NAD and NADP. The mutated *Tb*XDH gene mediated xylose growth when coexpressed with the XR gene in *S. cerevisiae*; however, bioethanol production from xylose was not reported (Metzger and Hollenberg, 1995). Engineering of a matched pair of XR and XDH for xylose fermentation by *S. cerevisiae* was achieved; however, only 50% reduction in glycerol yield was declared with no increase in ethanol production during xylose fermentation (Krahulec et al., 2009). This result may be due to the use of a *Galactocandida mastotermitis* XDH.

A counterintuitive route toward NADPH/$NADP^+$ also has been in development, and was begun by a complete reversal of coenzyme specificity of XDH from NAD^+ to $NADP^+$ by site-directed mutagenesis. The quadruple mutant ARSdR (D207A/I208R/F209S/N211R) showed more than 4500-fold higher values in *kcat/Km* with $NADP^+$ than in the wild-type enzyme, reaching values comparable with *kcat/Km* with NAD^+ of the wild-type enzyme (Watanabe et al., 2005). The significant enhancement of that method was clearly reflected in the 86% decrease in unfavorable xylitol excretion, with the 41% increase in ethanol production by a recombinant *S. cerevisiae* strain; the strain harbored *Ps*XR and a quadruple mutant ARSdR of *Ps*XDH (Watanabe et al., 2007c). A noteworthy accomplishment in xylose-fermenting yeast research is the chromosomal integration of the gene for $NADP^+$-dependent *Ps*XDH, a quadruple mutant ARSdR, along with *Ps*XR and *S. cerevisiae* endogenous xylulokinase genes (*Sc*XK) (Matsushika et al., 2008). The authors successfully reported an increase in ethanol production and a decrease in xylitol excretion, compared to the reference strain. The new construct expressed wild-type XDH when fermenting, not only xylose, but also mixed sugars containing glucose and xylose. Furthermore, the recombinant strain produced a high yield of ethanol from a nonsulfuric acid hydrolysate of wood chips (Matsushika et al., 2008). A novel, strictly NADPH-dependent *Ps*XR, constructed by site-directed mutagenesis with a double mutation (E223A/S271A), showed strict NADPH dependency with 106% increase in activity over the wild-type. Another double mutant, E223D/S271A, showed a 1.27-fold increase in activity compared to the wild-type XR with NADPH, and almost negligible activity with NADH (Khattab et al., 2011). This was done to evaluate the effects of exclusively recycling NADPH/$NADP^+$ cofactors between *Ps*XR and *Ps*XDH on xylitol excretion and bioethanol production (Fig. 10.8). Chromosomal integration of double-mutated *Ps*XRs (E223A/S271A, E223D/S271A, and E223G/S271A) was accomplished with the gene for the $NADP^+$-dependent quadruple mutant ARSdR XDH, and the *Sc*XK genes. All mutated strains demonstrated a reduction in xylitol accumulation, ranging from 34.4% to 54.7% compared to the control strain. Moreover, 20% and 10% improvement in bioethanol production was observed compared to the control strain that had chromosomal integration of wild-type *Ps*XR, ARSdR *Ps*XDH, and XK. A few positive points for using this route include the decrease in the *Km* of xylose with NADPH, the wild-type *Km* values for NADPH-dependent *Ps*XR that were nearly maintained in the mutated genes, their roles in the acceleration and efficiency of fermentation reactions, as well as higher XDH: XR ratios by 6- and 10-folds (Khattab et al., 2013).

Efficient recycling of NADPH/$NADP^+$ has a great effect on bioethanol production by preventing the accumulation of xylitol through instant reoxidization of the produced $NADP^+$ and ARSdR (XDH) to xylulose. This was confirmed further with the overexpression of *Sc*XK and the native NADPH-dependent aldose reductase (GRE3), as shown in (Fig. 10.8). GRE3 plays a role in stress protection (Aguilera and Prieto, 2001) with the strict $NADP^+$-dependent quadruple mutant ARSdR *Ps*XDH, where researchers observed a considerably lower xylitol accumulation (eifhtfold lower) and a 21.4% higher increase in bioethanol production (Khattab and Kodaki, 2014). Later, this strain converted 77% of released sugars from pretreated bagasse to bioethanol (Khattab, 2015).

10.5.1.2 Optimization of the XR/XDH Ratio

Matching expression levels of oxidoreductase enzymes is one of the requirements for preventing xylitol excretion (Walfridsson et al., 1997). Strains with XR/XDH/XK ratios corresponding to $1 \geq 10 \geq 4$ achieve the most ethanol production more efficiently (Eliasson et al., 2001). Deletion of GRE3 gene, the endogenous NADPH-dependent aldose reductase in *S. cerevisiae*, decreased xylitol accumulation by 49% and increased the ethanol yield by 19% during fermentation of glucose/xylose in a mixed culture by a recombinant *S.*

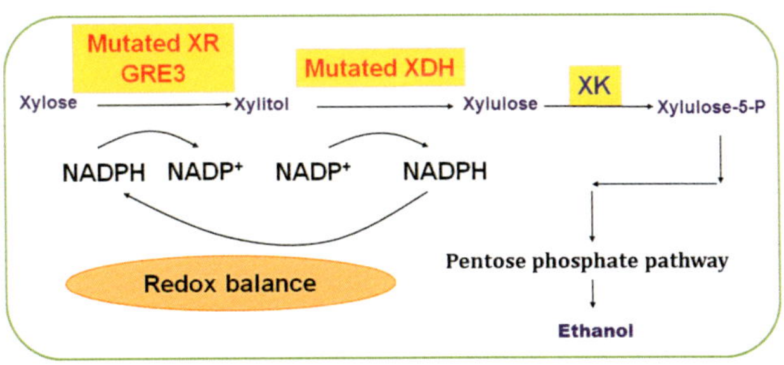

FIGURE 10.8 Schematic diagram for overcoming xylitol accumulation with mutations in both *Ps*XR and *Ps*XDH toward exclusively utilizing and recycling NADPH/$NADP^+$ during xylose fermentation in recombinant *S.cerevisiae*.

cerevisiae harbored xylose reductase *Ps*XR gene (Träff-Bjerre et al., 2004).

10.5.1.3 Increasing the NADPH Pool

Increasing the NADPH pool is another strategy for overcoming the problem of NADPH limitations. Glyceraldehyde 3-phosphate dehydrogenase (NADP-GAPDH), encoded by the GDP1gene from *Kluyveromyces lactis*, was overexpressed in *S. cerevisiae* to facilitate NADPH regeneration; therefore, the production of the unwanted byproduct xylitol decreased by more than 40% (Bera et al., 2011; Verho et al., 2003). Based on the idea of decreasing NADPH utilization during ammonium assimilation in *S. cerevisiae*, researchers are shifting assimilatory actions to depend on NADH. This has been accomplished by deleting the NADPH-dependent glutamate dehydrogenase (GDH1) and overexpressing the (GS-GOGAT) system, which consists of glutamate synthase (GLT1) and glutamine synthetase (GLN1). Results indicate that this will provide more available NADPH for the reduction of xylose. Additionally, overexpression of NADH-dependent glutamate dehydrogenase (GDH2) spares more NADPH for xylose reduction, which demonstrated an increase in the ethanol yield of ~15%, mainly due to a reduction in xylitol excretion by 44% (Roca et al., 2003).

10.5.1.4 IMPROVED FLUX IN THE OXIDATIVE PPP

Improved flux through the PPP in *S. cerevisiae*, including transaldolase (TAL1), transketolase (TKL1), D-ribose-5-phosphate ketol-isomerase (RKI1), and D-ribulose-5-phosphate 3-epimerase (RPE1) genes, is another method used to improve xylose consumption rates, also resulting in reduced xylitol accumulation and subsequent increased bioethanol production (Fig. 10.9). TKL1 and TAL1 genes were overexpressed in *S. cerevisiae*, showing considerably enhanced growth

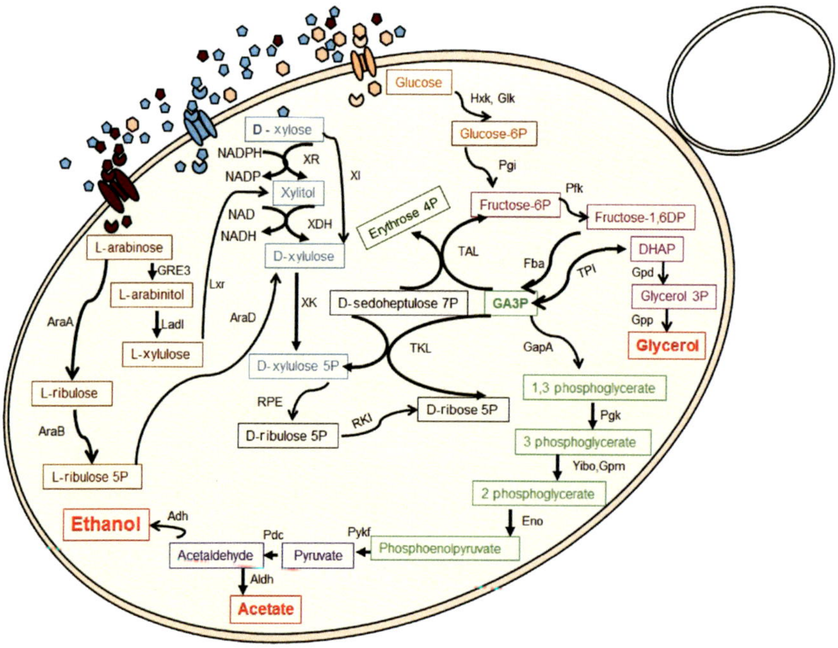

FIGURE 10.9 **Metabolic pathways of glucose, xylose, and arabinose in recombinant *S. cerevise*.**

on xylose compared with a strain that harbored only XR and XDH (Walfridsson et al., 1995). This point is confirmed by Karhumaa et al. (2007) who demonstrated that expression of high XR and XDH in the recombinant *S. cerevisiae* is governed by the rate of xylose fermentation, the overexpression of nonoxidative PPP, and the redirection of significant amounts of xylitol as a product to significant amounts of ethanol instead. On the contrary, reduced flux from the oxidative PPP was seen with disruptions in the GND1 gene, one of the isogenes of 6-phosphogluconate dehydrogenase, or the glucose 6-phosphate dehydrogenase that is encoded by the ZWF1 gene. Emerging from these disruptions were decreased phosphoglucose isomerase activity and an 84% lower xylose consumption rate, but the key results were the improvements in bioethanol production, as seen in these recombinant *S. cerevisiae* strains (Johansson and Hahn-Hägerdal, 2002).

10.5.1.5 Innovating the Integration of Recombinant and Evolutionary Strategies

Evolutionary engineering approaches for *S. cerevisiae* have provided promising strategies, despite the ambiguous and arduous reproduction. A novel evolutionary engineering approach has emerged for accelerated utilization of glucose, xylose, and arabinose mixtures by engineered *S. cerevisiae* strains. *S. cerevisiae* was engineered first to harbor structural genes with strong promotors for the overexpression of *Piromyces* XylA, *Lactobacillus plantarum* L-arabinose isomerase (araA), L-ribulokinase (araB), and L-ribulose-5-phosphate 4-epimerase (araD) genes. This was also true for the endogenous genes of the PPP for all enzymes involved in the conversion of xylulose to glycolytic intermediates: xylulokinase (*Ps*XK), ribulose 5-phosphate isomerase (RPI1), ribulose 5-phosphate epimerase (RPE1), transketolase (TKL1), and transaldolase (TAL1). In addition, the GRE3 gene encoding aldose reductase was deleted to further minimize xylitol production (Wisselink et al., 2007). Evolutionary strategies were based on a repetitive regimen of consecutive cultivation of the engineered strain, for utilization and fermentation of both xylose and arabinose (IMS0002). Three media were used with different compositions (glucose, xylose, and arabinose; xylose and arabinose; and only arabinose). *S. cerevisiae* was the first recombinant strain to ferment mixtures of glucose, xylose, and arabinose, resulting in a high ethanol yield (0.43 g/g of total sugar) without the formation of the side products xylitol or arabinitol. Moreover, this evolutionary strategy resulted in a 40% reduction in fermentation time (Wisselink et al., 2009). However, while there is no clear evidence of underlying genetic changes involved in the improvement in xylose and arabinose utilization by the evolved strain, potential changes would include mutations resulting in improved codon usage of introduced genes. Codon optimization for AraA, AraB, and AraD in engineered *S. cerevisiae* strains has been shown to result in improved arabinose conversion rates (Wiedemann and Boles, 2008).

Combining metabolic engineering and evolutionary engineering, with random mutagenesis has been successful for developing strains with improved D-xylose fermentation efficiencies (Liu and Hu, 2010; Zhou et al., 2012). Successfully exploiting a combined approach of consecutive metabolic engineering, mutation, genome shuffling, and evolutionary engineering developed a robust industrial *S. cerevisiae* strain exhibiting the highest yield of ethanol from D-xylose. An expression cassette of 13 genes from *Clostridium phytofermentans* was developed, including codon-optimized xylose isomerase (XI), codon-optimized yeast xylulokinase (XK), the yeast pentose/hexose transporter Hxt7, and all the enzymes of the nonoxidative part of the PPP: TKL1, TAL1, RPE1, and RKI1. Also included were the arabinose transporter (AraT) from *Scheffersomyces* (*Pichia*) *stipites*, codon-optimized arabinose isomerase (AraA) from *Bacillus licheniformis*, codon-optimized ribulokinase (AraB), and ribulose-5-p 4-epimerase (AraD) from *Escherichia coli* (Fig. 10.9), as well as codon-optimized yeast transketolase 2 (Tkl2) and transaldolase 2 (Nqm1). These were chromosomally integrated into PYK2 alleles of the ethanol red strain. Thereafter underwent to ethyl methanesulfonate mutagenesis, genome shuffling, and stress tolerance of inhibitors. The best-performing strain, GS1.11–26, showed a maximum specific D-xylose consumption rate of 1.1 g/g DW/h in synthetic medium, with complete attenuation of 35 g/L D-xylose in about 17 h. Cofermentation in rich Yeast peptone (YP) medium containing 36 g/L glucose and 37 g/L D-xylose resulted in nearly complete consumption of both glucose and D-xylose in about 13 h by the evolved strain, resulting in a high overall ethanol productivity of 1.4 g/g DW/h. In aseparate hydrolysis and fermentation of lignocellulosic hydrolysates of *Arundo donax* (giant reed), spruce, and a wheat straw/hay mixture, the maximum specific D-xylose consumption rates were 0.36, 0.23, and 1.1 g/g DW inoculum/h, respectively. The final ethanol titer was 4.2%, 3.9%, and 5.8% (v/v), respectively. This notable improvement in the xylose consumption accomplished by the synergistic effect of expressed XI and one or more mutations occurred in the genome (Demeke et al., 2013). Hitherto, the authors reported that this was the best xylose-fermenting strain, and it responded well to most research strategies, with the exception of a deletion of putative suppressor genes that are discussed below.

10.5.1.6 Deletion of Putative Suppressor Genes

New findings of genome-wide synthetic interactions reveal that more than 80 nonessential genes from a diverse range of cellular processes impact xylose utilization. Among those, four putative suppressor genes showed improved xylose utilization by *S. cerevisiae* S288C and CEN.PK when individually deleted. Those four genes, to date, have not been linked to pentose metabolism and include: the arginine transporter (ALP1), mitochondrial membrane–localized inositol phosphosphingolipid phospholipase C (ISC1), a component of the large (60S) ribosomal subunit (RPL20B), and component of the small ribosomal subunit (BUD21). Deleted BUD21 mutant cells produced ethanol, even in the absence of the exogenous xylose isomerase gene (XI) from *Piromyces* sp. E2. Ultimately, removed the suppressed genes, such as BUD21 retrieve the ability of *S. cerevisiae* to utilize and ferment xyloseas a sole carbon source (Usher et al., 2011).

10.6 CONCLUSIONS AND FUTURE PROSPECTS

Great efforts have been undertaken to diminish the complexity structure of lignocellulosic biomass (i.e., bagasse) using different pretreatment methods with the lowest amount of inhibitor production to be accessible by cellulase enzymes with a minimum saccharification dose, thereby releasing the highest sugar content. Engineered evolution of a robust yeast strain with high fermentative abilities has allowed higher efficiency conversions of a wide range of sugars, especially those dominant in lignocellulosic biomass. Considering the total production costs, we still need further developments in three parallel directions:

1. Pretreatments of lignocellulosic biomass exclusively based on microwave or ultrasonic wave treatments, assisted by organosolv and catalysts as seen in recent promising results.
2. Further development of cellulase enzymes with high catalytic activities at 30°C, and with hydrophobic lignin interactions to prevent adsorption.
3. Evolution of *S. cerevisiae* yeast with superfermenting abilities with all suppressor genes deleted that may have deleterious effects on fermentation. Moreover, overexpression of all fermenting metabolic pathways of glucose, xylose, and arabinose sugars with higher tolerances to inhibitors. At the same time, evolution of partially hydrolignocellulosic fermenting strains may provide synergetic effects between saccharfication and tolerance of inhibitors.

References

Abdul Khalil, H.P.S., Hossain, M.S., Rosamah, E., Azli, N.A., Saddon, N., Davoudpoura, Y., Islam, M.N., Dungani, R., 2015. The role of soil properties and its interaction towards quality plant fiber: a review. Renew. Sust. Energy Rev. 43, 1006–1015.

Aguilar, R., Ramírez, J.A., Garrote, G., Váazquez, M., 2002. Kinetic study of the acid hydrolysis of sugar cane bagasse. J. Food Eng. 55, 309–318.

Aguilera, J., Prieto, J.A., 2001. The *Saccharomyces cerevisiae* aldose reductase is implied in the metabolism of methylglyoxal in response to stress conditions. Curr. Genet. 39, 273–283.

Alves, L.A., Felipe, M.G.A., Silva, J.B.A., Silva, S.S., Prata, M.R., 1998. Pretreatment of sugarcane bagasse hydrolysate for xylitol production by *Candida guilliermondii*. Appl. Biochem. Biotechnol. 70–72, 89–98.

Alves, L.A., Vitolo, M., Felipe, M.G.A., Silva, J.B.A., 2002. Xylose reductase and xylitol dehydrogenase activities of *Candida guilliermondii* as a function of different treatments of sugarcane bagasse hydrolysate employing experimental design. Appl. Biochem. Biotechnol. 98–100, 403–413.

Amore, R., Kötter, P., Kuster, C., Ciriacy, M., Hollenberg, C.P., 1991. Cloning and expression in *Saccharomyces cerevisiae* of the NAD(P)H-dependent xylose reductase-encoding gene (XYL1) from the xylose-assimilating yeast *Pichia stipitis*. Gene 109, 89–97.

Avci, A., Saha, B.C., Dien, B.S., Kennedy, G.J., Cotta, M.A., 2013. Response surface optimization of corn stover pretreatment using dilute phosphoric acid for enzymatic hydrolysis and ethanol production. Bioresour. Technol. 130, 603–612.

Barber, J., 2009. Photosynthetic energy conversion: natural and artificial. Chem. Soc. Rev. 38, 185–196.

Benjamin, Y., Cheng, H., Görgens, J.F., 2013. Evaluation of bagasse from different varieties of sugarcane by dilute acid pretreatment and enzymatic hydrolysis. Ind. Crops Prod. 51, 7–18.

Bera, A.K., Ho, N.W.Y., Khan, A., Sedlak, M., 2011. A genetic overhaul of *Saccharomyces cerevisiae* 424A(LNH-ST) to improve xylose fermentation. J. Ind. Microbiol. Biotechnol. 38, 617–626.

Bidlack, J., Malone, M., Benson, R., 1992. Molecular structure and component integration of secondary cell wall in plants. Proc. Okla. Acad. Sci. 72, 51–56.

Blank, L.M., Lehmbeck, F., Sauer, U., 2005. Metabolic-flux and network analysis in fourteen hemiascomycetous yeasts. FEMS Yeast Res. 5, 545–558.

Brienzo, M., Siqueira, A.F., Milagres, A.M.F., 2009. Search for optimum conditions of sugarcane bagasse hemicellulose extraction. Biochem. Eng. J. 46, 199–204.

Brienzo, M., Tyhoda, L., Benjamin, Y., Görgens, J., 2015. Relationship between physicochemical properties and enzymatic hydrolysis of sugarcane bagasse varieties for bioethanol production. Nat. Biotechnol. 32, 253–262.

Brosse, N., 2012. In: Paterson, R.J. (Ed.), Lignin: Properties and Applications in Biotechnology and Bioenergy. Nova Science, New York, pp. 381–402.

Bruinenberg, P.M., de Bot, P.H.M., van Dijken, J.P., Scheffers, W.A., 1983a. The role of the redox balance in the anaerobic fermentation of xylose by yeasts. Eur. J. Appl. Microbiol. Biotechnol. 18, 287–292.

Bruinenberg, P.M., van Dijken, J.P., Scheffers, W.A., 1983b. An enzymic analysis of NADPH production and consumption in *Candida utilis*. J. Gen. Microbiol. 129, 965–971.

Buijs, N.A., Siewers, V., Nielsen, J., 2013. Advanced biofuel production by the yeast *Saccharomy cescerevisiae*. Curr. Opin. Chem. Biol. 17, 480–488.

Bustos, G., Ramírez, J.A., Garrote, G., Vázquez, M., 2003. Modeling of the hydrolysis of sugar cane bagasse with hydrochloric acid. Biotechnol Appl. Biochem. 104, 51–68.

Calvo-Flores, F.G., Dobado, J.A., 2010. Lignin as renewable raw material. Chem. Sust. Chem. 3, 1227–1235.

Canilha, L., Santos, V.T.O., Rocha, G.J.M., 2011. A study on the pretreatment of a sugarcane bagasse sample with dilute sulfuric acid. J. Ind. Microbiol. Biotechnol. 38, 1467–1475.

Chen, W., Tu, Y., Sheen, H., 2011a. Disruption of sugarcane bagasse lignocellulosic structure by means of dilute sulfuric acid pretreatment with microwave-assisted heating. Appl. Energy 88, 2726–2734.

Chen, W.H., Ye, S.C., Sheen, H.K., 2011b. Hydrolysis characteristics of sugarcane bagasse pretreated by dilute acid solution in a microwave irradiation environment. Appl. Energy 93, 237–244.

Cheng, K.K., Cai, B.Y., Zhang, J.A., Ling, H.Z., Zhou, Y.J., Ge, J.P., Xu, J.M., 2008. Sugarcane bagasses hemicellulose hydrolysate for ethanol production by acid recovery process. Biochem. Eng. J. 38, 105–109.

Chiang, V.L., Funaoka, M., 1990. The difference between guaiacyl and guaiacyl-gyringyl lignins in their responses to Kraft delignification. Holzforschung 44, 309–313.

da Silva, A.S., Inoue, H., Endo, T., Yano, S., Bon, E.P.S., 2010. Milling pretreatment of sugarcane bagasse and straw for enzymatic hydrolysis and ethanol fermentation. Bioresour. Technol. 101, 7402–7409.

da Silvaa, G.T., Chiarelloa, L.M., Limaa, E.M., Ramosa, L.P., 2016. Sono-assisted alkaline pretreatment of sugarcane bagasse for cellulosic ethanol production. Catal. Today 269, 21–28.

de Vasconcelos, S.M., Pinheiro Santos, A.M., Rocha, G.J.M., Souto-Maior, A.M., 2013. Diluted phosphoric acid pretreatment for production of fermentable sugars in a sugarcane-based biorefinery. Bioresour. Technol. 135, 46–52.

Demeke, M.M., Dietz, H., Li, Y., Foulquié-Moreno, M.R., Mutturi, S., Deprez, S., Abt, T.D., Bonini, B.M., Liden, G., Dumortier, F., Verplaetse, A., Boles, E., Thevelein, J.M.,

2013. Development of a D-xylose fermenting and inhibitor tolerant industrial *Saccharomyces cerevisiae* strain with high performance in lignocellulose hydrolysates using metabolic and evolutionary engineering. Biotechnol. Biofuels 6, 89.

Deng, X.X., Ho, N.W.Y., 1990. Xylulokinase activity in various yeasts including *Saccharornyces cerevisiae* containing the cloned xylulokinase gene. Appl. Biochem. Biotechnol. 24/25, 193–199.

Ding, S.Y., Himmel, M.E., 2006. The maize primary cell wall microfibril: a new model derived from direct visualization. J. Agric. Food Chem. 54, 597–606.

Eliasson, A., Hofmeyrc, J.S., Pedler, S., Hahn-Hägerdal, B., 2001. The xylose reductase/xylitol dehydrogenase/xylulokinase ratio affects product formation in recombinant xylose-utilising*Saccharomyces cerevisiae*. Enzyme Microb. Technol. 29, 288–297.

Fan, J.J., De Bruyn, M., Budarin, V.L., Gronnow, M.J., Shuttleworth, P.S., Breeden, S., et al., 2013. Direct microwave-assisted hydrothermal depolymerization of cellulose. J. Am. Chem. Soc. 135 (32), 11728–11731.

Fan, L.T., Gharpuray, M.M., Lee, Y.H., 1987. Cellulose Hydrolysisfirst ed. Springer-Verlag, New York.

Fengel, D., Wegener, G., 1984. Wood: Chemistry, Ultrastructure Reactions. W. De Gruyter, New York, p. 613.

Fox, D., Gray, P., Dunn, N., Marsden, W., 1987. Factors affecting the enzymic susceptibility of alkali and acid pretreated sugarcane bagasse. J. Chem. Technol. Biotechnol. 40, 117–132.

Frazer, J., 2013. Yeast: Making Food Great for 5,000 Years. But What Exactly Is it? Available from: https://blogs.scientificamerican.com/artful-amoeba/yeast-making-food-great-for-5000-years-but-what-exactly-is-it/.

Fulton, L., Howes, T., Hardy, J., 2004. Biofuels for Transport: An International Perspective. International Energy Agency, Paris, France., Available online at: http://www.iea.org/textbase/nppd/free/2004/biofuels2004.pdf.

Gámez, S., González-Cabriales, J.J., Ramírez, J.A., Garrote, G., Vázquez, M., 2006. Study of the hydrolysis of sugar cane bagasse using phosphoric acid. J. Food Eng. 74, 78–88.

Gárdonyi, M., Hahn-Hägerdal, B., 2003. The *Streptomyces rubiginosus* xylose isomerase is misfolded when expressed in *Saccharomyces cerevisiae*. Enzyme Microb. Technol. 32, 252–259.

Goffeau, A., Barrell, B.G., Bussey, B., et al., 1996. Life with 6000 genes. Science 274, 546–567.

Gong, C.S., Chen, C.S., Chen, L.F., 1993. Pretreatment of sugar cane bagasse hemicellulose hydrolyzate for ethanol production by yeast. Biotechnol. Appl. Biochem. 39/40, 83–88.

Guilherme, A.A., Dantas, P.V.F., Santos, E.S., Fernandes, F.A.N., Macedo, G.R., 2015. Evaluation of composition, characterization and enzymatic hydrolysis of pretreated sugar cane bagasse. Braz. J. Chem. Eng. 32, 23–33.

Gunsalus, I.C., Horecker, B.L., Wood, W.A., 1955. Pathways of carbohydrate metabolism in microorganisms. Bacteriol. Rev. 19, 79–128.

Harhangi, H.R., Akhmanova, A.S., Emmens, R., van der Drift, C., de Laat, W.T., van Dijken, J.P., Jetten, M.S., Pronk, J.T., Op den Camp, H.J., 2003. Xylose metabolism in the anaerobic fungus *Piromyces* sp. strain E2 follows the bacterial pathway. Arch. Microbiol. 180, 134–141.

Harrison, M.D., Zhang, Z., Shand, K., O'Hara, I.M., Doherty, W.O.S., Dale, J.L., 2013. Effect of pretreatment on saccharification of sugar cane bagasse by complex and simple enzyme mixtures. Bioresour. Technol. 148, 105–113.

Hasegawa, N., Mitani, T., Shinohara, N., Daidai, M., Katsura, Y., Sego, H., Watanabe, T., 2014. Pilot-plant scale 12 kW microwave irradiation reactor for woody biomass pretreatment. IEICE Trans. Electron. E97.C, 986–993.

Hatfield, R., Fukushima, R.S., 2005. Can lignin be accurately measured? Crop Sci. 45, 832–839.

Hector, R.E., Dien, B.S., Cotta, M.A., Jeffrey, A., Mertens, J.A., 2013. Growth and fermentation ofD-xylose by *Saccharomyces cerevisiae* expressing a novel D-xylose isomerase originating from the bacterium *Prevotella ruminicola* TC2-24. Biotechnol. Biofuels 6, 84.

Hernández-Salas, J.M., Villa-Ramírez, M.S., Veloz-Rendón, J.S., Rivera-Hernández, K.N., González-César, R.A., Plascencia-Espinosa, M.A., Trejo-Estrada, S.R., 2009. Comparative hydrolysis and fermentation of sugarcane and agave bagasse. Bioresour. Technol. 100, 1238–1245.

Himmel, M.E., Ding, S., Johnson, D.K., Adney, W.S., Nimlos, M.R., Brady, J.W., Foust, T.D., 2007. Biomass recalcitrance: engineering plants and enzymes for biofuels production. Science 315.

Ho, N.W.Y., Chen, Z., Brainard, A.P., 1998. Genetically engineered *Saccharomyces* yeast capable of effective cofermentation of glucose and xylose. Appl. Environ. Microbiol. 64, 1852–1859.

Hsiao, H.Y., Chiang, L.C., Chen, L.F., Tsao, G.T., 1982. Effects of borate on isomerization and yeast fermentation of high xylulose solution and acid hydrolysate of hemicellulose. Enzyme Microb. Technol. 4, 25–31.

Jeffries, T.W., Jin, Y.S., 2004. Metabolic engineering for improved fermentation of pentoses by yeasts. Appl. Microbiol. Biotechnol. 63, 495–509.

Jeffries, T.W., Shi, N.Q., 1999. Genetic engineering for improved xylose fermentation by yeasts. Adv. Biochem. Eng. Biotechnol. 65, 117–161.

Jeppsson, M., Bengtsson, O., Franke, K., Lee, H., Hahn-Hägerdal, B., Gorwa-Grauslund, M.F., 2006. The expression of a *Pichia stipites* xylose reductase mutant with higher K(M) for NADPH increases ethanol production from xylose in recombinant *Saccharomyces cerevisiae*. Biotechnol. Bioeng. 93, 665–673.

Jin, Y.S., Jeffries, T.W., 2003. Changing flux of xylose metabolites by altering expression of xylose reductase and xylitol dehydrogenase in recombinant *Saccharomyces cerevisiae*. Appl. Biochem. Biotechnol. 105/108, 277–285.

Jin, Y.S., Ni, H., Laplaza, J.M., Jeffries, T.W., 2003. Optimal growth and ethanol production from xylose by recombinant *Saccharomyces cerevisiae* require moderate D-xylulokinase activity. Appl. Environ. Microbiol. 69, 495–503.

Johansson, B., Christensson, C., Hobley, T., Hahn-Hägerdal, B., 2001. Xylulokinase overexpression in two strains of *Saccharomyces cerevisiae* also expressing xylose reductase and xylitol dehydrogenase and its effect on fermentation of xylose and lignocellulosic hydrolysate. Appl. Environ. Microbiol. 67, 4249–4255.

Johansson, B., Hahn-Hägerdal, B., 2002. The non-oxidative pentose phosphate pathway controls the fermentation rate of xylulose but not of xylose in *Saccharomyces cerevisiae*. FEMS Yeast Res. 2, 277–282.

Jørgensen, H., Kristensen, J.B.H., Felby, C., 2007. Enzymatic conversion of lignocellulose into fermentable sugars: challenges and opportunities. Biofuels Bioprod. Biorefin. 1, 119–134.

Karhumaa, K., Fromanger, R., Hahn-Hägerdal, B., Gorwa-Grauslund, M.F., 2007. High activity of xylose reductase and xylitol dehydrogenase improves xylose fermentation by recombinant *Saccharomyces cerevisiae*. Appl. Microbio. Biotechnol. 73, 1039–1046.

Khattab, S.M., Saimura, M., Kodaki, T., 2013. Boost in bioethanol production using recombinant *Saccharomyces cerevisiae* with mutated strictly NADPH-dependentxylose reductase and $NADP^+$-dependent xylitol dehydrogenase. J. Biotechnol. 165, 153–156.

Khattab, S.M., Watanabe, S., Saimura, M., Kodaki, T., 2011. A novel strictly NADPH dependent *Pichia stipitis* xylose reductase constructed by site-directed mutagenesis. Biochem. Biophys. Res. Commun. 404, 634–637.

Khattab, S.M.R., 2015. Production of bioethanol from sugarcane bagasse using recombinant *Saccharomyces cerevisiae* strain. J. Basic Appl. Mycol. 6, 17–25.

Khattab, S.M.R., Kodaki, T., 2014. Efficient bioethanol production by overexpression of endogenous *Saccharomyces cerevisiae* xylulokinase and NADPH-dependent aldose reductase with mutated strictly $NADP^+$-dependent *Pichia stipitis* xylitol dehydrogenase. Process Biochem. 49, 1838–1842.

Kim, S., Dale, B.E., 2004. Global potential bioethanol production from wasted crops and crop residues. Biomass Bioenergy 26, 361–375.

Klinke, H.B., Thomsen, A.B., Ahring, B.K., 2004. Inhibition of ethanol-producing yeast and bacteria by degradation products produced during pretreatment of biomass. Appl. Microbiol. Biotechnol. 66, 10–26.

Koshijima, T., Watanabe, T., 2003. Association between lignin and carbohydrates in wood and other plant tissues. In: Koshijima, T., Watanabe, T. (Eds.), Springer Series in Wood Science. Springer Verlag, Berlin, pp. 1–329.

Kötter, P., Amore, R., Hollenberg, C.P., Ciriacy, M., 1990. Isolation and characterization of the *Pichia stipitis* xylitol dehydrogenase gene, XYL2, and construction of a xylose-utilizing *Saccharomyces cerevisiae* transformant. Curr. Genet. 18, 493–500.

Krahulec, S., Klimacek, M., Nidetzky, B., 2009. Engineering of a matched pair of xylose reductase and xylitol dehydrogenase for xylose fermentation by *Saccharomyces cerevisiae*. Biotechnol. J. 4, 684–694.

Kuyper, M., Harhangi, H.R., Stave, A.K., Winkler, A.A., Jetten, M.S., de Laat, W.T., den Ridder, J.J., Op den Camp, H.J., van Dijken, J.P., Pronk, J.T., 2003. High-level functional expression of a fungal xylose isomerase: the key to efficient ethanolic fermentation of xylose by *Saccharomyces cerevisiae*? FEMS Yeast Res. 4, 69–78.

Kuyper, M., Hartog, M.M., Toirkens, M.J., Almering, M.J., Winkler, A.A., van Dijken, J.P., Pronk, J.T., 2005. Metabolic engineering of a xylose-isomerase-expressing *Saccharomyces cerevisiae* strain for rapid anaerobic xylose fermentation. FEMS Yeast Res. 5, 399–409.

Lancaster, M., 2002. Green Chemistry: An Introductory Text. Royal Society of Chemistry, Cambridge.

Lavarack, B.P., Griffin, G.J., 2002. The acid hydrolysis of sugarcane bagasse hemicellulose to produce xylose, arabinose, glucose and other products. Biomass Bioenergy 23, 367–380.

Lee, D., Yu, A.H.C., Wong, K.K.Y., Saddler, J.R., 1994. Evaluation of the enzymatic susceptibility of cellulosic substrates using specific hydrolysis rates and enzyme adsorption. Biotechnol. Appl. Biochem. 45/46, 407–415.

Liang, L., Zhang, J., Lin, Z., 2007. Altering coenzyme specificity of *Pichia stipitis* xylose reductase by the semi-rational approach CASTing. Microb. Cell Fact. 6, 36.

Liu, E., Hu, Y., 2010. Construction of a xylose-fermenting *Saccharomyces cerevisiae* strain by combined approaches of genetic engineering, chemical mutagenesis and evolutionary adaptation. Biochem. Eng. J. 48, 204–210.

Lönn, A., Gardonyi, M., van Zyl, W., Hahn-Hägerdal, B., Otero, R.C., 2002. Cold adaptation of xylose isomerase from *Thermus thermophilus* through random PCR mutagenesis: gene cloning and protein characterization. Eur. J. Biochem. 269, 157–163.

Lynd, L.R., 1996. Overview and evaluation of fuel ethanol production from cellulosic biomass: technology, economics, the environment and policy. Annu. Rev. Energy Environ. J. 21, 403–465.

Macquarrie, D.J., Clark, J.H., Fitzpatrick, E., 2012. The microwave pyrolysis of biomass. Biofuels Bioprod. Biorefin. 6, 549–560.

Martín, C., Galbe, M., Nilvebrant, N., Jönsson, J.L., 2002. Comparison of the fermentability of enzymatic hydrolyzates of sugarcane bagasse pretreated by steam explosion using different impregnating agents. Appl. Biochem. Biotechnol. 98-100, 699–716.

Martín, C., Puls, J., Saake, B., Schreiber, A., 2011. Effect of glycerol preatreatment on component recovery and enzymatic hydrolysis of sugarcane bagasse. Cellul. Chem. Technol. 45, 487–494.

Maryana, R., Marifatun, R., Wheni, A., Satriyo, K.W., Rizal, W.A., 2014. Alkaline pretreatment on sugarcane bagasse for bioethanol production. Energy Procedia 47, 250–254.

Matsushika, A., Sawayama, S., 2008. Efficient bioethanol production from xylose by recombinant *Saccharomyces cerevisiae* requires high activity of xylose reductase and moderate xylulokinase activity. J. Biosci. Bioeng. 106, 306–309.

Matsushika, A., Watanabe, S., Kodaki, T., Makino, K., Inoue, H., Murakami, K., Takimura, O., Sawayama, S., 2008. Expression of protein engineered $NADP^+$-dependent xylitol dehydrogenase increase ethanol production from xylose in recombinant *Saccharomyces cerevisiae*. Appl. Microbiol. Biotechnol. 81, 243–255.

Metzger, M.H., Hollenberg, C.P., 1995. Amino acid substitutions in the yeast *Pichia stipitis* xylitol dehydrogenase coenzyme-binding domain affect the coenzyme specificity. Eur. J. Biochem. 228 (1), 50–54.

Moes, C.J., Pretorius, I.S., van Zyl, W.H., 1996. Cloning and expression of the *Clostridium thermosulfurogenes* D-xylose isomerase gene (xylA) in *Saccharomyces cerevisiae*. Biotechnol. Lett. 18, 269–274.

Najafpour, M.M., Pashaei, B., 2012. Photosynthesis: how and why? in Chapter 1Advances in Photosynthesis: Fundamental Aspects. InTech.

Neureiter, M., Danner, H., Thomasser, C., Saidi, B., Braun, R., 2002. Dilute-acid hydrolysis of sugarcane bagasse at varying conditions. Appl. Biochem. Biotechnol. 98-100, 49–58.

Nissen, T.L., Anderlund, M., Nielsen, J., Villadsen, J., Kielland-Brandt, M.C., 2001. Expression of a cytoplasmic transhydrogenase in *Saccharomyces cerevisiae* results in formation of 2-oxoglutarate due to depletion of the NADPH pool. Yeast 18, 19–32.

Novo, L.P., Gurgel, L.V.A., Marabezi, K., Aprigio, A., Curvelo, S., 2011. Delignification of sugarcane bagasse using glycerol–water mixtures to produce pulps for saccharification. Bioresour. Technol. 102, 10040–10046.

Ohad, I., Danon, D., 1964. On the dimensions of cellulose microfibrils. J. Cell Biol. 22, 302–305.

Oshiro, S., Yamaguchi, A., Watanabe, T., 2017. Binding behaviour of a 12-mer peptide and its tandem dimer to gymnospermae andangiospermae lignins. RSC Adv. 7, 31338–31341.

Ostergaard, S., Olsson, L., Nielsen, J., 2000. Metabolic engineering of *Saccharomyces cerevisiae*. Microbiol. Mol. Biol. Rev. 64, 34–50.

Pandey, A., Soccol, C.R., Nigam, P., Soccol, V.T., 2000. Biotechnological potential of agro industries: sugarcane bagasse. Bioresour. Technol. 74, 69–80.

Pattra, S., Sangyoka, S., Boonmee, M., Reungsang, A., 2008. Bio-hydrogen production from the fermentation of sugarcane bagasse hydrolysate by *Clostridium butyricum*. Int. J. Hydrogen Energy 33, 5256–5265.

Peng, B., Huang, S., Liu, T., Geng, A., 2015. Bacterial xylose isomerases from the mammal gut bacteroidetes cluster function in *Saccharomyces cerevisiae* for effective xylose fermentation. Microb. Cell Fact. 14, 70.

Quintero, J.A., Montoya, M.I., Sánchez, O.J., Giraldo, O.H., Cardona, C.A., 2008. Fuel ethanol production from sugarcane and corn: comparative analysis for a Colombian case. Energy 33, 385–399.

Rabelo, S.C., Fonseca, N.A.A., Andrade, R.R., Filho, R.M., Costa, A.C., 2011. Ethanol production from enzymatic hydrolysis of sugarcane bagasse pretreated with lime and alkaline hydrogen peroxide. Biomass Bioenergy 35 (1), 2600–2607.

Ramadoss, G., Muthukumar, K., 2016. Mechanistic study on ultrasound assisted pretreatment of sugarcane bagasse using metal salt with hydrogen peroxide for bioethanol production. Ultrason. Sonochem. 28, 207–217.

Ramos, D.F., Fontana, J., 1996. Pretreatment of sugarcane bagasse for enhanced ruminal digestion. Appl. Biochem. Biotechnol. 57, 171–182.

Rizzi, M., Elrlemann, P., Bui-Thahn, N., Dellweg, H., 1988. Xylose fermentation by yeasts: purification and kinetic studies of xylose reductase from Pichia stipitis. Appl. Microbiol. Biotechnol. 29, 148–154.

Rizzi, M., Harwart, K., Bui-Thahn, N., Dellweg, H., 1989. A kinetic study of the NAD^+-xylitol dehydrogenase from the yeast *Pichia stipitis*. J. Ferment. Bioeng. 67, 25–30.

Roca, C., Nielsen, J., Olsson, L., 2003. Metabolic engineering of ammonium assimilation in xylose-fermenting *Saccharomyces cerevisiae* improves ethanol production. Appl. Environ. Microbiol. 69, 4732–4736.

Rocha, G.J.M., Gonçalves, A.R., Oliveira, B.R., Olivares, E.G., Rossell, C.E.V., 2012. Steam explosion pretreatment reproduction and alkaline delignification reactions performed on a pilot scale with sugarcane bagasse for bioethanol production. Ind. Crops Prod. 35, 274–279.

Rocha, G.J.M., Martin, C., Soares, I.B., Souto-Maior, A.M., Baudel, H.M., Abreu, C.A.M., 2011. Dilute mixed-acid pretreatment of sugarcane bagasse for ethanol production. Biomass Bioenergy 35, 663–670.

Rocha, G.J.M., Nascimento, V.M., Gonçalves, A.R., Silva, V.F.N., Martín, C., 2015. Influence of mixed sugarcane bagasse samples evaluated by elemental and physical–chemical composition. Ind. Crops Prod. 64, 52–58.

Rodríguez-Chong, A., Ramírez, J.A., Garrote, G., Vázquez, M., 2004. Hydrolysis of sugarcane bagasse using nitric acid: a kinetic assessment. J. Food Eng. 61, 143–152.

Rodriguez-Pena, J.M., Cid, V.J., Arroyo, J., Nombela, C., 1998. The YGR194c (XKS1) gene encodes the xylulokinase from the budding yeast *Saccharomyces cerevisiae*. FEMS Microbiol. Lett. 162, 15–160.

Saad, M.B.W., Oliveira, L.R.M., Candido, R.G., Quintana, G., Rocha, G.J.M., Gonçalves, A.R., 2008. Preliminary studies on fungal treatment of sugarcane straw for organosolv pulping. Enzyme Microb. Technol. 43, 220–225.

Saini, J.K., Saini, R., Lakshmi Tewari, L., 2015. Lignocellulosic agriculture wastes as biomass feedstocks for second-generation bioethanol production: concepts and recent developments. 3 Biotech 5, 337–353.

Santos, V.T., Esteves, P.J., Milagres, A.M., Carvalho, W., 2011. Characterization of commercial cellulases and their use in the saccharification of a sugarcane bagasse sample pretreated with dilute sulfuric acid. J. Ind. Microbiol. Biotechnol. 38, 1089–1098.

Sarkar, N., Ghosh, S.K., Bannerjee, S., Aikat, K., 2012. Bioethanol production from agricultural wastes: an overview. Renew. Energy 37, 19–27.

Sarthy, A.V., McConaughy, B.L., Lobo, Z., Sundstrom, J.A., Furlong, C.E., Hall, B.D., 1987. Expression of the *Escherichia coli* xylose isomerase gene in *Saccharomyces cerevisiae*. Appl. Environ. Microbiol. 53, 1996–2000.

Sasaki, C., Takada, R., Watanabe, T., Honda, Y., Karita, S., Nakamura, Y., Watanabe, T., 2011. Surface carbohydrate analysis and bioethanol production of sugarcane bagasse pretreated with the white rot fungus: *Ceriporiopsis subvermispora* and microwave hydrothermolysis. Bioresour. Technol. 102, 9942–9946.

Singh, A., Bishnoi, N.R., 2012. Enzymatic hydrolysis optimization of microwave alkali pretreated wheat straw and ethanol production by yeast. Bioresour. Technol. 108, 94–101.

Singh, B.R., Singh, O., 2012. Global trends of fossil fuel reserves and climate change in the 21st century. In: Khan, S. (Ed.), Fossil Fuel and the Environment. InTech, http://www.intechopen.com/books/fossil-fuel-and-the-environment/global-trends-of-fossil-fuelreserves-and-climate-change-in-the-21st-century.

Siqueira, G., Varnai, A., Ferraz, A., Milagres, A.M.F., 2013. Enhancement of cellulose hydrolysis in sugarcane bagasse by the selective removal of lignin with sodium chlorite. Appl. Energy 102, 399–402.

Soccol, C.R., Faraco, V., Karp, S., Vandenberghe, L.P.S., Thomaz-Soccol, V., Woiciechowski, A., Pandey, A., 2011. Lignocellulosic bioethanol: current status and future perspectives. In: Pandey, A., Larroche, C., Ricke, S.C., Dussap, C.G., Gnansounou, E. (Eds.), Biofuels: Alternative Feedstocks and Conversion Processes. Academic Press, San Diego, pp. 101–122.

Sugathapala, A.G.T., 2013. Technologies for Converting Waste Agricultural Biomass to Energy. Reading Material, compiled by United Nations Environmental Programme, Division of Technology, Industry and Economics, International Environmental Technology Centre, Osaka.

Sun, J.X., Sun, X.F., Sun, R.C., Su, Y.Q., 2004. Fractional extraction and structural characterization of sugarcane bagasse hemicelluloses. Carbohydr. Polym. 56, 195–204.

Suslick, K.S., 1989. The chemical effects of ultrasound. Sci. Am. 260, 80–86.

Taherzadeh, M.J., Karimi, K., 2008. Bioethanol: market and production processes. In: Nag, A. (Ed.), Biofuels Refining and Performance. McGraw-Hill, New York, NY, USA, pp. 69–106. doi: 10.1036/0071489703.

Tan, H., Yang, R., Sun, W., Wang, S., 2010. Peroxide-acetic acid pretreatment to remove bagasse lignin prior to enzymatic hydrolysis. Ind. Eng. Chem. Res. 49, 1473–1479.

Toivari, M.H., Aristidou, A., Ruohonen, L., Penttila, M., 2001. Conversion of xylose to ethanol by recombinant *Saccharomyces cerevisiae*: importance of xylulokinase (XKS1) and oxygen availability. Metab. Eng. 3, 236–249.

Träff-Bjerre, K.L., Jeppsson, M., Hahn-Hägerdal, B., Gorwa-Grauslund, M.F., 2004. Endogenous NADPH-dependent aldose reductase activity influences product formation during xylose consumption in recombinant *Saccharomyces cerevisiae*. Yeast 21, 141–150.

UN Energy Statistics Database, 2013. Available from: https://unstats.un.org/unsd/energy/edbase.htm

Usher, J., Balderas-Hernandez, V., Quon, P., Gold, N.D., Martin, V.J.J., Mahadevan, R., Baetzet, K., 2011. Chemical and synthetic genetic array analysis identifies genes that suppress xylose utilization and fermentation in *Saccharomyces cerevisiae*. Genes Genom. Genet. 1, 247–258.

van Maris, A.J.A., Abbott, D.A., Bellissimi, E., van den Brink, J., Kuyper, M., Luttik, M.A.H., Wisselink, H.W., Scheffers, W.A., van Dijken, J.P., Jack T. Pronk, J.P., 2006. Alcoholic fermentation of carbon sources in biomass hydrolysates by *Saccharomyces cerevisiae*: current status. Antonie Van Leeuwenhoek 90, 391–418.

Verho, R., Londesborough, J., Penttila, M., Richard, P., 2003. Engineering redox cofactor regeneration for improved pentose fermentation in *Saccharomyces cerevisiae*. Appl. Environ. Microbiol. 69, 5892–5897.

Wahlbom, C.F., Hahn-Hägerdal, B., 2002. Furfural, 5-hydroxymethyl furfural, and acetoin act as external electron acceptors during anaerobic fermentation of xylose in recombinant *Saccharomyces cerevisiae*. Biotechnol. Bioeng. 78, 172–178.

Walfridsson, M., Anderlund, M., Bao, X., Hahn-Hägerdal, B., 1997. Expression of different levels of enzymes from the *Pichia stipitis XYL1* and *XYL2* genes in *Saccharomyces cerevisiae* and its effects on product formation during xylose utilisation. Appl. Microbiol. Biotechnol. 48, 218–224.

Walfridsson, M., Bao, X., Anderlund, M., Lilius, G., Bulow, L., Hahn-Hägerdal, B., 1996. Ethanolic fermentation of xylose with *Saccharomyces cerevisiae* harboring the *Thermus thermophilus* xylA gene, which expresses an active xylose (glucose) isomerase. Appl. Environ. Microbiol. 62, 4648–4651.

Walfridsson, M., Hallborn, J., Penttilä, M., Keränen, S., Hahn-Hägerdal, B., 1995. Xylose-metabolizing *Saccharomyces cerevisiae* strains overexpressing the *TKL1* and

TAL1 genes encoding the pentose phosphate pathway enzymes transketolase and transaldolase. Appl. Environ. Microbiol. 61, 4148–4190.

Wang, K., Sun, R.C., 2010. In: Sun, R (Ed.), Cereal Straw as a Resource for Sustainable Biomaterials and Biofuels: Biorefinnery Straw for Bioethanol. Elsevier, New York, ISBN: 978-0-444-53234-3.

Wang, P.Y., Schneider, H., 1980. Growth of yeasts on D-xylulose. Can. J. Microbiol. 26, 1165–1168.

Watanabe, S., Abu Saleh, A., Pack, S.P., Annaluru, N., Kodaki, T., Makino, K., 2007b. Ethanol production from xylose by recombinant *Saccharomyces cerevisiae* expressing protein-engineered NADH-preferring xylose reductase from *Pichia stipitis*. Microbiology 153, 3044–3054.

Watanabe, S., Kodaki, T., Makino, K., 2005. Complete reversal of coenzyme specificity of xylitol dehydrogenase and increase of thermostability by the introduction of structural zinc. J. Biol. Chem. 280, 10340–10349.

Watanabe, S., Pack, S.P., Saleh, A.A., Annaluru, N., Kodaki, T., Makino, K., 2007a. Positive effect of the decreased NADPH-preferring activity of xylose reductase from *Pichia stipitis* on ethanol production using xylose-fermenting recombinant *Saccharomyces cerevisiae*. Biosci. Biotechnol. Biochem. 71, 1365–1369.

Watanabe, S., Saleh, A.A., Pack, S.P., Annaluru, N., Kodaki, T., Makino, K., 2007c. Ethanol production from xylose by recombinant *Saccharomyces cerevisiae* expressing protein engineered $NADP^+$-dependent xylitol dehydrogenase. J. Biotechnol. 130, 316–319.

Watanabe, T., 2013. In: Vincenza, F. (Ed.), Lignocellulose Conversion Introduction: Potential of Cellulosic Ethanol Enzymatic and Microbial Tools for Bioethanol Production. Springer, Heidelberg, New York, Dordrecht, Londondoi: 10.1007/978-3-642-37861-4.

Watanabe, T., Mitani, T., 2012. Microwave technology for lignocellulosic biorefinery. In: Xie, H., Gathergood, N. (Eds.), The Role of Green Chemistry in Biomass Processing and Conversion. John Wiley & Sons, Hoboken, NJ, pp. 281–291.

Wiedemann, B., Boles, E., 2008. Codon-optimized bacterial genes improve L-arabinose fermentation in recombinant *Saccharomyces cerevisiae*. Appl. Environ. Microbiol. 74, 2043–2050.

Wisselink, H.W., Toirkens, M.J., Berriel, M.R.F., Winkler, A.A., Van Dijken, J.P., Pronk, J.T., Van Maris, A.J.A., 2007. Engineering of *Saccharomyces cerevisiae* for efficient anaerobic alcoholic fermentation of L-arabinose. Appl. Environ. Microbiol. 73, 4881–4891.

Wisselink, H.W., Toirkens, M.J., Wu, Q., Pronk, J.T., Van Maris, A.J., 2009. Novel evolutionary engineering approach for accelerated utilization of glucose, xylose, and arabinose mixtures by engineered *Saccharomyces cerevisiae* strains. Appl. Environ. Microbiol. 75, 90–914.

Wyman, C.E., Hinman, N.D., 1990. Ethanol: fundamentals of production from renewable feedstocks and use as transportation fuel. Appl. Biochem. Biotechnol. 24, 735–775.

Zeng, Q., Du, H., Wang, J., Wei, D., Wang, X., Li, Y., Lin, Y., 2009. Reversal of coenzyme specificity and improvement of catalytic efficiency of *Pichia stipitis* xylose reductase by rational site-directed mutagenesis. Biotechnol. Lett. 31, 1025–1029.

Zhang, Z., Harrison, M.D., Rackemann, D.W., Doherty, W.O.S., O'Hara, I.M., 2016. Organosolv pretreatment of plant biomass for enhanced enzymatic saccharification. Green Chem. 18, 360–381.

Zhang, Z., Wongb, H.H., Peter, L., Albertson, P.L., Doherty, W.O.S., O'Hara, I.M., 2013. Laboratory and pilot scale pretreatment of sugarcane bagasse by acidified aqueous glycerol solutions. Bioresour. Technol. 138, 14–21.

Zhao, X., Peng, F., Cheng, K., Liu, D., 2009. Enhancement of the enzymatic digestibility of sugarcane bagasse by alkali-peracetic acid pretreatment. Enzyme Microb. Technol. 44, 17–23.

Zhou, H., Cheng, J.S., Wang, B.L., Fink, G.R., Stephanopoulos, G., 2012. Xylose isomerase overexpression along with engineering of the pentose phosphate pathway and evolutionary engineering enable rapid xylose utilization and ethanol production by *Saccharomyces cerevisiae*. Metab. Eng. 14, 611–622.

Zhu, Z., Rezende, C.A., Simister, R., McQueen-Mason, S.J., Macquarrie, D.J., Polikarpovd, I., Gomez, L.D., 2016. Efficient sugar production from sugarcane bagasse by microwave assisted acid and alkali pretreatment. Biomass Bioenergy 93, 269–278.

CHAPTER

11

Bioethanol Production From Rice- and Wheat Straw: An Overview

Manas R. Swain, Ajit Singh, Ajay K. Sharma, Deepak K. Tuli

DBT-IOC Centre for Advance Bio-energy Research Centre, Faridabad, India

11.1 INTRODUCTION

Biofuels are predominantly produced from biomass resources due to the following advantages: (1) it is a renewable resource that can be sustainably developed in the future, (2) it appears to have effectively positive environmental impacts resulting in no net release of carbon dioxide and very low sulfur content, and (3) it appears to have significant economic potential provided that fossil fuel prices increase in the future. Biofuels are liquid or gaseous fuels made from plant matter and residues, such as agricultural crops, municipal wastes, and agricultural and forestry by-products. Biofuels have been shown to reduce carbon emissions and help to increase energy security. They can be classified broadly as biodiesel and bioethanol, and may be subdivided into conventional or advanced fuels (Gnansounou and Dauriat 2010; Hendriks and Zeeman 2009).

The focus toward renewable sources, such as bioethanol for power production in various countries of the world has been shifted due to depletion of crude oil reserves. Brazil and USA are the leading countries that share 26.72% and 56.72%, respectively, of the global ethanol production (Baker, 2014). Conventionally, ethanol is mostly produced from starchy grains and sugar crops. However, there is debate on ethanol from food grains versus food security. Alternately, Kim and Dale (2004) reported that 442 billion liters of bioethanol can be produced from lignocellulosic biomass, and total crop residues and waste crops can produce 491 billion liters of bioethanol. In particular, rice straw is potentially the most favorable feedstock as compared to wheat straw, corn stover, and sugarcane bagasse in terms of quantity of biomass available. The most favorable area is Asia, which can produce 291GL of bioethanol because of biomass availability (Kim and Dale, 2004). This chapter focuses on the bioethanol production from rice and wheat straw, scopes, technoeconomic feasibility, bottlenecks in commercial production, and future perspectives.

Bioethanol Production From Food Crops
http://dx.doi.org/10.1016/B978-0-12-813766-6.00011-4

11.2 AVAILABILITY OF RICE STRAW

Annual global rice production is about 730 million metric tons on about 165 million ha. More than 50 countries contribute to this sum. Grain to straw ratio in rice varies from 1:1.3 to 1:3.

11.2.1 Rice Straw

The straw yields depend on factors like variety, season, location, plant height, soil texture and fertility, plant density, available water, fertilizer, weeds and their control, harvesting stage and methods (Binod et al., 2010). Rice straw is one of the most abundant lignocellulosic crop residues in the world because half of the world population uses this cereal as a principal food. Its annual production is about 731 million tons, which is distributed in Africa, Asia, Europe, and America. In Asia it is a major field-based residue that is produced in large amounts (667.59 million metric tons). In fact, the total amount equaling 668 million metric tons of straw can produce theoretically 282 billion liters of ethanol if appropriate technology is available. However, an increasing proportion of rice straw undergoes field burning. This waste of energy seems inept, given the high fuel prices and the great demand for reducing greenhouse gases (GHG) emissions as well as air pollution (Kim and Dale, 2004).

11.2.2 Wheat Straw

Similar to rice straw, wheat straw is considered a potential feedstock for the ethanol production. European Union, China, India, USA, and Canada are the leading wheat cultivating countries in the world. If residue and crop ration is 1.3, about 850 million metric tons wheat straw produced annually worldwide can be considered as a huge agricultural reside. As per the report from Otero et al. (2007) surplus wheat straw is able to produce 120 billion liters bioethanol annually, which can replace 93 billion liters gasoline. Similarly to the rice straw, wheat straw is also composed of cellulose, hemicellulose, and lignin in the range of 33%–40%, 20%–25%, and 15%–20% (w/w), respectively. Ash content in the wheat straw is almost three to four times lower than rice straw, which makes this substrate most suitable for the bioethanol production compared to rice straw (Prasad et al., 2007).

11.2.3 Straw Collection, Processing, and Transportation

Straw is collected using baler machines and farm transporting vehicles, such as tractors, power tillers, and trucks. At the time of harvest straw normally consists of high moisture content (60%–70%) on the weight basis. The straw is allowed to dry to bring down the moisture content to 25% by natural drying. Straw harvest, collection, and transportation play a major concern in ethanol production. If the bioethanol industry is located at the center of the circle, the relationship between distance and annual feedstock demand by an ethanol plant can be calculated as follows (Huang et al., 2009).

$$\text{Radius of the area } R(\text{km}) = \left(\frac{F}{\pi f_a f_{\text{lc}} Y}\right)^{\frac{1}{2}}$$

F = Annual feedstock demand

f_a = Fraction of total farmland from which feedstock can be collected or produced

f_{lc} = Fraction of surrounding farmland containing crops

Y = Biomass yield/unit area, t/km^2

Average transportation cost per dry feedstock can be calculated as per given formula:

$$\text{Annual transporatation cost} = \frac{2\text{Rf}_w\text{C}_1}{3(1-m)}[\$/\text{Mg}]$$

R = Radius of the area

f_w = Road-winding factor

C_1 = Hauling cost per unit weight distance

m = Biomass moisture in weight percent

The straw has low bulk density and transporting large volumes of straw is a major cost. The abrasive nature of straw contains higher wear and maintenance for processing equipment, resulting in high processing costs. Lack of proper straw harvesting technology, machinery for straw collection, and higher transportation cost encourage farmers to burn the straw at the field without doing any further processing. Accessibility of straw throughout the year, transportation cost, and infrastructural settings affect the bioethanol production.

11.2.4 Sustainable Straw Extraction

Straw is used as a source of solid nutrients and the extraction of straw from the field leads to humus imbalance and solid nutrient depletion (Singh et al., 2016). Sheehan et al. (2003) reported that 60% of the straw can be easily removed from the field without any loss of productivity and soil degradation. Koga and Tajima (2011) examined that there is more emissions of GHG when straw was incorporated to the soil. Straw removal from the field reduces the GHG emissions.

11.2.5 Potential of Rice Straw and Whet Straw for the Bioethanol Production

Rice and wheat straw are mainly composed of cellulose, hemicelluloses, lignin, silica, and ash content. The chemical composition of straw depends on crop varieties and growing season (Shen et al., 1998). The general compositions of rice and wheat straw are: cellulose (33%–47%), hemicelluloses (19%–27%), and lignin (5%–24%) (Saha 2003). But the silica content is higher in the case of rice straw (12%) than in wheat straw. The fundamental structural units in straw cell walls are microfibrils. In the cell walls, the cellulose chains are joined by hydrogen bonds to form crystalline microfibrils (Xu et al., 2009).

Burning of crop residues obtained from rice–wheat cropping systems causes GHG emissions (70% CO_2, 7% CO, 0.66% CH_3, and 2.09% N_2O) that ultimately lead to climate change. The exploitation of straw for bioethanol generation emits fewer amounts of GHG and also lignin, unreacted carbohydrates, and other organic products during the process. These unreacted by-products (lignin, residual carbohydrates, and organic products) are used as energy sources for distillation of ethanol and electricity generation (Singh et al., 2016).

11.3 ETHANOL AS BIOENERGY RESOURCES

Bioenergy concept is a sustainable technology that encompasses three domains, that is, economics, environmental, and social.

11.3.1 Bioethanol Production From Rice/Wheat Straw

Mostly the enzymatic hydrolysis is preferred for the conversion of the complex form of the carbohydrates in straw to simpler form. However, the process is limited by several factors, such as crystallinity of cellulose, degree of polymerization (DP), moisture content, available surface area, and lignin content (Laureano-Perez et al., 2005). Therefore, the production of ethanol from straw strongly needs pretreatment prior to enzymatic hydrolysis and fermentation.

In the next section the pretreatment process followed for rice and wheat straw is discussed (Fig. 11.1).

11.3.2 Pretreatment of Rice and Wheat Straw

Broadly, pretreatment methods can be divided into different categories: physical (milling and grinding), physicochemical (steam pretreatment/autohydrolysis, hydrothermolysis, and wet oxidation), chemical (alkali, dilute acid, oxidizing agents, and organic solvents), biological, electrical, or a combination of these. The details

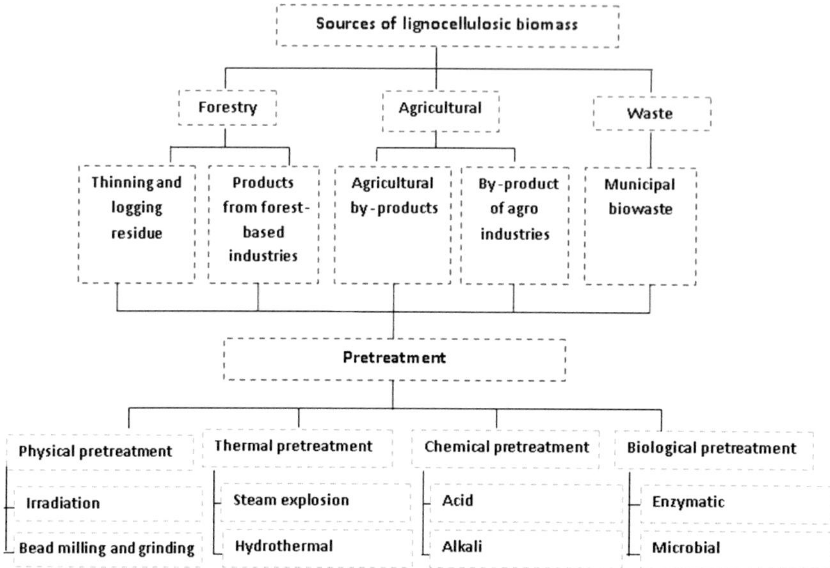

FIGURE 11.1 **Different sources of lignocellulosic biomass and mode of pretreatment for the bioethanol production.**

of the pretreatment conditions and their impacts on the biomass are documented in Table 11.1.

11.3.3 Physical Pretreatment

Physical pretreatment includes grinding and milling, irradiation, steaming, temperature, and pressure that increases the accessible surface area and size of pores, and decreases the crystallinity and DP of cellulose. Microwave irradiation could change the ultrastructure of cellulose (Zhu et al., 2005), degrade lignin and hemicelluloses in lignocellulosic materials, and increase the enzymatic susceptibility of lignocellulosic materials (Abedinifar et al., 2009). Hideno et al. (2009) found that in terms of glucose recovery as well as energy saving, wet disk milling is better than ball milling.

11.3.3.1 Milling

Milling results in reducing particle size of straw that leads to an increase in available specific surface and reduction of the DP (Chang and Holtzapple, 2000; Palmowski and Muller, 1999). For mechanical pretreatment, factors like capital costs, operating costs, scale-up possibilities, and depreciation cost of equipment are very important. The increase in specific surface area and the shearing are important factors that increase the total hydrolysis yield of the lignocellulose in most cases by 5%–25% (depending on kind of biomass, kind, and duration of milling), but also reduces the technical digestion time by 23%–59% (thus an increase in hydrolysis rate) (Delgenés et al., 2002).

Wet disk milling is an economical, practical pretreatment for the enzymatic hydrolysis of herbaceous biomass, such as rice straw (Hideno et al., 2009). The energy consumption of wet disk milling was found lower than that of other milling practices (ball milling) (Hideno et al., 2009). Ball milling is also considered as an effective pretreatment with superior glucose and carbohydrate yield compared to steam explosion pretreatment in wheat straw.

11.3.3.2 High-Energy Irradiation

Various studies have shown that irradiation reduces the DP and the crystallinity of lignocellulose (Bouchard et al., 2006) increases the accessibility of the cellulose crystalline regions (Stepanik et al., 1998). The application of electron

TABLE 11.1 Pretreatment Conditions for the Rice and Wheat Straw for Bioethanol Production

Reagents	Conditions followed for the pretreatment	Advantages of pretreatment	Glucan recovery	Xylan recovery	References
Pleurotus ostreatus	Rice straw (water content of 60%) was pretreated with *Pl. ostreatus* for 60 days	Removal of lignin up to 41%	83%	52%	Taniguchi et al. (2005)
Acid	1% (w/w) sulfuric acid at 1:10 ratio of rice straw and acid reaction time of 1–5 min at 160°C or 180°C for 1 min residence time	Solubilized more than 90% xylan	57%	4.7%	Hsu et al. (2010)
Alkali and microwave	Microwave frequency was 2450 MHz and the power could be set at 700, 500, and 300 W, respectively followed by 1% NaOH solution taken at 1:8 solid and liquid ratio ranging from 15 min to 2 h	Removing more hemicellulose and lignin and increasing its accessibility to hydrolytic enzymes	69.2%	10.2%	Zhu et al. (2005)
Phanerochaete chrysosporium	*Ph. chrysosporium* cultivated on rice straw for 15 days in the media optimized by response surface methodology	White rot fungi produces ligninolytic enzyme, which remove lignin from rice straw naturally	37.7%	10.9%	Bak et al. (2009)
Dilute acid hydrolysis	Rice straw soaked in the 0.5% sulfuric acid at 1:6.66 solid liquid ratio for 20 h the mixture was then introduced into the reactor, steam-heated for 1.5 min to achieve 15 bar pressure, and remained at this pressure for 10 min to carry out the hydrolysis	Maximum removal of hemicelluloses (99%), which helps in enzymatic hydrolysis	55%	1%	Abedinifar et al. (2009)
Alkali (calcium capturing by carbonation)	Rice and line mixed in different proportion with water keeping solid and liquid ratio 1:20 under 120°C for 1 h	Solid–liquid-separation step nor a washing step of the solid portion is needed	95.2%	81.5%	Park et al. (2010)
Organosolv	Rice straw mixed with at solid-to-liquid ratio of 1:8 of 75% (v/v) aqueous ethanol containing 1% (w/w) sulfuric acid as a catalyst. Treatments were carried out in a high-pressure stainless steel vessel at 180°C for 60 min	Organosolv pretreatment can be applied for efficient production of the solvents from rice straw	62.3%	22.4%	Amiri et al. (2014)
Ionic liquids	Rice straw mixed with ionic liquid at 1:20 solid liquid ratio and incubated in an oven at 130°C for 24 h	Glucan and xylan recovery rate is quite high	79%	76%	Nguyen et al. (2010)
Alkali	Rice straw mixed with 10% (v/v) ammonia solution at 1:10 solid liquid ratio and incubated at 100°C for 6 h	Glucan and xylan recovery rate is quite high	84%	52%	Nguyen et al. (2010)

(Continued)

TABLE 11.1 Pretreatment Conditions for the Rice and Wheat Straw for Bioethanol Production (*cont.*)

Reagents	Conditions followed for the pretreatment	Advantages of pretreatment	Glucan recovery	Xylan recovery	References
Acid	Rice straw at substrate concentration of 10% (w/v) subjected to dilute acid hydrolysis with sulfuric acid at 2% (w/v) at 132°C for 30 min under pressure	Reduces the crystallinity property of the rice straw and remove the xylan, which makes it more suitable for the enzymatic hydrolysis	50.7%	10.8%	Oberoi et al. (2010)
Alkali	Rice straw at 10% (w/v) treated with 1%–5% (w/v) NaOH concentrations and incubated in an incubator shaker at 150 rpm, 40°C for 1 h, and thereafter autoclave-sterilized at 121°C for 30 min	Decreased lignin by about 47% and increased glucan by about 50%.	57.78%	11.84%	Oberoi et al. (2012)
Microwave alkali	Rice straw 8%r (w/v) pretreated with 1% NaOH aqueous solution at boiling condition for 30 min and followed by microwave frequency (2450 MHz)t at 700 W for 30 min	Decrease lignin about 76% due to alkaline pretreatment and changes in the crystallinity of the rice straw	69.2%	10.2%	Zhu et al. (2005)
Microwave alkali	Wheat straw was pretreated with alkali (NaOH) concentrations of 2.75% (w/v) at a solid loading of 10% (w/v) and microwave for residence time 22.50 min and during treatment temperature reached up to 100°C	Decrease lignin about 60.8% due to alkaline pretreatment and microwave treatment at high temperature	51.0%	31.0%	Singh and Bishnoi (2012)
Beam radiation	Rice straw was irradiated with accelerated electrons by using a linear electron accelerator. Prior to the irradiation, straw was soaked in mineral water overnight in order to enhance the effects of pretreatment	Improve the enzymatic digestibility	—	—	Bak (2014)
Microwave organic acid	Rice straw was mixed with a certain concentration of acetic acid and radiated at 100–700 W for 2–5 min in the microwave oven	Removal ratio of lignin was 51.54%	—	—	Gong et al. (2010)

beam irradiation pretreatment has been reported to improve enzymatic hydrolysis of rice straw (Bak et al., 2009).

Jin et al. (2009) carried out physical pretreatment of milled dry rice straw using electron beam irradiation with accelerated electrons by a linear electron accelerator that had the capacity to produce electron beams. Enzymatic hydrolysis of electron beam irradiated and untreated rice straw resulted in glucose yield of 22.6% and the electron beam irradiated sample produced a glucose yield of 52.1% after hydrolysis for 132 h. Bak (2014) reported electron beam irradiation pretreatment for water soaking rice straw. Aqueous ammonia soaked rice straw was treated with proton beam irradiation at a beam energy of 45 MeV, which improved glucose conversion up to 90%.

11.3.4 Microwave Pretreatment

Microwave irradiation could change the ultrastructure of cellulose, degrade lignin, and hemicelluloses as well as increase the enzymatic susceptibility of rice straw (Kitchaiya et al., 2003; Xiong et al., 2000). Combined pretreatment of rice straw using microwave and alkali affects the weight loss (41.5%–44.6%) due to the removal of lignin and hemicellulose. Microwave/alkali had a higher hydrolysis rate and glucose content in the hydrolysate in comparison with the one by alkali alone (Zhu et al., 2005).

Similarly Zhu et al. (2006) reported microwave effect on alkali pretreated wheat straw resulting in higher ethanol concentration and yield than from alkali pretreated alone. The impact of microwave-assisted alkali pretreatment on wheat straw for the saccharide and lignin-releasing potential was studied. More than 80% of the hemicellulose and 90% of the lignin could be removed from the solid wheat straw without excessive saccharide degradation or solubilizing high amounts of cellulose (Janker-Obermeier et al., 2012; Singh and Bishnoi, 2012).

11.3.5 Chemical Pretreatments

Chemical pretreatment methods [i.e., alkali, acids, organic acids, peroxides ionic liquid (IL)] are normally coupled with physical methods, such as steam explosion, sulfur dioxide explosion, ammonia fiber explosion (AFEX) for pretreatment of biomass (Badiei et al., 2014). Silverstein et al. (2007) reported that alkali pretreatment of rice straw has resulted in minor cellulose and hemicellulose solubilization than acid or hydrothermal process.

11.3.6 Alkali Pretreatment

Alkaline pretreatment of chopped rice straw with 2% NaOH with 20% solids loading at 85°C for 1 h decreased lignin by 36% (Zhang and Cai, 2008). The separated and fully exposed microfibrils increased the external surface area and the porosity of the rice straw, thus facilitating enzymatic hydrolysis (Tarkov and Feist, 1969). Cheng et al. (2010) reported optimum reaction temperature at 95°C and 55°C for lime and NaOH pretreatment, remove lignin (23%–27%) from freshly harvested rice straw.

Alkaline wet oxidation with Na_2CO_3 is an efficient pretreatment of wheat straw that resulted in solid fractions with high cellulose recovery (96%) and high enzymatic convertibility to glucose (67%) (Klinke et al., 2002). The entire process was carried out at low temperature (25–40°C) using a low concentration (0.2%–0.5%) of chemicals, resulting in a relatively low chemical cost and low waste liquors. Recovery of cellulose after the double pretreatment reached 90% of that contained in the starting material, with a concomitant 81% degradation of lignin (Curreli et al., 1997).

11.3.6.1 Ammonia Treatment

Ammonia is an effective swelling reagent for lignocellulosic materials and has high selectivity for reactions with lignin which cleavage of C–O–C bonds in lignin as well as ether and ester bonds in the lignin–carbohydrate complex (Kim and Lee, 2007). A flow-through process called ammonia recycle percolation was developed for pretreatment. In this process, ammonia is pumped through a bed of biomass maintained at 170°C. By this process, up to 85% delignification and almost theoretical yield (75%) of glucose in enzyme hydrolysis can be achieved (Drapcho et al., 2008). Soaking in aqueous ammonia pretreatment at temperatures (40–90°C) for longer reaction times has been used to preserve most of the glucan and xylan in the samples, which is subsequently fermented using the simultaneous saccharification and cofermentation (SSCF) process (Kim and Lee, 2007; Kim et al., 2008). Rice straw pretreated using aqueous-ammonia solution [21% (w/w)] at moderate temperatures (69°C) for 10 h enabled the maximum enzymatic digestibility (71.1%). The sequential fermentation of sugar hydrolysate of ammonia pretreated rice straw would be a promising process for production of ethanol and xylitol (Swain and Chandraraj, 2015).

The AFEX process uses anhydrous ammonia. In this treatment, there is no loss of any carbohydrate fraction and no need for pH adjustment (Ko et al., 2009). Enzyme hydrolysis of AFEX-treated biomass can produce glucose with greater than 90% theoretical yield and xylose with up to 80% theoretical yield. AFEX is reported as an effective pretreatment process for rice straw as it resulted in only 3% sugar loss during pretreatment (Zhong et al., 2009). Ferrer et al. (1997) carried out pretreatment of rice straw by a process called ammonia pressurization and depressurization using a laboratory-scale ammonia reactor unit consisting of a 4-liter reactor with appropriate support equipment. Pretreatment followed by enzymatic hydrolysis resulted in a significant increase in sugar yield.

11.3.7 Acid Pretreatment

Literature regarding dilute acid hydrolysis of rice straw is limited because of the inability of the process to remove lignin and low sugar yield (Sumphanwanich et al., 2008). Dilute acid (0.75%) pretreatment of wheat straw have better enzymatic saccharification. Detoxification of the acid- and enzyme-treated wheat straw hydrolysate by overliming reduced the fermentation time from 118 to 39 h in the case of separate hydrolysis and fermentation (SHF) (35°C, pH 6.5), and increased the ethanol yield from 13 ± 2 to 17 ± 0 g/L and decreased the fermentation time from 136 to 112 h in the case of simultaneous saccharification and fermentation (SSF) (35°C, pH 6.0) (Saha et al., 2005). Steam exploded wheat straw soaked in acetic acid is a suitable raw material for ethanol production, because it has high cellulose content compared to steam explosion alone (Zabihi et al., 2010). Kootstra et al. (2009) reported impact of dilute maleic acid pretreatment of wheat straw by optimizing pretreatment time, temperature, and maleic acid concentration. Furfural formation was limited and on subsequent enzymatic hydrolysis, almost complete conversion to glucan and xylan was achieved. Dilute fumaric or maleic acid treatment can be alternative to dilute sulfuric acid pretreatment for pretreatment of wheat straw (Kootstra et al., 2009).

11.3.8 Pretreatment With Oxidizing Agent

In some of the pretreatment process oxidizing compound, like hydrogen peroxide or peracetic acid is added to the biomass, which is called oxidative pretreatment. The pretreatment removes hemicellulose and lignin to increase the accessibility of the cellulose. During oxidative pretreatment, major reactions take place, like electrophilic substitution, displacement of side chains, cleavage of alkyl aryl ether linkages, or the oxidative cleavage of aromatic nuclei (Hon and Shiraishi, 2001). Hydrogen peroxide pretreatment did not cause detectable changes in the structure of highly polymerized cellulose but solubilized the lignin and loosened the lignocellulosic matrix thus improved the enzyme digestibility (Martel and Gould, 1990). Wei and Cheng (1985) evaluated changes in the lignin content, weight loss, accessibility for Cadoxen solvent, water-holding capacity, and crystallinity of straw were measured during pretreatment to express the modification of the lignocellulosic structure of straw. Pretreatment at 60°C for 5 h in a solution with 1% (w/w) H_2O_2 and NaOH resulted in 60% delignification, 40% weight loss, a 5-fold increase in the accessibility for Cadoxen, 1-fold increase in the water-holding capacity, and only a slight decrease in crystallinity as compared with that of the untreated straw. Reports are there for employing acetic acid for the pretreatment of rice straw (Taniguchi et al., 1982). The acetic acid treatments caused little or no breakdown of the crystalline structure of cellulose in the straw. The degree of enzymatic solubilization relative to the amount of residual straw was 42% after treatment with 20% per-acetic acid. The maximum monomeric sugars were produced during the enzymatic hydrolysis of wheat straw [8.6% (w/v)] by alkaline peroxide pretreatment (Saha and Cotta, 2006). Jahan et al. (2006) reported acetic acid or formic acid pretreatment of rice straw. Maximum pentosan dissolution was observed in 80% acetic acid with 0.6% H_2SO_4 catalyst at 80°C for 120 min. Acetic acid dissolved pentosan more slowly than formic acid.

11.3.9 Organosolv Pretreatment

Organosolv pretreatment mostly does the delignification and hemicellulose removal leaving a cellulose-rich residue, which maximizes enzymatic hydrolysis at high rates. The change of cellulose crystallinity during organosolv pretreatment is not clear yet, but it has been found that the swelling of cellulose in organic solvent strongly depends on the species of organic solvents, solvent concentration, and temperature (Koo et al., 2011). The organosolv process uses hot organic solvents, such as ethanol at acidic pH to fractionate biomass components. High temperature (185–210°C) organosolv pretreatment processes did not require any acid addition during the process (Sun and Cheng, 2002). Jamshid et al. (2005) reported rice straw pulping using diethylene glycol, mixture of diethylene glycol and ethylene glycol at atmospheric pressure (Xuebing et al., 2009). In addition, mild pretreatment temperature and pressure and a neutral pH condition reduced carbohydrate degradation into undesired furfural and hydroxymethylfurfural (Sun and Cheng, 2002). The autocatalytic organosolv pretreatment-crude glycerol enabled wheat straw to achieve with reasonable enzymatic hydrolysis yields 75% for wet substrate and 63% for dried ones. Lignin recondensation and/or lignin–carbohydrate complex helped explain why dried pretreated wheat straw had a low enzymatic hydrolysis yield (Sun and Chen, 2008). Sindhu et al. (2010) reported acetone pretreated rice straw can be used as a good feedstock for bioethanol production based on the fermentable sugar yield.

11.3.10 IL Pretreatment

ILs are molten salts under 100°C, which comprise organic cations and organic or inorganic anions; and they were considered "eco-friendly" because of negligible vapor pressure, nonflammability, good thermal and chemical stability (Hallett and Welton 2011). The IL pretreatment results in significant decrystallization of cellulose and/or extensive extraction of lignin from lignocellulosic biomass, thus improving the subsequent enzymatic degradation (Lee et al., 2009). Efforts have been made to synthesize novel ILs from natural and renewable biomaterials for biomass pretreatment (Zhang et al., 2012). It has been demonstrated that the IL anions play a dominant role in the dissolution of cellulose and lignocellulose (Wang et al., 2012). As compared to extensive studies on cellulose-dissolving ILs, there is limited work about lignin-dissolving ILs and the influence of IL structures on the dissolution of lignin (Hossain and Aldous, 2012). Casas et al. (2012) explored lignin and cellulose solvation in ILs by using pinoresinol and glucose as models for lignin and cellulose, respectively, through computer simulation, and found that like cellulose dissolution in ILs, the anion also plays a major role in the dissolution process of lignin.

Poornejad et al. (2014) reported that rice straw was pretreated by an IL, 1-ethyl-3-methylimidazolium acetate ([Emim][OAc]), prior to enzymatic hydrolysis; glucose yield was significantly increased from 25.7% for the untreated straw to more than 75% for the treated straw. Similarly rice straw pretreated with cholinium lysine IL significantly improved the surface area and pore volume; this substantially increased polysaccharides accessibility to enzymes and enhanced polysaccharides digestion. The sugar yields of 81% for glucose and 48% for xylose were achieved in the enzymatic hydrolysis of pretreated rice straw. Wheat straw pretreated with 1-ethyl-3-methylimidazolium acetate ([Emim][CH_3COO]) completely dissolved biomass led to cellulose and hemicellulose-rich fractions which further showed a very high digestibility of the cellulose samples and confirmed a high glucose yield for the optimized pretreatment methodology (da Costa Lopes et al., 2013).

11.3.11 Biological Pretreatment

White-rot fungi, that is, *Phanerochaete chrysosporium*, *Ceriporiopsis subvermispora*, *Phlebia subserialis*, and *Pleurotus ostreatus* can metabolize lignin (Keller et al., 2003; Taniguchi et al., 2005). Likewise, several actinomycetes produce enzymes involved in lignocelluloses degradation. Zhang et al. (2016) reported biological pretreatment of rice straw with *Streptomyces griseorubens* JSD-1. After 10 days incubation, cellulolytic enzyme by *St. griseorubens* JSD-1 favored the conversion of cellulose to reducing sugars with an optimum saccharification efficiency of 88.13% was obtained. Zeng et al. (2011) study suggested that the supplementation of inorganic salts in the solid-state fermentation of wheat straw significantly enhanced the biomass degradation rate by *Ph. chrysosporium*. Mild alkali treated wheat straw was further pretreated with ligninolytic fungi (*Pleurotus eryngii* and *Irpex lacteus*) showed maximum digestibility after 21 days of pretreatment (López-Abelairas et al., 2013). Five different fungi obtained from screening were evaluated for pretreatment of wheat straw (Patel et al., 2006). Pretreatment with *Aspergillus niger* and *Aspergillus awamori* showed the best results regarding to yields of total sugars and ethanol after fermentation.

11.3.12 Combined Pretreatment

Kun et al. (2009) reported pretreatment of rice straw with alkali assisted by photocatalysis, which efficiently changed the physical properties and microstructure of rice straw and decreased the lignin content. Alkali treatment of rice straw in the absence of H_2O_2 favored solubilization of the small molecular size of hemicelluloses, which are rich in glucose, probably originating from α-glucan, while the second-stage treatment by alkaline peroxide enhanced dissolution of larger molecular size hemicelluloses rich in xylose (Sun et al., 2002). Kim et al. (2013) reported two-step pretreatment with dilute sulfuric acid and aqueous ammonia for rice straw to obtain fermentable sugar. The combined pretreatment using dilute sulfuric acid and aqueous ammonia on rice straw efficiently yielded fermentable sugar and achieved almost the same crystallinity index as that of α-cellulose. Zhu et al. (2006) reported several combinations of microwave pretreatment of rice straw along with acid and alkali, which removes hemicelluloses and lignin. Yang and Fang (2014) reported impact of ultrasonic irradiation and ILs for pretreatment of rice straw under different processes. ILs and ultrasound pretreated rice straw showed highest enzymatic efficiency among the tested hydrolysis. Jin and Chen (2006) studied a combination of steam explosion and superfine grinding of rice straw and its enzymatic hydrolysis. Superfine grinding were combined with low severity steam explosion for treating rice straw to shorten the grinding time, save the energy cost, avoid the inhibitors, and obtain high enzymatic hydrolysis. Superfine grinding was conducted after rice straw was steam exploded at low Ro (steam explosion severity factor) to avoid excessive decomposition of hemicellulose and side products generation from sugars and lignin. It showed differences in enzymatic hydrolysis, chemical compositions, fiber characteristics, and cells contents of the superfine ground steam exploded rice straw product and the ground steam exploded rice straw residue.

11.3.13 Enzymatic Hydrolysis

Enzymatic hydrolysis is the second step in the production of ethanol from lignocellulosic materials. It involves cleaving the polymers of cellulose and hemicellulose using cellulytic enzyme (Zhang and Lynd 2004). The cellulose usually contains only glucans, whereas hemicellulose contains polymers of several sugars, such as xylan, glucan, galactan, mannan, and arabinan. Consequently, the main hydrolysis product of cellulose is glucose, whereas the hemicellulose gives rise to several pentoses and hexoses (Taherzadeh and Niklasson, 2004). However, high lignin content blocks enzyme accessibility, causes end-product inhibition, and reduces the rate and yield of hydrolysis. In addition to lignin, cellobiose and glucose also act as strong inhibitors of cellulases (Knauf and Moniruzzaman, 2004). Enzymatic hydrolysis is usually carried out under mild conditions, that is, low pressure and long retention time in connection to the hydrolysis of hemicellulose. The pretreatment of the straw with dilute sulfuric acid resulted in 0.72 g/g sugar yield during 48 h enzymatic hydrolysis, which was higher than steam-pretreated (0.60 g/g) and untreated straw (0.46 g/g) (Abedinifar et al., 2009). The kinetics of glucose production from rice straw by *A. niger* was studied by Aderemi et al. (2008). The study showed that the concentration and rate of glucose production was depended on pretreatment of rice straw, substrate concentration, and cell loading. Enzymatic hydrolysis of alkali assisted photocatalysis of rice straw resulted 2.56 times higher hydrolysis rate than that of alkali process (Kun et al., 2009) whereas, ammonia treated rice straw resulted an increase of monomeric sugars from 11% in the untreated to 61% (Sulbaran-de-Ferrer et al., 2003). Hydrolysis efficiency of lignocellulosic biomass increases when combination of enzymes, such as cellulase, xylanases, and pectinases are employed rather than only cellulase (Sulbaran-de-Ferrer et al., 2003).

The yield of reducing sugars from wheat straw pretreated with these ILs at 130°C for 30 min reached 54.8% after being enzymatically hydrolyzed for 12 h. The atmospheric aqueous glycerol autocatalytic organosolv pretreated wheat straw was achieved with ~90% of enzymatic hydrolysis yield after 48 h. Ultimately this process enhanced the enzymatic hydrolysis of lignocellulosic biomass by removing the chemically compositional barrier and altering the physically structural impediment (Sun and Cheng, 2002). Zhu et al. (2006) reported that the microwave-assisted alkali pretreatment removed more lignin and hemicellulose from wheat straw and enhanced the enzymatic hydrolysis. But in the other study mechanical crushed irradiate pretreated wheat straw also improve enzymatic hydrolysis (Yang et al., 2008).

11.4 FERMENTATION

Fermentation is one significant process in bioethanol production, where the ethanol is directly produced from the metabolic activity of the fermentation agent. In the lignocellulosic hydrolysate, the fermentation process is more critical due to the complex nature of the hydrolysate (Fig. 11.2).

Hydrolysate is mostly composed of pentose, hexose sugar, and inhibitors produced during the process of pretreatment of biomass. It is quite

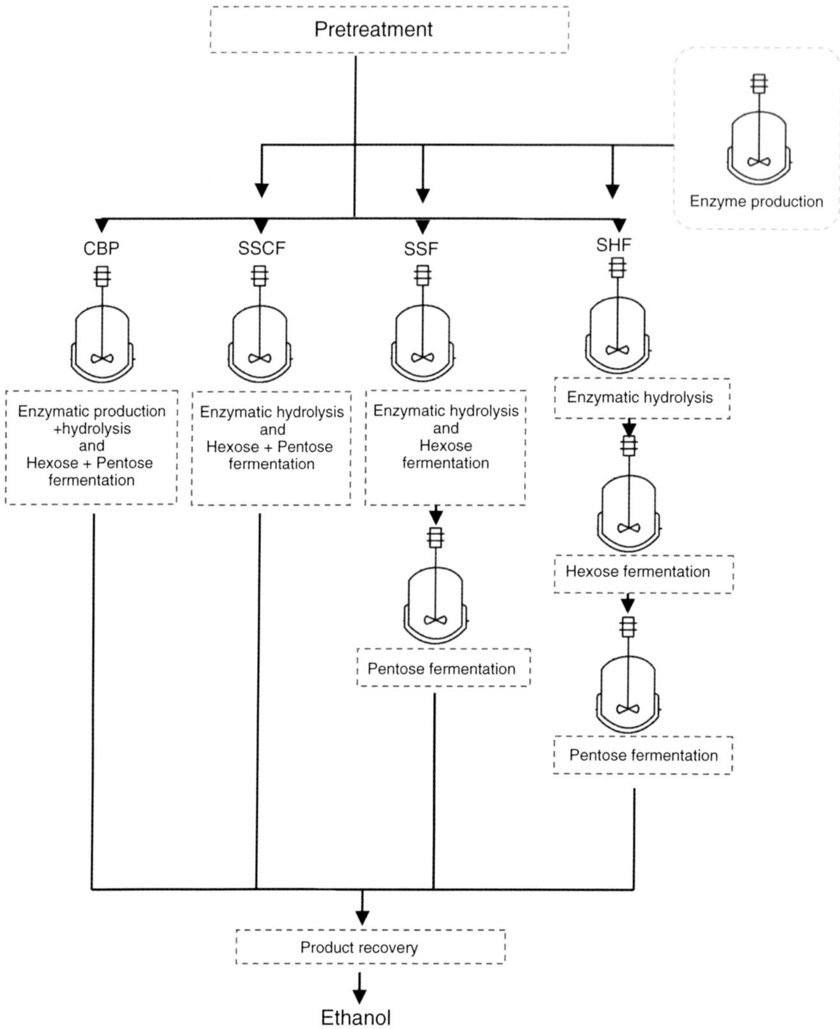

FIGURE 11.2 **Schematic diagrams of different available methods of ethanol production from lignocellulosic biomass.**

easy to ferment the hexose sugars by normal yeast. But for pentose sugar utilization it needs a specific fermenting agent (yeast or bacteria) according to the suitability to digest the pentose sugar. Theoretically, each kilogram of glucose and xylose can produce 0.49 kg carbon dioxide and 0.51 kg of ethanol (Hamelinck et al., 2005). Fermentation requires particular conditions for the microbes used, namely temperature and pH range (Demirbas, 2004). As mesophilic organisms, most of fermentation agents are comfortable within 30–35°C temperature (Hettenhaus, 1998). For bacteria types of fermentation microbes, pHs of 6.5–7.5 are typically essential to sustain the bacteria's growth (Aminifarshidmehr, 1996). Fungi are capable to resist more acidic environment of pH 3.5–5.0 than bacteria (Balat et al., 2008). Additional factors that play important roles in the fermentation process are the microbes' growth rate and genetic stability; tolerance of inhibitors, osmosis and alcohol; productivity and the yield of ethanol (Demirbas, 2004). There are numbers of discovered bacteria suitable as fermentation microorganisms.

Zymomonas mobilis widely used as fermenting bacteria, which converts glucose, sucrose, and fructose to ethanol. Like *Z. mobilis*, *Saccharomyces cerevisiae* is naturally consumes hexose sugars (e.g., glucose, fructose). Although known as the most commonly employed microorganisms, both *Z. mobilis* and *S. cerevisiae* are incapable to ferment pentose sugars. *Candida shehatae*, *Pichia stipitis*, and *Pachysolen tannophilus* are recognized for their ability to convert pentose sugars (xylose). The limitations of the natural type bacteria and yeasts drive many researches to modify the behavior and function genetically. Through metabolic engineering, additional fermenting capability can be achieved in yeasts to ferment more than one type of sugars. Buaban et al. (2010) reported that by using engineered *Pi. stipitis* BCC15191 as the fermentation agent it resulted 8.4 g/L of ethanol after 24 h of fermentation of hydrolyzed sugarcane bagasse. The study also reported that engineered *Pi. stipitis* BCC15191 was able to ferment both xylose and glucose, whereas the wild type of *Pi. stipitis* was only capable in fermenting xylose sugar.

The engineered microorganisms employed in the effort to enhance ethanol yield are listed in Table 11.2 (Aditiya et al., 2016).

11.4.1 Separate Hydrolysis and Fermentation

The SHF offers various processing advantage and opportunities. It enables enzymes to operate at higher temperature for increased performance and fermentation organisms to operate at moderate temperatures, optimizing the utilization of sugars (Wingren et al., 2003).

11.4.2 Simultaneous Saccharification and Fermentation

SSF is claimed to result higher ethanol yield since the hydrolysis inhibitions are solved by fermentation process, which makes this approach more desirable (Gupta and Verma, 2015; Sun and Chen, 2008). Watanabe et al. (2012) extracted ethanol from rice straw frond by SSF method, and the experiment resulted 38 g/L ethanol or about 84.7% ethanol yield. However, it is a challenge to control the solution environment since enzymes and microorganisms are employable only within the preference temperature and pH. Most enzymes can still become suitable with mild acid environment (pH 5) and resistant to more than 31°C of temperature, although those conditions threaten the microorganisms' sustainability (Huang et al., 2005).

11.4.3 Simultaneous Saccharification and Cofermentation

In SSF process though fermentation and hydrolysis continue simultaneously but mostly hexose sugar is targeted for fermentation, leaving pentose sugar untouched. SSCF approach allows the mixed-culture microbes to start the continuous process without sugars separation, to use various materials as the substrate without sterilization (Kleerebezem and van Loosdrecht, 2007). SSCF process has been proven effective in ethanol production for rice straw, wheat straw, and other biomass (Moreno et al., 2013). Moreno et al. (2013) reported 22 g/L ethanol from 20% solid loading of laccases mediated acid pretreated wheat straw. Similarly Olofsson et al. (2010) reported 35 g/L ethanol concentration from the 11% solid loading of steam pretreated wheat straw. SSCF of mechanically milled and hydrothermally pretreated rice straw produced 52 g/L ethanol from 25% of solid loading (Sasaki et al., 2015).

11.4.4 Consolidated Bioprocessing (CBP)

Another alternative is CBP. In ethanol production from cellulosic materials, this approach performs self-cellulase production, hydrolysis of the substrates and fermentation of hexose as well as pentose sugars within the same reactor by utilizing the ability of cocktail of microorganisms to perform the set of tasks (van Zyl et al., 2007). In comparison to SSF and SSCF, CBP

TABLE 11.2 List of Engineered Microorganisms Employed in the Improvement of Ethanol Production (Aditiya et al., 2016)

Employed microorganism	Microorganism type	Features	Ethanol yield
C. shehatae NCL3501	Yeast	Co ferment xylose and glucose	0.45 g/g of sugar by auto hydrolysis 0.5 g/g of sugar by immobilized cells
Clostridium thermocellum DSM1313	Bacterium	Improved ethanol yield	0.8 g/L
Cl. thermocellum YD01	Bacterium	Improved ethanol yield	1.33 mol-ethanol/mole-glucose equivalent (three times than resulted from the wild type)
Cl. thermocellum YD02	Bacterium	Bacterium improved ethanol yield	1.28 mol-ethanol/mole-glucose equivalent (three times than resulted from the wild type)
Escherichia coli KO11	Bacterium	Ferment both xylose and glucose	31.5 g/L or theoretically 91.5% after 48 h fermentation
E. coli FBR5	Bacterium	Ferment xylose and arabinose, the process was bioabated by *Coniochaeta ligniaria* NRRL30616	2.25% (w/v)
Pi. stipitis A	Yeast	Bacterium adapted at hydrolysate increased concentration	0.41 g product/gram substrate
Pi. stipitis NRRL Y-7124	Yeast	Adapted at hydrolysate in creased concentration	0.35 g product/gram substrate
Pi. stipitis BCC15191	Yeast	Ferment both xylose and glucose	8.4 g/L after 24 h fermentation
S. cerevisiae D5a	Yeast	Improved ethanol yield	0.58% (w/v) or 100% theoretical yield
S. cerevisiae RWB217	Yeast	Ferment glucose and xylose	0.43 g/g of sugars
Z. mobilis ZM4(pZB5)	Bacterium	Ferment both xylose and glucose	28 g/L with supplementation with 5 g/L yeast extract
Z. mobilis AX101	Bacterium	Ferment glucose, xylose, and arabinose	3.54 g/L/h (no presence of acetic acid); 1.17 g/L/h (with presence of acetic acid)
Thermoanaerobacterium saccharolyticum ALK2	Bacterium	Improved ethanol yield, able to ferment glucose, xylose, mannose, and arabinose	37 g/L
Thermoanaerobacter mathranii BG1L1	Bacterium	Improved ethanol yield	0.39–0.42 g/g sugars

method offers better benefits including lower production cost due to lesser steps required and no additional purchase of enzymes; better efficiency of conversion; and lesser energy required to run the production system (Carere et al., 2008). It is uncommon for wild type of microorganisms to work on CBP method, but several studies have reported the employment of specific microorganism in CBP method. *Trichoderma reesei* is a known cellulolytic fungi secreting cellulases to degrade cellulose in biomass (Xu et al., 2009). In addition, several fungi, namely *Fusarium oxysporum*, *Neurospora crassa*, and *Paecilomyces* sp. and *Clostridium thermocellum* are said to be practicable for CBP (Sarkar et al., 2012). However, the development of CBP method is encountered with low rate of ethanol conversion process and considerably low ethanol yield (Xu et al., 2009).

The paths of modification in CBP are exclusively classified into two terms:

1. Modification of strains of cellulolytic microbes (cellulase producer) to produce ethanol as well (ethanologen). Amore and Faraco (2012) described the potential of several fungi towards this type of strain modification. These include *T. reesei*, *Aspergillus* spp. (mostly *A. sojae*, *A. niger*, *A. oryzae*, and *A. terreus*), *F. oxysporum*, and *Rhizopus* spp. (*R. koji*, *R. oryzae*, and *R. stolonifera*).
2. Modification of genes of naturally ethanologen microbes to equip cellulolytic property. Various bacteria, namely, *Escherichia coli*, *Klebsiella oxytoca*, and *Z. mobilis* are reported to have been modified as well as yeasts, including *C. shehatae*, *P. tannophilus*, *S. cerevisiae*, and *Pi. stipitis* (Aditiya et al., 2016).

11.5 TECHNOECONOMICAL FEASIBILITY OF BIOETHANOL PRODUCTION

The production cost of straw-based ethanol is sensitive to key parameters, such as the type, composition, and price of the feedstock, the size of ethanol plant, the conversion efficiency, and the level of investment costs. The net production cost of ethanol is divided into (1) investment costs, (2) fixed operating costs (including salaries, general overhead, insurance, taxes, and maintenance), (3) variable operating costs (including purchase of consumables and sales of excess electricity), and (4) feedstock costs. Feedstock costs are separated from variable operating costs due to their large share of the net production cost. The choice of the production capacity has an effect not only on investment costs, but also on feedstock transport costs and fixed operating costs; salaries and maintenance costs depend on the size of the plant, but not linearly. The larger the ethanol plant, the lower the production cost. It can be considered, due to the relatively low contribution of operating costs to the total production cost, that the effect of plant size on operating costs is almost negligible. The optimal size of an ethanol plant therefore largely depends on regional conditions and on the availability of feedstock. The latter will have an effect on feedstock transport costs, but may also have some on feedstock nontransport costs depending on local conditions. A doubling of the production capacity (from 200 to 400 mL/year) results in a 10% reduction of the net production cost (from US$0.73/L to US$0.67/L). A halving of the production capacity (from 200 to 100 mL/year) results in a 15% increase of the net production cost (from US$0.73/L to US$0.84/L). The trade-off between plant size and transport distance in favor of plant size in terms of production cost may be largely different when considering the environmental impact of ethanol production.

11.6 CONCLUSION

The utilization of rice and wheat straw for bioethanol production requires the production technology to be cost-effective and environmentally sustainable. Biological conversion of straw into fermentable sugars, employing hydrolyzing enzymes is at present the most attractive alternative and requires the involvement of the government by issuing supporting energy policies. Different approaches in both process engineering and strain engineering have to be carried out to circumvent the difficulties of xylose and glucose cofermentation and to improve the system efficiency. A very balanced and intelligent combination of pretreatment, hydrolysis and fermentation process has to be selected for maximum efficacy of the process.

References

Abedinifar, S., Karimi, K., Khanahmadi, M., Taherzadeh, M.J., 2009. Ethanol production by *Mucor indicus* and *Rhizopus oryzae* from rice straw by separate hydrolysis and fermentation. Biomass Bioenergy 33, 828–833.

Aderemi, B.O., Abu, E., Highina, B.K., 2008. The kinetics of glucose production from rice straw by *Aspergillus niger*. Afr. J. Biotechnol. 7, 1745–1752.

Aditiya, H.B., Mahlia, T.M.I., Chong, W.T., Nur, H., Sebayang, A.H., 2016. Second generation bioethanol production: a critical review. Renew. Sustain. Energy Rev. 66, 631–653.

Aminifarshidmehr, N., 1996. The management of chronic suppurative otitis media with acid media solution. Am. J. Otolaryngol. 17, 24–25.

Amiri, H., Karimi, K., Zilouei, H., 2014. Organosolv pretreatment of rice straw for efficient acetone, butanol, and ethanol production. Bioresour. Technol. 152, 450–456.

Amore, A., Faraco, V., 2012. Potential of fungi as category I consolidated bioprocessing organisms for cellulosic ethanol production. Renew. Sustain. Energy Rev. 16, 3286–3301.

Badiei, M., Asim, N., Jahim, J.M., Sopian, K., 2014. Comparison of chemical pretreatment methods for cellulosic biomass. APCBEE Procedia 9, 170–174.

Bak, J.S., 2014. Electron beam irradiation enhances the digestibility and fermentation yield of water-soaked lignocellulosic biomass. Biotechnol. Rep. 4, 30–33.

Bak, J.S., Ko, J.K., Han, Y.H., Lee, B.C., Choi, I.G., Kim, K.H., 2009. Improved enzymatic hydrolysis yield of rice straw using electron beam irradiation pretreatment. Bioresour. Technol. 100, 1285–1290.

Baker, B., 2014. Global renewable fuel alliance. Available from: http://globalrfa.org/news-media/global-ethanol-production-will-rise-to-over-90-billion-litres-in.

Balat, M., Balat, H., Oz, C., 2008. Progress in bio-ethanol processing. Prog. Energy Combust. Sci. 34, 551–573.

Binod, P., Sindhu, R., Singhania, R.R., Vikram, S., Devi, L., Nagalakshmi, S., Kurien, N., Sukumaran, R.K., Pandey, A., 2010. Bio ethanol production from rice straw: an overview. Bioresour. Technol. 101, 4767–4774.

Bouchard, J., Méthot, M., Jordan, B., 2006. The effects of ionizing radiation on the cellulose of wood free paper. Cellulose 13, 601–610.

Buaban, B., Inoue, H., Yano, S., Tanapongpipat, S., Ruanglek, V., Champreda, V., Pichyangkura, R., Rengpipat, S., Eurwilaichitr, L., 2010. Bioethanol production from ball milled bagasse using an on-site produced fungal enzyme cocktail and xylose-fermenting *Pichia stipitis*. J. Biosci. Bioeng. 110, 18–25.

Carere, C., Sparling, R., Cicek, N., Levin, D., 2008. Third generation biofuels via direct cellulose fermentation. Int. J. Mol. Sci. 9, 1342–1360.

Casas, A., Palomar, J., Alonso, M.V., Oliet, M., Omar, S., Rodriguez, F., 2012. Comparison of lignin and cellulose solubilities in ionic liquids by COSMO-RS analysis and experimental validation. Ind. Crops Prod. 37, 155–163.

Chang, V.S., Holtzapple, M.T., 2000. Fundamental factors affecting biomass enzymatic reactivity. Appl. Biochem. Biotechnol. 84–86 (1), 5–37.

Cheng, Y.S., Zheng, Y., Yu, C.W., Dooley, T.M., Jenkins, B.M., Vander Gheynst, J.S., 2010. Evaluation of high solids alkaline pretreatment of rice straw. Appl. Biochem. Biotechnol. 162, 1768.

Curreli, N., Fadda, M.B., Rescigno, A., Rinaldi, A.C., Soddu, G., Sollai, F., Vaccargiu, S., Sanjust, E., Rinaldi, A., 1997. Mild alkaline/oxidative pretreatment of wheat straw. Process Biochem. 32, 665–670.

da Costa Lopes, A.M., Joao, K.G., Rubik, D.F., Łukasik, E.B., Duarte, L.C., Andreaus, J., Bogel-Łukasik, R., 2013. Pretreatment of lignocellulosic biomass using ionic liquids: wheat straw fractionation. Bioresour. Technol. 142, 198–208.

Delgenés, J.P., Penaud, V., Moletta, R., 2002. Pretreatments for the enhancement of anaerobic digestion of solid wastes, Chapter 8. Biomethanization of the Organic Fraction of Municipal Solid Wastes. IWA Publishing, London, pp. 201–228.

Demirbas, A., 2004. Ethanol from cellulosic biomass resources. J. Green Energy 1, 79–87.

Drapcho, C.M., Nhuan, N.P., Walker, T.H., 2008. Biofuels Engineering Process Technology. McGraw Hill Companies, Inc., New York.

Ferrer, S., et al., 1997. Sugar production from rice straw. Arch. Latinoam. Prod. Anim. 5, 112–114.

Gnansounou, E., Dauriat, A., 2010. Techno-economic analysis of lignocellulosic ethanol: a review. Bioresour. Technol. 101, 4980–4991.

Gong, G., Liu, D., Huang, Y., 2010. Microwave-assisted organic acid pretreatment for enzymatic hydrolysis of rice straw. Biosyst. Eng. 107, 67–73.

Gupta, A., Verma, J.P., 2015. Sustainable bio-ethanol production from agro-residues: a review. Renew. Sustain. Energy Rev. 41, 550–567.

Hallett, J.P., Welton, T., 2011. Room-temperature ionic liquids: solvents for synthesis and catalysis. 2. Chem. Rev. 111 (5), 3508–3576.

Hamelinck, C., Hooijdonk, G., Faaij, A., 2005. Ethanol from lignocellulosic biomass: techno economic performance in short, middle and long term. Biomass Bioenergy 28, 384–410.

Hendriks, A.T.W.M., Zeeman, G., 2009. Pretreatments to enhance the digestibility of lignocellulosic biomass. Bioresour. Technol. 100, 10–18.

Hettenhaus, J., 1998. Ethanol fermentation strains: present and future requirements for biomass to ethanol commercialization. United States Department of Energy and National Renewable Energy Laboratory.

Hideno, A., Inoue, H., Tsukahara, K., Fujimoto, S., Minowa, T., Inoue, S., Endo, T., Sawayama, S., 2009. Wet disk milling pretreatment without sulfuric acid for enzymatic hydrolysis of rice straw. Bioresour. Technol. 100 (10), 2706–2711.

Hon, D.N.S., Shiraishi, N., 2001. Wood and cellulose chemistry, second ed Marcel Dekker Inc, New York.

Hossain, M.M., Aldous, L., 2012. Ionic liquids for lignin processing: dissolution, isolation, and conversion. Aust. J. Chem. 65 (11), 1465–1477.

Hsu, T.C., Guo, G.L., Chen, W.H., Hwang, W.S., 2010. Effect of dilute acid pretreatment of rice straw on structural properties and enzymatic hydrolysis. Bioresour. Technol. 101, 4907–4913.

Huang, H.I., Ramaswamy, S.A., Dajani, W., Tschirner, U., Cairncross, R.A., 2009. Effect of biomass species and plant size on cellulosic ethanol: a comparative process and economic analysis. Biomass Bioenergy 33, 234–246.

Huang, L., Jin, B., Lant, P., Zhou, J., 2005. Simultaneous saccharification and fermentation of potato starch waste water to lactic acid by *Rhizopus oryzae* and *Rhizopus arrhizus*. Biochem. Eng. J. 23, 265–276.

Jahan, M.S., Lee, Z.Z., Jin, Y., 2006. Organic acid pulping of rice straw. I: cooking. Truck. J. Agric. For. 30, 231–239.

Jamshid, M.R., Alireza, T., Pejman, R.C., 2005. Pulping of rice straw by high boiling solvents in atmospheric pressure. Iran. Polym. J. 14, 223–227.

Janker-Obermeier, I., Sieber, V., Faulstich, M., Schieder, D., 2012. Solubilization of hemicellulose and lignin from wheat straw through microwave-assisted alkali treatment. Ind. Crops Prod. 39, 198–203.

Jin, S., Chen, H., 2006. Superfine grinding of steam-exploded rice straw and its enzymatic hydrolysis. Biochem. Eng. J. 30, 225–230.

Jin, S.B., Ja, K.K., Young, H.H., Byung, C.L., In-Geol, C., Heon, K.K., 2009. Improved enzymatic hydrolysis yield of rice straw using electron beam irradiation pretreatment. Bioresour. Technol. 100, 1285–1290.

Keller, F.A., Hamilton, J.E., Nguyen, Q.A., 2003. Microbial pretreatment of biomass: potential for reducing severity of thermo-chemical biomass pretreatment. Appl. Biochem. Biotechnol., 105–108, 27–41.

Kim, S., Dale, B.E., 2004. Global potential bioethanol production from wasted crops and crop residues. Biomass Bioenergy 26, 361–375.

Kim, T.H., Lee, Y.Y., 2007. Pretreatment of corn stover by soaking in aqueous ammonia at moderate temperatures. Appl. Biochem. Biotechnol. 137–140 (1–12), 81–92.

Kim, S.B., Lee, S.J., Lee, J.H., Jung, Y.R., Thapa, L.P., Kim, J.S., Um, Y., Park, C., Kim, S.W., 2013. Pretreatment of rice straw with combined process using dilute sulfuric acid and aqueous ammonia. Biotechnol. Biofuels 6, 109.

Kim, T.H., Taylor, F., Hicks, K.B., 2008. Bioethanol production from barley hull using SAA (soaking in aqueous ammonia) pretreatment. Bioresour. Technol. 99, 5694–5702.

Kitchaiya, P., Intanakul, P., Krairish, M., 2003. Enhancement of enzymatic hydrolysis of lignocellulosic wastes by microwave pretreatment under atmospheric pressure. J. Wood Chem. Technol. 23 (2), 217–225.

Kleerebezem, R., van Loosdrecht, M.C.M., 2007. Mixed culture biotechnology for bioenergy production. Curr. Opin. Biotechnol. 18, 207–212.

Klinke, H.B., Ahring, B.K., Schmidt, A.S., Thomsen, A.B., 2002. Characterization of degradation products from alkaline wet oxidation of wheat straw. Bioresour. Technol. 82, 15–26.

Knauf, M., Moniruzzaman, M., 2004. Lignocellulosic biomass processing: a perspective. Int. Sugar J. 106, 147–150.

Ko, J.K., Bak, J.S., Jung, M.W., Lee, H.J., Choi, I.G., Kim, T.H., Kim, K.H., 2009. Ethanol production from rice straw using optimized aqueous-ammonia soaking pretreatment and simultaneous saccharification and fermentation processes. Bioresour. Technol. 100, 4374–4380.

Koga, N., Tajima, R., 2011. Assessing energy efficiencies and green house gas emissions under bioethanol-oriented paddy rice production in northern Japan. J. Environ. Manage. 92, 967–973.

Koo, B.W., Kim, H.Y., Park, N., Lee, S.M., Yeo, H., Choi, I.G., 2011. Organosolv pretreatment of *Liriodendron tulipifera* and simultaneous saccharification and fermentation for bioethanol production. Biomass Bioenergy 35, 1833–1840.

Kootstra, A.M.J., Beeftink, H.H., Scott, E.L., Sanders, J.P.M., 2009. Optimization of the dilute maleic acid pretreatment of wheat straw. Biotechnol. Biofuels 2, 31.

Kun, N., Peng, C., Xu, Z., Wen-Song, T., 2009. Enhanced enzymatic hydrolysis of rice straw pretreated by alkali assisted with photocatalysis technology. J. Chem. Technol. Biotechnol. 84, 1240–1245.

Laureano-Perez, L., Teymouri, F., Alizadeh, H., Dale, B.E., 2005. Understanding factors that limit enzymatic hydrolysis of biomass. Appl. Biochem. Biotechnol. 124, 1081–1099.

Lee, J.W., Rodrigues, R.C., Jeffries, T.W., 2009. Simultaneous saccharification and ethanol fermentation of oxalic acid pretreated corncob assessed with response surface methodology. Bioresour. Technol. 100, 6307–6311.

López-Abelairas, M., Álvarez Pallín, M., Salvachúa, D., Lú-Chau, T., Martínez, M.J., Lema, J.M., 2013. Optimization of the biological pretreatment of wheat straw with white-rot fungi for ethanol production. Bioprocess Biosyst. Eng. 36 (9), 1251–1260.

Martel, P., Gould, J.M., 1990. Cellulose stability and delignification after alkaline hydrogen peroxide treatment of straw. J. Appl. Polym. Sci. 39, 707–714.

Moreno, A.D., Tomás-Pejó, E., Ibarra, D., Ballesteros, M., Olsson, L., 2013. In situ laccase treatment enhances the fermentability of steam-exploded wheat straw in SSCF processes at high dry matter consistencies. Bioresour. Technol. 143, 337–343.

Nguyen, T.A., Kim, K.R., Han, S.J., Cho, H.Y., Kim, J.W., Park, S.M., Park, J.C., Sim, S.J., 2010. Pretreatment of rice straw with ammonia and ionic liquid for lignocellulose conversion to fermentable sugars. Bioresour. Technol. 101 (19), 7432–7438.

Oberoi, H.S., Babbar, N., Sandhu, S.K., Dhaliwal, S.S., Kaur, U., Chadha, B.S., Bhargav, V.K., 2012. Ethanol production from alkali-treated rice straw via simultaneous saccharification and fermentation using newly isolated thermotolerant *Pichia kudriavzevii* HOP-1. J. Ind. Microbiol. Biotechnol. 39 (4), 557–566.

Oberoi, H.S., Vadlania, P.V., Brijwania, K., Bhargavb, V.K., Patil, R.T., 2010. Enhanced ethanol production via fermentation of rice straw with hydrolysate-adapted *Candida tropicalis* ATCC 13803. Process Biochem. 45, 1299–1306.

Olofsson, K., Palmqvist, B., Lidén, G., 2010. Improving simultaneous saccharification and co-fermentation of pretreated wheat straw using both enzyme and substrate feeding. Biotechnol. Biofuels 3, 17.

Otero, J., Panagiotou, G., Olsson, L., 2007. Fueling industrial biotechnology growth with bioethanol. Biofuels 108, 1–40.

Palmowski, L., Muller, J., 1999. Influence of the size reduction of organic waste on their anaerobic digestion. In: Proceedings of the II International Symposium on Anaerobic Digestion of Solid Waste, Barcelona 15–17 June, pp. 137–144.

Park, J., Shiroma, R., Haq, M., Zhang, Y., Ike, M., Sanoh, Y.A., Ida, A., Kondo, M., Tokuyasu, K., 2010. A novel lime pretreatment for subsequent bioethanol production from rice straw—calcium capturing by carbonation (CaCCO) process. Bioresour. Technol. 101 (17), 6805–6811.

Patel, S.J., Onkarappa, R., Shobha, K., 2006. Study of ethanol production from fungal pretreated wheat and rice straw. Int. J. Microbiol. 4.

Poornejad, N., Karimi, K., Behzad, T., 2014. Ionic liquid pretreatment of rice straw to enhance saccharification and bioethanol production. J. Biomass Biofuel 1, 8–15.

Prasad, S., Singh, A., Joshi, H.C., 2007. Ethanol as an alternative fuel from agricultural, industrial and urban residues. Resour. Conserv. Recycl. 50, 1–39.

Saha, B.C., Cotta, M.A., 2006. Ethanol production from alkaline peroxide pretreated enzymatically saccharified wheat straw. Biotechnol. Prog. 22, 449–453.

Saha, B.C., Iten, L.B., Cotta, M.A., Wu, Y.V., 2005. Dilute acid pretreatment, enzymatic saccharification and fermentation of wheat straw to ethanol. Process Biochem. 40, 3693–3700.

Saha, B.C., 2003. Hemicellulose bioconversion. Ind. Microbiol. Biotechnol. 30, 279–291.

Sarkar, N., Ghosh, S., Bannerjee, S., Aikat, K., 2012. Bioethanol production from agricultural wastes: an overview. Renew. Energy 37, 19–27.

Sasaki, K., Tsuge, Y., Sasaki, D., Teramura, H., Inokuma, K., Hasunuma, T., Ogino, C., Kondo, A., 2015. Mechanical milling and membrane separation for increased ethanol production during simultaneous saccharification and co-fermentation of rice straw by xylose-fermenting *Saccharomyces cerevisiae*. Bioresour. Technol. 185, 263–268.

Sheehan, J., Aden, A., Paustian, K., Killian, K., Walsh, M., Nelson, R., 2003. Energy and environmental aspects of using corn stover for fuel ethanol. J. Ind. Ecol. 7, 117–146.

Shen, H.S., Ni, D.B., Sundstol, F., 1998. Studies on untreated and urea-treated rice straw from three cultivation seasons: a physical and chemical measurements in straw and straw fractions. Anim. Feed Sci. Technol. 73, 243–261.

Silverstein, R.A., Chen, Y., Sharma-Shivappa, R.R., Boyette, M.D., Osborne, J., 2007. A comparison of chemical pretreatment methods for improving saccharification of cotton stalks. Bioresour. Technol. 98, 3000–3011.

Sindhu, R., Binod, P., Nagalakshmi, S., Janu, K.U., Sajna, K.V., Kurien, N., Sukumaran, R.K., Pandey, A., 2010. Formic acid as a potential pretreatment agent for conversion of sugarcane bagasse to ethanol. Appl. Biochem. Biotechnol. 162, 2313–2323.

Singh, A., Bishnoi, N.R., 2012. Enzymatic hydrolysis optimization of microwave alkali pretreated wheat straw and ethanol production by yeast. Bioresour. Technol. 108, 94–101.

Singh, R., Srivastava, M., Shukla, A., 2016. Environmental sustainability of bio-ethanol production from rice straw in India: a review. Renew. Sustain. Energy Rev. 54, 202–216.

Stepanik, T.M., Rajagopal, S., Ewing, D., Whitehouse, R., 1998. Electron processing technology: a promising application for the viscose industry. Radiat. Phys. Chem. 52, 505–509.

Sulbaran-de-Ferrer, B., Aristiguieta, M., Dale, B.E., Ferrer, A., Ojeda-de-Rodriguez, G., 2003. Enzymatic hydrolysis of ammonia-treated rice straw. Appl. Biochem. Biotechnol. 105–108, 155–164.

Sumphanwanich, J., Leepipatpiboon, N., Srinorakutara, T., Akaracharanya, A., 2008. Evaluation of dilute-acid pretreated bagasse corn cob and rice straw for ethanol fermentation by *Saccharomyces cerevisiae*. Ann. Microbiol. 58, 219–225.

Sun, Y., Cheng, J., 2002. Hydrolysis of lignocellulosic materials for ethanol production: a review. Bioresour. Technol. 83, 1–11.

Sun, F., Chen, H., 2008. Organosolv pretreatment by crude glycerol from oleochemicals industry for enzymatic hydrolysis of wheat straw. Bioresour. Technol. 99, 5474–5479.

Sun, F., Chen, H., 2008. Enhanced enzymatic hydrolysis of wheat straw by aqueous glycerol pretreatment. Bioresour. Technol. 99, 6156–6161.

Sun, R.C., Tomkinson, J., Ma, P.L., Liang, S.F., 2002. Comparative study of hemicelluloses from rice straw by alkali and hydrogen peroxide treatments. Carbohydr. Polym. 42, 111–122.

Swain, M.R., Chandraraj, K., 2015. Improved conversion of rice straw to ethanol and xylitol by combination of moderate temperature ammonia pretreatment and sequential fermentation using *Candida tropicalis*. Ind. Crops Prod. 77, 1039–1046.

Taherzadeh, M.J., Niklasson, C., 2004. Ethanol from lignocellulosic materials: pretreatment, acid and enzymatic hydrolyses, and fermentation. ACS Symp. Ser. 889, 49–68.

Taniguchi, M., Tanaka, M., Matsuno, R., Kamikubo, T., 1982. Evaluation of chemical pretreatment for enzymatic solubilization of rice straw. Eur. J. Appl. Microbiol. Biotechnol. 14, 35–39.

Taniguchi, M., Suzuki, H., Watanabe, D., Sakai, K., Hoshino, K., Tanaka, T., 2005. Evaluation of pretreatment with *Pleurotus ostreatus* for enzymatic hydrolysis of rice straw. J. Biosci. Bioeng. 100 (6), 637–643.

Tarkov, H., Feist, W.C., 1969. A mechanism for improving the digestibility of lignocellulosic materials with dilute alkali and liquid ammonia. Adv. Chem. 95, 197–218.

van Zyl, W., Lynd, L., Den Haan, R., Mc Bride, J., 2007. Consolidated bio-processing for bioethanol production using Saccharomyces cerevisiae. Adv. Biochem. Eng. Biotechnol. 108, 205–235.

Wang, C.Z., He, H., Wang, X., Yuan, C.S., 2012. Trends in scientific publications of chinese medicine. Am. J. Chin. Med. 40, 1099–1108.

Watanabe, I., Miyata, N., Ando, A., Shiroma, R., Tokuyasu, K., Nakamura, T., 2012. Ethanol production by repeated-batch simultaneous saccharification and fermentation (SSF) of alkali-treated rice straw using immobilized *Saccharomyces cerevisiae* cells. Bioresour. Technol. 123, 695–698.

Wei, C.J., Cheng, C.Y., 1985. Effect of hydrogen peroxide pretreatment on the structural features and the enzymatic hydrolysis of rice straw. Biotechnol. Bioeng. 27, 1418–1426.

Wingren, A., Galbe, M., Zacchi, G., 2003. Techno-economic evaluation of producing ethanol from softwood: comparison of SSF and SHF and identification of bottlenecks. Biotechnol. Prog. 19 (4), 1109–1117.

Xiong, J., Ye, J., Liang, W.Z., Fan, P.M., 2000. Influence of microwave on the ultrastructure of cellulose I. J. South China Union Technol. 28 (3), 84–89.

Xu, Q., Singh, A., Himmel, M., 2009. Perspectives and new directions for the production of bioethanol using consolidated bioprocessing of lignocellulose. Curr. Opin. Biotechnol. 20, 364–371.

Xuebing, Z., Keke, C., Dehua, L., 2009. Organosolv pretreatment of lignocellulosic biomass for enzymatic hydrolysis. Appl. Microbiol. Biotechnol. 82, 815–827.

Yang, C., Shen, Z., Yu, G., Wang, J., 2008. Effect and after effect of γ radiation pretreatment on enzymatic hydrolysis of wheat straw. Bioresour. Technol. 99, 6240–6245.

Yang, C.Y., Fang, T.J., 2014. Combination of ultrasonic irradiation with ionic liquid pretreatment for enzymatic hydrolysis of rice straw. Bioresour. Technol. 164, 198–202.

Zabihi, S., Alinia, R., Esmaeilzadeh, F., Kalajahi, J.F., 2010. Pretreatment of wheat straw using steam, steam/acetic acid and steam/ethanol and its enzymatic hydrolysis for sugar production. Biosyst. Eng. 105, 288–297.

Zeng, J., Singh, D., Chen, S., 2011. Biological pretreatment of wheat straw by *Phanerochaete chrysosporium* supplemented with inorganic salts. Bioresour. Technol. 102, 3206–3214.

Zhang, Y.H., Lynd, L.R., 2004. Toward an aggregated understanding of enzymatic hydrolysis of cellulose: noncomplexed cellulase systems. Biotechnol. Bioeng. 88 (7), 797–824.

Zhang, Q.Z., Cai, W.M., 2008. Enzymatic hydrolysis of alkali-pretreated rice straw by *Trichoderma reesei* ZM4-F3. Biomass Bioenergy 32, 1130–1135.

Zhang, D., Luo, Y., Chu, S., Zhi, Y., Wang, B., Zhou, P., 2016. Biological pretreatment of rice straw with *Streptomyces griseorubens* JSD-1 and its optimized production of cellulase and xylanase for improved enzymatic saccharification efficiency. Prep. Biochem. Biotechnol. 17, 575–585.

Zhang, J., Tang, M., Viikari, L., 2012. Xylans inhibit enzymatic hydrolysis of lignocellulosic materials by cellulases. Bioresour. Technol. 121, 8–12.

Zhong, C., Lau, M.W., Balan, V., Dale, B.E., Yuan, Y.J., 2009. Optimization of enzymatic hydrolysis and ethanol fermentation from AFEX-treated rice straw. Appl. Microbiol. Biotechnol. 84, 667–676.

Zhu, S., Wu, Y., Yu, Z., Liao, J., Zhang, Y., 2005. Pretreatment by microwave/alkali of rice straw and its enzymic hydrolysis. Process Biochem. 40, 3082–3086.

Zhu, S., Wu, Y., Yu, Z., Zhang, X., Wang, C., Yu, F., Jin, S., 2006. Production of ethanol from microwave-assisted alkali pretreated wheat straw. Process Biochem. 41, 869–873.

Further Reading

Kim, I., Lee, B., Park, J.Y., Choi, S.A., Han, J.I., 2014. Effect of nitric acid on pretreatment and fermentation for enhancing ethanol production of rice straw. Carbohydr. Polym. 99 (2), 563–567.

CHAPTER

12

Forest Bioresources for Bioethanol and Biodiesel Production With Emphasis on Mohua (*Madhuca latifolia* L.) Flowers and Seeds

Sudhanshu S. Behera, Ramesh C. Ray***

*Department of Fisheries and Animal Resource Development, Government of Odisha, Cuttack, India

**ICAR-Central Tuber Crops Research Institute (Regional Centre), Bhubaneswar, India

ABBREVIATIONS

BHA	Butylated hydroxyl anisole
FAMEs	Fatty acid methyl esters
FPU	Filter paper unit
LHW	Liquid hot water
MESP	Minimum ethanol selling price
MOME	Mohua oil methyl ester
SE	Saccharification efficiency
SHF	Separate hydrolysis and fermentation
SmF	Submerged state fermentation
SSF	Simultaneous saccharification and fermentation
SsF	Solid state fermentation

12.1 INTRODUCTION

Global energy demand is increasing dramatically and an increase of 53% is expected by 2030 (Behera et al., 2017). In the past few decades, a major fraction of energy demand (88.6%) is met with fossil-based fuels [petroleum fuels (33.7%), natural gas (24.4%), and coal (30.5%)]. However, being nonrenewable in nature, fossil fuels are projected to be exhausted in the near future (Ashraful et al., 2014). It has been estimated that crude oil reserves will be exhausted by 2052 (at the current consumption rate of 4 billion tons/year), and coal deposits will last until 2088 (Ashraful et al., 2014). Therefore, more attention has currently been laid on biorenewable energy platforms, such as bioethanol, biobutanol, and biodiesel. The United States is pursuing several options in order to create biofuels from renewable biomass and has claimed to produce 90 billion gal (341 billion L) of biofuels in each year (Zhang et al., 2011).

Bioethanol Production From Food Crops
http://dx.doi.org/10.1016/B978-0-12-813766-6.00012-6

Various feedstocks, such as sugar, starch, and lignocelluloses, have been employed for bioethanol production. Forest biomass is a highly interesting feedstock as it is suitable for the production of biofuels used for transportation. There are several types of biofuels produced from forest biomass using various production techniques, such as fermentation of lignocellulose to ethanol or gasification of wood and other forest biomass to syngas that can be converted to diesel, methanol, methane, and other fuels. This chapter analyzes the potential use of forest biomass for the production of two major biofuels, bioethanol, and biodiesel, with Mohua (*Madhuca* spp.) flowers and seeds given as primary examples of raw materials.

The forest products, such as pine and eucalyptus, have been recently considered as potential bioenergy crops (Li et al., 2013). The increased use of forest biomass for bioenergy production can lead to the reduction of anthropogenic carbon emissions in the environment and assists forest-dependent communities achieving energy independence while generating jobs and improving the economy (Cambero and Sowlati, 2014). In Canada, after hydroenergy, the highest share of renewable energy production is contributed by forest biomass (2.9% in 2012) (Shabani et al., 2013). Table 12.1 summarizes the various types of biofuel production from forest biomass.

12.2 BIOPROSPECTING OF FOREST RESOURCES FOR HARNESSING BIOFUELS

Technology is rapidly advancing to utilize woody perennials and forest products (lignocellulosic biomass) for the production of biofuels (Shabani et al., 2013; Vohra et al., 2014). Forest biomass includes mainly woody materials, such as hardwoods and softwoods, while forest wastes are sawdust and pruning, and bark thinning residues, wood chips, and branches from dead trees (Zabed et al., 2016). Utilization of woody biomasses for energy production has been estimated roughly 370 million tons/year in the USA (Zhang et al., 2012). Other countries rich in forests are Canada, Russia, Brazil, and China that comprise together more than half of the total forest area of the world (Zhang et al., 2012). Softwood is derived from the evergreen species, such as pine, cedar, spruce, cypress, fir, hemlock, and redwood possess low densities and fast growth. Hardwoods are mainly found in the Northern hemisphere and include trees, such as poplar, willow, oak, cottonwood, and aspen (Zabed et al., 2017).

12.3 BIOREFINERY FROM FOREST WOODY BIOMASS

According to the taxonomical division of forest/woody biomass, there are two types of wood, softwood and hardwood. Softwoods are gymnosperms and originate from coniferous trees, including pines, spruces, and firs. Hardwoods are angiosperms and originate from deciduous trees, including oaks, maples, and birches. Forest biomass represents a valuable feedstock due to its composition, that is, more lignin and less ash content than agricultural residues (Choudhary et al., 2016). Forest wastes like wood chips, branches, and sawdusts have also been used as bioethanol feedstocks. Woody biomass, including wood, bark, and mixtures of forest residues are composed of cellulose, hemicellulose, lignin, and small amounts of other carbon moieties (Demirbaş, 2001; Naik et al., 2010). Approximately 368 million tons of woody biomass is annually produced in the United States, forming the basis of natural resource-based industries (e.g., energy and fiber) (Wang et al., 2012). In Sweden, Canada, and part of the USA, softwoods (pine and spruce) become the major renewable source of ethanol production,

TABLE 12.1 Biofuel Production From Forest Biomass

Substrate	Fermentation conditions	Saccharification efficiency/ biofuel yield	References
Pine (softwood)	Optimum extrusion conditions, screw speed 150 rpm, temperature 180°C and moisture content 25%	Maximum cellulose, hemicellulose, and total sugar recoveries 65.8%, 65.6%, and 66.1%, respectively	Karunanithy et al. (2012)
Yellow poplar (hardwood)	Temperature 152°C, acid concentration 1.6%, reaction time 16 min	Maximum ethanol: 42.8 g/L	Kim et al. (2015)
Bamboo	1. Formic acid at 100°C for 4 h 2. Aqueous solution (1% NaOH + 1% H_2O_2) at 80°C for 2 h	SE: 99.36%	Li et al. (2017)
Bamboo	SSF with *Clostridium beijerinckii* ATCC 55025-E604	SE: 72.44%; Butanol: 6.45 g/L	Kumar et al. (2017)
Mahula (*Madhuca latifolia* L.)	Yeast (*Saccharomyces cerevisiae*) cells immobilized in agar agar and calcium alginate	Bioethanol: 151.2, 154.5, and 149.1 g/kg flower	Behera et al. (2010a)
Mahula (*M. latifolia* L.)	Comparative study: Yeast (*S. cerevisiae*, CTCRI strain) with the bacterium (*Zymomonas mobilis*-MTCC 92)	Bioethanol: 149 and 122.9 g/kg flower	Behera et al. (2010b)
Mohua (*M. longifolia*) flowers	*S. cerevisiae* in acidic fermented media	Bioethanol: 170.03, 142.3, 127.7 mL/kg	Gedela et al. (2017)

SSF, Simultaneous saccharification and fermentation; *SE*, saccharification efficiency.

mitigating some of the environmental impacts of petroleum-based fuels (Monavari et al., 2009; Pasangulapati et al., 2012). Chapagain et al. (2009) studied the suitability of desert date (*Balanites aegyptiaca* L. Delile) as an oil crop in arid lands for large-scale sustainable industrial biodiesel production. Desert date kernels contain up to 46.7% oils. The four major fatty acids, [palmitic (16:0), stearic (18:0), oleic (18:1), and linoleic (18:2)] constitute 98%–100% of total fatty acids in the oils. Hundt et al. (2013) investigated a pulping process, using alkaline glycerol (polyol) and enzymatic hydrolysis aimed at achieving an effective saccharification of hardwood (beech). The conversion of polysaccharides in the pulp was up to 95% within 48 h of enzymatic digestion. *Shorea robusta* (sal) seeds have high content of carbohydrates, which makes them a potential source of bioethanol production (Choudhary et al., 2016). Choudhary et al. (2016) investigated the production of bioethanol from *S. robusta* seeds by *Zymomonas mobilis* MTCC 92. The effects of incubation period, temperature, pH, and nutrients were evaluated. An incubation time of 72 h, at a temperature of 37°C and pH 4, as well as supplementation with phosphorous and sulfur nutrients were found to enhance bioethanol production.

12.3.1 Processing Routes of Forest/ Woody Biomass to Bioethanol

Bioethanol production from forest biomass (lignocellulosic materials) involves the degradation of lignocelluloses to fermentable sugars followed by fermentation and distillation of the fermentation broth (Choudhary et al., 2016). The degradation of lignocelluloses (woody biomass) depends on upstream processes, such as pretreatment and saccharification for woody biomass and this biomass is distinctly different to nonwoody biomass (agricultural biomass). Woody biomass is especially recalcitrant to enzymatic decomposition due to its high lignin content (20%–30%) and is a major barrier to biological (enzymatic) saccharification of wood cellulose (Pasangulapati et al., 2012). Moreover, the enzymatically catalyzed conversion of cellulose to glucose is very slow unless the biomass has been subjected to pretreatment, as native cellulose is well protected by a matrix of hemicellulose and lignin. The woody biomass potentially possesses a lignocellulosic energy conversion factor of 16 compared to 1 and 8 in the cases of corn and sugarcane, respectively. Special attention needs to be given to factors, such as the pretreatment of wood cellulose, required to achieve complete saccharification and the energy consumption of wood size reduction to allow effective enzymatic saccharification. In general, the biochemical conversion approaches of woody biomass to biofuels consist of four stages: pretreatment, hydrolysis, fermentation, and distillation (Fig. 12.1).

12.3.1.1 Pretreatment (Delignification) of Woody Biomass

Feedstock pretreatment has been recognized as an essential upstream process to remove biomass recalcitrance for downstream microbial and enzymatic processing during cellulosic ethanol production (Rezania et al., 2017). During pretreatment, the woody biomass (predominately hemicellulose) is subjected to solubilization and produce oligomeric and monomeric sugars in liquid phase. The pretreatment process also offers a significant technology and provides an economic benefit being introduced prior to biological (treatment) means. The basic treatment methods include physical and thermochemical processes, which disrupt the recalcitrant biomass and enable the cellulose to undergo hydrolysis with higher efficiency and lower energy consumption. The agricultural biomass and the woody biomass are different in terms of physical properties and chemical composition. Unlike agricultural biomass, woody biomass requires a high amount of energy to achieve a size reduction that allows further enzymatic saccharification (Saini et al., 2015). An important study related to delignification/pretreatment of forest biomass was conducted through the utilization of chemicals (acids, alkalis, ionic liquids, etc.), physical (ball-milling, steam explosion, etc.), biological (enzymes, fungus, etc.) pretreatments and their combinations for destroying the rigid interactions between lignin and hemicelluloses (Hu et al., 2008; Li et al., 2017). In some other studies, chemical pretreatments, such as ionic liquid-mediated extraction (Lee et al., 2009), ozone delignification (García-Cubero et al., 2009), sodium chlorite delignification (Hubbell and Ragauskas, 2010), oxygen delignification (Koo et al., 2011), organosolv pretreatment (Romaní

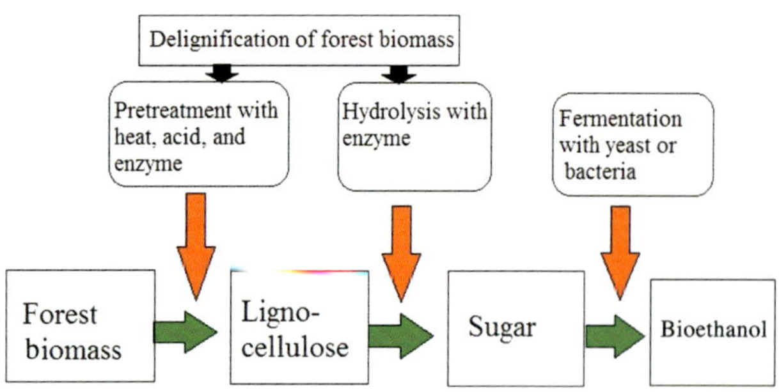

FIGURE 12.1 **Simplified process of bioethanol production from forest biomass.**

Kim, 2011), and lime pretreatment (Fuentes et al., 2011) were employed to improve the delignification of forest biomass (Lee et al., 2009). Li et al. (2017) developed a mild two-step pretreatment process using anhydrous formic acid and alkaline hydrogen peroxide, aqueous solution to fractionate lignocellulose to highly digestible cellulose for enzymatic saccharification. A high conversion rate of cellulose into glucose (99.36%) was achieved. The physical pretreatment processes, including ball milling and steam explosion (Asada et al., 2011) were found effective in decreasing lignin contents of forest biomass. Stream pretreatment/explosion is one of the most widely used method for pretreatment of lignocellulosic biomass. The woody biomass is first chipped and then treated with high-pressure saturated stream (at a temperature of 160–240°C and a pressure of 6–34 bar) for several seconds to few minutes, after which the pressure is released. The stream pretreatment/explosion can be improved by using an acid catalyst, such as H_2SO_4 or SO_2. Karunanithy et al. (2012) reported the use of an extrusion pretreatment method followed by enzymatic hydrolysis for the quantification of pine wood chips sugar and by-products. The pine wood chips pretreated at a screw speed of 150 rpm and a barrel temperature of 180°C with a moisture content of 25% resulted in a maximum recovery of total sugars of 66.1% (w/v) and 6.8-fold higher ethanol production compared to nonpretreated pine chips.

12.3.1.2 *Hydrolysis of Woody Biomass*

The performance of hydrolysis is highly associated with pretreatment operations of woody biomass. During hydrolysis, cellulose and hemicellulose fractions (woody biomass) are hydrolyzed into soluble components accessible for further conversion/fermentation to ethanol (Asada et al., 2011). There are two different types of hydrolysis that involve woody biomass: acidic and enzymatic (Devi et al., 2011). The application of dilute acids, such as sulfuric acid on woody biomass can be effective, providing a satisfactory cellulose conversion rate with certain hardwood species. However, the enzymatic process is considered to be the most attractive process and one of the critical processes for converting lignocellulosic biomass to sugars/glucose and bio-ethanol (Yu et al., 2011). Yu et al. (2011) studied the enzymatic (cellulase, xylanase, and β-glucosidase) hydrolysis of forest biomass on the effect of the hydrolysis process. The enzymatic hydrolysis increased by decreasing the lignin content (15%) and increasing accessible pore volume (0.30–0.35 g/g), especially in the case of soft wood of forest biomass. Several white-rot fungi, namely, *Ceriporia lacerata*, *Phanerochaete chrysoporium*, *Pleurotus ostreatus*, and *Cyathus stercoreus*, have been evaluated and showed high delignification efficiency when used for hydrolyzing lignocellulosic forest biomass (Kumar et al., 2009a).

12.3.1.3 *Fermentation of Woody Biomass to Bioethanol*

Fermentation is the following step. It is a microbial metabolic process that converts soluble sugars into alcohol (bioethanol). Some yeasts and bacteria can metabolize disaccharides (e.g., maltose and sucrose) and manosaccharides (e.g., glucose and fructose) in the absence of oxygen, which results in the production of ethanol and carbon dioxide (Zabed et al., 2017). Limayem and Ricke (2012) reported that yeast or bacterial cells work under high substrate concentration at the initial stage, while a high ethanol concentration is needed at the final stages. Several strains of yeast, including *S. cerevisiae*, *Z. mobilis*, *Pachysolen tannophilus*, *Pichia stipitis*, *Kluyveromyces marxianus*, and *Candida shehatae* have been reported to ferment sugars into ethanol. However, the tradtional strains of *S. cerevisiae* are among the most efficient, providing highest yields of bioethanol (12.0%–17.0% w/v; 90% of the theoretical yield) from sugars/glucose residues (Kumar et al., 2009b).

12.3.1.4 Distillation and Bioethanol Recovery

The fermentation woody biomass contains about 12%–18% (v/v) ethanol and it is thus necessary to separate the ethanol from the water (Devi et al., 2011). The secondary products, such as methanol, aldehydes, and esters are also intermingled with the water–ethanol binary system and are regulated in accordance with United States or European standards.

12.4 BIOREFINERY FROM MOHUA FLOWERS AND SEEDS

Mohua (*Madhuca* spp.) is a deciduous tree commonly found in the tropical rain forests of Asian and Australian continents (Gupta et al., 2012; Jadhav and Tiwari, 2011). There are three main species of *Madhuca* in India, that is, *M. latifolia*, *M. indica*, and *M. longifolia*. Mohua flowers are the cheapest source of carbohydrates (total sugar, 54.06%–72.9%) and are relatively unexploited having great potential of producing ethanol (148–193 g/kg) while 60 million tons per annum of nonedible oil can currently be obtained (Behera et al., 2016; Gupta et al., 2012; Swain et al., 2007).

12.4.1 Mohua as Bioethanol Crop

Mohua flower (corolla) (Fig. 12.2) is an important bioresource (US$ 35–40/ton) and a rich source (40%–47% dry weight basis) of fermentable sugars (glucose, sucrose, fructose, and maltose; 48%, w/w), next only to cane molasses. It is considered to be a potential source of sugar for bioethanol production (Behera et al., 2012; Mohanty et al., 2009). Moreover, Mohua flowers have been known for their use in the production of distilled liquor in rural areas of many Asian countries. Production of wine from Mohua flowers is uncommon. Soni and Dey (2013) claimed that improved the traditional brewing process to produce a fermented product with low alcohol content and rich in nutraceuticals.

FIGURE 12.2 (A) Mohua bud and flower (insert); (B) Mohua seeds.

12.4.2 Biodiesel From Mohua Seeds

Biodiesel is comprised of long-chain fatty acid methyl esters derived from vegetable oils or animal fats (Martínez et al., 2014). The biodiesel fuel has several advantages over diesel, being of domestic origin, biodegradable, nontoxic, characterized by a higher cetane number, and able to enhance engine lubricity (Martínez et al., 2014). Moreover, the biodiesel is 70% cleaner with 93% lower hydrocarbon, 50% lower carbon monoxide, and 45% lower particulate matter emissions in comparison to the conventional diesel fuel (Gupta et al., 2012).

Biodiesel is mainly produced throughout the world from edible oils of soybean, rapeseeds, and sunflower (Singh and Singh, 2010). However, vegetable oil esters may not be the right option to substitute diesel fuel because they are not self-sufficient (Ghosal et al., 2008). Recently, several attempts have been made to produce biodiesel from nonedible oil source, such as sal (*S. robusta* L.) (Vedaraman et al., 2012), neem (*Azadirachta indica* L.) (Dhar et al., 2012), karanja (*Millettia pinnata* L.) (Srivastava and Verma, 2008), and jatropha (*Jatropha curcas* L.) (Parawira, 2010), which have all great potential in the production of biodiesel for supplementing other conventional sources (Balat, 2011). Among the alternative fuels, vegetable oil esters (biodiesel) from Mohua seeds are more promising than petroleum fuel, used in compression ignition engines (Raheman and Ghadge, 2007). Mohua tree is cultivated in warm and humid regions for its flowers and oleaginous seeds and 20–200 kg seeds are produced annually per tree, depending on trees' maturity (Ashraful et al., 2014). Mohua seed has an oil content of 35–50 wt.% (v/w). Mohua oil contains approximately 41%–51% oleic acid. Some other fatty acids are also present in the oil, such as stearic (20.0%–25.1%, v/w), palmitic (16.0%–28.2%, v/w), and linoleic acids (8.9%–18.3%, v/w) (Ashraful et al., 2014). Ghosal et al. (2008) reported that Mohua-based methyl ester can be used as a substitute for diesel fuel in compression ignition engine with lower percentages of emission and engine wear.

12.4.3 Advance Technology for Production of Biorefinery Products From Mohua Plant Sources

Recent advances in the technology of the production of biorefinery products from forest biomass are mainly focused in achieving increased ethanol yield per unit of biomass, decreased processing time, and production cost, and more importantly, in minimizing process steps, particularly during higher lignin content of forest woody biomass (Halder et al., 2016).

12.4.3.1 Solid State Fermentation

Solid state fermentation (SsF) is a process of cultivation of microorganisms on a moist solid substrate and takes place in the absence or near absence of free water (Behera and Ray, 2016). Mohanty et al. (2009) investigated SsF for the production of ethanol from Mohua flowers using the ethanologenic yeast, *S. cerevisiae*. High ethanol productivity (3.13 g/kg flower/h) and ethanol yield (58.44 g/100 g flowers) were achieved with 77.1% fermentation efficiency.

12.4.3.2 Submerged Fermentation

Submerged fermentation (SmF) is the process in which growth and decomposition (anaerobic and/or partially anaerobic) of substrates (i.e., carbohydrates) is accomplished by microorganisms in the presence of plenty of free water (liquid medium) (Mussatto and Teixeira, 2010). SmF is the process of choice for industrial operations (Devi et al., 2011). Gedela et al. (2017) investigated the production of bioethanol from Mohua flowers using *S. cerevisiae* in acidic (pH 4–5.7) fermented medium (1 L) and an ethanol yield of 60/100 g flowers was obtained.

12.4.3.3 Cell Immobilization

Bioethanol can be produced using either free or immobilized cells. The use of immobilized cells is more efficient due to the easier separation of the cell mass from the bulk liquid, reduced risk of contamination, better operational stability and good cell viability for several cycles of operations (Wirawan et al., 2012). Among the different immobilization techniques, entrapment of microbial cells within polymeric matrices, such as agar agar, calcium alginate, gelatin, κ-carrageenan, have been studied widely (Nikolić et al., 2010). Two of the most suitable carriers for cell immobilization are entrapment in calcium alginate bead (Behera et al., 2010a) and agar agar cubes (Behera et al., 2010b). These techniques are simple, cost-effective, and nontoxic. Behera et al. (2010a) studied the batch fermentation of Mohua flowers using immobilized cells (in agar agar and calcium alginate) and free cells of *S. cerevisiae*. The ethanol yield was 152.1, 154.5, and149.1 g/kg of flowers using immobilized (in agar and calcium alginate) and free cells, respectively. Moreover, the immobilized cells remained physically active for at least three cycles of ethanol fermentation while no productivity loss was observed. Behera et al. (2010b) used batch fermentation of Mohua flowers with *Z. mobilis* MTCC 92 free cells and cells immobilized in a calcium alginate matrix, to produce ethanol. After 96 h of fermentation, 122.9 and 134.6 of ethanol/kg of flowers (on dry weight basis) were obtained using free and immobilized cells, respectively. The luffa (*Luffa cylindrical* L.) a cucurbitaceous crop, abundantly available in tropical and subtropical countries is believed to be a promising material for immobilizing microbial cells (Behera et al., 2011a). Behera et al. (2011b) reported the production of ethanol from Mohua flowers with immobilized cells of *S. cerevisiae* in *Luffa* sponge discs.

12.4.4 Thermooxidative Stability of (Mohua) Oil

Mohua oil has been identified as a potential alternative to diesel-engine fuels or biodiesel (Balat, 2011; Pradhan et al., 2016). High thermooxidative stability is very important for the maintenance of diesel engines or when uses in industrial lubricants are considered. However, the low thermooxidative stability of Mohua oil is an issue (Kumar and Sharma, 2011). This characteristic can be improved with the help of a suitable antioxidant (Singh and Singh, 2014). Numerous experimental studies have established the positive effect of antioxidants on the oxidative stability of Mohua oil. Functional groups containing oxygen, such as butylated hydroxyl anisole (BHA), alkyl amino piperazine, alkyl benzoate, alkyl citrate, and diphenyl amine are good antioxidants for boosting Mohua oil production (Singh and Singh, 2014).

12.4.5 Transesterification of (Mohua) Oil

The most common route for biodiesel production is the transesterification reaction, shown in Eq. 12.1 (Fig. 12.3). Transesterification refers to the catalyzed (KOH/NaOH) chemical reaction of oil/fat (triglycerides) with an alcohol (methanol/ethanol) to produce an ester of a fatty acid and glycerol (Chapagain et al., 2009; Jena et al., 2010). The properties of Mohua oil methyl ester (MOME) are close to those of diesel oil. Kapilan and Reddy (2008) prepared MOME through transesterification of Mohua oil using potassium hydroxide. MOME also exhibited a good performance and an emission rate close to diesel. Chapagain et al. (2009) investigated the transesterification (ester of methanol) of desert date palm (*B. aegyptiaca* L. Delile) for the production of biodiesel (methyl ester). The transesterification process was performed with methanol (6:1 molar ratio) and potassium hydroxide (catalyst) yielding biodiesel at an efficiency rate of 90% (weight basis).

12.4.6 Economics Mohua Biofuel

Economic analysis of bioethanol production from renewable resources (Mohua flower and seed) is one of the major areas that determine the overall cost of the process. Several studies have been carried out on this economic analysis using

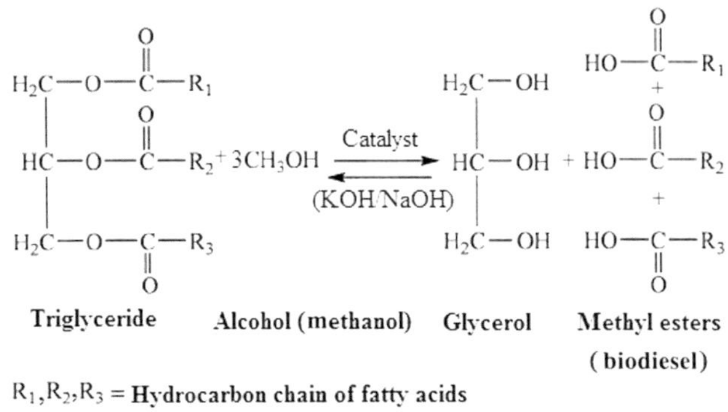

FIGURE 12.3 **Biodiesel production is the transesterification reaction.**

different feedstocks. However, no economic analysis of forest biomass, derived from Mohua flower and seeds has been carried out so far. The total production costs of bioethanol from Mohua flower and seeds depend on the sum of raw material (flower and seeds) costs, operating costs, labor, supplies, and overheads. Among the total capital costs, raw material costs are significantly lower in the case of forest/lignocellulosic biomass-based feedstocks compared to those of sugar and starch-based biomass. However, forest/lignocellulosic biomass requires the use of enzymes (during hydrolysis) that increase the overall capital costs (19%–22%) when compared with sugar- or starch-based biomass/feedstocks (10%–20%) (Gregg et al., 1998).

12.5 CONSOLIDATED BIOPROCESSING(CBP)/ ENGINEERED MICROORGANISM

The complex structure of lignocellulosic forest biomass requires a multistep process that is costly and time-consuming. In recent years, several bioprocessing approaches have been developed for the pretreatment, saccharification, and fermentation of forest biomass into biofuels production; however, these approaches are still limited as financially unfeasible. The cost constraint of these technologies can be overcome by designing and constructing robust biofuel-producing microbes/enzymes and using them in a CBP system (Salehi Jouzani and Taherzadeh, 2015). Horisawa et al. (2015) designed a CBP system of ethanol production from wood chips (polysaccharides) in a single reactor under anaerobic conditions using white rot fungus *Schizophyllum commune* NBRC 4928. The multifunctional fungus produced enzymes that degrade lignin, cellulose, and hemicelluloses with ethanol yields of 52.5%–80.5%.

In the past two decades, numerous microorganisms have been engineered to produce ethanol (Agarwal et al., 2013). All the major steps in a typical conversion process, and particularly the fermentation of sugars that is a common step for all types of biomass, are greatly influenced by the microorganisms used. Traditionally yeast, *S. cerevisiae*, and a bacterium, *Z. mobilis*, are genetically engineered and used to ferment ethanol (Zabed et al., 2017).

12.6 COMPARISON OF BIOREFINERY FROM FOREST BIOMASS WITH OTHER FEEDSTOCKS

The major differences between woody and nonwoody (agricultural) biomass includes: (1) woody biomass is physically and structurally stronger and denser (350–490 kg/m^3) than agricultural biomass (108–130 kg/m^3), (2) woody biomass has higher lignin content (23.9%–32.0%) than agricultural biomass (16.9%–23.1%), and (3) woody biomass is more resistant to microbial and enzymatic actions than agricultural biomass (Zhu and Pan, 2010). This suggests that future studies should focus on the thermochemical pretreatment of woody biomass to improve the cellulosic enzymatic saccharification and fermentable sugar yield. The upstream processing of woody biomass (particularly the wood size reduction and pretreatment processes) is essential to reduce recalcitrance in woody biomass and enhance microbial destruction of polysaccharides (Gaurav et al., 2017). A comparative analysis of feedstocks and technological characteristics governing the bioethanol production are given in Table 12.2.

12.7 TECHNOECONOMIC FEASIBILITY OF BIOETHANOL PRODUCTION FROM FOREST BIOMASS AND BOTTLENECKS

Optimizing biofuel inputs involves difficult obstacles that cover many aspects of primary production and transport, and will require region-specific design. Integrated bioprocessing and engineering strategies for the conversion of agricultural/forestry biomass to ethanol and other valued-added products are vital if the conversion of lignocellulosic biomass to ethanol is to be economically practicable. A technoeconomic assessment was used to evaluate the potential of producing bioethanol from forest biomass. Wingren et al. (2003) evaluated the production of biofuel (ethanol) from softwood from technological and economic points of view and subsequently analyzed how enzymatic processes can be involved. Two base configurations, that is, simultaneous saccharification and fermentation (SSF) and separate hydrolysis and fermentation (SHF) were evaluated and compared. The major economic improvements in both SSF and SHF could be achieved by lowering the energy consumption and recycling the process streams. Littlewood et al. (2013) investigated the use of bamboo. Liquid hot water (LHW) pretreatment was used under various pretreatment and saccharification conditions to produce bioethanol. A LHW pretreatment at 190°C for 10 min was selected as the optimal condition for maximizing sugar release, yielding 69% of the theoretical maximum amount, following 72 h of saccharification. Under these conditions a greater proportion of sugar was released during pretreatment in comparison to saccharification. However, enzymatic saccharification with five loadings (10–140 FPU/g glucan) of Cellic CTec2 led to a total sugar release ranging from 59%–76% of the theoretical maximum.

TABLE 12.2 Comparison Accounts of Biorefinery From Forest Biomass With Other Feedstocks

Substrates/ feedstocks	Cellulose content	Hemicellulose content	Lignin content	Ethanol yield	References
Perennial grasses	25%–40%	25%–50%	10%–40%	160–460 L/ton of biomass	Griffey et al. (2010)
Aquatic plants/ sea weeds	18%–35%	17%–44%	3.5%–19%	255 kg/ton of biomass	Saini et al. (2015)
Agricultural residues	30%–45%	20%–30%	10%–20%	235–450 L/ton of biomass	Rezania et al. (2017)
Municipal solid waste	33%–49%	9%–16%	10%–14%	154 L/ton of biomass	Saini et al. (2015)
Forest biomass and waste	20%–30%	35%–50%	20%–30%	220–285 L/ton of biomass	Kraan (2013); Zabed et al. (2017)

12.8 GLOBAL BIOFUEL ECONOMY AND WHERE FOREST BIOMASS STANDS

The global primary energy supply changes slowly and the biofuel share is expected to steadily increase until 2030. After 2030, the growth of energy supply from solar and wind technologies will abruptly change, while supply from biomass will increase slowly, hydroenergy will be more or less constant and fossil fuels will deplete completely (Berndes et al., 2003; Somerville et al., 2010). Moreover, ethanol production from sugar (Brazil) and cornstarch (USA) is unsustainable because it clashes with food production for human use and, more importantly, does not significantly diminish greenhouse gas emissions. These shortcomings could be addressed by producing ethanol from renewable lignocellulose feedstocks, such as agricultural and forest waste residues as well as energy-producing dedicated crops. According to renewable energy projections of the European Union, 56% of the renewable energy (biomass) will be estimated by 2020 (Ericsson and Nilsson, 2006). The annual demand of biomass for energy production is estimated to increase in the European Union from the current level of 5.7 EJ to 10.0 EJ in 2020 (Lamers and Junginger, 2013). However, the present forest or woody biomass shows the highest variation of energy production and is estimated between 0.8 and 6.0 EJ/year, whereas its production projected to 10.6 EJ/year in year 2050 (Bentsen and Felby, 2012).

12.9 CONCLUSION AND FUTURE PROSPECTS

Bioethanol is one of the most promising and environmentally friendly renewable sources of energy. In recent years, almost all required bioethanol is obtained from agricultural bioamass (at a commercial scale) usually corn and sugarcane feedstocks. However, these feedstocks are not adequate to meet the growing demand for bioethanol. Therefore, current research efforts have concentrated on alternative ethanol sources and particular attention is paid on the forest/woody biomass (lignocellulosic biomass). Considering the challenge of future resource scarcity, it is also important to understand the impacts of increasing the use of forest/woody biomass for bioethanol production. However, issues specifically associated with forest/woody biomass in the upstream processing need to be addressed for better harnessing of bioethanol. For instance, certain pretreatment methods have only been effective on few feedstocks and ineffective to forest biomass, and if possible/effective in certain cases difficulties have still been encountered during the large-scale production stages. Other challenges that forest/woody biomass utilization brings include high harvesting and transportation costs, overlapping demand for some forest biomass, and market promotion for new bioproducts. Key conversion technologies for forest/woody biomass are still studied and not yet viable to allow effective commercial production. Therefore, attention needs to be paid to challenges and public policies related to the development of bioenergy from forest-derived woody biomass as well as to processing, conversion technologies, and supportive public policies.

References

Agarwal, T., Quraishi, A., Tiwari, K., Jadhav, S., 2013. Ethanol production from Mohua (*Madhuca indica* JF Gmel) flowers by soil bacteria. Researcher 5, 102–106.

Asada, C., Kita, A., Sasaki, C., Nakamura, Y., 2011. Ethanol production from disposable aspen chopsticks using delignification pretreatments. Carbohyd. Polym. 85, 196–200.

Ashraful, A.M., Masjuki, H.H., Kalam, M.A., Fattah, I.R., Imtenan, S., Shahir, S.A., Mobarak, H.M., 2014. Production and comparison of fuel properties, engine performance, and emission characteristics of biodiesel from various non-edible vegetable oils: a review. Energ. Conserv. Manage. 80, 202–228.

Balat, M., 2011. Potential alternatives to edible oils for biodiesel production: a review of current work. Energ. Conserv. Manage. 52, 1479–1492.

Behera, S.S., Ray, R.C., 2016. Solid state fermentation for production of microbial cellulases: recent advances and improvement strategies. Int. J. Biol. Macromol. 86, 656–669.

Behera, S., Kar, S., Mohanty, R.C., Ray, R.C., 2010a. Comparative study of bio-ethanol production from mahula (*Madhuca latifolia* L.) flowers by *Saccharomyces cerevisiae* cells immobilized in agar agar and Ca-alginate matrices. Appl. Energy 87, 96–100.

Behera, S., Mohanty, R.C., Ray, R.C., 2010b. Comparative study of bio-ethanol production from mahula (*Madhuca latifolia* L.) flowers by *Saccharomyces cerevisiae* and *Zymomonas mobilis*. Appl. Energy 87, 2352–2355.

Behera, S., Mohanty, R., Ray, R., 2010c. Ethanol fermentation of mahula (*Madhuca latifolia*) flowers using free and immobilized bacteria *Zymomonas mobilis* MTCC 92. Biologia 65, 416–421.

Behera, S., Mohanty, R.C., Ray, R.C., 2011a. Ethanol production from mahula (*Madhuca latifolia* L.) flowers with immobilized cells of *Saccharomyces cerevisiae* in *Luffa cylindrica* L. sponge discs. Appl. Energy 88, 212–215.

Behera, S., Mohanty, R.C., Ray, R.C., 2011b. Ethanol production from mahula (*Madhuca latifolia* L.) flowers using free and immobilized (in *Luffa cylindrica* L. sponge discs) cells of *Zymomonas mobilis* MTCC 92. Ann. Microbiol. 61, 469–474.

Behera, S., Mohanty, R.C., Ray, R.C., 2012. Biochemistry of post-harvest spoilage of mahula (*Madhuca latifolia* L.) flowers: changes in total sugar, ascorbic acid, phenol and phenylalanine ammonia-lyase activity. Arch. Phytopathol. Plant Protect. 45, 846–855.

Behera, S., Ray, R.C., Mohanty, R.C., 2013. Bio-fuel agro-forestry industrial production system of mahula (*Madhuca latifolia* L.): process development and future perspectives. J. Sci. Ind. Res. (India) 72, 62–69.

Behera, S., Ray, R.C., Swain, M.R., Mohanty, R.C., Biswal, A.K., 2016. Traditional and current knowledge on the utilization of Mohua (*Madhuca latifolia* L.) flowers by Santhal tribe in Similipal biosphere reserve, Orissa, India. Ann. Trop. Res. 38, 94–104.

Behera, S.S., Ray, R.C., Ramachandran, S., 2017. Renewable energy from agro-industrial processing wastes: An overview. Elsevier, New York, ISBN: 978-967-14475-3-6; eISBN: 978-967-14475-2-9 322.

Bentsen, N.S., Felby, C., 2012. Biomass for energy in the European Union: a review of bioenergy resource assessments. Biotechnol. Biofuels 5, 25–45.

Berndes, G., Hoogwijk, M., Van den Broek, R., 2003. The contribution of biomass in the future global energy supply: a review of 17 studies. Biomass Bioenergy 25, 1–28.

Cambero, C., Sowlati, T., 2014. Assessment and optimization of forest biomass supply chains from economic, social and environmental perspectives: a review of literature. Renew. Sustain. Energ. Rev. 36, 62–73.

Chapagain, B.P., Yehoshua, Y., Wiesman, Z., 2009. Desert date (*Balanites aegyptiaca*) as an arid lands sustainable bioresource for biodiesel. Bioresour. Technol. 100, 1221–1226.

Choudhary, A., Tiwari, S., Jadhav, S.K., Tiwari, K.L., 2016. Bioethanol production from *Shorea robusta* (Sal) seeds using *Zymomonas mobilis* MTCC92. Silpakorn Univ. Sci. Technol. J. 10, 9–14.

De Meyer, A., Cattrysse, D., Rasinmäki, J., Van Orshoven, J., 2014. Methods to optimise the design and management of biomass-for-bioenergy supply chains: a review. Renew. Sustain. Energ. Rev. 31, 657–670.

Demirbaş, A., 2001. Biomass resource facilities and biomass conversion processing for fuels and chemicals. Energ. Convers. Manage. 42, 1357–1378.

Devi, S., Suhag, M., Dhaka, A., Singh, J., 2011. Biochemical conversion process of producing bioethanol from lignocellulosic biomass. Int. J. Microbial Resour. Technol. 1, 28–32.

Dhar, A., Kevin, R., Agarwal, A.K., 2012. Production of biodiesel from high-FFA neem oil and its performance, emission and combustion characterization in a single cylinder DICI engine. Fuel Process. Technol. 97, 118–129.

Ericsson, K., Nilsson, L.J., 2006. Assessment of the potential biomass supply in Europe using a resource-focused approach. Biomass Bioenergy 30, 1–15.

Fuentes, L.L., Rabelo, S.C., Maciel Filho, R., Costa, A.C., 2011. Kinetics of lime pretreatment of sugarcane bagasse to enhance enzymatic hydrolysis. Appl. Biochem. Biotechnol. 163, 612–625.

García-Cubero, M.T., González-Benito, G., Indacoechea, I., Coca, M., Bolado, S., 2009. Effect of ozonolysis pretreatment on enzymatic digestibility of wheat and rye straw. Bioresour. Technol. 100, 1608–1613.

Gaurav, N., Sivasankari, S., Kiran, G.S., Ninawe, A., Selvin, J., 2017. Utilization of bioresources for sustainable biofuels: a review. Renew. Sustain. Energ. Rev. 73, 205–214.

Gedela, R., Naidu, R.T., Rachakonda, S., Naidu, A., 2017. *Madhuca Longifolia* flowers for high yields of bio-ethanol feedstock production. Int. J. Appl. Sci. Biotechnol. 4, 525–528.

Ghosal, M.K., Das, D.K., Pradhan, S.C., Sahoo, N., 2008. Performance study of diesel engine by using Mohua methyl ester (biodiesel) and its blends with diesel fuel. Agric. Eng. Int. 10, 1–10.

Gregg, D., Boussaid, A., Saddler, J., 1998. Techno-economic evaluations of a generic wood to-ethanol process: effect of increased cellulose yields and enzyme recycle. Bioresour. Technol. 63, 7–12.

Griffey, C., Brooks, W., Kurantz, M., Thomason, W., Taylor, F., Obert, D., Moreau, R., Flores, R., Sohn, M., Hicks, K., 2010. Grain composition of Virginia winter barley and implications for use in feed, food, and biofuels production. J. Cereal Sci. 31, 41–49.

Gupta, A., Chaudhary, R., Sharma, S., 2012. Potential applications of Mohua (*Madhuca indica*) biomass. Waste Biomass Valoriz. 3, 175–189.

Halder, G., Dhawane, S.H., Dutta, D., Dey, S., Banerjee, S., Mukherjee, S., Mondal, M., 2016. Computational simulation and statistical analysis of bioethanol production from *Madhuca indica* by batch fermentation process using *Saccharomyces cerevisiae*. Sustain. Energ. Technol. Assess. 18, 16–33.

Horisawa, S., Ando, H., Ariga, O., Sakuma, Y., 2015. Direct ethanol production from cellulosic materials by consolidated biological processing using the wood rot fungus *Schizophyllum commune*. Bioresour. Technol. 197, 37–41.

Hu, G., Heitmann, J.A., Rojas, O.J., 2008. Feedstock pretreatment strategies for producing ethanol from wood, bark, and forest residues. BioResources 3, 270–294.

Hubbell, C.A., Ragauskas, A.J., 2010. Effect of acid-chlorite delignification on cellulose degree of polymerization. Bioresour. Technol. 101, 7410–7415.

Hundt, M., Schnitzlein, K., Schnitzlein, M.G., 2013. Alkaline polyol pulping and enzymatic hydrolysis of hardwood: effect of pulping severity and pulp composition on cellulase activity and overall sugar yield. Bioresour. Technol. 136, 672–679.

Jadhav, S.K., Tiwari, K.L., 2011. Renjini production of bioethanol form Mohua flower. Bioethanol: A Bioenergy Resource. LAP Lambert Academic Publishing, pp. 1–76.

Jena, P.C., Raheman, H., Kumar, G.P., Machavaram, R., 2010. Biodiesel production from mixture of Mohua and simarouba oils with high free fatty acids. Biomass Bioenergy 34, 1108–1116.

Kapilan, N., Reddy, R.P., 2008. Methyl esters of Mohua oil as an ecofriendly fuel in heavy duty vehicles. Asian J. Chem. 20, 1245–1250.

Karunanithy, C., Muthukumarappan, K., Gibbons, W.R., 2012. Extrusion pretreatment of pine wood chips. Appl. Biochem. Biotechnol. 167, 81–99.

Kim, H.Y., Hong, C.Y., Kim, S.H., Yeo, H.Y., Choi, I.G., 2015. Optimization of the organosolv pretreatment of yellow poplar for bioethanol production by response surface methodology. J. Korean Wood Sci. Technol. 43, 600–612.

Koo, B.W., Treasure, T.H., Jameel, H., Phillips, R.B., Chang, H.M., Park, S., 2011. Reduction of enzyme dosage by oxygen delignification and mechanical refining for enzymatic hydrolysis of green liquor-pretreated hardwood. Appl. Biochem. Biotechnol. 165, 832–844.

Kraan, S., 2013. Mass-cultivation of carbohydrate rich macroalgae, a possible solution for sustainable biofuel production. Mitig. Adapt. Strat. Glob. Change 18, 27–46.

Kumar, A., Sharma, S., 2011. Potential non-edible oil resources as biodiesel feedstock: an Indian perspective. Renew. Sustain. Energ. Rev. 15, 1791–1800.

Kumar, P., Barrett, D.M., Delwiche, M.J., Stroeve, P., 2009a. Methods for pretreatment of lignocellulosic biomass for efficient hydrolysis and biofuel production. Ind. Eng. Chem. Res. 48, 3713–3729.

Kumar, S., Gujjala, L.K., Banerjee, R., 2017. Simultaneous pretreatment and saccharification of bamboo for biobutanol production. Ind. Crop Prod. 101, 21–28.

Kumar, S., Singh, S.P., Mishra, I.M., Adhikari, D.K., 2009b. Recent advances in production of bioethanol from lignocellulosic biomass. Chem. Eng. Technol. 32, 517–526.

Lamers, P., Junginger, M., 2013. The 'debt'is in the detail: a synthesis of recent temporal forest carbon analyses on woody biomass for energy. Biofuels Bioprod. Biorefin. 7, 373–385.

Lee, S.H., Doherty, T.V., Linhardt, R.J., Dordick, J.S., 2009. Ionic liquid-mediated selective extraction of lignin from wood leading to enhanced enzymatic cellulose hydrolysis. Biotechnol. Bioeng. 102, 1368–1376.

Lehto, J., Alén, R., 2013. Alkaline pre-treatment of hardwood chips prior to delignification. J. Wood Chem. Technol. 33, 77–91.

Li, X., Kim, T.H., 2011. Low-liquid pretreatment of corn stover with aqueous ammonia. Bioresour. Technol. 102, 4779–4786.

Li, C., Sun, L., Simmons, B.A., Singh, S., 2013. Comparing the recalcitrance of eucalyptus, pine, and switchgrass using ionic liquid and dilute acid pretreatments. BioEnergy Res. 6, 14–23.

Li, M.F., Yu, P., Li, S.X., Wu, X.F., Xiao, X., Bian, J., 2017. Sequential two-step fractionation of lignocellulose with formic acid organosolv followed by alkaline hydrogen peroxide under mild conditions to prepare easily saccharified cellulose and value-added lignin. Energ. Convers. Manage. 148, 1426–1437.

Limayem, A., Ricke, S.C., 2012. Lignocellulosic biomass for bioethanol production: current perspectives, potential issues and future prospects. Prog. Energ. Combust. 38, 449–467.

Littlewood, J., Wang, L., Turnbull, C., Murphy, R.J., 2013. Techno-economic potential of bioethanol from bamboo in China. Biotechnol. Biofuels 6, 173–179.

Martínez, G., Sánchez, N., Encinar, J.M., González, J.F., 2014. Fuel properties of biodiesel from vegetable oils and oil mixtures: influence of methyl esters distribution. Biomass Bioenergy 63, 22–32.

McKendry, P., 2002. Energy production from biomass (part 1): overview of biomass. Bioresour. Technol. 83, 37–46.

Mohanty, S.K., Behera, S., Swain, M.R., Ray, R.C., 2009. Bioethanol production from mahula (*Madhuca latifolia* L.) flowers by solid-state fermentation. Appl. Energy 86, 640–644.

Monavari, S., Galbe, M., Zacchi, G., 2009. The influence of solid/liquid separation techniques on the sugar yield in two-step dilute acid hydrolysis of softwood followed by enzymatic hydrolysis. Biotechnol. Biofuels 2, 6–15.

Mussatto, S.I., Teixeira, J.A., 2010. Lignocellulose as raw material in fermentation processes. Méndez-Vilas, A. (Ed.), Current Research, Technology and Education Topics in Applied Microbiology and Microbial Biotechnology, vol. 2, Formatex Research Center, Badajoz, Spain, pp. 897–907.

Naik, S.N., Goud, V.V., Rout, P.K., Dalai, A.K., 2010. Production of first and second generation biofuels: a comprehensive review. Renew. Sustain. Energ. Rev. 14, 578–597.

Nikolić, S., Mojović, L., Pejin, D., Rakin, M., Vukašinović, M., 2010. Production of bioethanol from corn meal hydrolyzates by free and immobilized cells of *Saccharomyces cerevisiae* var. *ellipsoideus*. Biomass Bioenergy 34, 1449–1456.

Parawira, W., 2010. Biodiesel production from *Jatropha curcas*: a review. Sci. Res. Essays 5, 1796–1808.

Pasangulapati, V., Ramachandriya, K.D., Kumar, A., Wilkins, M.R., Jones, C.L., Huhnke, R.L., 2012. Effects of cellulose, hemicellulose and lignin on thermochemical conversion characteristics of the selected biomass. Bioresour. Technol. 114, 663–669.

Pradhan, D., Singh, R.K., Bendu, H., Mund, R., 2016. Pyrolysis of Mohua seed (*Madhuca indica*):production of biofuel and its characterization. Energ. Convers. Manage. 108, 529–538.

Puhan, S., Vedaraman, N., Ram, B.V., Sankarnarayanan, G., Jeychandran, K., 2005. Mohua oil (*Madhuca Indica* seed oil) methyl ester as biodiesel-preparation and emission characterstics. Biomass Bioenergy 28, 87–93.

Raheman, H., Ghadge, S.V., 2007. Performance of compression ignition engine with Mohua. Biodiesel Fuel 86, 2568–2573.

Rezania, S., Din, M.F., Mohamad, S.E., Sohaili, J., Taib, S.M., Yusof, M.B., Kamyab, H., Darajeh, N., Ahsan, A., 2017. Review on pretreatment methods and ethanol production from cellulosic water hyacinth. BioResources 24, 1024–2108.

Romaní, A., Garrote, G., López, F., Parajó, J.C., 2011. Eucalyptus globulus wood fractionation by autohydrolysis and organosolv delignification. Bioresour. Technol. 102, 5896–5904.

Saini, J.K., Saini, R., Tewari, L., 2015. Lignocellulosic agriculture wastes as biomass feedstocks for second-generation bioethanol production: concepts and recent developments. 3 Biotech 5, 337–353.

Salehi Jouzani, G., Taherzadeh, M.J., 2015. Advances in consolidated bioprocessing systems for bioethanol and butanol production from biomass: a comprehensive review. Biofuel Res. J. 2, 152–195.

Shabani, N., Akhtari, S., Sowlati, T., 2013. Value chain optimization of forest biomass for bioenergy production: a review. Renew. Sust. Energ. Rev. 23, 299–311.

Singh, R.K., Singh, A.K., 2014. Abilities of some compounds to stabilize mahwa oil from high temperature oxidative degradation for biolubricant applications. Waste Biomass Valoriz. 5, 847–855.

Singh, S.P., Singh, D., 2010. Biodiesel production through the use of different sources and characterization of oils and their esters as the substitute of diesel: a review. Renew. Sustain. Energ. Rev. 14, 200–216.

Somerville, C., Youngs, H., Taylor, C., Davis, S.C., Long, S.P., 2010. Feedstocks for lignocellulosic biofuels. Science 329, 790–792.

Soni, S., Dey, G., 2013. Studies on value-added fermentation of *Madhuca latifolia* flower and its potential as a nutrabeverage. Int. J. Biotechnol. Bioeng. Res. 4, 215–226.

Srivastava, P.K., Verma, M., 2008. Methyl ester of karanja oil as an alternative renewable source energy. Fuel 87, 1673–1677.

Swain, M.R., Kar, S., Sahoo, A.K., Ray, R.C., 2007. Ethanol fermentation of mahula (*Madhuca latifolia* L.) flowers using free and immobilized yeast *Saccharomyces cerevisiae*. Microbiol. Res. 162, 93–98.

Vedaraman, N., Puhan, S., Nagarajan, G., Ramabrahmam, B.V., Velappan, K.C., 2012. Methyl ester of Sal oil (*Shorea robusta*) as a substitute to diesel fuel—a study on its preparation, performance and emissions in direct injection diesel engine. Ind. Crop Prod. 36, 282–288.

Vohra, M., Manwar, J., Manmode, R., Padgilwar, S., Patil, S., 2014. Bioethanol production: feedstock and current technologies. J. Environ. Chem. Eng. 2, 573–584.

Wang, Z.J., Zhu, J.Y., Zalesny, R.S., Chen, K.F., 2012. Ethanol production from poplar wood through enzymatic saccharification and fermentation by dilute acid and SPORL pretreatments. Fuel 95, 606–614.

Wingren, A., Galbe, M., Zacchi, G., 2003. Techno-economic evaluation of producing ethanol from softwood: comparison of SSF and SHF and identification of bottlenecks. Biotechnol. Prog. 19, 1109–1117.

Wirawan, F., Cheng, C.L., Kao, W.C., Lee, D.J., Chang, J.S., 2012. Cellulosic ethanol production performance with SSF and SHF processes using immobilized *Zymomonas mobilis*. Appl. Energy 100, 19–26.

Yu, Z., Jameel, H., Chang, H.M., Park, S., 2011. The effect of delignification of forest biomass on enzymatic hydrolysis. Bioresour. Technol. 102, 9083–9089.

Yue, D., You, F., Snyder, S.W., 2014. Biomass-to-bioenergy and biofuel supply chain optimization: overview, key issues and challenges. Comput. Chem. Eng. 66, 36–56.

Zabed, H., Sahu, J.N., Boyce, A.N., Faruq, G., 2016. Fuel ethanol production from lignocellulosic biomass: an overview on feedstocks and technological approaches. Renew. Sustain. Energ. Rev. 66, 751–774.

Zabed, H., Sahu, J.N., Suely, A., Boyce, A.N., Faruq, G., 2017. Bioethanol production from renewable sources: current perspectives and technological progress. Renew. Sustain. Energ. Rev. 71, 475–501.

Zhang, F., Johnson, D.M., Johnson, M.A., 2012. Development of a simulation model of biomass supply chain for biofuel production. Renew. Energ. 44, 380–391.

Zhang, F., Johnson, D.M., Sutherland, J.W., 2011. A GIS-based method for identifying the optimal location for a facility to convert forest biomass to biofuel. Biomass Bioenergy 35, 3951–3961.

Zhu, J.Y., Pan, X.J., 2010. Woody biomass pretreatment for cellulosic ethanol production: technology and energy consumption evaluation. Bioresour. Technol. 101, 4992–5002.

CHAPTER

13

Microbial Enzyme Applications in Bioethanol Producing Biorefineries: Overview

Francisco J. Ríos-Fránquez, Óscar A. Rojas-Rejón**, Carlos Escamilla-Alvarado†*

*Tecnológico Nacional de México, Torreón, Mexico

**Instituto Tecnológico de Estudios Superiores de Occidente, Tlaquepaque, Mexico

†Universidad Autónoma de Nuevo León, San Nicolás de los Garza, Mexico

ABBREVIATIONS

1stGB	First-generation bioethanol
2ndGB	Second-generation bioethanol
BG	β-glucosidase
CBD	Cellulose binding domain
CBH	Cellobiohydrolase
CBM	Carbohydrate binding module
CBP	Consolidated bioprocessing
CMC	Carboxymethylcellulose
DM	Dry mass
EG	Endoglucanase
GH	Glycoside hydrolase
GHG	Greenhouse gases
LCA	Life cycle assessment
LCB	Lignocellulosic biomass
LiP	Lignin peroxidase
MnP	Manganese peroxidase
PLCB	Pretreated lignocellulose biomass
SHF	Separate hydrolysis and fermentation
SSF	Simultaneous saccharification and fermentation
VP	Versatile peroxidases

13.1 INTRODUCTION

From the enzymatic pretreatments to the fermentation of sugars by yeasts, enzymes are the core part of every biotechnological process, including those comprised within biorefineries. The most important enzymes for bioethanol biorefineries are amylases, cellulases, and xylanases, yet expansin-like proteins (loosenins), ligninases, and pectinases are also of potential interest (Cannella and Jørgensen, 2014; Cinelli et al., 2015; Favalo et al., 2015; Knežević et al., 2013; Quiroz-Castañeda et al., 2011). These enzymes are substrate-specific and should be chosen adequately depending on the substrate and processes used in the biorefinery.

Different approaches have been proposed to develop rational use of enzymes in bioethanol production. Separated hydrolysis and fermentation

Bioethanol Production From Food Crops
http://dx.doi.org/10.1016/B978-0-12-813766-6.00013-8

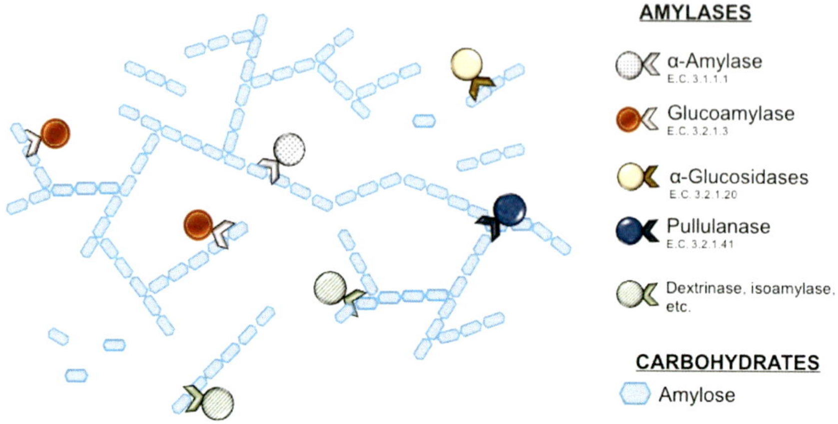

FIGURE 13.1 **Enzymatic hydrolysis of starch.**

(SHF) and simultaneous saccharification and fermentation (SSF) are the two main applications to achieve high-yield bioconversion of sugars into bioethanol. Recently, a newer configuration named consolidated bioprocessing (CBP) has profited from the advances in genetic engineering to design microorganisms that can perform cellulose hydrolysis and fermentation themselves or in coculture (Escamilla-Alvarado et al., 2017). Other developments include efficient immobilization, synergistic enzyme cocktails, and microbial engineering (Sheehan and Himmel, 1999).

13.2 BASICS OF ENZYME TECHNOLOGY

The main enzymes for carbohydrate hydrolysis in bioethanol biorefineries are amylases and cellulases. Cellulases and most hemicellulases belong to the glycosyl hydrolases group, containing up to 115 families (Dashtban et al., 2009). Recently, lytic polysaccharide monooxygenases (LPMO) have gained attention, yet their application is still a pending matter (Table 13.1).

13.2.1 Amylases

Starch is actually the most commonly used substrate for industrial bioethanol production (Favalo et al., 2015). In 2012, the United States produced 52.5×10^9 L of bioethanol, obtained mainly from starchy crops, which has led to an outstanding use of amylases.

Starch is a branched homopolymer of α-1,4-glucosidic bond linked glucose residues. Starch from crops is a mixture of two naturally occurring polymers: amylopectin (in greater proportion) that is constituted of D-glucose units with α-1,4 and α-1,6 glucosidic linkages, and amylose, which only has α-1,4 glucosidic bonds. Its hydrolysis is relatively easy and can be performed under mild conditions using amylases (Jönsson and Martín, 2016; Fig. 13.1).

Amylases hydrolyze starch, glycogen, and related polysaccharides (soluble oligosaccharides) by random cleavage; many of the maltodextrins released retain the α configuration of the anomeric carbon. During initial stages of enzymatic hydrolysis of starch and glucans, reducing groups are liberated yielding a reduced molecular weight polysaccharide (Brandam et al., 2003). After soluble oligosaccharide accumulation saccharification occurs with the accumulation of monosaccharides and disaccharides.

Amylases used in industrial starch hydrolysis are mainly α-amylase (α-1,4-glucan 4-glucanohydrolase) (E.C. 3.2.1.1) and glucoamylase

TABLE 13.1 Main Enzymes for Bioethanol Production

Technology	Enzyme	Function	References
1stGB	α-amylase	Hydrolyzes inner links of starch to maltose, glucose, and maltotriose	Pandey et al. (2000)
1stGB	Glucoamylase	Releases glucose from amylose and amylopectin	Pandey et al. (2000)
2ndGB	Endoglucanase	Hydrolyzes amorphous cellulose to cellooligosaccharides	Lynd et al. (2002)
2ndGB	Exoglucanase	Attacks reducing and nonreducing ends of cellulose releasing cellobiose	Lynd et al. (2002)
2ndGB	β-glucosidases	Release glucose by hydrolyzing the glycosidic bond of cellobiose	Lynd et al. (2002)
2ndGB	Endoxylanases	Hydrolyze xylans to xylooligosaccharides	Bastawde (1992)
2ndGB	β-xylosidases	Release xylose from xylooligosaccharides	Bastawde (1992)
2ndGB	Glucoronidases and acetylxylanesterases	Hydrolyze hemicellulose residue to arabinose and galactose, glucuronic, galacturonic, ferulic, and coumaric acids	Bastawde (1992)
2ndGB	Monooxygenases	Hydrolyzes crystalline cellulose and chitin to cellooligosaccharides	Payne et al. (2015)
2ndGB	Ligninases	Release phenolic compounds, oxidized products, 4-ethoxy-3-methoxybenzaldehyde, guaiacol, vanillic and protocatechuic acids from lignin, aromatics, veratrylalcohol and methoxybenzenes	Plácido and Capareda (2015)

(α-1,4-glucan glucohydrolase) (E.C. 3.2.1.3). Starch hydrolysis by amylases is dependent on starch composition (amylose/amylopectin ratio); α-amylases cleave α-1,4-glucosidic links randomly at the interior of amylose fraction either by the retaining or the inverting reaction mechanism (Davies and Henrissat, 1995; McCarter and Withers, 1994), and glucoamylases cleave the exterior of amylose and amylopectin at, α-1,3-, α-1,4-, and α-1,6-glucosidic linkages (Schusterb et al., 2000).

It has been reported that α-amylase and glucoamylase are not the only starch degrading enzymes, α-glucosidase (E.C. 3.2.1.20), dextran glucosidase (E.C. 3.2.1.70), isoamylase (E.C. 3.2.1.68), pullulanase (E.C. 3.2.1.41), amylopullulanase (E.C. 3.2.1.1/41), neopullulanase (E.C. 3.2.1.135), and cyclodextrin glucanotransferase (E.C. 2.4.1.19) are some of the exo-acting enzymes that depolymerize starch (Henrissat, 1991; Janecek et al., 2014). The enzymatic hydrolysis of starchy substrates requires α-exo-amylases, endo-amylases, and debranching enzymes with synergistic action. The hydrolysis begins with the cleaving of α-1,4 glucosidic bonds of the inner fraction of amylopectin or amylose chain. Linear and soluble oligosaccharides as well as α-limit dextrin are formed by the action of α-amylase. The external residues of amylose or amylopectin are hydrolyzed by exo-amylases (glucoamylases) and α-glucosidades, yielding glucose and small soluble dextrins.

Exo-amylase cleave both α-1,4 and α-1,6 glucosidic bonds, whereas β-amylase (E.C. 3.2.1.2) cleave exclusively α-1,4 glucosidic bonds and produce disaccharides (maltose) and β-limit dextrin. Pullulanase and starch debranching enzymes are the major responsible for complete starch hydrolysis to glucose (Hii et al., 2012).

Amylases and other enzymes related to the hydrolysis of starch are obtained from various microorganisms, such as *Bacillus acidopullulyticus*, *Klebsiella planticola*, *Bacillus deramificans*, thermophilic *Bacillus* sp. AN-7, *Bacillus cereus* FDTA-13, and *Geobacillus stearothermophilus*, *Bacillus amyloliquefaciens* as well as *Kleibsella aerogenes* (Henrissat, 1991; Hii et al., 2012).

13.2.2 Cellulases

Cellulose, the most abundant natural polymer, is constituted of glucose residues linked linearly by β-1,4-glucosidic bonds, which in order to be utilized by living beings, first needs to be hydrolyzed. Naturally, cellulolytic fungi are able to produce large amounts of extracellular cellulolytic enzymes, whereas bacterial and a few anaerobic fungal strains mostly produce cellulolytic enzymes in a cell wall–associated complex called cellulosome (Dashtban et al., 2009; Escamilla-Alvarado et al., 2017).

The mechanism for complete enzymatic cellulose hydrolysis is a synergetic biochemical process between endoglucanases or carboxymethylcellulases (EC 3.2.1.4), cellobiohydrolases (CBHs) or exoglucanases (EC 3.2.1.91, EC 3.2.1.74), and β-glucosidases (EC 3.2.1.21) (Fig. 13.2). The β-1,4-endoglucanases (EG) hydrolyze randomly β-1,4-glucosidic bonds of cellulose to produce open ends, which are recognized and attacked either by β-1,4-exoglucanases (CBH), which progressively hydrolyze β-1,4 links to release cellobiose, or by β-1,4-glucan glucohydrolases (EC 3.2.1.74) that liberates glucose, whereas β-glucosidases (BG) specialize on the release of glucose from cellobiose and other soluble cellodextrins using the retaining mechanism (Dashtban et al., 2009; Sheehan and Himmel, 1999; Viikari et al., 2012). BG, which may be found in bacteria, fungi, plant, and even mammalians, present the most variable structure (monomeric, dimeric, or trimeric) and localization (intracellular, cell wall–associated, and extracellular) amid the cellulolytic enzymes (Dashtban et al., 2009).

Endoglucanases and CBHs from bacteria and fungi may have a small independently folded carbohydrate binding module (CBM) that is connected to the catalytic domain by a flexible linker rich in proline, serine, and threonine. The CBM, present at the *N* or *C*-terminus of the catalytic domain after the linker, is responsible for binding the enzyme to the crystalline cellulose and enhances its activity (Dashtban et al., 2009) and presents a flat surface with exposed aromatic side-chains that mediate the binding to the hydrophobic cellulose surface (Shimon et al., 2000).

Even though the enzymatic step performed by endoglucanases and CBHs is the rate-limiting

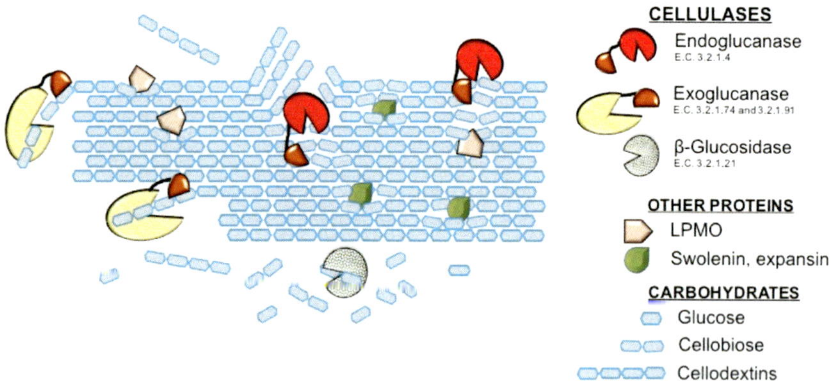

FIGURE 13.2 **Enzymatic hydrolysis of cellulose.**

step for the cellulose hydrolysis process, the following hydrolysis of cellobiose to glucose by β-glucosidases is relevant to avoid competitive from cellobiose (Viikari et al., 2012). In turn, glucose should be removed to avoid inhibition of β-glucosidases (Sheehan and Himmel, 1999). Indeed, this synergism is not only observed in the enzyme machinery of a single species of microorganism, but also between cellulases from different microorganisms (Sheehan and Himmel, 1999). One of the most common preparations is to use endo- and exoglucanases from *Trichoderma reesei*, and glucosidases from *Aspergillus niger* (Sheehan and Himmel, 1999).

The cellulose hydrolysis by cellulases involves the following steps (Bansal et al., 2009; Behera and Ray, 2016): (1) adsorption of cellulases onto the substrate via binding domain; (2) location of a bond susceptible of hydrolysis; (3) formation of enzyme-substrate complex; (4) hydrolysis of β-glucosidic bond and simultaneous forward sliding of the enzyme along the cellulose chain; (5) desorption of cellulases from substrate or repetition of step 2, 3, or 4 if binding domain is not detached from chain; and (6) hydrolysis of cellobiose to glucose by BG. Most enzymatic reactions benefit from the Arrhenius relationship, thus higher operating temperatures translate into increased diffusion and thermodynamics of catalysis (Sheehan and Himmel, 1999). Cellulases kinetic models in lignocellulose substrates are difficult to assess for several reasons: heterogeneous reaction, substrate is not in excess in contrast to enzymes (accessible cellulose is found in a very low fraction, 0.002–0.04). Still, classical homogenous enzyme catalysis is modeled by Michaelis–Menten kinetics, whereas heterogeneous catalysis on a catalyst support is modeled by Langmuir–Hinshelwood kinetics. Moreover, conversion of cellobiose to glucose by BG may indeed be modeled by Michaelis–Menten kinetics as it is a homogeneous reaction (Bansal et al., 2009).

13.2.3 Hemicellulases

Hemicellulases are a set of enzymes that hydrolyze hemicellulose, which commonly comprise 20%–30% of lignocellulose (Dhiman et al., 2008). Xylanases are the most important kind of hemicellulases due to the major abundance of the heteropolysaccharide xylan. Mannan, galactan, and arabinan are other structural polysaccharides of hemicellulose (Beg et al., 2001) (Fig. 13.3). Xylan is mainly constituted of D-xylose, but acetyl, arabinosyl, and glucuronyl residues may also be present (Bastawde, 1992). In hardwoods, O-acetyl-4-O-methylglucurono-D-xylans are the most relevant heteropolysaccharides, whereas for softwood O-acetylgalactoglucomannans and arabino-4-O-methylglucurono-D-xylans (Jönsson and Martín, 2016). In annual plants the

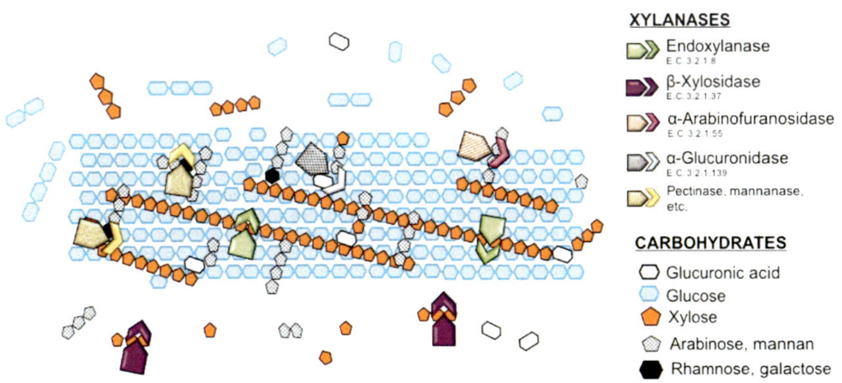

FIGURE 13.3 **Enzymatic hydrolysis of hemicellulose.**

most important kind of hemicelluloses are arabino-(O-acetyl-4-O-methylglucurono)-D-xylans, with *P*-coumaric and ferulic acids attached to the arabinose residues.

Xylanases hydrolyze β-1,4 xylopyranosyl linkages in xylan. The β-1,4-endoxylanases (EC 3.2.1.8) cleave the internal linkages of xylans in hemicellulose whereas β-1,4-xylosidases (EC 3.2.1.37) hydrolyze short xylooligosaccharides to D-xylose (Bastawde, 1992).

Accessory enzymes that hydrolyze the other substituents (glucuronic acid, arabinofuranose, ferulic, and coumaric acids) are α-L-arabinofuranosidases (EC 3.2.1.55), α-glucoronidases (EC 3.2.1.139), and acetylxylan esterases (EC 3.1.1.72). From a saccharification perspective, these are responsible for breaking the side chains and opening sites in the hemicellulose matrix that allows the attack of xylanases (Bastawde, 1992; Dhiman et al., 2008). Similar to cellulases, hemicellulases may have CBM, and in the case of anaerobic microorganisms it may be present in cellulosome association (Dashtban et al., 2009).

Hemicellulases are important to bioethanol production since any improvement in pentoses generation will have a direct impact in more substrate conversion (Tabka et al., 2006) and because the synergic relation with cellulases has been demonstrated (Gottschalk et al., 2010).

13.2.4 Accessory Enzymes and Others

Enzymes with a relatively minor participation in major hydrolysis effects of the substrate are known as accessory enzymes. The main interest in bioethanol biorefineries for these enzymes is its potential in allowing pore widening that leads to further cellulose and xylan conversion. Other enzymes, such as ligninases, would be helpful in increasing biomass digestibility.

13.2.4.1 Pectinases

Pectin is a gel forming branched heteropolysaccharide comprised a copolymer of 1,4-α-galacturonic acid and 1,2-rhamnose, which branches in 1,4-β-D-galactose or 1,5-α-L-arabinose (Ciriminna et al., 2015). Pectinases can be used for ethanol production but these are highly substrate dependent to pectin-rich biomass, such as banana and papaya. Moreover, ethanol has inhibitory effects on pectinases (Kashyap, 2001), which hence would limit its application in SSF and CBP processes.

13.2.4.2 Loosenins

Expansins (expansin-like proteins or loosenins) are proteins showing loosening activity of plant cell wall. They are expressed in plants as well as in microorganisms, and can be useful for lignocellulose saccharification. It is recognized that expansin-like and related proteins lack of true lytic activity; however, these have shown the ability to modify plant cell wall by the loosening or disruption of hydrogen bonds between cellulose networks (Cosgrove, 2000; Georgelis et al., 2014). Recently, loosenins have attracted attention in lignocellulose hydrolysis processes for bioethanol production because the addition of expansin-like proteins in lignocellulose saccharification processes increases biological conversion of cellulose to fermentable sugars and hence to ethanol (Nakatani et al., 2013; Quiroz-Castañeda et al., 2011), because disruption of hydrogen bonds from adjacent cellulose chains provides spacing for cellulase access leading to further lignocellulose hydrolysis.

13.2.4.3 Ligninases

Lignin is a complex aromatic polymer composed of phenylpropanoid units. In softwood, lignin is comprised primarily by guaiacyl units, whereas in hardwood it contains mainly syringyl units and important amounts of guaiacyl units. Along with guaiacyl and syringyl units, lignin in annual plants contains also *P*-hydroxyphenyl units. Phenylpropanoid units are linked through a complex network of ether and carbon–carbon bonds. Other phenolic compounds in lignocellulose are *P*-coumaric, ferulic and diferulic acid, which are considered to be crosslinking agents by esterification to arabinoxylans and by ether- or ester-links to lignin (Jönsson and Martín, 2016).

Lignin inhibits saccharification in diverse manners: by adsorption of cellulases through hydrophobic, electrostatic interactions, or hydrogen bonding, by blocking accessible surface of carbohydrates, and by inhibition of cellulases by its derivates (Agrawal et al., 2017); therefore, removal of lignin by ligninases would increase cellulases and hemicellulases effectiveness (Singh et al., 2014). The enzymes that remove this compound from its covalent union with hemicellulose are the laccases (EC 1.10.3.3), lignin peroxidases (LiP, EC 1.11.1.14), manganese peroxidases (MnP, EC 1.11.1.13), and versatile peroxidases (VP, EC 1.11.1.16). Laccases, which use oxygen as cofactor and oxidizing agent, are the most important ligninases; whereas LiP oxidizes aromatic rings moderately activated by electron donating substitutes and degrades several aromatic structures, such as veratryl alcohol (3,4-dimethoxybenzyl) and methoxybenzenes; VP on the other hand have a broader substrate preference.

13.2.4.4 Lytic Polysaccharide Monooxygenases

The LPMO, formerly known as glycoside hydrolase family 61 (GH61) and now belonging to the auxiliary activity family 9 proteins (AA9), are copper-dependent oxidases that have been proposed to act synergistically with cellulases, hemicellulases, amylases, and even chitinases (Escamilla-Alvarado et al., 2017; Kim et al., 2014). LPMOs exist as a single domain or contain a CBM frequently found in cellulolytic fungi (Kim et al., 2014).

LPMO are produced by fungi and bacteria, such as *Myceliophtora thermophila*, *Neurospora crassa*, *Phanerochaete chrysosporium*, *T. reesei*, *Bacillus licheniformis*, *Bacillus thuringiensis*, *Serratia marcescens*, and *Aspergillus oryzae* (Hemsworth et al., 2015).

No full molecular functions have been addressed to this family, the only recognized function is endoglucanase activity by the cleaving of β-glucosidic bonds; its oxidative effect has been shown in reducing and nonreducing ends. Oxidation at C1 and C4 generates oligosaccharides in the form of aldonic acid, 4-ketoaldose (Kim et al., 2014) and gluconic acid (Cannella et al., 2012). Indeed, sugars are released by LPMO alone in xylan or cellulose, yet concentrations are really low, as 1000-fold below common EG. It has been hypothesized that LPMOs attack the inaccessible highly crystalline cellulose, which creates more accessible sites for cellulases, enhancing cellulases activity (Kim et al., 2014; Quinlan et al., 2011). Oxygen is fundamental for oxidizing reaction of LPMO and its deficit reduces synergism with cellulases (Cannella and Jørgensen, 2014).

13.2.5 Inhibition

Pretreatment and hydrolysis of lignocellulose lead to the formation of different acids, such as uronic, acetic, and phenolic (Jönsson and Martín, 2016), which along with other solvents need to be separated because of their inhibitory risks to enzymes and fermentative microorganisms (Alvira et al., 2010). Enzyme inhibition may occur in two ways: by enzyme adsorption onto particulate lignin and by interactions with soluble particulates (Berlin et al., 2006). Cellulases are much affected by lignin moieties, specially by those containing hydrophobic groups, such as aliphatic hydroxy and carboxy; xylanases and glucosidases are affected to a much lesser extent (Berlin et al., 2006). Polyphenols have been studied for inhibition and deactivation of cellulases and BG. Tannic, gallic, hydroxy-cinnamic, and 4-hydroxybenzoic acids are produced during pretreatments and act as cellulose deactivators (Ximenes et al., 2010a,b).

In addition to inhibitors generated during pretreatment and hydrolysis, enzymes are subjected to feedback inhibition due to the accumulation of soluble sugars, such as cellobiose first, and glucose in second place (Hsieh et al., 2014). This limitation is perhaps one of the most important because improvements in catalytic efficiencies are needed to yield high sugar syrups to obtain high ethanol yields. An ethanol concentrated fermentation broth is key for downstream product recovery and purification (Sánchez et al., 2013, 2015). Kumar and Wyman (2014) reported that xylan and mannan strongly inhibit fungalcellulases (>50% drop in relative conversion), even at a small concentration (0.1 g/L). *T. reesei* expresses insufficiently BG activity to catalyze hydrolysis of the accumulated cellobiose, mainly because a major part of the BG activity is bound to the fungal mycelium and hence is not recovered during industrial cellulase production. Thus additional BG is required to achieve sufficiently high glucose yields (Tsai and Meyer, 2014). Indeed, *T. reesei* cellulases were found to be inhibited by glucose, cellobiose, ethanol, and butanol as noncompetitive inhibitors, δ-gluconolactone as a mixed inhibitor, and acetone as a noncompetitive activator. Holtzapple et al. (1990) found that the conversion of cellobiose into glucose reduces the effective inhibitor binding constant 6-fold, whereas its conversion to ethanol reduces it 16-fold.

13.3 ENZYME INDUSTRIAL APPLICATIONS AND NEW INSIGHTS

Bioethanol production has been achieved at pilot and industrial scales for first- and second-generation biomass, both consuming high volumes of enzymes. Whereas for first-generation biomass (starchy crops), α-amylase and glucoamylase dosages are typically of 0.36 and 0.61 kg enzyme/ton DM substrate, respectively, for second-generation biomass (lignocellulosic biomass, LCB) cellulase dosages range 10–100 g enzyme/ton DM substrate (Dunn et al., 2012). According to Dunn et al. (2012) and based on data from MacLean and Spatari (2009), energy consumption from amylose, glucoamylose, and cellulase production is 23, 100, and 45 kJ/kg enzyme, respectively.

In the decision of which type of cellulases is best for bioethanol biorefineries some other consideratios should be reckoned, such as the life cycle assessment (LCA). LCA is handy tool for identifying opportunity areas, mainly used to calculate contributions to environmental impacts from a given process or service. In the case of enzymes for bioethanol, it has been demonstrated a direct relationship between environmental burdens and costs, specifically in enzymatic hydrolysis, which was assessed to be one of the main impacting processes due to electricity use and ingredients consumptionin enzyme production (Nielsen et al., 2007; Sebastião et al., 2016).

13.3.1 Enzymes in First-Generation Bioethanol

First-generation biofuels (1stGB) are commonly produced from agricultural materials containing sugar or starch (Kunamneni and Singh, 2005). Indeed, Vohra et al. (2014) have estimated that between 2009 and 2010 the world total ethanol production was about 100×10^9 L; almost 40% of this production corresponded to starchy or saccharose substrates.

Agricultural grains, such as corn, wheat, oat, and barley are mainly composed of starch (ca. 60%–70% for corn) (Kunamneni and Singh, 2005; Naik et al., 2010; Vohra et al., 2014).

Starch is subjected to chemical or enzymatic hydrolysis prior fermentation by techniques that include mild acid hydrolysis with H_2SO_4 or HCl at 45–70°C (Hoover, 2000; Miao et al., 2011; Wang et al., 2003), enzymatic hydrolysis with amylases (Cinelli et al., 2015; Henrissat and Davies, 1997; Hii et al., 2012) and a combination of both mineral acid and enzymatic preparations (Khawla et al., 2014; Stevnebo et al., 2006).

Most of the enzymes used in 1stGB production are classified into family 13 of glycoside hydrolase (GH) (Henrissat and Davies, 1997). Accessory enzymes play a significant role in starch degradation through the amylase and pullulanase complexes. There have been efforts in the development of a 1stGB-CBP with the production of an amylolytic *S. cerevisiae* suitable for industrial bioethanol production (Favalo et al., 2015).

1stGB have advantages, such as simplicity at industrial scale, but there is still debate on the use of food for energy production. Moreover, there is no concrete evidence on the possibility of generating a zero-carbon balance (in contrast to cellulosic ethanol). An argument in favor of 1stGB is that, according to MacLean and Spatari (2009), chemical and enzyme inputs contribute only 3% of fossil energy use and greenhouse gas (GHG) emissions for corn ethanol, whereas for the near-term improved lignocellulosic technologies these are responsible for 30%–40% of fossil energy use and 30%–35% of GHG. However, fossil energy consumption and GHG emissions from corn ethanol biorefineries (0.9 MJ/MJ ethanol and 86 gCO_2 eq./MJ ethanol, respectively) are ca. 3-fold higher than switchgrass ethanol biorefineries under different pretreatments.

13.3.2 Enzymes in Second-Generation Bioethanol

There are recent technologies to overcome drawbacks generated by 1stGB, such as the fuel–food and carbon dioxide emissions into the atmosphere debates. Lignocellulosic biomass has been presented as an alternative source of materials for bioethanol production in the so-called second-generation biofuels (2ndGB). LCB and lignocellulosic wastes are generated after primary grains and crops are harvested. The 2ndGB has several advantages, among which there are the possibility of reducing and reverting CO_2 production, no competition with food crops, and under biorefinery concept, production of a wide variety of biofuels (bioethanol, biogas, hydrogen, methane, and biodiesel) and high value-added bioproducts.

LCB are composed primarily by cellulose, hemicellulose, and lignin, comprising plant biomass and residues of forest management or food crop production (Naik et al., 2010; Vohra et al., 2014). Structure and components proportion of LCB is diverse due to factors, such as plant biodiversity, physiological status of the plant, geographic location, season, nutrient availability, and age (Lynd et al., 2002).

LCB presents such a recalcitrance and resistance to chemical, physical, or biological damage (the protection of plant cell), which improvements in chemical, physical, and biological treatments are being required to yield access to the complex polysaccharides in LCB (Naik et al., 2010).

Considering that pretreatment technologies must be designed according to LCB origin and composition, several steps must be taken to effectively depolymerize LCB: particle size reduction, fiber damage with mineral acids and alkali, thermochemical procedures to expose lignocellulose fibers (steam explosion), and biological degradations (Rojas-Rejón and Sánchez, 2014; Sánchez et al., 2013, 2015).

Pretreated lignocellulose biomass (PLCB) is more easily degraded by cellulases and hemicellulases. Similarly to amylases, hemicellulases, and cellulases present synergism. Branched hemicellulose hydrolysates are removed to give a free pass to cellulases for complete PLCB hydrolysis (saccharification). Although it has been recognized that the hydrolysis of PLCB is carried out by cellulase and xylanase complexes, some bifunctional enzymes, such as that reported by Pérez-Avalos et al. (2008), from *Cellulomonas flavigena* could be useful.

Enzymatic hydrolysis of PLCB has been extensively studied at laboratory scale with highly controlled variables while less work has been done at pilot and industrial scale conditions (Jørgensen et al., 2007). The main challenges in enzymatic hydrolysis of PLCB are related to enzymes, such as feedback enzymatic inhibition, mass transfer issues, processivity, hydrolysis of "ideal substrates," and inherent complex matrix of LCB. One of the main concerns of 2ndGB production is the high costs of enzymes used in saccharification processes. Higher enzyme dosage is needed to obtain high sugar concentrations and thus high bioethanol yields through fermentation and distillation. However, to achieve high glucose–content syrups, enzymatic processes must be carried out at high solids loadings (Kristensen et al., 2009), but conversion yields remain a challenge (Modenbach and Nokes, 2013; Rojas-Rejón and Sánchez, 2014). Efforts to develop efficient technologies in the production of 2ndGB are focusing on achieving efficient pretreatment and hydrolysis with recombinant enzymes and enhanced catalytic activity.

The enzymes commonly used for industrial applications are Celluclast 1.5L (Novozymes) and Spezyme (Genencor International). Interestingly, fungal cellulase preparations are commercially available yet no bacterial cellulase preparations have yet been achieved industrially (Teter et al., 2006).

Actually, ethanol cost varies 0.24–0.57 EUR/L depending on the region of production (Olofsson et al., 2017; Viikari et al., 2012), and it has been stated that in order to be competitive with fossil fuels, the cost of enzyme should be reduced below 0.02 EUR/L of ethanol (Viikari et al., 2012); however, this cost is still 0.03–0.07 EUR/L ethanol, considering that the purchase price of enzyme is 3.55 EUR per million FPU and that enzyme dosage is 12.4–30.4 g enzyme cocktail/kg dry mass (DM), at enzyme cocktail density of 1.1 g/mL, enzyme activity of 213 FPU/mL enzyme cocktail, and yield of 270 L ethanol/t DM spruce (Olofsson et al., 2017).

13.3.3 Advances and Insights on Enzyme Technology

Innovations for industrial enzymes are being developed at laboratory scale. Although production of advanced biofuels from lignocellulose is much less mature than starch-based production, the high level of attention recently given to industrial application of biomass degrading enzymes has induced advances in our understanding of such enzymes (Harris et al., 2014), especially in the areas of recombinant cellulase production, immobilization, recycling, synergy, and tailored enzyme cocktails. These techniques are mostly based on the reasonable use of enzymes coming from recent developments and knowledge continuously gained in biochemistry, enzymology, and other disciplines.

13.3.3.1 Recombinant Enzymes

With the currently available molecular tools, it is possible to create strains that express a variety of enzymes for improving bioethanol biorefineries. Future work may yield industrially useful strains that produce diferent recombinant enzymes that synergistically degrade different types of biomass. A combination of recombinant cellulase production, cell surface engineering, and intracellular metabolic engineering is a very effective approach to develop cells with novel fermentation ability for industrial applications, which would be applicable to the development of efficient CBP systems. These advances are discussed in depth in other chapters in this book.

Among the common process configurations, CBP has been increasingly recognized for its potential for low-cost bioethanol, as it combines enzyme production, saccharification of cellulose and hemicellulose, and fermentation of hydrolysates in a single process (Van Zyl et al., 2007). However, native microorganisms with the capability for efficient enzyme production, lignocellulose saccharification, and ethanol production are not naturally available (Viikari et al., 2012). Therefore, CBP requires genetically engineered microorganisms to be developed for different process-specific characteristics (Hasunuma and Kondo, 2012).

Some enzyme-related modifications include expression of recombinant cellulases, of laccases to increase the resistance to phenolic inhibitors (Larsson et al., 2001), and of genes that allow the use of xylose and arabinose (Zaldivar et al., 2001).

Bacteria and yeast are the most widely used recombinant cellulase expression platforms so far, because of their ease of manipulation and genetic transformation, high growth rates and productivity of recombinant enzymes.

Until now, *S. cerevisiae* is the most studied and commonly used yeast for recombinant cellulase expression (Hasunuma and Kondo, 2012), and despite that cellulases of fungal origin have proven to be more effectively expressed, an extent of bacterial cellulases have also been assayed recombinantly (Haan et al., 2007). Many attempts have been made to engineer *S. cerevisiae* strains to secrete cellulases efficiently for the hydrolysis of cellulose (Van Zyl et al., 2007); however, when cellulase is expressed solubly and extracellularly, the enzymes remain in the culture supernatant, which makes it difficult to reuse them. An alternative to immobilize the recombinant cellulases and facilitate their reuse is the cell surface engineering to display the cellulolytic enzymes on the yeast cell surface.

The development of recombinant yeast strains displaying cellulases on the cell surface represents significant progress toward realization of CBP. A promising approach to cost-efficiently convert LCB into fermentable sugars for biofuel production is the use of minicellulosomes, which are surface displayed multicellulase-containing complexes present in cellulolytic *Clostridium* and *Ruminococcus* species (Fontes and Gilbert, 2010).

13.3.3.2 Immobilization and Enzyme Reuse

Enzymes recycling is an alternative if economic viability is desired. Recycling and reuse of enzymes may be achieved through several strategies, such as: washing off from the residue, by contacting the whole hydrolysate with fresh substrate, ultrafiltration recovery, and immobilization (Escamilla-Alvarado et al., 2015; Singh et al., 1991). Cellulases in particular present a high adsorption affinity on cellulose and lignin that should be carefully considered when recovery processes are being pondered (Gomes et al., 2015). Indeed, after strong adsorption of the EG and CBH to cellulose occurs, desorption is not carried out until the substrate has been degraded. Binding of CBH substrates occurs through either the catalytic dominion or the cellulose binding domain (CBD). Cellulases bound to cellulose do not return to the liquid fraction until hydrolysis has been performed, while in contrast, cellulases bound to ligning remain adsorbed in a nonproductive way (Gomes et al., 2015). Recovery of these enzymes could be performed applying surfactants as Tween 20 or 80 (Chylenski et al., 2012), or even bovine serum albumin (BSA) at 1% concentration before adding cellulases (Yang and Wyman, 2006). The binding of BSA to lignin is stronger than that of cellulases mainly due to its nonspecific competitive irreversible adsorption.

Free cellulases in media can be recovered by membrane ultrafiltration. Ultrafiltration follows centrifugation of reaction broth to remove insoluble material, and thus concentrate enzymes, in a similar method as enzymes are first recovered and concentered after production by cellulolytic microorganisms. However, a simpler reuse method is possible as was performed by Escamilla-Alvarado et al. (2015), who demonstrated that postcentrifuged enzyme broth from hydrolysis of fermented organic waste, retained 70%–86% of enzyme activity when assayed by filter paper hydrolysis. Enzyme residual activity was higher when saccharification broth received lower enzyme loads. Indeed, enzyme adsorption proportion depends on enzyme to substrate ratio, hydrolysis efficiency, and lignin content (Singh et al., 1991). Another interesting and simple method is the recycling of hydrolases by contacting the hydrolysate with fresh substrate, with the inconvenience that insoluble solids are carried along that increase the total solid concentrations and lignin content in the broth, which could be detrimental to the process (Weiss et al., 2013).

Concerning immobilization, this technique allows the recovery and recycling of enzymes, adding the opportunity of continuous operation. Besides, it promotes stability from thermal and pH stress (Abraham et al., 2014; Kar and Ray, 2008). Immobilization leads to mass transfer issues and may decrease enzyme-substrate affinity due to the interactions that change the final configuration of the enzyme (Vaz et al., 2016).

Immobilization requires the interaction of a carrier and the enzyme. Important features needed in the carrier are adequate and large surface area to interact with the enzyme, as well as chemical and mechanical stability (Hanefeld et al., 2009). It may be performed randomly or by design. Random immobilization assays trial and error to get a satisfactory system. When design immobilization is used, enzymes are less sensitive to environmental conditions, the control of the reaction is improved, product contamination with the enzyme is avoided, and activity is enhanced (Contesini et al., 2013). Enzyme immobilization techniques may be roughly grouped in (Contesini et al., 2013; Hanefeld et al., 2009; Roy and Gupta, 2004): (1) noncovalent adsorption and deposition, (2) covalent tethering, (3) encapsulation or entrapment, and (4) crosslinking of enzyme aggregates with bifunctional reagents to produce macroparticles. However, other possibilities and divisions may arise since combinations of two or more different techniques can be used, for instance crosslinking with entrapment and adsorption with covalent bonding.

Generally, immobilization promotes new interactions between the enzyme and support. Enzyme supports belong to three categories (Vaz et al., 2016): synthetic organic polymers (e.g., acrylic resins), biopolymers (e.g., cellulose, agarose, and chitosan), and inorganic solids (e.g., silica, alumina). Nanoparticles have recently attracted the attention of researchers due to their high superficial area to size ratio. Some studies use as immobilization nanomaterials gold, magnetic nanoparticles, carbon nanotubes, and carbon derivatives. Nanosized magnetic particles are easily separated from broth, present high surface area, and reduced toxicity (Abraham et al., 2014). Nanoporous materials are also of interest due to their catalytic, diffusion, and adsorptive properties (Vaz et al., 2016).

One of the most recurrent strategies for immobilization is based in enzyme entrapment. Immobilization through this technology, increments thermostability, pH, and kinetics parameters, such as Michaelis–Menten constant (Konsoula and Liakopoulou-Kyriakides, 2006; Reshmi et al., 2007; Roy and Gupta, 2004).

Although it has been proposed that immobilization of EGs and CBHs limits the enzyme catalysis due to mass transfer restrictions, as a consequence of slow or no diffusion of the insoluble substrate into the particles (Tsai and Meyer, 2014), some immobilization studies have shown interesting hydrolysis features insoluble substrates, such as carboxymethylcellulose (CMC) derivates (Vaz et al., 2016) and in insoluble, too, such as CMC, filter paper, and hemp hurds (Abraham et al., 2014; Cho et al., 2012). The latter has been achieved through man-sized materials by strong crosslinking covalent bonds to CBH and BG through glutaraldehyde.

It has been discussed that BG is more appropriate for immobilization considering that its substrate, cellobiose, is water soluble (Tsai and Meyer, 2014). Among the immobilization technologies investigated for BG (polyacrylamide, soil humates, and silica), calcium alginate entrapment is simple and inexpensive, whereas its recycling is simple done by recovery from reaction slurry. Crosslinking prior to immobilization in calcium alginate beads may be used to avoid detachment of enzyme from the calcium alginate matrix. Glutaraldehyde is a well-known agent for intra- and intermolecular crosslinking, which may be applied to BG to form large enzyme aggregates (Tsai and Meyer, 2014).

13.3.3.3 Enzymatic Synergy and Design of Enzyme Cocktails

The reduction of the enzyme loadings for saccharification of lignocellulose is desirable for an economic feasible production of bioethanol. In this context, for a desirable efficient enzymatic hydrolysis of lignocellulose, a more rational use of enzymes is required. This could be attained by profiting of the synergy between cellulases and other enzymes and proteins (Kim et al., 2014).

Synergistic effect of cellulases-xylanases has been demonstrated by summing the monosaccharides released using only xylanases to those sugars released by only cellulases, which did not achieve the same amounts when using both enzymes (Song et al., 2016), this effect has been attributed to the increase of swelling and porosity by xylanases that led to a better accession of cellulose into the microfibril pores of cellulose. Moreover, it has been suggested that removal of xylan even by pretreatment methods has more impact on cellulose chain reduction than lignin removal (Alvira et al., 2010).

According to Viikari et al. (2012), the optimal ratios for hydrolysis of lignocellulosic materials using common hydrolases are roughly (based on protein loading): CBHI 50%–60%, CBHII 10%–30%, EGs 10%–30%, xylanases 5%–10%, and BG (usually less than 1%). However, each substrate and the pretreatment method used requires its own evaluation of optimal ratios, which has led to the development of the minimal dosage enzyme mixture concept (Meyer et al., 2009), that is, the identification of minimal efficient enzyme mixtures that minimize the amounts of key enzymes.

Other synergistic proteins are CBMs, plant expansins, expansin-like proteins, and LPMO or AA9 proteins (Kim et al., 2014). Among these, LPMO are the most promising due to its commercial feasibility and elevated synergism with cellulases, which effectively reduce enzyme loadings in at least 2-fold. Indeed, the formulation of Cellic CTec2 (Novozymes), which is ca. 2-fold higher than Celluclast 1.5 L includes these kind of proteins (Cannella et al., 2012). *S. cerevisiae* is unable to ferment gluconic acid produced by LPMO, yet it profits from the previously enhanced hexose production (Cannella et al., 2012). This suggests that LPMO dosage should be carefully formulated, as an excess of gluconic acid could be detrimental for bioethanol production. Regarding expansins, a protein from corn (Zea h) was used with cellulose (*T. reesei*, Ningxia xiasheng Co. Ltd, China) on filter paper, proving synergistic interaction through an improved adsorption of cellulase onto substrate and the reduction of the hydrogen-bond intensity and crystallinity index, for which Zea h was ascribed for (Han and Chen, 2007). Furthermore, Han and Chen (2010) asserted that synergistic proteins have strong hydrophobic characterization, which is helpful in reducing unspecific cellulase adsorption onto noncellulose component of lignocelluloses during hydrolysis. This effect was more significant for CBH and BG.

Surfactants, such as Triton X100, Tween-80, PEG-6000, and PEG-1000 also have shown synergistic effect when used with cellulases (*T. viride*, Shanghai Baiao Biotech. Corp., China) by reducing the nonspecific binding of enzyme to the hydrophobic sites of the lignin (Han and Chen, 2010). Other surfactants with cellulase improvement on wheat straw were PEG-400 and polyalkylene glycols (Agrawal et al., 2017). Hydrothermally pretreated wheat straw also presented higher sugars and consequently higher bioethanol yield (over 40% increase) when PEG-3000 was added to an LPMO containing enzyme preparation (Cannella and Jørgensen, 2014).

Synergism between cellulases (*A. niger*, Sigma-Aldrich, USA) and xylanases (*Thermomyces lanuginosus*, Sigma-Aldrich, USA) has been demonstrated on unpretrated corncob, corn stover, and rice straw (Song et al., 2016); when compared to the treatment with only cellulases, the mixture improved the reducing sugar concentrations by 133%, 164%, and 545%, respectively, whereas conversion yields were 43.9%, 48.5%, and 40.2%, respectively. The synergy ratios were 1.31, 1.43, and 1.26, for each substrate.

Laccase pretreatment has also enhanced the cellulose conversion rate of wheat straw (Qiu and Chen, 2012). Breakage of feruloyl ester linkages between hemicellulose and lignin accelerated the hydrolysis of rice straw by destructing covalent linkages (Chen and Fu, 2016).

13.4 CONCLUSION AND FUTURE PERSPECTIVES

Whereas amylases are the main enzymes in first-generation bioethanol, cellulases are for second-generation bioethanol. Improvements in the production of second-generation bioethanol have been focused on achieving fast and efficient saccharification and fermentation of the carbohydrate fractions from LCB. Moreover, it is a concerted prediction that through heterologous expression of hydrolases, enzyme recycling and immobilization, and complementation of enzyme cocktails with hemicellulases and accessory enzymes, second-generation bioethanol production will come closer to a full industrial scale reality.

Acknowledgments

CE-A is grateful to Dr. Pasiano Rivas-García for his assistance in reviewing Sections 13.1–13.3.

References

Abraham, R.E., Verma, M.L., Barrow, C.J., Puri, M., 2014. Suitability of magnetic nanoparticle immobilised cellulases in enhancing enzymatic saccharification of pretreated hemp biomass. Biotechnol. Biofuels 7 (1), 2–24.

Agrawal, R., Satlewal, A., Kapoor, M., Mondal, S., Basu, B., 2017. Investigating the enzyme-lignin binding with surfactants for improved saccharification of pilot scale pretreated wheat straw. Bioresour. Technol. 224, 411–418.

Alvira, P., Tomás-Pejó, E., Ballesteros, M., Negro, M.J., 2010. Pretreatment technologies for an efficient bioethanol production process based on enzymatic hydrolysis: a review. Bioresour. Technol. 101, 4851–4861.

Bansal, P., Hall, M.l., Realff, M.J., Lee, J.H., Bommarius, A.S., 2009. Modeling cellulase kinetics on lignocellulosic substrates. Biotechnol. Adv. 27 (6), 833–848.

Bastawde, K.B., 1992. Xylan structure, microbial xylanases, and their mode of action. World J. Microbiol. Biotechnol. 8, 353–368.

Beg, Q.K., Kapoor, M., Mahajan, L., Hoondal, G.S., 2001. Microbial xylanases and their industrial applications: a review. Appl. Microbiol. Biotechnol. 56 (3–4), 326–338.

Behera, S.S., Ray, R.C., 2016. Solid state fermentation for production of microbial cellulases: recent advances and improvement strategies. Int. J. Biol. Macromol. 86, 656–669.

Berlin, A., Balakshin, M., Gilkes, N., Kadla, J., Maximenko, V., Kubo, S., et al., 2006. Inhibition of cellulase, xylanase and β-glucosidase activities by softwood lignin preparations. J. Biotechnol. 125 (2), 198–209.

Brandam, C., Meyer, X.M., Proth, J., Strehaiano, P., Pingaud, H., 2003. An original kinetic model for the enzymatic hydrolysis of starch during mashing. Biochem. Eng. J. 13, 43–52.

Cannella, D., Hsieh, C.-W.C., Felby, C., Jørgensen, H., 2012. Production and effect of aldonic acids during enzymatic hydrolysis of lignocellulose at high dry matter content. Biotechnol. Biofuels 5 (1), 26.

Cannella, D., Jørgensen, H., 2014. Do new cellulolytic enzyme preparations affect the industrial strategies for high solids lignocellulosic ethanol production? Biotechnol. Bioeng. 111 (1), 59–68.

Chen, H., Fu, X., 2016. Industrial technologies for bioethanol production from lignocellulosic biomass. Renew. Sust. Energ. Rev. 57, 468–478.

Cho, E.J., Jung, S., Kim, H.J., Lee, Y.G., Nam, K.C., Lee, H.-J., et al., 2012. Co-immobilization of three cellulases on Au-doped magnetic silica nanoparticles for the degradation of cellulose. Chem. Commun. 48 (6), 886–888.

Chylenski, P., Felby, C., Østergaard Haven, M., Gama, M., Selig, M.J., 2012. Precipitation of *Trichoderma reesei* commercial cellulase preparations under standard enzymatic hydrolysis conditions for lignocelluloses. Biotechnol. Lett. 34 (8), 1475–1482.

Cinelli, B., Castilho, L.R., Freire, D.M.G., Castro, A.M., 2015. A brief review on the emerging technology of ethanol production by cold hydrolysis of raw starch. Fuel 150, 721–729.

Ciriminna, R., Chavarría-Hernández, N., InésRodríguezHernández, A., Pagliaro, M., 2015. Pect: a new perspective from the biorefinery standpoint. Biofuel. Bioprod. Bioresour. 9 (4), 368–377.

Contesini, F.J., Figueira, J.D.B., Carvalho, P.D.O., Nascimento, M.D.G., et al., 2013. Potential applications of carbohydrases immobilization in the food industry. Int. J. Mol. Sci. 14 (1), 1335–1369.

Cosgrove, D.J., 2000. Loosening of plant cell walls by expansins. Nature 407, 321–326.

Dashtban, M., Schraft, H., Qin, W., 2009. Fungal bioconversion of lignocellulosic residues; opportunities and perspectives. Int. J. Biol. Sci. 5 (6), 579–595.

Davies, G., Henrissat, B., 1995. Structures and mechanisms of glycosyl hydrolases. Structure 3 (9), 853–859.

Dhiman, S., Sharma, J., Battan, B., 2008. Industrial application and future prospects of microbial xylanases: a review. Bioresource 3 (4), 1377–1402.

Dunn, J.B., Mueller, S., Wang, M., Han, J., 2012. Energy consumption and greenhouse gas emissions from enzyme and yeast manufacture for corn and cellulosic ethanol production. Biotechnol. Lett. 34 (12), 2259–2263.

Escamilla-Alvarado, C., Poggi-Varaldo, H.M., Ponce-Noyola, T., Ríos-Leal, E., Robles-Gonzalez, I., Rinderknecht-Seijas, N., 2015. Saccharification of fermented residues as integral part in a conceptual hydrogen-producing biorefinery. Int. J. Hydrogen Energy 40 (48), 17200–17211.

Escamilla-Alvarado, C., Pérez-Pimienta, J.A., Ponce-Noyola, T., Poggi-Varaldo, H.M., 2017. An overview of the enzyme potential in bioenergy-producing biorefineries. J. Chem. Technol. Biotechnol. 92 (5), 906–924.

Favalo, L., Viktor, M.J., Rose, S.H., Viljoen-Bloom, M., van Zyl, W.H., Basaglia, M., Cagnin, L., Casella, S., 2015. Consolidated bioprocessing of starchy substrates into ethanol by industrial *Saccharomyces cerevisiae* strains secreting fungal amylases. Biotechnol. Bioeng. 112 (9), 1751–1760.

Fontes, C.M., Gilbert, H.J., 2010. Cellulosomes: highly efficient nano machines designed to deconstruct plant cell-wall complex carbohydrates. Annu. Rev. Biochem. 79, 655–681.

Georgelis, N., Nikolaidis, N., Cosgrove, D.J., 2014. Biochemical analysis of expansin-like proteins from microbes. Carbohydr. Polym. 100, 17–23.

Gomes, D., Rodrigues, A.C., Domingues, L., Gama, M., 2015. Cellulase recycling in biorefineries: is it possible? Appl. Microbiol. Biotechnol. 99 (10), 4131–4143.

Gottschalk, L.M.F., Oliveira, R.A., Bon, E.P.D.S., 2010. Cellulases, xylanases β-glucosidase and ferulic acid esterase produced by *Trichoderma and Aspergillus* act synergistically in the hydrolysis of sugarcane bagasse. Biochem. Eng. J. 51 (1), 72–78.

Haan, R.D., Mcbride, J.E., laGrange, D.C., Lynd, L.R., van Zyl, J.M., 2007. Functional expression of cellobiohydrolases in *Saccharomyces cerevisiae*. Enzyme Microb. Technol. 40, 1291–1299.

Han, Y., Chen, H., 2007. Synergism between corn stover protein and cellulase. Enzyme Microb. Technol. 41 (5), 638–645.

Han, Y., Chen, H., 2010. Synergism between hydrophobic proteins of corn stover and cellulase in lignocellulose hydrolysis. Biochem. Eng. J. 48 (2), 218–224.

Hanefeld, U., Gardossi, L., Magner, E., 2009. Understanding enzyme immobilization. Chem. Soc. Rev. 38 (2), 453–468.

Harris, P.V., Xu, F., Kreel, N.E., Kang, C., Fukuyama, S., 2014. New enzyme insights drive advances in commercial ethanol production. Curr. Opin. Chem. Biol. 19, 162–170.

Hasunuma, T., Kondo, A., 2012. Development of yeast cell factories for consolidated bioprocessing of lignocellulose to bioethanol through cell surface engineering. Biotechnol. Adv. 30, 1207–1218.

Hemsworth, G.R., Johnston, E.M., Davies, G.J., Walton, P.H., 2015. Lytic polysaccharide monooxygenases in biomass conversion. Trends Biotechnol. 33 (12), 747–761.

Henrissat, B., 1991. A classification of glycosyl hydrolases based on amino acid sequence similarities. Biochem. J. 280 (2), 309–316.

Henrissat, B., Davies, G., 1997. Structural and sequence-based classification of glycoside hydrolases. Curr. Opin. Struct. Biol. 7 (5), 637–644.

Hii, S.L., Tan, J.S., Ling, T.C., Ariff, A.B., 2012. Pullulanase: role in starch hydrolysis and potential industrial applications. Enzyme Res., 1–14.

Holtzapple, M., Cognata, M., Shu, Y., Hendrickson, C., 1990. Inhibition of *Trichoderma reesei* cellulase by sugars and solvents. Biotechnol. Bioeng. 36 (3), 275–287.

Hoover, R., 2000. Acid-treated starches. Food Rev. Int. 16 (3), 369–392.

Hsieh, C., Cannella, D., Jørgensen, H., Felby, C., Thygesen, L., 2014. Cellulase inhibition by high concentrations of monosaccharides. J. Agric. Food Chem. 62 (17), 3800–3805.

Janecek, S., Svensson, B., MacGregor, E.A., 2014. α-Amylase: an enzyme specifity found in various families of glycoside hydrolases. Cell Mol. Life Sci. 71 (7), 1149–1170.

Jönsson, L.J., Martín, C., 2016. Pretreatment of lignocellulose: formation of inhibitory by-products and strategies for minimizing their effects. Bioresour. Technol. 199, 103–112.

Jørgensen, H., Kristensen, J., Felby, C., 2007. Enzymatic conversion of lignocellulose into fermentable sugars: challenges and opportunities. Biofuels Bioprod. Bior. 1 (2), 119–134.

Kar, S., Ray, R.C., 2008. Statistical optimization of α-amylase production by *Streptomyces erumpens* MTCC 7317 cells in calcium alginate beads using response surface methodology. Pol. J. Microbiol. 57 (1), 49–57.

Kashyap, D., 2001. Applications of pectinases in the commercial sector: a review. Bioresour. Technol. 77, 215–227.

Khawla, B.J., Sameh, M., Imen, G., Donyes, F., Dhouha, G., Raoudha, E.G., Qumema, N.E., 2014. Potato peel as feedstock for bioethanol production: a comparison of acidic and enzymatic hydrolysis. Ind. Crops Prod. 52, 144–149.

Kim, I.J., Lee, H.J., Choi, I.-G., Kim, K.H., 2014. Synergistic proteins for the enhanced enzymatic hydrolysis of cellulose by cellulase. Appl. Microbiol. Biotechnol. 98 (20), 8469–8480.

Knežević, A., Milovanović, I., Stajić, M., Lončar, N., Brčeski, I., Vukojević, J., et al., 2013. Lignin degradation by selected fungal species. Bioresour. Technol. 138, 117–123.

Konsoula, Z., Liakopoulou-Kyriakides, M., 2006. Starch hydrolysis by the action of an entrapped in a ligninate capsules α-amylase from *Bacillus subtilis*. Process Biochem. 41 (2), 343–349.

Kristensen, J.B., Felby, C., Jørgensen, H., 2009. Yield-determining factors in high-solids enzymatic hydrolysis of lignocellulose. Biotechnol. Biofuel. 2, 11.

Kumar, R., Wyman, C.E., 2014. Strong cellulase inhibition by mannan polysaccharides in cellulose conversion to sugars. Biotechnol. Bioeng. 111 (7), 1341–1353.

Kunamneni, A., Singh, S., 2005. Response surface optimization of enzymatic hydrolysis of maize starch for higher glucose production. Biochem. Eng. J. 27, 179–190.

Larsson, S., Cassland, P., Jonsson, L.J., 2001. Development of a *Saccharomyces cerevisiae* strain with enhanced resistance to phenolic fermentation inhibitors in lignocellulose hydrolysates by heterologous expression of laccase. Appl. Environ. Microbiol. 67, 1163–1170.

Lynd, L.R., Weimer, P.J., vanZyl, W.H., Pretorius, I., 2002. Microbial cellulose utilization: fundamentals and biotechnology. Microbiol. Mol. Biol. Rev. 66 (3), 506–577.

MacLean, H.L., Spatari, S., 2009. The contribution of enzymes and process chemicals to the life cycle of ethanol. Environ. Res. Lett. 4 (1), 014001.

McCarter, J.D., Withers, S., 1994. Mechanisms of enzymatic glycoside hydrolysis. Curr. Opin. Struct. Biol. 4 (6), 885–892.

Meyer, A.S., Rosgaard, L., Sørensen, H.R., 2009. The minimal enzyme cocktail concept for biomass processing. J. Cer. Sci. 50 (3), 337–344.

Miao, M., Jiang, B., Zhang, T., Jin, Z., Mu, W., 2011. Impact of mild acid hydrolysis on structure and digestion properties of waxy maize starch. Food Chem. 126, 506–513.

Modenbach, A.A., Nokes, S.E., 2013. Enzymatic hydrolysis of biomass at high-solids loadings: a review. Biomass Bioenergy 56, 526–544.

Naik, S.N., Goud, V.V., Rout, P.K., Dalai, A.K., 2010. Production of first and second generation biofuels: a comprehensive review. Renew. Sust. Energ. Rev. 14, 578–597.

Nakatani, Y., Yamada, R., Ogino, C., Kondo, A., 2013. Synergistic effect of yeast cell surface expression of cellulase and expansin-like protein on direct ethanol production from cellulose. Microb. Cell Fact. 12, 66.

Nielsen, P.H., Oxenbøll, K.M., Wenzel, H., 2007. Cradle-to-gate environmental assessment of enzyme products produced industrially in Denmark by novozymesA/S. Int. J. LCA 12 (6), 432–438.

Olofsson, J., Barta, Z., Börjesson, P., Wallberg, O., 2017. Integrating enzyme fermentation in lignocellulosic ethanol production: life-cycle assessment and techno-economic analysis. Biotechnol. Biofuel. 10 (1), 51.

Pandey, A., Soccol, C.R., Mitchell, D., 2000. New developments in solid state fermentation: I-bioprocesses and products. Process Biochem. 35, 1153–1169.

Payne, C.M., Knott, B.C., Mayes, H.B., Hansson, H., Himmel, M.E., Sandgren, M., et al., 2015. Fungal cellulases. Chem. Rev. 115, 1308–1448.

Pérez-Avalos, O., Sánchez-Herrera, L.M., Salgado, L.M., Ponce-Noyola, T., 2008. A bifunctional endoglucanase/endoxylanase from *Cellulomonas flavigena* with potential use in industrial processes at different pH. Curr. Microbiol. 57 (1), 39–44.

Plácido, J., Capareda, S., 2015. Ligninolytic enzymes: a biotechnological alternative for bioethanol production. Bioresour. Bioproc. 2, 23.

Qiu, W., Chen, H., 2012. Enhanced the enzymatic hydrolysis efficiency of wheat straw after combined steam explosion and laccase pretreatment. Bioresour. Technol. 118, 8–12.

Quinlan, R.J., Sweeney, M.D., LoLeggio, L., Otten, H., Poulsen, J.-C.N., Johansen, K.S., et al., 2011. Insights in to the oxidative degradation of cellulose by a copper metalloenzyme that exploits biomass components. Proc. Natl. Acad. Sci. 108 (37), 15079–15084.

Quiroz-Castañeda, E.R., Martínez-Anaya, C., Cuervo-Soto, L.I., Segovia, L., Folch-Mallol, J.L., 2011. Loosenin, a novel protein with cellulose-disrupting activity from *Bjerkandera adusta*. Microb. Cell Fact. 10, 8.

Reshmi, R., Sanjay, G., Sugunan, S., 2007. Immobilization of α-amylase on zirconia: a heterogeneous biocatalyst for starch hydrolysis. Catal. Commun. 8 (3), 393–399.

Rojas-Rejón, O.A., Sánchez, A., 2014. The impact of particle size and initial solid loading on thermo-chemical pretreatment of wheat straw improving sugar recovery. Bioproc. Biosyst. Eng. 37 (7), 1427–1436.

Roy, I., Gupta, M.N., 2004. Hydrolysis of starch by a mixture of glucoamylases and pullulanase entrapped individually in calcium alginate beads. Enzyme Microb. Technol. 34 (1), 26–32.

Sánchez, A., Sevilla-Güitrón, V., Magaña, G., Gutierrez, L., 2013. Parametric analysis of total costs and energy efficiency of 2G enzymatic ethanol production. Fuel 113, 165–179.

Sánchez, A., Gil, J.C., Rojas-Rejón, O.A., deAlba, A.P., Medina, A., Flores, R., Puente, R., 2015. Sequential pretreatment strategies under mild conditions for efficient enzymatic hydrolysis of wheat straw. Bioproc. Biosyst. Eng. 38 (6), 1127–1141.

Schusterb, K.C., Ehmosera, H., Gapesb, J.R., Lendla, B., 2000. On-line FT-Raman spectroscopic monitoring of starch gelatinization and enzyme catalyzed starch hydrolysis. Vib. Spectro. 22, 181–190.

Sebastião, D., Gonçalves, M.S., Marques, S., Fonseca, C., Gírio, F., Oliveira, A.C., et al., 2016. Life cycle assessment

of advanced bioethanol production from pulp and paper sludge. Bioresour. Technol. 208, 100–109.
Sheehan, J., Himmel, M., 1999. Enzymes, energy, and the environment: a strategic perspective on the U.S. Department of Energy's research and development activities for bioethanol. Biotechnol. Prog. 15, 817–827.
Shimon, L.J., Pagès, S., Belaich, A., Belaich, J.P., Bayer, E.A., Lamed, R., et al., 2000. Structure of a family IIIa scaffoldin CBD from the cellulosome of *Clostridium cellulolyticum* at 2.2 A resolution. Acta Crystallogr. D 56, 1560–1568.
Singh, A., Kumar, P.K.R., Schügerl, K., 1991. Adsorption and reuse of cellulases during saccharification of cellulosic materials. J. Biotechnol. 18, 205–212.
Singh, R., Shukla, A., Tiwari, S., Srivastava, M., 2014. A review on delignification of lignocellulosic biomass for enhancement of ethanol production potential. Renew. Sustain. Energy Rev. 32, 713–728.
Song, H.-T., Gao, Y., Yang, Y.-M., Xiao, W.-J., Liu, S.-H., Xia, W.-C., et al., 2016. Synergistic effect of cellulase and xylan as e during hydrolysis of natural lignocellulosic substrates. Bioresour. Technol. 219, 710–715.
Stevnebo, A., Sahlström, S., Svihus, B., 2006. Starch structure and degree of starch hydrolysis of small and large starch granules from barley varieties with varying amylose content. Anim. Feed Sci. Technol. 130, 23–38.
Tabka, M.G., Herpoël-Gimbert, I., Monod, F., Asther, M., Sigoillot, J.C., 2006. Enzymatic saccharification of wheat straw for bioethanol production by a combined cellulase xylanase and feruloylesterase treatment. Enzyme Microb. Technol. 39 (4), 897–902.
Teter, S.A., Xu, F., Nedwin, G.E., Cherry, J.R., 2006. Enzymes for biorefineries. In: Kamm, B., Gruber, P.R., Kamm, M. (Eds.), Biorefineries—Industrial Processes and Products. Status Quo and Future Directions. Wiley-VCH Verlag GmbH & Co, Weinheim, Germany.
Tsai, C.-T., Meyer, A.S., 2014. Enzymatic cellulose hydrolysis: enzyme reusability and visualization of β-glucosidase immobilized in calcium alginate. Molecules 19 (12), 19390–19406.
Van Zyl, W.H., Lynd, L.R., denHaan, R., McBride, J.E., 2007. Consolidated bioprocessing for bioethanol production using *Saccharomyces cerevisiae*. Adv. Biochem. Eng. Biotechnol. 108, 205–235.
Vaz, R.P., de Souza Moreira, L.R., Ferreira Filho, E.X., 2016. An overview of holocellulose-degrading enzyme immobilization for use in bioethanol production. J. Molec. Catal. B 133, 127–135.
Viikari, L., Vehmaanperä, J., Koivula, A., 2012. Lignocellulosic ethanol: from science to industry. Biomass Bioenergy 46, 13–24.
Vohra, M., Manwar, J., Manmode, R., Padgilwar, S., Patil, S., 2014. Bioethanol production: feedstock and current technologies. J. Environ. Chem. Eng. 2 (1), 573–584.
Wang, Y., Truong, V., Wang, L., 2003. Structures and rheological properties of corn starch as affected by acid hydrolysis. Carbohydr. Polym. 52, 327–333.
Weiss, N., Börjesson, J., Pedersen, L.S., Meyer, A.S., 2013. Enzymatic lignocellulose hydrolysis: improved cellulase productivity by insoluble solids recycling. Biotechnol. Biofuels 6 (1), 5.
Ximenes, E., Kim, Y., Mosier, N., Dien, B., Ladisch, M., 2010a. Inhibition of cellulases by phenols. Enzyme Microb. Technol. 46 (3–4), 170–176.
Ximenes, E., Kim, Y., Mosier, N., Dien, B., Ladisch, M., 2010b. Deactivation of cellulases by phenols. Enzyme Microb. Technol. 48 (1–5), 54–60.
Yang, B., Wyman, C.E., 2006. BSA treatment to enhance enzymatic hydrolysis of cellulose in lignin containing substrates. Biotechnol. Bioeng. 94 (4), 611–617.
Zaldivar, J., Nielsen, J., Olsson, L., 2001. Mini-review fuel ethanol production from lignocellulose: a challenge for metabolic engineering and process integration. Appl. Microbiol. Biotechnol. 56, 17–34.

Further Readings

Alberghina, L., 2003. Protein Engineering for Industrial Biotechnology. Hardwood Academic Publishers, CRC Press, Boca Raton, FL.
Alper, H., Stephanopoulos, G., 2009. Engineering for biofuels: exploiting innate microbial capacity or importing biosynthetic potential. Nat. Rev. Microbiol. 7, 715–723.
FitzPatrick, M., Champagne, P., Cunningham, M.F., Whitney, R.A., 2010. A biorefinery processing perspective: treatment of lignocellulosic materials for the production of value-added products. Bioresour. Technol. 101 (23), 8915–8922.
Hatakeyama, H., Hatakeyama, T., 2010. Lignin structure, properties, and applications. Biopolymers 232, 1–63.
Kim, Y., Ximenes, E., Mosier, S.N., Ladisch, M.R., 2011. Soluble inhibitors/deactivators of cellulase enzymes from lignocellulosic biomass. Enzyme Microb. Technol. 48 (4–5), 408–415.
Lambertz, C., Garvey, M., Klinger, J., Heesel, D., Klose, H., Fischer, R., Commandeur, U., 2014. Challenges and advances in the heterologous expression of cellulolytic enzymes: a review. Biotechnol. Biofuel. 7, 135.
Li, S.Z., 2006. Engineered Bacteria for Ethanol Production From Lignocellulose Center for Biomass. Engineering China Agricultural University.
Mathews, C.K., vanHolde, K.E., Ahern, K.G., 2002. Biochemistry, third ed. Addison-Wesley, Boston, MA.
Puls, J., 1997. Chemistry and biochemistry of hemicelluloses: relationship between hemicellulose structure and enzymes required for hydrolysis. Macromol. Symp. 120 (1), 183–196.

Sims, R.E.H., Mabee, W., Saddler, J.N., Tylor, M., 2010. An overview of second generation biofuels technologies. Biores. Technol. 101 (6), 1570–1580.

Tester, R.F., Karkalas, J., Qi, X., 2004. Starch structure and digestibility enzyme-substrate relationship. Worlds Poult. Sci. J. 60 (2), 186–195.

Toon, S.T., Philippidis, G.P., Ho, N.W.Y., Chen, Z., Brainard, A., Lumpkin, R.E., Riley, C.J., 1997. Enhanced cofermentation of glucose and xylose by recombinant *Saccharomyces* yeast strains in batch and continuous operating modes. Appl. Biochem. Biotechnol. 65, 243–255.

CHAPTER

14

Application of Fungal Pretreatment in the Production of Ethanol From Crop Residues

Thelmo A. Lu-Chau, María García-Torreiro*, María López-Abelairas*, Natalia A. Gómez-Vanegas**, Beatriz Gullón*, Juan M. Lema*, Gemma Eibes**

*Department of Chemical Engineering, Institute of Technology, Universidade de Santiago de Compostela, Santiago de Compostela, Spain

**Department de Ingeniería Química, Universidad de Antioquia, Medellín, Colombia

ABBREVIATIONS

AAO	Aryl alcohol oxidase
BRF	Brown-rot fungi
CBP	Consolidated bioprocessing
Lac	Laccase
LCB	Lignocellulosic biomass
LiP	Lignin peroxidase
MnP	Manganese peroxidase
SmF	Submerged fermentation
SsF	Solid state fermentation
SsF	Simultaneous saccharification and fermentation
SHF	Separate hydrolysis and fermentation
VP	Versatile peroxidase
WRF	White-rot fungi

14.1 INTRODUCTION

For the last two decades, there has been a growing interest in the search for alternative energy sources due to the fast depletion of fossil fuels as well as the increasing of greenhouse gas emissions (Gao et al., 2017). In this framework, the production of bioethanol from lignocellulosic feedstocks, such as crop residues is considered one of the most promising alternatives to fossil fuels. However, the complex and recalcitrant structure of lignocellulosic biomass requires a severe pretreatment to improve their

Bioethanol Production From Food Crops
http://dx.doi.org/10.1016/B978-0-12-813766-6.00014-X

conversion into fermentable sugars, being this step the main bottleneck for the economically viable bioethanol production (Liu et al., 2015).

Several pretreatments have been extensively investigated for this purpose, including physical (milling, grinding, pyrolysis, microwave), chemical (acid, alkaline), and physicochemical (ammonia fiber explosion, steam explosion) techniques (Liu et al., 2015; Sarkar et al., 2012). However, these pretreatments require high consumption of energy and chemicals, which often lead to the formation of inhibitory compounds that affect adversely the fermentation (Alvira et al., 2010).

In this context, the biological pretreatment has been considered as a low-cost, safe, and environmentally friendly alternative (Gao et al., 2017; Salvachúa et al., 2011), which can be used to modify the lignocellulose matrix and improve the action of hydrolytic enzymes for the subsequent fermentation to ethanol (López-Abelairas et al., 2013a; Martín-Sampedro et al., 2015). Thus, the biological pretreatment could be an alternative to degrade lignocellulose selectively and produce fewer inhibitors than steam explosion (Salvachúa et al., 2011).

The fungal treatment has long been investigated for biopulping applications (Gao et al., 2017; Singh et al., 2010). With the renewed interest in biofuels, white-rot fungi (WRF) are garnering attention once again. However, instead of modifying wood for pulp and paper production, WRF are being explored to condition a variety of biomass relevant to biofuel production (Canam et al., 2013). This interest is evidenced by the recent publication of many articles reviewing the application of fungal pretreatment for producing ethanol using lignocellulosic biomass from different origin (Canam et al., 2013; Millati et al., 2011; Moreno et al., 2015; Sindhu et al., 2016; Tian et al., 2012; Wan and Li, 2012).

Biological pretreatment is based on the capacity of some microorganisms including white-rot, brown-rot, and soft-rot fungi to use extracellular ligninolytic enzymes and other oxidative systems to degrade lignin and modify lignocellulose structure (Gao et al., 2017; Martín-Sampedro et al., 2015). The evolution of lignin decay capabilities in white-rot fungi caused a sharp decline in the rate of organic carbon burial at the end of the Permo-Carboniferous (Floudas et al., 2012). Thus, WRF evolved to have the biological function of deconstructing lignocellulosic biomass to use it for growth. The use of WRF as a pretreatment step in a biorefinery scheme needs to be adapted to the special characteristics of these microorganisms, to take full advantage of its great potential and to overcome the inherent drawbacks.

The main disadvantage of fungal pretreatment is the long incubation time necessary to reach yields similar to those obtained with current physicochemical pretreatments. Other disadvantages are the consumption of cellulose and/or hemicellulose for fungal growth and the requirement of sterile conditions (García-Torreiro et al., 2016a). Several strategies have been proposed to overcome these drawbacks, as, for example, the combination of the biological treatment with a physical or chemical pretreatment (García-Torreiro et al., 2016b; Salvachúa et al., 2011), or the in vitro use of fungal ligninolytic enzymes (Moreno et al., 2015).

In the present chapter, the application of fungal pretreatment has been focused on the use of crop residues as substrate for producing ethanol. This study presents different aspects to consider when using the fungal pretreatment in a biorefinery concept, which include the mechanism of action of the WRF, its implementation at lab scale, the use of different strategies for enhancing the performance of the pretreatment, and finally, its integration into the subsequent fermentation step.

14.2 MECHANISM INVOLVED IN THE PRETREATMENT OF CROP RESIDUES BY LIGNINOLYTIC FUNGI

White-rot fungi are the sole organisms in nature that can efficiently degrade and even mineralize lignin to water and carbon dioxide. Over the past three decades, the ligninolytic enzyme consortium has been extensively studied to elucidate the role of the enzymes involved on lignin degradation (Martínez et al., 2009; Sánchez, 2009).

During their growth in nature, WRF use two extracellular enzymatic systems to degrade lignocellulosic biomass: (1) a hydrolytic system, which produces cellulases and hemicellulases that degrade polysaccharides; and (2) an oxidative ligninolytic system, which degrades lignin and opens phenyl rings (Sánchez, 2009). Furthermore, lignocellulose degradation also involves a nonenzymatic system (Fig. 14.1). Although many WRF can degrade simultaneously all the plant cell wall components, some of them, such as *Ceriporiopsis subvermispora*, selectively decompose lignin in advance of cellulose and hemicellulose degradation.

The key enzymes of the ligninolytic enzyme consortium are lignin peroxidase (LiP), manganese peroxidase (MnP), versatile peroxidase (VP), and laccase (Lac), although other accessory enzymes, are also involved in lignin degradation (Alcalde, 2015; Dashtban et al., 2010; Martínez et al., 2009). Whereas some WRF produce several different kinds of ligninolytic enzymes, others produce only one or two of them (Hatakka and Hammel, 2010; Sánchez, 2009).

Laccases (EC 1.10.3.2) are blue-multicopper oxidoreductases, capable of oxidizing

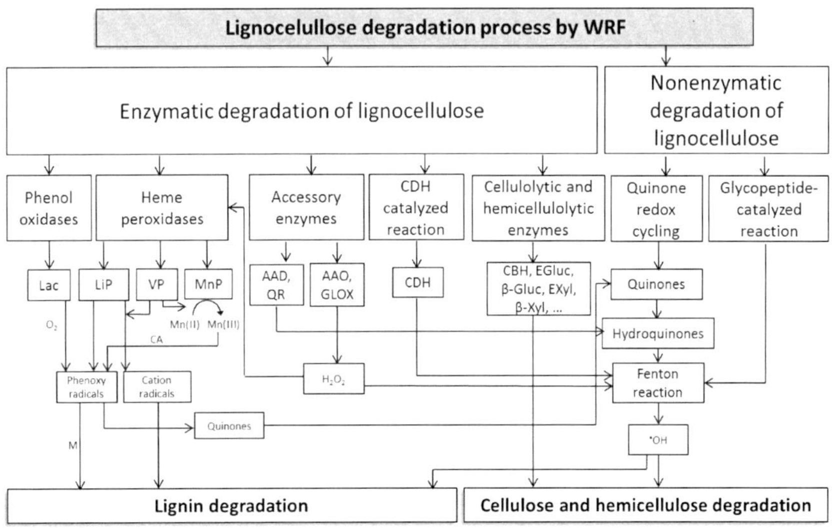

FIGURE 14.1 **Schematic diagram of lignocellulose degradation by white rot fungi.** *β-Gluc*, β-Glucosidases; *β-Xyl*, β-xylosidases; *AAD*, ary-alcohol dehydrogenase; *AOO*, aryl-alcohol oxidase; *CA*, chelating agent; *CBH*, cellobiohydrolase; *CDH*, cellobiose dehydrogenase; *EGluc*, endoglucanases; *Exyl*, endoxylanases; *GLOX*, glyoxal oxidase; *Lac*, laccase; *LiP*, lignin peroxidase; *M*, mediator; *MnP*, manganese peroxidase; *QR*, quinone reductase; *VP*, versatile peroxidase. *Source: Adapted from Dashtban, M., Schraft, H., Syed, T.A., Qin, W., 2010. Fungal biodegradation and enzymatic modification of lignin. Int. J. Biochem. Mol. Biol. 1(1), 36.*

phenols, polyphenols, aromatic amines, and many other compounds, using molecular oxygen and releasing water as the only byproduct. Laccase can oxidize low redox potential phenolic lignin structures. However, phenolic subunits often comprise less than 15% of the lignin polymer (Eggert et al., 1996), and as consequence, laccases are incapable of depolymerizing lignin. Nevertheless, in the presence of small compounds forming stable radicals that act as redox mediators, laccases oxidize lignin (Alcalde, 2015). Certain phenolic compounds derived from the degradation of lignocellulose by WRF, as for instance acetosyringone, syringaldehyde, vanillin, and *P*-hydroxycinnamic acids, act as natural mediators of laccase (Camarero et al., 2014).

Heme peroxidases show high redox potential, and are represented by LiP (EC 1.11.1.14), MnP (EC 1.11.1.13) and VP (EC 1.11.1.16). LiP is characterized by its ability to catalyze the H_2O_2-dependent oxidative depolymerization of nonphenolic lignin compounds, forming aromatic cations, and a wide range of phenolic compounds (e.g., guaiacol, vanillyl alcohol, catechol). It has been also shown that veratryl alcohol contributes to LiP oxidation of nonphenolic aromatic compounds (Martínez, 2002). On the other hand, MnP catalyzes the H_2O_2-dependent oxidation of Mn(II) to Mn(III), which is then released from the enzyme surface forming a complex with a chelating agent, such as oxalic or other organic acids. Chelated Mn(III) complex is a stable and strong oxidizing agent and acts as a diffusible redox-mediator of phenolic substrates. Other studies expanded the role of MnP in lignin biodegradation via lipid peroxidation. In this sense, chelated Mn(III) generates reactive intermediates (peroxy radicals) from unsaturated fatty acids, such as linoleic acid and their derivatives (lipids), which are able to oxidize a variety of nonphenolic aromatic compounds (Martínez, 2002). Finally, VP shows a dual oxidative ability to oxidize Mn(II) and phenolic and nonphenolic aromatic compounds. The presence of low molecular-weight redox mediators is particularly important during the initial steps of lignin degradation, since the compact molecular structure of the intact plant cell wall impedes the direct contact of the enzymes with lignin (Martínez, 2002).

Accessory enzymes involved in lignin degradation, include hydrogen peroxide generating oxidases (such as aryl-alcohol oxidase, AAO, and glyoxal oxidase, GLOX), aryl-alcohol dehydrogenase (AAD), which is also involved in the H_2O_2 during cyclic redox reactions, and quinone reductase (QR), which is a ferric-ion-reducing enzyme (Dashtban et al., 2010; Martínez et al., 2005). Moreover, cellobiose dehydrogenase (CDH), which is produced by many different fungi under cellulolytic conditions, is also involved in lignocellulose degradation through the reduction of quinones (Dashtban et al., 2010).

For the hydrolysis of cellulose, basidiomycetes utilize a set of hydrolytic enzymes typically composed of endoglucanase, cellobiohydrolase, and β-glucosidase (Baldrian, 2008). The hemicellulolytic system is more complex, and involves among others endoxylanases, β-xylosidases, mannanases, and mannosidases (Heidorne et al., 2006).

The degradation of lignocellulose by wood-rotting basidiomycetes also involves a nonenzymatic system based on Fenton reaction ($H_2O_2 + Fe^{2+} + H^+ \rightarrow H_2O + Fe^{3+} + {}^{\bullet}OH$). Hydrogen peroxide is produced by enzymes, such as AAO and GLOX in WRF, while the reduction of Fe^{3+} can be achieved through the quinone redox cycling, the glycopeptide catalyzed reaction or by the enzyme CDH (Baldrian, 2008). Hydroxyl radicals generated by Fenton chemistry attack polysaccharides as well as lignin in a nonspecific manner creating some cleavage sites, which facilitates the penetration of ligninolytic and hydrolytic enzymes on the cell wall (Dashtban et al., 2010).

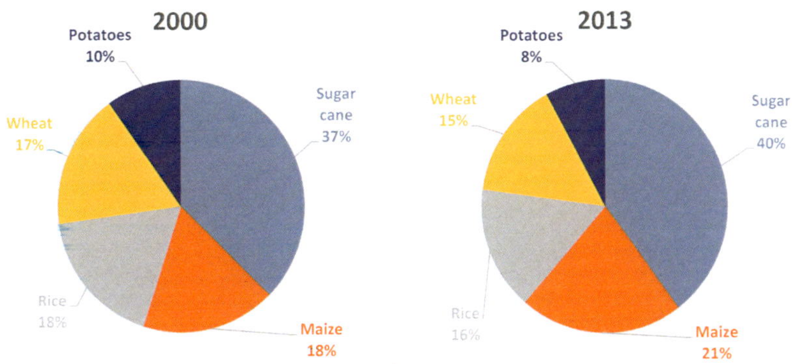

FIGURE 14.2 **Distribution of main crops in the world in 2000 and 2013.** *Source: Data taken from FAOSTAT, 2015. FAO Statistical Pocketbook. Available from: http://www.fao.org/3/a-i4691e.pdf.*

14.3 AGRICULTURAL RESIDUES AS POTENTIAL SUBSTRATE TO BE FUNGAL PRETREATED

According to statistical data collected from FAO (FAOSTAT, 2015) sugarcane, maize, rice, wheat, and potatoes represent the main crops grown in the world in the years 2000 and 2013 (Fig. 14.2). Nevertheless, cereals continue to be one of the most relevant sources for food worldwide, wherein wheat, rice, barley, maize, rye, oats, and millet represent most of the production within the farming sector. In addition, cereals continue to be the most important food source in human diets all over the world.

However, new tendencies in the resettlement of the population within urban areas, as well as the rising salaries, are causing diets to change towards the consumption of higher quantities of protein, fat, and sugar. Besides, the land is being shared with other industries, such as livestock and biofuel, which are growing at a faster rate than crop farming. This trend is causing a change in the use of soil, which is moving away from traditional crops (wheat and rice) into coarse grains and oilseeds to cope the rising demand for food, feed, and biofuel.

This new approach in the use of soil generates a high amount of lignocellulosic residues that can be used in the production of ethanol. This raw material does not compete in food markets, like other substrates derived from plants, such as cereals or sugarcane, used as well in the bioethanol industry (García-Torreiro et al., 2016a). Besides, the local availability of the raw material is an important factor to consider for biofuel production from an economical and environmental perspective. To produce lignocellulosic ethanol at commercial scale the production cost will be determined not only by the performance of the conversion process, but also by the entire supply-chain from feedstock production to consumption (Slade et al., 2009).

To be a feasible alternative, the fungal pretreatment should be evaluated using different feedstocks to demonstrate its flexibility (García-Torreiro et al., 2016a). WRF degrade larger amounts of lignin relative to cellulose. However, selectivity varies among fungal species and with pretreatment time. In addition, fungal performance on degradation and the resulting digestibility varies with different feedstocks (Wan and Li, 2012). Recently, the use of agricultural residues for producing ethanol has been extensively reviewed by Gupta and Verma (2015).

TABLE 14.1 Chemical Composition of Representative Crop Residues

	Composition (%w/w)		
Crop residue	**Cellulose**	**Hemicellulose**	**Lignin**
Barley hull	34	36	19
Barley straw	36–43	24–33	6.3–9.8
Corn cob	32.3–45.6	39.8	6.7–13.9
Corn stover	35.1–39.5	20.7–24.6	11–19.1
Cotton stalk	31	11	30
Coffee pulp	33.7–36.9	44.2–47.5	15.6–19.1
Empty fruit bunch palm	49.95–39.8	18.91–17.3	25.17–28.8
Rice straw	29.2–34.7	23–25.9	17–19
Rice husk	28.7–35.6	11.96–29.3	15.4–20
Wheat straw	35–39	22–30	12.0–16
Sugarcane bagasse	25–45	28–32	15–25
Oat straw	31–35	20–26	10.0–15
Sorghum straw	32–35	24–27	15–21

Source: Data compiled from Menon, V., Rao, M., 2012. Trends in bioconversion of lignocellulose: biofuels, platform chemicals and biorefinery concept. Prog. Energy Combust. Sci. 38(4), 522–550; Kim, S., Park, J.M., Seo, J.W., Kim, C.H., 2012. Sequential acid-/alkali-pretreatment of empty palm fruit bunch fiber. Bioresour. Technol. 109, 229–233; and Piñeros-Castro, Y., Velásquez-Lozano, M., 2014. Biodegradation kinetics of oil palm empty fruit bunches by white rot fungi. Int. Biodeterior. Biodegrad. 91, 24–28.

The potential of a crop residue to be used as a substrate for ethanol production, as well as its availability for being fungal pretreated depends on its composition. A summary of some of the most common agricultural residues along with their cellulose, hemicellulose, and lignin composition is shown in Table 14.1. The highest fraction is represented by cellulose and hemicellulose, from which sugars can be obtained to produce ethanol.

14.4 DEVELOPMENT OF FUNGAL PRETREATMENT

14.4.1 Solid State Fermentation (SsF)

Filamentous fungi, such as WRF, can grow in the absence of free water, using the bound water of the substrate. In nature, WRF develop in terrestrial habitats on wet substrates. On lignocellulosic materials, fungi must first colonize the substrate by adhesion and, then, spread between substrate particles by branching (Tengerdy and Szakacs, 2003). Consequently, SsF is a suitable cultivation method to perform fungal pretreatment of lignocellulosic substrates.

The term SsF describes the cultivation of microorganisms on solid and moist substrates, in the absence of a free aqueous phase (Hölker and Lenz, 2005). Ligninolytic and hydrolytic enzymes produced by mycelium diffuse to the solid matrix and catalyze the degradation of lignin and polysaccharides into smaller units.

SsF shows some advantages over traditional submerged fermentation (SmF). Among them, it is noteworthy its higher economic efficiency due to the lower cost of substrates, such as crop residues, in addition to the low requirements for instrumentation and equipment. Fungal enzymes produced in SsF are less prone to substrate inhibition and present higher temperature or pH stability than those produced in SmF. In addition, SsF provides higher volumetric productivities and less degradation of enzymes by proteases (Hölker and Lenz, 2005).

On the other hand, the main drawbacks of SsF are of operational and technical nature. The scale-up of SsF represents an important bottleneck due to the presence of different gradients (humidity, temperature, substrate concentration). There are few studies reporting the fungal pretreatment of crop residues at bioreactor scale (Planinić et al., 2016). Other SsF processes at bench and pilot reactor scale have been carried out not as a pretreatment, but for producing enzymes using agricultural residues as substrate (Mohseni et al., 2012; Pitol et al., 2016).

14.4.2 Fungal Pretreatment Carried Out at Lab Scale

The typical stages of fungal pretreatment at lab scale are shown in Fig. 14.3. Fungal pretreatment is conducted in static cultures typically carried out in Erlenmeyer flasks (Fig. 14.4). This fungal pretreatment scheme was proposed by López-Abelairas (2014) and has been used as a based protocol in the studies of López-Abelairas et al. (2013a,b), and García-Torreiro et al. (2016b) using wheat straw as substrate, and on palm oil EFB (Gómez Vanegas, 2016) and other crop residues, such as barley straw, corncob, and corn stover (García-Torreiro et al., 2016a). This scheme includes the mechanical and thermal treatments for substrate conditioning that precede the biological pretreatment as well as the subsequent stages for measuring its effect on the crop residue.

Other protocols used for fungal pretreatment of crop residues have been described by Zeng et al. (2011), and by Zhou et al. (2015), who applied a high-throughput methodology based on the use of miniaturized SsF in 24-well plates.

14.5 PARAMETERS AFFECTING THE PERFORMANCE OF FUNGAL PRETREATMENT AND THEIR OPTIMIZATION

The optimization of fungal pretreatment parameters is key to attain an efficient operation. Numerous parameters have influence on the biological pretreatment (Rouches et al., 2016; Sindhu et al., 2016). The most important factors are time, temperature, moisture content, inoculum concentration, and particle size. Moreover, the optimal conditions are substrate and fungal strain dependent (Sindhu et al., 2016).

14.5.1 Time

Incubation time depends on the composition and recalcitrance of the biomass and the fungal strain used (Rouches et al., 2016; Sindhu et al., 2016). During the colonization of the substrate, the sugars digestibility decreases, while it tends to increase during fungus maturation (Rouches et al., 2016). This fact can be explained by the early consumption of the sugars solubilized in water, previous to the decrease of substrate recalcitrance by fungal action (Zadražil, 1977). However, after weeks of treatment, the net losses of glucan and xylan could reach 40%–60% for some strains (Sindhu et al., 2016). Thus, it is necessary to achieve a balance between the increased efficiency in enzymatic saccharification and the consumption of polysaccharides (Du et al., 2011). The attainment of such equilibrium implies the study and optimization of pretreatment time.

Salvachúa et al. (2011) studied the evolution of the digestibility of biologically pretreated wheat straw. The highest ligninolytic activity was detected between 7 and 14 days, while the highest recovery of glucose after hydrolysis (62% for *Irpex lacteus*, the most promising strain) occurred after 21 days. At that time, the loss of glucose was around 17%. Gao et al. (2017) used WRF for the deconstruction of wheat

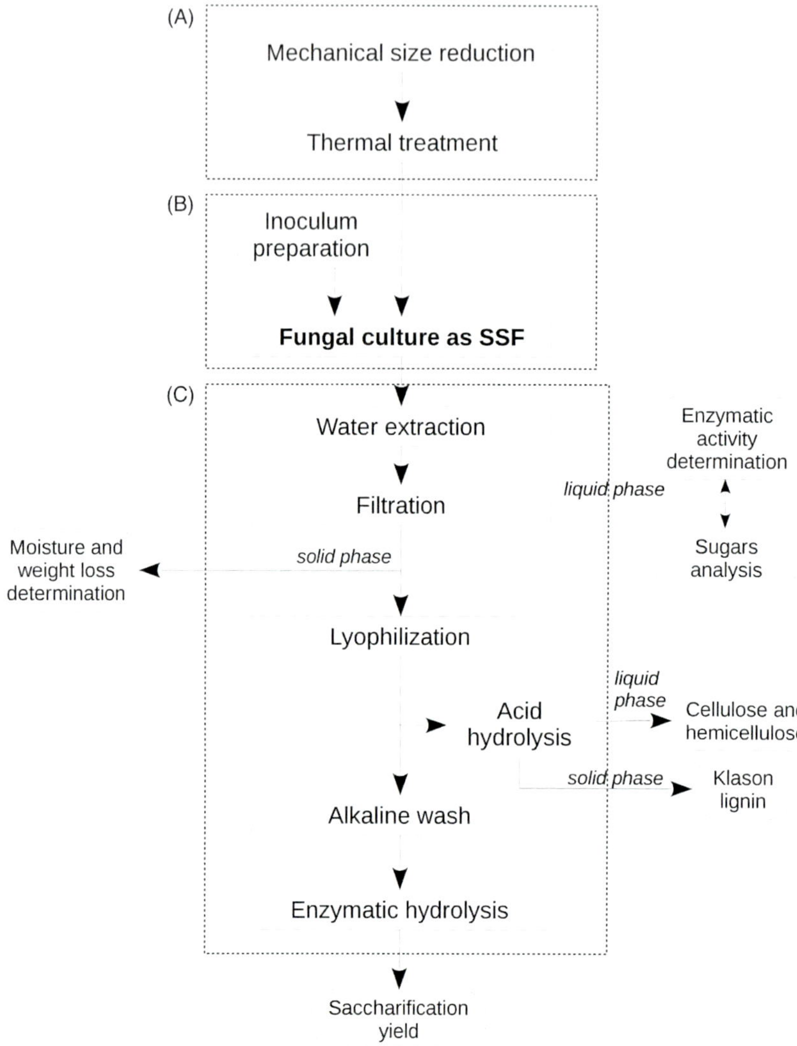

FIGURE 14.3 Scheme of the experimental implementation of fungal pretreatment showing the stages of (A) substrate conditioning, (B) biological pretreatment, and (C) fractioning and analysis of the pretreated crop residue.

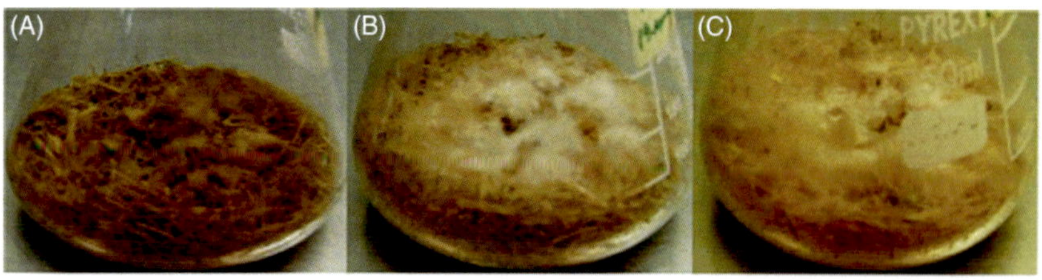

FIGURE 14.4 Evolution of the fungal pretreatment of wheat straw by action of the white-rot fungi *Irpex lacteus* after (A) 0, (B) 7, and (C) 14 days of solid state fermentation.

straw. They evaluated the optimal fermentation time and found that the longest (35 days) and shortest (21 days) times were achieved with *Phanerochaete chrysosporium* and *Trametes versicolor*, respectively.

14.5.2 Temperature

The optimum temperature largely depends on the type of microorganism employed. Most of the white-rot ascomycetes grow optimally at 39°C, while the white-rot basidiomycetes do it at 25–30°C (Sindhu et al., 2016). Papadopoulu et al. (2015) have found that the extent of colonization and the propagation velocity mainly depends on the pretreatment temperature, followed by the moisture of the substrate.

14.5.3 Moisture

SsF allows high substrate concentrations, which favors the economic viability of the biological pretreatment. On the other hand, initial moisture content is crucial for the establishment of microbial growth since it affects the fungal growth and enzyme production and, hence, significantly affects lignin degradation.

Camarero (1995) has reported that the optimum water activity for straw delignification by fungi is between 60% and 80%. Besides, it must be considered that in SsF the water volume in medium should not exceed the highest retention capacity of the substrate. López-Abelairas (2014) has studied the percent of solids that satisfies this requirement in wheat straw. They observed that free water is not present at solid concentrations above 7.5%, and that a solid concentration within the range 20%–40% is suitable for SsF in wheat straw. The study conducted by Xu et al. (2001) revealed that a lower solid–liquid ratio is more beneficial to produce LiP and MnP. Moreover, Shi et al. (2008) reported the biological pretreatment of cotton stalks using *Penicillium chrysogenum* where higher moisture content (75%–80%) led to more lignin degradation than lower moisture content (65%).

14.5.4 Aeration

Oxygen is essential for both WRF growth and the oxidative process of delignification. A high oxygenation rate often enhances ligninolytic enzyme production (Millati et al., 2011) and, thus, delignification (Kamra and Zadražil, 1986). Aeration rate can become a limiting factor during SsF. Depending on the compaction of the substrate, oxygen can penetrate in it to a greater or lesser extent.

Otherwise, fungal metabolism generates heat and leads to temperature gradients in solid-state media. A good aeration of all the zones of substrate would contribute to regulate the temperature and reduce temperature gradients.

14.5.5 Mycelium Content

Inoculum has an important role in biological pretreatment since the time required for the colonization of the substrate is clearly influenced by the type and amount of inoculum (van Kuijk et al., 2015a). Filamentous fungi are morphologically complex microorganisms and could exhibit different structural forms depending on the medium characteristics and culture conditions. The basic structural growth on solid substrates is as mycelium, while in submerged cultures, they adopt different forms, from dispersed mycelia to pellets (Papagianni, 2004).

The adequate concentration of inoculum must be determined to ensure good fungal growth and substrate colonization (Yoon et al., 2014). López-Abelairas (2014) has studied the effect of inoculum concentration on the degradation of wheat straw by *I. lacteus*. Within the studied range (0.5–8 mL inoculum/g straw), it was observed that an increase in inoculum concentration did not improve substrate degradation by the fungus, although it had a positive effect on growth rate.

14.5.6 Particle Size

The particle size of the substrate also affects the fungal pretreatment (van Kuijk et al., 2015b). The use of large particles limits the penetration of fungi into the biomass and the diffusion of air, water and metabolites. Small particles reduce the size of interparticle channels, which will adversely affect the gas circulation. Hence, the use of an optimum particle size is essential for effective biological pretreatment.

A similar conclusion was obtained by López-Abelairas et al. (2013a). They found the highest digestibility with the smallest particle size, which may be related to larger specific surfaces, a reduction of the crystallinity index of cellulose and better accessibility to cellulose and hemicellulose by hydrolytic enzymes (Sun and Cheng, 2002).

14.6 STRATEGIES TO IMPROVE THE ACTION OF FUNGAL PRETREATMENT

Despite fungal pretreatment is an environmentally friendly and energy efficient process compared to thermal and chemical pretreatments, its most significant drawback is the long time required for the process. Different strategies have been investigated to overcome this and other drawbacks of the fungal pretreatment. These strategies have been focused on enhancing the lignin degradation ability of the fungi, combining the fungal action with another type of pretreatment, the use of alternative ethanol production schemes, or the application in vitro of the ligninolytic enzymes produced by WRF.

14.6.1 Screening of Ligninolytic Fungi

The performance of the fungal pretreatment greatly depends on the white-rot fungus used. In nature, the colonization of lignocellulosic biomass is commonly carried out by a specific type of WRF. For example, in the case of wood, different fungal species have varying capacities for its decomposition (Valmaseda et al., 1990) and are dependently related to specific tree species (Rajala et al., 2010).

Two relevant parameters for selecting a WRF are the growth rate and the selectivity towards lignin degradation. For example, *Phanerochaete chrysosporium*, a WRF that has been used as a model microorganism in ligninolytic enzyme production and lignin degradation (Kersten and Cullen, 2007), has been reported to have low selectivity toward lignin degradation and minimal effect on increasing LCB digestibility (Wan and Li, 2012). Fungal metabolism can cause cellulose loss and it is therefore necessary to use the appropriate fungal strain-biomass-type combination (Cianchetta et al., 2014).

Taking this into account, one approach that has been used to improve fungal pretreatment has been the selection of the most adequate ligninolytic fungus on the target substrate. Table 14.2 shows some of the fungal screening studies carried out on different crop residues.

In one of the first studies describing the use of WRF for pretreating residual crops, Hatakka (1983) carried out a screening of 19 WRF on wheat straw. They observed that four of these WRF increased the enzymatic saccharification of wheat straw. After five weeks of pretreatment with *Pycnoporus cinnabarinus*, 54.6% of the residue was converted to reducing sugars in the enzymatic saccharification.

Salvachúa et al. (2011) screened 21 fungi to evaluate the potential of a fungal pretreatment to replace or complement physicochemical methods for ethanol production. They evaluated the changes in wheat straw composition, secretion of ligninolytic enzymes, and enzymatic hydrolysis efficiency after 7, 14, and 21 days of SsF. Most fungi degraded lignin, although only eight of them improved sugar recovery. They obtained the best results with *Poria subvermispora* and *I. lacteus* reaching sugars yields from cellulose of 69% and 66%, respectively.

TABLE 14.2 Screening of Fungal Strains to Be Used in the Pretreatment of Different Crop Residues

Substrate	Number of fungal strains assayed	Selected fungus	Treatment time (D)	Max. sugar digestibility (%)	References
Corn stover	26 WRF	*Cyathus stercoreus* NRRL-6573	30	52	Saha et al. (2016)
Cotton stalks	3 WRF	*Daedalea flavida* MTCC 145	20–40	83	Meehnian and Jana (2016)
Rice straw	3 fungal strains	*Fusarium equiseti* TWRF-10, *Penicillium simplicissimum* TRF-27	8	NA	Chaudhari et al. (2015)
Sugarcane bagasse	2 WRF, 1 BRF	*Ceriporiopsis subvermispora*	60	47	da Silva Machado and Ferraz (2017)
Wheat straw	19 WRF	*Pycnoporus cinnabarinus* 115	35	55	Hatakka (1983)
Wheat straw	20 WRF, 1 BRF	*Poria subvermispora, Irpex lacteus*	21	66–69	Salvachúa et al. (2011)
Wheat straw	5 WRF	*C. subvermispora*	70	44–60	Cianchetta et al. (2014)
Wheat straw, miscanthus	53 WRF, 10 BRF	17 WRF + 1 BRF	12	62	Zhou et al. (2015)

NA, Not assayed.

Using a different methodology, Zhou et al. (2015) evaluated 53 WRF and 10 BRF in 12d small-scale SsF experiments using 24-well plates. Based on hierarchical clustering, according to sugars released, they grouped the fungi in efficient, no-effect, and detrimental-effect species. The efficient group contained 17 WRF and one BRF. They observed that the yield of sugar released increased significantly (up to 62%) compared with noninoculated controls for both substrates.

In some screening studies, the fungi used in the pretreatment were isolated from nature. For example, Chaudhari et al. (2015) evaluated eight fungi isolated from compost and forest litter, during SsF with rice straw. They found that *Fusarium equiseti* TWRF-10 showed the highest endoglucanase, exoglucanase, and endoxylanase activities, whereas *Penicillium simplicissimum* TRF-27 presented the highest laccase and MnP activities. Besides, *F. equiseti* TRF-27 caused a 65% reduction in cellulose crystallinity.

14.6.2 Combination of Fungal and Physicochemical Pretreatments

The combination of the fungal pretreatment with other methods might reduce the pretreatment time, and it can also improve its effectiveness in terms of products yield and cost (Shirkavand et al., 2016). A short-term fungal pretreatment can modify the biomass cell wall before evident degradation of cellulose and/or hemicellulose takes place, while the required severity of the physico-chemical pretreatment can be significantly reduced (Wan and Li, 2012).

Physical pretreatment, such as size reduction by milling, grinding, or freezing, is already commonly combined with other kind of pretreatments. Size reduction increases the surface area of the biomass, while the production of toxic materials is negligible. Small particle size enhances the penetration of the fungi into the biomass, however, the size of the interparticle channels for the diffusion of air is reduced, hampering the transfer of oxygen, and thus delignification rate is not necessarily increased. As can be observed in Table 14.3, the effect of particle size on fungal performance have been studied applying different substrates and fungal species (Membrillo et al., 2008; Sarikaya and Ladisch, 1999; Wan and Li, 2010). In all the cases, an optimal particle size to enhance and maximize fungal growth or lignin degradation was found. In general terms, a smaller size improves fungal performance rather than a bigger one. However, the optimum size is a compromise between the substrate surface, for maximize fungal attack, and the space between particles, for gas transfer.

The chemical pretreatment includes acid, alkali, organic solvent, and ionic liquid pretreatments. In this case a chemical agent or solvent is applied to break down the lignocellulosic components. These treatments are usually highly effective at deconstruction level. However, the production of toxic compounds, the loss of carbohydrates, and their high cost are the main disadvantages. The severity of these methods could be reduced if the biomass was previously fungal pretreated. There is a synergic effect between both pretreatments, dealing with an enhanced sugar yield or lignin removal.

Yu et al. (2010) studied the effect of mild alkali treatment on corn stalks. They reported that the treatment with *I. lacteus* modified the lignin structure and significantly facilitated lignin degradation and xylan removal during mild alkali pretreatment. This synergy effect was found to be highly dependent on the severity of the chemical treatment. This effect was also observed by García-Torreiro et al. (2016b) with wheat straw, after the fungal pretreatment with *I. lacteus*. Applying an alkali treatment with lower severity (3.9) it was possible to reduce the enzyme dose in 33%, and to obtain the same sugars yield than with the harsher alkali treatment (5.5 of severity).

Similar results were found using other chemicals, such as acid (Gui et al., 2014), ethanol (Baba et al., 2011), or hydrogen peroxide (Yu et al., 2009). Besides the severity, the sequence of the treatments is also important. In all the previous examples, fungus was applied prior to the chemical treatment. The application of a chemical treatment before the fungal growth could present negative impact on lignin degradation (Kadimaliev et al., 2003). The chemical treatment of the biomass partially degraded the polysaccharides, thus, resulted in a higher sugars consumption and lower fungal lignin degradation.

More complex pretreatments allow the removal of lignin and hemicellulose, plus the disruption of the cellulose fraction, by abruptly changing the operating conditions, such as temperature and/or pressure, in the presence or not of a chemical compound. Namely, steam explosion, ammonia fiber explosion (AFEX), carbon dioxide explosion, and wet oxidation are the most known physicochemical pretreatments. Steam explosion can cause toxic compounds and condensation of acid insoluble lignin with other carbohydrates. These newly formed compounds can negatively interfere the enzymatic hydrolysis and fermentation (Sawada et al., 1995). Then, to obtain a better synergistic effect with the fungal pretreatment, slightly moderate conditions for the physicochemical pretreatment are required. The performance of the steam explosion prior to fungal treatment can break down lignin-hemicellulose bonds that may help the fungal action; however, the release of toxic compounds could have the opposite effect on fungal growth. Sawada et al. (1995) reported beneficial effects on sugar yields by the fungal treatment prior to steam explosion. While, Taniguchi et al. (2010) and Li and Chen (2014) reported as well beneficial effects of the fungal pretreatment when this is applied after steam explosion (Table 14.3).

In general terms, fungal treatment combined with physic, chemical, or physicochemical pretreatments enhance global performance in terms of sugar yield, digestibility, or fermentation yields. However, the severity of physical and

TABLE 14.3 Examples of the Application of Fungal Pretreatment Combined With Physical, Chemical, and Physico-Chemical Pretreatments to Different Crop Residues

Fungus	Substrate	Pretreatment	Results	References
PHYSICAL PRETREATMENTS				
Pleurotus ostreatus	Sugar cane bagasse	Milling	Higher lignocellulolytic enzyme production at 2.9 mm particle size. Test on different particle size: 0.92, 1.68, and 2.9 mm.	Membrillo et al. (2008)
P. ostreatus	Rapeseed	Grinding	Particle size between 0.42 and 0.84 allowed the highest lignin reduction (36%). With a particle size of 1 cm no lignin was reduced.	Sarikaya and Ladisch (1999)
C. subvermispora	Corn stover	Grinding	Lower lignin degradation was obtained with 15 mm (12.8–24.9%) compared to particle sizes between 5 and10 mm (19.5%–31.6%).	Wan and Li (2010)
CHEMICAL PRETREATMENTS				
Phanerochaete chrysosporium	*Glycyrrhiza uralensis* residue	Acetic acid 2 M	The reducing sugars were 1.62 times higher when biomass was cotreated than of solely acid pretreatment.	Gui et al. (2014)
I. lacteus	Corn stalks	1.5% NaOH, 30–75°C for 15–120 min	Fungal pretreatment improved delignification and xylan removal during alkali treatment.	Yu et al. (2010)
C. subvermispora	Japanese cedar wood	Ethanolysis	The combined treatment increased 7 times the sugar yield over the ethanolysis alone.	Baba et al. (2011)
P. ostreatus	Rice hull	H_2O_2	The combined pretreatment led to significant increases in lignin degradation and higher sugar yields.	Yu et al. (2009)
Panus tigrinus	Beech wood and pine sawdust	NH_4OH or Sulfuric acid Both at 2.5%, 165°C, 10 min	Both treatments applied prior fungus reduced lignin consumption.	Kadimaliev et al. (2003)
PHYSICOCHEMICAL PRETREATMENTS				
Phanerochaete chrysosporium	Beech wood meal	Steam explosion	The fungus treatment prior to steam explosion enhanced the saccharification.	Sawada et al. (1995)
P. ostreatus	Rice straw	Steam explosion	Steam explosion prior to fungal pretreatment decreased the treatment time from 60 to 36 days.	Taniguchi et al. (2010)
Phellinus baumii	Corn stalk	Steam explosion	The combined pretreatment improved the sugar yield 1.32 times.	Li and Chen (2014)

chemical pretreatment strongly influences the fungi performance; being optimal the milder conditions to favor the fungal pretreatment.

14.6.3 Coculture of WRF

The coculture of WRF has been proposed as a strategy to improve the fungal pretreatment of crop residues (Asiegbu et al., 1996; Ma and Ruan, 2015; Meehnian et al., 2017). Several studies showed that fungal consortia perform better and faster degradability of biomass when compared to monoculture. The complementation of the different ligninolytic activities provided by each WRF in the coculture has been suggested as an explanation of the improvement of coculture over monoculture.

Asiegbu et al. (1996) studied the delignification of spruce saw dust using *Phanerocheate chrysogenum*, *Tinea versicolor*, and *Pleurotus sajor-kaju*. When pure cultures were used, the delignification rate reached 0%–5%, while the consortium showed 16% of lignin removal.

In another study, Meehnian et al. (2017) observed an increase in the production of ligninolytic enzyme when *Daedalea flavida* and *Phlebia radiata* were cocultured during the pretreatment of cotton stalks in SsF. The pretreatment time was reduced to half in the coculture, and the glucose released after the enzymatic hydrolysis of 20-d pretreated cotton stalks was significantly higher than the glucose released by the WRF in monoculture.

14.6.4 Supplementation of Nutrients and Inducers of Ligninolytic Activity

Different types of supplements, including (1) ligninolytic enzymes inducers, such as salts ($MnSO_4$, $CuSO_4$, and $FeSO_4$) or phenolic compounds, (2) external carbon sources (glucose, malt extract, stillage), and (3) nitrogen sources (peptone, malt extract) have been evaluated for enhancing the fungal pretreatment of crop residues. Culture media, such as nutrients, buffers, and inducers play important roles in growth and ligninolytic enzyme production of WRF (Rüttimann-Johnson et al., 1993). The rationale behind this strategy is that increasing ligninolytic activity by providing exogenous inducers or nutrients the degradation of lignin by WRF is increased, as well as the digestibility of the fungal pretreated crop residues. The optimal combination of supplements used to enhance the production of multiple ligninolytic enzymes that act synergistically could give maximum lignin degradation and selectivity (Mishra et al., 2017).

Salvachúa et al. (2013) evaluated the effect of organic (peptone and wheat thin stillage) and inorganic ($MnSO_4$, $CuSO_4$, and $FeSO_4$) supplements on sugar yield of wheat straw during SsF of *I. lacteus*. Supplementation of SsF with 0.3 mM Mn(II) for 21 days resulted in glucose yields 10% and 106% higher than those obtained with the nonsupplemented and the untreated raw material, respectively. They correlated lignin loss, wheat straw digestibility, peroxidase activity, and fungal biomass with sugar yields. Mn(II) supplementation probably affects the induction of MnP, thus increasing the efficiency of the process (Camarero et al., 1996).

Mishra et al. (2017) used metallic salts ($MnSO_4$ and $CuSO_4$) and phenolic compounds as supplements in the pretreatment of sweet sorghum bagasse (SSB) by *Coriolus versicolor*. SsF supplemented with syringic acid increased the activities of LiP, AAO, and laccase; gallic acid increased MnP; and $CuSO_4$ increased laccase and polyphenol oxidase (PPO). The addition of these supplements improved lignin degradation and enhanced the enzymatic hydrolysis of pretreated SSB, yielding ~2.43 times the fermentable sugar obtained in the nonsupplemented SsF.

Wan and Li (2011) evaluated the supplementation of exogenous carbon and nitrogen sources on laccase activity and the delignification of wheat straw by *C. subvermispora* after 21 days of fermentation. The addition of glucose and malt extract increased the sugar yield of wheat straw by 10% over the control.

14.6.5 Alternative Strategies

The previous strategies were based on improving the performance of the fungal strain, but always considering as the final objective to replace one step of physicochemical pretreatment by an equivalent fungal-based pretreatment. However, a conceptually different approach is the integration of the fungal pretreatment in an alternative ethanol production scheme. Some strategies that take into account this approach are the consolidated bioprocessing, decentralized pretreatment, integration of process for obtaining valuable by-products, or a combination of some of them.

14.6.5.1 Consolidated Bioprocessing

Consolidated bioprocessing (CBP) is an alternative method that combines steps of biomass conversion to reduce the cost of ethanol. CBP combines enzyme production, polysaccharide hydrolysis, and sugar fermentation into one operation unit performed by one or several organisms (Lynd et al., 2005; Schuster and Chinn, 2013).

WRF can grow on plant biomass directly transforming lignocellulose into fermentable sugars, thus enabling CBP and single-step processes for ethanol production (Kamei et al., 2014; Mattila et al., 2017). Although fungal pretreatment and CBP comprise challenges to overcome, they hold yet unexplored potential in bioconversion and fermentation of agricultural and industrial waste lignocelluloses (Liguori and Faraco, 2016; Sindhu et al., 2016).

Based on the CBP concept, Kamei et al. (2014) applied the WRF *Phlebia* sp. MG-60 to the fermentation of high-solid loadings of unbleached hardwood kraft pulp (UHKP) without the addition of commercial cellulase. They obtained an ethanol production equivalent to 61.7% of the theoretical maximum.

The same WRF was applied to sugarcane bagasse through integrated fungal fermentation, involving a unified process for biological delignification with CBP (Kondo et al., 2014). After 4 weeks of aerobic incubation with *Phlebia* sp. MG-60, the lignin-to-glucan ratio decreased from 0.57 to 0.43, and then, after 10 days of semi-aerobic conditions an ethanol yield of 44% was achieved.

Ligninolytic fungi can degrade lignin at aerobic conditions and produce ethanol when cultivated at oxygen-limited or anaerobic conditions (Ali et al., 2012; Mattila et al., 2017; Shrestha et al., 2008). This characteristic has been evaluated in different studies to implement CBP configurations based on the use of these fungi. For example, Shrestha et al. (2008) developed a fungal process for ethanol production based on this approach. After two days of aerobic SsF of corn fiber with *P. chrysosporium*, the subsequent anaerobic static SmF for 6 days resulted in 1.7 g of ethanol/100 g of corn fiber.

14.6.5.2 Decentralized Pretreatment

Regional biomass processing depots (RBPD) have been proposed to address the logistical problems of a centralized biorefinery (Eranki et al., 2011). RBPD that pretreats and concentrates biomass before shipping it to a biorefinery for final processing into ethanol can potentially reduce logistical costs (Bals and Dale, 2012). In this production scheme, the pretreatment is geographically isolated from the biorefinery and is located near feedstock production areas (Kim and Dale, 2015). Moreover, RBPDs provides rural employment by establishing low capital biomass processing facilities owned by the community to process the feedstock produced by regional farmers (Kurian et al., 2013). To be implemented in a RBPD scheme, a pretreatment should be simple, have a low capital cost, simple catalyst recovery, create few if any harmful degradation products, and produce a stable intermediate that could be upgraded to valuable products (Eranki et al., 2011; Kurian et al., 2013).

Most of these characteristics of an ideal RBPD are met by applying fungal pretreatment.

Thompson et al. (2013) considered biological treatment as a good low-severity preconversion system for a depot setting, accounting that they are low cost, need low maintenance, and there is sufficient land area around the depot to perform the treatments. This approach has the additional advantage that the long fungal pretreatment time would not be considered as part of the ethanol processing time in the biorefinery, because the pretreatment would be previously carried out in the RBPD.

14.6.5.3 Integration of the Fungal Pretreatment for Obtaining Different By-products

The incorporation of fungal pretreatment in an integrated processing scheme may contribute to increasing the economical valorization of the crop residues. Typical by-products that are obtained during fungal pretreatment of lignocellulosic biomass are ligninolytic and cellulolytic enzymes. These enzymes could be subsequently used for the enzymatic hydrolysis of fermented solids (Chang and Webb, 2017) or used in vitro for different biotechnological applications.

One example of process integration is the enzyme-assisted ensiling of sweet sorghum (Tengerdy et al., 1996). In this process, freshly harvested sweet sorghum was ensiled with enzymes produced in situ on extracted pulp by SsF with the fungus *Gliocladium* sp. TUB-F-498. A small part of the fermented pulp (2%–3%) was used as an enzyme source, and the rest was valorized as enzyme-enriched animal feed. Applying this approach, the cost of the enzyme was reduced from 9.0 to 0.5US$ per MT of sweet sorghum.

Another interesting approach could be to integrate mushroom cultivation and fungal pretreatment. Wheat straw and other lignocellulosic materials are common substrates to cultivate edible mushrooms. Then, the spent mushroom substrate (SMS) is a potential substrate to produce biofuels. Worldwide production of mushrooms is greater than 6 Mt per year, which leads to an annual SMS production of almost 12 Mt (Rinker, 2002).

The most common use of SMS is as substrate for obtaining compost. The SMS is spread onto land and it remains there for several years. Consequently, salts and nitrates leach from the spent material with the subsequent environmental concerns. However, some efforts have been focused on valorizing SMS to produce fermentable sugars by enzymatic hydrolysis after the application of a pretreatment at soft conditions (Balan et al., 2008; Kapu et al., 2012; Teoh et al., 2014).

14.6.6 In Vitro Application of Fungal Ligninolytic Enzymes

Owing to the strong ability of ligninolytic enzymes to degrade lignin, they might be used in aqueous solutions for the in vitro pretreatment of lignocellulosic biomass. The main advantages of this strategy compared to in vivo pretreatment are (1) the substrate specificity, with no carbohydrate consumption, and (2) the possibility to increase the reaction rates and delignification efficiency, reducing the process time from weeks to hours (Moreno et al., 2015). However, since the full spectrum of effectors involved in the complete degradation of lignin remains unknown, their performance is not as good as that of solid-state fermentation of WRF (Tian et al., 2012).

Table 14.4 shows several studies reporting the enzymatic pretreatment of crops residues, mostly using crude ligninolytic enzymes, but also using concentrated or purified enzymes (Lu et al., 2010). The culture supernatants offer two main advantages over the use of purified enzymes: a more cost-efficient process and a higher stabilization of crude enzymes by proteins and other factors present in the medium (Call and Mücke, 1997).

In enzymatic reactions, the addition of mediators or cofactors is sometimes required. As, for

TABLE 14.4 Use of Fungal Ligninolytic Enzymes for Enzymatic Delignification of Crop Residues

Enzyme	Crop residue	Conditions of enzymatic pretreatment	Effect of the treatment	References
Laccase (*P. sanguineus*) + violuric acid	Extracted wheat straw and corn stover	66.6 U/L; 72 h; 28°C; 2 mM violuric acid	Klason lignin removal (86%–93%); TRS 1.9–2.2-fold higher than control	Lu et al. (2010)
Mixture of crude LE (*Pleurotus. florida*, *Pleur. sajor-caju*, *P. eryngii*)	Corn cob	10 days; 22°C; SLR: 1:2.8	5.2% lignin loss, also accompanied by 13.6% hemicellulose loss and 9.2% cellulose loss	Naraian et al. (2010)
Laccase (*Ganoderma lucidum*)	Alkali pretreated *Brassica campestris* straw	Not specified	Diameters and depths of holes on the surface larger than those with alkali alone; TRS 69% higher than with alkali alone	Yang et al. (2011)
Laccase (*Trametes versicolor*) + HBT	Ensiled corn stover	4000 U/g; 48 h; 40°C; SLR: 1:20; 5% HBT (w:w)	Lignin structural changes; 7% increased hydrolysis compared to control	Chen et al. (2012)
Crude LE (*Trametes hirsuta*)	Alkali pretreated corn straw	6 h RT, 5 g/70 mL	Porosity and surface area increased significantly; reducing sugar yields doubled	Li et al. (2012)
Laccase (*Sclerotium* sp.)	Steam-exploded wheat straw	0.55 U/g; simultaneous lac + cellulase; 48 h; 50°C; SLR: 1:20	Formation of micropores; reducing sugars increased 16.8% compared to control	Qiu and Chen (2012)
Crude LE (*Pleur. ostreatus*)	Sugarcane bagasse	48 h; 35°C; SLR: 1:6.7	33.6% delignification; 72.9% cellulose hydrolysis, higher than the obtained with alkali pretreatment	Asgher et al. (2013)
Crude LE (*P. chrysosporium* and *Coriolus versicolor*)	Corn stover	24 h; 45°C; SLR: 1:10	Lignin partially removed; hydrolysis increased 50% compared to control	Wang et al. (2013)
Laccase (*Pycn. cinnabarinus*) + synthetic mediators	Steam-exploded wheat straw	10 U/g; 3 h; 50°C; SLR: 1:20; 1.5% mediator (w:w)	Polymerization and grafting reactions instead of lignin degradation; lower glucose recoveries compared to control	Moreno et al. (2016)
Crude LE (*P. sanguineus*) + synthetic mediators + Mn(II)	Oil palm empty fruit bunch	3 days; 50°C; 10% substrate loading; 2 mM Mn(II); 1.5% HBT or 4% ABTS	4% removal of lignin; hydrolysis yield increased 17% compared to control	Zanirun et al. (2015)

ABTS, 2,2′-Azino-bis-(3-ethylbenzthiazoline-6-sulfonic acid); *HBT*, 1-hydroxybenzotriazole; *LE*, ligninolytic enzymes; *SLR*, solid to liquid ratio (w:v).

instance, synthetic mediators of laccase, such as ABTS, HBT, or violuric acid have been used to pretreat crop residues (Chen et al., 2012; Lu et al., 2010; Moreno et al., 2015), but also Mn(II) plays important roles in modification and partial removal of lignin (Zanirun et al., 2015).

In the in vitro pretreatment, the optimal pH and temperature can have wider ranges compared to in vivo cultures (pH 3–8 and 25–80°C) (Moreno et al., 2015). Special attention should be paid to enzymatic loading, which is quite variable from one study to another (from 0.55 to 4000 U/g substrate, Table 14.4), and solid:liquid rate (1:2.7–1:20, Table 14.4). Furthermore, the operational strategy should be carefully considered. In this sense, Qiu and Chen (2012) observed that the simultaneous laccase and cellulase treatment led to the highest cellulose conversion rate of steam-exploded wheat straw. However, Wang et al. (2013) concluded that ligninolytic enzymes could inhibit the activity of cellulase on corn stover and suggest removing the lignin-degrading enzyme from the reaction system after biopretreatment.

In some cases, ligninolytic enzymes had no significant effects on lignin removal, but enhanced subsequent enzymatic saccharification. Li et al. (2012) suggested as the possible mechanism, their complementary effects in processing time, detoxification of inhibitory compounds and lignocellulose depolymerization. Interestingly, Moreno et al. (2016) observed a lignin increase after the laccase treatment of steam-exploded wheat straw. This was ascribed to laccase-catalyzed polymerization and grafting reactions of lignin derived phenols. In contrast, this effect resulted in a reduction of soluble phenols content, and consequently promoted the detoxification of steam-exploded material, improving the ethanol yields.

14.7 FERMENTATION OF FUNGAL PRETREATED CROP RESIDUES

Several processes have been proposed for the biotechnological production of ethanol from lignocellulose (Fig. 14.5). They include separate hydrolysis and fermentation (SHF), simultaneous saccharification and fermentation (SSF) and consolidated bioprocessing (CBP) (Li et al., 2014). In the SHF the pretreated biomass is previously hydrolyzed using enzyme cocktails to obtain a stream rich in sugars that is used in a subsequent stage to produce ethanol by the action of microorganisms. In SSF configuration, saccharification of the pretreated lignocellulose is carried out simultaneously in the same reactor avoiding the inhibition of enzymes that occurs in SHF. In addition, the material cost is reduced

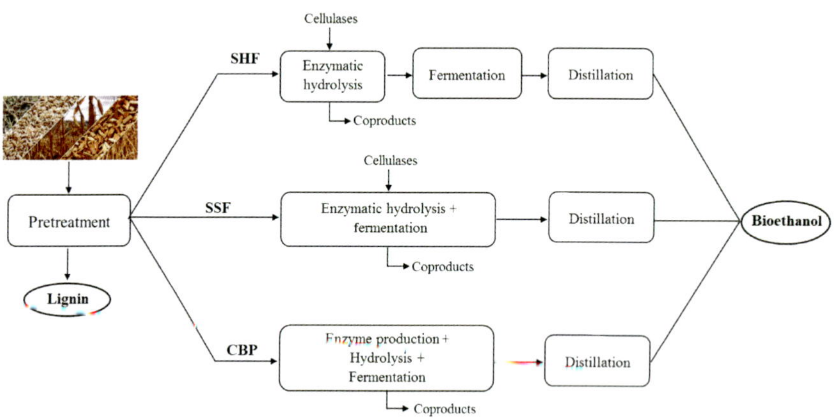

FIGURE 14.5 **Alternative systems for the saccharification and fermentation of fungal pretreated crop residues.**

since only one unit is required. This configuration has been applied successfully for bioethanol production from lignocellulosic feedstocks (Olofsson et al., 2008). In both configurations, it is possible to ferment the hemicellulosic sugars at the same time (Xu et al., 2009).

After the fungal pretreatment, the lignin content of LCB is reduced, and therefore the composition of cellulose and hemicellulose fraction is increased. Contrary to the application of other pretreatments like steam explosion, in which cellulose remain as a solid fraction, and most of the hemicellulose and lignin are separated in a liquid stream, after the fungal pretreatment the three components remain in the pretreated substrate. Consequently, a mixture of cellulases and hemicellulases is required to saccharify the biopretreated LCB into a stream rich in hexoses and pentoses, and free of inhibitory compounds (López-Abelairas et al., 2013b).

There is abundant literature reported on the ethanol production using biological pretreated feedstocks. A comprehensive review on different feedstocks, process configurations, microorganisms, and ethanol yields is presented in Table 14.5. Some of these studies are described later.

Bak et al. (2009) evaluated the ethanol production from rice straw pretreated by *P. chrysosporium*. The enzymatic digestibility after 5 days of biological pretreatment reached 64.9%. The ethanol production of the treated solid was performed via SSF using *Saccharomyces cerevisiae* D5A. After 24 h 9.5 g/L of ethanol was reached (58.2% of the maximum theoretical value).

In a related study, López-Abelairas et al. (2013a) examined the potential of *I. lacteus* to pretreat wheat straw and enhance the enzymatic saccharification as the previous step to produce bioethanol. As a working strategy, the authors combined a 14-day fungal pretreatment with a mild alkaline wash to improve the enzymatic hydrolysis. In the next stage, the hydrolysate was fermented with *S. cerevisiae* reaching an ethanol yield of 0.151 g/g straw (74% of the maximum theoretical yield). As a continuation of this work, the authors evaluated different process configurations and six microorganisms to produce ethanol (López-Abelairas et al., 2013b). The higher yield of ethanol (163 mg/g of raw wheat straw) was achieved using sequential pentoses and hexoses processing (H5-SSF) and the yeast *Pachysolen tannophilus*.

Wan and Li (2010) evaluated the feasibility of *C. subvermispora* to degrade lignin from corn stover and improve the enzymatic digestibility. After 35 days of biological pretreatment, they studied the ethanol production by *S. cerevisiae* through SSF obtaining an ethanol yield of 57.8%.

14.8 CONCLUDING REMARKS

To be considered as a feasible alternative, the fungal pretreatment needs to be evaluated from both economic and environmental perspectives. In a recent study, Baral and Shah (2017) evaluated the cost of four different pretreatments, compared them technoeconomically, and concluded that the cost of sugar yield obtained by applying the biological pretreatment was the highest. However, this analysis was based on process modeling that used as input experimental data obtained at small lab scale, as most of the studies presented in this chapter.

One aspect that needs to be carried out to evaluate more precisely the feasibility of the fungal pretreatment is the scale-up of the process from small flasks to bench or pilot bioreactors. This will require solving another important aspect of fungal pretreatment, such as the culture of WRF at nonaseptic conditions. Biomass decontamination prior to fungus inoculation is a challenge in biological pretreatment (Salvachúa et al., 2011; da Silva Machado and Ferraz, 2017), despite some studies have reported successful biopretreatments under nonaseptic conditions (Thomsen et al., 2016; Vasco-Correa et al., 2016).

Finally, the environmental impacts of the fungal pretreatments need to be evaluated. One of

TABLE 14.5 Fermentation of Fungal Pretreated Crop Residues Using Different Configurations

Raw material	Microorganism used in the fungal pretreatment	Saccharification and fermentation strategy	Fermentative microorganism	EtOH yield	References
Corn fiber	*Gloeophyllum trabeum*	SHF	*Saccharomyces cerevisiae*	0.04 g/g fiber	Rasmussen et al. (2010)
Corn fiber	*G. trabeum*	CBP	*G. trabeum*	0.033 g/g fiber	Rasmussen et al. (2010)
Corn stover	*I. lacteus* (with manganese supplement)	SSF	*S. cerevisiae*	0.144 g/g corn stover	Song et al. (2013)
Corn stover	Mild NaOH + *Phanerochaete chrysosporium*	SSF	*Escherichia coli* K011	0.05 g/g corn stover	Vincent et al. (2014)
Corn stover	Mild NaOH + *G. trabeum*	SSF	*E. coli* K011	0.067 g/g corn stover	Vincent et al. (2014)
Corn stover	Mild NaOH + *P. chrysosporium*	SSF	*S. cerevisiae*	0.029 g/g corn stover	Vincent et al. (2014)
Corn stover	Mild NaOH + *G. trabeum*	SSF	*S. cerevisiae*	0.031 g/g corn stover	Vincent et al. (2014)
Corn stover	*Phlebia brevispora*	SSF	Recombinant *E. coli* strain FBR5	0.26 g/pretreated corn stover	Saha et al. (2017)
Corn stover	*P. brevispora*	SSF	Industrial *S. cerevisiae* D5A	0.16 g/pretreated corn stover	Saha et al. (2017)
Corn stover	*P. brevispora*	SSF	Recombinant *S. cerevisiae* YHR400	0.19 g/pretreated corn stover	Saha et al. (2017)
Rice straw	*Dichomitus squalens*	SSF	*S. cerevisiae* D5A	0.30 g/of glucan	Bak et al. (2010)
Sorghum bagasse	*Neurospora crassa* DSM 1129	SSF	*Neur. crassa* DSM 1129	0.047 g/g of sorghum bagasse	Dogaris et al. (2009)
Sugarcane bagasse (SCB)	*C. subvermispora* + microwave hydrothermolysis	SSF	*S. cerevisiae* NBRC 0233	0.31 g/g pretreated SCB	Sasaki et al. (2011)
Wheat straw	*I. lacteus*	SHF	*S. cerevisiae*	0.097 g/g of wheat straw	López-Abelairas et al. (2013a)
Wheat straw	Ensiling with *Lactobacillus buchneri* + *C. subvermispora*	SSF	*S. cerevisiae*	0.131 g/g wheat straw	Thomsen et al. (2016)

the advantages most frequently argued in favor of fungal pretreatment is its low environmental impact. However, this point still needs to be evaluated using standard tools, such as life cycle assessment, which considers all the factors involved in the process.

The application of the strategies presented in this chapter, alone or combined, may be a starting point for improving the performance of the fungal pretreatment to integrate it in a feasible way into an ethanol biorefinery based on the use of crop residues.

Acknowledgments

This research was supported by the Spanish Government (Ministry of Economy and Competitiveness) through the 2G-Enzymes project (PCIN-2015-031). B. Gullón thanks the postdoctoral program *Juan de la Cierva* from the Spanish Government (MINECO) for the economic support provided during the development of this work (Grant Ref. IJCI-2015-25305). The authors belong to the Galician Competitive Research Group GRC2013-032 and to the CRETUS Strategic Partnership (AGRUP2015/02). All these programs are cofunded by FEDER (EU).

References

Alcalde, M., 2015. Engineering the ligninolytic enzyme consortium. Trends Biotechnol. 33 (3), 155–162.

Ali, S.S., Khan, M., Fagan, B., Mullins, E., Doohan, F.M., 2012. Exploiting the inter-strain divergence of *Fusarium oxysporum* for microbial bioprocessing of lignocellulose to bioethanol. AMB Express 2 (1), 16.

Alvira, P., Tomás-Pejó, E., Ballesteros, M., Negro, M.J., 2010. Pretreatment technologies for an efficient bioethanol production process based on enzymatic hydrolysis: a review. Bioresour. Technol. 101, 4851–4861.

Asgher, M., Ahmad, Z., Iqbal, H.M.N., 2013. Alkali and enzymatic delignification of sugarcane bagasse to expose cellulose polymers for saccharification and bio-ethanol production. Ind. Crops Prod. 44, 488–495.

Asiegbu, F.O., Paterson, A., Smith, J.E., 1996. The effects of co-fungal cultures and supplementation with carbohydrate adjuncts on lignin biodegradation and substrate digestibility. World J. Microbiol. Biotechnol. 12 (3), 273–279.

Baba, Y., Tanabe, T., Shirai, N., Watanabe, T., Honda, Y., Watanabe, T., 2011. Pretreatment of Japanese cedar wood by white rot fungi and ethanolysis for bioethanol production. Biomass Bioenergy 35 (1), 320–324.

Bak, J.S., Kim, M.D., Choi, I.G., Kim, K.H., 2010. Biological pretreatment of rice straw by fermenting with *Dichomitus squalens*. New Biotechnol. 27 (4), 424–434.

Bak, J.S., Ko, J.K., Choi, I.G., Park, Y.C., Seo, J.H., Kim, K.H., 2009. Fungal pretreatment of lignocellulose by *Phanerochaete chrysosporium* to produce ethanol from rice straw. Biotechnol. Bioeng. 104 (3), 471–482.

Balan, V., da Costa Sousa, L., Chundawat, S.P., Vismeh, R., Jones, A.D., Dale, B.E., 2008. Mushroom spent straw: a potential substrate for an ethanol-based biorefinery. J. Ind. Microbiol. Biotechnol. 35 (5), 293–301.

Baldrian, P., 2008. Enzymes of saprotrophic basidiomycetes. British Mycological Society Symposia Series, 28, Academic Press, Oxford, UK, pp. 19–41.

Bals, B.D., Dale, B.E., 2012. Developing a model for assessing biomass processing technologies within a local biomass processing depot. Bioresour. Technol. 106, 161–169.

Baral, N.R., Shah, A., 2017. Comparative techno-economic analysis of steam explosion, dilute sulfuric acid, ammonia fiber explosion and biological pretreatments of corn stover. Bioresour. Technol. 232, 331–343.

Call, H.P., Mücke, I., 1997. History, overview and applications of mediated lignolytic systems, especially laccase-mediator-systems (Lignozym-process). J. Biotechnol. 53 (2), 163–202.

Camarero, S., 1995. Estudio de la deslignificación de la paja de trigo con *Pleurotus* para la fabricación de papel (Doctoral dissertation). Facultad de Ciencia Biológicas. Universidad Complutense, Madrid (Spain).

Camarero, S., Bockle, B., Martinez, M.J., Martinez, A.T., 1996. Manganese-mediated lignin degradation by *Pleurotus pulmonarius*. Appl. Environ. Microbiol. 62 (3), 1070–1072.

Camarero, S., Martínez, M.J., Martínez, A.T., 2014. Understanding lignin biodegradation for the improved utilization of plant biomass in modern biorefineries. Biofuel. Bioprod. Biorefin. 8 (5), 615–625.

Canam, T., Town, J., Iroba, K., Tabil, L., Dumonceaux, T., 2013. Pretreatment of lignocellulosic biomass using microorganisms: Approaches, advantages, and limitations. InTech, 181–206. doi: 10.5772/55088.

Chang, C.W., Webb, C., 2017. Production of a generic microbial feedstock for lignocellulose biorefineries through sequential bioprocessing. Bioresour. Technol. 227, 35–43.

Chaudhari, Y.B., Talukdar, N.C., Adhikary, N.C., Kalita, M.C., Khan, M.R., 2015. Rice straw based evaluation of lignolytic and cellulolytic capabilities of novel strains of saprophytic fungi from Indo-Burma biodiversity hotspot. Energy Fuels 29 (2), 784–792.

Chen, Q., Marshall, M.N., Geib, S.M., Tien, M., Richard, T.L., 2012. Effects of laccase on lignin depolymerization and enzymatic hydrolysis of ensiled corn stover. Bioresour. Technol. 117, 186–192.

Cianchetta, S., Di Maggio, B., Burzi, P.L., Galletti, S., 2014. Evaluation of selected white-rot fungal isolates for improving the sugar yield from wheat straw. Appl. Biochem. Biotechnol. 173 (2), 609–623.

da Silva Machado, A., Ferraz, A., 2017. Biological pretreatment of sugarcane bagasse with basidiomycetes producing varied patterns of biodegradation. Bioresour. Technol. 225, 17–22.

Dashtban, M., Schraft, H., Syed, T.A., Qin, W., 2010. Fungal biodegradation and enzymatic modification of lignin. Int. J. Biochem. Mol. Biol. 1 (1), 36.

Dogaris, I., Vakontios, G., Kalogeris, E., Mamma, D., Kekos, D., 2009. Induction of cellulases and hemicellulases from *Neurospora crassa* under solid-state cultivation for bioconversion of sorghum bagasse into ethanol. Ind. Crops Prod. 29 (2), 404–411.

Du, W., Yu, H., Song, L., Zhang, J., Weng, C., Ma, F., Zhang, X., 2011. The promoting effect of byproducts from *Irpex lacteus* on subsequent enzymatic hydrolysis of bio-pretreated cornstalks. Biotechnol. Biofuel. 4 (1), 37.

Eggert, C., Temp, U., Dean, J.F., Eriksson, K.E.L., 1996. A fungal metabolite mediates degradation of non-phenolic lignin structures and synthetic lignin by laccase. FEBS Lett. 391 (1-2), 144–148.

Eranki, P.L., Bals, B.D., Dale, B.E., 2011. Advanced regional biomass processing depots: a key to the logistical challenges of the cellulosic biofuel industry. Biofuel. Bioprod. Biorefin. 5 (6), 621–630.

FAOSTAT, 2015. FAO Statistical Pocketbook. Available from: http://www.fao.org/3/a-i4691e.pdf

Floudas, D., Binder, M., Riley, R., Barry, K., Blanchette, R.A., Henrissat, B., Martínez, A.T., Otillar, R., Spatafora, J.W., Yadav, J.S., Aerts, A., 2012. The Paleozoic origin of enzymatic lignin decomposition reconstructed from 31 fungal genomes. Science 336 (6089), 1715–1719.

Gao, W., Tabil, L.G., Dumonceaux, T., Ríos, S.E., Zhao, R., 2017. Optimization of biological pretreatment to enhance the quality of wheat straw pellets. Biomass Bioenergy 97, 77–89.

García-Torreiro, M., López-Abelairas, M., Lu-Chau, T.A., Lema, J.M., 2016a. Fungal pretreatment of agricultural residues for bioethanol production. Ind. Crops Prod. 89, 486–492.

García-Torreiro, M., Pallín, M.Á., López-Abelairas, M., Lu-Chau, T.A., Lema, J.M., 2016b. Alkali treatment of fungal pretreated wheat straw for bioethanol production. Bioethanol 2 (1), 32–43.

Gómez Vanegas, N.A., 2016. Aprovechamiento del subproducto lignocelulósico raquis, procedente de la extracción de aceite de palma para la producción de bioetanol. Doctoral Thesis. Universidad de Antioquia, Colombia.

Gui, X., Wang, G., Li, X., Yan, Y., 2014. Fungus-assisted mild acid pretreatment of *Glycyrrhiza uralensis* residues to enhance enzymatic hydrolysis and oil production by green microalga *Chlorella protothecoides*. Ind. Crops Prod. 62, 466–473.

Gupta, A., Verma, J.P., 2015. Sustainable bio-ethanol production from agro-residues: a review. Renew. Sustain. Energy Rev. 41, 550–567.

Hatakka, A., Hammel, K.E., 2010. In: The Mycota, X., Hofrichter, M. (Eds.), Fungal Biodegradation of Lignocelluloses Industrial Applications. Springer-Verlag, Berlin Heidelberg, Germany.

Hatakka, A.I., 1983. Pretreatment of wheat straw by white-rot fungi for enzymic saccharification of cellulose. Appl. Microbiol. Biotechnol. 18 (6), 350–357.

Heidorne, F.O., Magalhães, P.O., Ferraz, A.L., Milagres, A.M., 2006. Characterization of hemicellulases and cellulases produced by *Ceriporiopsis subvermispora* grown on wood under biopulping conditions. Enzyme Microb. Technol. 38 (3), 436–442.

Hölker, U., Lenz, J., 2005. Solid-state fermentation–are there any biotechnological advantages? Curr. Opin. Microbiol. 8 (3), 301–306.

Kadimaliev, D.A., Revin, V.V., Atykyan, N.A., Samuilov, V.D., 2003. Effect of wood modification on lignin consumption and synthesis of lignolytic enzymes by the fungus *Panus (Lentinus) tigrinus*. Appl. Biochem. Microbiol. 39 (5), 488–492.

Kamei, I., Hirota, Y., Meguro, S., 2014. Direct fungal production of ethanol from high-solids pulps by the ethanol-fermenting white-rot fungus *Phlebia* sp MG-60. BioResources 9 (3), 5114–5124.

Kamra, D.N., Zadražil, F., 1986. Influence of gaseous phase, light and substrate pretreatment on fruit-body formation, lignin degradation and in vitro digestibility of wheat straw fermented with *Pleurotus* spp. Agr. Wastes 18 (1), 1–17.

Kapu, N.U.S., Manning, M., Hurley, T.B., Voigt, J., Cosgrove, D.J., Romaine, C.P., 2012. Surfactant-assisted pretreatment and enzymatic hydrolysis of spent mushroom compost for the production of sugars. Bioresour. Technol. 114, 399–405.

Kersten, P., Cullen, D., 2007. Extracellular oxidative systems of the lignin-degrading basidiomycete *Phanerochaete chrysosporium*. Fungal Genet. Biol. 44 (2), 77–87.

Kim, S., Dale, B.E., 2015. Comparing alternative cellulosic biomass biorefining systems: centralized versus distributed processing systems. Biomass Bioenergy 74, 135–147.

Kim, S., Park, J.M., Seo, J.W., Kim, C.H., 2012. Sequential acid-/alkali-pretreatment of empty palm fruit bunch fiber. Bioresour. Technol. 109, 229–233.

Kondo, R., De Leon, R., Anh, T.K., Meguro, S., Shimizu, K., Kamei, I., 2014. Effect of chemical factors on integrated fungal fermentation of sugarcane bagasse for ethanol production by a white-rot fungus *Phlebia* sp. MG-60. Bioresour. Technol. 167, 33–40.

Kurian, J.K., Nair, C.R., Hussain, A., Raghavan, G.V., 2013. Feedstocks, logistics and pre-treatment processes for sustainable lignocellulosic biorefineries: a comprehensive review. Renew. Sustain. Energy Rev. 25, 205–219.

Li, G., Chen, H., 2014. Synergistic mechanism of steam explosion combined with fungal treatment by *Phellinus baumii* for the pretreatment of corn stalk. Biomass Bioenergy 67, 1–7.

Li, J., Lin, J., Zhou, P., Wu, K., Liu, H., Xiong, C., Gong, Y., Xiao, W., Liu, Z., 2014. One-pot simultaneous saccharification and fermentation: a preliminary study of a novel configuration for cellulosic ethanol production. Bioresour. Technol. 161, 171–178.

Li, J., Sun, F., Li, X., Yan, Z., Yuan, Y., Liu, X., 2012. Enhanced saccharification of corn straw pretreated by alkali combining crude ligninolytic enzymes. J. Chem. Technol. Biotechnol. 87 (12), 1687–1693.

Liguori, R., Faraco, V., 2016. Biological processes for advancing lignocellulosic waste biorefinery by advocating circular economy. Bioresour. Technol. 215, 13–20.

Liu, J., Sidhu, S.S., Wang, M.L., Tonnis, B., Habteselassie, M., Mao, J., Huang, Q., 2015. Evaluation of various fungal pretreatment of switchgrass for enhanced saccharification and simultaneous enzyme production. J. Cleaner Prod. 104, 480–488.

López-Abelairas, M., Lu-Chau, T.A., Lema, J.M., 2013b. Fermentation of biologically pretreated wheat straw for ethanol production: comparison of fermentative microorganisms and process configurations. Appl. Biochem. Biotechnol. 170 (8), 1838–1852.

López-Abelairas, M., Pallín, M.A., Salvachúa, D., Lú-Chau, T., Martínez, M.J., Lema, J.M., 2013a. Optimisation of the biological pretreatment of wheat straw with white-rot fungi for ethanol production. Bioprocess Biosyst. Eng. 36 (9), 1251–1260.

López-Abelairas, M., 2014. Bioprocesses for production of Fuels. Chemicals and Polymers From Lignocellulosic Materials. University of Santiago de Compostela, Spain, Doctoral thesis, Santiago de Compostela.

Lu, C., Wang, H., Luo, Y., Guo, L., 2010. An efficient system for pre-delignification of gramineous biofuel feedstock in vitro: application of a laccase from *Pycnoporus sanguineus* H275. Process Biochem. 45 (7), 1141–1147.

Lynd, L.R., Van Zyl, W.H., McBride, J.E., Laser, M., 2005. Consolidated bioprocessing of cellulosic biomass: an update. Curr. Opin. Biotechnol. 16 (5), 577–583.

Ma, K., Ruan, Z., 2015. Production of a lignocellulolytic enzyme system for simultaneous bio-delignification and saccharification of corn stover employing co-culture of fungi. Bioresour. Technol. 175, 586–593.

Martín-Sampedro, R., Fillat, Ú., Ibarra, D., Eugenio, M.E., 2015. Use of new endophytic fungi as pretreatment to enhance enzymatic saccharification of *Eucalyptus globulus*. Bioresour. Technol. 196, 383–390.

Martínez, A.T., 2002. Molecular biology and structure-function of lignin-degrading heme peroxidases. Enzyme Microb. Technol. 30 (4), 425–444.

Martínez, A.T., Ruiz-Dueñas, F.J., Martínez, M.J., del Río, J.C., Gutierrez, A., 2009. Enzymatic delignification of plant cell wall: from nature to mill. Curr. Opin. Biotechnol. 20 (3), 348–357.

Martínez, A.T., Speranza, M., Ruiz-Dueñas, F.J., Ferreira, P., Camarero, S., Guillén, F., Martínez, M.J., Gutiérrez, A., Del Rio, J.C., 2005. Biodegradation of lignocellulosics: microbial, chemical, and enzymatic aspects of the fungal attack of lignin. Int. Microbiol. 8, 195–204.

Mattila, H., Kuuskeri, J., Lundell, T., 2017. Single-step, single-organism bioethanol production and bioconversion of lignocellulose waste materials by phlebioid fungal species. Bioresour. Technol. 225, 254–261.

Meehnian, H., Jana, A.K., 2016. Cotton stalk pretreatment using *Daedalea flavida*, *Phlebia radiata*, and *Flavodon flavus*: lignin degradation, cellulose recovery, and enzymatic saccharification. Appl. Biochem. Biotechnol. 181, 1–20.

Meehnian, H., Jana, A.K., Jana, M.M., 2017. Pretreatment of cotton stalks by synergistic interaction of *Daedalea flavida* and *Phlebia radiata* in co-culture for improvement in delignification and saccharification. Int. Biodeterior. Biodegrad. 117, 68–77.

Membrillo, I., Sánchez, C., Meneses, M., Favela, E., Loera, O., 2008. Effect of substrate particle size and additional nitrogen source on production of lignocellulolytic enzymes by *Pleurotus ostreatus* strains. Bioresour. Technol. 99 (16), 7842–7847.

Menon, V., Rao, M., 2012. Trends in bioconversion of lignocellulose: biofuels, platform chemicals and biorefinery concept. Prog. Energy Combust. Sci. 38 (4), 522–550.

Millati, I.R., Syamsiah, S., Niklasson, C., Cahyanto, M.N., Ludquist, K., Taherzadeh, M.J., 2011. Biological pretreatment of lignocelluloses with white-rot fungi and its applications: a review. BioResources 6 (4), 5224–5259.

Mishra, V., Jana, A.K., Jana, M.M., Gupta, A., 2017. Enhancement in multiple lignolytic enzymes production for optimized lignin degradation and selectivity in fungal pretreatment of sweet sorghum bagasse. Bioresour. Technol. 236, 49–59.

Mohseni, S., Najafpour, G.D., Vaseghi, Z., Mahjoub, S., 2012. Solid state fermentation of agricultural residues for lipase production in a tray-bioreactor. World Appl. Sci. J. 16 (7), 1034–1039.

Moreno, A.D., Ibarra, D., Alvira, P., Tomás-Pejó, E., Ballesteros, M., 2015. A review of biological delignification and detoxification methods for lignocellulosic bioethanol production. Crit. Rev. Biotechnol. 35 (3), 342–354.

Moreno, A.D., Ibarra, D., Alvira, P., Tomás Pejó, E., Ballesteros, M., 2016. Exploring laccase and mediators behavior during saccharification and fermentation of steam exploded wheat straw for bioethanol production. J. Chem. Technol. Biotechnol. 91 (6), 1816–1825.

Naraian, R., Singh, D., Verma, A., Garg, S.K., 2010. Studies on in vitro degradability of mixed crude enzyme extracts produced from *Pleurotus* spp. J. Environ. Biol. 31 (6), 945–951.

Olofsson, K., Bertilsson, M., Lidén, G., 2008. A short review on SSF: an interesting process option for ethanol production from lignocellulosic feedstocks. Biotechnol. Biofuel. 1 (1), 7.

Papadopoulou, K., Dimitropoulos, V., Rigas, F., 2015. Assessment of *Pleurotus ostreatus* mediated degradation of agro residues by using design of experiments methodologies. Environ. Prog. Sustain. Energy 34 (6), 1705–1713.

Papagianni, M., 2004. Fungal morphology and metabolite production in submerged mycelial processes. Biotechnol. Adv. 22 (3), 189–259.

Piñeros-Castro, Y., Velásquez-Lozano, M., 2014. Biodegradation kinetics of oil palm empty fruit bunches by white rot fungi. Int. Biodeterior. Biodegrad. 91, 24–28.

Pitol, L.O., Biz, A., Mallmann, E., Krieger, N., Mitchell, D.A., 2016. Production of pectinases by solid-state fermentation in a pilot-scale packed-bed bioreactor. Chem. Eng. J. 283, 1009–1018.

Planinić, M., Zelić, B., Čubel, I., Bucić-Kojić, A., Tišma, M., 2016. Corn forage biological pretreatment by *Trametes versicolor* in a tray bioreactor. Waste Manage. Res. 34 (8), 802–809.

Qiu, W., Chen, H., 2012. Enhanced the enzymatic hydrolysis efficiency of wheat straw after combined steam explosion and laccase pretreatment. Bioresour. Technol. 118, 8–12.

Rajala, T., Peltoniemi, M., Pennanen, T., Mäkipää, R., 2010. Relationship between wood-inhabiting fungi determined by molecular analysis (denaturing gradient gel electrophoresis) and quality of decaying logs. Can. J. For. Res. 40 (12), 2384–2397.

Rasmussen, M.L., Shrestha, P., Khanal, S.K., Pometto, A.L., van Leeuwen, J.H., 2010. Sequential saccharification of corn fiber and ethanol production by the brown rot fungus *Gloeophyllum trabeum*. Bioresour. Technol. 101 (10), 3526–3533.

Rinker, D.L., 2002. Handling and using "spent" mushroom substrate around the world. Mushroom Biology and Mushroom Products. UAEM, Cuernavaca, Mexico.

Rouches, E., Herpoël-Gimbert, I., Steyer, J.P., Carrere, H., 2016. Improvement of anaerobic degradation by white-rot fungi pretreatment of lignocellulosic biomass: a review. Renew. Sustain. Energy Rev. 59, 179–198.

Rüttimann-Johnson, C., Salas, L., Vicuña, R., Kirk, T.K., 1993. Extracellular enzyme production and synthetic lignin mineralization by *Ceriporiopsis subvermispora*. Appl. Environ. Microbiol. 59 (6), 1792–1797.

Saha, B.C., Kennedy, G.J., Qureshi, N., Cotta, M.A., 2017. Biological pretreatment of corn stover with *Phlebia brevispora* NRRL 13108 for enhanced enzymatic hydrolysis and efficient ethanol production. Biotechnol. Prog. 33 (2), 365–374.

Saha, B.C., Qureshi, N., Kennedy, G.J., Cotta, M.A., 2016. Biological pretreatment of corn stover with white-rot fungus for improved enzymatic hydrolysis. Int. Biodeterior. Biodegrad. 109, 29–35.

Salvachúa, D., Prieto, A., López-Abelairas, M., Lu-Chau, T., Martínez, Á.T., Martínez, M.J., 2011. Fungal pretreatment: an alternative in second-generation ethanol from wheat straw. Bioresour. Technol. 102 (16), 7500–7506.

Salvachúa, D., Prieto, A., Vaquero, M.E., Martínez, Á.T., Martínez, M.J., 2013. Sugar recoveries from wheat straw following treatments with the fungus *Irpex lacteus*. Bioresour. Technol. 131, 218–225.

Sánchez, C., 2009. Lignocellulosic residues: biodegradation and bioconversion by fungi. Biotechnol. Adv. 27 (2), 185–194.

Sarikaya, A., Ladisch, M.R., 1999. Solid-state fermentation of lignocellulosic plant residues from *Brassica napus* by *Pleurotus ostreatus*. Appl. Biochem. Biotechnol. 82 (1), 1–15.

Sarkar, N., Ghosh, S.K., Bannerjee, S., Aikat, K., 2012. Bioethanol production from agricultural wastes: an overview. Renew. Energy 37 (1), 19–27.

Sasaki, C., Takada, R., Watanabe, T., Honda, Y., Karita, S., Nakamura, Y., Watanabe, T., 2011. Surface carbohydrate analysis and bioethanol production of sugarcane bagasse pretreated with the white rot fungus *Ceriporiopsis subvermispora* and microwave hydrothermolysis. Bioresour. Technol. 102 (21), 9942–9946.

Sawada, T., Nakamura, Y., Kobayashi, F., Kuwahara, M., Watanabe, T., 1995. Effects of fungal pretreatment and steam explosion pretreatment on enzymatic saccharification of plant biomass. Biotechnol. Bioeng. 48 (6), 719–724.

Schuster, B.G., Chinn, M.S., 2013. Consolidated bioprocessing of lignocellulosic feedstocks for ethanol fuel production. BioEnergy Res. 6 (2), 416–435.

Shi, J., Chinn, M.S., Sharma-Shivappa, R.R., 2008. Microbial pretreatment of cotton stalks by solid state cultivation of *Phanerochaete chrysosporium*. Bioresour. Technol. 99 (14), 6556–6564.

Shirkavand, E., Baroutian, S., Gapes, D.J., Young, B.R., 2016. Combination of fungal and physic chemical processes for lignocellulosic biomass pretreatment : a review. Renew. Sustain. Energy Rev. 54, 217–234.

Shrestha, P., Rasmussen, M., Khanal, S.K., Pometto Iii, A.L., van Leeuwen, J., 2008. Solid-substrate fermentation of corn fiber by *Phanerochaete chrysosporium* and subsequent fermentation of hydrolysate into ethanol. J. Agr. Food Chem. 56 (11), 3918–3924.

Sindhu, R., Binod, P., Pandey, A., 2016. Biological pretreatment of lignocellulosic biomass: an overview. Bioresour. Technol. 199, 76–82.

Singh, P., Sulaiman, O., Hashim, R., Rupani, P.F., Peng, L.C., 2010. Biopulping of lignocellulosic material using different fungal species: a review. Rev. Environ. Sci. Biotechnol. 9 (2), 141–151.

Slade, R., Bauen, A., Shah, N., 2009. The commercial performance of cellulosic ethanol supply-chains in Europe. Biotechnol. Biofuel. 2 (1), 3.

Song, L., Ma, F., Zeng, Y., Zhang, X., Yu, H., 2013. The promoting effects of manganese on biological pretreatment with *Irpex lacteus* and enzymatic hydrolysis of corn stover. Bioresour. Technol. 135, 89–92.

Sun, Y., Cheng, J., 2002. Hydrolysis of lignocellulosic materials for ethanol production: a review. Bioresour. Technol. 83 (1), 1–11.

Taniguchi, M., Takahashi, D., Watanabe, D., Sakai, K., Hoshino, K., Kouya, T., Tanaka, T., 2010. Effect of steam explosion pretreatment on treatment with *Pleurotus ostreatus* for the enzymatic hydrolysis of rice straw. J. Biosci. Bioeng. 110 (4), 449–452.

Tengerdy, R.P., Szakacs, G., 2003. Bioconversion of lignocellulose in solid substrate fermentation. Biochem. Eng. J. 13 (2), 169–179.

Tengerdy, R.P., Szakacs, G., Sipocz, J., 1996. Bioprocessing of sweet sorghum within situ-produced enzymes. Appl. Biochem. Biotechnol. 57 (1), 563–569.

Teoh, T.C., Soo, C.S., Heidelberg, T., Zainudin, A., 2014. Bioethanol from spent mushroom sawdust waste by perchloric acid: easy sugars separation and hydrolysis without charring. Chiang Mai J. Sci. 41 (3), 513–523.

Thompson, D.N., Campbell, T., Bals, B., Runge, T., Teymouri, F., Ovard, L.P., 2013. Chemical preconversion: application of low-severity pretreatment chemistries for commoditization of lignocellulosic feedstock. Biofuels 4 (3), 323–340.

Thomsen, S.T., Londoño, J.E., Ambye-Jensen, M., Heiske, S., Kádár, Z., Meyer, A.S., 2016. Combination of ensiling and fungal delignification as effective wheat straw pretreatment. Biotechnol. Biofuel. 9 (1), 16.

Tian, X.F., Fang, Z., Guo, F., 2012. Impact and prospective of fungal pre-treatment of lignocellulosic biomass for enzymatic hydrolysis. Biofuel. Bioprod. Biorefin. 6 (3), 335–350.

Valmaseda, M., Almendros, G., Martínez, A.T., 1990. Substrate-dependent degradation patterns in the decay of wheat straw and beech wood by ligninolytic fungi. Appl. Microbiol. Biotechnol. 33 (4), 481–484.

van Kuijk, S.J., Sonnenberg, A.S., Baars, J.J., Hendriks, W.H., Cone, J.W., 2015a. Fungal treatment of lignocellulosic biomass: importance of fungal species, colonization and time on chemical composition and in vitro rumen degradability. Anim. Feed Sci. Technol. 209, 40–50.

van Kuijk, S.J.A., Sonnenberg, A.S.M., Baars, J.J.P., Hendriks, W.H., Cone, J.W., 2015b. Fungal treated lignocellulosic biomass as ruminant feed ingredient: a review. Biotechnol. Adv. 33, 191–202.

Vasco-Correa, J., Ge, X., Li, Y., 2016. Fungal pretreatment of non-sterile miscanthus for enhanced enzymatic hydrolysis. Bioresour. Technol. 203, 118–123.

Vincent, M., Pometto, A.L., van Leeuwen, J.H., 2014. Ethanol production via simultaneous saccharification and fermentation of sodium hydroxide treated corn stover using *Phanerochaete chrysosporium* and *Gloeophyllum trabeum*. Bioresour. Technol. 158, 1–6.

Wan, C., Li, Y., 2010. Microbial pretreatment of corn stover with *Ceriporiopsis subvermispora* for enzymatic hydrolysis and ethanol production. Bioresour. Technol. 101 (16), 6398–6403.

Wan, C., Li, Y., 2011. Effectiveness of microbial pretreatment by *Ceriporiopsis subvermispora* on different biomass feedstocks. Bioresour. Technol. 102 (16), 7507–7512.

Wan, C., Li, Y., 2012. Fungal pretreatment of lignocellulosic biomass. Biotechnol. Adv. 30 (6), 1447–1457.

Wang, F.Q., Xie, H., Chen, W., Wang, E.T., Du, F.G., Song, A.D., 2013. Biological pretreatment of corn stover with ligninolytic enzyme for high efficient enzymatic hydrolysis. Bioresour. Technol. 144, 572–578.

Xu, F., Chen, H., Li, Z., 2001. Solid-state production of lignin peroxidase (LiP) and manganese peroxidase (MnP) by *Phanerochaete chrysosporium* using steam-exploded straw as substrate. Bioresour. Technol. 80 (2), 149–151.

Xu, Q., Singh, A., Himmel, M.E., 2009. Perspectives and new directions for the production of bioethanol using consolidated bioprocessing of lignocellulose. Curr. Opin. Biotechnol. 20 (3), 364–371.

Yang, P., Jiang, S., Zheng, Z., Luo, S., Pan, L., 2011. Effect of alkali and laccase pretreatment of Brassica campestris straw: architecture, crystallisation, and saccharification. Polymers Renew. Resour. 2 (1), 21.

Yoon, L.W., Ang, T.N., Ngoh, G.C., Chua, A.S.M., 2014. Fungal solid-state fermentation and various methods of enhancement in cellulase production. Biomass Bioenergy 67, 319–338.

Yu, H., Du, W., Zhang, J., Ma, F., Zhang, X., Zhong, W., 2010. Fungal treatment of cornstalks enhances the delignification and xylan loss during mild alkaline pretreatment and enzymatic digestibility of glucan. Bioresour. Technol. 101 (17), 6728–6734.

Yu, J., Zhang, J., He, J., Liu, Z., Yu, Z., 2009. Combinations of mild physical or chemical pretreatment with biological pretreatment for enzymatic hydrolysis of rice hull. Bioresour. Technol. 100 (2), 903–908.

Zadražil, F., 1977. The conversion of straw into feed by basidiomycetes. Appl. Microbiol. Biotechnol. 4 (4), 273–281.

Zanirun, Z., Bahrin, E.K., Lai-Yee, P., Hassan, M.A., Abd-Aziz, S., 2015. Enhancement of fermentable sugars production from oil palm empty fruit bunch by ligninolytic enzymes mediator system. Int. Biodeterior. Biodegrad. 105, 13–20.

Zeng, J., Singh, D., Chen, S., 2011. Biological pretreatment of wheat straw by *Phanerochaete chrysosporium* supplemented with inorganic salts. Bioresour. Technol. 102 (3), 3206–3214.

Zhou, S., Raouche, S., Grisel, S., Navarro, D., Sigoillot, J.C., Herpoël-Gimbert, I., 2015. Solid-state fermentation in multi-well plates to assess pretreatment efficiency of rot fungi on lignocellulose biomass. Microb. Biotechnol. 8 (6), 940–949.

CHAPTER

15

Currently Used Microbes and Advantages of Using Genetically Modified Microbes for Ethanol Production

*Bishnu Joshi**, *Jarina Joshi**, *Tribikram Bhattarai**, *Lakshmaiah Sreerama***

*Tribhuvan University, Kirtipur, Nepal

**Qatar University, Doha, Qatar

15.1 INTRODUCTION

Ethanol is a high-octane liquid transportation fuel that can be produced either by fermentation or via synthetic routes. Ethanol is produced presently by fermentation of sugar rich biomass, mainly starchy grains (mainly corn), tubers (cassava, potato), and/or saccharides (sugarcane/sugar beet), whereas in synthetic processes ethanol is produced by catalytic hydration of ethylene. Ethanol (ethyl alcohol, CH_3–CH_2–OH) is an oxygenated fuel with high octane and low cetane number than that of petroleum fuels (Lynd et al., 1991). Cetane index is a value derived from fuel density and its volatility that will give a reasonably close approximation to cetane number. Besides high octane and low cetane value, ethanol possesses high heat of evaporation, higher flammable temperature, which consequently increases the overall performance of engines with high compression ratios (Wheals et al., 1999). Bioethanol qualifies to be an alternative fuel, given it is environmentally friendly and mitigates the challenges associated with rising energy demands and cost of procuring them (Perlack et al., 2005). Ethanol is blended with petroleum in certain percentage (5%–15%), which will significantly decreases in petroleum product usage and at the same time reduces the overall greenhouse gas emissions.

Bioethanol Production From Food Crops
http://dx.doi.org/10.1016/B978-0-12-813766-6.00015-1

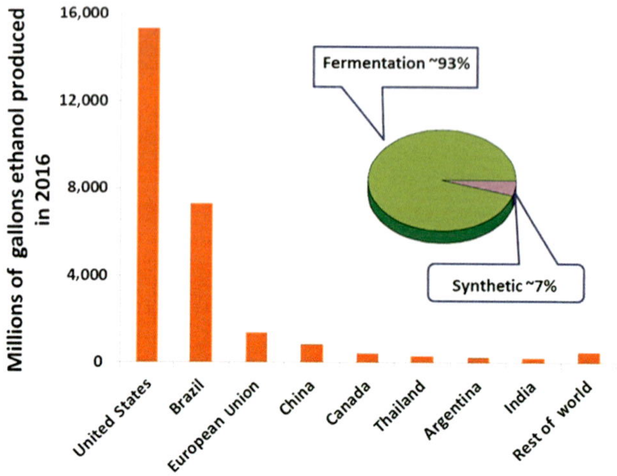

FIGURE 15.1 **World's leading ethanol producers.** *Source: Renewable Fuel Association. Available from: http://www.ethanolrfa.org/resources/industry/statistics.*

15.2 WORLD FUEL ETHANOL PRODUCTION AND CHALLENGES

As per 2016 statistics, the annual worldwide ethanol production (Fig. 15.1) was about 32 billion L (RFA, 2017). About 7% of it corresponds to the synthetic alcohol derived from ethylene or coal, and the rest is bioethanol (Smeets et al., 2005). Brazil and USA are the leading countries in ethanol production and they use sugarcane and corn as the raw material, respectively. Sucrose from sugar beet is also used by growing ethanol industries of Western Europe (Fig. 15.1). The advantages of using sugar beet include a shorter crop production cycle, higher yield, and higher tolerance to wide range of climatic variations, and low water and fertilizer requirement.

Owing to global industrialization, increasing population and other factors, the demand of transportation fuels is increasing every year. In this scenario, bioethanol production by first-generation methodologies as well as synthetic processes is not enough to meet the global demand. Further, ethanol produced by first-generation technologies directly competes with food security and its use in biofuel production is often criticized.

Starch-rich feedstocks are a common source of sugar needed for ethanol production. Starch is branched homopolymer of D-glucose units that is broken down to D-glucose using a variety of enzymes, that is, α-amylases break drown starch to limited dextrin, β-amylases break down starch and limited dextrin to maltose and glucoamylase breaks down starch, dextrin, and maltose to D-glucose. Debranching enzymes play an important role in breaking α (1–6) linkages in starch and dextrin. This process generates fermentable sugars. The factors that affect the enzymatic hydrolysis of starch include substrate used, enzyme activity, and the reaction conditions (temperature, pH, and other parameters). Dilute hydrochloric acid (~0.5 M) may also be used for breakdown of starch, however, the enzymatic process is most commonly used.

The starch- and sugarcane-based industry for bioethanol production has been commercially viable for decades. However, the cost of bioethanol production is higher in first-generation technologies. Yeast cannot utilize starchy materials directly and the processing requires a large amount of amylolytic enzymes. The process also needs the starchy material to be cooked at higher temperature (140–180°C) (Bai et al., 2008; Balat et al., 2008).

Current ongoing research in the area of bioethanol production is focused on utilizing lignocellulosic biomass (LCB) as source of sugar for fermentation. In this process (also known as second-generation technology), LCB is used as cheap feedstock to produce ethanol. LCB is cheap, renewable, and widely available (Lynd et al., 1996; Mosier et al., 2005; Sun and Cheng, 2002). In the past decade, tremendous progress has been made to utilize LCB for the commercial production of liquid biofuels, including bioethanol. This process of ethanol production generally involves pretreatment and hydrolysis of LCB to sugars followed by fermentation of such sugars to ethanol (Fig. 15.2) (Joshi et al., 2011).

The main bottleneck in commercial ethanol production from second generation technologies is mainly due to the cost associated with chemical and/or enzymatic pretreatment processes involved to generate fermentable sugars. The commonly used pretreatment processes, and their advantages and disadvantages are summarized in Table 15.1.

The main reason for pretreatment processes is to assist in lignin removal, cellulose/hemicellulose decrystalization and partial depolymerization, and to increase the surface area to make cellulose/hemicellulose accessible for the enzymatic hydrolysis. Biological pretreatment methods are more ecofriendly than chemical pretreatment methods because nature has bestowed us with tremendous microorganisms that are a potential source of valuable enzymes, which could be utilized rationally to hydrolyze recalcitrant biomass, such as cellulose and lignin (Sindhu et al., 2016). In this process, mostly wood-feeding termites have been explored for isolating cellulolytic microbes. The most well-studied species in this regard include *Amycolatopsis*, *Acinetobacter*, *Pseudomonas*, *Staphylococcus*, and various species of *Enterobacteriaceae* and *Bacillaceae* families (Pourramezan et al., 2012; Sharma et al., 2015). Filamentous fungi (*Trichoderma reesei*, *Trich. viride*, *Aspergillus niger*, *A. oryzae*, and *Fusarium oxysporum*) and *basidiomycetes* (*Phanerochaete chrysosporium*, *Fomitopsis* sp.) have also shown promising cellulase production abilities under different cultivation conditions using a number of carbon sources (Chandel et al., 2012; Mathew et al., 2008). To date, *Phan. chrysosporium* and *Gloeophyllum trabeum* are considered as very competent cellulase producers. Similarly, *Trich. reesei* and their mutant strain lineages have been reported to produce 40–100 g cellulase enzyme per liter of culture broth (Vitikainen et al., 2010).

Cellulase system of many fungi, including *Trich. reesei* is deficient in β-glucosidase and sensitive to feedback inhibition by glucose. Considering this aspect, attempts have been made to increase the copy number and introduce glucose tolerant β-glucosidase gene in *Trich. reesei* (Nakari-Setälä and Penttilä, 1995). The most productive cellulolytic bacteria produce only a few grams cellulase per liter (Xu et al., 2009). Highly cellulolytic anaerobic hyperthermophiles are found in the genera *Thermotoga* and *Caldicellulosiruptor bescii* (formerly *Anaerocellum thermophilum*, isolated from a geothermally heated pool) and some of its relatives in the same genus, such as *Caldicellulosiruptor saccharolyticus* (Bok et al., 1998; Rainey et al., 1994). From the genome sequences of *Caldicellulosiruptor* sp., it has been known that ~20% of the total open reading frames (ORF) are involved in the degradation, transport, and metabolism of carbohydrates, such as cellulose, xylan, and starch (Lee et al., 2017). Hydrolases of *Caldicellulosiruptor* are considered as "megazymes" due to their multidomain functions and multicatalytic

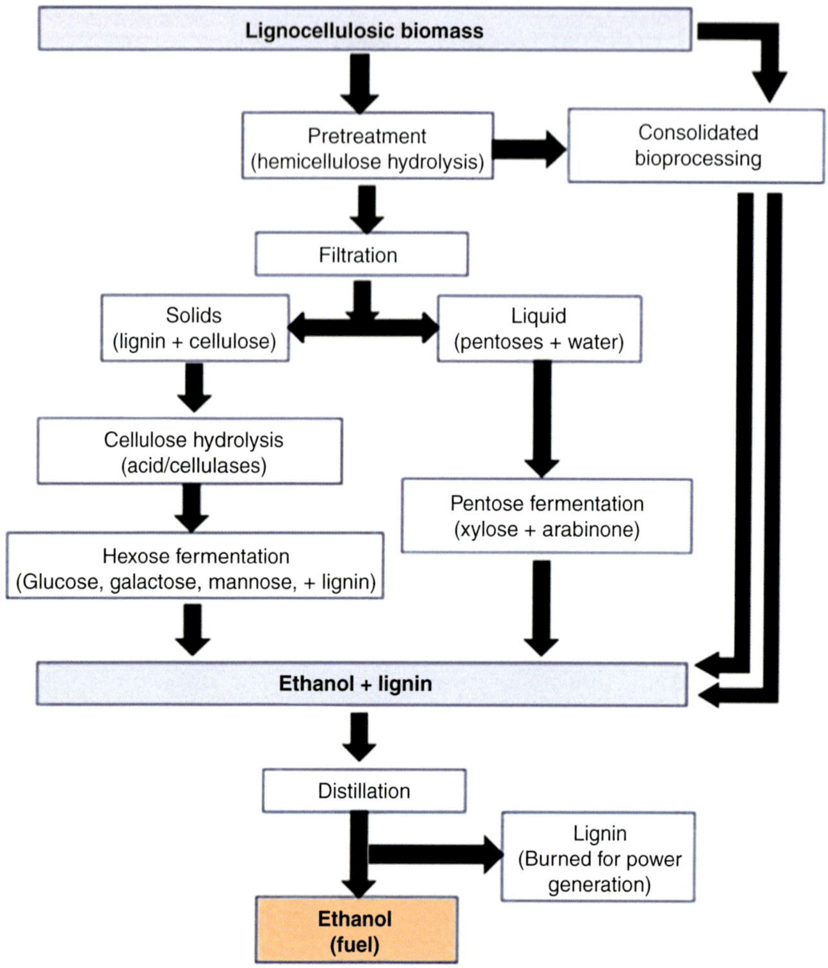

FIGURE 15.2 **Schematic of various processes used to utilize lignocellulosic biomass to produce bioethanol.** *Source: Adapted with permission from Joshi, B., Bhatt, M.R., Sharma, D., Joshi, J., Malla, R., Sreerama, L., 2011. Lignocellulosic ethanol production: current practices and recent developments. Biotechnol. Mol. Biol. Rev. 6, 172–182.*

nature. Many of these megazymes contain five or more domains, which can include a variety of cellulases, hemicellulases, and carbohydrate binding modules (Bergquist et al., 1999). One of the enzymes for cellulose degradation termed CelA from *Cald. bescii* has been characterized (Zverlov et al., 1998), however its physiochemical properties, such as cellulose and hemicellulose degrading ability at elevated temperatures were only reported very recently (Kanafusa-Shinkai et al., 2013). *Cald. bescii* has the capability of not only solubilizing cellulose and hemicellulose in unpretreated biomass but also much of the lignin in unpretreated switchgrass (Kataeva et al., 2013).

Likewise, fungi belonging to Ascomycota and Basidiomycota, such as brown-rot and white-rot fungi, respectively (Dashtban et al., 2010), are also efficient degraders of LCB. From a productive perspective, brown-rot fungi appear

TABLE 15.1 Biomass Pretreatment Methods: Advantages and Disadvantages

Pretreatment method	Process	Advantages	Disadvantages	Sources
Physical pretreatment methods	Milling • Ball milling • Two row milling • Colloid milling	Reduces cellulose crystallinity Increases surface area for enzymatic hydrolysis	High power and energy consumption	Sun and Cheng (2002)
	Irradiation • Gamma-ray irradiation • Microwave irradiation	Decreases cellulose crystallinity and decrease degree of polymerization	Lignin removal is inefficient	Saini et al. (2015)
	Others • Extrusion • Pyrolysis	80%–85% conversion of cellulose to reducing sugars with more than 50% glucose	Temperature >300°C is required	Taherzadeh and Karimi (2008)
Chemical pretreatment methods	Acids • Sulfuric acid • Hydrochloric acid • Phosphoric acid Dilute vs. Concentrated acids	Partial or nearly complete delignification High glucose yields Ambient temperature Less corrosion problem than concentrated acid and dilute acids forms lower concentrations of inhibitors, for example, furfural	High cost of acids and need to be recovered/ neutralization Reactor corrosion problems Acid pretreatment leads to formation of degradation products	Mosier et al. (2005), Sun and Cheng (2002), Yang and Wyman (2008)
	Alkalis Organic solvents	Effective lignin removal and hemicellulose solubilization Causes lignin and hemicellulose hydrolysis	Long exposure times and requires alkali removal High cost; solvent needs to be drained and recycled	Taherzadeh and Karimi (2008)
Physico chemical methods	Ammonia fiber expansion (AFEX) Stem explosion	Increase accessible surface area Minimal inhibitor formation Causes lignin transformation and Hemicellulose solubilization	Not efficient for raw material with high lignin content. High cost and requires large amount of ammonia Generation of toxic compounds	Chiaramonti et al. (2012)
Biological methods	Use of lignin degrading microorganisms, mainly rot fungi and actinomycetes Use of cellulolytic organisms from termite gut and soil	Degrade lignin and hemicelluloses Low energy consumption Minimal environmental impact	Requires long incubation time Requires careful control of growth condition	Ali et al. (2016)
Ionic liquid pretreatment methods	Biomass hydrolysis in ionic liquids with different mineral acids as catalysts	Achieves ~ 81% liberation of total reducing sugars initially present in the biomass with 1-N-butyl-3-methylimidazolium chloride and hydrochloric acid	Acids are corrosive	Li et al. (2008)

to contribute very little toward lignin degradation. For example, Hatakka (1983) have shown that 35% of the straw was converted to sugars by *Pleurotus ostreatus*. On the contrary, white-rot fungi are very efficient for lignin breakdown and some of them can even exhibit high selectivity, thus leaving the cellulose and hemicellulose fractions almost intact (Mäkelä et al., 2014). The lignin degradation is due to the presence of efficient laccase enzymes in these fungi. In our earlier studies, we have successfully purified laccase enzyme from *Ganoderma* sp. having molecular mass of 43 kDa, exhibiting optimal activity at 30°C and pH 5.0. The isolated laccase was thermally stable (up to 70°C for 1 h) and exhibited broad pH stability (Shrestha et al., 2016). The microorganisms used in the fermentation process to date are inefficient/incompatible in the cofermenting the variety of sugars released from LCB. To economize the process of ethanol production, there is an urgent need to identify robust microoraganisms that can ferment both hexose (C_6) and pentose (C_5) sugars, as well as compatible with process ingeration, for example, consodilated bioprocessing (CBP) in which pretreatment, saccharification, and fermentation are all carried out in a single bioreactor using genetically modified microorganisms.

15.3 HETEROGENITY IN CARBOHYDRATE COMPOSITION OF LIGNOCELLULOSIC BIOMASS

Many countries and laboratories are working toward the development of technologies to exploit the potential of LCB as a substrate for the production of bioethanol. This process generally involves pretreatment and hydrolysis of LCB to fermentable sugars (see earlier) followed by fermentation of such sugars to ethanol. Achieving fermentable levels of sugars from LCB requires relatively harsh pretreatment processes, costly enzymes and design of different processing tanks for the saccharification, hydrolysis and fermentation processes (Banerjee et al., 2010). All these steps make the cost of production of ethanol from LCB to be prohibitive, for example, pretreatment processes takes up to 33% of the overall cost of producing lignocellulosic bioethanol (Behera et al., 2014). Given this, the current research focus is on the genetic manipulation of the lignin biosynthetic pathway in plants in order to improve bioethanol yields from LCB (Fu et al., 2011). An alternative strategy could be CBP scheme (Fig. 15.2) in which lignin degradation, cellulase/hemicellulase production, substrate hydrolysis, and fermentation are accomplished in a single step and in one reactor (Lynd et al., 2017). It is important to note that in CBP only one microbial consortium is employed for simultaneous biodegradation, fermentation, and distillation, a three-in-one process, which eventually help in lowering ethanol production costs due to simpler LCB processing, lower energy inputs and higher conversion efficiencies than simultaneous saccharification fermentation (SSF) or simultaneous saccharification cofermentation (SSCF) based processes. Efforts are being made to engineer ethanol producing yeast (*Saccharomyces cerevisiae*) capable of producing lacase/cellulases/hemicellulases and lignocellulose degrading bacteria (*Clostridium thermocellum*) to be efficient ethanol producers (Chen, 2017; Khatun et al., 2017; Kricka et al., 2014; Lynd et al., 2017).

In nature, LCB is in abundance and it is renewable. LCB is a complex matrix with three main components; lignin, cellulose (polymer of β-glucose units), and hemicellulose (polymer of C5/C6 sugars). More than 70% of the dry mass LCB is made of sugar polymers. If these sugar polymers can be efficiently hydrolyzed and fermented, bioethanol production from LCB holds tremendous potential in meeting energy needs of the world providing many environmental benefits. Suitable substrates for LCB includes corn stover, wheat straw, sugarcane bagasse, rice straw, rice hull, corncob, oat hull, corn

TABLE 15.2 Compositional Analysis of Carbohydrate Present in Various Weed Species and Their Carbohydrate Composition

Biomass source	Cellulose (%)	Hemicellulose (%)	Lignin (%)	Moisture (%)	Ash (%)	Sources
Eichhornia crassipes	36.5	22.0	15.2	92.8	NR	Mishima et al. (2008)
E. crassipes	18.2	48.7	3.5	NR	NR	Kumar et al. (2009)
E. crassipes	25.61	18.4	9.9	NR	NR	Poddar et al. (1991)
E. crassipes	33.97	18.0	26.4	NR	NR	Chanakya et al. (1993)
E. crassipes	36.5	22.0	5.2	NR	NR	Mukherjee and Nandi (2004)
Saccharum spontaneum	45.1	22.75	24.5	4.9	2.5	Chandel et al. (2011)
Sacchar. spontaneum	36.8	2.2 – 21.5	20.1	5.4	NR	Scordia et al. (2010)
Lantana camara	44.1	17.0	32.3	4.4	2.3	Kuhad et al. (2010)
L. camara	45.1	17.0	27.3	6.5	2.3	Pasha et al. (2007)

NR, Not reported.

fiber, wood chips, saw dust, and cotton stalk, energy crops, such as switch grass and alfa alfa, and various agricultural and forest weeds, for example, *Saccharum spontaneum*, *Lantana camara*, *Eichhornia crassipes* (water hyacinth) (Chandel and Singh, 2011; Joshi et al., 2011). The amount of biomass (cellulose, hemicellulose, and lignin) present in these feedstocks also depends on the region from where the plant has been harvested. Studies have shown of obtaining varying amounts of cellulose from the same species obtained from different places/regions, accordingly various weeds have been evaluated for their carbohydrate content with goal of using them for bioethanol production (Table 15.2).

15.4 INHIBITORS GENERATED DURING PRETREATMENT/ HYDROLYSIS AND THEIR EFFECTS ON FERMENTING MICROORGANISMS

Use of concentrated acids and higher temperatures in the pretreatment process generates not only large amounts of fermentable sugars but also generates intermediates that act as inhibitors later in the fermentation process. These inhibitors either inhibit the enzymes responsible for saccharification and fermentation and/or growth of the organisms used in pretreatment/fermentation process. These fermentation inhibitors (Fig. 15.3 and Table 15.3) can be divided into three major groups: (1) organic acids, such as acetic, formic, and levulinic acids, (2) furan derivatives, such as furfural and 5-hydroxymethylfurfural (5-HMF), and (3) phenolic compounds (Chandel et al., 2010). These intermediates slow down or even completely inhibit fermentation process (Frazer and McCaskey, 1989; Larsson et al., 1999). Among the inhibitors produced furfural (a pentose sugar byproduct) and hydroxy methyl furfural (a hexose sugar by-product) are the most potent toxic compounds/inhibitors among compounds listed earlier (Taherzadeh et al., 1999). The effects of these compounds may be overcome via (1) overliming fermentation broths, (2) use of ion exchange resins to selectively remove them from the broth, and (3) use enzymes to mitigate the effect of toxins/inhibitors by converting them to nontoxic products. More recently, the researchers (Chong et al., 2013; Yuan et al., 2017; Zhang et al., 2017; Zhu et al., 2013) are moving toward

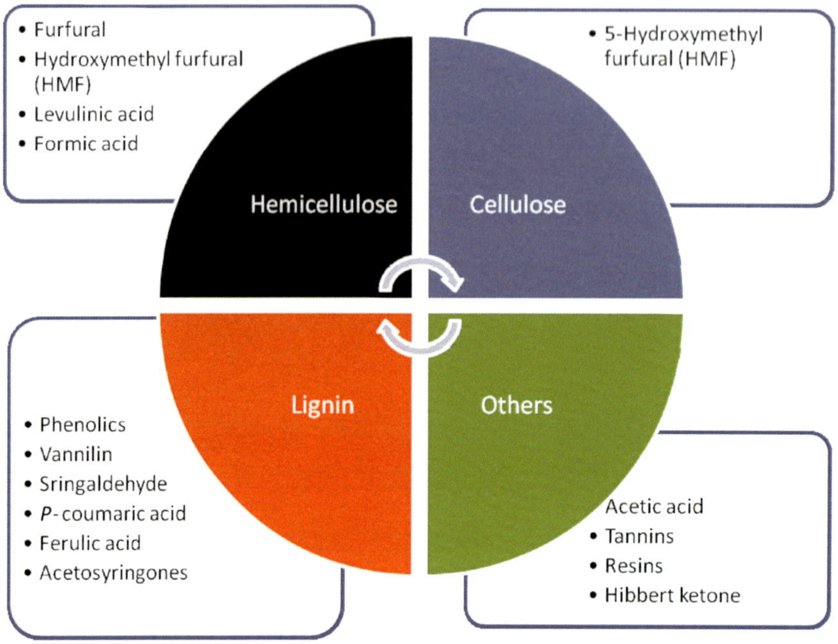

FIGURE 15.3 **Inhibitors of fermentation process generated during pretreatment process.**

TABLE 15.3 Effects of Fermentation Inhibitors Aliphatic Acid, Furan Derivative, and Aromatics

Fermentation inhibitors	Effect	Detoxification	Sources
Aliphatic acids • Acetic acid • Levulinic acid • Formic acid	Uncoupling and intracellular anion accumulation Inhibit the growth and metabolism allowing the higher permeation of cell membrane in microorganisms.	Activated charcoal	Zaldivar and Ingram (1999)
Furan derivatives • Furfural • 5-HMF	ROS accumulation causes damage to mitochondria and vacuole membrane Cell wall damage, DNA breakdown, inhibition of protein and RNA synthesis	Use of calcium hydroxide	Sanchez and Bautista (1988)
Aromatics • Vanillin • 4 hydroxy-benzaldehyde	Loss of integrity of biological membrane, diminished cell growth and sugar assimilation	Use of stress (inhibitor) tolerant yeast	Palmqvist et al. (1999)

metabolic engineering of microbes (yeast/bacteria) that can tolerate high concentration of alcohol as well as inhibitors. Tolerant ethanologenic yeast strains have been found to convert furfural and 5-(hydroxymethyl)-2-furaldehyde to less toxic compounds, such as furan methanol (FM) and furan-2,5-dimethanol (FDM; 2,5-bis-hydroxymethylfuran) and at the same time produce normal yields of ethanol (Liu et al., 2004; Liu and Slininger, 2005). The enzymes that have shown to catalyze detoxification of fermentation inhibitors (earlier) include alcohol dehydrogenases (ADH1, ALD4, ADH6, and ADH7), and methylglyoxal reductases (GRE2 and GRE3) (Liu, 2011, 2012). Systems biology approaches, including disruptome screening, transcriptomics, and metabolomics have been recently exploited to gain insight into the molecular and genetic traits involved in tolerance and adaptation to the fermentation inhibitors in the above organisms (Cheng et al., 2017; Kasavi et al., 2016).

15.5 ENGINEERING BACTERIA FOR ETHANOL PRODUCTION

The following group of microorganisms are genetically engineered for higher ethanol productivity.

15.5.1 Escherichia coli

Escherichia coli is considered as an attractive host for the conversion of LCB into ethanol because of the useful characteristics, such as (1) metabolism of a wide variety of substrates (pentose and hexose sugars as well as uronic acid from in pectin) found in pretreated LCB, (2) sustained high glycolytic flux (both aerobic and anaerobic), and (3) high tolerance to ethanol (up to 50 g/L). Other advantages of using *E. coli* for ethanol production are the extensive understanding of its physiology and metabolism and the ease of genetic manipulation. Therefore, *E. coli* has been an obvious choice for metabolic engineering towards ethanol production. An *E.coli* strain, ATCC 11303 (pLOI297), is reported to be the most promising strain for xylose fermentation under a wide variety of culture conditions. Under the optimum fermentation conditions (≤42°C, pH > 6, xylose 80 g/L), xylose fermenting cultures showed maximum ethanol tolerance (53–56 g/L) and an average productivity 0.72 g/L convert the pyruvate to ethanol has been unsuccessful due to low level of ADH expression in native *E. coli*. To overcome this shortcoming, *E. coli* W strain KO11 was constructed by introducing *pdc* (pyruvate decarboxylase) and *adhB* (ADH) genes isolated from *Zymomonas mobilis* under the control PET operon and deletion of *frd* gene (fumarate reductase) to prevent succinate production (Ingram et al., 1987). In this strain, the PET operon was stably integrated into the chromosome at the *pfl* locus along with an antibiotic resistance marker. Spontaneous mutants exhibiting high ADH activity and high antibiotic resistance were selected to ensure high PET operon activity. The resulting strain KO11 produced ethanol at a yield of 95% in complex media (Ohta et al., 1991a). However, recombinant *E. coli* has several drawbacks as it requires high nutrient supply, weak ability of secreting active cellulase, and sensitivity to end product inhibition. Recombinant *E. coli* W KO11 has been used for production of ethanol from biomass using multiple substrates including, rice hull (Moniruzzaman and Ingram, 1998), sugarcane bagasse, corncobs, hull and agricultural residues (Asghari et al., 1996), pectin-rich beet pulp (Doran et al., 2000), and sweet whey (Leite et al., 2000). Similarly, feedback resistant strains have been developed using *pfl* and *ldh* mutant strains in which the conversion of pyruvate to lactate and recycling of the NADH generated from glycolysis was blocked. Transforming the strains with plasmid pLOI297, which encodes the PET operon, restored fermentative growth. The transformed strains selectively produced ethanol from arabinose, glucose and/or xylose. These strains have been used to ferment hydrolysates prepared from corn hull and germ meal. Ethanol production by the best strain, FBR5, was 0.46–0.51 g/g, and fermentation was complete within 36–60 h (Dien et al., 2000).

Recombinant *E. coli* DH5α (pEglABglA) expressing endophytic endoglucanase A (*EglA*) from *Bacillus pumilus* (Lima et al., 2005) and hyperthermophilic β-glucosidase A (*BglA*) from *Fervidobacterium* sp. (Lima et al., 2009) has been shown to grow in carboxymethyl cellulose (CMC) as sole source of carbon and energy. The expression of the *Bgl*A and *Egl*A genes enabled

E. coli to partially degrade CMC. Among the five strains of *E. coli* transfected with pEglABglA, when cultured in CMC media, specific growth rates of recombinant *E. coli* strains BL21, JM101, and Top 10 were higher than those of DH5α and DH10B strains. The recombinant *E. coli* DH5α reduced the viscosity of the CMC medium by 5-fold and increased the reducing sugar released from CMC by 30-fold. The pEglABglA plasmid can be used as backbone for further cellulase gene addition, which may enhance the *E. coli*'s cellulolytic capacity even more (Rodrigues et al., 2010).

The gene for cellulase from *Ruminococcus albus* F-40 when cloned in *E. coli* HB101 with pBR322 vector and cultured in LB broth the cellulase production was remarkably enhanced (up to ~170 U/L after 8 h). Further, maintaining pH at ~6.5 and NaCl concentration at ~ 80 mM lead to stable cellulase expression for several hours (Ohmiya et al., 1988).

The success of using any microorganism for industrial production of fuels depends on its ability to quickly convert renewable raw material into fuel with high productivity at a low price without being toxic to the organism itself. Availability of genetic and molecular tools to engineer existing native pathways or to create a synthetic new pathway has made *E. coli* as the microorganism of best choice in order to produce biofuels from renewable energy sources. Although significant work has been done, some challenges still exist to consider using *E. coli* as a cost-efficient strategy for commercial production of bioethanol, higher chain alcohols, and biodiesel.

To date, the state-of-the-art bioethanol-producing *E. coli* strains have shown titers in the range of 40–55 g/L and yield of ~100% theoretical maximum from various cellulosic and hemicellulosic feed sources and are similar to bioethanol produced through *Sacch. cerevisiae*. However, it is important to note that, although the cellulosic and hemicellulosic materials are cheap, their hydrolysates contain toxic compounds, such as organic acids, furan derivatives, and phenolic compounds that inhibit the growth of *E. coli* more as compared to *Sacch. cerevisiae*. Improved pretreatment technology and genetic engineering approaches to improve tolerance to these compounds are useful to mitigate the challenges stated earlier.

15.5.2 Ethanol Production Using Actinomycetes

Among the actinomycetes, *Thermomonospora fusca* (Deng and Fong, 2010), *Streptomyces thermodiastaticus* (Crawford and McCoy, 1972), *Thermomonospora curvata*, *Streptomyces viridosporus*, *Streptomyces setonii*, and other strains (Godden et al., 1989) have been shown to contain cellulolytic activity. Arora et al. (2005) studied the role of *Streptomyces griseus* B1 isolated from leaf litter in ethanol production. These studies used hardwood as well as softwood as substrates in fermentation at 37°C and *Strep. griseus* B1 caused much higher loss of Klason lignin content compared to the control. The control experiments showed hardly any loss of Klason lignin when treated with 0.1 M alkali for 2 h. *Strep. griseus* B1 preferred hardwood substrates (23.4% lignin loss) as compared to softwood substrates (10.5% lignin loss). Lee et al. (2013) have recently heterologously expressed *pdc* and *Adh II* genes in the Gram-positive *Strep. lividans* TK24. The recombinant *Strep. lividans* TK24 produced ethanol from glucose with a yield of 23.7%. The recombinant *Strep. lividans* TK24 was also able to produce ethanol from xylose, L-arabinose, mannose, L-rhamnose, galactose, ribose, and cellobiose with yields of 16.0, 25.6, 21.5, 33.6, 30.6, 14.6, and 33.3%, respectively.

15.5.3 Thermophilic Bacteria

Production of ethanol from thermophilic organisms have distinctive advantages over mesophiles in terms of their ability to (1) increase solubility of substrates, (2) improved mass transfer due to decreased viscosity, (3) increased diffusion rates, (4) high bioconversion rates, (5) their ability to use variety of inexpensive biomass feedstocks, (6) low risk of contamination, (7) reduced cost of bioreactor operation (avoid cooling costs of bioreactors), and (8) facilitated product recovery (high vapor pressure of the volatile compounds). Given the aforementioned, thermophilic bacteria are currently of interest in bioethanol production because they have the ability to ferment biomass to ethanol without the addition of external hydrolytic enzymes, a substantial cost advantage. These bacteria also metabolize both pentose and hexose sugars found in lignocellulosic hydrolysates (Lynd et al., 2005).

Several anaerobic thermophiles have been shown to utilize cellulose, hemicellulose, and starch. These organisms include *Clost. thermocellum*, *Clost. raminisolvens*, *Clost. stercorarium*, *Cald. saccharolyticus*, and *Cald. obsidiansis* (Freier et al., 1988; Hamilton-Brehm et al., 2010; Wiegel and Dykstra, 1984). The noncellulolytic thermophiles, such as *Clost. thermohydrosulphuricum*, *Clost. themosaccharolyticus* and *Thermoanaerobacter ethanolicus* utilize xylose and other pentose sugars at comparable levels to that of hexose. On the other hand, cellulolytic thermophile *Clost. thermocellum* does not utilize pentoses (Lynd et al., 2005).

Thermo. saccharolyticum JW/SL-YS485, a thermophilic anaerobic bacterium, grows in a temperature range of 45–65°C and a pH range of 4.0–6.5 and is able to ferment hemicellulose and xylan polymers directly. It also ferments primary sugars found in cellulosic biomass, including cellobiose, glucose, xylose, mannose, galactose, and arabinose (Shaw et al., 2008).

Therm. ethanolicus is a nonspore-forming extreme thermophilic anaerobic bacterium that produces ethanol at 69°C (Kannan and Mutharasan, 1985). The thermophilic anaerobic bacterial strain *Thermoanaerobacter* BG1L1 was able to ferment undetoxified corn stover hydrolysate in a continuous immobilized reactor system at 70°C with ethanol yield of 0.39–0.42 g/g sugars consumed. Nearly 89%–98% xylose in corn stover hydrolysate was utilized to produce ethanol by this bacterium (Angelidaki et al., 2012). Researchers have explored the coculture of *Clost. thermocellum* paired with *Clost. thermosaccharolyticus*, *Clost. themohydrosulphuricum* for simultaneous utilization of pentose and hexose sugars to produce ethanol and these studies do show that the coculture yield higher ethanol concentrations (Angelidaki et al., 2012).

Clost. thermocellum exhibits a high growth rate on crystalline cellulose (Zhang and Lynd, 2005). Specific growth rates reported for thermophiles, such as *Clost. thermohydrosulfuricum*, *Clost. theromocellum*, *Clost. thermosaccharolyticum*, and *Thermo. ethanolicus* on soluble substrates generally range between 0.4 and 0.6/h (Zeikus et al., 1981), which are comparable to 0.43/h to 0.46/h for the yeast (*Sacch. cerevisiae*) growing on glucose (Laopaiboon et al., 2008). *Clost. thermocellum* has been reported to directly ferment cellulosic biomass without pretreatment process and produce 0.28–0.29 g ethanol/g glucose fermented (Ram and Seenayya, 1989). It is not exploited industrially due to low ethanol tolerance [tolerates only up to 1.5% (v/v) ethanol]. Research to improve ethanol tolerance by these organisms is ongoing. Rani and Seenayya (1999) have isolated *Clost. thermocellum* strain SS21 and SS22, producing high yields of ethanol, that are tolerant to 4.0% and 5.0% (v/v) ethanol, respectively. These strains are also tolerant to various solvents and acetic acid. Similarly, Brown et al. (2011) have reported that mutated bifunctional acetaldehyde-CoA/alcohol dehydrogenase gene (adhE), leads to improved ethanol tolerance in *Clost. thermocellum*. Extremely thermophilic cellulolytic *Cald. saccharolyticus* can coutilize glucose and xylose and its close relative *Cald. bescii* DSM 6725 has been found to degrade xylan and xylose (Yang et al., 2010).

Several mesophilic *Clostridium* strains have also been reported to utilize both cellulose and xylan, including strains of *Clost. phytofermentans* and *Clost. cellulovorans* (Warnick et al., 2002). *Clost. phytofermentans* is an obligate anaerobe, mesophilic, Gram-negative, cellulolytic, motile, straight rod, spherical terminal spore former that can utilize cellulose, pectin, polygalacturonic acid, starch, xylan, arabinose, cellobiose, fructose, galactose, glucose, lactose, maltose, mannose, ribose, and xylose in growth medium in the temperature range of 35–37°C. The major end products of this fermentation are ethanol, acetate, CO_2, and H_2 and minor products include formate and lactate (Warnick et al., 2002). The genome of *Clost. phytofermentans* (ATCC 700394) encodes for the highest number of enzymes for degradation of lignocellulosic biomass among the sequenced *Clostridia* genomes (Weber et al., 2010). In one of our laboratories, this species has been used to ferment LCB with and without pretreatment and observed production of ethanol under laboratory conditions. Others have also observed that this species can ferment LCB directly without any pretreatment of biomass and the efficiency of ethanol production remains the same with or without pretreatment of biomass. Thus, it has become a promising native anaerobic CBP microorganism. Recently published work shows that it secretes individual enzymes, which not only degrade cellulose and hemicellulose to fermentable sugars but also can consume almost all the sugars present in LCB and produce ethanol and acetate as the major product (Warnick et al., 2002; Weber et al., 2010). In a CBP scheme, *Clost. phytofermentans* can be the best microorganism for the production of ethanol in a cost-effective and viable way. In this regard, Jin et al. (2011) have evaluated the performance of *Clost. phytofermentans* (ATCC 700394) on ammonia fiber expansion-treated corn stover (AFEX-CS). Under optimal conditions with 0.5% (w/w) glucan load of AFEX-CS, *Clost. phytofermentans* hydrolyzed 76% of glucan and 88.6% of xylan in 10 days. These values reached 87% and 102% of those obtained by SSCF using commercial enzymes and *Sacch. cerevisiae* 424A. Ethanol titer for CBP was found to be 2.8 g/L which was about 72% of that yielded by SSCF (3.9 g/L) procedure.

Z. mobilis is an anaerobic, Gram-negative bacterium, which produces ethanol from glucose via the Entner–Doudoroff (ED) pathway in conjunction with the enzymes PDC and ADH (Conway, 1992). It is reported that the ethanol yield in *Z. mobilis* cultures could be as high as 97% of the theoretical yield of ethanol from glucose (Sprenger, 1996). When compared to yeast, *Z. mobilis* exhibits 5%–10% higher yields and up to 5-fold greater volumetric productivity while tolerating upto 120 g/L ethanol. Also as a consequence of the low ATP yield, *Z. mobilis* maintains a higher glucose metabolic flux, and correspondingly, higher ethanol productivity (3–5-fold higher as compared to *Sacch. cerevisiae*) (Gunasekaran and Raj, 1999; Sprenger, 1996). *Z. mobilis* also has simple nutritional needs with the exception that some strains require the adition of pantothenate and biotin for their growth (Rogers et al., 1982). However, it is not exploited industrially because of narrow substrate range, that is, it can only utilize three sugars: D-glucose, D-fructose, and sucrose, and is unable to ferment widely available pentose sugars. Second, although *Z. mobilis* is generally regarded as safe (GRAS organism) (Lin and Tanaka, 2006), the postfermentation biomass is not acceptable to be used as animal feed. The later inevitably generates the problem for biomass disposal if it replaces *Sacch. cerevisiae* in the industrial ethanol production. Finally, the continuous ethanol fermentation with *Z. mobilis* tends to be oscillatory (McLellan et al., 1999).

15.5.4 Engineering *Z. mobilis* for Ethanol Production

Z. mobilis is the first recombinant strain engineered to ferment xylose (Zhang et al., 1995a). To achieve this, four *E. coli* genes: xylose isomerase (xylA), xylulose kinase (xylB), transketolase (tktA), and transaldolase (talB) were introduced to this strain (Zhang et al., 1995a). Xylose isomerase and xylulose kinase convert xylose into xylulose-5-phosphate, an important pentose that can enter the pentose phosphate pathway. Xylulose-5-phosphate is next converted to intermediates of the ED pathway by transketolase and transaldolase. The genes clonned into a plasmid (pZB5) using strong constitutive promoters of either enolase or glyceraldhyde-3-phosphate dehydrogenase from *Z. mobilis*. The transformed strain CP4 with pZB5 grew on xylose, and the ethanol yield was 86% of theoretical yield. This strain also simultaneously fermented glucose and xylose. Xylose uptake depends on the native glucose permease; *Z. mobilis* does not have active sugar transport systems (Parker et al., 1995).

For arabinose fermentation, pZB206 plasmid containing five genes isolated from *E. coli* coding for (1) L-arabinose isomerase (*araA*), (2) L-ribulose kinase (*araB*), (3) L-ribulose-5-phosphate-4-epimerase (*araD*), (4) transketolase (*tktA*), and (5) transaldolase (*talB*) has been contructed and transforend into *Z. mobilis*. The first three enzymes are responsible for converting arabinose to xylulose-5-phosphate, which is converted to ED pathway intermediate xylulose-5-phosphate via transketolase and transaldolase. The *Z. mobilis* strain transformed, namely, ATCC 39676, with the genes coding for the five enzymes earlier, successfully fermented arabinose to ethanol (25 g/L) and displayed a very high yield (98% of theoretical). But the rate of arabinose fermentation was much lower compared to that observed for the xylose. This may be due to low affinity of the glucose permease transporter for arabinose (Parker et al., 1995). Yet another strain of *Z. mobilis*, ATCC 39767, carrying pZB5 containing cellulase genes was able to convert cellulose substrate to ethanol efficiently (Zhang et al., 1995b), however, this strain was sensitive to inhibitors. The strain has previously been used for industrial scale trails (up to 586,000 L) for fermenting ground milo (sorghum), corn or wheat (Doelle and Doelle, 1990).

Lawford and Rousseau (1999) successfully adapted xylose-fermenting ATCC 39767 strain to tolerate higher concentrations of acetic acid, as well as other inhibitors, by culturing the strain continuously in higher levels of lignocellulosic hydrolysates. The continuous culture was run for 149 days and the level of hydrolysate in the medium increased from 10% to 50% (v/v). Isolates recovered at the end of the fermentation demonstrated significantly improved ethanol productivity in the presence of acetic acid compared to the unadapted strain. The adapted strain was subsequently evaluated for converting poplar wood hydrolysate to ethanol in SSF. The hydrolysate was prepared by steam explosion of poplar wood chips followed by overliming to reduce inhibitor levels. The fermentation was performed at 34°C and pH 5.5, which represents a compromise between the optimal conditions for the cellulase enzymes and typical culture conditions for *Z. mobilis*. After 7 days, the fermentation reached an ethanol concentration of 30 g/L, a yield of 54% based on total initial carbohydrate content (McMillan et al., 1999).

Recently, Zhang et al. (2017) have made a number improvements to their *Z. mobilis* strains. The new stain (AX101, parental strain ATCC 39676) ferments both arbinose and xylose and carries the seven important heterologous genes. When AX101 was used to ferment a mixture of sugars (40 g/L glucose, 40 g/L xylose, and 20 g/L arbinose) it fermented all of the glucose and xylose, and 75% of the arabinose in 50 h (Lawford and Rousseau, 2002; Mohagheghi et al., 2002). Overall, ethanol yield was 0.43 to 0.46 g/g with by product of xylitol (3.35 g/L), lactic acid (0.21 g/L), and acetic acid (0.84 g/L). Lactic acid production was reduced in this strain compared to previous pentose utilizing strains, possibly because the arabinose fermentation genes were integrated at the site of the putative *ldh* gene.

15.5.5 Engineering *Klebsiella* for Ethanol Production

Klebsiella oxytoca is an enteric bacterium found growing in paper and pulp streams as well as around other sources of wood. This microorganism is capable of growing at pH as low as 5.0 and temperatures as warm as 35°C. *K. oxytoca* can utilize a wide variety of sugars, including hexoses and pentoses, as well as cellobiose and cellotriose (Doran and Ingram, 1993). The strain is especially appealing for cellulose fermentation by eliminating the need for β-glucosidadase and xylosidases. Another difference between *K. oxytoca* and *E. coli* is ability of *K. oxytoca* to use urea as a sole source of nitrogen. Urea costs roughly half of the ammonium hydroxide on an equivalent nitrogen basis. Urea metabolism does not contribute to media acidification (rather reduces it) and therefore reduces the production costs (Jarboe et al., 2007). In mineral salt medium, *K. oxytoca* starin BW21 produces more than 40 g/L of ethanol from 90 g/L of glucose within 48 h. As in *E. coli*, the PET operon has also been expressed in *K. oxytoca* (Ohta et al., 1991a) and later chromosomally integrated resulting in strain P2 (Wood and Ingram, 1992). *K. oxytoca* strain BW21 derived from strain P2 by eliminating butanediol pathway, produces over 40 g/liter ethanol in 48 h in optimized urea medium, a medium designed specifically for *K. oxytoca*. In addition to the urea as the sole nitrogen source, this medium contains corn steep liquor, mineral salts, and glucose (Wood et al., 2005).

K. oxytoca P2 strains in which *Z. mobilis* ethanol genes have been integrated into the chromosome, have some distinct characteristics (Ohta et al., 1991a). They have advantage over *E. coli* based constructs, for example, (1) they contain transport systems and enzymes that allow the intracellular metabolism of short oligosaccharides from cellulose and xylan (Burchhardt and Ingram, 1992), eliminating the need for β-glucosidase and xylosidases, (2) contain chromosomally integrated *pdc* and *adh* genes (Wood and Ingram, 1992). These strains are designed specifically for their use in SSF process, but their ethanol tolerance is inferior to that of most other yeasts (Doran and Ingram, 1993), however, they rapidly produce ethanol at high yield (30 g/L in 71 h) from both glucose and cellobiose (Golias et al., 2002; Wood and Ingram, 1992).

K. oxytoca strains expressing two endoglucanases from *Erwiniachrys anthemi* and genes for ethanol production from *Z. mobilis* (Zhou and Ingram, 2001) also have high demand for bioethanol production. A native strain of *K. oxytoca* THLC0409 was isolated from a lignocelluloses-degrading microflora. Preliminary investigations were conducted to obtain the optimal growth conditions for this strain. Batch operations were applied for producing ethanol from various lignocellulose sources via direct microbial conversion. *K. oxytoca* THLC0409 has been shown to utilize cellulose (avicel, α-cellulose) as well as natural LCB (corncob, Napier grass, purified bamboo, raw bamboo, rice straw) as substrates. On the other hand, under the optimal conditions, ethanol yields on corncob and Napier grass were 0.0623 and 0.0475 g/g, respectively. This was far higher when compared to other substrates, such as avicel (0.019 g/g), α-cellulose (0.02 g/g), purified bamboo (0.02 g/g), raw bamboo (0.018 g/g), and rice straw (0.016 g/g). Ethanol yields on oat-extracted xylan and corncob-extracted xylan were 0.015 and 0.04 g/g, respectively (Tran et al., 2011).

15.5.6 Engineering *Bacillus* for Ethanol Production

Recombinant cellulolytic *Bacillus subtilis* strains are under construction these days, because of their numerous advantages, for example, (1) it is industrially safe (GRAS microorganism by the Food and Drug Administration), (2) has a very high-protein-secreting capability, and (3) fast growing with very low nutrient requirements. It has native properties of utilizing soluble pentose, hexose sugars including glucose, xylose, mannose, cellobiose, and so on, with tolerance to very high concentrations of salts and solvents. Further, its genome has been sequenced. Given it is an industrially safe microorganism, the biomass after fermentation can be used as animal feed (Bien et al., 2014). Further, *B. subtilis* is considered a perfect host for the production of secretary protein (Yamane et al., 2004). It can naturally produce several polysaccharide degrading enzymes. Besides carbohydrate active enzymes, *B. subtilis* can also produce metabolites, which can enhance the enzymatic hydrolysis of cellulose (Kim et al., 2009).

A family of 48 cellobiohydrolases from *Clost. phytofermentans* have been successfully expressed and secreted as major extracellular protein in *B. subtilis* using high-level expression and secretion system (Zhang et al., 2010). Doi (2008) has constructed a cellulosome producing *B. substilis* strain from a protease deficient *B. subtilis*. However, for making good cellulolytic strain, coexpression of many cellulases in the same strain remains to be challenging. Plasmid and chromosome integration expression has been proposed to be a better alternative for minimizing complex expression (Nguyen et al., 2005) as the chromosome of *B. subtilis* has a large number of dispensable regions (Westers et al., 2003). Cellulase productivity of *B. subtilis* may be enhanced by reducing its genome size by eliminating spore-formation-related genes, protease genes, sigma factors, genes related to catabolite inhibition, glucose metabolism, and intercellular signaling mechanism (Ara et al., 2007). In this regard, Morimoto et al. (2008) have enhanced recombinant protein productivity by genome reduction by depleting of 874 kb (20%) of genomic sequence and named the new stain as *B. subtilis* MBG 874. One-step production of ethanol or lactate from cellulose or pretreated biomass from recombinant *B. subtilis* is also possible (Zhang and Zhang, 2010).

15.5.7 Yeasts and Their Role in Fermentation

Sacch. cerevisiae is a facultative anaerobe that can ferment glucose into ethanol efficiently. It can tolerate temperatures up to 40°C with an optimum temperature of about 30–35°C. It is a preferred organism in the industrial scale for ethanol production from sugar or starch rich substrates because fermentation of sugar and starch generally reaches 90%–95% of the theoretical yield, equivalent to 0.45–0.48 g/g sugar in the raw material. The theoretical specific productivity for *Sacch. cerevisiae* is around 2 g ethanol/g cells/h. This makes it a much faster ethanol producing organism compared to filamentous fungi (Skoog and Hahn-Hägerdal, 1988). Lately *Sacch. cerevisiae* is also being employed in ethanol production from LCB because it can tolerate ethanol up to 105 g/liter (Busche et al., 1992). This organism can also tolerate lignocellulose derived metabolic inhibitors (organic acids, phenolics and furan derivatives) (Hahn-Hägerdal et al., 2006) as well as a wide range of pH (although acidic pH preferred to avoid the problem of contamination by pathogenic bacteria). The main challenge associated with ethanol production from lignocellulosic hydrolysate using *Sacch. cerevisiae* is its inability to ferment a wide range of sugars (mono-, di-, and trisaccharides) derived from cellulose and hemicellulose thus making the ethanol production cost-prohibitive. Neither it can readily utilize starch, cellulose, and hemicellulose, nor the disaccharides cellobiose and xylobiose (Lynd et al., 2002; Zhang and Lynd, 2005)]. Further, *Sacch. cerevisiae* is also unable to ferment pentose sugars (xylose and arabinose). Given the earlier challenges, extensive research has been focused on engineering *Sacch. cerevisiae* for efficient utilization of sugars in commercial settings.

Yeasts, other than *Sacch. cerevisiae*, also referred to as nonconventional yeast, are important in alcohol fermentation. They include the species of *Zygosaccharomyces* and *Hanensula polymorpha*. These species can utilize pentose sugars, such as xylose and arabinose, as well as starch and lignocellulose derived di- and trisaccharides (Skoog and Hahn-Hägerdal, 1988). However, their use in industrial settings to produce bioethanol from LCB has been unsuccessful due to lack of robustness.

So far, more than 2000 yeast species have been described and some of these could provide a solution to these limitations mostly encountered by native *Saach. cerevisiae*, that is, unable to ferment pentose sugars (xylose and arabinose). Utilization of pentose sugars is a predominant cost-reducing factor in second-generation bioethanol reactors. The nonconventional yeast species showing unusual tolerance to stresses include *Zygosaccharomyces rouxii* (osmotolerance), *Kluyveromyces marxianus* (Ogataea; methylotrophic yeast), and *H. polymorpha* (thermotolerance), *Dekkera bruxellensis* (ethanol tolerance), *Pichia kudriavzevii* (furan derivatives tolerance), and *Z. bailii* (acetic acid tolerance). These organisms could be potentially used as model organisms to study the molecular basis for the tolerances as well as to further develop robust *Sacch. cerevisia*e (Radecka et al., 2015).

The hemicellulosic hydrolysate consists pentose sugar mainly xylose and arabinose, which consistitute about 25% of the total ferentable sugar present in the lignocellulosic hydrolysates. However, depending on the raw material the arabinose fraction may be substantial. The natural xylose fermenting organisms are very limited; for example, *Scheffersomyces stipitis* (*Pichia*) and its anamorphs *Candida shehatae* and *Pachysolen tanophilus* have been isolated that can ferment xylose to ethanol under oxygen-limited conditions. *Pich. stipitis* is also able to produce hemicellulolytic enzymes (Njoku et al., 2013) and its unique ability to ferment xylose with less production of xyliol (a poor substrate for pentose sugar fermentation) (Nigam, 2001). Furthermore, it will not require vitamins for growth. It also possesses both high-affinity and low-affinity proton symport that operate simultaneously. These yeasts are less tolerant to pH, ethanol, and hydrolysate inhibitors when compared to *Sacch. cerevisiae* (Koti et al., 2016; Ma et al., 2013). Nonetheless, they can be utilized as the source of genes to be introduced into *Sacch. cerevisiae* for improvement of ethanol production in industrial processes.

15.5.8 Metabolic Engineering of *Sacch. cerevisiae* for Pentose Fermentation

Sacch. cerevisiae cannot utilize xylose but it can use its keto isomer xylulose and ferment it to ethanol (Chiang et al., 1981). Although it harbors all the genes (xylose reductase, xylose dehydrogenase, xylulokinase) required for the xylose utilization (Deng and Ho, 1990), it cannot grow in xylose medium or ferment it to ethanol. It might be due to low level of expression of these genes. Attempts to upregulate endogenous xylose utilization pathways through an adaptation protocol have been positive, the rate of xylose utilization and its fermentation to ethanol by such strains have not reached the industrially competitive levels yet (Attfield and Bell, 2006).

Initial attempts to introduce the xylose isomerization pathway genes (Fig. 15.4) into *Sacch. cerevisiae* employing bacterial xylose isomerization pathway genes have been unsuccessful due to the difficulties in expressing these genes functionally in yeast (Amore et al., 1989; Moes et al., 1996; Sarthy et al., 1987). There have been attempts to identify novel xylose isomerization pathway genes from soil metagenomics libraries based on protein sequences and activities in *E. coli*, however, the identified xylose isomerization pathway genes could not perform efficiently in *Sacch. cerevisiae* (Parachin and Gorwa-Grauslund, 2011). Discovery and application of eukaryotic xylose isomerization coding genes from anaerobic fungi is ongoing (Harhangi et al., 2003; Madhavan et al., 2009). Kuyper et al. (2003) have reported that *Sacch. cerevisiae* can utilize xylose when heterologous genes XYL1 and XYL2 encoding xylose reductase and xylitol

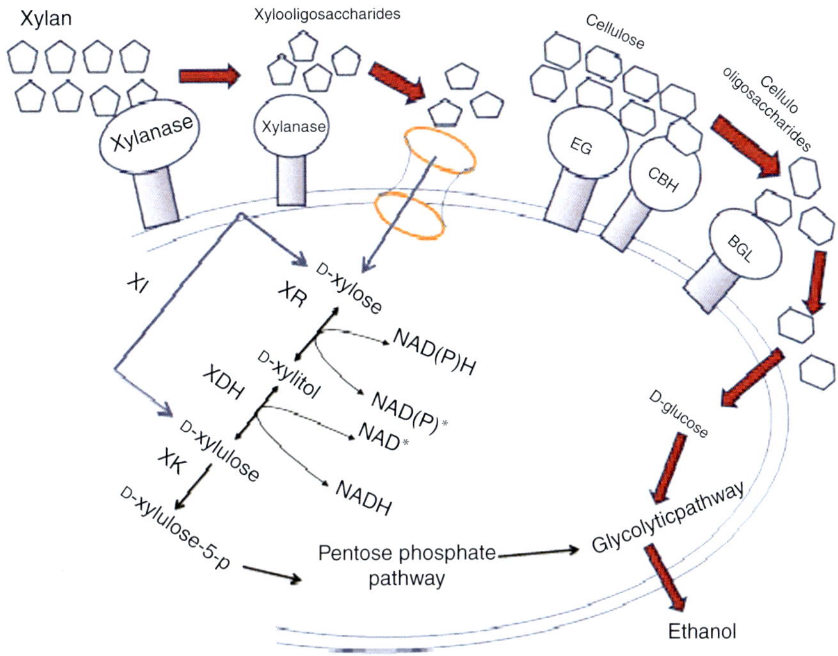

FIGURE 15.4 **Schematic representation of xylose utilization pathway.** BGL: β-Glucosidases, CBH: cellobiohydrolase EG: Endogluconase, *XDH*, Xylose dehydrogenase; *XI*, xylose isomerase; *XK*, xylose kinase; *XR*, xylose reductase.

dehydrogenase from *P. stipites* were introduced and overexpressed in *Sacch. cerevisiae*. The ethanol yields obtained in such engineered *Sacch. cerevisiae* was found to be low due to cofactor imbalance. Xylose reductase has a higher specificity for NADPH than for NADH (Km = 3.2 μmol./L for NADPH and Km = 40 μmol./L for NADH) and xylose dehydrogenase uses only NAD. In yeast, there is a disparity between the amount and accessibility to intracellular NADPH and NADH for the xylose pathway enzymes (Fernandes et al., 2009), an impairment in the redox balance resulting from the different coenzyme specificities of xylose reductase and xylitol dehydrogenase. This disparity could be due to competition for NAD by other endogenous metabolic enzymes or inefficiency of other xylose pathway enzymes (Fernandes et al., 2009; Krahulec et al., 2010). A severe cofactor imbalance can also lead to poor cell growth or little ethanol production even though the enzymes are actively expressed (Bruinenberg et al., 1983; Hahn-Hägerdal et al., 2007). Various approaches to alleviate the cofactor imbalance have been reported and these efforts include (1) controlling xylose reductase/xylose dehydrogenase expression ratio (Walfridsson et al., 1996), (2) mutations to reduce the affinity of xylose reductase for NADPH (Jeppsson et al., 2006), (3) reduce the affinity of xylose dehydrogenase for NAD (Metzger and Hollenberg, 1995), and (4) shifting the cofactor specificity of xylose dehydrogenase from NAD to NADP (Hou et al., 2007). Similarly, to avoid xylitol accumulation and cofactor imbalances xylose isomerization pathway genes that directly converts D-xylose into D-xylulose without redox reactions have also been attempted (Jeffries, 2006).

Heterologous expression of bacterial xylose isomeraization genes, which was initially from bacterial strains (Amore et al., 1989), anaerobic fungi assimilating xylose, for example,

Piromyces (Kuyper et al., 2003) and *Orpinomyces* (Madhavan et al., 2009) have also been introduced into *Sacch. cerevisiae* but expression of an active prokaryotic xylose isomerase in *Sacch. cerevisiae* has been a challenge. Efforts to express an active prokaryotic xylose isomerase from *E. coli* (Sarthy et al., 1987), *B. subtilis* (Amore et al., 1989), and *Clost. thermosulfurogens* (Moes et al., 1996) in *Sacch. cerevisiae* have failed. The first successful attempt at expression of *xylaA* gene encoding for xylose isomerase from thermophilic bacterium *Thermus thermophilus* into *Sacch. cerevisiae* was performed by Walfridsson et al. (1996), however, the activity of thermophilic enzymes (temperature optima of 85°C) was very low at 30°C. More recently, a xylose isomerase from obligate anaerobic rumen fungus, that is, *Piromyces*, has been expressed in *Saach. cerevisiae* with an activity of ~1 U/mg protein at 30°C and produced a high yield of ethanol (0.42 g/g xylose). One of the disadvantages has been that the xylose isomerase is strongly inhibited by xylitol. Bacterial xylose isomerase with high sequence similarity to the *Piromyces* xylose isomerase, such as those from *Bacteroides thetaiotaomicron* and *Xanthomonas campestris*, have also expressed in *Sacch. cerevisiae*, but the enzyme expression levels in *Sacch. cerevisiae* are lower than that of *Piromyces*. Similarly, Brat et al. (2009) have reported successful cloning and expression of xylose isomerase from *Clost. phytofermentans* in *Sacch. cerevisiae* with ethanol yield of 0.42 g/g of xylose consumed.

15.6 CONCLUSION AND FUTURE PERSPECTIVES

In conclusion, much research has gone into identifying many organisms that are capable of producing industrially important enzymes for ethanol production. Isolation of genes, heterologous cloning of these genes into suitable organisms and metabolic engineering has been largely successful. Some of recombinant organisms have been very robust, however, challenges to use them in commercial settings remain to be addressed, although the potential of using these organisms in commercial settings for ethanol production remains high and optimistic.

References

Ali, S.S., Nugent, B., Mullins, E., Doohan, F.M., 2016. Fungal-mediated consolidated bioprocessing: the potential of *Fusarium oxysporum* for the lignocellulosic ethanol industry. AMB Express 1, 1–13.

Amore, R., Wilhelm, M., Hollenberg, C.P., 1989. The fermentation of xylose—an analysis of the expression of *Bacillus* and *Actinoplanes* xylose isomerase genes in yeast. Appl. Microbiol. Biotechnol. 30, 351–357.

Angelidaki, I., Tomás, A.F., Karakashev, D., 2012. DSMZ 24726 for second generation bioethanol production. Patent number WO/2012/059105.

Ara, K., Ozaki, K., Nakamura, K., Yamane, K., Sekiguchi, J., Ogasawara, N., 2007. *Bacillus* minimum genome factory: effective utilization of microbial genome information. Biotechnol. Appl. Biochem. 46, 169–178.

Arora, A., Nain, L., Gupta, J., 2005. Solid-state fermentation of wood residues by *Streptomyces griseus* B1, a soil isolate, and solubilization of lignins. World J. Microbiol. Biotechnol. 21, 303–308.

Asghari, A., Bothast, R., Doran, J., Ingram, L., 1996. Ethanol production from hemicellulose hydrolysates of agricultural residues using genetically engineered *Escherichia coli* strain KO11. J. Ind. Microbiol. 16, 42–47.

Attfield, P.V., Bell, P.J., 2006. Use of population genetics to derive non recombinant *Saccharomyces cerevisiae* strains that grow using xylose as a sole carbon source. FEMS Yeast Res. 6, 862–868.

Bai, F., Anderson, W., Moo-Young, M., 2008. Ethanol fermentation technologies from sugar and starch feedstocks. Biotechnol. Adv. 26, 89–105.

Balat, M., Balat, H., Öz, C., 2008. Progress in bioethanol processing. Progr. Energy Combust. Sci. 34, 551–573.

Banerjee, S., Mudliar, S., Sen, R., Giri, B., Satpute, D., Chakrabarti, T., Pandey, R., 2010. Commercializing lignocellulosic bioethanol: technology bottlenecks and possible remedies. Biofuel. Bioprod. Biorefin. 4, 77–93.

Behera, S., Arora, R., Nandhagopal, N., Kumar, S., 2014. Importance of chemical pretreatment for bioconversion of lignocellulosic biomass. Renew. Sustain. Energy Rev. 36, 91–106.

Bergquist, P.L., Gibbs, M.D., Morris, D.D., Te'o, V.J., Saul, D.J., Morgan, H.W., 1999. Molecular diversity of thermophilic cellulolytic and hemicellulolytic bacteria. FEMS Microbiol. Ecol. 28, 99–110.

Bien, T.L.T., Tsuji, S., Tanaka, K., Takenaka, S., Yoshida, K.-i., 2014. Secretion of heterologous thermostable cellulases in *Bacillus subtilis*. J. Gen. Appl. Microbiol. 60, 175–182.

Bok, J.D., Yernool, D.A., Eveleigh, D.E., 1998. Purification, characterization, and molecular analysis of thermostable cellulases CelA and CelB from *Thermotoga neapolitana*. Appl. Environ. Microbiol. 64, 4774–4781.

Brat, D., Boles, E., Wiedemann, B., 2009. Functional expression of a bacterial xylose isomerase in *Saccharomyces cerevisiae*. Appl. Environ. Microbiol. 75, 2304–2311.

Brown, S.D., Guss, A.M., Karpinets, T.V., Parks, J.M., Smolin, N., Yang, S., Land, M.L., Klingeman, D.M., Bhandiwad, A., Rodriguez, M., 2011. Mutant alcohol dehydrogenase leads to improved ethanol tolerance in *Clostridium thermocellum*. Proc. Natl. Acad. Sci. 108, 13752–13757.

Bruinenberg, P.M., de Bot, P.H., van Dijken, J.P., Scheffers, W.A., 1983. The role of redox balances in the anaerobic fermentation of xylose by yeasts. Europ. J. Appl. Microbiol. Biotechnol. 18, 287–292.

Burchhardt, G., Ingram, L., 1992. Conversion of xylan to ethanol by ethanologenic strains of *Escherichia coli* and *Klebsiella oxytoca*. Appl. Environ. Microbiol. 58, 1128–1133.

Busche, R.M., Scott, C.D., Davison, B.H., Lynd, L.R., 1992. Ethanol, the ultimate feedstock. Appl. Biochem. Biotechnol. 34, 395.

Chanakya, H., Borgaonkar, S., Meena, G., Jagadish, K., 1993. Solid-phase biogas production with garbage or water hyacinth. Bioresour. Technol. 46, 227–231.

Chandel, A.K., Chandrasekhar, G., Silva, M.B., Silvério da Silva, S., 2012. The realm of cellulases in biorefinery development. Crit. Rev. Biotechnol. 32, 187–202.

Chandel, A.K., Singh, O.V., 2011. Weedy lignocellulosic feedstock and microbial metabolic engineering: advancing the generation of "Biofuel". Appl. Microbiol. Biotechnol. 89, 1289–1303.

Chandel, A.K., Singh, O.V., Rao, L.V., 2010. Biotechnological applications of hemicellulosic derived sugars: state-of-the-art. Sustainable Biotechnology. Springer, Dordrecht, Netherlands, 63-81.

Chandel, A.K., Singh, O.V., Rao, L.V., Chandrasekhar, G., Narasu, M.L., 2011. Bioconversion of novel substrate *Saccharum spontaneum*, a weedy material, into ethanol by *Pichia stipitis* NCIM3498. Bioresour. Technol. 102, 1709–1714.

Chen, X., 2017. Yeast cell surface display: an efficient strategy for improvement of bioethanol fermentation performance. Bioengineered 8, 115–119.

Cheng, C., Zhang, M., Xue, C., Bai, F., Zhao, X., 2017. Development of stress tolerant *Saccharomyces cerevisiae* strains by metabolic engineering: new aspects from cell flocculation and zinc supplementation. J. Biosci. Bioeng. 123, 141–146.

Chiang, L.-C., Gong, C.-S., Chen, L.-F., Tsao, G.T., 1981. D-Xylulose fermentation to ethanol by *Saccharomyces cerevisiae*. Appl. Environ. Microbiol. 42, 284–289.

Chiaramonti, D., Prussi, M., Ferrero, S., Oriani, L., Ottonello, P., Torre, P., Cherchi, F., 2012. Review of pretreatment processes for lignocellulosic ethanol production, and development of an innovative method. Biomass Bioenergy 46, 25–35.

Chong, H., Huang, L., Yeow, J., Wang, I., Zhang, H., Song, H., Jiang, R., 2013. Improving ethanol tolerance of *Escherichia coli* by rewiring its global regulator cAMP receptor protein (CRP). PLoS ONE 8, e57628.

Conway, T., 1992. The Entner-Doudoroff pathway: history, physiology and molecular biology. FEMS Microbiol. Rev. 103, 1–27.

Crawford, D.L., McCoy, E., 1972. Cellulases of *Thermomonospora fusca* and *Streptomyces thermodiastaticus*. Appl. Microbiol. 24, 150–152.

Dashtban, M., Schraft, H., Syed, T.A., Qin, W., 2010. Fungal biodegradation and enzymatic modification of lignin. Int. J. Biochem. Mol. Biol. 1, 36–50.

Deng, X.X., Ho, N.W., 1990. Xylulokinase activity in various yeasts including *Saccharomyces cerevisiae* containing the cloned xylulokinase gene. Appl. Biochem. Biotechnol. 24, 193–199.

Deng, Y., Fong, S.S., 2010. Influence of culture aeration on the cellulase activity of *Thermobifida fusca*. Appl. Microbiol. Biotechnol. 85, 965–974.

Dien, B.S., Nichols, N.N., O'bryan, P.J., Bothast, R.J., 2000. Development of new ethanologenic *Escherichia coli* strains for fermentation of lignocellulosic biomass. Appl. Biochem. Biotechnol. 84, 181–196.

Doelle, M.B., Doelle, H.W., 1990. Sugar-cane molasses fermentation by *Zymomonas mobilis*. Appl. Microbiol. Biotechnol. 33, 31–35.

Doi, R., 2008. Cellulases of mesophilic microorganisms. Annal. New York Acad. Sci. 1125 (1), 267–270.

Doran, J.B., Cripe, J., Sutton, M., Foster, B., 2000. Fermentations of pectin-rich biomass with recombinant bacteria to produce fuel ethanol. Appl. Biochem. Biotechnol. 84, 141–152.

Doran, J.B., Ingram, L., 1993. Fermentation of crystalline cellulose to ethanol by *Klebsiella oxytoca* containing chromosomally integrated *Zymomonas mobilis* genes. Biotechnol. Prog. 9, 533–538.

Fernandes, S., Tuohy, M.G., Murray, P.G., 2009. Xylose reductase from the thermophilic fungus *Talaromycese mersonii*: cloning and heterologous expression of the native gene (Texr) and a double mutant (Texr K271R+ N273D) with altered coenzyme specificity. J. Biosci. 34, 881–890.

Frazer, F., McCaskey, T., 1989. Wood hydrolyzate treatments for improved fermentation of wood sugars to 2,3-butanediol. Biomass 18, 31–42.

Freier, D., Mothershed, C.P., Wiegel, J., 1988. Characterization of *Clostridium thermocellum* JW20. Appl. Environ. Microbiol. 54, 204–211.

Fu, C., Mielenz, J.R., Xiao, X., Ge, Y., Hamilton, C.Y., Rodriguez, M., Chen, F., Foston, M., Ragauskas, A., Bouton, J., 2011. Genetic manipulation of lignin reduces recalcitrance and improves ethanol production from switchgrass. Proc. Natl. Acad. Sci. 108, 3803–3808.

Godden, B., Legon, T., Helvenstein, P., Penninckx, M., 1989. Regulation of the production of hemicellulolytic and cellulolytic enzymes by a *Streptomyces* sp. growing on lignocellulose. J. Gen. Microbiol. 135, 285–292.

Golias, H., Dumsday, G.J., Stanley, G.A., Pamment, N.B., 2002. Evaluation of a recombinant *Klebsiella oxytoca* strain for ethanol production from cellulose by simultaneous saccharification and fermentation: comparison with native cellobiose-utilising yeast strains and performance in co-culture with thermotolerant yeast and *Zymomonas mobilis*. J. Biotechnol. 96, 155–168.

Gunasekaran, P., Raj, K.C., 1999. Ethanol fermentation technology–*Zymomonas mobilis*. Curr. Sci., 56–68.

Hahn-Hägerdal, B., Galbe, M., Gorwa-Grauslund, M.-F., Lidén, G., Zacchi, G., 2006. Bio-ethanol: the fuel of tomorrow from the residues of today. Trends Biotechnol. 24, 549–556.

Hahn-Hägerdal, B., Karhumaa, K., Fonseca, C., Spencer-Martins, I., Gorwa-Grauslund, M.F., 2007. Towards industrial pentose-fermenting yeast strains. Appl. Microbiol. Biotechnol. 74, 937–953.

Hamilton-Brehm, S.D., Mosher, J.J., Vishnivetskaya, T., Podar, M., Carroll, S., Allman, S., Phelps, T.J., Keller, M., Elkins, J.G., 2010. *Caldicellulosiruptor obsidiansis* sp. nov., an anaerobic, extremely thermophilic, cellulolytic bacterium isolated from Obsidian Pool, Yellowstone National Park. Appl. Environ. Microbiol. 76, 1014–1020.

Harhangi, H.R., Akhmanova, A.S., Emmens, R., van der Drift, C., de Laat, W.T., van Dijken, J.P., Jetten, M.S., Pronk, J.T., den Camp, H.J.O., 2003. Xylose metabolism in the anaerobic fungus *Piromyces* sp. strain E2 follows the bacterial pathway. Arch. Microbiol. 180, 134–141.

Hatakka, A.I., 1983. Pretreatment of wheat straw by white-rot fungi for enzymic saccharification of cellulose. Appl. Microbiol. Biotechnol. 18, 350–357.

Hou, J., Shen, Y., Li, X., Bao, X., 2007. Effect of the reversal of coenzyme specificity by expression of mutated *Pichia stipitis* xylitol dehydrogenase in recombinant *Saccharomyces cerevisiae*. Lett. Appl. Microbiol. 45, 184–189.

Ingram, L., Conway, T., Clark, D., Sewell, G., Preston, J., 1987. Genetic engineering of ethanol production in *Escherichia coli*. Appl. Environ. Microbiol. 53, 2420–2425.

Jarboe, L.R., Grabar, T.B., Yomano, L.P., Shanmugan, K.T., Ingram, L.O., 2007. Development of ethanologenic bacteria. Biofuels. Springer, Berlin, Heidelberg, Germany, 237-261.

Jeffries, T.W., 2006. Engineering yeasts for xylose metabolism. Curr. Opin. Biotechnol. 17, 320–326.

Jeppsson, M., Bengtsson, O., Franke, K., Lee, H., Hahn Hägerdal, B., Gorwa Grauslund, M.F., 2006. The expression of a *Pichia stipitis* xylose reductase mutant with higher KM for NADPH increases ethanol production from xylose in recombinant *Saccharomyces cerevisiae*. Biotechnol. Bioeng. 93, 665–673.

Jin, M., Balan, V., Gunawan, C., Dale, B.E., 2011. Consolidated bioprocessing (CBP) performance of *Clostridium phytofermentans* on AFEX-treated corn stover for ethanol production. Biotechnol. Bioeng. 108, 1290–1297.

Joshi, B., Bhatt, M.R., Sharma, D., Joshi, J., Malla, R., Sreerama, L., 2011. Lignocellulosic ethanol production: current practices and recent developments. Biotechnol. Mol. Biol. Rev. 6, 172–182.

Kanafusa-Shinkai, S., Wakayama, J.I., Tsukamoto, K., Hayashi, N., Miyazaki, Y., Ohmori, H., Tajima, K., Yokoyama, H., 2013. Degradation of microcrystalline cellulose and non-pretreated plant biomass by a cell-free extracellular cellulase/hemicellulase system from the extreme thermophilic bacterium *Caldicellulosiruptor bescii*. J. Biosci. Bioeng. 115, 64–70.

Kannan, V., Mutharasan, R., 1985. Ethanol fermentation characteristics of *Thermoanaerobacter ethanolicus*. Enzyme Microbiol. Technol. 7, 87–89.

Kasavi, C., Eraslan, S., Oner, E.T., Kirdar, B., 2016. An integrative analysis of transcriptomic response of ethanol tolerant strains to ethanol in *Saccharomyces cerevisiae*. Mol. Biosyst. 12, 464–476.

Kataeva, I., Foston, M.B., Yang, S.-J., Pattathil, S., Biswal, A.K., Poole II, F.L., Basen, M., Rhaesa, A.M., Thomas, T.P., Azadi, P., 2013. Carbohydrate and lignin are simultaneously solubilized from unpretreated switch grass by microbial action at high temperature. Energ. Env. Sci. 6, 2186–2195.

Khatun, M.M., Yu, X., Kondo, A., Bai, F., Zhao, X., 2017. Improved ethanol production at high temperature by consolidated bioprocessing using *Saccharomyces cerevisiae* strain engineered with artificial zinc finger protein. Bioresoures Technology. Springer, New York, NY, United States.

Kim, E.S., Lee, H.J., Bang, W.G., Choi, I.G., Kim, K.H., 2009. Functional characterization of a bacterial expansion from *Bacillus subtilis* for enhanced enzymatic hydrolysis of cellulose. Biotechnol. Bioeng. 102, 1342–1353.

Koti, S., Govumoni, S.P., Gentela, J., Rao, L.V., 2016. Enhanced bioethanol production from wheat straw hemicellulose by mutant strains of pentose fermenting organisms *Pichia stipites* and *Candida shehatae*. Springerplus 5, 1–9.

Krahulec, S., Petschacher, B., Wallner, M., Longus, K., Klimacek, M., Nidetzky, B., 2010. Fermentation of mixed glucose-xylose substrates by engineered strains of *Saccharomyces cerevisiae:* role of the coenzyme specificity of xylose reductase, and effect of glucose on xylose utilization. Microb. Cell Fact. 9, 1–14.

Kricka, W., Fitzpatrick, J., Bond, U., 2014. Metabolic engineering of yeasts by heterologous enzyme production for degradation of cellulose and hemicellulose from biomass: a perspective. Front Microbiol. 5, 1–11.

Kuhad, R.C., Gupta, R., Khasa, Y.P., Singh, A., 2010. Bioethanol production from *Lantana camara* (red sage): pretreatment, saccharification and fermentation. Bioresour. Technol. 101, 8348–8354.

Kumar, A., Singh, L., Ghosh, S., 2009. Bioconversion of lignocellulosic fraction of water- hyacinth (*Eichhornia crassipes*) hemicellulose acid hydrolysate to ethanol by *Pichia stipitis*. Bioresour. Technol. 100, 3293–3297.

Kuyper, M., Harhangi, H.R., Stave, A.K., Winkler, A.A., Jetten, M.S., de Laat, W.T., den Ridder, J.J., Op den Camp, H.J., van Dijken, J.P., Pronk, J.T., 2003. High-level functional expression of a fungal xylose isomerase: the key to efficient ethanolic fermentation of xylose by *Saccharomyces cerevisiae*. FEMS Yeast Res. 4, 69–78.

Laopaiboon, L., Nuanpeng, S., Srinophakun, P., Klanrit, P., Laopaiboon, P., 2008. Selection of *Saccharomyces cerevisiae* and investigation of its performance for very high gravity ethanol fermentation. Biotechnology 7, 493–498.

Larsson, S., Palmqvist, E., Hahn-Hägerdal, B., Tengborg, C., Stenberg, K., Zacchi, G., Nilvebrant, N.-O., 1999. The generation of fermentation inhibitors during dilute acid hydrolysis of softwood. Enzyme. Microbial. Technol. 24, 151–159.

Lawford, H.G., Rousseau, J.D., 1999. The effect of glucose on high-level xylose fermentations by recombinant *Zymomonas* in batch and fed-batch fermentations. Appl. Biochem. Biotechnol. 77, 235–249.

Lawford, H.G., Rousseau, J.D., 2002. Performance testing of *Zymomonas mobilis* metabolically engineered for cofermentation of glucose, xylose, and arabinose. Biotechnology for Fuels and Chemicals. Humana Press, New York, NY, United States, 429-448.

Lee, A., Choi, K.-H., Yoon, D., Kim, S., Cha, J., 2017. Characterization of a thermostable glycoside hydrolase family 36 α-galactosidase from *Caldicellulosiruptor bescii*. J. Biosci. Bioeng. 124, 289–295.

Lee, J.S., Chi, W.-J., Hong, S.-K., Yang, J.-W., Chang, Y.K., 2013. Bioethanol production by heterologous expression of Pdc and AdhII in *Streptomyces lividans*. Appl. Microbiol. Biotechnol. 97, 6089–6097.

Leite, A.R., Guimarães, W.V., Araújo, E.F.D., Silva, D.O., 2000. Fermentation of sweet whey by recombinant *Escherichia coli* K011. Braz. J. Microbiol. 31, 211–214.

Li, C., Wang, Q., Zhao, Z.K., 2008. Acid in ionic liquid: an efficient system for hydrolysis of lignocellulose. Green Chem. 10, 177–182.

Lima, A.O., Davis, D.F., Swiatek, G., McCarthy, J.K., Yernool, D., Pizzirani-Kleiner, A.A., Eveleigh, D.E., 2009. Evaluation of GFP tag as a screening reporter in directed evolution of a hyperthermophilic β-glucosidase. Mol. Biotechnol. 42, 205–215.

Lima, A.O., Quecine, M.C., Fungaro, M.H., Andreote, F.D., Maccheroni, Jr., W., Araújo, W.L., Silva-Filho, M.C., Pizzirani-Kleiner, A.A., Azevedo, J.L., 2005. Molecular characterization of a β-1,4-endoglucanase from an endophytic *Bacillus pumilus* strain. Appl. Microbiol. Biotechnol. 68, 57–65.

Lin, Y., Tanaka, S., 2006. Ethanol fermentation from biomass resources: current state and prospects. Appl. Microbiol. Biotechnol. 69, 627–642.

Liu, Z., Slininger, P., Dien, B., Berhow, M., Kurtzman, C., Gorsich, S., 2004. Adaptive response of yeasts to furfural and 5-hydroxymethylfurfural and new chemical evidence for HMF conversion to 2,5-bis-hydroxymethylfuran. J. Ind. Microbiol. Biotechnol. 31, 345–352.

Liu, Z.L., 2011. Molecular mechanisms of yeast tolerance and in situ detoxification of lignocellulose hydrolysates. Appl. Microbiol Biotechnol. 90, 809–825.

Liu, Z.L., 2012. Genomics of yeast tolerance and in situ detoxification. Microbial Stress Tolerance for Biofuels. Springer, New York, NY, United States, 1-28.

Liu, Z.L., Slininger, P.J., 2005. Development of genetically engineered stress tolerant ethanologenic yeasts using integrated functional genomics for effective biomass conversion to ethanol. Agr. Prod. Consumer Energ. 1, 283–294.

Lynd, L.R., Cushman, J.H., Nichols, R.J., Wyman, C.E., 1991. Fuel ethanol from cellulosic biomass. Science 251, 1318–1323.

Lynd, L.R., Elamder, R.T., Wyman, C.E., 1996. Likely features and costs of mature biomass ethanol technology. Appl. Biochem. Biotechnol. 57, 741–761.

Lynd, L.R., Liang, X., Biddy, M.J., Allee, A., Cai, H., Foust, T., Himmel, M.E., Laser, M.S., Wang, M., Wyman, C.E., 2017. Cellulosic ethanol: status and innovation. Cur. Opin Biotechnol. 45, 202–211.

Lynd, L.R., Weimer, P.J., Van Zyl, W.H., Pretorius, I.S., 2002. Microbial cellulose utilization: fundamentals and biotechnology. Microbiol. Mol. Biol. Rev. 66, 506–577.

Lynd, L.R., Zyl, W.H.V., McBride, J.E., Laser, M., 2005. Consolidated bioprocessing of cellulosic biomass: an update. Cur. Opinion Biotechnol. 16, 577–583.

Ma, M., Wang, X., Zhang, X., Zhao, X., 2013. Alcohol dehydrogenases from *Scheffersomyces stipitis* involved in the detoxification of aldehyde inhibitors derived from lignocellulosic biomass conversion. Appl. Microbiol. Biotechnol. 97, 8411–8425.

Madhavan, A., Tamalampudi, S., Ushida, K., Kanai, D., Katahira, S., Srivastava, A., Fukuda, H., Bisaria, V.S., Kondo, A., 2009. Xylose isomerase from polycentric fungus *Orpinomyces*: gene sequencing, cloning, and expression in *Saccharomyces cerevisiae* for bioconversion of xylose to ethanol. Appl. Microbiol. Biotechnol. 82, 1067–1078.

Mäkelä, M.R., Donofrio, N., de Vries, R.P., 2014. Plant biomass degradation by fungi. Fungal Genet. Biol. 72, 2–9.

Mathew, G.M., Sukumaran, R.K., Singhania, R.R., Pandey, A., 2008. Progress in research on fungal cellulases for lignocellulose degradation 67, 898–907.

McLellan, P.J., Daugulis, A.J., Li, J., 1999. The incidence of oscillatory behavior in the continuous fermentation of *Zymomonas mobilis*. Biotechnol. Prog. 15, 667–680.

McMillan, J.D., Newman, M.M., Templeton, D.W., Mohagheghi, A., 1999. Simultaneous saccharification and cofermentation of dilute-acid pretreated yellow poplar hardwood to ethanol using xylose-fermenting *Zymomonas mobilis*. Appl. Biochem. Biotechnol. 79, 649–665.

Metzger, M.H., Hollenberg, C.P., 1995. Amino acid substitutions in the yeast *Pichia stipitis* xylitol dehydrogenase coenzyme binding domain affect the coenzyme specificity. FEBS J. 228, 50–54.

Mishima, D., Kuniki, M., Sei, K., Soda, S., Ike, M., Fujita, M., 2008. Ethanol production from candidate energy crops: water hyacinth (*Eichhornia crassipes*) and water lettuce (*Pistia stratiotes* L.). Bioresour. Technol. 99, 2495–2500.

Moes, C.J., Pretorius, I.S., van Zyl, W.H., 1996. Cloning and expression of the *Clostridium* thermosulfurogenes D-xylose isomerase gene (xylA) in *Saccharomyces cerevisiae*. Biotechnol. Lett. 18, 269–274.

Mohagheghi, A., Evans, K., Chou, Y.-C., Zhang, M., 2002. Cofermentation of glucose, xylose, and arabinose by genomic DNA-integrated xylose/arabinose fermenting strain of *Zymomonas mobilis* AX101. Appl. Biochem. Biotechnol. 98, 1–9.

Moniruzzaman, M., Ingram, L.O., 1998. Ethanol production from dilute acid hydrolysate of rice hulls using genetically engineered *Escherichia coli*. Biotechnol. Lett. 20, 943–947.

Morimoto, T., Kadoya, R., Endo, K., Tohata, M., Sawada, K., Liu, S., Ozawa, T., Kodama, T., Kakeshita, H., Kageyama, Y., 2008. Enhanced recombinant protein productivity by genome reduction in *Bacillus subtilis*. DNA Res. 15, 73–81.

Mosier, N., Wyman, C., Dale, B., Elander, R., Lee, Y., Holtzapple, M., Ladisch, M., 2005. Features of promising technologies for pretreatment of lignocellulosic biomass. Bioresour. Technol. 96, 673–686.

Mukherjee, R., Nandi, B., 2004. Improvement of in vitro digestibility through biological treatment of water hyacinth biomass by two *Pleurotus* species. Int. Biodeterior. Biodegrad. 53, 7–12.

Nakari-Setälä, T., Penttilä, M., 1995. Production of *Trichoderma reesei* cellulases on glucose-containing media. Appl. Environ. Microbiol. 61, 3650–3655.

Nguyen, H.D., Nguyen, Q.A., Ferreira, R.C., Ferreira, L.C., Tran, L.T., Schumann, W., 2005. Construction of plasmid-based expression vectors for *Bacillus subtilis* exhibiting full structural stability. Plasmid 54, 241–248.

Nigam, J., 2001. Ethanol production from wheat straw hemicellulose hydrolysate by *Pichia stipitis*. J. Biotechnol. 87, 17–27.

Njoku, S.I., Iversen, J.A., Uellendahl, H., Ahring, B.K., 2013. Production of ethanol from hemicellulose fraction of cocksfoot grass using *Pichia stipitis*. Sustain. Chem. Proc. 1, 1–7.

Ohmiya, K., Nagashima, K., Kajino, T., Goto, E., Tsukada, A., Shimizu, S., 1988. Cloning of the cellulase gene from *Ruminococcus albus* and its expression in *Escherichia coli*. Appl. Environ. Microbiol. 54, 1511–1515.

Ohta, K., Beall, D., Mejia, J., Shanmugam, K., Ingram, L., 1991a. Genetic improvement of *Escherichia coli* for ethanol production: chromosomal integration of *Zymomonas mobilis* genes encoding pyruvate decarboxylase and alcohol dehydrogenase II. Appl. Environ. Microbiol. 57, 893–900.

Palmqvist, E., Almeida, J.S., Hahn-Hägerdal, B., 1999. Influence of furfural on anaerobic glycolytic kinetics of *Saccharomyces cerevisiae* in batch culture. Biotechnol. Bioeng. 62, 447–454.

Parachin, N.S., Gorwa-Grauslund, M.F., 2011. Isolation of xylose isomerases by sequence-and function-based screening from a soil metagenomic library. Biotechnol. Biofuel. 4, 1–10.

Parker, C., Barnell, W.O., Snoep, J.L., Ingram, L.O., Conway, T., 1995. Characterization of the *Zymomonas mobilis* glucose facilitator gene product (glf) in recombinant *Escherichia coli:* examination of transport mechanism, kinetics and the role of glucokinase in glucose transport. Mol. Microbiol. 15, 795–802.

Pasha, C., Nagavalli, M., Venkateswar Rao, L., 2007. *Lantana camara* for fuel ethanol production using thermotolerant yeast. Lett. Appl. Microbiol. 44, 666–672.

Perlack, R.D., Wright, L.L., Turhollow, A.F., Graham, R.L., Stokes, B.J., Erbach, D.C., 2005. Biomass as feedstock for a bioenergy and bioproducts industry: the technical feasibility of a billion-ton annual supply. DTIC Doc. U.S. Department of Agriculture, U.S. Department of Energy, Washington, DC, United States.

Poddar, K., Mandal, L., Banerjee, G., 1991. Studies on water hyacinth (*Eichhornia crassipes*): chemical composition of the plant and water from different habitats. Ind. Vet. J. 68, 833–837.

Pourramezan, Z., Ghezelbash, G., Romani, B., Ziaei, S., Hedayatkhah, A., 2012. Screening and identification of newly isolated cellulose-degrading bacteria from the gut of xylophagous termite *Microcerotermes diversus* (Silvestri). Microbiologiia 81, 736–742.

Radecka, D., Mukherjee, V., Mateo, R.Q., Stojiljkovic, M., Foulquié-Moreno, M.R., Thevelein, J.M., 2015. Looking beyond *Saccharomyces*: the potential of non-conventional yeast species for desirable traits in bioethanol fermentation. FEMS Yeast. Res. 15, 1–13.

Rainey, F., Donnison, A., Janssen, P., Saul, D., Rodrigo, A., Bergquist, P.L., Daniel, R.M., Stackebrandt, E., Morgan, H.W., 1994. Description of *Caldicellulosiruptor saccharolyticus* gen. nov., sp. nov: an obligately anaerobic, extremely thermophilic, cellulolytic bacterium. FEMS Microbiol. Lett. 120, 263–266.

Ram, M.S., Seenayya, G., 1989. Ethanol production by *Clostridium thermocellum* SS8, a newly isolated thermophilic bacterium. Biotechnol. Lett. 11, 589–592.

Rani, K.S., Seenayya, G., 1999. High ethanol tolerance of new isolates of *Clostridium thermocellum* strains SS21 and SS22. World J. Microbiol. Biotechnol. 15, 173–178.

Renewable fuels associations, Building Partnerships Growing Markets 2017 Ethanol Industry Outlook, 2017

Rodrigues, A.L., Cavalett, A., Lima, A.O., 2010. Enhancement of *Escherichia coli* cellulolytic activity by co-production of beta-glucosidase and endoglucanase enzymes. Elect. J. Biotechnol. 13, 5–6.

Rogers, P., Lee, K., Skotnicki, M., Tribe, D., 1982. Ethanol production by *Zymomonas mobilis*. Microbial Reactions. Springer, Berlin, Heidelberg, Germany, 37-84.

Saini, A., Aggarwal, N.K., Sharma, A., Yadav, A., 2015. Prospects for irradiation in cellulosic ethanol production. Biotechnol. Res. Int. 11, 1–13.

Sanchez, B., Bautista, J., 1988. Effects of furfural and 5-hydroxymethylfurfural on the fermentation of *Saccharomyces cerevisiae* and biomass production from *Candida guilliermondii*. Enzyme. Microbiol. Technol. 10, 315–318.

Sarthy, A., McConaughy, B., Lobo, Z., Sundstrom, J., Furlong, C., Hall, B., 1987. Expression of the *Escherichia coli* xylose isomerase gene in *Saccharomyces cerevisiae*. Appl. Environ. Microbiol. 53, 1996–2000.

Scordia, D., Cosentino, S.L., Jeffries, T.W., 2010. Second generation bioethanol production from *Saccharum spontaneum* L. ssp. aegyptiacum (Willd.) Hack. Bioresour. Technol. 101, 5358–5365.

Sharma, D., Joshi, B., Bhatt, M.R., Joshi, J., Malla, R., Bhattarai, T., Sreerama, L., 2015. Isolation of cellulolytic organisms from the gut contents of termites native to Nepal and their utility in saccharification and fermentation of lignocellulosic biomass. J. Biomass Biofuel. 2, 11–20.

Shaw, A.J., Podkaminer, K.K., Desai, S.G., Bardsley, J.S., Rogers, S.R., Thorne, P.G., Hogsett, D.A., Lynd, L.R., 2008. Metabolic engineering of a thermophilic bacterium to produce ethanol at high yield. Proc. Natl. Acad. Sci., 105, pp. 13769–13774.

Shrestha, P., Joshi, B., Joshi, J., Malla, R., Sreerama, L., 2016. Isolation and physicochemical characterization of laccase from *Ganoderma lucidum*-CDBT1 isolated from its native habitat in Nepal. BioMed. Res. Int. 10, 1–10.

Sindhu, R., Binod, P., Pandey, A., 2016. Biological pretreatment of lignocellulosic biomass: an overview. Bioresour. Technol. 199, 76–82.

Skoog, K., Hahn-Hägerdal, B., 1988. Xylose fermentation. Enzyme. Microb. Technol. 10, 66–80.

Smeets, E., Junginger, H., Faaij, A., 2005. Supportive study for the OECD on alternative developments in biofuel production across the world. Report NWS-E-2005-141. Copernicus Institute for Sustainable Development, Utrecht University, Utrecht, Netherlands.

Sprenger, G.A., 1996. Carbohydrate metabolism in *Zymomonas mobilis*: a catabolic highway with some scenic routes. FEMS Microbiol. Lett 145, 301–307.

Sun, Y., Cheng, J., 2002. Hydrolysis of lignocellulosic materials for ethanol production: a review. Bioresour. Technol. 83, 1–11.

Taherzadeh, M.J., Gustafsson, L., Niklasson, C., Lidén, G., 1999. Conversion of furfural in aerobic and anaerobic batch fermentation of glucose by *Saccharomyces cerevisiae*. J. Biosci. Bioeng. 87, 169–174.

Taherzadeh, M.J., Karimi, K., 2008. Pretreatment of lignocellulosic wastes to improve ethanol and biogas production: a review. Int. J. Mol. Sci. 9, 1621–1651.

Tran, D.-T., Lin, C.-W., Lai, C.-Y., Wu, C.-H., 2011. Ethanol production from lignocelluloses by native strain *Klebsiella oxytoca* THLC0409. Waste Biomass Valoriz. 2, 389–396.

Vitikainen, M., Arvas, M., Pakula, T., Oja, M., Penttilä, M., Saloheimo, M., 2010. Array comparative genomic hybridization analysis of *Trichoderma reesei* strains with enhanced cellulase production properties. BMC Genomics 11, 1–16.

Walfridsson, M., Bao, X., Anderlund, M., Lilius, G., Bülow, L., Hahn-Hägerdal, B., 1996. Ethanolic fermentation of xylose with *Saccharomyces cerevisiae* harboring the *Thermus thermophilus* xylA gene, which expresses an active xylose (glucose) isomerase. Appl. Environ. Microbiol. 62, 4648–4651.

Warnick, T.A., Methé, B.A., Leschine, S.B., 2002. *Clostridium phytofermentans* sp. nov., a cellulolytic mesophile from forest soil. Int. J. Syst. Evol. Microbiol. 52, 1155–1160.

Weber, C., Farwick, A., Benisch, F., Brat, D., Dietz, H., Subtil, T., Boles, E., 2010. Trends and challenges in the microbial production of lignocellulosic bioalcohol fuels. Appl. Microbiol. Biotechnol. 87, 1303–1315.

Westers, H., Dorenbos, R., van Dijl, J.M., Kabel, J., Flanagan, T., Devine, K.M., Jude, F., Séror, S.J., Beekman, A.C., Darmon, E., 2003. Genome engineering reveals large dispensable regions in *Bacillus subtilis*. Mol. Biol. Evol. 20, 2076–2090.

Wheals, A.E., Basso, L.C., Alves, D.M., Amorim, H.V., 1999. Fuel ethanol after 25 years. Trends Biotechnol. 17, 482–487.

Wiegel, J., Dykstra, M., 1984. *Clostridium thermocellum*: adhesion and sporulation while adhered to cellulose and hemicellulose. Appl. Microbiol. Biotechnol. 20, 59–65.

Wood, B.E., Ingram, L., 1992. Ethanol production from cellobiose, amorphous cellulose, and crystalline cellulose by recombinant *Klebsiella oxytoca* containing chromosomally integrated *Zymomonas mobilis* genes for ethanol production and plasmids expressing thermostable cellulase genes from *Clostridium thermocellum*. Appl. Env. Microbiol. 58, 2103–2110.

Wood, B.E., Yomano, L., York, S., Ingram, L., 2005. Development of industrial medium required elimination of the 2,3-butanediol fermentation pathway to maintain ethanol yield in an ethanologenic strain of klebsiella oxytoca. Biotechnol. Prog. 21, 1366–1372.

Xu, Q., Singh, A., Himmel, M.E., 2009. Perspectives and new directions for the production of bioethanol using consolidated bioprocessing of lignocellulose. Curr. Opin. Biotechnol. 20, 364–371.

Yamane, K., Bunai, K., Kakeshita, H., 2004. Protein traffic for secretion and related machinery of *Bacillus subtilis*. Biosci. Biotechnol. Biochem. 68, 2007–2023.

Yang, B., Wyman, C.E., 2008. Pretreatment: the key to unlocking low cost cellulosic ethanol. Biofuels, Bioprod. Bioref. 2, 26–40.

Yang, S.-J., Kataeva, I., Wiegel, J., Yin, Y., Dam, P., Xu, Y., Westpheling, J., Adams, M.W., 2010. Classification of *Anaerocellum thermophilum* strain DSM 6725 as *Caldicellulosiruptor bescii* sp. nov. Int. J. Syst. Evol. Microbiol. 60, 2011–2015.

Yuan, S.F., Guo, G.L., Hwang, W.S., 2017. Ethanol production from dilute acid steam exploded lignocellulosic feedstocks using an isolated multistress tolerant *Pichia kudriavzevii* strain. Microb. Biotechnol 10, 1581–1590.

Zaldivar, J., Ingram, L.O., 1999. Effect of organic acids on the growth and fermentation of ethanologenic *Escherichia coli* LY01. Biotechnol. Bioeng. 66, 203–210.

Zeikus, J., Ben-Bassat, A., Ng, T.K., Lamed, R.J., 1981. Thermophilic ethanol fermentations. Trends in the Biology of Fermentations for Fuels and Chemicals. Springer, New York, NY, United States, 441-461.

Zhang, K., Fang, Y.-H., Gao, K.-H., Sui, Y., Zheng, D.-Q., Wu, X.-C., 2017. Effects of genome duplication on phenotypes and industrial applications of *Saccharomyces cerevisiae* strains. Appl. Microbiol. Biotechnol. 101, 5405–5414.

Zhang, M., Eddy, C., Deanda, K., Finkelstein, M., Picataggio, S., 1995a. Metabolic engineering of a pentose metabolism pathway in ethanologenic *Zymomonas mobilis*. Science 267, 240.

Zhang, M., Franden, M., Newman, M., McMillan, J., Finkelstein, M., Picataggio, S., 1995b. Promising ethanologens for xylose fermentation. Appl. Biochem. Biotechnol. 51, 527–536.

Zhang, X.Z., Zhang, Y.H.P., 2010. One-step production of biocommodities from lignocellulosic biomass by recombinant cellulolytic *Bacillus subtilis*: opportunities and challenges. Eng. Life Sci. 10, 398–406.

Zhang, X.-Z., Zhang, Z., Zhu, Z., Sathitsuksanoh, N., Yang, Y., Zhang, Y.-H.P., 2010. The noncellulosomal family 48 cellobiohydrolase from *Clostridium phytofermentans* ISDg: heterologous expression, characterization, and processivity. Appl. Microbiol. Biotechnol. 86, 525–533.

Zhang, Y.-H.P., Lynd, L.R., 2005. Cellulose utilization by *Clostridium thermocellum*: bioenergetics and hydrolysis product assimilation. Proc. Natl. Acad. Sci. 102, 7321–7325.

Zhou, S., Ingram, L., 2001. Simultaneous saccharification and fermentation of amorphous cellulose to ethanol by recombinant *Klebsiella oxytoca* SZ21 without supplemental cellulase. Biotechnol. Lett. 23, 1455–1462.

Zhu, X., Cui, J., Feng, Y., Fa, Y., Zhang, J., Cui, Q., 2013. Metabolic adaption of ethanol-tolerant *Clostridium thermocellum*. PLoS ONE 8 (7), e70631.

Zverlov, V., Mahr, S., Riedel, K., Bronnenmeier, K., 1998. Properties and gene structure of a bifunctional cellulolytic enzyme (CelA) from the extreme thermophile *Anaerocellum thermophilum* with separate glycosyl hydrolase family 9 and 48 catalytic domains. Microbiology 144, 457–465.

Further Readings

Ingram, L., Gomez, P., Lai, X., Moniruzzaman, M., Wood, B., Yomano, L., York, S., 1998. Metabolic engineering of bacteria for ethanol production. Biotechnol. Bioeng. 58, 204–214.

Markou, G., Angelidaki, I., Georgakakis, D., 2012. Microalgal carbohydrates: an overview of the factors influencing carbohydrates production, and of main bioconversion technologies for production of biofuels. Appl. Microbiol. Biotechnol. 96, 631–645.

Ohta, K., Beall, D., Mejia, J., Shanmugam, K., Ingram, L., 1991b. Metabolic engineering of *Klebsiellao xytoca* M5A1 for ethanol production from xylose and glucose. Appl. Environ. Microbiol. 57, 2810–2815.

SECTION IV

BIOETHANOL FROM ALGAE

16 *Biorefinery Approach for Ethanol Production From Bagasse* 319

17 *Biorefinery as a Promising Approach to Promote Ethanol Industry From Microalgae and Cyanobacteria* 343

18 *Role of Genetic Engineering in Bioethanol Production From Algae* 361

CHAPTER

16

Biorefinery Approach for Ethanol Production From Bagasse

Stavros E. Michailos, Colin Webb***

*University of Sheffield, Sheffield, United Kingdom

**University of Manchester, Manchester, United Kingdom

16.1 INTRODUCTION

16.1.1 Ethanol from Bagasse: Motivation and Background

The constant increase in greenhouse gas (GHG) emissions [Fig. 16.1 (WRI, 2014)] coupled with the restricted supply of fossil fuels, has brought forth considerations relating to the sustainability of our world in terms of attaining a less polluted environment. As a result of CO_2 emissions, typical global temperatures are rising, with potentially disastrous consequences for our world. In view of this and the related, inevitable, depletion of fossil reserves, the biorefinery concept has recently emerged. The focal aim of the biorefineries is the integration of biomass conversion processes for the sustainable production of biofuels and chemicals with the aim of substituting petroleum derived products, such as gasoline, diesel, kerosene, and plastics. Biomass is the sole renewable energy source that can produce liquid fuels with properties similar to fossil fuels.

Furthermore, because of oil price volatility as well as social concerns (e.g., energy independency), significant research has been focused on the sustainable production of lignocellulose-based fuels and chemicals with the aim of substituting existing petroleum-derived products. In developed countries, the transport sector represents approximately 35% of the total energy consumption, which even now is still highly dependent on fossil sources (~98%) (He and Zhang, 2011). However, due to the inevitable depletion of fossil reserves, the manufacture of biobased fuels from renewable sources is crucial for meeting increasing world demand for transportation fuels as an essential element of the long term energy mix (Naik et al., 2010). Production of biofuels commenced in the late 19th century, when corn ethanol was produced for the very first time and Rudolf Diesel's first engine utilized peanut oil. Until the 1940s, biofuels were considered as feasible transportation fuels, but the decrease in fossil fuel prices halted their further expansion (Antizar-Ladislao and Turrion-Gomez, 2008). It was not until the

Bioethanol Production From Food Crops
http://dx.doi.org/10.1016/B978-0-12-813766-6.00016-3

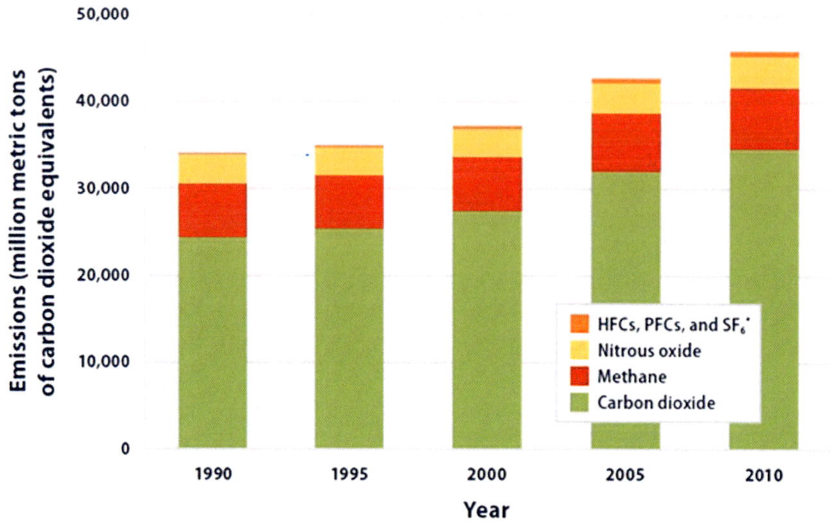

FIGURE 16.1 **Global greenhouse gas emissions by gas, 1990–2010.** *Source: Adapted from World Resources Institute.*

mid-1970s that significant interest in commercial manufacture of biofuels rose again. At that time, ethanol from sugar cane in Brazil and corn in the USA were efficiently produced. However, in most regions of the world the sharpest growth in biofuel manufacture has been observed during the past 20 years, mainly due to government policies (Mohr and Raman, 2013). Nowadays, bioethanol is the chief biofuel, and is swiftly escalating as a fuel additive along with its utilization as a chemical raw material. At present, it is, in principle, exploited in gasoline blends at 10% with the potential to reach 85% in flexible fuel cars (Gupta and Verma, 2015). Even though the production of first-generation ethanol has raised concerns over its sustainability, the exploitation of lignocellulosic biomass derived from forestry or agricultural residues, including bagasse, can positively contribute to the renewable production of building block chemicals and biofuels without competing for land and with the food supply chain (Saini et al., 2015). As a result, alternative sources of energy that have the potential to provide sustainability are being explored. The refocusing of the "Global Energy Economy" toward a greener pathway will, however, not be an easy task, at least not until the utilization of renewable raw materials is commonplace. To achieve this, a successful alternative energy source should meet the following feasibility criteria (Saini et al., 2015):

1. It should be readily available
2. It should be cheap
3. It should be GHG neutral
4. It should not offer a threat to food and land availability

A promising biomass feedstock that meets the aforementioned requirements is sugarcane bagasse. It is a major agroindustrial residue in many countries [such as Brazil, India, China, and Thailand (De Aquino et al., 2017)] and, despite its suitability for fuels and/or chemicals production (Pandey et al., 2000), it is mainly utilized to raise steam and generate electricity; nevertheless, there is still some excess. Furthermore, if more efficient combustion boilers are implemented, it has been estimated that less bagasse would be incinerated (approximately 36% reduction) (Zanin et al., 2000). To this direction, research has focused on exploiting bagasse in biorefining technologies that are capable of producing saleable fuels from sugars, pyrolysis, and syngas (Bozell et al., 2014). Sugar cane milling processes for ethanol or sugar production leave a solid residue called bagasse; one ton of sugarcane produces approximately 250–300 kg of bagasse (Chauhan et al., 2011). In addition, the expected escalation in cane production in order to satisfy the increase of global ethanol demand (from 70 to 130 million m^3 between 2009 and 2025), will also give rise to further bagasse availability (Balat and Balat, 2009). The composition of bagasse chiefly comprises two polysaccharides (cellulose and hemicellulose) as well as a polyphenolic macromolecule (lignin). Cellulose consists of a linear chain of numerous highly structured β (1→4)-linked D-glucose units that creates crystalline districts and as a result it is more resistant to hydrolysis compared to hemicellulose. The latter component is made of heteroxylans, with a prevalence of xylose subunits, and along with lignin (possesses a complex structure derived from aromatic alcohols polymerization), cover the cellulosic matrix providing a barrier to enzymatic and chemical degradation (Ververis et al., 2007). It has been reported that sugarcane bagasse possesses the greatest positive net energy balance among the prevalent raw materials considered nowadays (Wyman et al., 2005). Presently, the energy generated from sugarcane is seven times more compared to that consumed in its creation. And when bagasse is counted in the equation, it is projected that the number might rise to as much as 15 times (Isikgor and Becer, 2015). It is, also, believed that if bagasse is converted to ethanol the overall fuel yield can be increased by 30% (mass basis) without the requirement of using extra land and with the additional advantage of further decreasing the carbon footprint (Mosier et al., 2005). Back in 2012 the global production of sugarcane was 670 million tons (Kiatkittipong et al., 2009), which left as residue approximately 180 million tons of bagasse, resulting in a disposal issue. Thus it is sensible to consider bagasse as a potential feedstock for a sustainable biofuel sector. Its exploitation, in this way, would provide a significant solution to overcome waste problems. Moreover, as a waste material, it will be considerably less expensive than other biomass material, such as wood. Another advantage is that there is no additional competition with the accessibility and usage of food and land. Essentially, it precludes the tightness of food supply (Basu, 2010) and averts the conflict with the cultivation of farmland (Rabelo et al., 2011). Similarly to various energy crops, it manages neutral carbon footprint and its composition (high holocellulose content) makes it appropriate for a number of conversion procedures.

16.2 ETHANOL FROM BAGASSE: TECHNOLOGICAL STATUS

As previously mentioned, three are the major technologies that are able to transform lignocellulosics into ethanol, that is, biochemical, thermochemical, and hybrid (Srivastava et al., 2015). The biochemical route is established in three steps: (1) pretreatment, (2) saccharification, and (3) sugars fermentation to ethanol. This route has received much attention from the research community mostly because of its similarity with the existing production technology of ethanol and the anticipated lower capital costs (Liguori et al., 2016). The main disadvantage of this route is that lignin cannot be decomposed and as such does not contribute to ethanol production along with the high enzymes cost. The first step includes pretreatment that aims at cracking down the lignin structure of lignin and cellulose hydrolysis to C_6 sugars. The most common pretreatment methods are briefly presented (Moreira, 2000; Sun and Cheng, 2002):

- Mechanical: Chemical composition and structure is not affected, no production of toxic inhibitors, energy intensive method.
- Steam explosion: Solubilizes the majority of xylan (the principal component of hemicellulose) leaving cellulose, lignin, and xylan leftovers in the solid phase.

- Ozonolysis: Utilized to eradicate lignin structure effectively; hemicellulose is to a certain degree influenced and cellulose negligibly. Also, in this process no inhibitors are produced and the reactions take place under moderate conditions (temperature and pressure). However, the necessity of enormous quantities of ozone makes this process excessively costly.
- Dilute acid hydrolysis: Accomplishes substantially significant conversion of xylan to xylose but generation of inhibitors is inevitable. Prior to pentose fermentation this liquid stream must be detoxified by engaging various techniques, such as neutralization, over liming and adsorption.
- Liquid hot water (LHW): It is a thermal pretreatment that employs pressurized water at high temperatures in order to crack down the lignin structure. Xylan to xylose conversion factor is relative high, no generation of toxic waste but requires high energy loads.

Hydrolysis of cellulose to C_6 sugars can be catalyzed either by acids or enzymes. Undoubtedly enzymatic hydrolysis is the dominant one due to high sugar yields, low operating costs, and avoidance of inhibitors production. The main products of enzymatic hydrolysis of cellulose are glucose and cellobiose and it takes place under mild conditions; temperature of 50°C and pH of 5. The production of inhibitors from the pretreatment mostly depends on the feedstock and the pretreatment system employed. The most common inhibiting compounds are phenols and other aromatics, aliphatic acids, and furan aldehydes. Many different approaches have been proposed recently for the detoxification of hydrolysates, including chemical biological and physical methods. These include overliming treatment, liquid–liquid extraction, liquid–solid extraction, heating and evaporation, and microbial catalyzed procedures (Ravagnani et al., 2010). Different approaches have been proposed for the biochemical conversion route mainly related to alternative pretreatment and fermentation sections (Fig. 16.2). The main trends in the fermentation section can be classified based on whether hydrolysis and fermentation take place in a single operation or separately and whether C_5 and C_6 sugars are cofermented or not. Hence, we have four alternative pathways: separate hydrolysis and separate C_5 and C_6 fermentations (SHF); separate hydrolysis with cofermentation of C_5 and C_6 sugars (SHcF); simultaneous saccharification and fermentation of C_6 sugars with separate fermentation of C5 sugars (SSF); and simultaneous saccharification with cofermentation of C_5 and

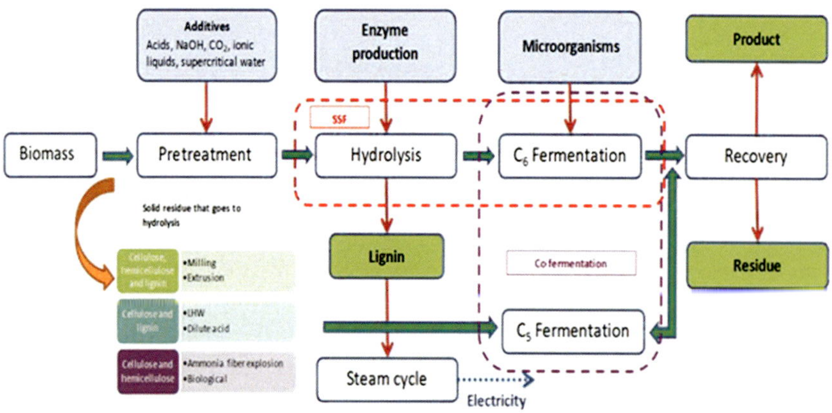

FIGURE 16.2 **Biochemical conversion route for biomass to fermentation products.**

C_6 sugars (SScF) (Olofsson et al., 2008). Recently, it has been proposed that the SScF process can be further integrated so as to include enzyme production in the same stage, known as consolidated bioprocessing (CBP). The traditional method for anhydrous ethanol purification from fermentation broth is a three-stage procedure (Ravagnani et al., 2010):

- Broth distillation in order to reach the ethanol-water azeotrope (~95 wt.% ethanol (Kumar et al., 2010).
- Azeotropic, extractive distillation, or adsorption with molecular sieves employing a solvent, such as cyclohexane or benzene so as to break the azeotrope and eradicate the residual water.
- Distillation to recover the solvent from the water, in order to be reused.

The thermochemical route consists of the transformation of biomass into syngas via gasification followed by mixed alcohol catalytic synthesis. The procedure is established with five steps: biomass pretreatment, gasification, syngas cleaning and conditioning, alcohol synthesis, and alcohol separation. Pretreatment is essential because the gasifier requires a fuel, which is relatively dry and has a small particle size and thereby it includes process units, such as grinding, screening, and drying. Gasification is a similar process to combustion but with different operating conditions (lower air: biomass ratio and temperature). The main goal is to produce syngas (a mixture of carbon dioxide and hydrogen), which subsequently is subjected to various treatments in order to give a wide range of products. As oxidant, air, purified oxygen, steam, or even a mixture of these could be utilized. Air gasification generates a producer gas with energy content of 5 MJ/m^3 (STP), which contains a large amount of nitrogen (approximately 50% v/v). This results in a more intensive downstream process, reducing the overall economic and thermodynamic efficiency of the plant. The lower heating value of the producer gas can be increased to 9–13 MJ/m^3(STP) if oxygen is used as medium (Bridgwater, 2003). Nevertheless, the requirement for an air separation unit affects negatively the investment cost. In theory steam gasification can augment the syngas energy content to around 13–18 MJ/m^3 (STP) mainly because of the increased yields of methane and lower hydrocarbons (Solantausta et al., 1996). However, if the desired product is liquid fuels, the reforming process becomes more complex and costly. The producer gas from the gasification procedures is mainly composed of hydrogen, carbon monoxide, carbon dioxide, and methane, but typically contains numerous contaminants, such as tar, alkali compounds, hydrogen sulfide, ammonia, and hydrochloric acid, which have to be removed effectively because the fuel synthesis catalysts are highly sensitive to slight quantities of such impurities. Some common methods are adsorption, scrubbing, and catalytical cracking. The cleaned and conditioned syngas is, then, catalytically converted to alcohols in a fixed bed reactor. Rhodium-based catalysts are among the most selective catalysts reported in the literature (Spivey and Egbebi, 2007). The overall stoichiometric reaction for alcohol synthesis can be summarized as (He and Zhang, 2011).

$$nCO + 2nH_2 \rightarrow C_nH2_{n+1}OH + (n-1)H_2O \quad (16.1)$$

The product gas is subsequently cooled, allowing the alcohols to condense and separate from the unconverted syngas. The liquid alcohols are then sent to alcohol separation and purification whereby common separation units, such as distillation and adsorption, are utilized. The drawbacks of this alternative are the high-energy requirements and low ethanol selectivity (Fang et al., 2009). The hybrid route is, in fact, a combination of the other two technologies. Biomass is gasified and then syngas is fermented by acetogenic bacteria to ethanol. Several advantages accompany this technology including high reaction specificity, low energy demands, syngas composition versatility, and high resistance to impurities (Acharya et al., 2014). In addition, it exploits the lignin content of the raw material and the use of expensive enzymes is avoided. However, the hybrid route is associated with low ethanol yields due to low solubility and mass transfer limitations of the CO and H_2 gaseous substrates (Piccolo and Bezzo, 2009).

16.3 PROCESS SIMULATION AS A TOOL FOR ROUTE SELECTION

Process simulation stands for the model-based depiction and design of chemical, physical, biological, and other technical processes and unit operations by executing the appropriate mass and energy balances. Basic preconditions are an in-depth familiarity with the chemical and physical properties of pure components and mixtures, of reaction kinetics and mechanisms, and of mathematical modeling techniques, which, jointly, enable the software-based design of a process.

Process simulation history is intimately connected with the progression of computer science and hardware as well as the development of programming languages. It was not until the 1970s that proper software and hardware became available and the first applications of chemical process simulation were implemented. However, the modeling of chemical properties was possible long before the existence of such software. The cubic equations of state and the Antoine equation, for example, were available during the previous century. The main interface for modern process simulation software is, generally, a representative flow diagram where the user is prompted to position and connect with streams (energy or material) the various unit operations. The software solves mass and energy balance equations, usually in their steady state form, to compute a stationary operating point. At all times, process simulations utilize models that are associated with approximations and assumptions but gives the opportunity to the user to describe a property within a broad range of conditions (such as temperature and pressure), which are not necessarily found in the literature.

Process simulation is frequently carried out during the design phase or in advance of the full operation of a plant so as to test how changes in unit operation specifications can affect a procedure throughout the duration of its life cycle. By utilizing process simulation software, the user can set up and compare alternative but equivalent process routes and thereby select the preferred option on the basis of cost, energy consumption, carbon footprint, or other criteria.

There are several simulation packages used by process industries for simulating, analyzing, and optimizing their various processes, in order to achieve efficient operations and maximize their profits. Simulation packages are used for generating high-accuracy predictive information, which will aid decision making and supports decisions regarding the product and process innovation, design, and operation. Among these packages are Aspen Plus (by AspenTech), Aspen Hysys (by AspenTech), Pro/II (by SimSci), ProSimPlus (by ProSim), SuperPro (by Intelligen, Inc.), Unisim (Honeywell), ChemCAD (from Chemstations, Inc.), gPROMS (from Process Systems Enterprise Limited [PSE]), and so on, have been successfully used by process engineers and scientists in simulating and optimizing complex conventional processes. Some of these simulators are designed specifically for a type of reaction, for example, simulators designed specifically for bioprocesses that include BioProcess Simulator (by AspenTech), Biotechnology Design Simulator (BDS developed by Life Sciences International Philadelphia, PA), SuperPro Designer (developed by Intelligen, Inc.), and BATCHES (from Batch Process Technologies) have been found; all designed for bioprocesses.

TABLE 16.1 Typical Bagasse Composition

Proximate analysis		Ultimate analysis	
Parameters	**Mass fraction (%)**	**Element**	**Dry weight (%)**
Moisture	50 (wb)	C	45.38
Ash	3.2 (db)	H	5.96
Volatile matter	83.65 (db)	O	45.21
Fixed carbon	13.15 (db)	N	0.15
Chemical composition			
Component		Dry weight (%)	
Cellulose		45	
Hemicellulose		25	
Lignin		20	
Extractives		6.8	

Source: Gao, Y., Xu, J., Zhang, Y., Yu, Q., Yuan, Z., Liu, Y., 2013. Effects of different pretreatment methods on chemical composition of sugarcane bagasse and enzymatic hydrolysis. Biores. Technol. 144, 396–400; De Medeiros, E.M., Posada, J.A., Noorman, H., Osseweijer, P., Filho, R.M., 2017. Hydrous bioethanol production from sugarcane bagasse via energy self-sufficient gasification-fermentation hybrid route: simulation and financial analysis. J. Cleaner Prod. 168, 1625–1635.

16.4 PROCESS DESIGN AND MODELING

The purpose of this section is to perform fundamental mass and energy calculations of the renewable ethanol production processes with the aim of estimating overall mass and energy efficiencies as well as defining the key heuristics of each route. For this purpose, Aspen Plus software was employed to conduct mass and energy balances in conjunction with Excel spreadsheets. The basis for each biorefinery is 100 dry tons per h of bagasse with 50% moisture content (dry basis). The design starts with feedstock pretreatment and ends with fuels storage. Bagasse consists of cellulose, hemicellulose, and lignin; it was assumed that cellulose and hemicellulose consist only of glucan and xylan, respectively. For the thermochemical procedures, bagasse was defined in terms of the elements in the proximate and ultimate analysis, whereas for the biochemical process it was defined by its chemical composition. Bagasse compositions vary for different species of sugarcane, however, for the sake of conducting simulations typical ultimate and proximate analyses as well as chemical composition were utilized and those are illustrated in Table 16.1 (De Medeiros et al., 2017; Gao et al., 2013). The physical properties of the conventional components have been estimated by using the Redlich-Kwong-Soave cubic equation of state with Boston-Mathias alpha function (RKS-BM) (Aspen Technology, 2012). This method is suitable for gas-processing, refinery and petrochemical applications, such as gas plants, crude towers, and ethylene plants. Using the RKS-BM model, rational and reliable outcomes can be anticipated at all temperatures and pressures. The SOLIDS property option was employed for the biomass crushing and drying units as it is recommended for solids processing unit operations (Aspen Technology, 2013) and typical steam tables for the CHP units.

16.4.1 Biochemical Production of Ethanol

The bioprocessing platform consists of: (1) pretreatment, (2) cellulose hydrolysis, (3) sugars fermentation, (4) ethanol recovery, and (5) CHP unit. Bagasse is initially introduced to a primary grinder and then a secondary grinder, which decreases wood to the proper size for pretreatment (5 mm). Chopped bagasse is then sent to the pretreatment reactor. The main goal is to break down the lignin structure to render the holocellulose accessible to the enzymes. The hemicellulose fraction is degraded to mainly xylose along with other materials that act as inhibitors for the subsequent fermentations, such as acetic acid and furfural. Also, some portion of the cellulose is converted into glucose. Dilute acid hydrolysis with H_2SO_4 solution (2% w/w) was selected as it can achieve a high conversion of hemicellulose to xylose. A stirred batch reactor was employed operating at 130°C and 2 bar. The reaction time is equal to 15 min (Michailos et al., 2016a). The key reactions along with conversion factors utilized in the simulations are illustrated in Table 16.2. Then, a flash separator operating at 105°C vaporizes a large amount of water, a portion of the acetic acid, and much of the furfural. Solids (cellulose and lignin) are separated from liquids via filtration. Afterward, hemicellulose hydrolysates have to be detoxified mainly from the sulfuric acid, which is partially eliminated after reacting with $Ca(OH)_2$. A conversion reactor was employed assuming 100% conversion of $Ca(OH)_2$ to gypsum. The produced gypsum is then separated from the liquid via centrifugation and sold as by-products. A 2% sugars loss is realized during detoxification. The next step is the hydrolysis of the cellulose to hexoses. Cellulose hydrolysis is a batch process with reaction times within the range of 24–72 h (Gupta and Verma, 2015) (45 h was selected for the simulations) and operating temperature of 50°C. The required enzymes for the hydrolysis step were obtained by an external supplier and it was assumed that 0.02 kg of cellulases can hydrolyze 1 kg of cellulose. Table 16.2 contains reactions involved and the respective conversion factors. A typical batch reactor is used to simulate the hydrolysis step. After hydrolysis, lignin is removed from the liquid product and is sent to the CHP unit. The latter is a typical Rankine cycle process consisting of feed boiler water pump, economizer, evaporator, superheater, steam turbine, and condenser. The hexoses and the pentoses are mixed and are fed to the fermentation section.

TABLE 16.2 Reactions Involved in Bagasse Pretreatment and Cellulose Hydrolysis

Reactions	Conversion to products
Pretreatment	
Hemicellulose + water → xylose	90%
Hemicellulose → furfural + 2water	5%
Cellulose + water → glucose	7%
Acetate → acetic acid	100%
Cellulose hydrolysis	
Cellulose + water → glucose	90%
Cellulose + 0.5water — 0.5cellobiose	1.2%
Cellobiose + water → 2glucose	95%

Source: Alvarado-Morales, M., Terra, J., Gernaey, K.V., Woodley, J.M., Gani, R., 2009. Biorefining: computer aided tools for sustainable design and analysis of bioethanol production. Chem. Eng. Res. Des. 87, 1171–1183; Aden, A., 2002. National Renewable Energy Laboratory (U.S.). Lignocellulosic biomass to ethanol process design and economics utilizing co-current dilute acid prehydrolysis and enzymatic hydrolysis for corn stover. Technical report NREL/TP-510-32438. Golden, Colo.: National Renewable Energy Laboratory.

Several types of yeast strains have been used in fermentation for ethanol production, including some recombinants, which are capable of cofermenting hexoses and pentoses (Mohd Azhar et al., 2017). A substantial metabolic engineering effort has been directed toward development of strains of *Saccharomyces cerevisiae* (Krahulec et al., 2010) with increased xylose conversion (Kricka et al., 2015). Fermentation takes place in a common batch reactor with reaction time equal to 30 h and operating temperature of 35°C. Yeast is recovered from the fermentation broth via filtration and is recycled to the fermenter. Table 16.3 provides a summary of the participant reaction schemes along with

TABLE 16.3 Ethanol Fermentation Reaction Data

Reactions	Conversion to products
Glucose	
Glucose → 2EtOH + $2CO_2$	90%
Glucose + 2water → 2glycerol + O_2	0.5%
Glucose + $2CO_2$ → 2succinic acid + O_2	0.6%
Glucose → 3acetic acid	1.5%
0.55 Glucose → 0.33yeast + 1.11water + $0.45CO_2$	2%
Xylose	
3Xylose → 5EtOH + $5CO_2$	85%
3Xylose + 5water → 5glycerol + $2.5O_2$	0.3%
Xylose + water → xylitol + $0.5O_2$	4.6%
3Xylose +$5CO_2$ → 5succinic acid + $2.5O_2$	0.9%
2Xylose → 5acetic acid	1.4%
0.66Xylose → 0.33yeast + 1.11water + $0.45CO_2$	4%

Source: Kumar, D., Murthy, G.S., 2011. Impact of pretreatment and downstream processing technologies on economics and energy in cellulosic ethanol production. Biotechnol. Biofuels, 4, 27; Walter, A., Ensinas, A.V., 2010. Combined production of second-generation biofuels and electricity from sugarcane residues. Energy 35, 874–879.

the conversion factors. Finally ethanol is recovered via extractive distillation whereby the fermentation broth enters a series of two distillation columns, which give as output distillate with approximately 95% wt. of ethanol—water mixture (azeotrope). The bottom stream of the first column is sent to a multievaporator and after water evaporation enters the CHP unit. The bottom stream from the second column is recycled to the first column. Extractive ethanol recovery comprises two distillation columns: the extractive distillation unit and the solvent recovery unit where the entrainer is obtained. The entrainer used in this technology should be nonvolatile and have a high boiling point and essentially does not form any azeotrope with either component of the original azeotropic mixture, such as ethylene-glycol. The distillate of the first column (extractive) is pure ethanol and the bottom stream a mixture of water and the entrainer. The latter stream is sent to the second column where pure ethylene-glycol is recovered from the bottom and is returned to the extractive column. The main advantage of extractive distillation compared to azeotropic is that the necessary amount of the entrainer is much lower, which leads to lower energy consumption and lower capital costs (due to lower volume of the distillation units). A simplified process flow diagram (PFD) for the bioethanol case is depicted in Fig. 16.3.

16.4.2 Thermochemical Production of Ethanol

The thermochemical conversion route to ethanol is established in six steps: (1) bagasse pretreatment, (2) gasification, (3) syngas cleaning and conditioning, (4) mixed alcohols synthesis (MAS) reactor, (5) product separation, and (6) CHP unit. Two Aspen Plus crusher blocks have been employed to simulate a gyratory crusher, which chops bagasse to a final particle size of 2 mm. An electrical requirement of 50 kWh per ton of dry feed was assumed (Ringer et al., 2006). Afterward, bagasse enters a dryer in order to reduce its moisture content to 10%. The high initial moisture content of bagasse makes it necessary to employ a dryer so as to reduce subsequent heat losses in the gasifier unit. The fractional conversion of wood to water has been estimated by using a FORTRAN statement. A RSTOICH reactor was used to model the drying of bagasse. Even if bagasse drying is not generally considered a chemical reaction, the RSTOICH module is capable of converting a portion of the wood to form water. Eq. (16.2) presents the chemical reaction for bagasse drying:

$$\text{Bagasse(wet)} \rightarrow 0.0555\text{H}_2\text{O} \tag{16.2}$$

Aspen Plus treats all nonconventional components as if they have a molecular weight of 1.0. The reaction indicates that 1 mole (or 1 kg) of wood reacts to form 0.0555 mole (or 1 kg)

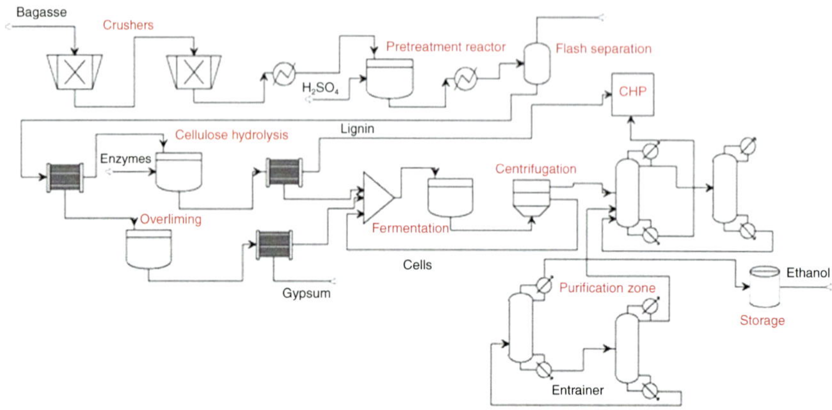

FIGURE 16.3 **PFD for bioethanol production from bagasse.**

of water. Then, a calculator block was used (includes FORTRAN statements) to specify the moisture content of the dried wood (10%) and calculates the corresponding conversion of wood to water. More details about this technique can be found online in a report published by AspenTech regarding solids processing (Aspen Technology, 2013). The energy required for the drying process is provided by heat produced within the processes (e.g., exhausted flue gas from the CHP unit). Dried and chopped biomass is then sent to the gasifier. A steam blown circulated fluidized bed (CFB) reactor operating at 950°C and 20 bar was considered. A CFB gasifier is suitable for large-scale biorefinery applications and the utilization of steam, instead of air, reduces the capital costs from smaller equipment sizes and more effective processes (Bridgwater, 2009). Pressurized gasifiers are better suited for fuels production since the pressure of product syngas will be sufficient or will need to be slightly increased to be fed directly to synthesis reactors. A common technique for modeling biomass gasification is by estimating thermodynamic equilibrium composition through Gibbs free energy minimization calculations for the C, H, and O atoms of the fuel and the gasification agent mixture. It was also assumed that the nitrogen and sulfur content of bagasse convert to NH_3 and H_2S, respectively. This approach has been used before in several studies (Sudiro and Bertucco, 2009; Trippe et al., 2011) and it is suitable for feasibility studies but not for reactor design (Sreejith et al., 2013). For simulation purposes, the gasifier was divided in three steps. The first step employs a RYIELD reactor, which decomposes bagasse to its constituent elements according to the ultimate analysis. Afterward, the elemental composition is fed to the (RSTOIC) reactor where CH_4 and tar are formed (Panopoulos et al., 2006). The extent of methane formation was set to use 10% of the elemental carbon in the biomass fuel and that of tar 0.1%. The purpose of the addition of this step is that methane and tar compositions are underestimated under equilibrium conditions. The remaining elemental biomass along with steam (steam to biomass ratio equal to 0.4) are fed to an (RGIBBS) reactor model where according to Gibbs free energy minimization calculations (for the specified temperature and pressure), the equilibrium composition is evaluated, considering the components CH_4, CO, CO_2, H_2, H_2O, and C(s). The equilibrium outlet is mixed with the hydrocarbons streams (methane and tar) to form the final syngas outlet stream. Furthermore, to avoid the inclusion of a huge variety of tar compounds, toluene was selected to represent this group (Liu et al., 2017). Afterward, two cyclones were employed, the

TABLE 16.4 Alcohols Product Distribution

Product	Distribution (wt.%)
Methanol	4.83
Ethanol	67.13
Propanol	9.48
Butanol	1.17
Pentanol	0.15
Others (CH_4, C_2H_6, C_4H_{10})	14.58
Water	2.66

Source: Dutta, A., Phillips, S., 2009. National Renewable Energy Laboratory (U.S.). Thermochemical ethanol via direct gasification and mixed alcohol synthesis of lignocellulosic biomass. Nrel/Tp 510-45913. Golden, CO: National Renewable Energy Laboratory.

first one separates the unreacted char from the producer gas and recycles it to the gasifier and the second distinguishes the ash from the gas mixture. Light hydrocarbons (mainly methane) and tar are converted to syngas in a catalytic tar reformer which operates at 800°C and pressure of 12 bar. Hot gases are then cooled down, in a series of heat exchangers, to a final temperature of 110°C (ideal for the subsequent cleaning process). This can be achieved by raising medium pressure saturated steam (5 bar) that serves as the gasifier's oxidizing agent. Subsequently, the acid gases H_2S and CO_2 are removed by using a solution of monoethanolamine (MEA) in an absorber. The entrainer is recovered in a stripper and recycled to the absorber. Clean syngas is cooled down to 50°C and water is removed in a flash drum. Syngas is, then, compressed and heated to meet the operating conditions of the MAS reactor, that is, 68 bar and 300°C. The syngas reacts to form an alcohol mixture across a fixed bed catalyst. The product gas is successively cooled, allowing the alcohols to condense and separate from the unconverted syngas. The liquid alcohols are then sent to alcohol separation and purification. The residual gas stream is recycled back to the MAS reactor with a small purge to fuel combustion (10%). The product distribution is presented in Table 16.4. The CO per pass conversion is equal to 40% and the total alcohol selectivity is 80% (the balance goes to CO_2). A molybdenum-based catalyst was considered and the reactor was modeled using a RYIELD reactor.

The product stream is separated from water via adsorption in molecular sieves by a combination of depressurization and flushing with methanol. Afterward, two distillation units were utilized in order to recover ethanol. The first separates ethanol and methanol from the heavier alcohols while the bottom stream of the second one contains high-purity ethanol. Methanol from the top stream head to the molecular sieves unit for the regeneration step. Fig. 16.4 provides the PFD for the thermochemical production of ethanol.

16.4.3 Ethanol Production via Syngas Fermentation

The design of the hybrid route to ethanol incorporates: (1) bagasse pretreatment, (2) gasification, (3) syngas fermentation to ethanol, (4) ethanol recovery, and (5) CHP unit (Fig. 16.5). The first two steps are identical to the MAS case. Unlike other biofuel processes based on gasification (e.g., Fischer-Tropsch), the adjustment of syngas ratio is not essential for the syngas fermentation process. On top of this, removal of CO_2 is not necessary as it acts as carbon source for ethanol synthesis and the concentration of H_2S in the producer gas (120 ppm) is assumed to be too low to contaminate the fermentation (Drzyzga et al., 2015). Therefore, there is no need to add to the design any water–gas–shift reactors or acid gases removal units. The next step of the process is the syngas fermentation. The modeling of the bioreactor is based on the experiments conducted by Arora et al. (1997). As so, syngas fermentation takes place in a CSTR with operating conditions of temperature equal to 38°C and pressure of 1.5 bar. Fermentation is catalyzed by bacterium *Clostridium ljungdahlii* and apart from ethanol, acetic acid is also produced. In

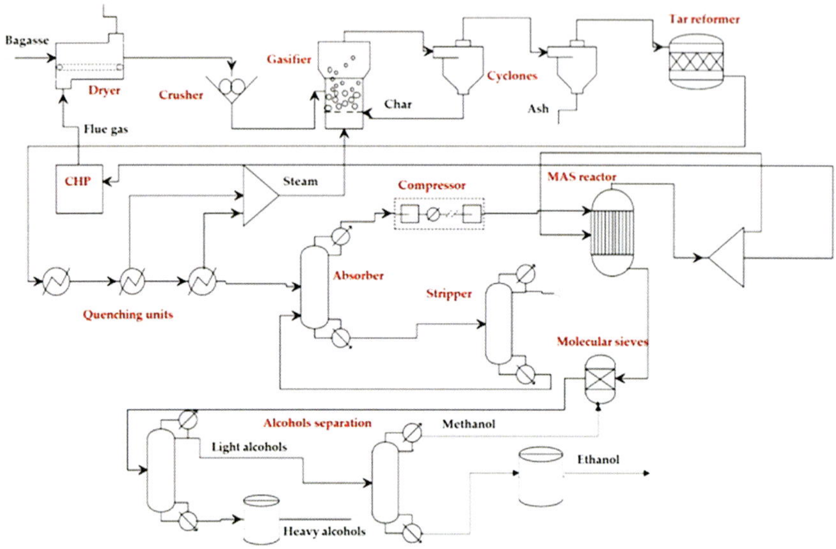

FIGURE 16.4 **PFD for thermochemical ethanol production from bagasse.**

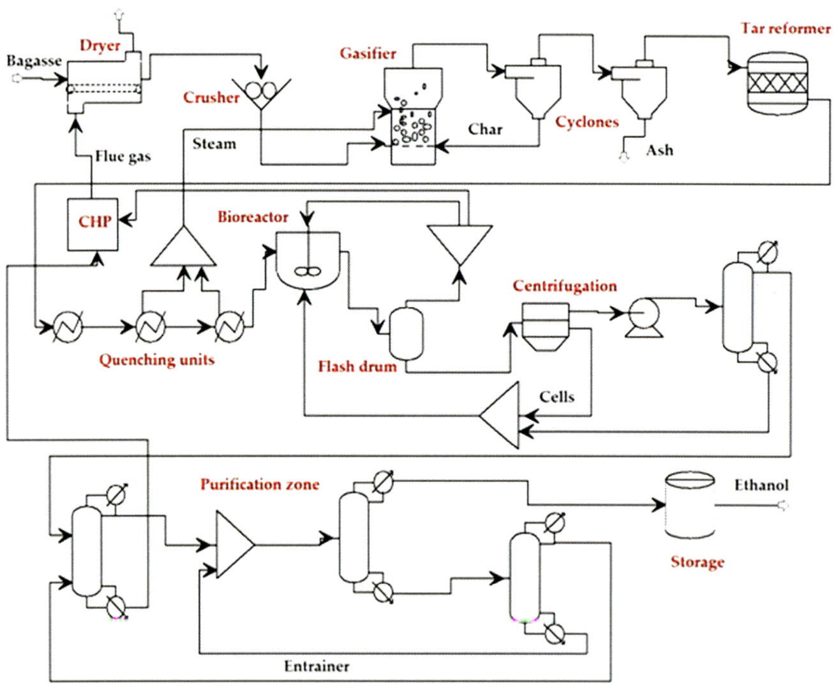

FIGURE 16.5 **PFD for ethanol via bagasse-derived syngas fermentation.**

the current model, the reactions that were taken under consideration are:

$$6CO + 3H_2O \rightarrow CH_3CH_2OH + 4CO_2 \quad (16.3)$$

$$2CO_2 + 6H_2 \rightarrow CH_3CH_2OH + 3H_2O \quad (16.4)$$

$$4CO + 2H_2O \rightarrow CH_3COOH + 2CO_2 \quad (16.5)$$

$$2CO_2 + 4H_2 \rightarrow CH_3COOH + 2H_2O \quad (16.6)$$

A conversion reactor block was used to simulate the bioreactor considering that 70% of CO is converted to ethanol and 5% to acetic acid, while 50% of H_2 is converted to ethanol and 2% to acetic acid (Arora et al., 1997). Cells are recycled to the bioreactor after they are separated in a centrifugation unit. The fermenter broth contains a significant amount of unreacted gases, which are separated from the liquid products in a flash drum separator. This gas stream is recycled to the bioreactor but in order to avoid accumulation a small percentage (~15%) is split and heads to a combined gas–steam turbine system to raise electricity. A minor proportion of the unreacted gases still remain in the liquid mixture and will separate in the first distillation column condenser using a pressure relief valve. Also, the bottom stream is recycled to the fermenter. Then, the liquid fermentation product enters the purification zone, which is the same as described for the biochemical ethanol scenario.

16.5 FEASIBILITY ANALYSIS

16.5.1 Mass and Energy Efficiency

The term mass conversion efficiency refers to the ratio of kg of biofuels produced over kg of feedstock input on a dry basis. It is a significant performance indicator especially when same product processes are compared, as in this case (ethanol). The biochemical route achieves the highest value, that is, almost 27%, due to the high efficiency of cellulose hydrolysis and sugars fermentation. The major bottleneck is that lignin cannot be hydrolyzed but this is partially offset by generating excess of electricity. The thermochemical route attains a 19.1% ethanol mass efficiency and if we take into account the by-products this value increases to around 25%. The process suffers from the low CO conversion across the MAS reactor and from the many process steps involved in the production line. A significant advantage is that lignin is gasified to syngas and as a result it contributes to fuel production. At the current technological status, the low solubility and mass transfer limitations of the CO and H_2 gaseous substrates prevent the syngas fermentation technology from surpassing the biochemical route. Nevertheless, the value of approximately 24% indicates that syngas is better utilized biochemically rather than chemically when ethanol is the target product.

In any chemical process, energy quantitative evaluation can be carried out by utilizing the first law of thermodynamics. Energy analysis can also be used to compare components or systems to help make informed design decisions. Energy efficiency can be calculated by utilizing the following equation:

$$\eta = \frac{m_{fuel} LHV_{fuels} + Q_{out} + W_{out}}{m_{feedstock} LHV_{feedstock} + Q_{in} + W_{in}} \quad (16.7)$$

Where m_{fuel} and $m_{feedstock}$ are the mass flow rates of feedstock and produced fuels, respectively, and subscripts *in* and *out* stand for produced energy flows and external energy flows respectively. The LHV of bagasse is equal to 16.4 MJ/kg (Michailos et al., 2016b) while for the values for ethanol (29.7 MJ/kg) and by-products can be found in several textbooks. Based on the simulations and the given process design, the energy duties and outputs of each alterative were calculated. For the bioethanol case and hybrid route, the energy hot spots are traced in the purification zone since ethanol concentration in fermentation broth is low (~11 wt.%). The severe operating conditions (elevated temperatures and pressures) associated

TABLE 16.5 Simulation Results for the Investigated Ethanol Production Routes From Bagasse

	Bioethanol	Thermochemical ethanol	Syngas fermentation
Ethanol (ton/h)	26.8	19.1	23.7
By-products (ton/h)	—	5.8	—
Heating duties (MW)	47	38	42
Cooling duties (MW)	45	35	40
Electricity demand (MW)	10	25	9
Electricity output (MW)	18	7	13
Mass efficiency (%)	26.8	24.9	23.7
Energy efficiency (%)	43.1	39.7	38.4

with the thermochemical ethanol scenario are responsible for the majority of the energy demand (~82%) while the distillation units count for the balance. Electricity imports, necessary to pressurize syngas, pose as the major process barrier of the thermochemical ethanol. Table 16.5 presents the relevant data. It can be observed that biochemical ethanol outplays the other candidates followed by thermochemical ethanol.

16.5.2 Economic Evaluation

There are several methods for calculating the equipment costs, including cost estimation based on historic data, functional unit estimating methods, which associate plant cost with number of processing stages, the factorial method of cost estimation, via which the total capital cost is estimated as the sum of the equipment cost multiplied by an installation factor (known as Lang factor) and cost estimating based on recent data on actual prices paid for similar equipment. The latter method is the most accurate one but access to large amounts of high-quality data is required, which is possible only for international companies that carry out projects globally. Thus, for this study cost estimation based on historic data was used by utilizing the following equation (Towler and Sinnott, 2013a):

$$C = C_0\left(\frac{S}{S_0}\right)f \quad (16.8)$$

Where C is the estimated actual cost of the unit, C_0 the base cost of the unit, S the actual size or capacity of the unit (extracted from simulations), S_0 the base or capacity, and f an empirical scaling factor. The reference year of the study is 2014 and the currency is USD. A scaling factor f is applied to estimate the cost of a system based on the known cost of the system for a different size. This factor is typically 0.8 to 0.9 for processes that use a lot of mechanical work or gas compression (e.g., CHP units, solids-handling plants). For typical petrochemical processes f is usually about 0.7. For small-scale, highly instrumented processes, such as specialty chemical or pharmaceuticals manufacture, f is in the range 0.4 to 0.5. Averaged across the whole chemical industry, f is about 0.6, and hence it is commonly referred to as the "six-tenths rule" (Towler and Sinnott, 2013a). Costs were adopted from relevant literature. Where costs were reported at different currency than USD the value was converted to USD by using exchange rates of the respective reference year and were subsequently brought up to date (2014). The latter can be attained by utilizing proper cost indices. These incorporate changes in the value of money due to inflation and deflation of raw materials, products, labor, material, and energy costs. To get the best estimate, each job should be broken down into its components and separate indices should be used for labor and materials. It is often more convenient to use the composite indices (as here) published for various industries in the trade journals. These are weighted average indices combining the various components of costs in proportions considered typical for the particular industry. A composite index for the United States process plant industry is published monthly in the journal *Chemical Engineering*; this is the chemical engineering plant cost index (CEPCI). These indices do not include learning effects, location factors, or the effect of supply and demand on prices (Sadhukhan et al., 2014). Furthermore, Lang factors were utilized to calculate installed and indirect costs as well as total capital investment (TCI), that is, 2.47, 0.89, and 4.64, respectively. In this study, it is assumed that all factors apply to delivered-equipment costs. The full methodology for the calculation of TCI is illustrated in Table 16.6 (Peters et al., 2003).

Subsequent to the estimation of total capital investment, operating costs (OC), comprising feedstock price, utilities costs, general overhead, catalysts cost, insurance, maintenance, and labor expenditures, were also calculated. For the labor requirement the following correlation was utilized (Towler and Sinnott, 2013b):

$$N_{OL} = (6.29 + 31.7P + 0.23N)^{0.5} \tag{16.9}$$

Where P the number of the solids handling steps and N number of other processing steps. For each of the N_{OL} operators per 8-h shift, approximately five operators must be hired for a plant that runs 24 h per day, to account for the three shifts per day and the 3 weeks of leave typically taken by each operator per year. In addition, because the assumed average salary listed is not fully loaded (i.e., do not include benefits), a general overhead factor was used. Regarding the catalysts costs, data from the literature was utilized in order to correlate the annual catalyst costs with the annual fuels productivity and the resultant equations were implemented to the present data. Typical values for the operating costs are presented in Table 16.7. Prices reported in different currency were converted to USD at the exchange rate of the reported year and updated to 2014 (an annual average inflation rate of 2.5% was considered).

Afterward, it is possible to calculate a crucial economic factor, the cost of production. The production cost of a product is a significant index especially when comparing the financial feasibility between different conversion pathways. It is extremely useful when the value of a product cannot be determined clearly, for instance when a known product is produced from a nonconventional feedstock (as in the bagasse case), Eq. (16.10).

$$\text{Production Cost} = \frac{\text{Total annual cost}}{\text{Production rate}} \tag{16.10}$$

TABLE 16.6 Total Capital Investment (TCI) Estimation Methodology

Cost component	Lang factor
Purchased equipment	1
Purchased equipment installation	0.39
Instrumentation and controls	0.26
Piping	0.31
Electrical systems	0.1
Buildings (including services)	0.29
Yard improvements	0.12
Direct Costs (DC)	**2.47**
Engineering and supervision	0.32
Construction expenses	0.34
Legal expenses	0.04
Contractor's fee	0.19
Indirect Costs (IC)	**0.89**
Contingency	0.2 × (DC + IC)
Fixed Capital Investment (FCI)	**Contingency + IC + DC**
Working capital (WC)	0.15 × FCI
Total Capital Investment (TCI)	**FCI + WC**

Source: Peters, M., Timmerhaus, K., West, R., 2003. Plant Design and Economics for Chemical Engineers, McGraw-Hill Education.

TABLE 16.7 General Operating Costs and By-Product Prices

Parameter	Price	Unit	References
Bagasse	44	\$/dt	Gubicza et al. (2016)
Process steam	0.027	\$/kWh	Vlysidis et al. (2011)
Cooling water	0.00135	\$/kWh	Vlysidis et al. (2011)
Feed boiler water	0.225	\$/m^3	Vlysidis et al. (2011)
Electricity	0.08	\$/kWh	Assumption
Maintenance	0.03 × TCI	\$	Towler and Sinnott (2013b)
Insurance and general	0.015 × TCI	\$	Towler and Sinnott (2013b)
Ash disposal	25.8	\$/t	Caputo et al. (2005)
Labor	60,000	\$	Assumption
General overhead	0.7 × labor	\$	Towler and Sinnott (2013b)
Enzymes	5.07	\$/kg	Kazi (2010)

The total annual cost (TAC) of each project derives from the sum of the annualized capital cost (ACC) and the operating costs. The former is calculated as:

$$ACC = TCI \times \frac{i_c \times (1+i_c)^n}{-1+(1+i_c)^n} \tag{16.11}$$

Where i_c is the cost of capital (assumed to be equal to 10%) and n the lifetime of the project, that is, 25 years. Fig. 16.6 presents the values for equipment cost (EC), TCI, OC, and production costs for each alternative.

As depicted in Fig. 16.6, capital costs are higher for the thermochemical production of ethanol and they are mainly associated with the gasification island and the alcohols synthesis reactor (~60%). Hybrid and bioethanol come at similar capital costs but with significant difference in operating costs, which in sequence affect the operating costs, that is, 0.76 and 0.67 \$/kg, respectively. The latter is attributed to the high enzymes costs, which count for approximately 33% of the bioethanol production costs. A detailed breakdown of the operating costs for each scenario is illustrated in Fig. 16.7. Feedstock price is the main contributor to the OC for all cases. This issue can be eliminated if the second-generation ethanol plant is integrated with a sugarcane ethanol plant and bagasse is treated as an on-site by-product. Other benefits from this approach are the integration of the ethanol recovery units and the fermentation of hexose sugars to ethanol (in the case of bioethanol). Reported reductions in ethanol production costs for an integrated first- and second-generation ethanol plant are in the range of 20%–30%.

16.5.3 Environmental Sustainability Metrics

Biofuels are recognized as a form of renewable energy and so far they are commonly treated as naturally carbon neutral sources, in the sense that carbon dioxide emitted from their combustion should not be reckoned since it is counterbalanced by CO_2 uptake during photosynthesis process. However, this approach does not take into account emissions related with energy imported (e.g., pumping water, heat duties for the biomass conversion, fuel consumption for transport purposes) at any stage of the biofuel life cycle. Hence, sustainability is a cross-cutting element for the establishment of novel

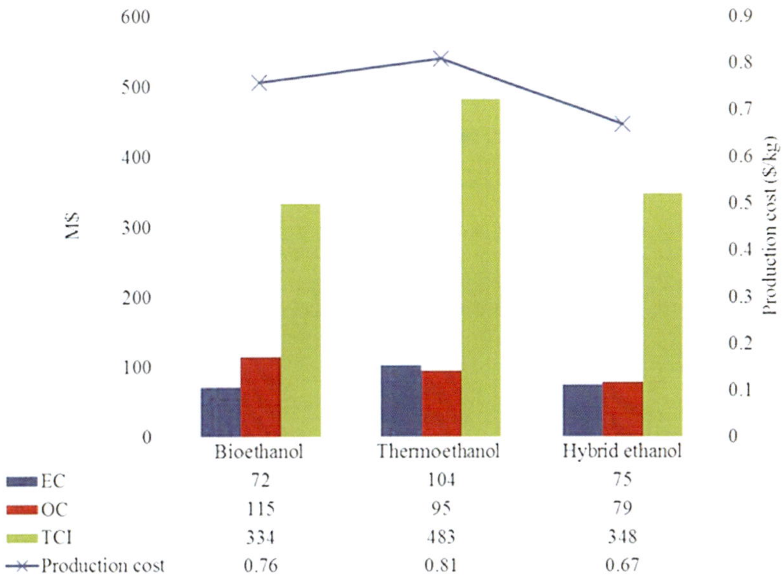

FIGURE 16.6 **Cost data for the investigated ethanol production alternatives.**

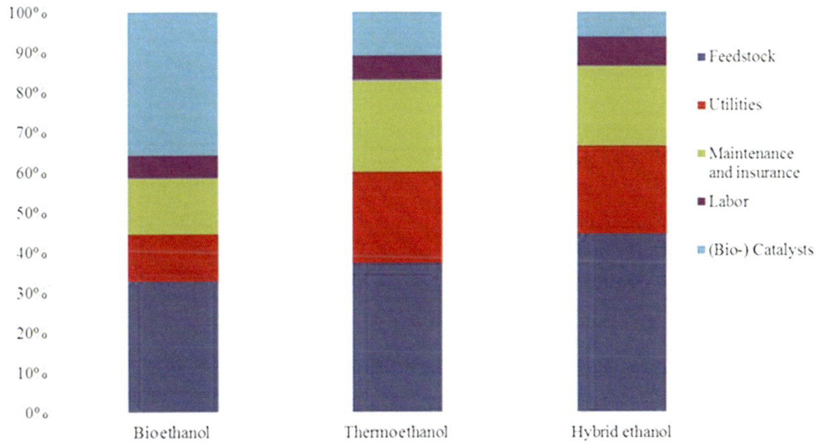

FIGURE 16.7 **Breakdown of the OC for each alternative.**

biorefineries systems. In this chapter, the emissions related to the conversion step of bagasse were calculated (gate-to-gate analysis) based on the simulations and it is sensible to consider that technologies with lower emissions will have a higher contribution to the development of a sustainable biofuel sector. The Bioenergy Technologies Office (BETO) has developed an initial set of significant sustainability metrics for the conversion stage of the biofuel life cycle: CO_2 emissions, fossil energy consumption, fuel yield and biomass carbon-to-fuel efficiency (Jones et al., 2013). Fossil consumption can be quantified by means of fossil fuel energy ratio (FER)

TABLE 16.8 Sustainability Metrics for the Investigated Biojet Fuel Routes

Processes	CO_2 emissions (gCO_2/MJ fuel)	FER	Total fuel yield (kg/dt of feedstock)	Biomass carbon-to-fuel efficiency (%)
Bioethanol	11.9	3.88	268	30.9
Thermoethanol	22.7	3.36	249	29.7
Hybrid ethanol	12.1	3.83	237	27.4

that is the ratio of biofuel energy, in terms of low heating value (LHV), over the fossil energy input into the system, Eq. (16.12).

$$FER = \frac{E_{fuel}}{E_{fossil}} \tag{16.12}$$

In the case that FER is less than 1, then the fuel is categorized as nonrenewable, such as more fossil energy is needed to manufacture it than the energy contained in the final biofuel product. Table 16.8 illustrates the calculated values for the aforementioned metrics for the investigated procedure. It should be noted here that, in the calculations, the emissions are related with producing steam and electricity from conventional sources while CO_2 emissions derived from bagasse conversion steps were considered biogenic and neglected. Typical values are 0.201 kg-CO_2/kWh for raising steam and 0.537 kg-CO_2/kWh for generating electricity (Vlysidis et al., 2011). Due to high electricity imports, the thermochemical route is the more inefficient process while the environmental effect of the other two is very similar.

16.5.4 Multicriteria Decision Analysis

During the last few years, the necessity to determine the best option among the applicable alternatives and classify these alternatives according to their importance for a specific purpose, as well as the need to comparatively evaluate sophisticated technological and social–economic procedures, the use of quantitative multicriteria techniques has been extensively proposed and implemented. The key idea behind the quantitative assessment techniques is to efficiently integrate values of the criteria outlining a specific concept and their significance (known as weights) into a single magnitude. Depending on the project and selecting process, for some certain criteria the largest value is the desirable one, whereas for others the smallest value is the preferred. The scope of decision making process should be to (Osyczka and Montusiewicz, 1988): (1) to identify goals, (2) to identify alternatives for accomplishing these goals, (3) to identify the factors to be used to compare the alternatives, (4) to analyze of the alternatives, and (5) to make choices. The appraisal of several engineering development projects involves consideration of different decision criteria. The multicriteria framework allows such factors to be presented in a comprehensive and consistent format (Thayer, 2016). Although it is a methodology derived from the cost–benefit framework, it has evolved into a decision system that allows all types of attributes to be assessed on an equal footing, possessing many of the characteristics of an effective multicriteria model. Some of the most popular methods are: (1) simple "noncompensatory" methods, (2) simple additive weighting (SAW) method, (3) analytic hierarchy process (AHP), and (4) TOPSIS technique. In this research, TOPSIS method was utilized.

TOPSIS (technique for order preference by similarity to an ideal solution) method is a popular approach to MADM and has been widely used in the literature. TOPSIS was first

TABLE 16.9 Structure of the Alternative Performance

	Criterion 1	**Criterion 2**	**......**	**Criterion n**
Alternative 1	X_{11}	X_{12}		X_{1n}
Alternative 2	X_{21}	X_{22}		X_{2n}
	.	.		.
.	.	.		.
.	.	.		.
Alternative m	X_{m1}	X_{m2}		X_{mn}
	W_1	W_2	......	W_n

developed by Hwang and Yoon (1981) for solving a MADM problem. TOPSIS simultaneously considers the distances to the ideal solution and negative ideal solution regarding each alternative and selects the most relative closeness to the ideal solution as the best alternative. That is, the best alternative is the nearest one to the ideal solution and the farthest one from the negative ideal solution. A relative advantage of TOPSIS is the ability to identify the best alternative quickly. The structure of the alternative performance matrix is expressed as shown in Table 16.9.

Where x_{ij} is the rating of alternative i with respect to criterion j, w_j is the weight of criterion j. The normalized importance weight for each criterion can be calculated using Eq. (16.13):

$$W_i = \frac{n - x_i + 1}{\sum_{i=1}^{n} n - x_i + 1} \quad (16.13)$$

The procedure of TOPSIS is summarized as:

1. Calculation of normalized decision matrix: The normalized value n_{ij} is calculated as

$$n_{ij} = \frac{x_{ij} + 1}{\sqrt{\sum_{j=1}^{m} x_{ij^2}}} \; j = 1, \ldots., m, i = 1\ldots..n \quad (16.14)$$

2. Calculation of the weighted normalized decision matrix: The weighted normalized value v_{ij} is calculated as:

$$v_{ij} = n_{ij} w_{ij}, j = 1, \ldots., m, i = 1\ldots..n \quad (16.15)$$

3. Determination of the positive ideal and negative ideal solution:

$$A^+ = \{v_1^+, \ldots., v_n^+\} = \{(\max v_{ij}) \mid i \in I\}, \{(\min v_{ij}) \mid i \in J\} \quad (16.16)$$

$$A^- = \{v_1^-, \ldots., v_n^-\} = \{(\min v_{ij}) \mid i \in I\}, \{(\max v_{ij}) \mid i \in J\} \quad (16.17)$$

Where I is associated with benefit criteria, and J is associated with cost criteria.

4. Calculation of the separation of each alternative from the positive ideal (d^+) and negative ideal (d^-) solution measures, using the n-dimensional Euclidean distance:

$$d_j^+ = \sqrt{\sum_{j=1}^{n} \left(v_{ij} - v_1^+\right)^2}, j = 1, \ldots., m \quad (16.18)$$

$$d_j^- = \sqrt{\sum_{j=1}^{n} \left(v_{ij} - v_1^-\right)^2}, j = 1, \ldots., m \quad (16.19)$$

5. Calculation of the relative closeness to the ideal solution:

$$S_j = \frac{d_j^-}{d_j^+ + d_j^-} \; j = 1, \ldots., m \quad (16.20)$$

6. Ranking the preference order: The closer the S_j is to 1 implies the higher priority of the jth alternative.

The criteria used are energy efficiency (rank 1), production cost (rank 2), mass efficiency

TABLE 16.10 Computation of Normalized Weights Using Ranking Method

Criteria	Rank Position	Rank Score R_i	$N-R_i+1$	W_i
Energy efficiency	First	1	4	0.4
Production cost	Second	2	3	0.3
Mass efficiency	Third	3	2	0.2
FER	Fourth	4	1	0.1

(rank 3), and fossil energy input (rank 4). Energy efficiency determinations extracted by rigorous thermodynamic analysis and assessment of the biofuel process, would provide a more reliable, rational and real basis for evaluation and should therefore be considered as the foremost crucial and vital decision making factor. If the thermodynamic evaluation suggests that the tested pathway is inefficient and unappealing then supplementary elements will not enhance its feasibility by a great deal. As economic criterion production cost was selected as it considers both capital and operating costs and by taking into consideration that the FER metric accounts for all fossil energy inputs to the system the latter was selected as environmental factor. Table 16.10 presents the values of the weighting factor derived from Eq. (16.13).

After defining the weight of each criterion it was possible to calculate the score, S_j, attributed to each conversion route. Table 16.11 presents the separation measures from the positive ideal solution and the negative ideal solution as well as the final overall scores for each alternative. As expected from the discussion developed throughout the whole chapter, the biochemical route is the most sustainable way to produce ethanol. The hybrid route comes close mainly favored from the low production costs and the higher percentage difference observed for this criterion compared to the rest performance factors. Thermochemical ethanol is a high energy intensive process and this affect both its economic and environmental efficiency, and consequently it attains a comparatively rather low score.

16.6 CONCLUDING REMARKS

The present chapter provides a holistic assessment of three biomass to ethanol conversion routes. It incorporates the design and analysis of chemical, biochemical, and hybrid biorefinery systems utilizing as feedstock sugarcane bagasse. Even though a recent analysis, conducted by the BP organization (2017), claims that there are sufficient oil supplies for the foreseeable future, crude oil is a finite resource and production is anticipated to ultimately deteriorate. Therefore, even without the pressing environmental motives for decreasing the "carbon footprin" in the transport sector, the ultimate depletion of fossil fuel sources triggers

TABLE 16.11 TOPSIS Results for the Investigated Routes

	d_j^+	d_j^-	Sj (%)	Result – Rank
Bioethanol	0.020	0.204	91	1
Thermoethanol	0.223	0.009	4	3
Hybrid ethanol	0.030	0.222	88	2

the research on exploring alternative sources of energy, such as biomass and in particular agricultural and/or municipal wastes. The investigated pathways are established as follow: (1) saccharification of bagasse followed by sugars fermentation, (2) gasification followed by mixed alcohols synthesis, and (3) gasification followed by syngas fermentation. Robust and thermodynamically rigorous simulations of the constituent processes of these biofuel-integrated conversion options were built in process simulation software, that is, Aspen Plus. Based on the quantification and assessment of the yields, mass and energy balances of the constituent processes and overall thermodynamic energy, economic and environmental efficiencies were calculated for each option. In order to take into consideration all the criteria studied and thus make a reliable comparison possible, multicriteria analysis was applied where each alternative was issued a score. The methodology utilized combines the most important feasibility factors and builds a reliable decision-making tool that can be useful in assessing and comparing the performance of the currently biomass conversion routes to ethanol. Due to limited biomass availability, the establishment of optimal route biofuel value chains (one conversion route for each feedstock) is a key prerequisite for a feasible bioenergy sector and as so the methodology is developed in this way to suggest only one process as the best option.

The analysis showed that the most feasible pathway is the biochemical route as it is favored for the high theoretical yields of cellulose hydrolysis and sugar fermentation steps. The process can be, in theory, further optimized via heat integration of the ethanol recovery stage. Nevertheless, the chief drawback of this process is not directly associated with the design but with market conditions and in particular the enzymes price. The hybrid route seems to be a promising route combining benefits from both the mother processes, such as lignin exploitation for fuel production (from thermochemical) and low equipment costs (fermentor are typically cheaper than chemical synthesis reactors). However, this process is yet to achieve commercial success due to low ethanol productivities. Nevertheless, recent advances, in this direction, include two-stage continuous fermentation (Richter et al., 2013) that can achieve yields of 300–330 litre/ton of lignocellulosic biomass. The thermochemical route suffers from the many process steps involved in the production line and the complexity associated with these units. The low conversion of syngas to alcohols along with the elevated operating pressures were identified as the main bottleneck of this pathway and consequently increased electricity imports are required. This, in effect, negatively affects the performance of the process. Furthermore, if ethanol is the target product then the thermochemical pathway is the least desired due to the synthesis of by-products. Overall, at the current technological status and based on the examined criteria the bioprocessing of bagasse to ethanol is the more efficient technology. In fact, a more thorough decision analysis depends on numerous aspects including criteria aside from those calculated here, such as the market demand, possible social concerns and the location of the plant. However, since we deal with a single feedstock and a single product those criteria would not change the final outcome.

References

Acharya, B., Roy, P., Dutta, A., 2014. Review of syngas fermentation processes for bioethanol. Biofuels 5, 551–564.

Aden, A., 2002. National Renewable Energy Laboratory (U.S.): Lignocellulosic biomass to ethanol process design and economics utilizing co-current dilute acid prehydrolysis and enzymatic hydrolysis for corn stover. Technical Report NREL/TP-510-32438. National Renewable Energy Laboratory, Golden, CO, United States.

Alvarado-Morales, M., Terra, J., Gernaey, K.V., Woodley, J.M., Gani, R., 2009. Biorefining: computer aided tools for sustainable design and analysis of bioethanol production. Chem. Eng. Res. Des. 87, 1171–1183.

Antizar-Ladislao, B., Turrion-Gomez, J.L., 2008. Second-generation biofuels and local bioenergy systems. Biofuel. Bioprod. Bioref. 2, 455–469.

Arora, D., Basu, R., Breshears, F.S., Gaines, L. D., Hays, K. S. Phillips, J. R., Wikstrom, C. V., Clausen, E. C., Gaddy, J.L., 1997. United States Department of Energy, Office of Energy Efficiency and Renewable Energy, Albuquerque Operations Office, Office of Scientific and Technical Information. Production of ethanol from refinery waste gases. Final report, April 1994–July 1997. United States Department of Energy, Office of Energy Efficiency and Renewable Energy; distributed by the Office of Scientific and Technical Information, U.S. Department of Energy. http://www.osti.gov/servlets/purl/565441-UPx17r/webviewable/

2012. Aspen Physical Property System: Physical Property Models. Aspen Technology, Burlington, MA.

2013. Aspen Plus, Getting Started Modeling Processes with Solids. Aspen Technology, Burlington, MA.

Balat, M., Balat, H., 2009. Recent trends in global production and utilization of bio-ethanol fuel. Appl. Energy 86, 2273–2282.

Basu, P., 2010. Biomass Gasification and Pyrolysis: Practical Design and Theory. Academic Press, Boston, MA, United States.

Bozell, J.J., Astner, A., Baker, D., Biannic, B., Cedeno, D., Elder, T., Hosseinaei, O., Delbeck, L., Kim, J.-W., O'lenick, C.J., Young, T., 2014. Integrating separation and conversion—conversion of biorefinery process streams to bio-based chemicals and fuels. BioEnergy Res. 7, 856–866.

Bridgwater, A.V., 2003. Renewable fuels and chemicals by thermal processing of biomass. Chem. Eng. J. 91, 87–102.

Bridgwater, A.V., 2009. Technical and economic assessment of thermal processes for biofuels. NNFCC project 08/018.

Caputo, A.C., Palumbo, M., Pelagagge, P.M., Scacchia, F., 2005. Economics of biomass energy utilization in combustion and gasification plants: effects of logistic variables. Biomass Bioenergy 28, 35–51.

Chauhan, M.K., Chaudhary, V.S., Kumar, S., Samar, S., 2011. Life cycle assessment of sugar industry: a review. Renew. Sustain. Energy Rev. 15, 3445–3453.

De Aquino, G.S., De Conti Medina, C., Da Costa, D.C., Shahab, M., Santiago, A.D., 2017. Sugarcane straw management and its impact on production and development of ratoons. Industrial Crop. Prod. 102, 58–64.

De Medeiros, E.M., Posada, J.A., Noorman, H., Osseweijer, P., Filho, R.M., 2017. Hydrous bioethanol production from sugarcane bagasse via energy self-sufficient gasification-fermentation hybrid route: simulation and financial analysis. J. Clean. Prod. 168, 1625–1635.

Drzyzga, O., Revelles, O., Durante-Rodríguez, G., Díaz, E., García, J.L., Prieto, A., 2015. New challenges for syngas fermentation: towards production of biopolymers. J. Chem. Technol. Biotechnol. 90, 1735–1751.

Dutta, A., Phillips, S., 2009. National renewable energy laboratory (U.S.) 2009. Thermochemical Ethanol Via Direct Gasification and Mixed Alcohol Synthesis of Lignocellulosic Biomass. National Renewable Energy Laboratory, Golden, CO, United States, Nrel/Tp 510-45913.

Fang, K., Li, D., Lin, M., Xiang, M., Wel, W., Sun, Y., 2009. A short review of heterogeneous catalytic process for mixed alcohols synthesis via syngas. Catalysis Today 147, 133–138.

Gao, Y., Xu, J., Zhang, Y., Yu, Q., Yuan, Z., Liu, Y., 2013. Effects of different pretreatment methods on chemical composition of sugarcane bagasse and enzymatic hydrolysis. Biores. Technol. 144, 396–400.

Gubicza, K., Nieves, I.U., Sagues, W.J., Barta, Z., Shanmugam, K.T., Ingram, L.O., 2016. Techno-economic analysis of ethanol production from sugarcane bagasse using a liquefaction plus simultaneous saccharification and co-fermentation process. Biores. Technol. 208, 42–48.

Gupta, A., Verma, J.P., 2015. Sustainable bio-ethanol production from agro-residues: a review. Renew. Sustain. Energy Rev. 41, 550–567.

He, J., Zhang, W., 2011. Techno-economic evaluation of thermo-chemical biomass-to-ethanol. Appl. Energy 88, 1224–1232.

Hwang, C.L., Yoon, K., 1981. Multiple Attribute Decision Making: Methods and Applications: A State-of-the-Art Survey. Springer-Verlag, Berlin, Germany.

Isikgor, F.H., Becer, C.R., 2015. Lignocellulosic biomass: a sustainable platform for the production of bio-based chemicals and polymers. Polymer Chem. 6, 4497–4559.

Jones, S., Meyer, P., Snowden-Swan, L., Padmaperum, A., Tan, E., Dutta, A., Jacobson, J., Cafferty, K., 2013. Process design and economics for the conversion of lignocellulosic biomass to hydrocarbon fuels fast pyrolysis and hydrotreating bio-oil pathway. Pacific Northwest National Laboratory (U.S.), Department of Energy, Office of Scientific and Technical Information.

Kamm, B., Gruber, P.R., Kamm, M., 2010. Biorefineries-Industrial processes and products: Status quo and future directions Wiley-VCH, Weinheim.

Kazi, F.K., 2010. National Renewable Energy Laboratory (U.S.) 2010. Techno-Economic Analysis of Biochemical Scenarios for Production of Cellulosic Ethanol. National Renewable Energy Laboratory, Golden, CO, United States, Technical report NREL/TP 6A2-46588.

Krahulec, S., Petschacher, B., Wallner, M., Longus, K., Klimacek, M., Nidetzky, B., 2010. Fermentation of mixed glucose-xylose substrates by engineered strains of *Saccharomyces cerevisiae*: role of the coenzyme specificity of xylose reductase, and effect of glucose on xylose utilization. Microb. Cell Factor. 9, 16.

Kricka, W., Fitzpatrick, J., Bond, U., 2015. Challenges for the production of bioethanol from biomass using recombinant yeasts. Adv. Appl. Microbiol. 92, 89–125.

Kumar, D., Murthy, G.S., 2011. Impact of pretreatment and downstream processing technologies on economics and energy in cellulosic ethanol production. Biotechnol. Biofuels 4, 27.

Kumar, S., Singh, N., Prasad, R., 2010. Anhydrous ethanol: a renewable source of energy. Renew. Sustain. Energy Rev. 14, 1830–1844.

Liguori, R., Ventorino, V., Pepe, O., Faraco, V., 2016. Bioreactors for lignocellulose conversion into fermentable sugars for production of high added value products. Appl. Microbiol. Biotechnol. 100, 597–611.

Liu, S., Mel, D., Wang, L., Tu, X., 2017. Steam reforming of toluene as biomass tar model compound in a gliding arc discharge reactor. Chem. Eng. J. 307, 793–802.

Michailos, S., Parker, D., Webb, C., 2016a. Simulation studies on ethanol production from sugar cane residues. Ind. Eng. Chem. Res. 55, 5173–5179.

Michailos, S., Parker, D., Webb, C., 2016b. A multicriteria comparison of utilizing sugar cane bagasse for methanol to gasoline and butanol production. Biomass Bioenergy 95, 436–448.

Mohd Azhar, S.H., Abdulla, R., Jambo, S.A., Marbawi, H., Gansau, J.A., Mohd Faik, A.A., Rodrigues, K.F., 2017. Yeasts in sustainable bioethanol production: a review. Biochem. Biophys. Rep. 10, 52–61.

Mohr, A., Raman, S., 2013. Lessons from first generation biofuels and implications for the sustainability appraisal of second generation biofuels. Energy Policy 63, 114–122.

Moreira, J.R., 2000. Sugarcane for energy: recent results and progress in Brazil. Energy Sustain. Develop. 4, 43–54.

Mosier, N., Wyman, C., Dale, B., Elander, R., Lee, Y.Y., Holtzapple, M., Ladisch, M., 2005. Features of promising technologies for pretreatment of lignocellulosic biomass. Bioresour. Technol. 96, 673–686.

Naik, S.N., Goud, V.V., Rout, P.K., Dalai, A.K., 2010. Production of first and second generation biofuels: a comprehensive review. Renew. Sustain. Energy Rev. 14, 578–597.

Olofsson, K., Bertilsson, M., Lidén, G., 2008. A short review on SSF: an interesting process option for ethanol production from lignocellulosic feedstocks. Biotechnol. Biofuel. 1, 7.

Osyczka, A., Montusiewicz, J., 1988. A multicriteria approach to optimum design of vehicle brakes. Int. J. Vehicle Des. 9, 438–446.

Pandey, A., Soccol, C.R., Nigam, P., Soccol, V.T., 2000. Biotechnological potential of agro-industrial residues I: sugarcane bagasse. Biores. Technol. 74, 69–80.

Panopoulos, K.D., Fryda, L.E., Karl, J., Poulou, S., Kakaras, E., 2006. High temperature solid oxide fuel cell integrated with novel allothermal biomass gasification. J. Power Sources 159, 570–585.

Peters, M., Timmerhaus, K., West, R., 2003. Plant Design and Economics for Chemical Engineers. McGraw-Hill Education, New York, NY, United States.

Piccolo, C., Bezzo, F., 2009. A techno-economic comparison between two technologies for bioethanol production from lignocellulose. Biomass Bioenergy 33, 478–491.

Rabelo, S.C., Carrere, H., Maciel Filho, R., Costa, A.C., 2011. Production of bioethanol, methane and heat from sugarcane bagasse in a biorefinery concept. Biores. Technol. 102, 7887–7895.

Ravagnani, M.A.S.S., Reis, M.H.M., Filho, R.M., Wolf-Maciel, M.R., 2010. Anhydrous ethanol production by extractive distillation: a solvent case study. Proc. Saf. Environ. Protec. 88, 67–73.

Richter, H., Martin, M., Angenent, L., 2013. A two-stage continuous fermentation system for conversion of syngas into ethanol. Energies 6, 3987.

Ringer, M., Putsche, V., Scahill, J., 2006. Large-scale pyrolysis oil production: a technology assessment and economic analysis. Technical Report NREL/TP-510-37779. National Renewable Energy Laboratory, Golden, CO, United States.

Sadhukhan, J., Ng, K.S., Hernandez, E.M., 2014. Economic analysis. Biorefineries and Chemical Processes. Wiley, Hoboken, NJ.

Saini, J.K., Saini, R., Tewari, L., 2015. Lignocellulosic agriculture wastes as biomass feedstocks for second-generation bioethanol production: concepts and recent developments. Biotech 5, 337–353.

Solantausta, Y., Bridgwater, A.V., Beckman, D., 1996. Electricity production by advanced biomass power systems. Report VTT. Technical Research Centre, Finland.

Spivey, J.J., Egbebi, A., 2007. Heterogeneous catalytic synthesis of ethanol from biomass-derived syngas. Chem. Soc. Rev. 36, 1514–1528.

Sreejith, C.C., Muraleedharan, C., Arun, P., 2013. Thermochemical analysis of biomass gasification by gibbs free energy minimization model: Part II (Optimization of biomass feed and steam to biomass ratio). Int. J. Green Energy 10, 610–639.

Srivastava, N., Rawat, R., Singh Oberoi, H., Ramteke, P.W., 2015. A review on fuel ethanol production from lignocellulosic biomass. Int. J. Green Energy 12, 949–960.

Sudiro, M., Bertucco, A., 2009. Production of synthetic gasoline and diesel fuel by alternative processes using natural gas and coal: process simulation and optimization. Energy 34, 2206–2214.

Sun, Y., Cheng, J., 2002. Hydrolysis of lignocellulosic materials for ethanol production: a review. Biores. Technol. 83, 1–11.

Thayer, M., 2016. Renewable energy and energy efficiency: assessment of projects and policies. Appl. Econ. J. 23, 92–100.

Towler, G., Sinnott, R., 2013a. Capital cost estimating, 2nd ed. Chemical Engineering Design. Butterworth-Heinemann, Boston, MA, United States.

Towler, G., Sinnott, R., 2013b. Economic evaluation of projects, 2nd ed. Chemical Engineering Design. Butterworth-Heinemann, Boston, MA, United States.

Trippe, F., Fröhling, M., Schultmann, F., Stahl, R., Henrich, E., 2011. Techno-economic assessment of gasification as a process step within biomass-to-liquid (BtL) fuel and chemicals production. Fuel Process. Technol. 92, 2169–2184.

Ververis, C., Georghiou, K., Danielidis, D., Hatzinikolaou, D.G., Santas, P., Santas, R., Corleti, V., 2007. Cellulose, hemicelluloses, lignin and ash content of some organic materials and their suitability for use as paper pulp supplements. Biores. Technol. 98, 296–301.

Vlysidis, A., Binns, M., Webb, C., Theodoropoulos, C., 2011. A techno-economic analysis of biodiesel biorefineries: assessment of integrated designs for the co-production of fuels and chemicals. Energy 36, 4671–4683.

Walter, A., Ensinas, A.V., 2010. Combined production of second-generation biofuels and electricity from sugarcane residues. Energy 35, 874–879.

WRI, 2014. Climate Analysis Indicators Tool (CAIT) 2.0: WRI's Climate Data Explorer. World Resources Institute, 2014, http://cait.wri.org.

Wyman, C.E., Dale, B.E., Elander, R.T., Holtzapple, M., Ladisch, M.R., Lee, Y.Y., 2005. Coordinated development of leading biomass pretreatment technologies. Biores. Technol. 96, 1959–1966.

Zanin, G.M., Santana, C.C., Bon, E.P.S., Giordano, R.C.L., De Moraes, F.F., Andrietta, S.R., Neto, C.C.D.C., Macedo, I.C., Lahr Fo, D., Ramos, L.P., Fontana, J.D., 2000. Brazilian bioethanol program. Appl. Biochem. Biotechnol. 84, 1147–1161.

Further Reading

BP Global, 2017. Oil Reserves. Available from: http://www.bp.com/en/global/corporate/energy-economics/statistical-review-of-world-energy/oil/oil-reserves.html.

CHAPTER

17

Biorefinery as a Promising Approach to Promote Ethanol Industry From Microalgae and Cyanobacteria

Carlos E. de Farias Silva,**, Elena Barbera*, Alberto Bertucco**

*Department of Industrial Engineering DII, University of Padova, Padova, Italy

**Institute of Chemistry and Biotechnology, Federal University of Alagoas, Maceió, Brazil

17.1 INTRODUCTION

In both developed and developing countries, exploiting natural resources in economic activities to provide clean technologies, especially in the industrial and agricultural sectors, has numerous advantages, including those related to resource saving, recycling, and reuse of wastes. Therefore, exploitation of natural resources in industrial processes is essential to achieve a more sustainable world (Uenojo and Pastore, 2007).

In general, increased and improved global strategies for energy security and mitigation of CO_2 emissions from energy production processes are required, especially those aimed at maximizing the energy efficiency by expanding the use of clean energy. This means using fuels that promote the carbon cycle without changing the atmospheric balance (renewable fuels), and developing energy resources in CO_2 neutral systems (Brennan and Owende, 2010).

In particular, biofuels play an important role in reducing global climate change. Their impact will depend on several aspects related to the selection of new technologies, legal restrictions, international trade, land use, choice of raw materials, and management techniques (Worldwatch Institute, 2007). This is one of the main goals of what is known as a biorefinery approach.

17.1.1 The Biorefinery and Bioethanol Production

Among other definitions, according to the International Energy Agency (IEA Bioenergy)

Bioethanol Production From Food Crops
http://dx.doi.org/10.1016/B978-0-12-813766-6.00017-5

biorefining is "the sustainable processing of biomass into a spectrum of marketable products and energy" (IEA, 2008). The National Renewable Energy Laboratory (NREL) comments that "biorefinery is a facility that integrates biomass conversion processes and equipment to produce fuels, power, and chemicals from biomass" (NREL, 2015). As a rule, biorefinery is the integration of a given biomass, with all its components separated, to produce energy and chemicals. Biorefineries enlarge and spread the concept of biofuels, energy, and chemicals from a renewable source and promote the concept of carbon cycle, aiming at sustainability, while helping in the reduction of production costs.

Ethanol is the main biofuel produced worldwide. Between 2007 and 2015, bioethanol throughput practically doubled, reaching 25 billion gal/year, even though after 2010 the production was stagnant due the limitations of arable land for biomass cultivation ("Food vs. Fuel") (AFDC, 2016). The United States and Brazil are the main contributors in the field, using corn and sugarcane, respectively.

Sugarcane ensures the lowest bioethanol production costs. In spite of its significant advantage, it is not a viable option for all the regions of the planet. Consequently, countries of the northern hemisphere have been incessantly looking for new technological routes that permit the efficient production of biofuels while respecting environmental and economic sustainability issues (BNDS, 2008), and "new" generations of biomass-to-ethanol are proposed. In fact, based on the type of biomass, bioethanol production is classified as first (raw material saccharine or starch-based); second (lignocellulosic materials); third (microalgal/macroalgal biomass); and fourth (genetically modified cyanobacteria) generation.

Fig. 17.1 represents a scheme of the four generations and the advantages/disadvantages of each one of them. The main concerns are about the land use, food versus fuel competition, recalcitrance of the lignocellulosics/macroalgae, high cultivation costs of microalgae, and little information in literature about genetically modified cyanobacteria.

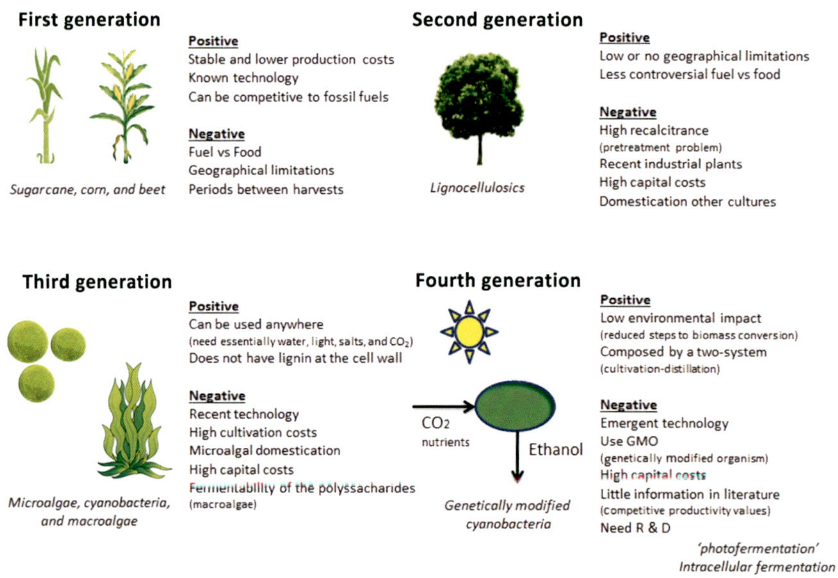

FIGURE 17.1 **Biomass generations aimed to bioethanol production.**

Low costs are proper to first-generation bioethanol (with the exception of corn-based one), which has a well-established and economically sustainable technology, while second-generation still requires a decrease in production costs to become competitive, with pretreatment and hydrolysis processes needing to be more effective and largely scalable (Gupta and Verna, 2015). On the other hand, micro- and macroalgae have not reached an industrial scale operating plant yet. In the case of third- and fourth-generation bioethanol, further studies are required to develop a competitive and consolidated technology, taking into account also issues other than technological ones.

In addition, bioethanol based on lignocellulosic biomass (second generation) has greater saccharification difficulties, due to hemicellulose–cellulose–lignin tridimensional structure. In particular, lignin is a recalcitrant and nonfermentable component. High production rates associated with violent pretreatments and the use of a number of enzymes, in order to improve the productivity, make this process difficult for large-scale consolidation with acceptable production costs, even if new lignocellulosic-based industrial plants in the last decade have been trying to stimulate this approach and to increase significantly the ethanol production worldwide.

In third-generation bioethanol, microalgae/macroalgae biomass is used, which does not have lignin in its cellular structure, and is cultivated with higher growth rates when compared to higher plants (Harun and Danquah, 2011; Ho et al., 2013; Hong et al., 2014; John et al., 2011; Wei et al., 2013). As for this biomass a suitable process is not available yet, and the related costs cannot be properly estimated, researchers are trying:

- For microalgae: to optimize microalgal productivity and cultivation conditions, as this represents the highest production costs, considering that hydrolysis and fermentation are instead easier compared with lignocellulosics and macroalgae (Silva and Bertucco, 2016).
- For macroalgae: to increase the polysaccharide fermentability, considering that when the entire biomass (and not only the cellulose-rich residue remaining after hydrocolloid extraction) is saccharified, some residual sugars, such as mannitol, are not fermented by *Saccharomyces cerevisiae* (Daroch et al., 2013; Enquist-Newman, 2014; Wei et al., 2013). Genetically modified microorganisms are proposed to this aim (Wei et al., 2013).

At last, fourth-generation bioethanol is still an emerging technology and refers to the process in which ethanol is produced without need of biomass breakdown, that is, using genetically modified cyanobacteria able to capture sunlight, water, and nutrients and convert them directly into ethanol by "photofermentation." It has a limited use protected by patents in the USA issued by Algenol and Joule Unlimited and little information is shared, which makes a real validation of the potentiality for these microorganisms quite difficult (Silva and Bertucco, 2016). Data available in the literature report a production of 0.025 mg/L in 6 days (*Synechococcus* PCC 7942) and 0.55 g/L after 6.25 days (*Synechocystis* PCC 6803) (Angermayr et al., 2009; Deng and Coleman, 1999; Tan et al., 2011), values indeed much lower than those divulgated by the companies (Silva and Bertucco, 2016).

Because of their high growth rate, and relatively simple biochemical composition (partitioned among carbohydrates, lipids, and proteins), microalgae are acknowledged as very promising feedstock for bioethanol production, and are the focus of this chapter. However, in order for this process to be economically and environmentally sustainable, it is essential to fully exploit the microagal biomass, by carefully recovering from it all the possible products as well as energy. Accordingly, in this chapter, we suggest the idea that it should be more correct to address the issue of microalgal bioethanol production from a biorefinery standpoint, with the main focus directed toward maximizing production of biofuels (ethanol/biodiesel) and energy (power generation), while at the same time promoting nutrients (nitrogen, phosphorous, and other micronutrients) and CO_2 recovery/recycle integrated approach.

This concept can be applied to bioethanol production processes from microalgal biomass as proposed in the block-flow diagram of Fig. 17.2. The different units of this flowsheet are discussed separately in the following sections. In our opinion, the main challenges of microalgae-based ethanol biorefinery are related to: cultivation and carbohydrate production, hydrolysis and fermentation, CO_2/water/salts recycling, and nutrient recovery by anaerobic digestion (AD). Fine chemicals were excluded from the currently discussed process because they generally use carbohydrate or lipid fractions, and we focus on the biofuel-based (mainly bioethanol and biorefinery).

17.2 MICROALGAL BIOMASS CULTIVATION AND CARBOHYDRATE PRODUCTION

Microalgae (eukaryote) and cyanobacteria (prokaryote) are photosynthetic microorganisms, exploited for their biochemical plasticity, that is, the ability to change the biochemical composition as an adaptation to environmental and nutritional changes, and consequently to produce energy reserves (carbohydrates and lipids). Even though applications as foods, food additives, animal feed, fertilizers, and biochemicals are under investigation, this chapter is focused on microalgal biomass for biofuels. *Scenedesmus*, *Chlorella*, *Chlamydomonas*, *Tetraselmis*, and *Nannochloropsis* are examples of industrial microalgae for carbohydrate/lipid production. As to cyanobacteria species, *Spirulina*, *Chlorococcus*, *Gloeocapsa*, *Synechocystis*, and *Synechococcus* can be cited.

Photosynthesis represents a unique process of sunlight energy conversion. In this process,

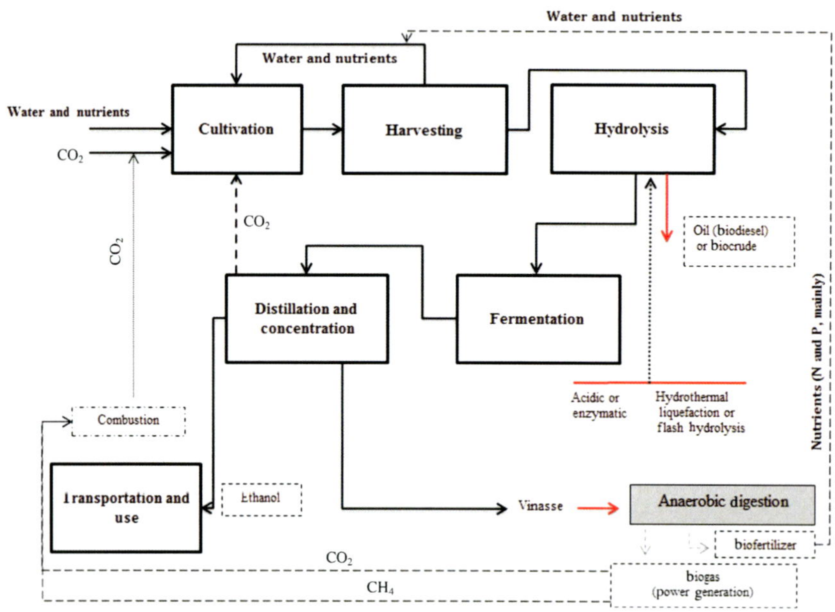

FIGURE 17.2 **Flow chart of bioethanol production from microalgae biomass within the concept of biorefinery.**

inorganic compounds and sunlight energy are converted to organic matter by photoautotrophs. Oxygenic photosynthesis can be expressed as a redox reaction driven by light energy, in which carbon dioxide and water are converted to carbohydrates (not complex molecules) and oxygen.

When nitrogen limitation is used, according to the main method currently applied, it shifts the carbon fixed by Calvin's Cycle to produce substances other than nitrogen-based compounds (proteins), that is, carbohydrates (polysaccharides) and lipids. Starch and glycogen are examples of polysaccharides accumulated as energetic reserves. Most of the green algae accumulate starch as primary energy and carbon reserve, whereas lipids serve as a secondary storage, while cyanobacteria preferentially produce glycogen (Gonzalez-Fernandez and Ballesteros, 2012; Vitovà et al., 2015). On the other hand, some species, such as chrysolaminarin producers (Heterokotophyta, Bacillariophyceae) and some oleaginous algae (*Nannochloropsis*, *Trachydiscus*) are apparently unable to produce polyglucans, thus they are not applicable in the context of this chapter (Vitovà et al., 2015).

Microalgae must ensure a high production rate of carbohydrates if bioethanol from fermentation is desired. Namely, nutritional techniques (nutrient starvation and carbon source), saline stress, light intensity, and temperature (Chen et al., 2013; Silva and Bertucco, 2016) can be applied to this purpose. For instance, these are efficiently used for *Scenedesmus*, *Chlorella*, *Chlamydomonas*, and *Tetraselmis*. Under proper conditions, microalgae/cyanobacteria species can accumulate carbohydrates with a fraction of dry cell weight up to 60%.

Cultivation can be carried out in batch, fed-batch, or continuous systems. These operating modes, with their characteristics and procedure to accumulate carbohydrates, including pros and cons, are represented in Fig. 17.3.

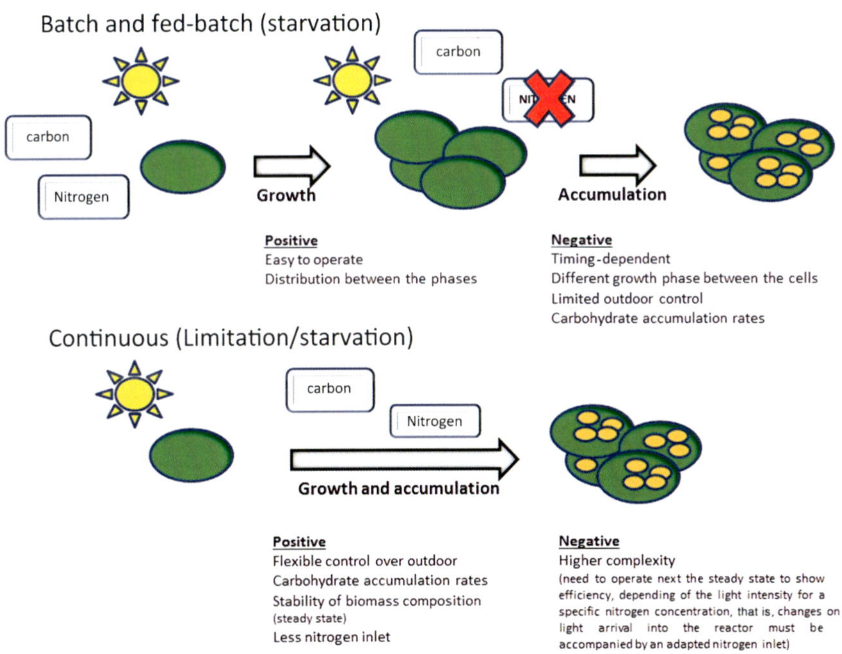

FIGURE 17.3 **Mechanism of carbohydrate accumulation under nitrogen starvation/limitation at different operating systems.**

In batch mode the main concern is related to biomass productivity, as a two-step operation is required: a first one (excess of nutrients), where biomass grows until nutrient starvation, and a second one (accumulation) where carbon continues to be fixed but it is converted in energy reserves, like carbohydrates and lipids. In this second step, biomass is not produced anymore and, although the carbohydrate productivity slightly increases, the biomass productivity considerably decreases (generally 40%–60%). When continuous mode is carried out, using lower nitrogen concentration in the feed it is possible to obtain growth and carbohydrate accumulation simultaneously, thus achieving a steady state with 2–3 times higher carbohydrate productivity, even though biomass productivity decreases by 0–20% in comparison with a steady state performed with excess nutrients. When the stress is moderate, carbohydrate is generally accumulated, but if it is strong, lipids are preferred, because these metabolisms (carbohydrate and lipids) are interconnected to the environmental/nutritional stresses.

17.3 HARVESTING AND WATER/ NUTRIENT RECYCLE STEPS

In microalgae and cyanobacteria cultivation units a biomass concentration of not more than a few grams per liter can be achieved. In order to remove water, the most applied harvesting methods comprise flocculation (both pH induced and bioflocculation), coagulation, gravitational sedimentation, electric-based processes, filtration, and centrifugation (Barros et al., 2015). Gravitational sedimentation, flocculation, and centrifugation usually have higher yields and lower costs with respect to the others. It was shown that a carbohydrate-rich microalgal biomass (N stressed) increases its settling velocity up to 10 times in comparison with a control condition (N replete) (Depraetere, 2015).

Because the amounts of fertilizers and carbon dioxide required to microalgal cultivation represents a significant portion of production costs (Acién et al., 2012; Slade and Bauen, 2013), it is mandatory to minimize the waste of nutrients due to process purges. With respect to water/nutrient/CO_2 recycling, two process streams are discussed in this chapter (Fig. 17.2): the culture medium after cultivation and the exhausted/solubilized biomass/organic matter at the end of ethanolic production step (the latter will be specifically discussed in later sections).

Water and nutrient recycling after cultivation is an essential stage to make biofuels processes environmentally and economically sustainable, as it warrants a better use of available resources, with an increase in the process yield.

Microalgae biomass production has been estimated to consume water in a ratio 200–1000 kg_{water}/kg of dry biomass, with closed systems clearly losing less water than open raceways (Béchet et al., 2016; Farooq et al., 2015). For this reason, a reliable low-cost water supply is critical to the success of biofuel production from microalgae (Slade and Bauen, 2013). In particular, freshwater needs to be added to raceway ponds systems to compensate evaporation, while some PBR designs require high water consumption for heating/cooling. In addition, cleaning and purge water are examples of water use in an industrial plant and need to be accurately quantified. A water balance of the system can be written according to (Guieysse et al., 2013):

$$V_{\text{total,input}} = V_{\text{fill}}F_{\text{req}} + V_{\text{evap}} + V_{\text{blowdown}} + V_{\text{photo}} + V_{\text{harvest}} + V_{\text{drying}} + V_{\text{biomass}} + V_{\text{gray water}} \quad (17.1)$$

where

$V_{\text{total,input}}$ = water volume that needs to be added to the cultivation system annually ($m^3/(m^2$ year));
V_{fill} = nominal water required to fill the cultivation system (m^3/m^2);
F_{req} = frequency of emptying the cultivation system for cleaning (per year);

V_{evap} = water loss due to evaporation ($m^3/(m^2$ year));
$V_{blowndown}$ = water removed during cultivation to maintain stable salinity and culture control ($m^3/(m^2$ year));
V_{photo} = water loss due the photosynthesis ($m^3/(m^2$ year));
$V_{harvest}$ = water removal during the harvesting process ($m^3/(m^2$ year));
V_{drying} = water removed if the biomass is dried ($m^3/(m^2$ year));
$V_{biomass}$ = water in biomass (moisture content) ($m^3/(m^2$ year));
$V_{gray\ water}$ = water with nutrients or contaminants nonrecycled in the process of cultivation ($m^3/(m^2$ year));
V_{photo} is a less amount (photosynthesis) and $V_{gray\ water}$ if the water is recycled can be neglected. The values of evaporation and water supply cost are geographic-dependent.

The water requirement of the system can be significantly reduced by recycling the cultivation medium after harvesting. For example, it was reported that the water consumption to produce 1 L of biodiesel with *Chlorella vulgaris* would be reduced 1500–3700 L to a range of 500–800 L (Farooq et al., 2015; Yang et al., 2011). Quantitative data related to specific water consumption for bioethanol in literature is lacking, but clearly recycling water after cultivation would allow significant savings also in this case.

Water recycling is not an issue for microalgal cultivation. For instance, *Chlorella zoofigiensis* was cultivated twice in water harvested and totally recycled, being adjusted the nitrogen and phosphorous content only (Zhu et al., 2013).

In relation to nutrients, it is necessary especially to control the correct use of macronutrients, such as nitrogen, phosphorus, magnesium, and sulfur, that is, to quantify their concentrations in the effluent after cultivation, together with those of a number of micronutrients, such as, Fe, Al, and Cu. Nutrient recycling is the only alternative to promote their maximum utilization.

However, water/nutrient recycling after cultivation can cause a number of problems:

- Salinity of the medium increases as a result of water evaporation (mainly Na, Ca, Mg, and Cl) and strategies to monitorate them are necessary, in view of an economic and ecological viability, because saline stress is a problem for biomass productivity in microalgal systems;
- Aluminum and ferric salts used as bioflocculant deteriorate the water quality, because they can show inhibitory effects, which are species-dependent;
- The presence of organic molecules/particulate released from microalgae (extracellular compounds or from died cells) can inhibit the growth, even though the productivity can be improved due to mixotrophy;
- Higher bacterial, fungi, and virus contamination is likely to occur (Farooq et al., 2015; Slade and Bauen, 2013).

Reverse osmosis and dilution are methods proposed to control water salinity, the latter being more feasible. To prevent contamination some techniques are suggested, such as filtration, ozonation, chlorination, treatment with hydrogen peroxide, or heating. Also centrifugation reduces significantly the content of contaminant microorganisms (Farooq et al., 2015).

Eventually, it is noteworthy that, when nitrogen limitation/starvation is applied, this nutrient is used efficiently, maximizing the carbohydrate/lipid production. Phosphorous content in the biomass can increase in nitrogen limitation/starvation conditions. An example is presented in Fig. 17.4, which shows the remarkable difference between biomass production under nitrogen excess and limited/starved conditions, corresponding to around 50 g N/kg of microalgal biomass for nitrogen limited, instead of 130 g N/kg in nitrogen excess condition, thus reducing significantly the requirement of this fertilizer.

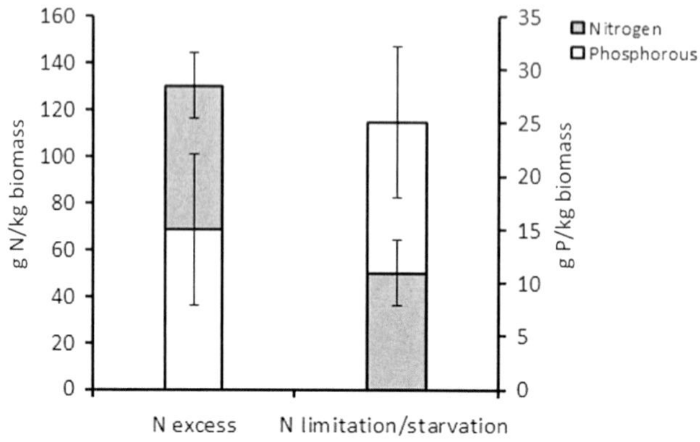

FIGURE 17.4 **Difference between the nitrogen and phosphorous supplied to cultivate microalgae under excess and limitation/starvation of nitrogen.** *Source: Data from Silva, C.E.F., Sforza, E., 2016. Carbohydrate productivity in continuous reactor under nitrogen limitation: effect of light intensity and residence time on nutrient uptake in* Chlorella vulgaris. *Process Biochem. 51, 2112–2118.*

17.4 HYDROLYSIS AND FERMENTATION

Once biomass is harvested and a concentration of about 50–200 g/L is achieved, the carbohydrate fraction needs to be converted into simple sugars and fermented to produce bioethanol. The conventional saccharification–fermentation process is described in the next section, followed by possible alternative processing routes aiming at a more strategic fractionation and recovery of products from the biomass.

17.4.1 Traditional Methods

The main routes to saccharify microalgae are chemical (thermochemical) or enzymatic hydrolysis, each one with its own advantages. In any case, microalgae and cyanobacteria present lower recalcitrance in comparison to lignocellulosics/macroalgae, so that no are simple pretreatment is needed, and hydrolysis can be performed in one step.

Concerning the thermochemical process, sulfuric, hydrochloric, and nitric acid are effective in chemical saccharification, although more aggressive conditions than the enzymatic process are required for hydrolysis, generally ranging between 120 and 140°C of temperature and between 15 and 30 min of reaction time, if >80% saccharification and >80% of theoretical fermentation yield are desired. If acid is not added, a hydrothermal process can be performed with the natural acidity of biomass but it requires higher temperature, 160–220°C (see earlier sections).

A few examples of acidic hydrolysis on microalgal biomass reported in the literature include:

- *Scenedesmus obliquus* when hydrolyzed at 120°C using 2–3N sulfuric acid for 30 min (biomass concentration 20–500 g/L) yielded an almost total transformation of the carbohydrate content in reducing sugars (71%–97% of hydrolysis extension), of which 65% was made of glucose (Miranda et al., 2012).
- *Scenedesmus bijugatus* with 26% carbohydrate content after lipid extraction, after acid hydrolysis [H_2SO_4 (0.36–1.08N) at 130°C, 45 min, and 20 g/L solid concentration] achieved 84% of sugars solubilization and was 70% converted in bioethanol (Ashokkumar et al., 2015).
- Sulfuric acid hydrolysis was more efficient than enzymatic treatment with pectinases, amylases, and cellulases (Kim et al., 2014)

for *C. vulgaris* FSP-E. In fact, Ho et al. (2013) reported that this acidic process, with H_2SO_4 (0.036–1.8N) at 121°C for 20 min and 10–80 g/L of biomass concentration, ensured 95% of glucose saccharification, and reached approximately 90% of the theoretical fermentation yield in 12 h with *Zymomonas mobilis* ATCC 29191.

- *Arthrospira platensis* was hydrolyzed using different acids (H_2SO_4, HNO_3, HCl, and H_3PO_4, alone or combined) at a concentration of 0.25–2.5N, and a temperature 60–100°C. The best result was 80% saccharification and 55% fermentation yield using *S. cerevisiae* MV 92081 (Markou et al., 2013).

Enzymatic hydrolysis features mild temperatures and lower degradation risks. Enzymes used for the saccharification of microalgal biomass normally include amylases (glucoamylases and α-amylases), cellulases (endo-β-glucanase, exo-β-glucanase, and β-glucosidase) and pectinases, although hemicellulase can be required (separate or together). In this case, cell disruption is a crucial step. However, the major drawbacks of this treatment are related to the cost of enzymes, together with the necessity to control pH (generally using buffers) and to optimize temperature, according to the optimum conditions required by the specific enzyme. The performances obtained with enzymatic hydrolysis on different microalgal/cyanobacterial species can be summarized as follows:

- *C. vulgaris* was subjected to different methods of cell disruption, with beadbeating combined with pectinase (from *Aspergillus aculeatus*), resulting in sugars extraction increased between 45% and 70% and a subsequent fermentation yield of 89% after 12 h with *S. cerevisiae* KCTC 7906. For this species, pectinase appears as a more effective enzyme if compared with cellulases, amylases, and xylanases (Kim et al., 2014).
- *Chlamydomonas reinhardtii* (50 g/L of biomass concentration and 59.7% of carbohydrate content) treated by separated hydrolysis and fermentation (SHF) using amylases (0.005% α-amylase from *Bacillus licheniformis* at 90°C and 30 min to liquefaction and 0.2% glucoamylase from *Aspergillus niger* at 55°C and 30 min to saccharification, pH 4.5) resulted in 94% hydrolysis efficiency and 60% of fermentation yield with *S. cerevisiae* S288C (Choi et al., 2010).
- *Synechococcus* sp. PCC 7002 (with 3 g/L of biomass concentration and 60% of carbohydrate content), hydrolyzed 80% of sugars after enzymatic treatment (lysozyme, and α-glucanases Liquozyme SC DS and Spirizyme fuel). After fermentation with *S. cerevisiae*, an ethanol yield of 86% with respect to the theoretical maximum rate was reached (Mollers et al., 2014).

17.4.2 Alternatives Processes

As an alternative to conventional acid or enzymatic hydrolysis, different processes have been proposed in the literature to fractionate the algal biomass by exploiting the solvent-like, tunable properties of water under hydrothermal conditions (high T and P). These processes [namely, hydrothermal liquefaction and flash hydrolysis (FH)] have been originally developed with the aim of converting microalgal biomass into biocrude, hence exploiting mostly the oil fraction. They are particularly suitable for microalgae/cyanobacteria a system, considering that this type of biomass brings along high water content and its slurry is relatively easy to pump in continuous-flow systems.

Typically, hydrothermal liquefaction (HTL) is carried out in a temperature range between 220 and 374°C (i.e., subcritical water conditions), and the residence time is in the order of a few minutes (10–30 min) (Lopez-Barreiro et al., 2013). It leads to the production of gaseous, aqueous, and solid by-products. Under such conditions, sugars are degraded to furfurals, hydroxymethylfurfural (HMF), and other complex aromatic compounds

(Martinez-Fernandez and Chen, 2017). However, by properly acting on the operating variables (i.e., temperature and residence time) carbohydrates can be extracted/hydrolyzed without being degraded. In addition, with respect to HTL applied to lignoscellulosic biomass, microalgal carbohydrates are easier to extract/hydrolyze, being present mostly as starch and without the presence of lignin.

As an example, a two-step sequential hydrothermal liquefaction (SHTL) process has been proposed, which consists in a first extraction of polysaccharides under milder temperatures (140–200°C), followed by a second step of bio-oil production under typical HTL conditions (Chakraborty et al., 2013; Martinez-Fernandez and Chen, 2017; Miao et al., 2012). It has been reported that, under a temperature $T = 160°C$ and 20 min of residence time, a maximum yield of 32% wt. polysaccharides could be extracted from *Chlorella sorokiniana* biomass into the aqueous phase (Miao et al., 2012). That was characterized to be mainly a linear 1–4 linked glucan with only minor branching, which can be easily hydrolyzed to simple sugars (97.5% glucose) (Chakraborty et al., 2013). Similar results have been reported by Yuan et al. (2016), who obtained a recovery of sugars in the aqueous phase of 84.15% at 200°C and 60 min with *Scenedesmus* sp., even though only 1% was present as glucose, the rest being mostly oligosaccharides. By slightly increasing the temperature, polysaccharides could be, at least partially, directly hydrolyzed to monosaccharides (i.e., glucose), even though the formation of protein–sugar interaction compounds might also start to occur, so the conditions need to be specifically optimized to maximize sugars recovery.

Alternatively, a process called FH is also promising in this regard. It consists in treating the wet algal biomass under temperatures similar to those of HTL (210–350°C), but with extremely short residence times (less than 10 s), in a continuous-flow system. Because of the fast reaction time, mainly proteins along with carbohydrates are extracted in the aqueous phase, while the rest ends up in a solid product (Garcia-Moscovo et al., 2013). This process looks very promising, considering that for crystalline cellulose treated at 335°C and 4.7 s, a 66.8% yield of hydrolysis products (mainly glucose and some oligomers) is obtained (Kumar and Gupta, 2008). In addition, a recovery of reducing sugars between 4% and 40% is reported for *Scenedesmus* sp. treated under temperatures 205–325°C and 9–10 s of residence time (with the highest value reported for the highest temperature) (Garcia-Moscovo et al., 2013). Controlling the residence time and ensuring rapid cooling is a key aspect to minimize degradation of glucose into undesired products, such as aldehydes, furfurals. A great advantage of FH is related to the rapid residence time, which allows easily working in continuous-flow systems, so that it appears more suitable for industrial scale plants.

One of the main advantages of these two hydrothermal processes is that they allow fractionation of the whole algal biomass with simultaneous recovery of different coproducts: most importantly, the lipid fraction is converted into fuel products (see above sections). Additionally, nutrients (mainly N, P, but also other inorganics, such as S, Mg, and K) are also extracted in the aqueous phase (Martinez-Fernandez and Chen, 2017; Teymouri and Kumar, 2017), which can therefore be suitable for the subsequent fermentation step, by providing nutrients for yeasts growth. Alternatively, other high-value bioproducts might be recovered by inducing precipitation of minerals from the aqueous phase: for example, Barbera et al. (2017) showed how magnesium ammonium phosphate ($MgNH_4PO_4$) could be precipitated from the hydrolysate after FH and recycled as a slow-release fertilizer for algae cultivation or other applications. The hydrolysate itself has shown to be a good substrate for algal growth, suggesting that part of it could be recycled to the cultivation system. For instance, *S. obliquus* was grown in the hydrolysate obtained from the same species for FH at 280°C and 9 s of residence time. This species ensured better performance in batch cultivation than the standard medium for autotrophic growth, thanks to the combination with mixotrophy. In continuous mode, the productivities ranged 0.62–0.72 g/L/day, showing satisfactory performances (Barbera et al., 2016).

Clearly, a solid residue is also left over by both of these processes: such biochar, which has applications in heavy metals removal from polluted water or as soil amendment (Abdel-Fattah et al., 2015). The removal of heavy metals, such as copper, cadmium, chromium, and lead has already been proven feasible with biochar obtained from switchgrass and other terrestrial biomasses (Abdel-Fattah et al., 2015; Regmi et al., 2012). Alternatively, even though it is not the preferable solution, the solid char could be combusted as a low-value fuel to supply a portion of the heat energy required by the process (Martinez-Fernandez and Chen, 2017).

The main disadvantage of hydrothermal processes is the high energy duty, caused by the rather strong temperature and pressure operating conditions. These might hinder the convenience of applying such a technology to exploit the useful fraction of microalgae. However, as water acts as both reaction medium and reactant, they allow avoiding the costs associated with chemicals (acids) or enzymes, provided that the conditions and fermentable sugars recovery yields are optimized. A comparison between the operating process conditions and main limitations are presented in Table 17.1.

17.4.3 Lipids Utilization

In view of a biorefinery concept, it has been stressed how the importance to recover products from the algal biomass needs to be maximized. In particular, the lipid fraction should also be exploited for the production of additional liquid biofuels. When considering the traditional saccharification process, this is particularly important because chemical hydrolysis methods may also facilitate the subsequent solvent extraction of lipids, thereby recovering both fermentable sugars and lipids from the microalgal biomass.

TABLE 17.1 Comparison Between the Main Discussed Process for Microalgal Biomass Hydrolysis

Process of hydrolysis	Range of temperature (°C)	Operation time	Limitations/particularities
Acidic	110–140	5–60 min	Neutralization after hydrolysis (increase the salinity of the broth); additional cost with chemicals
Enzymatic	50–80	24–72 h	Cellular rupture is necessary; necessity to find an average value of pH and temperature (enzymatic mix: each enzyme has an optimum value); accurately control of pH and temperature (sensibility of enzymes); use of buffers to control the pH; cost of enzymes
HTL	160–200°C	10–30 min	High energy duty; incomplete hydrolysis of polysaccharides;
Flash hydrolysis	250–350	Less than 10 s	New technology (scale-up to pilot scale is needed to obtain more reliable results).

Nannochloropsis gaditana, *C. sorokiniana* and *Phaeodactylum tricornutum* have been treated by steam explosion with sulfuric acid (H_2SO_4, 0–3.6N at 120–150°C for 5 min), and approximately 96% of the sugar content was hydrolyzed using 0.6N of the acid at 150°C. The acid hydrolysis of these microalgae biomasses also increased the efficiency of lipid extraction (Lorente et al., 2015). Wang et al. (2014) reported a 25% increase in lipid contents obtained before and after hydrolysis of the microalgae *Tribonema* sp. with H_2SO_4 1N. The carbohydrate content was hydrolyzed to 80% for a biomass concentration of 50 g/L in suspension at 121°C and 45 min, and 70% of the theoretical yield was achieved after fermentation with *S. cerevisiae*. Recent studies on wet lipid extraction from algal biomass are also promising (Naghdi et al., 2016), encouraging the recovery of oil, to be then converted into biodiesel via transesterification, by eliminating the extremely high cost otherwise associated with biomass drying.

The FH process described in the previous section also allows recovering the lipid fraction from the starting microalgal feedstock, which could be subsequently converted into biodiesel. In fact, from experiments carried out at 280°C and 9 s residence time, using different algal species (e.g., *Scenedesmus* sp. and *N. gaditana*) it was found that roughly 90% of initial lipids are preserved in the solid fraction (Garcia-Moscovo et al., 2013, 2015; Teymouri and Kumar, 2017). In particular, Teymouri and Kumar (2017) verified that the fatty acids methyl esters (FAME) profile of the lipids retained in the solid was consistent with that of the starting algal biomass. Moreover, thanks to the fact that the solid product is already partially degraded, and that its lipid content can be as high as 70% (Teymouri and Kumar, 2017), lipids extraction will also be facilitated.

If the SHTL approach is followed, instead, the residual biomass remaining after the first step is then converted into biocrude. Compared to direct HTL, the biocrude produced with the two-step approach is reported to have lower oxygen, nitrogen, and sulfur contents, thanks to the prior extraction of proteins along with polysaccharides. In addition, the production of solid char residues is also reduced, while biocrude yields of roughly 30% and higher heating values (HHV) between 33 and 37 MJ/kg are reported for different *Chlorella* species (Martinez-Fernandez and Chen, 2017).

17.5 NUTRIENTS RECOVERY

The possibility of nutrients recovery is of tremendous importance when large-scale biomass cultivation is desired for biofuels production. In fact, standard mediums, which are recipes providing an optimized mixture of nutrients to support microalgae growth, are very expensive. Thus, in a biorefinery approach, recycling carbon, nitrogen, phosphorous, and other micronutrients is mandatory, and could definitely increase the sustainability of the proposed ethanol process from photosynthetic microorganisms (Barreiro et al., 2015). While the recycling of cultivation medium after harvesting was discussed in an earlier section, nutrients can also be recovered from the residual biomass, because only little amounts of them end up in the biofuels products. In this regard, the recycling of nutrients (N and P) and CO_2 by AD and from fermentation is discussed in the following.

17.5.1 Anaerobic Digestion of Vinasse

After the operations of sugars fermentation and subsequent bioethanol distillation, all the remaining organic matter from biomass and yeasts together with large amounts of macro- and micronutrients (especially K, N, and P) end up in a wastewater effluent, the so-called vinasse. Vinasse produced from sugarcane or sugar beet biorefineries is typically employed as field fertilizer for crop production, or as an additive animal feed (Moraes et al., 2017). However, considering the high organic content, this application does not appear as a good one in terms of environmental impacts (high toxicity to the soil and water bodies, it need to be carefully applied).

In agreement with the concept of biorefinery, treating this effluent by AD looks like a promising option (Fig. 17.2). Through AD, in fact, the organic matter can be converted into biogas (i.e., a mixture of mainly CH_4 and CO_2, with volumetric amounts of roughly 55%–70% and 45%–30%, respectively). The biogas produced can then be used for combined heat and power generation, to be integrated in any of the energy consuming operations within the industrial process. The CO_2 obtained could, on the other hand, be reused in loco for the cultivation of the microalgae/cyanobacteria (Thiansathit et al., 2015), together with that derived from the fermentation process (see next section).

In addition to biogas, AD also produces a liquid digestate in which nutrients are mineralized: for example, organic nitrogen is converted into ammonium (i.e., the most preferred form for microalgae to uptake) (Markou et al., 2014), while phosphorus into phosphate salts. Specifically, for stillage, which has a good C/N for AD, after the digestion nitrogen and phosphorous content in liquid phase are higher, because these fractions are not left in gaseous phase (Moraes et al., 2015; Silva and Abud, 2016).

Such a digestate can indeed be employed as a nutrient source for microalgal cultivation: for instance, *C. vulgaris* was successfully cultivated in anaerobically digested sugarcane stillage, showing good consumptions of N and P from the effluent (Marques et al., 2013). Thanks to nutrients recycling into the microalgal cultivation system by AD of vinasse, the supply of fresh fertilizers required would be greatly reduced, thus increasing the sustainability of the process, as well as its economicity.

So far, AD of vinasse from microalgal bioethanol production has not been investigated much. However, a few reports carried out on AD of sugarcane vinasse, which is starting to be considered a promising approach in the Brazilian area, show that promising results can be obtained, provided that the operating parameters are properly considered (Koyama et al., 2016). Among the aspects to be investigated are the likely better suitability of employing thermophilic rather than mesophilic conditions (given the temperature of the effluent from the distillation process); also, the high amount of proteinaceous material (after the extraction of both carbohydrates and lipids from the microalgal biomass), may lead to low C/N ratios in the vinasse, which could hinder the AD process, so that codigestion with other C-richer substrates may be advisable (Moraes et al., 2017). Overall, given all the promising aspects related to nutrients and CO_2 recovery together with energy generation, the possibility of integrating AD of vinasse in microlagal/cyanobacterial bioethanol production certainly deserves further attention and investigation.

17.5.2 CO_2 Recycling

Because in the future it might not be permissible for industrial applications to emit CO_2 in large amounts at ground level (Slade and Bauen, 2013), recycling carbon streams in the process is important, especially in microalgal systems.

The reuse of carbon dioxide in ethanol biorefineries is a must (Moncada et al., 2014). Generally, 1 kg of dry algal biomass requires at least 1.83 kg of CO_2. In practical terms, the efficiency of CO_2 fixation in open raceways may be less than 10%; for thin layer cultivation it can achieve 35%. In closed tubular photobioreactors (PBRs) CO_2 fixation efficiencies of around 75% are reported (Slade and Bauen, 2013).

Almost half of initial sugars mass during the fermentation is "lost" as carbon dioxide (CO_2) (glucose → ethanol + $2CO_2$). This carbon has a market value and can be used in several processes, such as the food industry or chemical synthesis, or be stored as value-added product and softener the environmental impact (CO_2

emissions), thus decreasing energy losses and process costs (Kheshgi and Prince, 2005; Xu et al., 2010).

In this sense, the development of a technology able to capture carbon dioxide becomes important, being essential in view a sustainable/economic/environmental aspect of the whole process (Mollersten et al., 2003; Rhodes and Keith, 2005; Xu et al., 2010).

Another application is to use this carbon dioxide to cultivate microalgae, because at industrial level supply and maintenance of this gas can correspond up to 50% of production costs (Chisti, 2013; Silva et al., 2016). As an example, *Dunaliella teriolecta* was cultivated with CO_2 from beer fermentation, reaching a growth rate of 0.58/day and biomass productivity 0.18 g/L/day (18 klx, 28°C and airflow rate of 1 L/min), which can represent a valuable contribution even to integrate first- and second- with third-ethanol generation (Chagas et al., 2015).

17.6 POTENTIAL BIOETHANOL PRODUCTIVITY FROM MICROALGAE

Thanks to the higher growth rates and, consequently, biomass productivities, microalgae/cyanobacteria have a good potential for bioethanol extensive production. Maximum ethanol throughput is estimated to be 46–140,000 L bioethanol/(ha year) (Cheryl, 2008). Actually, in continuous mode for *C. vulgaris* and *S. obliquus* under constant and day–night light cycles ethanol productivities of 11,000 and 30,000 L bioethanol/(ha year) were calculated (Silva et al., 2017), which are 2–5 times higher than sugarcane [average of 6000 L bioethanol/(ha year)].

An economic/energetic analysis exclusively for an ethanol biorefinery from microalgae is lacking in the literature, and real figures of the feasibility of this process cannot be provided, so that it is presently not possible to compare with a biodiesel plant (first objective of microalgae exploitation for biofuels). However, cultivating microalgae for biofuels is still expensive and new developments are necessary.

17.7 CONCLUDING REMARKS AND FUTURE OUTLOOK

Research on microalgae and cyanobacteria cultivation, hydrolysis, and fermentation are, as a rule, the object of many investigations even though they are not currently consolidated within a continuous process, as cultivation costs are still high (Tercero et al., 2016). The stage of water and nutrient recycles and the reuse of lipids in the biomass saccharification process of sugars/lipid extraction, AD and energy-economic analysis of viability still lack technical information. Further investigations should be undertaken to consolidate and guarantee the viability for the production of bioethanol from microalgae, but they are more likely to be successful if addressed within a biorefinery approach. Thus, a valuable analysis for this adapted technology is necessary, because most of the data available are related to other processes, such as microalgal biomass for food or biodiesel.

Acknowledgments

The authors thank CNPq—Brazil (National Research Council of Brazil)—Process number 249182/2013-0, for resources and fellowship.

References

Abdel-Fattah, T.M., Mahmoud, M.E., Ahmed, S.B., Huff, M.D., Lee, J.W., Kumar, S., 2015. Biochar from woody biomass for removing metal contaminants and carbon sequestration. J. Ind. Eng. Chem. 22, 103–109.

Acién, F.G., Fernandez, J.M., Magan, J.J., Molina, E., 2012. Production cost of a real microalgae production plant and strategies to reduce it. Biotechnol. Adv. 30, 1344–1353.

ADFC, 2016. Ethanol production in the world. Available from: http://www.afdc.energy.gov/data/.

Angermayr, S.A., Hellingwerf, K.J., Lindblad, P., Mattos, M.J.T., 2009. Energy biotechnology with cianobactéria. Curr. Opin. Biotech. 20, 257–263.

Ashokkumar, V., Salam, Z., Tiwari, O.N., Chinnasamy, S., Mohammed, S., Ani, F.N., 2015. An integrated approach for biodiesel and bioethanol production from *Scenedesmus bijugatus* cultivated in a vertical tubular photobioreactor. Energy Convers. Manage. 101, 778–786.

Barbera, E., Sforza, E., Kumar, S., Morosinotto, T., Bertucco, A., 2016. Cultivation of *Scenedesmus obliquus* in liquid hydrolysate from flash hydrolysis for nutrient recycling. Bioresour. Technol. 207, 59–66.

Barbera, E., Teymouri, A., Bertucco, A., Stuart, B.J., Kumar, S., 2017. Recycling minerals in microalgae cultivation through a combined flash hydrolysis–precipitation process. ACS Sustain. Chem. Eng. 5, 929–935.

Barreiro, D.L., Bauer, M., Hornung, U., Posten, C., Kruse, A., Prins, W., 2015. Cultivation of microalgae with recovered nutrients after hydrothermal liquefaction. Algal Res. 9, 99–106.

Barros, A.I., Gonçalves, A.L., Simoes, M., Pires, J.C.M., 2015. Harvesting techniques applied to microalgae: a review. Renew. Sust. Energ. Rev. 41, 1489–1500.

Béchet, Q., Shilton, A., Guieysse, B., 2016. Maximizing productivity and reducing environmental impacts of full-scale algal production through optimization of open pond depth and hydraulic retention time. Environ. Sci. 50, 4102–4110.

BNDS, 2008. Banco Nacional de Desenvolvimento Econômico e Social. Bioetanol de cana-de-açúcar: Energia para o desenvolvimento sustentável. BNDS, Rio de Janeiro, Brazil.

Brennan, L., Owende, P., 2010. Biofuels from microalgae: a review of technologies for production, processing, and extractions of biofuels and co-products. Renew. Sustain. Energ. Rev. 14 (2), 557–577.

Chagas, A.L., Rios, A.O., Jarenkow, A., Marcilio, N.R., Ayub, M.A.Z., Rech, R., 2015. Production of carotenoids and lipids by *Dunaliella tertuiolecta* using CO_2 from beer fermentation. Process Biochem. 50, 981–988.

Chakraborty, M., McDonald, A.G., Nindo, C., Chen, S., 2013. An α-glucan isolated as a co-product of biofuel by hydrothermal liquefaction of *Chlorella sorokiniana* biomass. Algal Res. 2, 230–236.

Chen, C., Zhao, X., Yen, H., Ho, S., Cheng, C., Lee, D., Bai, F., Chang, J., 2013. Microalgae-based carbohydrates for biofuel production. Biochem. Eng. J. 78, 1–10.

Cheryl, 2008. Algae Becoming the New Biofuel of Choice. http://duelingfuels.com/biofuels/non-food-biofuels/algae-biofuel.php.

Chisti, Y., 2013. Constraints to commercialization of algal fuels. J. Biotechnol. 167, 201–214.

Choi, S.P., Nguyen, M.T., Sim, S.J., 2010. Enzymatic pretreatment of *Chlamydomonas reinhardtii* biomass for ethanol production. Bioresour. Technol. 101, 5330–5336.

Daroch, M., Geng, S., Wang, G., 2013. Recent advances in liquid biofuel production from algal feedstocks. Appl. Energy 102, 1371–1381.

Deng, M., Coleman, J.R., 1999. Ethanol synthesis by genetic engineering in cyanobacteria. Appl. Environ. Microb. 65 (2), 523–528.

Depraetere, O., et al., 2015. Harvesting carbohydrate-rich *Arthrospira platensis* by spontaneous settling. Bioresour. Technol. 180, 16–21.

Enquist-Newman, M., et al., 2014. Efficient ethanol production from brown macroalgae sugars by a synthetic yeast platform. Nature 505, 239–243.

Farooq, W., Suh, W.I., Park, M.S., Yang, J., 2015. Water use and its recycling in microalgae cultivation for biofuel application. Bioresour. Technol. 184, 73–81.

Garcia-Moscovo, J.L., Obeid, W., Kumar, S., Hatcher, P.G., 2013. Flash hydrolysis of microalgae (*Scenedesmus* sp.) for protein extraction and production of biofuels intermediates. J. Supercrit. Fluids 82, 183–190.

Garcia-Moscovo, J.L., Teymouri, A., Kumar, S., 2015. Kinetics of peptides and arginine production from microalgae (*Scenedesmus* sp.) by flash hydrolysis. Ind. Eng. Chem. 54, 2048–2058.

Gonzalez-Fernandez, C., Ballesteros, M., 2012. Linking microalgae and cyanobacteria culture conditions and key-enzymes for carbohydrate accumulation. Biotechnol. Adv. 30, 1655–1661.

Guieysse, B., Béchet, Q., Shilton, A., 2013. Variability and uncertainty in water demand and water footprint assessment of fresh algae cultivation based on case studies from five climatic regions. Bioresour. Technol. 128, 317–323.

Gupta, A., Verna, J.P., 2015. Sustainable bio-ethanol production from agroresidues: a review. Renew. Sust. Energ. Rev. 41, 550–567.

Harun, R., Danquah, M.K., 2011. Enzymatic hydrolysis of microalgal biomass for bioethanol production. Chem. Eng. J. 168, 1079–1084.

Ho, S., Huang, S., Chen, C., Hasunuma, T., Kondo, A., Chang, J., 2013. Bioethanol production using carbohydrate-rich microalgae biomass as feedstock. Bioresour. Technol. 135, 191–198.

Hong, I.K., Jeon, H., Lee, S.B., 2014. Comparison of red, brown and green seaweeds on enzymatic saccharification process. J. Ind. Eng. Chem. 20, 2687–2691.

IEA, 2008. IEA Bioenergy Task 42 on Biorefineries: Co-production of Fuels, Chemicals, Power and Materials from Biomass. IEA Bioenergy, Copenhagen, Denmark, March 25–26, 2007.

Jonh, R.P., Anisha, G.S., Nampoothiri, K.M., Pandey, A., 2011. Micro and macroalgal biomass: a renewable source for bioethanol. Bioresour. Technol. 102, 186–193.

Kheshgi, H., Prince, R.C., 2005. Sequestration of fermentation CO_2 from ethanol production. Energy 30, 1865–1871.

Kim, K.H., Choi, I.S., Kim, H.M., Wi, S.G., Bae, H., 2014. Bioethanol production from the nutrient stress-induced microalga *Chlorella vulgaris* by enzymatic hydrolysis and immobilized yeast fermentation. Bioresour. Technol. 153, 47–54.

Koyama, M.H., Junior, M.M.A., Zaiat, M., Junior, A.D.N.F., 2016. Kinetics of thermophilic acidogenesis of typical Brazilian sugarcane vinasse. Energy 116, 1097–1103.

Kumar, S., Gupta, R.B., 2008. Hydrolysis of microcrystalline cellulose in subcritical and supercritical water in a continuous flow reactor. Ind. Eng. Chem. Res. 47 (23), 9321–9329.

Lopez-Barreiro, D., Prins, W., Ronsee, F., Brilman, W., 2013. Hydrothermal lioquefaction (HTL) of microalgae for biofuel production: state of the art review and future prospects. Biomass Bioenergy 53, 113–127.

Lorente, E., Farriol, X., Salvado, J., 2015. Steam explosion as a fractionation step in biofuel production. Fuel Process. Technol. 131, 93–98.

Markou, G., Angelidaki, I., Nerantzis, E., Georgakakis, D., 2013. Bioethanol production by carbohydrate-enriched biomass of *Antrospira* (*Spirulina*) *platensis*. Energies 6, 3937–3950.

Markou, G., Vandamme, D., Muylaert, K., 2014. Microalgal and cyanobacterial cultivation: the supply of nutrients. Water Res. 65, 186–202.

Marques, S.S.I., Nascimento, I.A., Almeida, P.F., Chinalaia, F.A., 2013. Growth of *Chlorella vulgaris* on sugarcane vinasse: the effect of anaerobic digestion pretreatment. Appl. Biochem. Biotechnol. 171, 1933–1943.

Martinez-Fernandez, J.S., Chen, S., 2017. Sequential hydrothermal liquefaction characterization and nutrient recovery assessment. Algal Res. 25, 274–284.

Miao, C., Chakraborty, M., Chen, S., 2012. Impact of reaction conditions on the simultaneous production of polysaccharides and bio-oil from heterotrophically grown by a unique sequential hydrothermal liquefaction process. Bioresour. Technol. 110, 617–627.

Miranda, J.R., Passarinho, P.C., Gouveia, L., 2012. Pre-treatment optimization of *Scenedesmus obliquus* microalga for bioethanol production. Bioresour. Technol. 104, 342–348.

Mollers, K.B., Cannella, D., Jorgensen, H., Frigaard, N., 2014. Cyanobacterial biomass as carbohydrate and nutrient feedstock for bioethanol production by yeast fermentation. Biotechnol. Biofuels 7, 1–11.

Mollersten, K., Yan, J., Moreira, J.R., 2003. Potential market niche for biomass energy with CO_2 capture and storage: opportunities for energy supply with negative CO_2 emissions. Biomass Bioenergy 25, 273–285.

Moncada, J., Tamayo, J.A., Cardona, C.A., 2014. Integrating first, second, and third generation biorefineries: incorporating microalgae into the sugarcane biorefinery. Chem. Eng. Sci. 118, 126–140.

Moraes, B.S., Zaiat, M., Bonomi, A., 2015. Anaerobic digestion of vinasse from sugarcane ethanol production in Brazil: challenges and perspectives. Renew. Sustain. Energ. Rev. 44, 888–903.

Moraes, B.S., Petersen, S.O., Zaiat, M., Sommer, S.G., Triolo, J.M., 2017. Reduction in greenhouse gas emissions from vinasse through anaerobic digestion. Appl. Energy 189, 21–30.

Naghdi, F.G., Gonzalez, L.M., Chan, W., Schernk, P.M., 2016. Progress on lipid extraction from wet algal biomass for biodiesel production. Microb. Biotechnol. 9 (6), 718–726.

NREL, 2015. What is a Biorefinary? U.S. Department of Bioenergy. http://www.nrel.gov/biomass/biorefinery.html.

Regimi, P., Moscovo, J.L.G., Kumar, S., Cao, X., MaO FJ., Schafran, G., 2012. Removal of copper and cadmium from aqueous solution using switchgrass biochar produced via hydrothermal carbonization process. J. Environ. Manage. 109, 61–69.

Rhodes, J.S., Keith, D.W., 2005. Engineering economic analysis of biomass IGCC with carbon capture and storage. Biomass and Bioenergy 29, 440–450.

Silva, C.E.F., Abud, A.K.S., 2016. Anaerobic biodigestion of sugarcane vinasse under mesophilic conditions using manure as inoculum. Revist. Ambient. Agua 11 (4), 763–777.

Silva, C.E.F., Bertucco, A., 2016. Bioethanol from microalgae and cyanobacteria. Process Biochem. 51, 1833–1842.

Silva, C.E.F., Sforza, E., 2016. Carbohydrate productivity in continuous reactor under nitrogen limitation: effect of light intensity and residence time on nutrient uptake in *Chlorella vulgaris*. Process Biochem. 51, 2112–2118.

Silva, C.E.F., Gris, B., Sforza, E., La Rocca, N., Bertucco, A., 2016. Effects of sodium bicarbonate on biomass and carbohydrate production in *Synechococcus* PCC 7002. Chem. Eng. Trans. 49, 241–246.

Silva, C.E.F., Sforza, E., Bertucco, A., 2017. Continuous cultivation of microalgae as an efficient method to improve carbohydrate productivity and biochemical stability. 26th European Biomass Conference and Exhibition. Stockholm, Sweden. .

Slade, R., Bauen, A., 2013. Micro-algae cultivation for biofuels: cost, energy balance, environmental impacts and future prospects. Biomass Bioenergy 53, 29–38.

Vitovà, M., Bisova, K., Kawano, S., Zachleder, V., 2015. Accumulation of energy reserves in algae: from cell cycles to biotechnological applications. Biotechnol. Adv. 6, 1204–1218.

Tan, X., Yao, L., Gao, Q., Wang, W., Qi, F., Lu, X., 2011. Photosynthesis driven conversion of carbon dioxide to fatty alcohols and hydrocarbons in cyanobacteria. Metabol. Eng. 13, 169–176.

Tercero, E.A.R., Domenicali, G., Bertucco, A., 2016. Autotrophic production of biodiesel from microalgae: an update process and economic analysis. Energy 76, 807–815.

Teymouri, A., Kumar, S., 2017. Integration of biofuels intermediates production and nutrients recycling in the processing of a marine algae. Alche J. 63 (5), 1494–1502.

Thiansathit, W., Keesser, T.C., Khang, S., Ratpukdi, T., Hovichitr, P., 2015. The kinetics of *Scenedesmus obliquus* microalgae growth utilizing carbon dioxide gas from biogas. Biomass Bioenergy 76, 79–85.

Uenojo, M., Pastore, G.M., 2007. Pectinases: applications and perspectives. Revist. Quím. Nova 30 (2), 388–394.

Wang, H., Ji, C., Bi, S., Zhou, P., Chen, L., Liu, T., 2014. Joint production of biodiesel and bioethanol from filamentous oleaginous microalgae *Tribonema* sp. Bioresour. Technol. 172, 169–173.

Wei, N., Quaterman, J., Jin, Y., 2013. Marine macroalgae: an untapped resource for producing fuels and chemicals. Trends Biotechnol. 31 (2), 70–77.

Worldwatch Institute, 2007. Biofuels for Transport. Earthscan, Washington, DC, United States.

Xu, Y., Isom, L., Hanna, M.A., 2010. Adding value to carbon dioxide from ethanol fermentations. Bioresour. Technol. 101, 3311–3319.

Yang, J., Xu, M., Zhang, X., Hu, Q., Sommerfeld, M., Chen, Y., 2011. Life-cycle analysis on biodiesel production from microalgae: water footprint and nutrients balance. Bioresour. Technol. 102, 159–165.

Yuan, T., Li, X., Xiao, S., Guo, Y., Zhou, W., Xu, J., Yuan, Z., 2016. Microalgae pretreatment with liquid hot water to enhance enzymatic hydrolysis efficiency. Bioresour. Technol. 220, 530–536.

Zhu, L.D., Takala, J., Hiltunen, E., Wang, Z.M., 2013. Recycling harvest water to cultivate *Chlorella zoofigiensis* under nutrient limitation for biodiesel production. Bioresour. Technol. 144, 14–20.

Further Reading

Tan, I.S., Lee, K.T., 2014. Enzymatic hydrolysis and fermentation of seaweed solid wastes for bioethanol production: an optimization study. Energy 78, 53–62.

CHAPTER

18

Role of Genetic Engineering in Bioethanol Production From Algae

*Duraiarasan Surendhiran**, *Abdul Razack Sirajunnisa***

*St. Joseph University, Dar es Salaam, United Republic of Tanzania

**Anna University, Chennai, India

18.1 INTRODUCTION

Biofuel has been recognized as a potential alternate fuel in future's transportation scenario. Biofuel could be a positive replacement of fossil fuel as it could mitigate the surplus demands of the latter in the near future and also could reduce the carbon emissions (Sirajunnisa and Surendhiran, 2016). Bioethanol is one among the various biofuel and it is potential alternate transport fuel. Bioethanol production is not a new concept; it was initiated during fuel crisis worldwide in 1970s and the capacity of ethanol production grew from less than 1 billion L in 1975, reached 39 billion L in 2006 and expected to reach 100 billion L in mid-2015 (Chaudhary et al., 2014; Nguyen et al., 2012) due to their wide applications in many sectors, particularly transport. Bioethanol production is categorized into first, second, and third generations based on various feedstocks, production technology, and their level of development (Dragone et al., 2010; Nguyen et al., 2012; Wu et al., 2014). Synthesis of each generation bioethanol and respective feedstocks are shown in Fig. 18.1. Ethanol produced from food crops like corn, wheat, barley, and sweet sorghum is called grain alcohol whereas ethanol produced from lignocellulosic biomass and agroresidue, such as rice straw, wheat straw, and grasses is known as biomass ethanol or bioethanol (Naik et al., 2010). Grain alcohol has been represented as the first generation bioethanol and biomass ethanol as the second-generation bioethanol. Recently, third-generation bioethanol has emerged whose feedstock is microalgae and seaweeds, which is further discussed in this chapter.

18.2 ALGAE

Algae have recently been considered as "third-generation feedstock" for bioethanol production (Abdallah et al., 2016; Ho et al., 2017).

Bioethanol Production From Food Crops
http://dx.doi.org/10.1016/B978-0-12-813766-6.00018-7

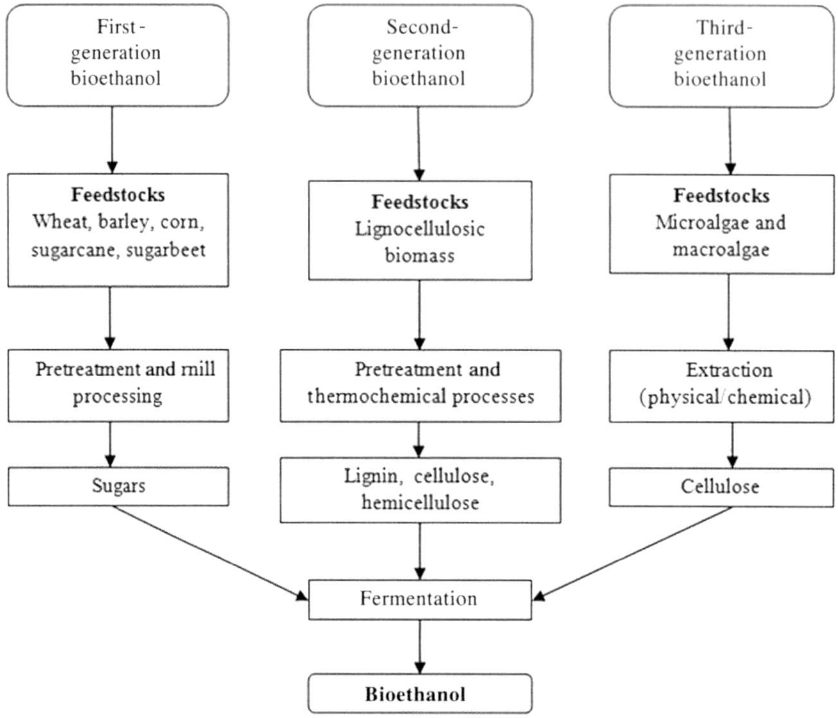

FIGURE 18.1 **Various generations of bioethanol production from different feedstocks.**

Algae are simple organisms containing chlorophyll and they use light for photosynthesis. Algae are the fastest growing plant in the world and generally divided into macroalgae and microalgae based upon morphology. The macroalgae, or "seaweeds," are larger, oceanic multicellular plants growing up to 60 m in length. Microalgae are microscopic, mostly existing as small cells of about 2–200 mm and inhabitants of fresh, sea, and even wastewater systems. They efficiently convert solar energy to biomass by photosynthetic process utilizing sunlight, CO_2, and nutrients from water (Sirajunnisa and Surendhiran, 2016).

The production of biofuel from algae has several advantages over the first- and second generation of biomass sources. First, algae serve as nonfood feedstock, which does not compromise our food security. Second, algae grow in aquatic habitats and thereby do not compete with food crops on agricultural land, or cause deforestation. Third, algal biomass can be used to produce two types of biofuel (bioethanol and biodiesel) because they accumulate high amounts of carbohydrates and lipids (Imran Khan et al., 2017; Radakovits et al., 2010). In addition, the freshwater requirement for algal growth is significantly lower than plant demands to produce the same volume of biofuel (Rajkumar et al., 2014). Finally, the absence of lignin in algae also makes the fermentation process much easier than those used with lignocellulosic biomass (Ho et al., 2017).

18.2.1 Microalgae

Microalgae, as their name implies, are prokaryotic or eukaryotic photosynthetic microorganisms. In general, microalgae are called phytoplankton by biologist, size ranging between 3 and 30 μm, very common (hundreds of thousands species exist), and occur both in fresh- and marine water where they form the basis for the aquatic food chain (Sirajunnisa and Surendhiran, 2016). Microalgae, documented as one of the ancient plant like organisms lacking roots, stems, and leaves and that have chlorophyll *a* as their primary photosynthetic pigment (Demirbas and Faith Demirbas, 2010). Microalgal cells are sunlight-driven cell factories that can convert carbon dioxide (CO_2) into raw materials for producing biofuels (e.g., biohydrogen, biodiesel, and bioethanol), animal food chemical feedstocks and high-value bioactive compounds (e.g., docosahexaenoic acid [DHA]). In particular, the ability of these cells to absorb CO_2 suggests microalgae cultivation as an attractive alternative for CO_2 sequestration that can be applied to fossil fuel power plant gas effluents to facilitate the reduction of greenhouse gas emissions (Klinthong et al., 2015).

18.2.2 Macroalgae

Macroalgae or "seaweeds," are multicellular plants growing in marine. They are often fast growing and can reach sizes of up to 60 m in length. Though there are about 9200 species of seaweeds, only 221 species are economically important. They are classified into three broad groups based on their pigmentation such as brown seaweed (*Phaeophyceae*), red seaweed (*Rhodophyceae*), and green seaweed (*Chlorophyceae*) (Sirajunnisa and Surendhiran, 2016). Generally, it can be cultivated in the vast expanse of the ocean with free sunlight and no need for nitrogen-based fertilizers (Karunakaran et al., 2013). On the contrary of their appearances, their features of morphologic and physiological and chemical compositions are different from terrestrial plants (Jung et al., 2013). Due to have high photosynthesis capability, they have sufficient carbon source for usage in biorefinery.

18.3 BIOFUELS FROM ALGAE

Algae have received a great deal of attention as a novel biomass source for the generation of renewable energy. Algae are both unicellular and multicellular autotrophic aquatic life forms. The unique feature of algae from the other entire microorganism community is that they contain chlorophylls (chlorophyll *a* and *b*), which are usually found in higher plants. Chlorophyll is an important feature for photosynthesis, which enables algae to absorb energy from light to fuel the manufacture of various biomasses (Saifuddin and Priatharsini, 2016). These biomasses can be utilized for production of various renewable fuels like biodiesel, biohydrogen, and bioethanol.

18.3.1 Biodiesel

Biodiesel is a green fuel generally referred to as fatty acid methyl esters (FAME) synthesized from animal fat or vegetable oil by the transesterification reaction (Surendhiran and Vijay, 2014; Yan et al., 2014). Nowadays, eminent concerns are on production of biodiesel from algae. Algae are most viable alternative oil source for biodiesel production due to many advantages over plant and animal sources. However, biodiesel has some disadvantages, such as long-term storage problem because of oxidation (Balat and Balat, 2010; Surendhiran and Vijay, 2014), produce nitrous oxide while burning (Balat and Balat, 2010), high viscosity, low energy content, and gets freezing properties during cold conditions (Srivatsava and Prasad, 2000).

18.3.2 Biohydrogen

Biohydrogen is an ideal alternate fuel produced from algae by the process called biophotolysis of water. It can be operated at ambient temperature and pressure with minimal energy consumption and are more environmental friendly. Using green as a means of producing biohydrogen is a very good alternative (Show et al., 2012). However, commercial production of biohydrogen from biological sources is still not feasible and faces a number of technical and economical barriers, such as due to the low biomass concentration, low yield and costly downstream processes than the current chemical system. Even though substantial progress is continuously being made, there are still many unknown aspects regarding hydrogen production mechanisms and how the efficiency can be improved (Saifuddin and Priatharsini, 2016). Moreover, hydrogen production in green algae is catalyzed by Fe hydrogenases, which are small, bidirectional enzymes, and very high sensitivity to oxygen (Vignais, 2008). Therefore, this technology is still in its infancy and there is a significant potential for technology development and improvement at every level.

18.3.3 Bioethanol

Bioethanol is considered as a clean and renewable fuel, simply produced by microbial fermentation process and have more advantages than the other biofuels. Bioethanol is a promising alternate source to fulfill the current requirement alternate green fuel for transportation sectors to minimize the carbon emissions and is being commercially produced in Brazil, Canada, and United States for transportation (Sirajunnisa and Surendhiran, 2016). Bioethanol is produced by several crops but algae is considered to be the best feedstock due to its ubiquitous in nature, high sugar content, and as the yield is significantly higher than that of other crops (Georgianna and Mayfield, 2012). Furthermore, they normally lack the recalcitrant cell wall structure and lignin that makes the hydrolysis difficult and the sugars for fermentation can be obtained without any expensive lignin removal process. Production of bioethanol from algae consists of following steps like pretreatment and hydrolysis, saccharification and fermentation (Wu et al., 2014).

18.4 BIOETHANOL FROM MICROALGAE

Microalgae exhibit several features that favor using them for industrial production of biofuel. First, they lack specialized tissues and structures, which simplify the cultivation and harvesting processes (Abdallah et al., 2016). Some species of microalgae have high amounts of carbohydrates as reserve polymers and such species are perfect candidates for carbohydrate extraction to produce fermentable sugars for the production of bioethanol (Nguyen et al., 2012). Because microalgae are the most abundant organisms on earth, the research has been a remarkable surge to investigate the utilization of microalgae as an advanced energy feedstock for bioethanol production and species like *Chlorella*, *Dunaliella*, *Chlamydomonas*, *Scenedesmus*, *Spirulina* are selected for ethanol production due to contains high amount of carbohydrate content (John et al., 2011). Additionally, other microalgae have been shown to be a potential source for third-generation biofuels due to their high oil and carbohydrates contents (Abdallah et al., 2016). Different carbohydrate content of various microalgal species are listed in Table 18.1.

Furthermore, most of the microalgae demonstrate fast growth rate under optimal conditions. For example, the replication time of some *Chlamydomanos* species is as short as 6 h (Chen et al., 2009); due to this quality, smaller in size the microalgae handling is often simpler in research than the energy crops. Besides biodiesel and bioethanol there are lots of high-value products and subproducts produced from microalgae, such as biogas, biobutanol, acetone, omega-3 oil, livestock feed, pharmaceuticals, and cosmetics. These subproducts are preferred for economic support of the main process (http://172.16.0.46:8081/FMSNEW/login.go).

Another important challenge of biofuel production from microalgae is to harvest and concentrate the cells because of their diminutive

TABLE 18.1 Various Microalgae and Their Different Concentration of Carbohydrate Content Used for Bioethanol Production

Microalgae	Carbohydrate content (%)	References
Microcystis aeruginosa	57	Imran Khan et al. (2017)
Scenedesmus obliquus CNW-N	45–50	Ho et al. (2017)
Chlorella vulgaris FSP-E	51	Ho et al. (2013)
Chlor.vulgaris	22.4	Kim et al. (2014)
Pavlova lutheri	28	Ryu et al. (2012)
Anabaena cylindrica	25–30	Spolaore et al. (2006)
Chlamydomonas rheinhardii	17	Spolaore et al. (2006)
Dunaliella salina	32	Spolaore et al. (2006)
Porphyridium cruentum	40–57	Spolaore et al. (2006); Harun et al. (2010)
Prymnesium parvum	25–33	Spolaore et al. (2006); Balat (2011)
Scenedesmus dimorphus	21–52	Spolaore et al. (2006); Balat (2011)
Spirogyra sp.	33–64	Harun et al. (2010)
Spirulina fusiformis	37.3–56.1	Rafiqul et al. (2003)
Arthrospira (Spirulina) platensis	16.3	Efremenko et al. (2012)
Nannochloropsis sp.	22	Efremenko et al. (2012)
Cosmarium sp.	19.5	Efremenko et al. (2012)
Oedigonium sp.	33.6	John et al. (2011)
Chlamydomonas reinhardtii UTEX 90	53.0	John et al. (2011)
Chlam.reinhardtii (UTEX2247)	45.0	John et al. (2011)
Chlorella sp. TISTR 8485	27.0	John et al. (2011)
C. emersonii	41	Illman et al. (2000)
Chlorococcum sp. TISTR8583	26.0	John et al. (2011)
Scenedesmus sp. TISTR 8579	20.4	John et al. (2011)
S. obliquus TISTR 8522	23.7	John et al. (2011)
Nostoc sp. TISTR 8873	32.9	John et al. (2011)
N. maculiforme TISTR 8406	30.1	John et al. (2011)
N. muscorum TISTR 8871	33.5	John et al. (2011)
N. paludosum TISTR 8978	32.1	John et al. (2011)
Oscillatoria sp. TISTR 8869	19.3	John et al. (2011)
Phormidium angustissimum TISTR 8979	28.5	John et al. (2011)

size. One of the difficulties in large-scale production of microalgae is the development of efficient techniques to separate cells, maintain their viability and bioactivity prior to use in the field. The harvesting step accounts almost 20%–30% of the total cost in the microalgal biodiesel production. In order to remove large amounts of water for separating microalgae, a suitable harvesting method may achieved by physical, chemical, and biological ways. Many harvesting methods are being reported for removing microalgae from culture medium, such as centrifugation, gravity sedimentation, filtration, flotation, immobilization, ultrasonic aggregation, electrolytic method, chemical flocculation, bioflocculation, and magnetic nanoparticle separation (Sirajunnisa and Surendhiran, 2016).

18.5 BIOETHANOL FROM MACROALGAE

Macroalgae is referred to as the ideal candidate for the bioethanol production than the microalgae because they don't contain high amounts of lipid as microalgae, instead they have high content of natural sugars and other carbohydrates, which can be fermented to produce alcohol-based fuels (www.fao.org/bioenergy/aquaticbiofuels, 2009). Different carbohydrate content of various seaweeds is shown in Table 18.2. Recently, preliminary investigation has been conducted in The Netherlands, USA, and Ireland for macroalgal bioethanol production. Macroalgae have been included in a recent report by the International Energy Agency (IEA) and the US Department of Energy's roadmap for algal biofuels. Pilot plant studies on the development of macroalgae as biofuel feedstock are now underway in Asia and Europe (Fasahati et al., 2015). It is well known that two thirds of the planet is covered by ocean so seaweed can be grown directly cultivated in open sea.

Moreover, cultivation and harvesting of seaweeds are very simple and large amounts of biomass could be attained than microalgae due to larger size. Hence, the current research is shifted to produce bioethanol from macroalgae. Unlike microalgae, cultivation of seaweed is easy and can be harvested by simple methods; while seaweed can be directly cultivated in open sea, microalgae has to be cultivated in a controlled, designed system. Generally, a part of the current seaweed production comes from harvesting natural population or collecting beach-cast seaweed. Besides the disturbance of the ecosystem by these practices, they are clearly unsustainable for application on a very large scale. Therefore, growing macroalgae in a dedicated cultivation system is worth considering. For seaweed cultivation several important factors, such as temperature, nutrient, and light source must be analyzed with other functions of the sea, distance from shore (or distance from a suitable harbor) as they imply energy and time spent on transportation (www.fao.org/bioenergy/aquaticbiofuels, 2009). Generally, seaweeds are cultivated in lagoons or sheltered bays so it attains nutrients directly from the seawater. Several methods are routinely applied for cultivation and harvesting of seaweeds, such as fixed off-bottom (postandline), long lines, rock-based farming (De San, 2012).

Seaweeds serve as an unexploited potential source for bioethanol production due to two facts. First, macroalgae combine high biomass productivity with low capital and operating costs owing to the fact that macroalgae are harvested from naturally occurring stocks or aquacultured sea farms. Such cultivation systems require capital and operating costs that are significantly lower than the microalgal open ponds, nevertheless they provide high biomass productivity (Rajkumar et al., 2014). Second, macroalgae are cultivated worldwide on a large scale for nonbiofuel purposes. The remainder of the biomass, which is rich in carbohydrates, can be hydrolyzed to produce ethanol. In fact, the worldwide biomass production from macroalgae greatly surpasses that of microalgae. For example, in 2010, approximately 9 million and 6.75 million wet metric tons (WMT) were harvested from red and brown macroalgae, respectively. In comparison, a total of only 6.2 thousand dry metric tons (DMT) were produced by major microalgal species, such as *Chlorella* sp., *Dunaliella*

TABLE 18.2 Different Carbohydrate Content of Various Seaweeds

Seaweeds	Carbohydrate content (%)	References
Monostroma sp.	63.9	Rajkumar et al. (2014)
Padina sp.	31.6	Rajkumar et al. (2014)
Alaria sp.	39.8	Rajkumar et al. (2014)
Ulva fasciata	43.0	Trivedi et al. (2013)
Gracilaria verrucosa	43 ± 2	Kumar et al. (2013)
Eucheuma cottonii	70.0	Fakhrudin et al. (2014)
Gracilaria sp.	76.67	Wu et al. (2014)
Gelidium amansii	66.0	Hong et al. (2014)
Laminaria japonica	51.5	Hong et al. (2014)
Codium fragile	58.7	Hong et al. (2014)
Capsosiphon fulvescens	48.1	Hong et al. (2014)
Enteromorpha prolifera	53.3	Hong et al. (2014)
Ulva lactuca	50.4	Hong et al. (2014)
Caulerpa lentillifera	45.5	Hong et al. (2014)
Porphyra sp.	45.5	Hong et al. (2014)
Gigartina tenella	42.2	Hong et al. (2014)
Hypnea charoides	57.3	Hong et al. (2014)
Carpopeltis cornea	60.7	Hong et al. (2014)
Undaria pinnatifida	43.2	Hong et al. (2014)
Hijikia fusiforme	47.5	Hong et al. (2014)
Eisenia bicyclis	72.7	Hong et al. (2014)
Ecklonia stolonifera	65.0	Hong et al. (2014)
Sargassum sp.	41.6	Jelynne et al. (2014)

salina, and *Haematococcus pluvialis*, in the same year. Additionally, 93% of the worldwide cultivated macroalgal biomass is produced from only four genera that belong to the brown algae (*Laminaria* 65.8% and *Undaria* 9.8%) and red algae (*Porphyra* 12.6% and *Gracilaria* 4.8%) (Abdallah et al., 2016).

In general, seaweeds can live variety of environmental conditions. There is a wide range of organisms that grow along the coastal areas. With the advancement of genetic engineering, it is now possible to develop a suitable species of macroalgae for bioethanol production. Genetically engineered macroalgae would need to be cultivated in enclosed bioreactors (Abdallah et al., 2016). These characters bring about high confidence for future development of macroalgae in the renewable energy area, such as bioethanol. Therefore, microalgae provide a promising bioethanol feedstock owing to their high biomass yield and high sugar content.

18.6 ALGAL POLYSACCHARIDES

Generally, an alga does not have lignin in their cell wall, which is a major advantage to convert their polysaccharides into fermentable sugars (Surendhiran and Vijay, 2014). Algal carbohydrates are the major products derived from photosynthesis and the carbon fixation metabolism (i.e., the Calvin cycle) (Chen et al., 2013). The process necessary for the production of bioethanol is different depending on the type of algal biomass and typically includes pretreatment and hydrolysis of the biomass, saccharification, fermentation, and recovery of the product. The pretreatment is an important process for the formation of the sugars used in the fermentation process (Alexandrov et al., 2015). Due to the rigid nature structural carbohydrates, the hydrolysis process of algal biomass is really challenging. Therefore, optimization of the hydrolysis process of structural carbohydrates carries the promise of maximizing ethanol yield. The pretreatment and hydrolysis may vary to the different types of algae contains polysaccharide that are used for bioethanol production.

Both micro- and macroalgae are excellent sources of polysaccharides, most of them being sulfated (de Jesus Raposo et al., 2015). Microalgae have large amounts of carbohydrates in the cell wall (mainly in the form of cellulose and soluble polysaccharides) and plastids (mainly in the form of starch) (Chen et al., 2013). Whereas, macroalgae produce a wide range of polysaccharides may vary species to species. This diverse collection of polysaccharides functions primarily as food reserves or structural material. Here, we describe the most economically important algal sugars, which have received considerable amount of research interest (Abdallah et al., 2016). Different algae groups and their different sugar compounds are shown in Table 18.3.

The majority of algae store their food reserves in the form of starch-type polysaccharides (such as starch, floridean starch, and glycogen) or laminarin-type polysaccharides (such as chrysolaminarin and laminarin) and brown algae accumulate large amounts of mannitol. Starch is a homopolysaccharide of glucose units that are linked by α-(1→4) glycosidic bonds to form a linear amylose; and via α-(1→6) bonds to form amylopectin (Abdallah et al., 2016). Several algal species have been reported to store relatively high concentrations of starch, reaching in some species to about 50% of the dry weight (John et al., 2011).

The laminarans, also known as laminarins, are reserve polysaccharide found in brown algae, such as *Laminaria digitata* (Hudson) Lamouroux, that grows in cold waters in the northern hemisphere. It is a linear homopolysaccharide with linkages β (1→3): β (1→6) in a ratio of 3:1. Another similar polysaccharide is the chrysolaminarin present in phytoplankton and considered one of the most abundant biopolymer in the nature. This water-soluble molecule is composed by glucose units with linkages β (1→3) and β (1→6) in a ratio of 11:1 (Stadnik and de Freitas, 2014).

Most of the microalgae contain cellulose as their cell wall materials (Surendhiran and Vijay, 2014). Cellulose is a homopolysaccharide made of a linear chain of D-glucose units, which are connected together by β-(1→4) glycosidic bonds. Cellulose chains aggregate together by intra- and intermolecular hydrogen bonds to form cellulose microfibrils. Microfibrils are packed together to form fibrils, which in turn aggregate to form cell wall fibers (Pu et al., 2013). Agarose is a nonsulfated, nonwater-soluble linear galactan that is composed of repeating disaccharide units of D-galactose and 3,6-anhydro-l-galactose. The repeating disaccharide unit is called agarobiose or neoagarobiose depending on (1) the position of each sugar in the disaccharide, (2) the bond that links the monomers within the disaccharide, and (3) the bond that links the disaccharides to form agarose (Abdallah et al., 2016). Carrageenan is a sulfated water-soluble linear galactan of carrabiose (or neocarrabiose) subunits. Carrabiose and neocarrabiose are similar in structure and linkage to agarobiose and neoagarobiose, respectively (Stadnik and de Freitas, 2014).

Mannitol is a sugar alcohol of the aldohexose D-mannose. In brown algae, it serves as a storage sugar, and an antioxidant, and protects against osmotic stress. It accumulates in the cell as a monosaccharide and as part of the

TABLE 18.3 Different Groups of Algae and Their Sugar Composition for Bioethanol Production

Algal group	Major carbohydrate	Composition	Glycosidic bonds
Chlorophyta Dinophyta Cryptophyta Glaucocystophyta	Starch (Amylose)	$Glucose_n$	α-(1→4)
Rhodophyta	Floridean starch	$Glucose_n$	α-(1→4) and α-(1→6)
Rhodophyta	Agarose	[Galactose and anhydro-L-galactose]$_n$	β-(1→4) and α-(1→3)
Rhodophyta	Carrageenan	[Galactose and anhydro-D-galactose]$_n$	β-(1→4) and α-(1→3)
Cyanophyta	Glycogen	$Glucose_n$	α-(1→4) and α-(1→6)
Phaeophyceae	Laminarin	$Glucose_n$	β-(1→3) and β-(1→6)
Bacillariophyceae Xanthophyceae Chrysophyceae Haptophyta Chlorarachniophyta	Chrysolaminarin	$Glucose_n$	β-(1→3) and β-(1→6)
Euglenophyta	Paramylon	$Glucose_n$	β-(1→3)
All algal groups except diatoms	Cellulose	$Glucose_n$	β-(1→4)
Phaeophyceae	Alginate	α-L-guluronaten and β-D-mannuronaten	β-(1→4) and α-(1→4)
Ulva and *Enteromorpha* sp. (green algae)	Ulvan	Sulfated rhamnose, glucuronic acid, iduronic acid, xylose, and sulfated xylose	β-(1→4) and α-(1→4)

Source: Adapted from Abdallah, Q.A., Tracy Nixon, B., Fortwendel, J.R., 2016. The enzymatic conversion of major algal and cyanobacterial carbohydrates to bioethanol. Front. Energy Res. 4, 36.

laminarin polysaccharide—forming the laminarin M-chain, where mannitol is at the reducing end. Mannitol is one of the major storage carbohydrates in brown algae, and some brown algae accumulate high amounts of mannitol reaching approximately 25% of *Laminaria hyperborea*, 30% of *Laminaria japonica*, and 33% of several *Sargassum* and *Turbinaria* dry weight (Abdallah et al., 2016).

Ulvan is a water-soluble cell wall polysaccharide, which is found in green seaweeds, such as *Ulva* and *Enteromorpha* sp. It is comprised of sulfated rhamnose, glucuronic acid, iduronic acid, xylose, and sulfated xylose. The ratio and linkage of ulvan constituent monosaccharides vary among species. Nevertheless, ulvan general structure is comprised of repeating disaccharide units that are linked by α-(1→) glycosidic bond. Ulvan disaccharides are formed from [glucuronic acid–sulfated rhamnose], [iduronic acid–sulfated rhamnose], [xylose–sulfated rhamnose], and [sulfated xylose–sulfated rhamnose] monomers, which are linked by either α- or β-(1→4) glycosidic bonds (Abdallah et al., 2016; Stadnik and de Freitas, 2014).

Fucoidan is a cell wall polysaccharide that is found in the members of family Laminariaceae of brown algae. The dry weight percentage of fucoidan ranges normally from 5% to 20%. Fucoidan structure is heterogeneous and varies among algal species. Nevertheless, it displays a general backbone consisting of predominantly α-(1→2)-linked sulfated l-fucose units and, in smaller amounts, α-(1→3)- and α-(1→4)-1inked sulfated fucose units (Abdallah et al., 2016; Stadnik and de Freitas, 2014).

18.7 ENHANCEMENT OF CARBOHYDRATE CONTENT FOR INCREASED BIOETHANOL PRODUCTION

To enhance the economic feasibility of using algal carbohydrates for biofuels production, the productivity is the key parameter that needs to be improved. Many parameters have been studied and reported on improvement of carbohydrate content in algae can be achieved by various techniques like optimization of medium components and environmental parameters and genetic engineering tools. The effects of these approaches are discussed in more detail in the following sections.

18.7.1 Optimization of Medium Components and Environmental Parameters

Generally, chemical composition of microalgae can change according to the cultivation type and cultivation conditions (Özçimen and nan, 2015). Various factors influence the algal growth and carbohydrate, lipid, pigment, and proteins content, such as light intensity, temperature, and nutrients and pH. Light and temperature mainly influence the growth of algae and other factors alter the biochemical pathways and increase the compositions of algae.

Light is a very essential factor, which determines the growth and photosynthetic rate and growth of algae and organisms use light energy to convert CO_2 into organic compounds like sugars (Sirajunnisa and Surendhiran, 2016). The light intensity and photoperiod are also highly involved in growth and metabolism. Microalga grown under varying light conditions exhibits remarkable changes in carbohydrate accumulation algae (Chen et al., 2013).

Previous studies reported that an increase in the light intensity resulted in an increase in the accumulation of carbohydrates. For example, a significant increase in starch content from 8.5% (dry weight basis) to 40% was observed when the mean light intensity was increased from 215 to 330 $\mu mol/m^2$ s (Cheng and He, 2014; Fernandes et al., 2010).

Temperature is the most important limiting factor, after light, for culturing algae in both closed and open systems. Generally, in optimal temperature range, a rise in temperature leads to improved microalgal biomass production and above optimum the growth declines despite which a high temperature stress increases the lipid content in microalgae (Sirajunnisa and Surendhiran, 2016). Similarly, many reports described that temperature stress could also influence the increasing carbohydrate content. For example, the carbohydrate content of *Spirulina* sp. was increased upto 50% when the temperature rose from 25 to 40°C (Ogbonda et al., 2007).

Nutrients stress conditions produces considerable variations in change of metabolic pathway and biochemical compositions in algae (Sirajunnisa and Surendhiran, 2016). As a result, the improved production of desired biofuels can be achieved by manipulating nutrient conditions. Nitrogen and phosphorous are the most important nutrients required for algae growth because the formation of proteins, amino acids, chloroplast, enzymes, coenzymes, and so on is associated with nitrogen assimilation (Chen et al., 2013). When under nitrogen-depletion conditions many microalgal strains could transform proteins or peptides to lipids or carbohydrates as energy reserve components (Sirajunnisa and Surendhiran, 2016). For example, carbohydrate content of *Tetraselmis suecica* was increased from 10% to 57% under nitrogen depleted conditions (D'Souza and Kelly, 2000). Likewise, 55% carbohydrate content was achieved for *Chlorella vulgaris* by cultivating on a low-nitrogen-containing medium (Illman et al., 2000).

Similar to the effects of nitrogen deficiency, phosphorus starvation reduces protein content thereby increasing the relative carbohydrate content in algal cells. Phosphorous is a very essential compound, which involved in many cellular metabolic processes (Cheng and He, 2014; Sirajunnisa and Surendhiran, 2016). It has been reported that the carbohydrate content in *Arthrospira (Spirulina) platensis* increased from 11% to 67% under phosphorus depleted growth (Markou et al., 2012). There are many other examples in the literature showing that limitation of nitrogen and phosphorous availability seems to be the most effective way of triggering the accumulation of carbohydrates in algae.

pH is an important environmental condition for the metabolism of microorganisms, affecting not only cell growth rate but also biochemical composition (Khalil et al., 2010). For example, the extracellular carbohydrate production of a marine diatom *Skeletonema costatum* was dramatically increased from 2.1% to 17.7% when it was increased to pH 9.4 from neutral pH (Taraldsvik and Myklestad, 2000). However, the changes in carbohydrate accumulation in algae are species specific (Chen et al., 2013). For example, *Dunaliella bardawil* produce their maximum carbohydrate at pH 7.5 and there is no higher accumulation during the higher or lower pH changes (Khalil et al., 2010).

18.7.2 Genetic Engineering in Ethanol Production

Unfortunately, medium components and environmental parameters do not meet the overall success rate of algal biorefinery due to following reasons:

- Accumulation of algal carbohydrates usually occurs when the algal cells face environmental stress (typically nutrients limitation), which often results in poor cell growth (or lower biomass productivity) (Chen et al., 2013). This would affect the overall process of algal bioethanol production, because the quantity of the algal product depends on the quantity of the biomass (Alexandrov et al., 2015).
- However, some studies indicated that there was a competition between lipid and carbohydrate synthesis under stress environments (e.g., nitrogen-deficient) because the metabolic pathways associated with synthesis and degradation of energy-rich compounds (e.g., lipids and carbohydrates) are closely linked. Precisely, starch biosynthesis of microalgae can directly proceed away from lipid synthesis. On the contrary, degradation of starch provides the metabolites for producing acetyl-CoA, which is the precursor of fatty acid synthesis. Thus, in some cases, decreasing starch degradation is necessary to block the synthetic pathway of lipids (Chen et al., 2013).
- The effect of temperature on the accumulation of microalgal carbohydrates is still inconclusive. Some researchers observed no significant differences in biochemical composition under the stress of temperature variations in some microalgal species, whereas some researchers pointed out that temperature is potentially able to change the biochemical composition of microalgae (Chen et al., 2013).
- Moreover, some reports exposed that there was no obvious positive correlation between light intensity and carbohydrate accumulation (Carvalho et al., 2009). Hence, these contradictory reports revealed that an optimal light intensity was required for favorable growth and productivity of the algae varying species to species (George et al., 2014).
- Besides, many studies have reported the possibility of successful cultivation of microalgae for commercial use including biofuel and food supplements in open raceway pond system due to low cost involvement (Sirajunnisa and Surendhiran, 2016). However, the manual changes of temperature and irradiance is impossible while cultivating microalgae in case of open system and cultivating seaweeds in marine regions. Therefore, how to enhance the carbohydrate content without compromising the cell growth rate is crucial to the success of carbohydrate production from algae.

18.7.2.1 Role of Genetic Modifications for Bioethanol Production

Genetic engineering of algae is considered necessary to overcome many of the drawbacks of algal biofuels (Alexandrov et al., 2015). Researchers have been studying algal fuels extensively since the 1970s, but no commercially viable strain has yet been isolated. In the past, some programs had to be shut down due to the higher cost of algal biofuels as compared to fossil fuels. The use of recombinant DNA techniques offer the greatest range of options to improve upon the performance of wild strains (Kumar, 2015). Significant advances in algal genomics and genetic engineering have been achieved during last decades and this is the another new-fangled way of improving algal biomass and specific product, such as lipid or carbohydrate (Ryan Georgianna and Mayfield, 2012). This was the assumption of John Craig Venter, a scientist known for his contribution to sequencing the human genome and later for creating the first living cell running on man-made DNA. He has now turned his attention to genetic manipulation of algae. His words at the New America Foundation in Washington, DC in 2011 are: "Nothing new has to be invented. We just have to combine genes in a way that nature has not done before. We're speeding up evolution by billions of years." However, modifying algae genetically is a daunting and a difficult task (Alexandrov et al., 2015).With the advent of next-generation sequencing technologies, our understanding of algal genomes, especially the genes involved in metabolic responses, has improved rapidly (Ryan Georgianna and Mayfield, 2012). Moreover, genetic engineering allows direct control over the algae's cellular machinery through mutagenesis or the introduction of transgenes (Singh et al., 2011). To meet the challenges on economic biofuels production from microalgae, it is important to integrate updated information of genomic sequences, transcriptomes, proteomes, and metabolomes data at systems level (Chen et al., 2013). With such technology, *Volvox carteri* was the first ever green algae to be carrying a hexose/H^+ symporter gene HupI from *Chlorella* and the transformation was confirmed by ^{14}C labeled glucose, glucosamine assimilation.

In addition, several nuclear genome sequencing projects have now been completed, including those for *Chlamydomonas reinhardtii*, *Phaeodactylum tricornutum*, *Thalassiosira pseudonana*, *Cyanidioschyzon merolae*, *Ostreococcus lucimarinus*, *Ostreococcus tauri*, and *Micromonas pusilla*. Currently, ongoing microalgal genome sequencing projects include those for *Fragilariopsis cylindrus*, *Pseudo-nitzschia*, *Thalassiosira rotula*, *Botryococcus braunii*, *Chlor. vulgaris*, *Dunaliella salina*, *Micro.pusilla*, *Galdieria sulphuraria*, *Porphyra purpurea*, *Vol. carteri*, and *Aureococcus anophageferrens* (Radakovits et al., 2010). Such a strategy will open a door for efficient carbohydrate metabolic regulation and genetic engineering of algae for biofuels production.

18.7.2.2 Carbohydrate Metabolism and Storage in Algae

Carbon metabolism in autotrophic algae starts with photosynthesis and carbon acquisition, followed by incorporation of carbon into different organic metabolites. Algae can make carbohydrates from CO_2 and water at the expense of energy furnished by the ATP and NADPH generated in photosynthetic electron transfer. Starch biosynthesis is unique to the phyla *Chlorophyta*, *Dinophyta*, *Glaucophyta* and *Rhodophyta* store glucans in linear α-1,4 and branched α-1.6 glycosidic linkages. In *Heterokontophyta*, *Phaeophyceae*, and *Bacillariophyceae*, water-soluble granules of laminarin and chrysolaminarin are synthesized, which are made up of β-1,3 linkages with branching at the C-2 and C-6 positions of glucose (Radakovits et al., 2010). Most cyanobacteria contain carboxysomes, and store carbohydrate as water-soluble glycogen (Hildebrand et al., 2013). However, the recent identification of new cyanobacterial species that make starch-like oligosaccharides with an intermediate type of chain length distribution between amylopectin and glycogen (Busi et al., 2014).In general, in green algae, starch is synthesized and stored within chloroplast, while it is stored in the cytoplasm *Glaucophyta* (Radakovits et al., 2010). Data suggest that cytoplasmic starch formation generated from ADP-glucose exported from the chloroplast resulted from the merging of bacterial and eukaryote pathways of storage polysaccharide metabolism. In the chlorophytes and related algae, gene duplications and enzyme retargeting resulted in starch synthesis being relocated to the chloroplast. Pyrenoid-associated starch in chlorophytes could play a role in carbon concentration. In rhodophytes, pyrenoids are not present in all species, and carbohydrate is stored in the cytoplasm as either glycogen or Floridean starch, which is a less crystalline starch form lacking amylase (Hildebrand et al., 2013). This indicates that there are substantial differences in the accessibility of different storage forms of carbohydrates that are less accessible energetically and produce bioethanol from algae.

18.7.2.3 Manipulation of Genetic Material for Higher Algal Ethanol

A potential alternate approach for increasing algal carbohydrate content is to alter the metabolic pathway by employing heterologous ones into the algal species. For example, introduce starch synthesizing enzyme coding gene into the cytosol. The cytosol would give more physical space for the starch granules to accumulate (Radakovits et al., 2010).

18.7.2.3.1 SELECTION OF ALGAL STRAINS

Screening and selection of algae is the most important parameter in genetic engineering for bioethanol production. There are more than 100,000 species of algae existing on earth (Demirbas and Faith Demirbas, 2010; Ullah et al., 2015), perhaps not all algae can be used for genetic modification because it is based on the essential factors like growth rate, photosynthetic yield, biomass productivity, carbohydrate, and lipid content and adaptation of different environmental conditions. In addition to these other advantages include robustness, survival of shear stress common in PBRs, ability to dominate wild strains in open pond production systems, high CO_2 sinking capacity, having limited nutrient requirements, tolerant to a wide range of temperatures resulting from the diurnal cycle and seasonal variations, providing valuable coproducts and displaying self-flocculation character (Sirajunnisa and Surendhiran, 2016). Genetically modified algae (GM) involved in fuel production should possess the tendency to withstand changes in temperature, light, salinity and pathogen load on land and also standardized strains have to tolerate the geographical regions and conditions. The GM algae would be successful only if it meets the aforementioned criteria along with excellent expression of heterologous genes from the nuclear genome of other algae species.

Among various algal species, *Chlam. reinhardtii* (Chlorophyta) has shown itself as a favorable photosynthetic organism for genetic studies because that has been studied for more than 30 years as a model for demonstrating great utility for the development of genetically modified organism (GMO). Moreover, this organism is often easier to study than higher plants due to its haploid genome, the short generation time and fast response to a changing environment (Radakovits et al., 2010). As such, it has been widely employed to pioneer genetic engineering for synthesis of desired goods.

18.7.2.3.2 TECHNIQUES TO DELIVER THE GENE OF INTEREST

Generally genetic manipulation algae could be done by chloroplast transformation system and DNA transformation. The DNA transformation system comprises of various techniques like generation of protoplasts, glass beads method (Doron et al., 2016), biolistic delivery (Chen et al., 2008), electroporation (Chen et al., 2008), and agrobacterium mediated gene transfer (Doron et al., 2016; Kumar et al., 2004; Radakovits et al., 2010). More than 30 different strains of microalgae have been transformed successfully to date (Table 18.4). In many cases, transformation resulted in stable expression of transgenes, from either the nucleus or the plastid, but in some cases only transient expression was observed. Methods developed primarily with *Chlam. reinhardtii* demonstrate that the stability of expression can be improved through proper codon usage, the use of strong endogenous promoters, and inclusion of species specific 5′, 3′, and intron sequences. The efficiency of transformation seems to be strongly species dependent, and the method of transformation has to be carefully selected and optimized for each alga (Radakovits et al., 2010).

Chloroplast transformation technique—It is a stable transformation system first successfully developed for *Chlam. reinhardtii* using biolistic delivery. However, still less transformation system that targets the chloroplast as compared to those targeting algal nucleus. One great advantage of chloroplast transformation system is that transgenes can be easily directed to integrate through homologous recombination, whereas nuclear transformation usually results in random integration events (Doron et al., 2016). Efficient isolation of genetic transformants is greatly facilitated by the use of selection markers, including antibiotic resistance and metabolic enzymes markers. Metabolic selection of transformed cells provides a great advantage over the use of genes encoding for resistance to antibiotics, as metabolic selection is considered to be more environmental friendly. Metabolic selection is common in yeast and animal cells, especially because many markers offers positive and negative selections. For example, acetylornithine aminotransferase (ARG9) is a key enzyme in the metabolic pathway of arginine. It is encoded in the nucleus of *Chlam. reinhardtii*, with protein translocating into the chloroplast. A mutation in ARG9 resulted in an auxotrophic phenotype, such that the algal cells could only grow if arginine was added to the medium (Nakamura et al., 2000). For selecting antibiotic resistance markers, a bacterial gene, *aphA6*, encoding aminoglycoside (3′) transferase that confers resistance to aminoglycoside antibiotics was used with *Chlam. reinhardtii*. The combination of *aadA* and *aphA6* extended the possibility for expressing multiple foreign genes in the chloroplast. For instance, *Dunaliella tertiolecta* was successfully modified via chloroplast transformation by inserting the gene that encodes both the gene of interest and erythromycin esterase gene, *ereB*, through particle bombardment (Doron et al., 2016). However, the chloroplast transformation technique is not suitable for all types of algae especially for diatoms.

Nuclear transformation techniques: It is widely used method having several advantages, including targeting of foreign proteins for expression in organelles, such as the chloroplast, and protein glycosylation and additional posttranslational modifications, as well as secretion (Jarvis and Brown, 1991). A collection of protocols for nuclear transformation, combined with the availability of several constitutive or inducible promoters and the availability of multiple selectable markers are facilitating the nuclear transformation. Many different antibiotic resistance genes have been successfully used for algal genetic modifications, including bleomycin, spectinomycin, streptomycin, nourseothricin (Radakovits et al., 2010), paromomycin (Sizova et al., 2001), G418 (Zaslavskaia et al., 2000), hygromycin and chloramphenicol (Radakovits et al., 2010).

However, all antibiotic resistance genes cannot be employed for the all species. This is due to that many algae are resistant to a wide range of antibiotics, the actual number of antibiotics that work with specific strain may be much more limited. Moreover, antibiotics like nourseothricin and G418 are much less effective in salt containing culture medium and not suitable for marine algae (Zaslavskaia et al., 2000).

Generation of protoplasts: In this method the foreign DNA is inserted through the protoplasm by cell loosening or disturbing cell wall. However, in many cases, generating protoplasts is a major bottleneck due to the diverse composition of the cell wall in different algae *species*, which, in many cases, remain poorly characterized (Doron et al., 2016). The *Chlamydomonas* cell wall is built from glycoproteins and contains little cellulose or chitin. As such, polysaccharide-degrading enzymes are ineffective, with protoplasts instead being generated by incubation with autolysins, namely hydroxylproline-specific proteases that are active during gametogenesis and mating. Unlike *Chlamydomonas*, the cell wall in *Chlorella* consists of sugar polymers that can be degraded by sugar digesting-enzymes (Surendhiran and Vijay, 2014; Takeda, 1991). Therefore, protoplasts of different *Chlorella* species can be generated by incubation with different sugar-degrading enzyme mixes (Doron et al., 2016). Protoplasts from red and brown algae have been prepared using numerous cell wall-digesting enzymes extracted from marine mollusks or echinoderms (Reddy et al., 2010).

Agitation with glass beads method—It was one of the first methods used to transform *Chlam. reinhardtii* and second for *Chlorella ellipsoidea* and it has been widely used in many laboratories because of its simplicity and reproducibility. It was then found that the addition of the membrane fusion agent polyethylene glycol (PEG) during agitation increased by 5–12-fold the number of transformants produced (Coll, 2006; Rivera et al., 2014). Agitation methods usually employ 30–100 million microalgae in a 0.4 ml volume, 1–20 μg of exogenous DNA, 300 mg of glass beads (500 μm of diameter), a 5%–13% solution of PEG (molecular weight 4,000–8,000 kDa), and possibly carrier DNA (25–50 μg) (Coll, 2006). The main disadvantage of this method is high quantities of DNA are damaged in the process and the viability of the cells is drastically reduced (Rivera et al., 2014). Moreover, before begin this process the cell wall might be removed to enhance the better DNA transfer. Transformation by agitation with glass beads is generally performed during logarithmic growth, when the number of cells undergoing mitosis, and therefore the absence of nuclear membranes, is maximal. This aids in the passage of DNA through the cell wall and membranes and, thus entry into the nucleus. After transfection, the selection of the few transformants produced is favored by axenic cloning for a few weeks in a semisolid medium (Rivera et al., 2014).

Biolistic Delivery—This is the most frequently used method of gene delivery is biolistic transformation, also referred to as "micro-projectile bombardment" (Doron et al., 2016) or "gene gun technique" (Rivera et al., 2014). Biolistics consists on the acceleration of high density carrier particles (usually made of gold, tungsten or platinum) covered with genes that pass through the cells, leaving the DNA inside by an adsorption mechanism. Cells, protoplasts, organized tissues like meristems, embryos or callus can be used as a target, and the introduction of multiple genes or chimeric DNA (DNA from two different species) is simple (Coll, 2006; Rivera et al., 2014). Successful transformation by this approach has been reported for *Chlam. reinhardtii*, *Dun. salina*, *Haem. pluvialis*, and *diatoms* (Doron et al., 2016). Disadvantage of this technique is an expensive technique, transformation parameters must be optimized to each bio-logical target employed and there is a risk of multiple copies of the introduced genes, which can lead to various undesirable side effects, such as gene silencing or altered gene expression (Sanford, 2000).

Electroporation—It is a most popular genetic transformation method due to less time consumption, low cost, and simplicity. Electroporation is based on the application of a strong electrical field to enhance the formation of pores on the cell membrane due to a polarity alteration, caused by the electrical field (alternated or pulsed) that induces a dipolar moment inside the cells, and a potential difference through the plasmatic membrane

(Rivera et al., 2014). This technique was used to transform algal protoplasts, cell wall-deficient mutants, and other thin walled algal cells. It was used to transform *Chlam. reinhardtii, Dun. salina, Chlor.vulgaris, Scenedesmus obliquus*, and *Nannochloropsis* sp. (Doron et al., 2016). The following factors, such as pulse length, type and duration of the electrical field, transmembrane potential created, extent of membrane permeation, duration of the permeated state, mode and duration of molecular flow, global and local (surface) concentrations of DNA, form of DNA, tolerance of cells to membrane permeation and the heterogeneity of the cell population have a strong effect on the transformation efficiency during electroporation (Rivera et al., 2014; Weaver and Chizmadzhev, 1996).

Agrobacterium-Based Transformation of Microalgae—Nowadays this technique is more popular for nuclear transformation of land plants using *Agrobacterium tumefaciens* and now employed for transformation of microalga, using the pCAMBIA transformation vector. Best example for *Agrobacterium tumefaciens*(T-DNA) based genetic modification of algae is *Chlam. reinhardtii*(~ 350 transformants per million transfected microalgae) (Coll, 2006). Other algal species were transformed by Agrobacterium, including *Haem. pluvialis, Schizochytrium, Isochrysis galbana*, and *Isochrysis* sp. (Doron et al., 2016).

18.7.2.3.3 STABILITY OF THE GENETICALLY MODIFIED ALGAL STRAINS

The stability of transformants remains a major challenge in genetic engineering. Confirmation of the presence of the exogenous DNA in the microalgae genome has been demonstrated in most reports by Southern (DNA), Northern (RNA) or Western (protein) blotting (Coll, 2006). GM algae-based biofuel projects are underway in many countries and are being supported by their respective governments and/or private agencies. A key aspect of making algal biofuel commercially viable is the requisite large-scale cultivation in open ponds, near oceans and in marine water, as opposed to sophisticated PBRs. Therefore, the introduction of GM algae into the environment and the associated biosafety issues should be addressed carefully and meticulously, especially the potential of unintended gene transfer of transgenic traits to compatible wild-type cells or others (Kumar, 2015). List of the GM algae and their stability is shown in Table 18.4 (Enzing and Nooijen, 2012). At the same time some reports indicated that the instability of GM and possibility of loss of exogenous gene expression in algae includes glass bead transformants of *Ch. ellipsoidea* (Jarvis and Brown, 1991) and *Chlor.vulgaris* (Hawkins and Nakamura, 1999), in electroporation transformants of *Chlor. vulgaris* (Chow and Tung, 1999), and in bombardment transformants of *Thalassiosira weissflogii* (Coll, 2006).

Another big challenge is risk of releasing GM algae into the environment and most of the environmental activists oppose this idea. During large-scale cultivation of GM algae in open ponds, individual cells have the potential to escape into the environment by various means, such as being carried by the wind, water, birds, other animals, natural disasters, or accidents. The transgenes may integrate into wild-type species via sexual reproduction and by horizontal gene transfer (HGT) (Kumar, 2015), which resulted in complete replacing of wild-type organisms. In addition, any antibiotic resistance marker genes used in the genetic transformation may themselves be a source of risk as there is the potential that they can be transferred to wild-type relatives and thus introduce novel antibiotic resistance. If necessary, they should be removed from the GM organism prior to commercial cultivation (Shao et al., 2014). Moreover, hazard assessment should be checked before the release of GM algae into industrial applications. In general, green algae are harmless to the environment except a few like *Prototheca wickerhami*, which causes protothecosis disease in dogs, cats, cattle, and humans. However, it is not an important candidate for biofuel production and most of the strains that have been genetically modified up until now are only model species (Kumar, 2015). Therefore, stability and potential risks involved in the mass cultivation of GM algae

TABLE 18.4 Some of the Genetically Modified Algal Species, Mode of Transformation and Their Stability

Organism	Stability	Mode of transformation
Chlamydomonas reinhardtii	Stable	Nucleus
Chlam.reinhardtii	Stable	Chloroplast
Volvox carteri	Stable	Nucleus
Dun.salina	Stable	Nucleus
Dunaliella viridis	Stable	Nucleus
Haematococcus pluvialis	Stable	Nucleus
Chlorella sorokiniana	Stable	Nucleus
Chlorella kessleri	Stable	Nucleus
Chlorella ellipsoidea	Stable	Nucleus
Chlor.vulgaris	Transient	Nucleus
Cyanidioschyzon merolae	Stable	Nucleus
Porphyra yezoensis	Stable/transient	Nucleus
Porphyra miniata	Transient	Nucleus
Kappaphycus alvarezii	Transient	Nucleus
Gracilaria changii	Transient	Nucleus
Gracilaria	Stable	Nucleus
Lam.japonica	Stable	Nucleus
Undaria pinnatifida	Stable	Nucleus
Phaeodactylum tricornutum	Stable	Nucleus
Navicula saprophila	Stable	Nucleus
Cylindrotheca fusiformis	Stable	Nucleus
Cyclotella cryptic	Stable	Nucleus
Thalassiosira weissflogii	Transient	Nucleus
Nannochloropsis sp.	Stable	Nucleus
Ostreococcus tauri	Stable	Nucleus
	Stable	Nucleus
Spirulina platensis (Arthrospira platensis)	Stable	Nucleus
Anabaena sp.	Stable	Nucleus
Synechocystis sp.	Stable	Nucleus
Synechococcus	Stable	Nucleus
Nosctoc muscorum	Stable	Nucleus
Euglena gracilis	Stable	Chloroplast
Porphyridium sp.	Stable	Chloroplast

Source: Adapted from Enzing, C., Nooijen, A., 2012 May. Algae and genetic modification, research, production and risks. www.technopolis-group.com.

need to be addressed in advance in order to commercialize algal-based bioethanol production.

18.8 PROSPECTS OF BIOETHANOL

Although bioethanol used as renewable fuel is one of the positive trends for sustainable development, there are some difficulties in production from algae due to which it has not yet been commercialized and still in its laboratory level. Both microalgae and macroalgae have their own advantages and disadvantages, which determine the commercialization of algal ethanol in near future. Because the price of a final product is directly related to the cost of feedstock, the price of algal biomass is an important factor in the economics of bioethanol production (Chen et al., 2015). Algae bioethanol production cost is connected with the overall process include cultivation, harvesting, and processing. Microalgae contains high amounts of carbohydrates especially starch in chloroplasts and cellulose on cell walls are easily converted into reducing sugars by cellulase enzyme, nevertheless due to its diminutive size facing many difficulties while harvesting of cells and yielded very low amount of biomass. Reports revealed that costs of algal bioethanol is higher than the corn-based production due to overall production costs of microalgae (Sirajunnisa and Surendhiran, 2016).

Unlike microalgae, macroalgae can be easily cultivated and harvested due to its bioavailability and cellular size (Sirajunnisa and Surendhiran, 2016). However, the production of bioethanol from seaweeds is a little difficult than from microalgae, because the main components of polysaccharides are alginate, laminaran, and mannitol. However, many organisms are able to convert mannitol and laminarin to glucose, which are a good substrate for bioethanol fermentation, but at present ethanol fermenting microbes cannot degrade alginate as a carbon source. Thus, multiple pretreatment, such as acid or enzyme lyses, were needed to decompose their structural polysaccharides to release monomer sugars from brown algae (Sirajunnisa and Surendhiran, 2016), which leads to additional cost involvement in overall production. Conversion of biomass from marine algae into ethanol could be economically feasible because some algae hydrolysates can contain more total carbohydrate and hexose sugars than some terrestrial, lignocellulosic biomass feedstock (Nguyen et al., 2012). Another important alternate to overcome these impediments is the development of a robust and high-yielding microbe is required. Recently, a variety of engineered microorganisms, by metabolic engineering integrated with genome engineering and synthetic biology, appear to be quite promising with improved yields of biofuel production (Sirajunnisa and Surendhiran, 2016).

18.9 CONCLUSION

Currently, many feedstocks are being experimented for bioethanol production. Though, algae are proved as one of the most promising sources of bioethanol due to its advantages over other sources. They are distributed worldwide and grow in different climates, under normal and extreme conditions bioethanol production using algae has a great possibility and significance in global sustainable development. However, some algae produces high amounts of biofuel feedstock in natural environments whereas the content would be declined while cultivating under controlled conditions especially PBRs. The artificial cultivation system consumes additional cost to the overall bioethanol production. Therefore, to fulfill all the needs to implement large scale algal bioethanol process, the natural strains have to be modified to produce high photosynthetic yield, high biomass, and high carbohydrate content at natural and artificial conditions. It is now possible since rapid advancements in genome sequencing, now routinely performed for all organisms, are present and will prove instrumental in advancing these goals in the near future. Nevertheless, the technology is in infancy stage still now and a lot needs to be done before commercial production of bioethanol by GM-algae. Finally, extra precautions are needed to prevent the escaping of GM algae while grown in open system and address the formal risk assessments of GM algae as per regulations for development of sustainable algal biofuel production.

References

Abdallah, Q.A., Tracy Nixon, B., Fortwendel, J.R., 2016. The enzymatic conversion of major algal and cyanobacterial carbohydrates to bioethanol. Front. Energy Res. 4 (36).

Alexandrov, S.I., Iliev, G., Gacheva, A., Kr umov, P., Pilarski, G., Petkov, 2015. Could algae be a real source of fuel? Genet. Plant Physiol. 5 (2), 105–122.

Balat, M., 2011. Production of bioethanol from lignocellulosic materials via the biochemical pathway: a review. Energy Convers. Manage. 52, 858–875.

Balat, M., Balat, H., 2010. Progress in biodiesel processing. Appl. Energy 87, 1815–1835.

Busi, M.V., Barchiesi, J., Martin, M., Gomez-Casati, D.F., 2014. Starch metabolism in green algae. Starch/Starke 66, 28–40.

Carvalho, A.P., Monteiro, C.M., Malcata, F.X., 2009. Simultaneous effect of irradiance and temperature on biochemical composition of the microalga *Pavlova lutheri*. J. Appl. Phycol. 21, 543–552.

Chaudhary, L., Pradhan, P., Soni, N., Singh, P., Tiwari, A., 2014. Algae as a feedstock for bioethanol production: new entrance in biofuel world. Int. J. Chem. Tech. Res. 6 (2), 1381–1389.

Chen, C.Y., Zhao, X.Q., Yen, H.W., Ho, S.H., Cheng, C.L., Lee, D.J., Bai, F.W., Chang, J.S., 2013. Microalgae-based carbohydrates for biofuel production. Biochem. Eng. J. 78, 1–10.

Chen, H.L., Li, S.S., Huang, R., Tsai, H.J., 2008. Conditional production of a functional fish growth hormone in the transgenic line of *Nannochloropsis oculata* (Eustigmatophyceae). J. Phycol. 44, 768–776.

Chen, J., Bai, J., Li, H., Chang, C., Fang, S., 2015. Prospects for bioethanol production from macroalgae. Trends Renew. Energy 1 (3), 185–197.

Chen, P., Min, M., Chen, Y., Wang, L., Li, Y., Chen, Q., Wang, C., Wan, Y., Wang, X., Cheng, Y., Deng, S., Hennessy, K., Lin, X., Liu, Y., Wang, Y., Martinez, B., Ruan, R., 2009. Review of the biological and engineering aspects of algae to fuels approach. Int. J. Agric. Biol. Eng. 2 (4), 1–30.

Cheng, D., He, Q., 2014. Assessment of environmental stresses for enhanced microalgal biofuel production: an overview. Front. Energy Res. 4 (2), 1–8.

Chisti, Y., 2016. Large-scale production of algal biomass: Raceway ponds. In: Bux, F., Chisti, Y. (Eds.), Algae Biotechnology: Products and Processes. Springer International Publishing, New York, pp. 21–40.

Chow, K.C., Tung, W.L., 1999. Electrotransformation of *Chlorella vulgaris*. Plant Cell Rep. 18, 778–780.

Coll, J.M., 2006. Review. Methodologies for transferring DNA into eukaryotic microalgae. Span. J. Agric. Res. 4 (4), 316–330.

D'Souza, F.M.L., Kelly, G.J., 2000. Effects of a diet of a nitrogen-limited alga (*Tetraselmis suecica*) on growth, survival and biochemical composition of tiger prawn (*Penaeus semisulcatus*) larvae. Aquacult 181, 311–329.

de Jesus Raposo, M.F., Bernardo de Morais, A.M., Costa de Morais, R.M.S., 2015. Marine polysaccharides from algae with potential biomedical applications. Mar. Drugs 13 (5), 2967–3028.

De San, M., 2012.The farming of seaweeds. Smart Fish program, March 2012.

Demirbas, A., Faith Demirbas, M., 2010. Algae Energy Algae as a New Source of Biodiesel. Springer-Verlag, London, e-ISBN 978-1-84996-050-2, 2010; 75.

Doron, L., Segal, N., Shapira, M., 2016. Transgene expression in microalgae: from tools to applications. Front. Plant. Sci. 7, 505. doi: 10.3389/fpls.2016.00505.

Dragone, G., Fernandes, B., Vicente, A.A., Teixeira, J.A., 2010. Third generation biofuels from microalgae. In: Mendez-Vilas, A. (Ed.), Current Research Technology and Education Topics in Applied Microbiology and Microbial Biotechnology. Formatex Research Center, Badajoz, Spain, pp. 1355–1366.

Efremenko, E.N., Nikolskaya, A.B., Lyagin, I.V., Senko, O.V., Makhlis, T.A., Stepanov, N.A., Maslova, O.V., Mamedova, F., Varfolomeev, S.D., 2012. Production of biofuels from pretreated microalgae biomass by anaerobic fermentation with immobilized *Clostridium acetobutylicum* cells. Bioresour. Technol. 114, 342–348.

Enzing, C., Nooijen, A., 2012. Algae and genetic modification: research, production and risks. CGM 2012-05, May; www.technopolis-group.com, Food & Biobased Research, Wageningen UR.

Fakhrudin, J., Setyaningsih, D., Rahayuningsih, M., 2014. Bioethanol Production from Seaweed *Eucheuma cottonii* by neutralization and detoxification of acidic catalyzed hydrolysate. Int. J. Environ. Sci. Develop. 5 (5), 455–458.

Fasahati, P., Woo, H.C., Liu, J.J., 2015. Industrial-scale bioethanol production from brown algae: effects of pretreatment processes on plant economics. Appl. Energy 139, 175–187.

Fernandes, B.D., Dragone, G.M., Teixeira, J.A., Vicente, A.A., 2010. Light regime characterization in an airlift photobioreactor for production of microalgae with high starch content. Appl. Biochem. Biotechnol. 161, 218–226.

George, B., Pancha, I., Desai, C., Chokshi, K., Paliwal, C., Ghosh, T., Mishra, S., 2014. Effects of different media composition, light intensity and photoperiod on morphology and physiology of freshwater microalgae *Ankistrodesmus falcatus*: a potential strain for bio-fuel production. Bioresour. Technol. 171, 367–374.

Georgianna, D.R., Mayfield, S.P., 2012. Exploiting diversity and synthetic biology for the production of algal biofuels. Nature 330 (488), 329–335.

Harun, R., Danquah, M.K., Forde, G.M., 2010. Microalgal biomass as a fermentation feedstock for bioethanol production. J. Chem. Technol. Biotechnol. 85 (2), 199–203.

Hawkins, R.L., Nakamura, M., 1999. Expression of human growth hormone by the eukaryotic alga *Chlorella*. Curr. Microbiol. 38, 335–341.

Hildebrand, M., Abbriano, R.M., Polle, J.E.W., Traller, J.C., Trentacoste, E.M., Smith, S.R., Davis, A.K., 2013. Metabolic and cellular organization in evolutionarily diverse microalgae as related to biofuels production. Curr. Opin. Chem. Biol. 17, 506–514.

Ho, S.H., Chen, Y.D., Chang, Y.U., Lai, Y.Y., Chen, C.Y., Kondo, A., Ren, N.Q., Chang, J.S., 2017. Feasibility of CO_2 mitigation and carbohydrate production by microalga *Scenedesmus obliquus* CNW-N used for bioethanol fermentation under outdoor conditions: effects of seasonal changes. Biotechnol. Biofuel. 10, 27.

Ho, S.H., Huang, S.W., Chen, C.Y., Hasunuma, T., Kondo, A., Chang, J.S., 2013. Bioethanol production using carbohydrate-rich microalgae biomass as feedstock. Bioresour. Technol. 135, 191–198.

Hong, I.W., Jeon, H., Lee, S.B., 2014. Comparison of red, brown and green seaweeds on enzymatic saccharification process. Ind. Eng. Chem. Res. 20, 2687–2691.

Illman, A.M., Scragg, A.H., Shales, S.W., 2000. Increase in *Chlorella* strains calorific values when grown in low nitrogen medium. Enzyme Microb. Technol. 27, 631–635.

Imran Khan, M., Lee, M.G., Shin, J.H., Kim, J.D., 2017. Pretreatment optimization of the biomass of *Microcystes aeruginosa* for efficient bioethanol production. AMB Express 7, 19.

Jarvis, E.E., Brown, L.M., 1991. Transient expression of firefly luciferase in protoplasts of the green alga *Chlorella ellipsoidea*. Curr. Genet. 19, 317–321.

Jelynne, P., Tamayo, Del Rosario, E.J., 2014. Chemical analysis and utilization of *Sargassum* sp. as substrate for ethanol production. Iran. J. Energy Environ. 5 (2), 202–208.

John, R.P., Anisha, G.S., Madhavan Nampoothiri, K., Pandey, A., 2011. Micro and macroalgal biomass: a renewable source for bioethanol. Bioresour. Technol. 102, 186–193.

Jung, K.A., Lim, S.R., Kim, Y., Park, J.M., 2013. Potentials of macroalgae as feedstocks for biorefinery. Bioresour. Technol. 135, 182–190.

Karunakaran, S., Gurusamy, R., Senbagam, D., Senthilkumar, B., 2013. Screening and optimization of fermentation media for increased production of ethanol from *Gracilaria* sp. using response surface methodology. J. Pure Appl. Microbiol. 7 (4), 2641–2652.

Khalil, Z.I., Asker, M.M.S., El-Sayed, S., Kobbia, I.A., 2010. Effect of pH on growth and biochemical responses of *Dunaliella bardawil* and *Chlorella ellipsoidea*. World J. Microbiol. Biotechnol. 26, 1225–1231.

Kim, K.H., Choi, I.S., Kim, H.M., Wi, S.G., Bae, H.J., 2014. Bioethanol production from the nutrient stress-induced microalga *Chlorella vulgaris* by enzymatic hydrolysis and immobilized yeast fermentation. Bioresour. Technol. 153, 47–54.

Klinthong, W., Yang, Y.H., Huang, C.H., Tan, C.S., 2015. A review: microalgae and their applications in CO_2 capture and renewable energy. Aerosol Air Qual. Res. 15, 712–742.

Kumar, S., 2015. GM Algae for biofuel production: biosafety and risk assessment. Collect. Biosaf. Rev. 9, 52–75.

Kumar, S., Gupta, R., Kumar, G., Sahoo, D., Kuhad, R.C., 2013. Bioethanol production from *Gracilaria verrucosa*, a red alga, in a biorefinery approach. Bioresour. Technol. 135, 150–156.

Kumar, S.V., Misquitta, R.W., Reddy, V.S., Rao, B.J., Rajam, M.V., 2004. Genetic transformation of the green alga—*Chlamydomonas reinhardtii* by *Agrobacterium tumefaciens*. Plant Sci. 166, 731–738.

Markou, G., Chatzipavlidis, I., Georgakakis, D., 2012. Carbohydrates production and bio-flocculation characteristics in cultures of *Arthrospira (Spirulina) platensis*: improvements through phosphorus limitation process. Bioenerg. Res. 5, 915–925.

Naik, S.N., Goud, V.V., Rout, P.K., Dalai, A.K., 2010. Production of first and second generation biofuels: a comprehensive review. Renew. Sust. Energy Rev. 14, 578–597.

Nakamura, Y., Gojobori, T., Ikemura, T., 2000. Codon usage tabulated from international DNA sequence databases: status for the year 2000. Nucleic Acids Res. 28, 292–1292.

Nguyen, Minh, T.H., Vu, Hanh, V., 2012. Bioethanol production from marine algae biomass: prospect and troubles. J. Viet. Env. 3 (1), 25–29.

Ogbonda, K.H., Aminigo, R.E., Abu, G.O., 2007. Influence of temperature and pH on biomass production and protein biosynthesis in a putative *Spirulina* sp. Bioresour. Technol. 98, 2207–2211.

Özçimen, D., nan, B., 2015. An overview of bioethanol production from algae. Biofuels—Status and Perspective. InTech, Chapter 7, 141-162.

Pu, Y., Hu, F., Huang, F., Davison, B.H., Ragauskas, A.J., 2013. Assessing the molecular structure basis for biomass recalcitrance during dilute acid and hydrothermal pretreatments. Biotechnol. Biofuels 6, 15. doi: 10.1186/1754-6834-6-15.

Radakovits, R., Jinkerson, R.E., Darzins, A., Posewitz, M.C., 2010. Genetic engineering of algae for enhanced biofuel production. Eukaryot. Cell. 9 (4), 486–501.

Rafiqul, I.M., Hassan, A., Sulebele, G., Orosco, C.A., Roustaian, P., Jalal, C.A., 2003. Salt stress culture of blue green algae *Spirulina fusiformis*. Pak. J. Biol. Sci. 6 (7), 648-L650.

Rajkumar, R., Zahira, B., Takriff, M.S., 2014. Potential of the micro and macro algae for biofuel production: a brief review. Bioresources 9 (1), 1606–1633.

Reddy, C., Gupta, V., Jha, B., 2010. Developments in biotechnology of red algae. In: Seckbachand, J., Chapman, D.J. (Eds.), Red Algae in the Genomic Age. Springer, Dordrecht, pp. 307–341.

Rivera, A.L., Gómez-Lim, M., Fernández, F., Loske, A.M., 2014. Genetic transformation of cells using physical methods. J. Genet. Syndr. Gene Ther. 5, 4.

Ryan Georgianna, D., Mayfield, S.P., 2012. Exploiting diversity and synthetic biology for the production of algal biofuels. Nature 488, 329–335.

Ryu, B.M., Kang, K.H., Ngo, D.H., Qian, J.Z., Kim, S.K., 2012. Statistical optimization of microalgae *Pavlova lutheri* cultivation conditions and its fermentation conditions by yeast. Candida rugopellicul. Bioresour. Technol. 107, 307–313.

Saifuddin, N., Priatharsini, P., 2016. Developments in biohydrogen production from algae: a review. Res. J. Appl. Sci. Eng. Technol. 12 (9), 968–982.

Sanford, J.C., 2000. The development of the biolistic process. In Vitro Cell Dev-P1. Springer, New York, 303-308.

Shao, M., Kumar, S., Thomson, J.G., 2014. Precise excision of plastid DNA by the large serine recombinase Bxb1. Plant Biotechnol. J. 12, 322–329.

Show, K.Y., Lee, D.J., Tay, J.H., Lin, C.Y., Chang, J.S., 2012. Biohydrogen production: current perspectives and the way forward. Int. J. Hydrogen Energ. 37 (20), 15616–15631.

Singh, A., Nigam, P.S., Murphy, J.D., 2011. Renewable fuels from algae: an answer to debatable land based fuels. Bioresour. Technol. 102, 10–16.

Sirajunnisa, A.R., Surendhiran, D., 2016. Algae: a quintessential and positive resource of bioethanol production: a comprehensive review. Renew. Sustain. Energy Rev. 66, 248–267.

Sizova, I., Fuhrmann, M., Hegemann, P., 2001. A *Streptomyces rimosus* aphVIII gene coding for a new type phosphotransferase provides stable antibiotic resistance to *Chlamydomonas reinhardtii*. Gene 277, 221–229.

Spolaore, P., Cassan, C.J., Duran, E., Isambert, A., 2006. Commercial applications of microalgae. J. Biosci. Bioeng. 101 (2), 87–96.

Srivatsava, A., Prasad, R., 2000. Triglycride-based derived fuels. Renew. Sustain. Energy Rev. 4 (2), 111-L113.

Stadnik, M.J., de Freitas, M.B., 2014. Algal polysaccharides as source of plant resistance inducers. Trop. Plant Pathol. 39 (2), 111–118.

Surendhiran, D., Vijay, M., 2014. Evaluation of various pretreatment process for extracting oil from *chlorella salina* under nitrogen rich and depleted conditions. Intern. J. Environ. Bioenergy 7 (3), 129–142.

Taraldsvik, M., Myklestad, S.M., 2000. The effect of pH on growth rate, biochemical composition and extracellular carbohydrate production of the marine diatom *Skeletonema costatum*. Eur. J. Pharmacol. 35, 189–194.

Trivedi, N., Gupta, V., Reddy, C.R., Jha, B., 2013. Enzymatic hydrolysis and production of bioethanol from common macrophytic green alga *Ulva fasciata* Delile. Bioresour Technol. 150, 106–112.

Ullah, K., Ahmad, M., Sofia, Vinod Kumar, S., Lu, P., Harvey, A., 2015. Assessing the potential of algal biomass opportunities for bioenergy industry: a review. Fuel 143, 414–423.

Vignais, P.M., 2008. Hydrogenases and $H^{(+)}$-reduction in primary energy conservation. Results Probl. Cell Differ. 45, 223–252.

Weaver, J.C., Chizmadzhev, Y.A., 1996. Theory of electroporation: a review. Bioelectroch. Bioener. 41, 135–160.

Wu, F.C., Wu, J.Y., Liao, Y.J., Wang, M.Y., Shih, I.L., 2014. Sequential acid and enzymatic hydrolysis in situ and bioethanol production from *Gracilaria* biomass. Bioresour. Technol. 156, 123–131, www.fao.org/bioenergy/aquaticbiofuels 2009.

Yan, J., Zheng, X., Li, S., 2014. A novel and robust recombinant *Pichia pastoris* yeast whole cell biocatalyst with intracellular overexpression of a *Thermomyces lanuginosus* lipase: preparation, characterization and application in biodiesel production. Bioresour. Technol. 151, 43–48.

Zaslavskaia, L.A., Lippmeier, J.C., Kroth, P.G., Grossman, A.R., Apt, A.E., 2000. Transformation of the diatom Phaeodactylum tricornutum (Bacillariophyceae) with a variety of selectable marker and reporter genes. J. Phycol. 36, 379–386.

Further Reading

Poulsen, N., Kroger, N., 2005. A new molecular tool for transgenic diatoms. FEBS J. 272, 3413–3423.

SECTION V

LIFE CYCLE ANALYSIS, ECONOMICS AND POLICY ISSUES

19 *Life Cycle Assessment (LCA) of Bioethanol Produced From Different Food Crops: Economic and Environmental Impacts* 385

20 *Upgrading Comparative and Competitive Advantages for Ethanol Fuel Production From Agroindustrial Crops in Developing Countries: Mexico as a Case Study* 401

21 *Bioethanol in Brazil: Status, Challenges and Perspectives to Improve the Production* 417

CHAPTER

19

Life Cycle Assessment (LCA) of Bioethanol Produced From Different Food Crops: Economic and Environmental Impacts

Poritosh Roy, Animesh Dutta
University of Guelph, Guelph, ON, Canada

19.1 INTRODUCTION

The global energy consumption was 13,583 × 10^6 ton of oil equivalent (tOe) in 2013 and emitted about 31,097 × 10^6 ton CO_2, which contributed to global warming; however, the global energy consumption is increasing remarkably at a rate of 2.3% since 2000 (Enerdata, 2014) and contributing to global warming. The energy sector is the main source of emissions in countries, such as 80% emissions are energy related in Canada (ECCC, 2017). Consequently, the growing concern about the climate change and its potential implications on the earth led countries to have ambitious targets to reduce their greenhouse gas (GHG) emissions (Canada: 26%–41%; Japan: 38%–48%; EU: 40% levels; US: 25%–30%; China: 19%–34%; India: 13%–17% from 1990 levels by 2030) to combat the growing challenges that humankind faces nowadays (Hof et al., 2015). The transportation sector is one of the major contributors to the global GHG emission, and is responsible for about one quarter of global energy related GHG emissions, and that share is rising (IPCC, 2001). For example, in Canada the transportation sector contributes 28% to total GHG emissions (ECCC, 2017). Liquid biofuels are identified as an alternative to fossil fuels to reduce the dependence on fossil fuels and abate GHG emissions.

In an effort to offset fossil fuel-related negative impacts on the environment, countries have set an ambitious target and enforced renewable energy regulations, which require fuel producers and importers to have a certain amount of biofuels of distillates that they produce or import yearly. For example, in Canada they are required to have renewables of at least 5% of gasoline (Environment Canada, 2010) and Japan committed to blend 3% bioethanol with

Bioethanol Production From Food Crops
http://dx.doi.org/10.1016/B978-0-12-813766-6.00019-9

gasoline (METI, 2008). The greater bioethanol blending has also been reported, such as 10% in the USA and EU, which may be raised over time (Lane, 2015; Worldwatch Institute, 2017), and 20% in Brazil (Benvenga et al., 2016). Consequently, bioethanol production has been increased significantly, especially from food crops (Table 19.1). This table shows that the USA is the main producer of bioethanol followed by Brazil. This rapid expansion affects virtually every aspect of the field crop sectors, ranging from domestic demand and exports to price and the allocation of land area among crops (Roy and Shiina, 2010). First generation biofuels (bioethanol and biodiesel) are produced from food crops (e.g., corn, sugarcane, soybeans, vegetable oil, wheat, sugar beets, rapeseed, and peanuts), thus, compete with food or feed, and contribute to higher food prices (Mueller et al., 2011; Yang et al., 2009). For example, in the USA most of the bioethanol is produced from corn (Worldwatch Institute, 2017).

The Energy Independence and Security Act (EISA) stipulated that conventional biofuels (usually produced from food crops, such as corn), advanced biofuels, and cellulosic biofuels must reduce 20%, 50%, and 60% GHG emissions, respectively, compared to conventional fossil fuels in USA (Mondou and Skogstad, 2012; Mueller, 2016). In the USA, corn bioethanol must meet a reduction target of 20% (Mueller, 2016). On the other hand, EU directive requires to reduce GHG emissions by 50% in 2017 and 60% in 2018 compared to fossil fuels for a sustainable biofuel (EU, 2009; Matías et al., 2015). However, all biofuels must reduce GHG emissions by 50% compared to fossil fuels (Biofuels International, 2016). Therefore, it is important to determine the environmental and economic viability of biofuels either produced from food crops or any other feedstock. Life cycle assessment (LCA) is a tool that can be used to evaluate the life cycle environmental impacts of a product, process, or activity. Although several studies have already discussed the environmental impacts (Gelfand et al., 2013) and cost of biofuels (Benvenga et al., 2016), there are no studies highlighting either environmental or economic

TABLE 19.1 Biofuel Production From Major Food Crops in Different Regions

Biofuels	Country/region	Major feedstock	Production (L) ($\times 10^9$)
Bioethanol	Australia	Sugarcane	0.5
	Brazil	Sugarcane	25
	China	Corn/cassava/rice	3
	Canada	Corn/wheat	2
	France	Wheat/sugarcane/sugar beet	1
	Germany	Sugarcane/molasses	0.75
	India	Sugarcane/molasses	1
	United States	Corn/wheat	40
Biodiesel	Australia	Soybean/jatropha/coconut/palm	1
	Europe	Canola/soybean/barley	7
	North America	Soybeans	3
	South America	Castor bean/sunflower	4

Data taken from BIOFUEL, 2010. Major Biofuel Producers by Region. Available from: http://biofuel.org.uk/major-producers-by-region.html.

sustainability of biofuels produced from food crops. Therefore, this study provides a synthesis on their potential as a viable economic and environmental alternative to fossil fuels.

19.2 LIFE CYCLE ASSESSMENT METHODOLOGY (AND DYNAMIC LIFE CYCLE ASSESSMENT)

LCA has several possible purposes, including the comparison of alternative products, processes, or services; the comparison of alternative life cycles for a certain product or service; and the identification of parts of the life cycle where the greatest improvement can be made. The main objective of an LCA is to provide as complete a portrait as possible of resources consumption, environmental viability and their rebound effects, hence enable effective planning for a sustainable society. The major steps in LCA are: goal definition and scoping, inventory analysis, impact assessment, and interpretation (Fig. 19.1).

Goal definition and scoping defines the purpose, system boundary, expected product, functional units, and assumptions of the study. The inventory data are normalized to the functional unit. The system boundary is often illustrated by a general input and output flow diagram representing all the activities and resources input (energy and materials) as well as output (product and coproduct/by-product) of the system. Although the ISO standards (ISO 2006) state that coproduct allocation should be avoided by system expansion, various allocation methods (such as physical, system expansion) have been used in the LCA of biofuels (Bessou et al., 2013; Chagas et al., 2016; Liu et al., 2013; Qin et al., 2016).

The inventory is the quantitative environmental profile of a system being examined, that is, inventory identifies and quantifies the material inputs, energy consumption, and environmental emissions as well as products associated with the system. The impact assessment stage of an LCA evaluates environmental impacts (i.e., global warming potential; photochemical ozone creation potential, eutrophication potential, acidification potential, human toxicity potentials, ozone layer depletion potential, and abiotic depletion potential) of the system based on the inventory analysis. Finally, based on the goal and scope of the study, the inventory and impacts assessment are discussed, hot spots of the system are identified for conclusions and recommendations. Interpretation may include both quantitative and qualitative measures of improvement, such as changes in product, process, and activity design; raw material use, industrial processing, consumer use, and waste management. The information

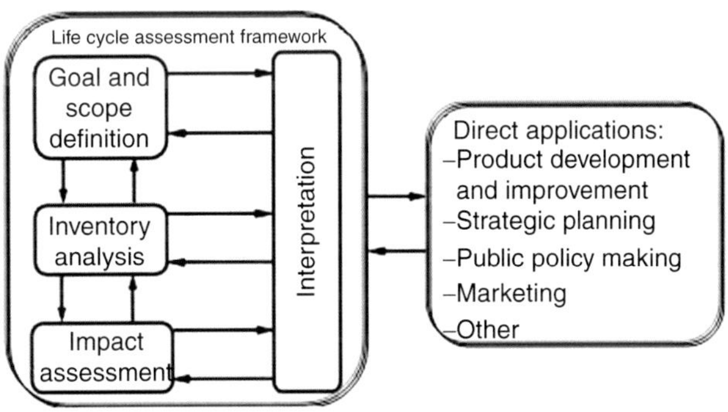

FIGURE 19.1 **Stages of an life cycle assessment.** *Modified from ISO (International Organization for Standardization), 2006. ISO 14040: 2006(E) Environmental management—Life cycle assessment—Principles and framework.*

generated from this assessment is usually used as a decision-support tool.

Furthermore, there is a growing recognition that GHGs released earlier have greater potential climate impacts compared with those released later (Levasseur et al., 2010; Tanaka et al., 2010), therefore the timing of GHG emissions is important for determining a project's life cycle emissions (Gaudreault and Miner, 2015; Levasseur et al., 2010), particularly when considering biomass. However, conventional LCA modeling is static, averaging emissions over a certain time period, therefore ignoring the implications of emission timing (Almeida et al., 2015; Giuntoli et al., 2016). Despite many studies have used LCA to demonstrate the GHG benefits for various biofuels, there are some concern related to the carbon debt of forest biomass, depending on how long it takes to be recaptured (Cherubini et al., 2011).

19.3 LIFE CYCLE ASSESSMENT STUDIES ON BIOETHANOL FROM FOOD CROPS

Due to renewable energy directive in different countries bioethanol production rapidly expanded mainly from food crops (corn, sugarcane, wheat, barley, etc.), which is known as first-generation bioethanol. Corn and sugarcane contributes to more than half and one third of global bioethanol production, respectively (REN 21, 2010). To date, corn-based bioethanol is the main contributor to total bioethanol production in Canada followed by wheat (23%) (Ethanol Producer Magazine, 2016). Food crops are also the main contributor to total bioethanol production in Europe. For example, in Spain soybeans, palm, rapeseed, and sunflower contribute 55%, 43%, 1.7%, and 0.1% to total biodiesel, respectively. On the other hand, contribution of corn, sugarcane, wheat, barley, and sugar beet to bioethanol production was 51%, 26%, 19%, 3%, and 2%, respectively (Garraín et al., 2016). However, sugarcane is the main source of bioethanol in Brazil (BIOFUEL, 2010).

Application of crop lands for biofuels could have either positive or negative environmental impacts depending on the types of crop, crop rotation, the way by-products are utilized (i.e., if by-products are considered as waste), thus affect indirect land use change (iLUC) (Garraín et al., 2016). The concern on iLUC forced EU to reshape her renewable energy policy. According to the current directive, EU member countries require to have biofuels at least 10% of transportation fuels by 2020 (20% of total energy); however, the first generation biofuels can contribute only up to 7% of distillates (Commission of the European Communities, 2007; Lane, 2015).

19.3.1 Grain Crops (Corn, Wheat, Rice, Barley, etc.)

Tillage and farming systems, crop rotation, and fertilizer application also affect the GHG emission from bioethanol produced from grain crops (Gelfand et al., 2013). No-till grain crop system (corn–soybean–wheat) noted to be environmentally beneficial compared with tilled system (Gelfand et al., 2013). Introduction of pea in crop rotation (oilseed, wheat, and peas) and better management of nitrogen fertilizer can reduce environmental impacts of grain crops, and improve ecoefficiency; however, environmental benefits depend on the design of crop rotation (Nemecek et al., 2015). The land use change has a significant impact (0.002–0.009 kg CO_2/MJ) on the life cycle of bioethanol produced from corn depending on the type of lands (e.g., conversion of forest land to cropland or conversion of cropland pasture to cropland as well as their conversion rates) (Qin et al., 2016). In contrast, corn butanol (produced in an acetone–butanol–bioethanol fermentation process) noted to have no GHG emissions savings compared to gasoline, where acetone was considered as a waste (Väisänen et al., 2016). The emission from the life cycle of butanol varied from 0.08–0.12 kg

CO_2 eq./MJ depending on the scenarios; thus, no emission saving was observed. Process energy consumption (i.e., steam production and fermentation) and corn farming are the major sources of emissions. On the other hand, GHG emission was reported to be 0.03 kg CO_2 eq./MJ of bioethanol produced from sorghum (García et al., 2017).

Although bioethanol produced from wheat can reduce about 9%–54% GWP (global warming potential) compared to fossil fuel (gasoline), it raises the land competition of about 95% compared with the gasoline (Gallejones et al., 2015). Similarly, biodiesel from rapeseed can reduce about 10%–64% GWP compared to diesel (Gallejones et al., 2015). However, emissions are noted to be dependent on the application of fertilizers. It is worthy to mention that acidification and eutrophication potential increased along with land competition in the case of biodiesel from rapeseed. However, both the biofuels (bioethanol and biodiesel) reduced the nonrenewable energy demand.

19.3.2 Tuber Crops (Sweet Beat, Cassava, and Sweet Potato)

Tuber crops have a significant contribution to total biofuels production in different countries for reducing GHG emissions (Bessou et al., 2013; Thatoi et al., 2016), and identified as a promising feedstock to meet rising bioethanol demand as well as reducing the dependency on fossil fuels (Thatoi et al., 2016). Bessou et al. (2013) reported that sugar beet bioethanol reduces environmental impacts for most of the categories (GWP, abiotic depletion, ozone layer depletion, and photochemical oxidation) compared with gasoline except acidification and eutrophication. For example, sugar beet bioethanol can reduce 28%–42% GHG emissions compared to gasoline; however, acidification and eutrophication potential rises because of the application of reactive nitrogen fertilizer on the arable land (Bessou et al., 2013).

On the other hand, planting mode affects the life cycle GHG emission of bioethanol produced from cassava. Intensive agriculture (e.g., high fertilizer application and highly mechanized harvesting) has lower GHG savings compared to extensive mode because mechanized harvesting reduces yield, and emission from fertilizer manufacturing and the use of fertilizer. Depending on the source of energy that is used in the conversion process and the mode of cultivation, GHG emissions savings varied from 26% to 34%. The conversion stage is reported to be the main contributor to total GHG emission followed by cultivation and biomass logistics (Liu et al., 2013).

The GHG emission from the life cycle of bioethanol also depends on the wastewater treatment and recovery of biogas, and the utilization of biogas as well as the sources of energy for steam production (Numjuncharoen et al., 2015). For example, emission varied from 0.55–0.61 kg CO_2 eq./L when biomass/biogas has been used for steam production (Numjuncharoen et al., 2015). On the other hand, it was 2.1 kg CO_2 eq./L when coal has been used for steam production (Moriizumi et al., 2012). Feedstock cultivation, harvesting, and processing are the main contributors to the life cycle of bioethanol from cassava if biomass/biogas is used for steam production (Numjuncharoen et al., 2015); however, steam production stage was the prime emitter when coal is used (Moriizumi et al., 2012). It is important to note that biomass is assumed to be carbon neutral in the preceding study.

19.3.3 Oil Seeds

GHG emission from the life cycle of renewable jet fuel produced from rapeseed that was grown in fallow period (e.g., wheat–wheat–rapeseed) reported to be dependent on the residues management and tillage systems. Increasing residue inputs improves soil carbon and the full tillage increase the loss of soil carbon. The emission was 0.004–0.07 kg CO_2 eq./L depending on the residues inputs for both crops (wheat and rapeseed). It can reduce 20%–96% of GHG emissions compared to the conventional jet fuels (Ukaew et al., 2015).

19.3.4 Herbaceous Crops (Sugarcane)

Sugarcane-based bioethanol has attracted a tremendous attention as an alternative to fossil fuels because of its prospects for producing alternate energy and reducing GHG emissions (Cavalett et al., 2012; Chagas et al., 2016; Seabra et al., 2011; Wang et al., 2012). The environmental impacts are dependent on the farming systems and cropping patterns. For example, better agricultural management (controlled traffic farming; crop rotation) allows to maximize the number of cuts in a sugarcane cycle, consequently abate GHG emissions, fossil fuel depletion and acidification of 43%, 24%, and 44%, respectively (Chagas et al., 2016). On the other hand, crop rotation improves soil quality, which helps in reducing fertilizer application and pest infestations, thus reducing environmental burden.

Nakashima and Ishikawa (2016) found that better mechanized (capital intensive) production systems (cultivation) increases GHG emissions. Fertilizer production is the main contributor to total GHG emissions followed by agricultural machinery, field operation, and transportation. The authors also noted that organic farming can reduce both energy consumption and GHG emissions. On the other hand, Väisänen et al. (2016) noted that side energy flow, that is, energy produced from bagasse helps biorefinery industry that produces butanol become self-sufficient as well as supply surplus electricity to the national grid offsetting some of the GHG emissions, consequently reduces GHG emissions (−0.055 to 0.018 kg CO_2 eq./MJ). In addition, García et al. (2017) found that bioethanol directly produced from sugarcane juice is environmentally beneficial, instead of producing bioethanol from molasses or sorghum. Producing bioethanol from juice not only reduces GHG emissions (28%) but also requires 10.6 times less land area and 52% blue water and create 11% more employment.

19.4 LAND, WATER, AND OTHER APPROACHES IN LIFE CYCLE ASSESSMENT OF BIOETHANOL

Life cycle emission of biofuels can be influenced by the land management change, thus, the effect of land use change and timing of emission from soil should be taken into account in agricultural LCAs (Goglio et al., 2015). Although biofuels can reduce GHG emissions, they require 95% more land area compared with fossil fuels (Gallejones et al., 2015). The land requirement also depends on the productivity of land, type of crops, and the bioethanol yield from feedstock (Thatoi et al., 2016). Water consumption in the life cycle of bioethanol depends on the feedstock used for bioethanol (Gheewala et al., 2013). For example, bioethanol produced from cassava root has the highest water footprint (2374–2841 L/L bioethanol); followed by molasses (1976–3105 L/L bioethanol) and sugarcane (1396–2196 L/L bioethanol); however, molasses has the highest fresh water demand (Gheewala et al., 2013). Feedstock cultivation stage is the main contributor to water footprint of bioethanol produced from these crops and only 2%–3% is contributed by industrial stage.

The sustainability of bioethanol from food crops also depends on the application of synthetic fertilizers and nitrogen fixation capacity of the crops that are cultivated for bioethanol. For example, legumes have the nitrogen fixation capability, thus reduce fertilizer application and improve the sustainability of bioethanol produced from legumes. The organic farming may avoid the application of synthetic fertilizer and improve the sustainability of agriculture (Bedoussac et al., 2015); however, it reduces the productivity of land, thus require greater land area and increase the land competition.

19.5 ENVIRONMENTAL BENEFITS OF BIOETHANOL

The use of bioethanol has significant potential to reduce GHG emissions, compared to the use of fossil fuels. Although LCA can consider various impacts that occur throughout the life cycle (i.e., global: global warming potential, ozone depletion potential; regional: acidification/eutrophication; local: ecotoxicity/human toxicity, particulate emissions), most studies focused mainly on the GWP of bioethanol. Consequently, we have summarized only the GWP of bioethanol produced from food crops (Table 19.2).

The ranges in results are because of differences in the system boundary, type feedstock, feedstock production process, industrial processes, and utilization of coproducts/by-products. A range of LCA modeling boundaries have been used in various studies, such as some using a cradle-to-gate approach (excluding several aspects of the life cycle), others using a well-to-pump/field to gate approach (which excludes the usage of the product), while others used a cradle-to-grave approach (which includes all operations from raw materials acquisition to waste disposal or use of final products). A range of approaches have also been used to handle coproducts, from physical allocation to system expansion.

TABLE 19.2 A Summary of the Feedstock, System Boundary, Allocation Method, Bioethanol Yield and Life Cycle GWP of Bioethanol

Authors	Feedstock	System boundary	Allocation method	Bioethanol yield (L/kg)	GWP (kg CO_2 eq/MJ)
Grains					
Wang et al. (2012)	Corn	Well to wheel	System expansion	0.425	0.053–0.062
Qin et al. (2016)	Corn	Field to wheel	System expansion	–	0.06–0.07
Gallejones et al. (2015)	Wheat	Well to tank	System expansion	0.36 kg/kg	0.047–0.076
Gallejones et al. (2015)	Rapeseed	Well to tank	System expansion	0.40 kg/kg	0.03–0.101
Ukaew et al. (2015)	Rapeseed	Cradle to grave	System expansion	–	0.004–0.07
Wang et al. (2012)	Sugarcane	Well to wheel	Energy	0.081	0.4–0.5 kg CO_2 eq./L
Chagas et al. (2016)	Sugarcane	Well to gate	Economic	0.085	0.42–0.51 kg CO_2 eq./L
Nakashima and Ishikawa (2016)	Sugarcane	Cradle to gate	System expansion	1.4–1.6 GJ/t cane (21.2 MJ/L)	0.617–0.667 kg CO_2 eq./L
Root crops					
Bessou et al., (2013)	Sugar beet	Well to tank	Mass	0.03–0.04 kg/kg	0.0578
Liu et al. (2013)	Cassava	Field to wheels	System expansion	0.36 kg/kg	0.02–0.07
[a]Numjuncharoen et al. (2015)	Cassava	Cradle to gate	System expansion	–	0.55–0.61 kg CO_2 eq./L

–, not specified; *AD*, anaerobic digestion; *DT*, Dry ton.

[a] *Without biogas utilization: 1.03 kg CO_2 eq./L.*

Consequently, the GWP benefits of bioethanol need to be considered on a case-by-case basis, and using the principles outlined by the ISO LCA standards (ISO, 2006). However, there are still a lot of gaps in understanding the industrial symbiosis/circular economy and permanence of carbon credit to different food crops, GWP needs to be interpreted and conveyed carefully. Furthermore, many of the studies have reported GWP benefits based on modeling or small-scale production, thus the benefits may not be compatible with commercial plants, Table 19.1 is a summary of the feedstock, system boundary, allocation method, bioethanol yield, and life cycle GWP of bioethanol.

19.6 ECONOMICS OF BIOETHANOL

Although GHG benefits of first generation biofuels are positive, the commercialization of biofuel technologies depends on the production cost and the competitiveness with their counterpart (i.e., conventional fossil fuels). The production cost depends on both fixed and variable costs. There is a variation in life cycle production costs of bioethanol (Table 19.3) because of a range of costs related to feedstock, product yield, life span of bioethanol plant, discount rates, and plant capacity. The production cost varies depending on how the feedstock is produced or sourced, the collection and processing of the feedstock, and the biomass logistics.

It is noted that production cost is dependent not only on the farming system but also on the crop rotation. For example, soybean cultivated as a rotation crop for sugarcane improved farm income, thus reduces production cost of bioethanol from sugarcane; however, the profitability depends on the type of crops grown as a rotation crop (Chagas et al., 2016). The configuration of cogeneration system (where bagasse is used for energy supply in the system) and the integration of the production of first- and second-generation bioethanol from sugarcane marginally affects the production cost (Dias et al., 2013). The study revealed that surplus bagasse can be used for second generation bioethanol in an integrated approach to improve economic viability of first generation bioethanol. Similarly, integration of

TABLE 19.3 Reported Production Costs of Bioethanol Produced From Food Crops

Authors	Feedstock type	Plant capacity (t/d)	Feed cost ($/ton)	Product yield (L/T)	System boundary	Products	Cost (US$/L)
[a]Chagas et al. (2016)	Sugarcane	20,000	26.2–29.5	84.8	Cradle to gate	Bioethanol	0.49–0.52
[a]Dias et al. (2013)	Sugarcane	12,000	23.24	82.3	Cradle to gate	Bioethanol	0.36
[b,c]Maung and Gustafson (2011)	Sugar beet	38–76 million L/year	63	–	Field to gate	Bioethanol	0.45–0.48
[b]Cobuloglu and Büyüktahtakın (2015)	Corn	95 million L/year	168	286	Field to gate	Bioethanol	0.28
[b]Benvenga et al. (2016)	Cassava	Lab scale	131	195	Field to gate	Bioethanol	0.62

[a] *Plant life span 25 years (167 days/year).*
[b] *Plant life span 10 years; t/d: ton/day.*
[c] *Interest rate 8.5%; – Not specified.*

first- and second-generation bioethanol from wheat grain and wheat straw found to be beneficial; however, it depends on the market price of biogas that produced from distiller dried grains (DDGS), and the production cost of bioethanol varied from 0.45 to 0.56 €/L (Joelsson et al., 2016).

Bioethanol production from tuber crops (especially from cassava) is increasing because of economic benefits compared with other crops (Jansson et al., 2009). Benvenga et al. (2016) optimized the starch hydrolysis conditions (23.5 g/L, 61.9°C and 111.0 min; which provided bioethanol yield of 88%) to determine the production cost of bioethanol from cassava starch with enzymatic hydrolysis process. The production cost of bioethanol was noted to be 0.62 US$/L. The profitability of bioethanol can also be improved by increasing the scale of production (Kuppens et al., 2015; López-González et al., 2015) because large-scale production and longer life span reduces fixed costs; however, it may increase transportation costs due to greater feedstock demand.

19.7 ONGOING EFFORTS TO IMPROVE THE SUSTAINABILITY OF BIOETHANOL

Efforts are underway to improve the sustainability of biofuels; thus, various approaches have been used to estimate the life cycle emissions of biofuels. Although several studies claimed that N_2O is greatly affected by the climatic conditions and fertilizer management (Flechard et al., 2007; Gerber et al., 2016); usually, the Intergovernmental Panel on Climate Change (IPCC) default emission factor is applied to evaluate the life cycle emissions. Various models have also been used to minimize the uncertainty in quantification of N_2O emission (e.g., Daily version of the CENTURY ecosystem model (DAYCENT): Parton et al., 1998; Simulateur Multidisciplinary pour Ies Cultures Standard (STICS): Brisson et al., 1998), which noted to have some limitations. Consequently, Gallejones et al. (2015) have used a new approach [simulation of nitrogen in cropping systems model ($SIMS_{NIC}$ model)] to capture some of the impacts specific to climate conditions and fertilizer management (improves the quantification of N_2O emission) to evaluate the life cycle bioethanol and biodiesel produced from wheat and rapeseed, respectively, and reported that about 40% GHG emissions can be reduced. On the other hand, Genetic Algorithms technique has been used to evaluate the environmental and economic viability of bioethanol produced from cassava root, which is claimed to be more viable method than other known techniques (Benvenga et al., 2016).

A mixed-integer optimization model has been introduced to determine the economic and environmental viability of bioethanol produced from switch grass and corn. The model can optimize the land allocations among crops to maximize the economic and environmental benefits (Cobuloglu and Büyüktahtakın, 2015). A linear optimization model (BioScope) has also been used to determine the optimal location for bioethanol industry to reduce transportation cost, thus the production cost, and improves the sustainability of bioethanol produced from sugarcane. On the other hand, environmental sustainability was determined with PCraster Land Use Change model (PLUC model) (Jonker et al., 2016). It is also reported that modeling approaches can be used to reduce water footprint of bioethanol. For example, application of linear programming approach reduced 53% annual water footprint of bioethanol (Pongpinyopap and Mungcharoen, 2015). Furthermore, agent-based model (ABM) (Bichraoui-Draper et al., 2015; Florent and Enrico 2015); geographic information systems (GIS) (Jäppinen et al., 2014); Biomass Inventory Mapping and Analysis Tool (BIMAT) (Agriculture and Agri-Food Canada: http://www.agr.gc.ca/atlas/bimat), Introductory Carbon Balance Model (ICBM) (Andren and Kätterer, 1997) and dynamic life cycle assessment (DLCA) (Levasseur et al., 2010) models have been proposed to reduce the uncertainty of emissions from the life cycle of biofuels. In addition to these models, Asia-Pacific integrated model (AIM), the multiregional, multisectoral, static, applied general equilibrium (MAGNET model), global change assessment model (GCAM), global biosphere management model (GLOBIOM); model of agricultural production and its impact on the environment (MAgPIE), and so on, have also been used to evaluate the sustainability of biofuels and to estimate carbon taxes or emission mitigation cost, especially for lignocellulosic biofuels (Lotze-Campen et al., 2014). In addition, biogeochemical process model (GIS based integrated climate model) has been introduced to determine the energy potential of cassava grown on the marginal land (Jiang et al., 2015).

Finally, industrial symbiosis/circular economic concepts may be useful in reducing the GHG emission from the life cycle of bioethanol and improve the sustainability as well as reduce production cost of bioethanol, thus improving the profitability. For example, sugar industry by-products, such as molasses and bagasse can be used for ethanol production or molasses can further refined or used as an animal feed. On the other hand bagasse can be used for second-generation bioethanol or for heat and power. Consequently, industrial symbiosis/circular economy concept is emerging as a viable option for improving the sustainability of bioethanol, as well as foster food and energy security (Concha et al., 2017; Hagman et al., 2017; Mouzakitis et al., 2017).

19.8 DISCUSSION

GHG emissions of various biofuels and their ability to reduce the GHG emissions vary widely, and dependent on the choice of feedstock, agricultural practices, land use changes, refining and conversion process, and finally the end use practices. A variety of production and utilization systems of available biofuels have different social, economic and environmental impacts. Effective and efficient recovery/utilization of coproducts improves the economy of the process (Liu and Barrows, 2013), and the profitability depends on the plant size (Rajendran et al., 2015) as well as the integration of the production of first- and second-generation bioethanol. For example, in the integrated biorefinery system two-thirds of energy from coproducts become excess and can be used for other industries (Mussoline and Wilkie, 2015), thus can offset some of the environmental and economic burden of bioethanol.

Although various efforts are underway to build consensus on the biofuel LCA till the biofuel LCA results are not directly comparable because of the differences in system boundary, allocation methods, assumptions, and integrated (either first- and second-generation ethanol or industrial symbiosis approaches) or nonintegrated production process. In addition, there is uncertainty toward the carbon neutrality and sustainability of biofuels. Although biofuels can create employment opportunity in the rural communities (Hondo and Moriizumi, 2017; Roy and Dias, 2017; tef nescu-Mih il , 2016), many studies argued that biofuels are responsible for food insecurity (Renzaho et al., 2017; Rulli et al., 2016), and loss of biodiversity (Hoekman and Broch, 2017) as well as put pressure on water (Gheewala et al., 2013). The economic cost is identified to be the main barrier for promotion of biofuels (Ji and Long, 2016) as well as to compete with their counterpart. The concerns over sustainability and perceptions about negative impacts of biofuels are growing and resulting in the closer scrutiny of policies designed to expand bioenergy use, especially from food crops.

Although LCA has emerged as an important tool to understand the environmental impacts of products and services (Martin et al., 2015), there are various challenges with the methodology because of the limited scope of sustainability assessment (Hellweg and Canals, 2014). Most of the LCA studies concluded that biofuels scenarios significantly reduce GHG emissions when these replace fossil fuels (Cherubini and Strømman, 2011; Roy et al., 2012); however, some LCA studies found significant increases in GHG emissions due to leakage arising from clearing of new land for food, which is known as iLUC is accounted in the later studies (Tonini et al., 2016; Wenzel et al., 2014). Consequential life cycle assessment (cLCA) linked to the market mechanism, that is, economic responses/market implications of the system are taken into account by expanding the system boundary (Martin et al., 2015; Meier et al., 2015; Weidema et al., 2009), even beyond the national boundary (Sevigné-Itoiz et al., 2015); however, there is no general framework for a cLCA (Rajagopal, 2017).

The implementation of revised renewable energy directive may help to reduce the use of food crops for bioethanol, thus reduce the uncertainty in food security. Although LCA accounts for immediate physical flows in determining the environmental impacts of a product, process, or activity, cLCA accounts for the changes in market demand and describe the changes in physical flows and their consequences (Garraín et al., 2016). As a result, application of cLCA is increasing to determine the sustainability of biofuel; however, multiple approaches (such as DLCA, cLCA, and LCC) need to be applied to determine the sustainability of biofuels from food crops to avoid any adverse impacts on the society and economy.

Food and energy security are two major issues that need to be addressed in order to meet the needs of the growing population and a better quality of life for the bottom of the pyramid. At the same time, food and energy production, and use are associated with severe environmental and resource degradation, such as the capacity of ecosystems, GWP, and biodiversity. Although many researchers and policy makers support the use of biofuel as a way to meet energy demand sustainably, this support has come with several controversies, including the food and fuel debate. Various researchers pointed out the severe consequences of inappropriate development of biomass resources for energy, including global agricultural production (Martin et al., 2015), land use change and food security (Nijsen et al., 2012; Vasile et al., 2016), and potential adverse impacts on the environment, such as increased GHG emissions (Searchinger et al., 2008).

Although initiatives are underway to minimize the use of food crops and crops land for bioethanol, the application of higher-quality land and food corps for bioethanol need to be strictly regulated so that adequate amount of food can be produced to meet the enormous demand of the growing population as well as ensure their food security. Carbon tax has been introduced by countries/provinces to abate/control GHG emissions and achieve the ambitious emission reduction target. Consequently, multiple approaches may need to apply to precisely determine the life cycle environmental impacts of bioethanol and their rebound effects on the society. Currently, EU has put a cap on the first-generation bioethanol to minimize the competition between food and fuels, which may need to follow by other countries on the earth to ensure adequate amount of food for the growing population.

19.9 CONCLUSIONS

A lot of efforts have been made to mitigate the GHG emissions from the life cycle of bioethanol to improve its sustainability. The ambitious target for biofuels affects virtually every aspect of the field crops sectors and the allocation of land area among crops; however, most of the earlier LCA studies failed to account the impact of land use change and emission timing. Moreover, due to different approaches have been used to determine the life cycle impacts of bioethanol;

the results are not comparable. Consequently, life cycle of bioethanol need to be evaluated by applying the cLCA to avoid any unwanted consequences from bioethanol. In addition, integrated/industrial symbiosis could be very important to improve the economic and environmental sustainability of bioethanol produced from food crops as well as may reduce food insecurity, and abate loss of biodiversity and ecosystem service.

Acknowledgments

This research is funded by the NSERC (Discovery Grant No. 400495), and the Ontario Ministry of Agriculture, Food and Rural Affairs (OMAFRA, Grant No. 200474).

References

Almeida, J., Degerickx, J., Achten, W.M., Muys, B., 2015. Greenhouse gas emission timing in life cycle assessment and the global warming potential of perennial energy crops. Carbon Manage. 6, 185–195.

Andren, O., Kätterer, T., 1997. ICBM: the introductory carbon balance model for exploration of soil carbon balances. Ecol. Appl. 7, 1226–1236.

Bedoussac, L., Journet, E.P., Hauggaard-Nielsen, H., Naudin, C., Corre-Hellou, G., Jensen, E.S., Prieur, L., Justes, E., 2015. Ecological principles underlying the increase of productivity achieved by cereal-grain legume intercrops in organic farming: a review. Agron. Sustain. Dev. 35, 911–935.

Benvenga, M.A.C., Librantz, A.F.H., Santana, J.C.C., Tambourgi, E.B., 2016. Genetic algorithm applied to study of the economic viability of alcohol production from Cassava root from 2002 to 2013. J. Clean Prod. 113, 483–494.

Bessou, C., Lehuger, S., Gabrielle, B., Mary, B., 2013. Using a crop model to account for the effects of local factors on the LCA of sugar beet in Picardy region, France. Int. J. Life Cycle Assess 18 (1), 24–36.

Bichraoui-Draper, N., Xu, M., Miller, S.A., Guillaume, B., 2015. Agent-based life cycle assessment for switchgrass-based bioenergy systems. Resour. Cons. Recycl. 103, 171–178.

BIOFUEL, 2010. Major biofuel producers by region. Available from: http://biofuel.org.uk/major-producers-by-region.html

Biofuels International, 2016. US corn ethanol may help Japan reach 50% GHG reduction target. Available from: http://biofuels-news.com/display_news/11233/US_corn_ethanol_may_help_Japan_reach_50__GHG_reduction_target__study_says/

Brisson, N., Mary, B., Ripoche, D., Jeuffroy, M.H., Ruget, F., Nicoullaud, B., Gate, P., Devienne-Barret, F., Antonioletti, R., Durr, C., Richard, G., 1998. STICS: a generic model for the simulation of crops and their water and nitrogen balances I: theory and parameterization applied to wheat and corn. Agronomie 18, 311–346.

Cavalett, O., Junqueira, T.L., Dias, M.O., Jesus, C.D., Mantelatto, P.E., Cunha, M.P., Franco, H.C., Cardoso, T.F., Maciel Filho, R., Rossell, C.E., Bonomi, A., 2012. Environmental and economic assessment of sugarcane first generation biorefineries in Brazil. Clean Technol. Environ. Policy 14, 399–410.

Chagas, M.F., Bordonal, R.O., Cavalett, O., Carvalho, J.L.N., Bonomi, A., La Scala, N., 2016. Environmental and economic impacts of different sugarcane production systems in the ethanol biorefinery. Biofuel. Bioprod. Bioref. 10, 89–106.

Cherubini, F., Strømman, A.H., 2011. Life cycle assessment of bioenergy systems: state of the art and future challenges. Bioresour. Technol. 102, 437–451.

Cherubini, F., Peters, G.P., Berntsen, T., StrØmman, A.H., Hertwich, E., 2011. CO_2 emissions from biomass combustion for bioenergy: atmospheric decay and contribution to global warming. GCB Bioenergy 3, 413–426.

Cobuloglu, H.I., Büyüktahtakın, E., 2015. Food vs. biofuel: an optimization approach to the spatio-temporal analysis of land-use competition and environmental impacts. Appl. Energy 140, 418–434.

Commission of the European Communities, 2007. Biofuel issues in the new legislation on the promotion of renewable energy public consultation exercise, April–May 2007. Energy and Transport Directorate-General.

Concha, D., Adams, M., Suárez, J., Faxas, R., 2017. Fostering food and energy security through by-product valorization within agricultural and agro-industrial networks: study of the province of Santiago de Cuba. Int. J. Sust. Dev. World Ecol. 24, 159–174.

Dias, M.O., Junqueira, T.L., Cavalett, O., Cunha, M.P., Jesus, C.D., Mantelatto, P.E., Rossell, C.E., Maciel Filho, R., Bonomi, A., 2013. Cogeneration in integrated first and second generation bioethanol from sugarcane. Chem. Eng. Res. Des. 91, 1411–1417.

ECCC (Environment Climate Change Canada) 2017. Available from: https://www.ec.gc.ca/ges-ghg/default.asp%3Flang(Enandn(02D095CB-1

Enerdata, 2014. Global Energy Statistical Yearbook. , Available from: https://yearbook.enerdata.net/#CO2-intensity-data.html.

Environment Canada, 2010. Renewable Fuels Regulations. Available from: http://www.ec.gc.ca/lcpe-cepa/eng/regulations/detailReg.cfm?intReg=186.

Ethanol Producer Magazine, 2016. Report: Canadian Ethanol Industry Operating at Full Capacity. , Available from: http://www.ethanolproducer.com/articles/13646/report-canadian-ethanol-industry-operating-at-full-capacity.
EU, 2009. Directive 2009/28/EC of the European Parliament and of the Council of 23 April 2009 on the promotion of the use of energy from renewable sources and amending and subsequently repealing Directives 2001/77/EC and 2003/30/EC. Off. J. Eur. Union 5, 2009.
Flechard, C.R., Ambus, P., Skiba, U., Rees, R.M., Hensen, A., Van Amstel, A., Van Den Pol-Van Dasselaar, A., Soussana, J.F., Jones, M., Clifton-Brown, J., Raschi, A., 2007. Effects of climate and management intensity on nitrous oxide emissions in grassland systems across Europe. Agric. Ecosyst. Environ. 121, 135–152.
Florent, Q., Enrico, B., 2015. Combining agent-based modeling and life cycle assessment for the evaluation of mobility policies. Environ. Sci. Technol. 49, 1744–1751.
Gallejones, P., Pardo, G., Aizpurua, A., Del Prado, A., 2015. Life cycle assessment of first-generation biofuels using a nitrogen crop model. Sci. Total Environ. 505, 1191–1201.
García, C.A., Manzini, F., Islas, J.M., 2017. Sustainability assessment of ethanol production from two crops in Mexico. Renew. Sustain. Energ. Rev. 72, 1199–1207.
Garraín, D., de la Rúa, C., Lechón, Y., 2016. Consequential effects of increased biofuel demand in Spain: global crop area and CO_2 emissions from indirect land use change. Biomass Bioenergy 85, 187–197.
Gaudreault, C., Miner, R., 2015. Temporal aspects in evaluating the greenhouse gas mitigation benefits of using residues from forest products manufacturing facilities for energy production. J. Ind. Ecol. 19, 994–1007.
Gelfand, I., Sahajpal, R., Zhang, X., Izaurralde, R.C., Gross, K.L., Robertson, G.P., 2013. Sustainable bioenergy production from marginal lands in the US Midwest. Nature 493, 514–517.
Gerber, J.S., Carlson, K.M., Makowski, D., Mueller, N.D., Garcia de Cortazar-Atauri, I., Havlík, P., Herrero, M., Launay, M., O'Connell, C.S., Smith, P., West, P.C., 2016. Spatially explicit estimates of N_2O emissions from croplands suggest climate mitigation opportunities from improved fertilizer management. Glob. Chang. Biol. 22, 3383–3394.
Gheewala, S.H., Silalertruksa, T., Nilsalab, P., Mungkung, R., Perret, S.R., Chaiyawannakarn, N., 2013. Implications of the biofuels policy mandate in Thailand on water: the case of bioethanol. Bioresour. Technol. 150, 457–465.
Giuntoli, J., Agostini, A., Caserini, S., Lugato, E., Baxter, D., Marelli, L., 2016. Climate change impacts of power generation from residual biomass. Biomass Bioenergy 89, 146–158.
Goglio, P., Smith, W.N., Grant, B.B., Desjardins, R.L., McConkey, B.G., Campbell, C.A., Nemecek, T., 2015. Accounting for soil carbon changes in agricultural life cycle assessment (LCA): a review. J. Clean. Prod. 104, 23–39.
Hagman, L., Blumenthal, A., Eklund, M., Svensson, N., 2017. The role of biogas solutions in sustainable biorefineries. J. Clean. Prod. 172, 3982–3989.
Hellweg, S., Canals, L.M., 2014. Emerging approaches, challenges and opportunities in life cycle assessment. Science 344, 1109–1113.
Hockman, S.K., Broch, A., 2017. Environmental implications of higher ethanol production and use in the US: a literature review. Part II—Biodiversity, land use change, GHG emissions, and sustainability. Renew. Sustain. Energ. Rev.doi: 10.1016/j.rser.2017.05.052.
Hof, A., Brink, C., Beltran, A.M., Elzen, den, 2015. Greenhouse Gas Emission Reduction Targets for 2030 Conditions for an EU Target of 40%. PBL Netherlands Environmental Assessment Agency, The Hague, Netherlands, Available from: http://www.pbl.nl/sites/default/files/cms/publicaties/PBL_2012_Greenhouse-gas-emission-reduction-targets-for-2030_500114023.pdf.
Hondo, H., Moriizumi, Y., 2017. Employment creation potential of renewable power generation technologies: a life cycle approach. Renew. Sustain. Energ. Rev. 79, 128–136.
IPCC, 2001. IPCC Third Assessment Report Climate change 2001: Mitigation. Cambridge University Press, Cambridge, United Kingdom.
ISO, 2006. Environmental Management—Life Cycle Assessment—Principles and Framework. International Organization for Standardization, Geneva, Switzerland, ISO 14040: 2006(E).
Jansson, C., Westerbergh, A., Zhang, J., Hu, X., Sun, C., 2009. Cassava, a potential biofuel crop in (the) People's Republic of China. Appl. Energy 86, S95–S99.
Jäppinen, E., Korpinen, O.J., Ranta, T., 2014. GHG emissions of forest-biomass supply chains to commercial-scale liquid-biofuel production plants in Finland. GCB Bioenergy 6, 290–299.
Ji, X., Long, X., 2016. A review of the ecological and socioeconomic effects of biofuel and energy policy recommendations. Renew. Sustain. Energ. Rev. 61, 41–52.
Jiang, D., Hao, M., Fu, J., Huang, Y., Liu, K., 2015. Evaluating the bioenergy potential of cassava on marginal land using a biogeochemical process model in GuangXi. China J. Appl. Remote Sens. 9, 097699.
Joelsson, E., Erdei, B., Galbe, M., Wallberg, O., 2016. Techno-economic evaluation of integrated first- and second-generation ethanol production from grain and straw. Biotechnol. Biofuels 9, 1.
Jonker, J.G.G., Junginger, H.M., Verstegen, J.A., Lin, T., Rodríguez, L.F., Ting, K.C., Faaij, A.P.C., van der Hilst, F., 2016. Supply chain optimization of sugarcane first generation and eucalyptus second generation ethanol production in Brazil. Appl. Energy 173, 494–510.

Kuppens, T., Van Dael, M., Vanreppelen, K., Thewys, T., Yperman, J., Carleer, R., Schreurs, S., Van Passel, S., 2015. Techno-economic assessment of fast pyrolysis for the valorization of short rotation coppice cultivated for phytoextraction. J. Clean. Prod. 88, 336–344.

Lane, J., 2015. EU reshapes its biofuels policy. Biofuels Digest, Available from: http://www.biofuelsdigest.com/bdigest/2015/04/16/eu-reshapes-its-biofuels-policy/.

Levasseur, A., Lesage, P., Margni, M., Deschenes, L., Samson, R., 2010. Considering time in LCA: dynamic LCA and its application to global warming impact assessments. Environ. Sci. Technol. 44, 3169–3174.

Liu, B., Wang, F., Zhang, B., Bi, J., 2013. Energy balance and GHG emissions of cassava-based fuel ethanol using different planting modes in China. Energy Policy 56, 210–220.

Liu, K., Barrows, F.T., 2013. Methods to recover value-added co-products from dry grind processing of grains into fuel ethanol. J. Agric. Food Chem. 61, 7325–7332.

López-González, D., Puig-Gamero, M., Acién, F.G., García-Cuadra, F., Valverde, J.L., Sanchez-Silva, L., 2015. Energetic, economic and environmental assessment of the pyrolysis and combustion of microalgae and their oils. Renew. Sustain. Energ. Rev. 51, 1752–1770.

Lotze-Campen, H., Lampe, M., Kyle, P., Fujimori, S., Havlik, P., Meijl, H., Hasegawa, T., Popp, A., Schmitz, C., Tabeau, A., Valin, H., 2014. Impacts of increased bioenergy demand on global food markets: an AgMIP economic model inter-comparison. Agric. Econ. 45, 103–116.

Martin, E.W., Chester, M.V., Vergara, S.E., 2015. Attributional and consequential life-cycle assessment in biofuels: a review of recent literature in the context of system boundaries. Curr. Sustain. Renew. Energ. Rep. 2, 82–89.

Matías, J., Encinar, J.M., González, J., González, J.F., 2015. Optimisation of ethanol fermentation of Jerusalem artichoke tuber juice using simple technology for a decentralised and sustainable ethanol production. Energy Sustainable Dev 25, 34–39.

Maung, T.A., Gustafson, C.R., 2011. The economic feasibility of sugar beet biofuel production in central North Dakota. Biomass Bioenergy 35, 3737–3747.

Meier, M.S., Stoessel, F., Jungbluth, N., Juraske, R., Schader, C., Stolze, M., 2015. Environmental impacts of organic and conventional agricultural products—are the differences captured by life cycle assessment? J. Environ. Manage. 149, 193–208.

METI, 2008. The Demonstration Project for an Area System of Eco-Fuel Utilization. Minister of Economy Trade and Industry, Japan, Available from: www.meti.go.jp/committee/materials2/downloadfiles/g81002b03j.pdf.

Mondou, M., Skogstad, G., 2012. The regulation of biofuels in the United States, European Union and Canada. Canadian Agricultural Innovation and Regulation Network (CAIRN). University of Saskatchewan, University of Toronto, Saskatoon, SK, Canada.

Moriizumi, Y., Suksri, P., Hondo, H., Wake, Y., 2012. Effect of biogas utilization and plant co-location on life-cycle greenhouse gas emissions of cassava ethanol production. J. Clean. Prod. 37, 326–334.

Mouzakitis, Y., Aminalragia-Giamini, R., Adamides, E.D., 2017. From the treatment of olive mills wastewater to its valorisation: towards a bio-economic industrial symbiosis. International Conference on Sustainable Design and Manufacturing. Springer, New York, NY, United States, pp. 267–276.

Mueller, S.A., Anderson, J.E., Wallington, T.J., 2011. Impact of biofuel production and other supply and demand factors on food price increases in 2008. Biomass Bioenergy 35, 1623–1632.

Mueller, S., 2016. Climate Benefits of US Produced Corn Ethanol. , Available from: http://www.erc.uic.edu/assets/pdf/uic_china_mexico_climate_ra.pdf.

Mussoline, W.A., Wilkie, A.C., 2015. Anaerobic digestion potential of co-products associated with ethanol production from sweet potato: a review. Ind. Biotechnol. 11, 113–126.

Nakashima, T., Ishikawa, S., 2016. Energy inputs and greenhouse gas emissions associated with small-scale farmer sugarcane cropping systems and subsequent bioethanol production in Japan. J. Life Sci. 76, 43–53.

Nemecek, T., Hayer, F., Bonnin, E., Carrouée, B., Schneider, A., Vivier, C., 2015. Designing eco-efficient crop rotations using life cycle assessment of crop combinations. Eur. J. Agron. 65, 40–51.

Nijsen, M., Smeets, E., Stehfest, E., Vuuren, D.P., 2012. An evaluation of the global potential of bioenergy production on degraded lands. Glob. Chang. Biol. Bioenergy 4, 130–147.

Numjuncharoen, T., Papong, S., Malakul, P., Mungcharoen, T., 2015. Life-Cycle GHG emissions of cassava-based bioethanol production. Energy Proceed. 79, 265–271.

Parton, W.J., Hartman, M., Ojima, D., Schimel, D., 1998. DAYCENT and its land surface submodel: description and testing. Glob. Planet. Chang. 19, 35–48.

Pongpinyopap, S., Mungcharoen, T., 2015. Bioethanol water footprint: life cycle optimization for water reduction. Water Sci. Technol. 15, 395–403.

Qin, Z., Dunn, J.B., Kwon, H., Mueller, S., Wander, M.M., 2016. Influence of spatially-dependent, modeled soil carbon emission factors on life-cycle greenhouse gas emissions of corn and cellulosic ethanol. Glob. Chang. Biol. Bioenergy 8, 1136–1149.

Rajagopal, D., 2017. A step towards a general framework for consequential life cycle assessment. J. Ind. Ecol. 21, 261–271.

Rajendran, K., Rajoli, S., Teichert, O., Taherzadeh, M.J., 2015. Impacts of retrofitting analysis on first generation ethanol production: process design and techno-economics. Bioproc. Biosyst. Eng. 38, 389–397.

REN 21, 2010. Renewables 2010 global status report. Available from: http://www.ren21.net/REN21ProductsServices/

Publications/GlobalStatusReport/Renewables2010GlobalStatusReport/tabid/5824/Default.aspxandgt.

Renzaho, A.M., Kamara, J.K., Toole, M., 2017. Biofuel production and its impact on food security in low and middle income countries: Implications for the post-2015 sustainable development goals. Renew. Sustain. Energ. Rev. 78, 503–516.

Roy, P., Dias, G., 2017. Prospects for pyrolysis technologies in the bioenergy sector: a review. Renew. Sustain. Energ. Rev. 77, 59–69.

Roy, P., Shiina, T., 2010. Global environment, biofuel: sustainable food production and distribution. Global Environmental Policies: Impact, Management and Effects. Nova Science Publishers, Happauge, NY, United States, 29–58.

Roy, P., Tokuyasu, K., Orikasa, T., Nakamura, N., Shiina, T., 2012. A techno-economic and environmental evaluation of the life cycle of bioethanol produced from rice straw by RT-CaCCO process. Biomass Bioenergy 37, 188–195.

Rulli, M.C., Bellomi, D., Cazzoli, A., De Carolis, G., D'Odorico, P., 2016. The water-land-food nexus of first-generation biofuels. Scientific Rep. 6, 22521.

Seabra, J.E., Macedo, I.C., Chum, H.L., Faroni, C.E., Sarto, C.A., 2011. Life cycle assessment of Brazilian sugarcane products: GHG emissions and energy use. Biofuels. Bioprod. Bioref. 5, 519–532.

Searchinger, T., Heimlich, R., Houghton, R.A., Dong, F., Elobeid, A., Fabiosa, J., Tokgoz, S., Hayes, D., Yu, T.H., 2008. Use of US croplands for biofuels increases greenhouse gases through emissions from land-use change. Science 319, 1238–1240.

Sevigné-Itoiz, E., Gasol, C.M., Rieradevall, J., Gabarrell, X., 2015. Methodology of supporting decision-making of waste management with material flow analysis (MFA) and consequential life cycle assessment (CLCA): case study of waste paper recycling. J. Clean. Prod. 105, 253–262.

tef nescu-Mih il , R.O., 2016. Rural economy and bioethanol production. Sustainability 8, 1148.

Tanaka, K., Peters, G.P., Fuglestvedt, J.S., 2010. Policy update: multi-component climate policy: why do emission metrics matter? Carbon Manage. 1, 191–197.

Thatoi, H., Dash, P.K., Mohapatra, S., Swain, M.R., 2016. Bioethanol production from tuber crops using fermentation technology: a review. Int. J. Sustain. Energy 35, 443–468.

Tonini, D., Hamelin, L., Alvarado-Morales, M., Astrup, T.F., 2016. GHG emission factors for bioelectricity, biomethane, and bioethanol quantified for 24 biomass substrates with consequential life-cycle assessment. Bioresour. Technol. 208, 123–133.

Ukaew, S., Beck, E., Archer, D.W., Shonnard, D.R., 2015. Estimation of soil carbon change from rotation cropping of rapeseed with wheat in the hydrotreated renewable jet life cycle. Int. J. LCA 20, 608–622.

Väisänen, S., Havukainen, J., Uusitalo, V., Havukainen, M., Soukka, R., Luoranen, M., 2016. Carbon footprint of biobutanol by ABE fermentation from corn and sugarcane. Renew. Energy 89, 401–410.

Vasile, A.J., Andreea, I.R., Popescu, G.H., Elvira, N., Marian, Z., 2016. Implications of agricultural bioenergy crop production and prices in changing the land use paradigm: the case of Romania. Land Use Policy 50, 399–407.

Wang, M., Han, J., Dunn, J.B., Cai, H., Elgowainy, A., 2012. Well-to-wheels energy use and greenhouse gas emissions of ethanol from corn, sugarcane and cellulosic biomass for US use. Environ. Res. Lett. 7, 045905.

Weidema, B.P., Ekvall, T., Heijungs, R., 2009. Guidelines for application of deepened and broadened LCA. Deliverable D18 Work Package 5, 17.

Wenzel, H., Høibye, L., Grandal, R.D., Hamelin, L., Bird, D.N., Olesen, A.S., 2014. Carbon Footprint of Bioenergy Pathways for the Future Danish Energy System. , Available from: http://www.ens.dk/sites/ens.dk/files/undergrund-forsyning/elnaturgas-varmeforsyning/Energianalyser/nyeste/carbon_footprint_of_bioeneergy_pathways_for_the_future_danish_energy_system_-_final_280314_l.pdf.

Worldwatch Institute, 2017. United States Considers Ethanol Blend Increase. , Available from: http://www.worldwatch.org/node/6017.

Yang, J., Huang, J., Qiu, H., Rozelle, S., Sombilla, M.A., 2009. Biofuels and the Greater Mekong sub-region: assessing the impact on prices, production and trade. Appl. Energy 86, S37–S46.

CHAPTER

20

Upgrading Comparative and Competitive Advantages for Ethanol Fuel Production From Agroindustrial Crops in Developing Countries: Mexico as a Case Study

*Noé Aguilar-Rivera**, Ricardo Serna-Lagunes*, Christian Michel-Cuello**, Armín Trujillo-Mata*[†]

*University of Veracruz, Cordoba Mexico

**Autonomous University of San Luis Potosí, Rioverde, Mexico

[†]Technology Institute of Orizaba, Orizaba, Mexico

20.1 INTRODUCTION

Ethanol (ethyl alcohol) or bioethanol, C_2H_5OH (EtOH) is an internationally traded commodity. Ethanol fuel can be used in internal combustion engines with spark ignition (better known as an Otto cycle engine), directly in vehicles designed to run on pure ethanol (hydrated ethanol, which usually contains about 5% water content), or by blending dehydrated (anhydrous) ethanol with gasoline (up to 25%). Ethanol can be used to increase octane and reduce pollution from additives in unleaded gasoline, substituting for chemical additives, such as methyl tertiary butyl ether (MTBE) (Sadeghinezhad et al., 2014). Bioethanol production generally utilizes derivatives from food crops, such as corn grain and sugarcane juice or molasses, but the limited supply of these crops can lead to competition between their use in bioethanol production and food provision.

Bioethanol Production From Food Crops
http://dx.doi.org/10.1016/B978-0-12-813766-6.00020-5

Along with sustained investment and secure policy frameworks, the biofuels industry can bring transformational technologies to market. These technologies not only deliver further reductions in carbon emissions, but also open up markets for a wide range of green chemicals and biobased coproducts, ethanol chemistry, and biorefineries. This will enable biofuel producers to diversify their feedstock and helps ensure a sustainable future for stakeholders worldwide (Kiran et al., 2014; Sara et al., 2016).

The situation changes considerably with increased domestic investments and research in biofuel conversion and crop productivity improvements. Mitigating potentially adverse impacts from aggressive increases in biofuel production requires a renewed focus on crop breeding and agroecological management practices for productivity improvement in wheat, maize, and sugar crops. For this reason, using agroindustrial by-products or energy crops in marginal lands could help this issue. While some crops may be more favorable from the perspective of profitability, they may encounter binding environmental constraints (Hildbrand et al., 2017; Oke et al., 2016; Zabed et al., 2017). Therefore, feedstock has been identified as a critical success factor in the intensive ethanol industry and the transition to a green economy (Msangi et al., 2007; Wang et al., 2015).

20.2 VALUE CHAIN ETHANOL FUEL

Bioethanol as a biofuel has been the subject of large-scale scientific and research development undertaken by academics at universities and industries throughout the value chain. Chan and Reiner (2011) mentioned that there are three markets along the value chain. The first is the market for feedstock, which growers and traders integrate downstream with ethanol manufacturing in order to create a stable business for their feedstock. It is also with the motivation of pursuing additional downstream rents. The other two markets—ethanol and retail fuel—are closely linked. The size of the ethanol market is linked to the size of the retail fuel market, particularly its ethanol component. There are mainly three stages: low blend (E5 or E10), midblend (E15–E25), and high blend (E85 or E100).

Blend mandates and the integration of the whole value chain (Fig. 20.1) and an orientation towards more value-added products could be a step forward in the direction of a more biobased economy through technology, innovation, research, and development. However, many authors argue that increasing *Cellulosic Ethanol Production* is a realistic option to avoid using food and feed crops in the production of biofuels, under current agricultural productivity conditions (Gregg et al., 2017; Kim and Dale, 2004), because agricultural activities produce significant amounts of waste and by-products that can be used in the production of fermentable sugars for second-generation ethanol (2G) (Koopmans et al., 1997; Koutinas et al., 2014).

Therefore, public policies and domestic strategic programs should consider actions and management practices to increase productivity of the main crops to fulfill the domestic requirements of food, feed, and fiber, as well as to have significant surpluses for the production of ethanol (1G and 2G) (Koutinas et al., 2014). Besides, scientific planning and stakeholder participation in the ethanol value chain are necessary to avoid increase in world commodity (i.e., sugarcane, maize, and sugar beet) prices related to ethanol production, and land expansion would affect the costs of food baskets around the world (Elobeid and Hart, 2007). Higher food costs will impact food security particularly in developing countries, because they tend to be not only net exporters of primary agricultural commodities, oil, or minerals but also net importers of food and fuel (gasoline).

Therefore, it is necessary to use different frameworks or methodological approaches, internationally evaluated models and/or develop

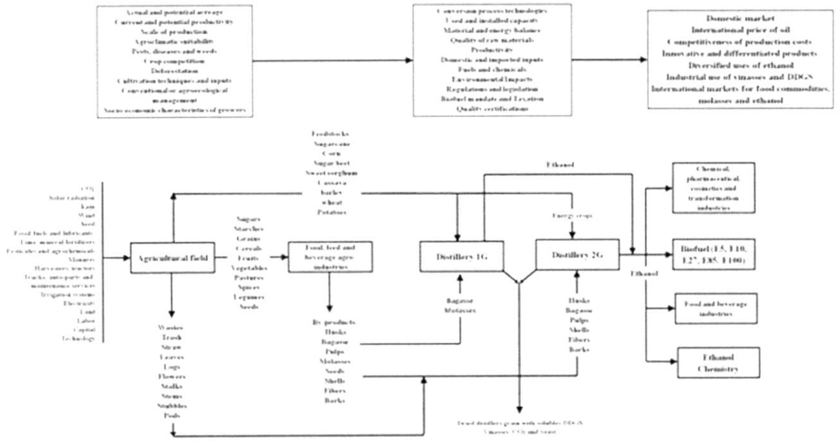

FIGURE 20.1 **Ethanol value chain.** *Source: Adapted from Aguilar-Rivera, N., Castillo-Moran, A., Enríquez-Ruvalcaba, V., Herrera-Solano, A., Milanés-Ramos, N., Rodríguez-Lagunes, D.A., 2017. Graduate education for sustainability of sugarcane biorefineries in Mexico. In: Kumar, A., Kim, D. (Eds.), Sustainability Practice and Education on University Campuses and Beyond. Bentham Science Publishers, United Arab Emirates, pp. 72–87.*

hybrid models of application to developing countries, according to the regional characteristics and constraints, which link various aspects of the biofuels, such as feedstock source location, supply chain, upstream and downstream logistics, as well as thermochemical, biochemical, processing and energy balances, sustainability, and competitiveness (Arodudu et al., 2017a,b; Friedemann, 2016; Hall et al., 2014; Kharrazi et al., 2014; Tilley, 2003). Likewise, Trujillo-Mata et al. (2016) concluded that the complexity of biofuel production nowadays forces managers to seek better techniques that can help them make more effective decisions. The last element that makes a decision hard is the existence of multiple decision makers or stakeholders (farmers, traders, government, sugar millers, distilleries, markets, flex fuel car manufacturers, and chemical industry), who have different perspectives about the same decision situation. It is important to mention that commonly two types of decisions are faced: *operating decisions* (production technology and management of raw materials and ethanol) and *strategic decisions* (stakeholder decisions, domestic public policies, and commodity geopolitics) (Fig. 20.2).

20.3 BIOFUELS FOR DEVELOPING COUNTRIES BOOST COMPARATIVE ADVANTAGES

In recent years, bioenergy and advances in technology have created numerous pathways to produce biochemicals and advanced biofuels through sugars present in agricultural crops and wastes. Moreover, the agroindustry may also provide energy security and new income to poor farmers and rural communities around the world, as production is expected to increase significantly over the next 10 years (Arodudu et al., 2017a,b; Liguori et al., 2013).

In developing countries, first-generation ethanol (fuel and nonfuel) appears unsustainable because of the potential stress that its production process places on food commodities and it largely depends on the type of feedstock and supply chain, where the crop has been cultivated and transported to a biorefinery. Food prices will be affected because increased biofuel production implies that the industry will compete with food crops for land use; however, the intensive production of biomass and conversion as feedstock to biofuel may create new jobs and

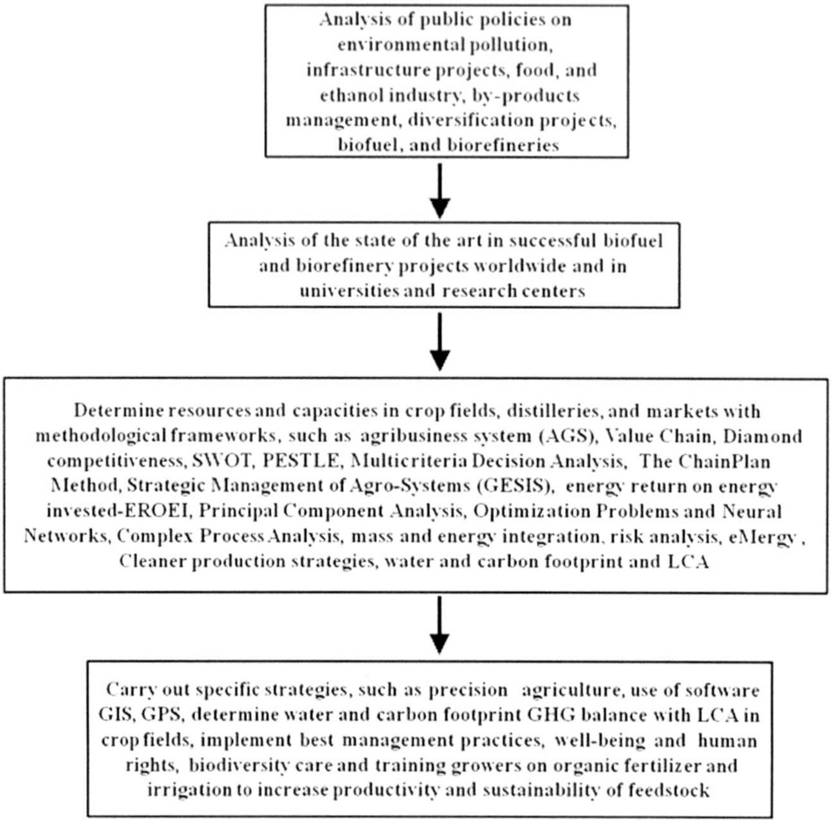

FIGURE 20.2 **Innovations for analysis of sustainably of the complexity of biofuels projects.**

increase revenue from the agricultural sector (Naqvi and Yan, 2015).

Use of agricultural feedstock in biofuels production and a sustainable supply chain remain important components of long-term demand for agricultural products that are holding prices at historically high levels. By 2022, world ethanol production is projected to increase by almost 70% compared to the 2010–12 average, which was 4% (OECD–FAO, 2016).

However, in the short term, the production of bioethanol as the most important vehicular biofuel is almost entirely dependent on the supply chain for starch and sugars from existing food and feed, but inputs from the economy (land, labor, capital, fertilizer, and fossil fuels) and the environment (water and nutrients) required to produce ethanol fuel from renewable resources at reasonable costs is reigniting heated controversies about food-versus-fuel concerns that are the drawbacks of producing bioethanol (Mathioudakis et al., 2017). But it remains too early to predict the future production of cellulosic ethanol, as the best sustainable way, because the conversion technology is still on the learning curve. Therefore, for the planning and implementation of a competitive and sustainable ethanol fuel production system, a large number of researchers using conventional and multidisciplinary methodologies have examined the impact of biogeophysical variables on site selection to increase the productivity of plantations for food, feed, and also feedstock for ethanol, in a sustainable way. Important factors include the supply chain, coproduct markets, utility costs and availability, government incentives, absence of operating ethanol plants, multimodal transportation, product markets, environmental constraints, soil erosion, nutrient loss, runoff, population, road and rail facilities, close proximity of large quantities of biomass and petroleum terminals, and the influence of political, labor, professional, and business organizations (Martinkus et al., 2017). The methodology adopted must start by gathering information regarding the feedstock availability using geographic information systems (GIS) in the region of interest. Further, understanding crop composition, variability during the growing season, harvesting, the supply chain are essential elements for evaluating the availability of convertible sugars and fiber (Aragon et al., 2013; de Jong et al., 2017).

20.4 LAND SUITABILITY FOR ETHANOL PRODUCTION IN MEXICO—A CASE STUDY

The various parameters for studying land suitability for ethanol production in Mexico are discussed in the following subsections.

20.4.1 Land and Climate as Competitive Advantage for Biofuel

Because changes in land use or an increase in arable land for ethanol purposes can have positive or negative environmental impacts (e.g., more greenhouse gases [GHG], indirect emissions, unsustainable harvesting, carbon sequestration, other land-use change effects, depleted soil nutrients, and increased water consumption). It is necessary to determine spatially and quantitatively the potential to expand arable land without jeopardizing food and feed production. Therefore, the application of successful methodological frameworks internationally recommended for this purpose is essential, because phenomena, such as soil erosion, loss of nutrients and organic matter, decrease energy efficiency, reduce biodiversity, cultural erosion and excessive use of agrochemicals, are causing this decline in soil quality (Huang et al., 2016).

Crop land has different quality conditions, which are reflected in different land suitability for agriculture. The term land suitability for agriculture is defined as the land's ability to reach potential yields in a number of crops. The levels of this classification are set as follows: prime land is able to produce 80% of the yields potentially reachable in its crops, good land is able to produce 40%–80% of its potential, and finally marginal land produces less than 40%. In order to assess the soil resources, the FAO has developed procedures, such as Agro-Ecological Zones (AEZ) for the quantification of agricultural land available, its evaluation according to quality and vocation, and finally the establishment of planning programs for the best use of these lands depending on global production trends, food demand, and feedstock for biofuel production.

The AEZ, as a tool, and novel technologies, such as Maximum Entropy Modeling (MaxEnt), the Soil and Water Assessment Tool, remote sensing, GIS, and precision agriculture, may contribute to achieving sustainability goals, supporting major strategic decisions to improve crop yields and the ethanol production value chain, and adding value through bioproducts, without affecting the production of food and feed (Dufossé et al., 2016). Analyzing climatic and edaphic information of the place in question and contrasting it with the environmental needs of the crop of interest has been used to improve the production and competitive advantages of various species worldwide.

20.4.2 Methodological Approach and Data

In Mexico, interest in producing biofuels emerged in the 2000s with the principal goal of mitigating GHG emissions in the transport sector, and achieving increased employment in the agricultural sector (Hennecke et al., 2016).

Therefore, the objective behind applying AEZ as a novel, MaxEnt-based technology is related to the classification, spatial representation, and mapping of Mexican agricultural land with respect to potential and suitability to expand productivity for feedstock for ethanol fuel from conventional main crops.

In this study, the authors have extracted geographical coordinate data from the Global Biodiversity Information Facility (GBIF, 2017) (www.gbif.org) database for the following worldwide raw material for ethanol and potential crops in Mexico: sugarcane (*Saccharum* spp.), corn (*Zea mays* L.), sugar beet (*Beta vulgaris* L.), sweet sorghum (*Sorghum bicolor* [L.] Moench), wheat (*Trilicvm aestivum* L.), cassava (*Manihot esculenta* Crantz), and barley (*Hordeum vulgare* L.).

Soil and climate conditions prevalent for the modeling, 19 climate variables, including one topographical variable (elevation) with a resolution of 30 arc sec or around 1 km^2, were used. All the variables were downloaded from the WorldClim Website (https://www.worldclim.org), where each cell contains information on each environmental layer during the last 67 years (Hijmans et al., 2005). These layers (soil, climate, and Digital Elevation Model) were cut out for the extreme coordinates: north limit: 33°, south limit 14°, east limit −86°, and west limit −119°. Additionally, nine layers were used related to Mexican soil properties at a scale of 1:1000000 (Cruz-Cárdenas et al., 2014). Also, nine bioclimatic parameters were used for the continental surface of Mexico, determined by season according to daily weather reports for the period 1902–2011, held by the National Meteorological Service. Based on ANUCLIM methodology; these layers were extracted from the Website (http://atlasclimatico.unam.mx/atlas/) of the Climate Atlas of Mexico (Fernández-Eguiarte et al., 2012).

Distribution modeling for each species using the MaxEnt software (MAXENT v. 3.4.1) was constructed, which produces a continuous binomial probability distribution representing habitat suitability according to the climate variables. MaxEnt executes log–log (clog–clog) transformation to estimate (calculate) probability of occurrence. The clog–log transform derives from the recently published interpretation of MaxEnt as an inhomogeneous Poisson process that gives it a stronger theoretical justification than the logistic transform which it replaces by default (Phillips et al., 2017).

A preliminary run with MaxEnt was performed in order to select the layers (environmental, topographic, and soil) that are the most important for the species' biological growth. That means the program was first ran with all 38 layers followed by a second run in which some variables were discriminated that did not provide any information to the ecological model, facilitating a better interpretation of the results (Soberón and Peterson, 2005). The model obtained with AUC (area under the curve) values >0.9 was evaluated, meaning that MaxEnt performed well by generating predicted niches for the presence of each species (Phillips et al., 2006). The model is a graphical output map in which the discrimination capacity of presence (sensitivity) versus the discrimination capacity of absence (specificity) is determined (Phillips et al., 2004). The color scale of the map signals the edaphoclimatic conditions for the growth of the crop. White indicates the best climate (environmental) conditions for the agroecological growth of the crop, and the probability of a suitable habitat for the crop is reduced as the color becomes more intense.

20.4.3 Competitive Advantages From Mexican Crop Lands for Ethanol Fuel

As parameters for the model, the environmental layers that have the greatest importance for the agroecological growth of the crop were selected, using an output format logistic prediction (Merow et al., 2013). In addition, the jackknife statistical test was applied to determine the relative importance of each variable in the model, as well as the requirements with the greatest agroecological importance (Phillips et al., 2006; Sokal and Rohlf, 1995). Additionally, the fraction of the areas that are predicted as the suitable habitat for the crop using the binomial probability test with its respective statistical significance of the prediction, were calculated (Elith and Leathwick, 2009)

Table 20.1 describes the characteristics of the model for each crop. The area under the curve (AUC) for all models is above 0.85 sensitivity and specificity for each species, being the receiver operating characteristic (ROC) curve for the data, which means the maximum value of the AUC is 1. The area under the ROC curve AUC was almost higher for MaxEnt, indicating better discrimination of suitable versus unsuitable areas for the species.

Table 20.2 shows the variables that have greater edaphoclimatic importance for the production of the crops those were analyzed in this study. The following tables give estimates of the relative contributions of the environmental variables to the MaxEnt model. To determine the first estimate, in each iteration of the training algorithm, the increase in regularized gain is added to the contribution of the corresponding variable, or subtracted from it if the change to the absolute value of lambda is negative. For the second estimate, for each environmental variable in turn, the values of that variable on training presence and background data are randomly permuted. The model is reevaluated on the permuted data, and the resulting drop in training AUC is shown in the table, normalized to percentages. As with the variable jackknife, variable contributions should be interpreted with caution when the predictor variables are correlated.

Figs. 20.3–20.9 are a representation of the MaxEnt model for each crop. The blue color shows a low potential of areas with conditions suitable for cultivation, while yellow represents intermediate conditions for the development of the crop, and red signifies the optimal environmental conditions for the establishment of the crop and utilization for food, feed, fiber, and fuels (4Fs).

Regarding the crop modeling, land suitability has a direct impact on crop productivity (Parthasarthy et al., 2007), and climatic variables are the most important factors of those determining the geographic distribution of a species, although soil variables are important, as well.

This study demonstrated that the habitat distribution patterns of bioenergy species in Mexico can be effectively modeled using occurrence data, environmental variables, and the edaphoclimatic requirements of species. The study presents the spatial distributions of seven bioenergy species for potential ethanol production, without affect-

TABLE 20.1 Characteristics of the Edaphoclimatic Zoning Model of Different Potential Crops for the Production of Biofuels

Crop	N	AUC	# of points used in the zoning	Regularized training gain	Coefficient of the model
Sugarcane	116	0.942	10116	1.682	7.54
Maize	721	0.910	10721	1.198	8.09
Beet	43	0.977	10043	2.748	6.46
Sorghum	270	0.865	10270	0.877	8.37
Wheat	17	0.971	10017	1.886	7.32
Cassava	186	0.942	10186	1.619	7.30
Barley	20	0.980	10020	2.818	7.33

TABLE 20.2 Relative Importance Values of the Layers of Soil and Climatic and Topographic Variables Used in Agroecological Zoning Models of Potential Crops for Biofuels

Variables	Sugarcane	Maize	Beet	Sorghum	Wheat	Cassava	Barley
Bio 1 Annual mean temperature	0	0	0	0.4	0	0.9	1.9
Bio 2 Mean diurnal range	0	0	0	15.5	5.4	0.3	4.3
Bio 3 Isothermality	7.3	0	0.3	1.4	25.8	0.1	2.8
Bio 4 Temperature seasonality	0	50	0	8.8	3.7	5	19.1
Bio 5 Max temperature of warmest month	0	5.9	11.1	0.4	0	0	0
Bio 6 Min temperature of coldest month	7.2	4.9	0	4.3	9.9	0.9	17.2
Bio 7 Temperature annual range	0	0	5.3	0.9	0	0.1	0
Bio 8 Mean temperature of wettest quarter	0	0	0	0.2	0	0.1	0
Bio 9 Mean temperature of driest quarter	6.1	4.9	5.2	1.8	0.1	1.8	0
Bio 10 Mean temperature of warmest quarter	0	0	5.2	0.7	0	0.7	0
Bio 11 Mean temperature of coldest quarter	0	0	1	0	0	1.4	0
Bio 12 Annual precipitation	22	17.8	0	0.2	3.3	0.5	0
Bio 13 Precipitation of wettest month	0	0	3.7	0.1	0	0.8	0
Bio 14 Precipitation of driest month	19.5	0	0.7	1.6	14.1	20.6	0.1
Bio 15 Precipitation seasonality	0	0	1.8	5	0	3	15.1
Bio 16 Precipitation of wettest quarter	15.6	0	0	0.3	2.6	0.1	0.6
Bio 17 Precipitation of driest quarter	7.5	1.7	0	0.1	0.2	1.4	0.3
Bio 18 Precipitation of warmest quarter	0	0	21.2	1.1	0	0.3	0
Bio 19 Precipitation of coldest quarter	0	3.1	2.3	1.6	2.1	0.9	0

DEM	6.2	2.8	9.9	4.6	20.3	4.1	12.1
K	0	2.6	1.8	3	4.4	0.8	1.7
Mg	0	0	20.9	2.2	1.8	0.3	1
Organic matter	0	0	0	3.2	0.1	1.3	0
Na	0	1.9	2	0.2	1.4	0.9	3.1
pH	1.4	4.3	0.6	0.1	3.6	1.3	0.5
CE	0	0	1.8	0	0	0.2	0
CO	0	0	0.7	2.8	0	0.8	0
Annual evaporation	0	0	0.6	1	0	46.5	0
Evaporation of wettest months	0	0	0	2.5	0	0.9	0
Evaporation of driest months	0	0	0	0	0.7	1	0.1
Annual precipitation	0	0	0	0	0	1	0
Precipitation of wettest months	0	0	0	35.5	0	0.2	0
Precipitation of driest months	0	0	0.5	0.3	0.7	0.1	0
Temperature of wettest months	0	0	1.7	0.1	0	1.1	20.3
Annual mean temperature	0	0	0	0	0	0	0
Temperature of driest months	0	0	0	0.1	0	0.5	0

DEM, Digital elevation model.

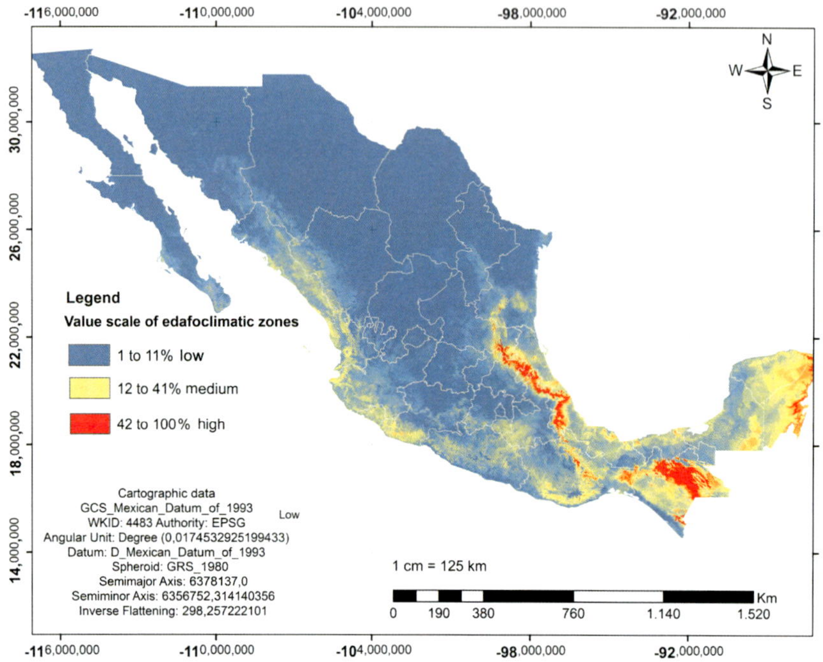

FIGURE 20.3 **Area with edaphoclimatic potential for the production of sugarcane.**

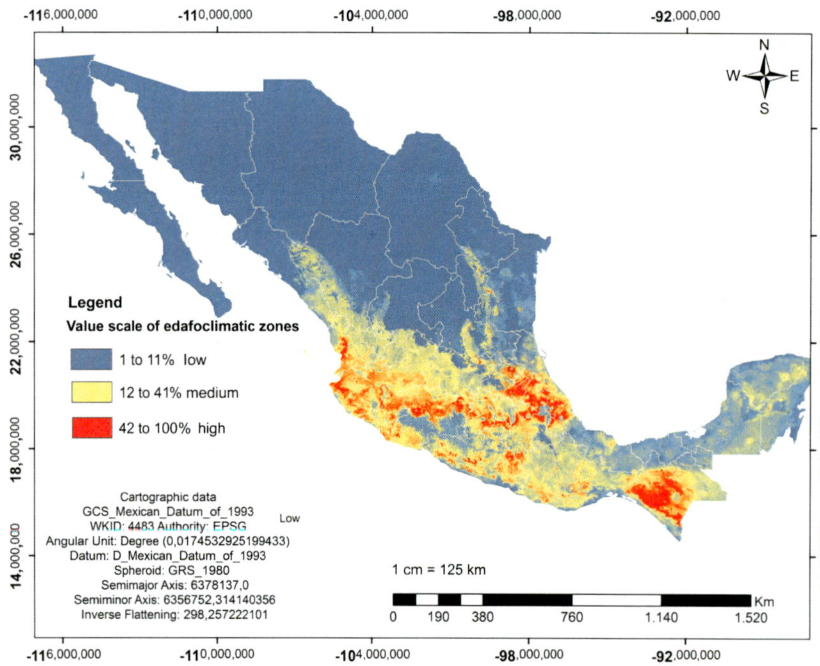

FIGURE 20.4 **Area with edaphoclimatic potential for the production of maize.**

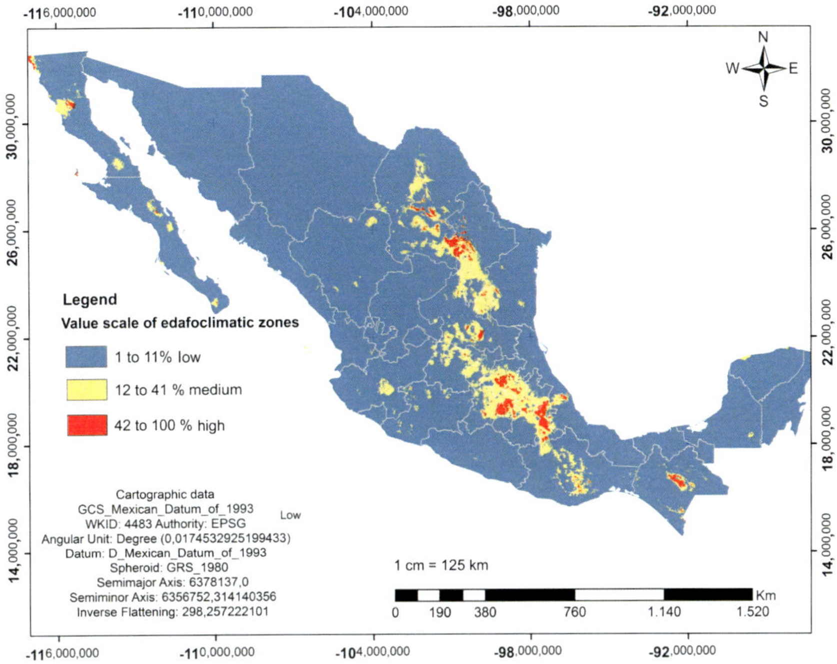

FIGURE 20.5 Area with edaphoclimatic potential for the production of sugar beet.

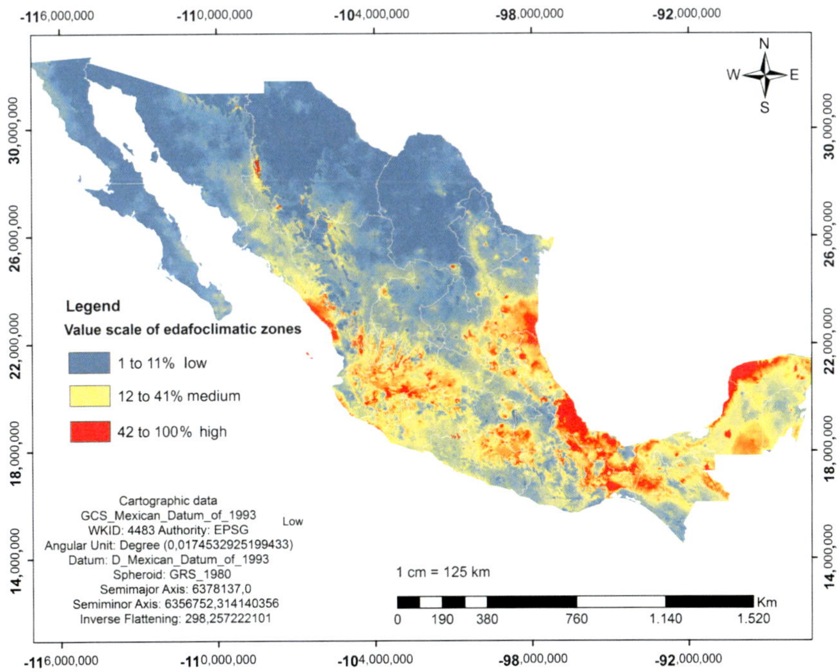

FIGURE 20.6 Area with edaphoclimatic potential for the production of sorghum.

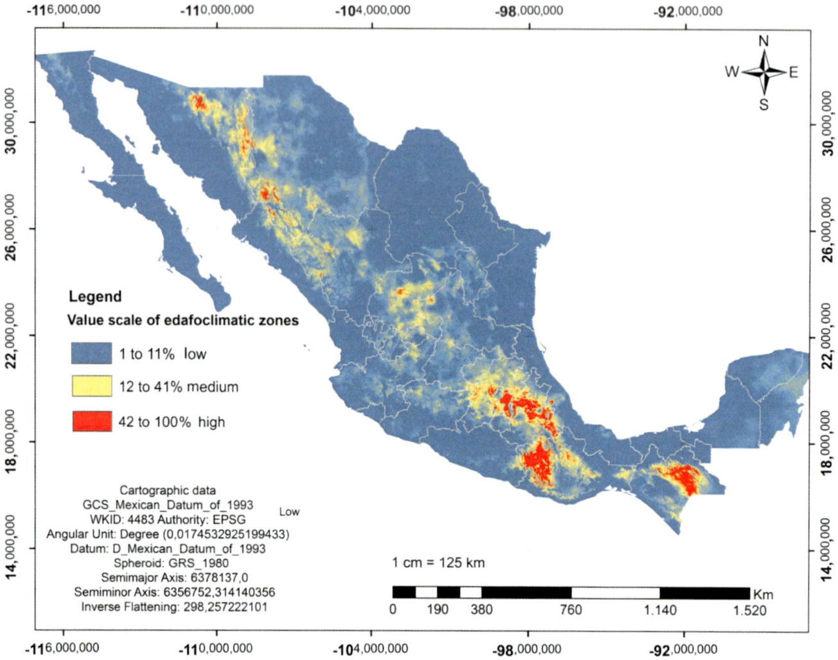

FIGURE 20.7 **Area with edaphoclimatic potential for the production of wheat.**

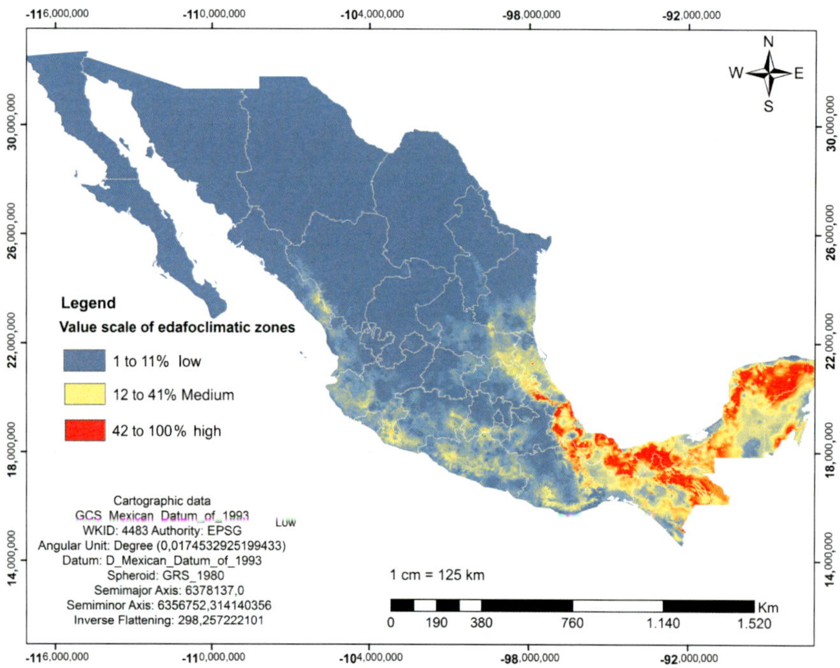

FIGURE 20.8 **Area with edaphoclimatic potential for the production of cassava.**

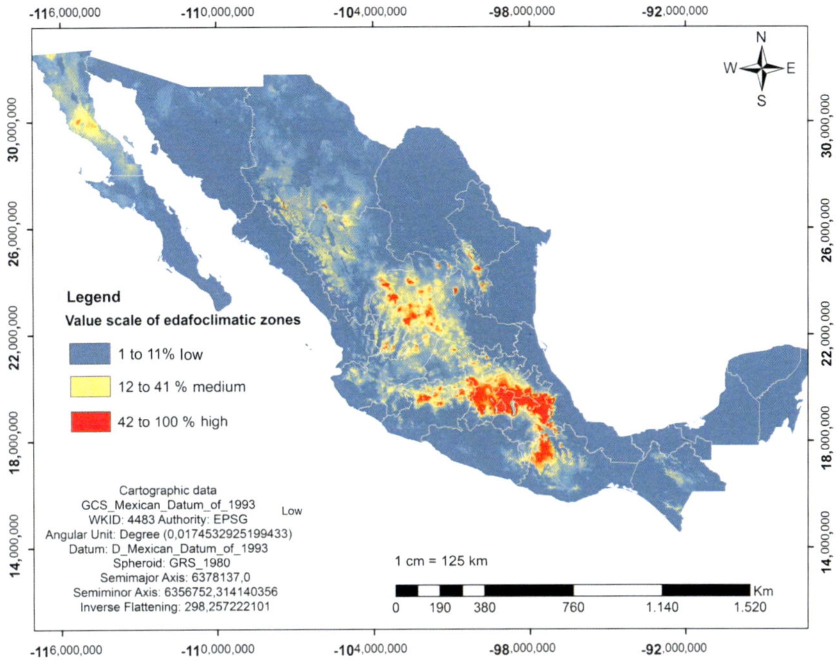

FIGURE 20.9 **Area with edaphoclimatic potential for the production of barley.**

ing current production of food and feed and food security; moreover, areas were identified and zoned to meet sustainability objectives based on the comparative advantages of climate and land and on their performance functions and independent explanatory variables. In addition, Mexico needs to encourage the introduction of the flexible fuel engines that run with any concentration of gasoline and ethanol blend. Automobiles that can run with variable fuel engines will be attractive to many customers due to their ability to choose between gasoline and ethanol fuel on the basis of price.

20.5 FINAL REMARKS

Mexico has great potential for ethanol fuel production based on designating the use of agricultural lands as a function of economic, ecological and social aspects, primordial factors in the pursuit of sustainable development. The proper application of this tool has allowed determining comparative advantages, such as addressing the proper application of land zoning of a crop to obtain the best yields for food, feed, fiber, and feedstock for biofuel and potential land expansion based on an assessment of the agroecological suitability. Because each plant crop has different requirements, the land classified as unsuitable may be reconverted to other uses, such as livestock grazing and energy crop cultivation. The results of this case study are promising, because any technology available to increase crop yield per unit land and decrease inputs, particularly fossil fuels, fertilizer, and water, should be of value to the future development of the biofuel industry. Bioenergy crops will steadily accumulate biomass over time, positively affecting soils and preserving the country's natural resources. Biomass is an abundant resource in Mexico, but the country strongly depends on imported energy sources (gasoline, diesel, MTBE, Ethyl tert-butyl ether (ETBE), natural gas, and petrochemicals), for transport, food production and industrialization despite being one of the world's top 10 oil producers. In a final remark, Mexico must aim to increase its use of biomass as an energy source.

References

Aguilar-Rivera, N., Castillo-Moran, A., Enríquez-Ruvalcaba, V., Herrera-Solano, A., Milanés-Ramos, N., Rodríguez-Lagunes, D.A., 2017. Graduate education for sustainability of sugarcane biorefineries in Mexico. In: Kumar, A., Kim, D. (Eds.), Sustainability Practice and Education on University Campuses and Beyond. Bentham Science Publishers, Emirate of Sharjah, United Arab Emirates, pp. 72–87.

Aragon, D., Suhr, M., Kochergin, V., 2013. Evaluation of energy cane and sweet sorghum as feedstocks for conversion into fuels and chemicals. Sugar Ind. Zucker. 138 (10), 651–655.

Arodudu, O.T., Helming, K., Voinov, A., Wiggering, H., 2017a. Integrating agronomic factors into energy efficiency assessment of agro-bioenergy production: a case study of ethanol and biogas production from maize feedstock. Appl. Energy 198, 426–439.

Arodudu, O., Helming, K., Wiggering, H., Voinov, A., 2017b. Towards a more holistic sustainability assessment framework for agro-bioenergy systems: a review. Environ. Impact Assess. Rev. 62, 61–75.

Chan, J.H., Reiner, D., 2011. Dynamics of Evolution in the Global Fuel-Ethanol Industry. Available from: https://www.repository.cam.ac.uk/handle/1810/242044.

Cruz-Cárdenas, G., López-Mata, L., Ortiz-Solorio, C.A., Villaseñor, J.L., Ortiz, E., Silva, J.T., Estrada-Godoy, F., 2014. Interpolation of Mexican soil properties at a scale of 1:1,000,000. Geoderma 213, 29–35.

de Jong, S., Hoefnagels, R., Wetterlund, E., Pettersson, K., Faaij, A., Junginger, M., 2017. Cost optimization of biofuel production: the impact of scale, integration, transport and supply chain configurations. Appl. Energy 195, 1055–1070.

Dufossé, K., Drouet, J.L., Gabrielle, B., 2016. Agro-ecosystem modeling can aid in the optimization of biomass feedstock supply. Environ. Model. Softw. 85, 139–155.

Elith, J., Leathwick, J.R., 2009. Species distribution models: ecological explanation and prediction across space and time. Annual review of ecology, evolution, and systematic. 40, 677–697.

Elobeid, A., Hart, C., 2007. Ethanol expansion in the food versus fuel debate: how will developing countries fare? J Agr. Food Ind. Org. 5 (2), 1–21, 6.

Fernández-Eguiarte, A., Zavala-Hidalgo, J., Romero-Centeno, R, 2012. Meteorological drought in the Digital Climatic Atlas of Mexico, ICAN. Newslett. Int. Coast. Atlas Netw. 1 (1), 13.

Friedemann, A.J., 2016. Post fossil fuels, if biomass is the "answer to everything," is there enough? When Trucks Stop Running. Springer, New York, pp. 41–44.

GBIF, 2017. www.gbif.org (June 3, 2017). GBIF Occurrence Download. Available from: http://doi.org/10.15468/dl.rbnuyg.

Gregg, J.S., Bolwig, S., Hansen, T., Solér, O., Ben Amer-Allam, S., Pladevall Viladecans, J., Fevolden, A., 2017. Value chain structures that define European cellulosic ethanol production. Sustainability 9 (1), 1–17, 118.

Hall, C.A., Lambert, J.G., Balogh, S.B., 2014. EROI of different fuels and the implications for society. Energy Policy 64, 141–152.

Hennecke, A.M., Mueller-Lindenlauf, M., García, C.A., Fuentes, A., Riegelhaupt, E., Hellweg, S., 2016. Optimizing the water, carbon, and land-use footprint of bioenergy production in Mexico—six case studies and the nationwide implications. Biofuels Bioprod. Biorefin. 10 (13), 222–239.

Hijmans, R.J., Cameron, S.E., Parra, J.L., Jones, P.G., Jarvis, A., 2005. Very high resolution interpolated climate surfaces for global land areas. Int. J. Climatol. 25, 1965–1978.

Hildbrand, S., Hildbrand, S., Bodhanya, S., Bodhanya, S., 2017. Exploring the complexity of sugarcane supply chains via systemic approaches. Kybernetes 46 (2), 310–329.

Huang, B., Li, J., Fang, W., Liu, P., Guo, M., Yan, D., Wang, Q., Cao, A., 2016. Effect of soil fumigation on degradation of pendimethalin and oxyfluorfen in laboratory and ginger field studies. J. Agric. Food Chem. 64, 8710–8721.

Kharrazi, A., Kraines, S., Hoang, L., Yarime, M., 2014. Advancing quantification methods of sustainability: a critical examination energy, exergy, ecological footprint, and ecological information-based approaches. Ecol. Indic. 37, 81–89.

Kim, S., Dale, B.E., 2004. Global potential bioethanol production from wasted crops and crop residues. Biomass Bioenergy 26 (4), 361–375.

Kiran, E.U., Trzcinski, A.P., Ng, W.J., Liu, Y., 2014. Bioconversion of food waste to energy: a review. Fuel 134, 389–399.

Koopmans, A., Koppejan, J., 1997. Agricultural and Forest Residues—Generation, Utilization and Availability. Available from: http://www.fao.org/docrep/006/AD576E/ad576e00.pdf.

Koutinas, A.A., Vlysidis, A., Pleissner, D., Kopsahelis, N., Garcia, I.L., Kookos, I.K., Lin, C.S.K., 2014. Valorization of industrial waste and by-product streams via fermentation for the production of chemicals and biopolymers. Chem. Soc. Rev. 43 (8), 2587–2627.

Liguori, R., Amore, A., Faraco, V., 2013. Waste valorization by biotechnological conversion into added value products. Appl. Microbiol. Biotechnol. 97 (14), 6129–6147.

Martinkus, N., Rijkhoff, S.A., Hoard, S.A., Shi, W., Smith, P., Gaffney, M., Wolcott, M., 2017. Biorefinery site selection using a stepwise biogeophysical and social analysis approach. Biomass Bioenergy 97, 139–148.

Mathioudakis, V., Gerbens-Leenes, P.W., Van der Meer, T.H., Hoekstra, A.Y., 2017. The water footprint of second-generation bioenergy: a comparison of biomass feedstocks and conversion techniques. J. Cleaner Prod. 148, 571–582.

Merow, C., Smith, M.J., Silander, J.A., 2013. A practical guide to MaxEnt for modeling species' distributions: what it does, and why inputs and settings matter. Ecography 36, 1058–1069.

Msangi, S., Sulser, T., Rosegrant, M., Valmonte-Santos, R., 2007. Global scenarios for biofuels: impacts and implications for food security and water use. 10th Annual Conference on Global Economic Analysis. Purdue University, West Lafayette, IN. p. 16.

Naqvi, M., Yan, J., 2015. First-Generation Biofuels Handbook of Clean Energy Systems. Wiley, Hoboken, NJ, pp. 1–18.

OECD–FAO, 2016. Agricultural Outlook Biofuel—OECD–FAO Agricultural Outlook 2016–2025. Available from http://stats.oecd.org/index.aspx?queryid=71231.

Oke, M.A., Annuar, M.S.M., Simarani, K., 2016. Mixed feedstock approach to lignocellulosic ethanol production: prospects and limitations. BioEnergy Res. 9 (4), 1189–1203.

Parthasarthy, U., Johny, A., Jayarajan, K., Parthasarathy, V., 2007. Site suitability for turmeric production in India—a GIS interpretation. Nat. Prod. Rad. 6, 142–147.

Phillips, S.J., Dudík, M., Schapire, R.E., 2004. A maximum entropy approach to species distribution modeling. In: Proceedings of the Twenty-First International Conference on Machine Learning. ACM. p. 83.

Phillips, S.J., Anderson, R.P., Schapire, R.E., 2006. Maximum entropy modeling of species geographic distributions. Ecol. Model 190 (3), 231–259.

Phillips, S.J., Anderson, R.P., Dudík, M., Schapire, R.E., Blair, M.E., 2017. Opening the black box: an open-source release of Maxent. Ecographydoi: 10.1111/ecog.03049.

Sadeghinezhad, E., Kazi, S.N., Badarudin, A., Togun, H., Zubir, M.N., Oon, C.S., Gharehkhani, S., 2014. Sustainability and environmental impact of ethanol as a biofuel. Rev. Chem. Eng. 30 (1), 51–72.

Sara, M., Rouissi, T., Brar, S.K., Blais, J.F., 2016. Life cycle analysis of potential substrates of sustainable biorefinery. In: Brar, S.K., Sarma, S.J., Rajan, Pakshi (Eds.), Platform Chemical Biorefinery. Elsevier, Amsterdam, Netherlands, pp. 55–76.

Soberón, J., Peterson, A.T., 2005. Interpretation of models of fundamental ecological niches and species' distributional areas. Biodiver. Inform. 2, 1–10.

Sokal, R.R., Rohlf, F.J., 1995. Biometry: The Principles and Practice of Statistics in Biological Research. State University of New York, Stony Brook, New York.

Tilley, D.R., 2003. Industrial ecology and ecological engineering. J. Ind. Ecol. 7, 13–32.

Trujillo-Mata, A., Cortés-Robles, G., Sánchez-Ramírez, C., Blanco-Fernández, J., Jiménez-Macías, E., 2016. An approach to estimate the water footprint of the bioethanol supply chain and its dynamic simulation. 4th International Workshop on Simulation for Energy, Sustainable Development and Environment, SESDE.

Wang, Z., Wang, Q., Wennersten, R., Sun, Q., 2015. Transitions to sustainable energy and material systems: outline of principles for scenarios. Energy Procedia 75, 2683–2693.

Zabed, H., Sahu, J.N., Suely, A., Boyce, A.N., Faruq, G., 2017. Bioethanol production from renewable sources: current perspectives and technological progress. Renew. Sustain. Energy Rev. 71, 475–501.

Further Readings

Beckman, J., Keeney, R., Tyner, W., 2011. Feed demands and co-product substitution in the biofuel era. Agribusiness 27 (1), 1–18.

FAOSTAT, 2017. Food and agriculture data. Food and Agriculture Organization of the United Nations. http://www.fao.org/faostat/en/#data.

Mohr, A., Raman, S., 2013. Lessons from first generation biofuels and implications for the sustainability appraisal of second generation biofuels. Energy Policy 63, 114–122.

RFA, 2017. World Fuel Ethanol Production. Renewable Fuels Association. Available from: http://www.ethanolrfa.org/resources/industry/statistics/#1454099788442-e48b2782-ea53.

CHAPTER

21

Bioethanol in Brazil: Status, Challenges and Perspectives to Improve the Production

Ana K. de Souza Abud, Carlos E. de Farias Silva***

*Department of Food Technology, Federal University of Sergipe, São Cristóvão, Brazil

**Institute of Chemistry and Biotechnology, Federal University of Alagoas, Maceió, Brazil

21.1 INTRODUCTION

The global energetic matrix currently consists of fossil carbon sources (about 87%), with about 33% of oil, 30% of coal, and 24% of natural gas. Renewable energy accounts for only 10% of the global energy. In particular, the growing need to expand the use of renewable energy sources (e.g., biofuels) in a sustainable manner has boosted biofuel production worldwide (Moraes et al., 2015).

Biofuel production is vigorously growing in the 21st century and the United States along with Brazil together account for about 70% of world production, usually reported as Mtoe (million tonnes of oil content). Thinking of a carbohydrate-rich biomass, bioethanol is the main current application due to its worldwide application, that is, it is the biofuel that is more produced. In fact, the global ethanol production is comprised 45–50 Mtoe (more than 100 million m^3–25 billion gal per year). The United States and Brazil hold hegemony in ethanol production, accounting for almost 85% (worldwide production), using corn and sugarcane as feedstock, respectively. However, this amount is less than 1% of the overall fuel consumed (Belicanta et al., 2016; Worldwatch Institute, 2007).

Bioethanol production is divided into four different generations: the 1G biofuels are produced from saccharine and starch feedstock (sugarcane, beetroot, and corn); 2G biofuels from lignocellulosic biomass; the 3G from the use of micro/macroalgae biomass; and, finally, the 4G, which produces biofuel from genetically modified cyanobacteria through a process denominated 'photofermentation' (direct conversion of light and carbon dioxide into ethanol) (Silva and Bertucco, 2016).

Bioethanol Production From Food Crops
http://dx.doi.org/10.1016/B978-0-12-813766-6.00021-7

In the case of Brazil, sugarcane is the main feedstock used, due to its lower production costs at an international level (Gupta and Verna, 2015), when compared to the other two principal alternative crops-corn (United States) and beetroot (Europe). Given the increased productivity of sugar and ethanol per unit area, both produce a significant volume of residual biomass, consisting of bagasse, straw, and spikes, which may be used as fuel in the generation of steam and electricity, making the units energetically self-sufficient and, even, energy exporters (Tolmasquim, 2016).

According to the *Brazilian Institute of Geography and Statistic*-Instituto Brasileiro de Geografia e Estatística (IBGE), in the 2016 harvest, the country reached a production of 74.77 tonnes of cane per acre. Moreover, the projections for 2G and 3G technologies, as well as of energy cane, have increased sharply, with several actions being taken to increase production and/or prospection in Brazil. The technology of 2G ethanol has only recently reached an industrial scale, with its energetic/economic validation still under development. Regarding micro and macroalgae, although there are many works emphasising the potential use of these biomasses, an adapted technology is yet to be developed (Hong et al., 2014; Silva and Bertucco, 2016; Silva and Sforza, 2016; Wei et al., 2013).

As for fourth-generation ethanol, little information is available regarding this type, and what is available is protected by international patents, with its use being limited to plants in the United States, which has suffered a great crisis in recent years. Therefore, little is known regarding a real perspective of industrial use for this class of modified microorganisms.

21.2 HISTORICAL BACKGROUND AND GEOGRAPHIC IMPOSITIONS TO SUGARCANE EXPLOITATION

In Brazil, sugarcane (*Saccharum* spp.) cultivation started straight after its discovery, approximately in 1532, by Martim Afonso de Souza, founder of Vila de São Vicente, in São Paulo, founding the first sugar mill in Brazil, named the São Jorge Mill. 20 years afterward, Brazil became the biggest world producer of sugar, exporting the product mainly to Europe, with its planting system mainly relying on slave labor (Rodrigues, 2012).

As fuel, sugarcane ethanol began to gain importance by the end of the 1920s, with the introduction of the first automotive vehicles in Brazil and the industrial revolution in full expansion, when the consumption of liquid fuels went through exponential growth, forcing the country to create management mechanisms in order to improve its ethanol production systems. The first state intervention in this productive segment was made by the *Institute of Sugar and Alcohol* – Instituto do Açúcar e Álcool (IAA) Foundation, in 1933, making it easier to obtain funds for the implementation of distilleries annexed to the sugarcane mills, as well as the creation of Decree-Law no. 737 (1938), which made the 5% (v/v) of anhydrous ethanol compulsory in commercial gasoline, as a replacement of the antidetonating additive tetraethyl lead [$Pb(C_2H_5)_4$], forbidden to be added to gasoline in Brazil and in the European Union. From March 2015, according to Decree no. 75 (2015), from the *Ministry of Agriculture, Livestock and Supply* – Ministério da Agricultura, Pecuária e Abastecimento (MAPA) and Resolution no. 1 (2015), from the *Inter-Ministerial Council for Sugar and Alcohol* – Conselho Interministerial do Açúcar e do Álcool (CIMA), the mandatory percentage of anhydrous ethanol fuel in regular gasoline, required elevate octane rating was set to 27%, and to 25% in premium gasoline.

Another governmental intervention of great impact in the country took place in the end of the 1970s, becoming an international reference in the market of biofuels, named the *National Alcohol Programme* – Programa Nacional do Álcool (Proálcool), consolidating and widening the national sugarcane industry. Considered a pioneer program of biofuels, Proálcool was mainly aimed at producing an alternative national fuel, given the oil crises of 1973 and 1979, considerably increasing its production and with a strong impact on the balance of payments (Tolmasquim, 2016). Based on government incentives, in 1989 the country had already roughly 4 million vehicles moved by ethanol-a third of the Brazilian fleet at the time.

The main species of cultivated sugarcane, *Saccharum officinarum*, has as predominant characteristics the high levels of sugar and low levels of fiber, with several varieties being sown throughout Brazil. In order to increase productivity, programs of genetic improvement of sugarcane have prioritized the rise in sucrose levels, with the development of regional varieties adapted to specific environments (Novacana, 2016a,b,c,d). Each variety, therefore, behaves according to the environmental conditions and climate of the cultivation area, as well as to possible outbreaks of pests and diseases and harvesting system. While in 1961 the average productivity of cane fields reached only 43 tonnes per acre, in 2016 it had already reached almost 75 tonnes of cane per acre, consequence of great technological development in this field.

In Brazil, genetic improvement and the creation of new varieties of cane, both by public and private entities, with a great interaction with the productive sector, have become even more frequent over the past years. The main research centres in this field are the *Agronomic Institute of Campinas* – Instituto Agronômico de Campinas (IAC), the *Center of Sugarcane Technology* – Centro de Tecnologia Canavieira (CTC) and the *Inter-university Network for the Development of the Sugar-alcohol Sector* – Rede Interuniversitária para o Desenvolvimento do Setor Sucroenergético (Ridesa), a group consisting of nine universities, as well as the company CanaVialis, acquired by the Monsanto group (www.monsanto.com) in 2008. Together, these institutions have provided the market between 2010 and 2013 with 30 new varieties of sugarcane (Cherubin, 2013).

The existence of a great variety of species is seen as an advantage, although it makes the choice of which one to use challenging, as it requires a greater knowledge from the rural producer regarding all available options. It is, therefore, important for the producer to sow a great variety of sugarcane types, thus reducing the possibility of outbreaks of pests or of diseases spreading over the cultivation area, which could cause great damage to production (Novacana, 2016a,b,c,d).

As for the production process of ethanol from sugarcane, the conventional system uses as feedstocks sugarcane juice and molasses, mixed in different proportions, with both being obtained after the processes of sugarcane milling, treatment, and boiling of the sugarcane juice. Some distilleries use only juice, while others only molasses, but the mixture is considered to be a better substrate, because the juice has some nutritional deficiencies, whereas molasses has inhibitory compounds for yeast fermentation (Basso et al., 2011). The process consists in three different main steps: fermentation, centrifugation, and distillation. The most frequent form of fermenting broth (conducted in 83% of sugar mills) is known as the Melle-Boinot process, with 8%–10% of ethanol being reached after an 8 h process, by means of centrifugation, to separate the wine from the yeast mixture. This is then recycled and sent to the fermenting vats for fermentation, while the wine obtained from centrifugation is sent to distillation columns for ethanol concentration. The remaining 17% of sugarcane mills use a continuous process, where the must/broth feed and withdrawal of fermented medium are processed continuously and under the same flow rate (Lopes et al., 2016). Recycling the cells enables the start process at higher concentrations of cells (8%–12% v/v), being the main reason why fermentation time periods in sugarcane mills in Brazil are much lower (6–12 h) in fermenting processes, when compared with the USA (54–72 h), where yeast cells are not recycled. In general, ethanol is obtained in its hydrous form, until its azeotropic point (95°GL), with the possibility of going through a dehydration process (molecular sieves or liquid–liquid process) for obtaining anhydrous ethanol (Basso et al., 2011; Lima, 2011; Lopes et al., 2016).

The production of ethanol from sugarcane relies mainly on the seasonality of the agricultural production and, being a culture that grows throughout the entire year, sugarcane crops suffer from all kinds of challenging climate conditions. It also relies on the characteristics of the plant harvested and, mainly on ethanol concentration contained in sugarcane, expressed in terms of total recoverable sugar/kg of sugarcane, which depends on the extraction process, steam generation capacity, as well as on energy consumption, linked to the production profile of the industrial plant (sugar, ethanol, and producing and exporting energy units). In general, large industrial plants are able to process, on average, 4.5–10.5 Mt of sugarcane per harvest season (Novacana, 2017).

Due to the great extension of the Brazilian territory, as well as to climate conditions and the maturation cycle of the plant (in general with five cuts and sown during one and a half years), sugarcane is harvested in the central–south region of the country, approximately between March and October, while in the north and northeastern regions, harvest occurs in the period between the off-season of the central–south region. Throughout the harvesting period, sugar content reaches maximum levels in stems and, even if not harvested, the plant will use this sugar storage to grow again (producing fibers), until the beginning of the following season (Tolmasquim, 2016).

In 2015, Brazil used roughly 9 million acres to plant sugarcane, resulting in a production of 658.4 Mt, generating 177.8 Mt of bagasse, derived from crushed sugarcane stalks, and 102.1 Mt of straws and spikes from harvest (Tolmasquim, 2016). In general, 50–70% of bagasse is used as fuel source for the boilers, in order to meet energy demands in plants, with the remaining bagasse (surplus bagasse) being commercialized or stocked to be used as fuel for sugar mills in the beginning of the subsequent harvest period. Part of the spikes and sugarcane straw should remain in the field, for agronomic purposes, to be used in soil conservation (Ferreira-Leitão et al., 2010); while the remaining parts, together with bagasse, are stored at an open patio, which can be taken to boilers to be burned as fuel source for steam and energy, given the residue's considerable calorific value (Saad, 2010). The use of straws and spikes for energy generation has become more significant over time, as the still common practice of burning those in cane fields to help manual harvest, culminating in great air pollution, is being increasingly inhibited under new legislations and deals established between public and private entities, therefore intensifying the use of mechanical harvesting.

Over the past years, driven by the development of 2G ethanol (second-generation ethanol), new generations of sugarcane, with higher fiber content, in detriment of sugar levels, have been developed based on another species of sugarcane, *Saccharum spontaneum*. The varieties developed from this species are denominated energy cane, opposed to sugarcane, appearing in the 1970s, as part of improvement programmes in countries like Barbados, India, Cuba, and Australia; in Brazil, energy cane only began to be explored in the 1980s, aimed at energy cogeneration. Nevertheless, only after 2002 the sector, facing a more solid market, grew interest in the development of other cane by-products, as is the case of 2G ethanol (Novacana, 2016a,b,c,d).

Vignis (www.vignis.com.br), originated from the company CanaVialis and founded in 2012, became a pioneer in the development of this product, conducting the biggest genetic improvement program of sugarcane in the world, focused on a greater production of fiber (bagasse) and sugar per acre. Ridesa and IAC began their research on energy cane between 2008 and 2009, respectively. GranBio, responsible for the first industrial plant of 2G ethanol in Brazil, developed the Vertix variety of energy cane, in partnership with IAC and Ridesa at the start of the improvement program (Cherubin, 2016).

Energy cane is an increasingly more productive variety, with the capacity of generating sugar and biomass at unprecedented amounts and costs ever seen in Brazil or elsewhere in the world. While for the traditional sugarcane, the multiplication ratio that generates new plants is of 1:10, reaching 1:5 in mechanical harvest, for energy cane this ratio reaches 1:30 or more. Besides that, the latter grows in diverse environments, even under improper growth conditions (dry, cold weather, etc.) (Cherubin, 2016).

According to the president director of the *Brazilian Sugarcane Industry Association* – União da Indústria de Cana-de-Açúcar (UNICA), Elizabeth Farina, the Brazilian proposal for Intended Nationally Determined Contributions, established during the 21st Conference of the Parties, which took place in December 2015, in Paris, was one of the boldest schemes; proposing to expand the use of electric energy from renewable sources to 18% until 2030, therefore producing 50 billion L of ethanol, in comparison with the current 30 million L produced, implying a reduction of 43% in the emission greenhouse gases (GHG) until 2030 (in comparison with the emissions from 2005) (UNICA, 2015a,b). The sugar–energy industry for the use of ethanol and sugarcane biomass in the country is currently responsible for 16% of the national energy matrix, representing 40% of the total renewable energy domestically offered, making use of only 0.5% of the Brazilian territory for the cultivation of sugarcane. Sugarcane straw and bagasse, by-products of this process, can also generate renewable and clean electric energy for the national system, which would therefore avoid an estimated 10 Mt of CO_2 emissions, a figure that would only be reached by cultivating 70 million native species over a period of 20 years (UNICA, 2016).

With this in mind, companies responsible for the production of 2G ethanol have used energy cane as feedstock, given its high biomass content per acre, associated with the flexible levels of sugar in stalks, making it more resilient to the continuous fluctuations in the prices of ethanol, sugar, and energy (Cherubin, 2016).

21.3 LIGNOCELLULOSIC ETHANOL IN BRAZIL: CURRENT STATUS, NEEDS, AND LIMITATIONS

Lignocellulosic materials are mainly used in 2G processes, being resistant to saccharification, especially due to lignin content, recalcitrant compound, nonfermentable, and difficult to be biologically degraded. Thus strong pretreatment processes and the use of several enzymes are needed, increasing production costs and lowering productivities, affecting a large-scale consolidation in the market. The costs of bioethanol based on lignocellulosic materials (USA, 0.43–0.93 US$/L) are still higher than 1G biofuels Sugarcane - Brazil, 0.16-0.22 US$/L, even considering corn (USA, 0.25–0.40 US$/L) and sugar beet (Europe, 0.43–0.73 US$/L) (Gupta and Verna, 2015).

The production of lignocellulosic ethanol is seen as an advantage mainly considering three aspects:

1. Directing agricultural, urban and forest residue for the production of bioethanol;
2. No competition with food supply;
3. A partial increase in productivity of ethanol for the same cultivated area.

Regarding the existing crops used in the production of ethanol, such as corn and sugarcane, the use of lignocellulosic residue complies with these three basic aspects, increasing its potential in biorefineries, which can even double the production capacity of ethanol (Santos et al., 2012).

The conversion of lignocellulosic material consists mainly of three different stages: pretreatment, enzymatic hydrolysis, and fermentation. The pretreatment stage is considered extremely vital for the biomass to be unstructured, given the recalcitrant structure of lignocellulose, removing hemicellulose, and reducing the negative influence/presence associated with lignin, improving the access of enzymes to cellulose. Therefore, pretreatment, besides its energy cost and use of chemical compounds (which must be reduced to a minimum), must exhibit the following essential functions:

- Breaks the bond between polymers (cellulose–hemicellulose–lignin);
- Reduces the crystallinity of cellulose;
- Avoids sugar degradation (degradation of hemicellulose and cellulose);

- Avoids fermentation-inhibiting compounds formation (usually products of sugar degradation or from breaking down lignin).

Thus, for the pretreatment step to be considered efficient, it must generate a substrate rich in cellulose, of easy hydrolysis, and with lower loss of soluble sugar in secondary reactions and lower generation of compounds potentially toxic for the bioprocess, being economically competitive when recovering hemicellulose and lignin for other processes. To improve the enzymatic hydrolysis process, strategies for increasing the velocity of product release for the same enzyme (turnover number), as well as to increase the stability of enzymes and to minimize/exclude the inhibition effects have been studied (Mosier et al., 2005; Rodrigues et al., 2013; Silva and Bertucco, 2016).

With the technologies currently available, the pretreatment is the most challenging step, as none of the methods used are universal, varying significantly between the types of biomass and other technological factors, with each method having its own advantages and disadvantages. The methods considered most promising are of dilute and alkaline acid and steam explosion.

In enzymatic hydrolysis, feedstock is pretreated and mixed to the enzymes (biocatalysts), which break the long chain of sugar molecules converting cellulose and hemicellulose in individual monomers. This is not a simple process, given that in order to carry out the total deconstruction of the biomass, a combination of different classes of enzymes is required (enzyme cocktail), in which each enzyme of this complex can demand a particular optimum pH condition, as well as an optimum temperature to reach maximum performance (Agostinho et al., 2015; Cunha et al., 2017; Rodrigues et al., 2013).

For the conversion of cellulose in glucose, for instance, an enzyme complex is required, originated from three different groups: endoglucanases, exoglucanases, and cellobioses (β-glucosidases). The endoglucanases break the cellulose chain, while exoglucanases allowing further crack down and form tetra, tri (cellotriose), or dissacharide (cellobiose), bonded glucose molecules. β-glucosidases are responsible for separating these molecules, generating the sugar available for fermentation (Rodrigues et al., 2013).

As a result of this, enzyme technology, a tool that enables making the process of 2G ethanol production more efficient, is an environmentally favorable process, with low energy consumption and high specificity, being a subject of many studies by several multinationals and research institutions worldwide. Once the group of potential enzymes (enzyme cocktails) is identified, the task of making one single microorganism to produce them more efficiently and at an industrial scale begins.

Regarding the fermentation step, the main restriction is in handling the sugar removed in the retreatment, mainly pentose originated from the hemicellulosic part, which is not naturally fermented by *Saccharomyces cerevisiae*, main industrial yeast (Silva and Bertucco, 2016).

In enzymatic hydrolysis and fermentation, the main difficulties in consolidating and minimizing costs in the process are due to:

- Prospection of more efficient hydrolytic enzyme in nature;
- Optimizing enzyme activity under the process conditions;
- Selecting the optimum enzyme cocktail;
- Reducing enzyme production costs;
- Prospection of microorganisms (bacteria and yeast) that enable the fermentation of hexoses and pentoses;
- Optimizing activities of the different microorganisms used [simultaneous saccharification and fermentation (SSF)];
- Genetic engineering of the organisms for cofermentation, thus reducing operational costs.

It is also important to highlight that the different feedstock types used in the production of 2G ethanol (sugarcane bagasse and sugarcane straw in Brazil, corn stove and fiber in the United States, wheat straw in Europe, as well as forest waste), and even residues from the same plant at different locations and under different forms of cultivation may influence the enzymes that act on the material, such as preliminary pretreatments.

The Danish company Novozymes (www.novozymes.com), which holds 48% of the global market share of enzymes for industrial application, directs 14% of its global revenue to research and developments in bioinnovation, providing enzyme technology to five 2G ethanol plants at an existing global commercial scale, two in Brazil, being one in São Miguel dos Campos, in the State of Alagoas (Bioflex 1, from GranBio) and another in the municipality of Piracicaba, in the State of São Paulo (Raízen), from sugarcane bagasse. The other two are located in the United States, using corn stover as feedstock (Poet/DSM and Abengoa) and another in Italy, Beta Renewables, which uses wheat straw. From the 1G of enzymes in pilot scale (2009) to the fourth- generation of enzymes at an industrial scale (2015), Novozymes managed to increase the power of the enzyme cocktail, enhancing performance, decreasing 50% in volume and reducing costs in roughly 30%. Besides Novozymes, DSM (www.dsm.com), DuPont (www.dupont.com), and Clariant (www.clariant.com) produce enzymes for the production of 2G ethanol. A summary of some of the industries that use lignocellulosic material in the production of biofuel is presented in Table 21.1.

21.3.1 Potential Biomasses

Brazil has one of the greatest biodiversities regarding biomass sources due to its agricultural nature. To make the processes of ethanol production more economically viable, the scientific community seeks to make use of the lignocellusic component of biomass, produced as by-products in several industrial sectors, as is the case of agricultural residues (sugarcane straw and bagasse, wheat straw and corn residues, among others) and forest residue (sawdust and wood residue), as well as grass (Santos et al., 2012). Elephant grass, sorghum biomass, and energy cane stand out as being potential cultures for the production of 2G ethanol, in terms of rapid growth and high biomass productivity, mainly during sugarcane off-season (Carlin and Rossetto, 2014).

21.3.1.1 Sugarcane Straw and Bagasse

Considered the most-efficient bioenergetic culture in tropical and subtropical regions worldwide, the continuous development reached by the sugarcane agroindustry, namely in the production of ethanol fuel, has generated incentives for making use of produced bagasse and to the total or partial recycling of sugarcane straw, as a way of increasing the amount of biomass available to be converted in ethanol. The complete use of the plant, as well as the increase in production without the need of expanding cultivation areas, leads to industrial energy self-sufficiency, contributing even further to the sustainability of the sugar–energy industry.

According to the *Brazilian Institute of Geography and Statistics* (IBGE), the sugarcane productivity in Brazil in the year of 2016 was of 74.77 tonnes per acre. Santos et al. (2012) states that for every tonne of sugarcane processed, 140 kg of straw and another 140 kg of bagasse on a dry base are generated, that is, 12 tonnes of straw and 12 tonnes of bagasse per acre.

Dias et al. (2011) also states that approximately 90% of sugarcane bagasse is used as fuel in steam and electricity production, with at least 10% of bagasse remaining available and at a very low cost for the production of 2G biofuels and bioproducts (Carvalho et al., 2014).

When considering the conversion of all glucose in ethanol, the total use of sugarcane (stalk, straw, and bagasse) may implicate on a significant increase on ethanol production per acre, reaching twice of the production, without the

TABLE 21.1 Cellulosic Biofuel Producers in 2016

Company name (location)	Feedstock	Fuel	Capacity (millions of liters/year)
Abengoa (Hungton, Kansas, USA)	Corn stover	Ethanol	94.6
Cool Planet (Alexandria, Louisiana, USA)	Wood waste	Gasoline	3.7
CNG/LNG producers (various locations)	Biogas	CNG/LNG	N/D
DuPont (Nevada, Iowa, USA)	Corn stover	Ethanol	113.5
Edeniq (various)	Corn stover	Ethanol	N/D
Ensyn (Renfrew, Ontario, Canada)	Wood waste	Heating oil	11.3
GranBio (Sâo Miguel dos Campos, São Paulo, Brazil)	Sugarcane bagasse	Ethanol	79.4
INEOS Bio (Vero Beach, Florida, USA)	Plant residues	Ethanol	30.2
Poet (Emmetsburg, Iowa, USA)	Corn stover	Ethanol	90.8
Qccp (Galva, Iowa, USA)	Corn stover	Ethanol	7.5

Source: EPA, Novacana 2016. The Raízen plant was not considered, for not taking part of data collection.

need of expanding the cultivation area. It is important to highlight that both bagasse and straw present adequate chemical compositions for the production of 2G ethanol, with high carbohydrate content (51%–78%) and at a low cost, relevant facts for the production of cellulose as well as for other industrial uses (Carvalho et al., 2014). However, due to the characteristics of the lignocellulosic biomass, the economic balance of the units of 2G ethanol still does not break even, with production being considered economically unsuitable.

21.3.1.2 Wheat and Rice Straw

Ferreira-Leitão et al. (2010) states that, in Brazil, wheat crops generate approximately 6 Mt of straw, in which 50% of those residues would be available for conversion, with great potential for the production of up to 600 million L of ethanol and 1 Mt of lignin, as this residue is among one of the main lignocellulosic materials.

Rice straw, due to physical processing, represents around 23% of rice weight, being considered one of the most abundant residual lignocellulosic materials in the world, where in Africa, Asia, and Europe its annual production reaches up to 731 million, with 20.9, 667.6,and 3.9 Mt, respectively (Tsunatu et al., 2017).

Rice straws include stalks, leaf blades, leaf sheaths, and remaining of panicle after threshing (Cheng and Timilsina, 2011), containing similar contents of cellulose (32%–47%), hemicellulose (19%–27%), and lignin (5%–24%) compared to other plant residues, which can produce up to 205 billion L of ethanol per year (Tsunatu et al., 2017).

21.3.1.3 Elephant Grass (*Pennisetum purpureum*)

Elephant grass, a semiperennial grass used in Brazil as forage, is seen as advantageous in corn, sugarcane, and soybean crops, given it presents a short production period (less than month) and at a lower cost, as well as not significantly interfering on food production (Marafon et al., 2012). Originated in Africa and introduced in Brazil around the 1920s, elephant grass is highly efficient in fixing atmospheric CO_2 from photosynthesis and is able to grow even in adverse conditions. Cultivated throughout the entire country, it is considered a biomass with a vital role to the national energy matrix, as a tool for decentralizing production, enabling energy generation, and biofuel production in places where the construction of hydroelectric or the cultivation of traditional biomasses is not possible (Chies, 2015). It is mostly cultivated in the State of Rio Grande do Sul, and can be harvested up to four times a year without great need of water and nutrients to grow, being possible to obtain 45 tonnes of dry biomass per acre (Grasel et al., 2017).

Switchgrass (*Panicum virgatum*), also part of the Miscanthus family, is a native species from North America, and considered of great interest in Europe as a source of cellulosic biomass (Cardoso et al., 2013).

Embrapa Agronergia, in the city of Brasilia, has a project conducted by the researcher Marcelo Ayres, from Embrapa Cerrados, carrying out studies in the production of 2G ethanol with this biomass, taking into account its high productivity (150–200 tonnes of fresh matter per acre, considering two annual cuts in order to obtain better performance both in the field and in the production of biofuel), even if it is a culture produced only in small areas and used as animal feed. The results obtained in the first experiments (high volume of recovered glucose and lower use of enzymes), made the research center sort for this crop, together with sugarcane bagasse, optimizing the production stages of 2G ethanol.

21.3.1.4 Sorghum

Sorghum (*Sorghum bicolor* [L.] *Moench*) is one of the main grains cultivated worldwide, presenting as main characteristics a wide genetic diversity and high tolerance to abiotic stresses (drought and heat), as well as requiring fewer fertilizers in comparison with corn and being tolerant to a wide range of conditions (Cardoso et al., 2013).

Sweet sorghum, cultivated in Brazil mainly for the production of grains and forage, although also used in the production of alcohol, began as a complementary technology to sugarcane, due to its low period of growth (120–130 days) and as being tolerant to drought, with a total mechanical production similar calorific value to sugarcane and to crops, such as eucalyptus and elephant grass, as well as high levels of cellulose, hemicellulose, and lignin; being considered a complementary biomass to 2G ethanol, thus becoming viable crop to be sown during the sugarcane off-season (Cardoso et al., 2013).

Among the companies that have invested in sorghum crops, NexSteppe (br.nexsteppe.com), founded in the United States, gained importance when launching Palo Alto sorghum hybrid biomass, with the first trials on Brazilian soil in the end of 2011, carried out in the farms of IAC – *Agronomic Institute of Campinas*, in the city of Ribeirão Preto (SP), in cooperation with Agência Paulista de Tecnologia dos Agronegócios (APTA) – *Paulista Agency of Technology and Agribusiness* (Vital, 2015).

Embrapa Corn and Sorghum, in Minas Gerais, launched its first sorghum hybrid biomass in 2014, the BRS 716, with an average productivity ranging from 120 to 150 tonnes of dry matter per acre, meeting the demands for alternative cultures in the sugarcane off-season and 2G ethanol.

Ceres (www.ceres.org.br), in partnership with Canaoeste (www.canaoeste.com.br), evaluates the best hybrid sorghum varieties (Blade saccharine sorghum production of ethanol and electricity from bagasse, and Blade high sorghum biomass production of electricity and heat), in reformed sugarcane fields in the Brazilian Central–South region. Saccharine sorghum uses the same harvesting, milling, and sugarcane manufacturing equipment, while ordinary sorghum is harvested with forage.

21.3.2 Research Centres and Lines of Research

21.3.2.1 Embrapa

The *Brazilian Agricultural Research Corporation* – Empresa Brasileira de Pesquisa Agropecuária (Embrapa) (www.embrapa.br) is a public entity, founded in April 1973 and linked to the *Ministry of Agriculture, Livestock and Supply* – MAPA, aimed at developing technologies, as well as technoscientific knowledge and information regarding agriculture and livestock in Brazil. It maintains bilateral agreements of technical cooperation with several countries and institutions around the world, as well as multilateral agreements with international organizations for joint research.

Regarding the sugar–energy sector, the company has been studying several sugarcane crops, as well as other feedstocks that may be useful in this sector, such as sorghum (in its sucrose and biomass varieties), grass, *Arundodonax* (giant cane), and starch products for 1G ethanol production (corn and sweet potato).

It receives investments from funding agencies, such as Financiadora de Estudos e Projetos (Finep), Innovation and Research, the *Brazilian National Council for Scientific and Technological Development*-Conselho Nacional de Desenvolvimento Científico e Tecnológico (CNPq), *São Paulo Research Foundation*-Fundo de Amparo à Pesquisa de São Paulo (FAPESP), and other state agencies, such as the *Brazilian Development Bank* – Banco Nacional do Desenvolvimento (BNDES), with funds, such as the *Brazilian Technology Fund* – Fundo Tecnológico (Funtec), among others, besides public–private partnerships.

Its current president, Guy de Capdeville, affirms that the restrictions of the agricultural, industrial, and those transversal to the sugar–energy sector have been studied in the various centers of the company, which instituted a research, development, and innovation portfolio, aimed at integrating and directing its research actions.

In the agricultural aspect, the main challenges are for: (1) increasing productivity, due to the low rate of renovation on the sugarcane field, expanding cultivation to other areas not appropriate for plantation, including in poor soil; (2) developing production systems of sugarcane with direct seeding, with less disruption to the soil, better edaphic environment for growing plants, less environmental risk, besides lower production costs; (3) developing more vertical and sustainable production systems, such as the production system of irrigated cane, as well as identifying water resources available for its sustainable use; (4) identifying, characterizing, and adopting new biomass sources, offering alternatives for the off-season; (5) developing or adopting new machineries and improving harvesting logistics, in order to improve operational performance and, consequently, the longevity of cultivation areas. The work carried out by Emprapa regarding sugarcane were also aimed at improving production systems by adapting cultivars developed by Ridesa and IAC in all Brazilian regions, as well as more efficient and economical fertilization strategies and the development of genetically modified cane crops, more resistant to hydric stress and to high levels of aluminium in the soil.

With regard to the industrial aspect, the challenges include: (1) reducing logistics costs for accessing and storing biomass, be it for energy cogeneration or for producing 2G ethanol, or even the distribution of refined products; (2) reducing idleness in current sugarcane mills during the off-season, which lowers production costs. Researchers have been carried out to ensure that sugarcane mills can operate under the biorefinery concept, enabling the use of multiple input feedstock types and the production of multiple output products (chemicals, fuels, new materials, etc.), considerably increasing its economic viability; (3) overcome the many technological challenges, including efficient and economically viable plantation and mechanical harvesting, extremely outdated plants and obsolete equipment; (4) developing and/or adapting new conversion processes for new focus-crops of the sugar–energy sector.

Thus far, enzymatic routes have exhibited significant advantages over chemical routes, when regarding bioethanol production from sugarcane bagasse. A great challenge consists in making this enzymatic process technical and economically viable. The viability of the enzymatic process requires the production of enzymes specific for pretreated sugarcane bagasse.

The physicochemical process of steam explosion in the presence/or not of a catalyst represents a promising technology for the pretreatment of cane bagasse in the short, medium, and long term, both for separate hydrolysis and fermentation and SSF or SSCF simultaneous saccharification and cofermentation.

Strategically, it is important to develop an efficient process of acid hydrolysis of cellulose from cane bagasse, even if with technical and economic limitations, as it might come as a technological alternative for preventing the uncertainties associated with the development of competitive enzymatic processes, as well as ensuring that enzymes are provided at an adequate cost.

As for the transversal aspects, the challenges include: (1) consider the reality of the country, regarding the use of international standards for describing the economic and environmental performance in the sector, requiring studies of the most different fields, as for instance the lifecycle analysis of the processes and technologies used in the sector; (2) the need to thoroughly understand the impact of emissions and the mitigation strategies for inserting the sector in the international market; (3) developing projects regarding the social impacts of sugar–energy production.

A more thorough understanding of the existing challenges in each of these aspects and the development of researches that can solve the issues raised are vital for ensuring economic, social, and environmental sustainability and viability for the sector. The economic viability of 2G ethanol will be reached by developing new fractioning processes of the different lignocellulosic biomasses, as well as of the logistics aspects regarding storage, manufacturing, and distribution of the products.

One of the projects led by Embrapa, the *Integrated Pluriannual Programme of Research, Development and Innovation on Sugarcane* – Programa Plurianual Integrado de Pesquisa, Desenvolvimento e Inovação (Pluricane), financed by Finep, gathers actions ranging from the introduction of plant quarantine to the conventional and assisted genetic improvement, production systems, and advanced sugarcane biology, seeking solutions for energy cogeneration with crops, such as *Arundodonax* (giant cane), elephant grass, green coconut shell, and saccharine sorghum. Actions in the production of saccharine sorghum and sorghum biomass are also contemplated in researches regarding biological nitrogen fixation and plant health of this crop, from sugarcane and elephant grass. The project also expects resources for expanding and maintaining experimental stations from the *Interuniversity Network for the Development of the Sugar-Energy Sector* – Ridesa and the *Agronomic Institute of Campinas* – IAC. This project is being run in almost all of Brazil by researchers from seven Embrapa units (Agrobiology, Agroenergy, Cerrados, Temperate Agriculture, Agricultural Informatics, Corn and Sorghum and Coastal Tablelands), as well as by Ridesa, IAC, the *State University of North Fluminense* – UENF, the *Federal University of Rio de Janeiro* – UFRJ, the *Federal University of Grande Dourados* – UFGD, and the *Agronomic Institute of the State of Paraná* – Instituto Agronômico do Paraná (IAPAR) (Collares, 2017).

21.3.2.2 CENPES—Petrobras

The *Research Centre Leopoldo Américo Miguez de Mello* – Centro de Pesquisa Leopoldo Américo Miguez de Mello (CENPES) (www.petrobras.com.br/pt/nossas-atividades/tecnologia-e-inovacao/), founded in 1963 in Rio de Janeiro, is the Petrobras unit responsible for research, development, and innovation activities (RD&I) for the production of biofuels, assisting the needs for exploring, producing, and refining oil and natural gas, as well as researches on renewable energies and sustainable development.

It maintains in its Pilot Plants park (Ilha do Fundão, Rio de Janeiro), since 2007, a pilot unit of enzymatic hydrolysis, developed together with the Brazilian company Albrecht EquipamentosIndustriais Ltda. (www.albrecht.com.br), for the production of 2G ethanol from sugarcane bagasse as a result of researches carried out with the *Federal University of Rio de Janeiro* – UFRJ, in partnership with the *Laboratory of Bioprocess Development* – Laboratório de Desenvolvimento de

Bioprocessos (Ladebio) (www.ladebio.org.br/) at Chemical School – EQ, since 2004.

Besides sugarcane biomass, Petrobras had invested until 2015 in seeking other potential lignocellulosic biomasses for the production of bioethanol, with great importance to the residues produced from palm oil (dendê) in the country, with the fruit bunches that were processed for removing the oil containing 40% of cellulose and 20% of hemicellulose, similar levels to sugarcane bagasse and favorable for the production of 2G ethanol. This biomass had already been analyzed in the laboratories in Rio de Janeiro, for a project developed by Petrobras Biofuels and by Belem Bioenergia Brasil, (http://www.belembioenergia.com.br/), in a joint venture between Petrobras Biofuels and Galp (www.galpenergia.com), from Portugal (Monteiro, 2015). However, according to the new business plan for 2017–21, Petrobras ceased its operations in the biofuels, fertilizers, and LPG sectors, concentrating its activities mainly in the production of petroleum. On the other hand, Belem Energia Brasil is expected to maintain its production of palm oil in the State of Pará, as well as its production of biodiesel in Portugal.

For the other lines of research regarding the production of 2G ethanol, CENPES has established interinstitutional agreements with the Chemical School of the *Federal University of Rio de Janeiro* – UFRJ, as well as with the *University of Brasilia* – UnB, the *University of São Paulo* – USP, the *Federal University of São Carlos* – UFSCar, the *Federal University of Rio Grande do Sul* – UFRGS and with the *Brazilian Bioethanol Science and Technology Laboratory* – Laboratório Nacional de Ciência e Tecnologia do Bioetanol (CTBE), in Campinas.

21.3.2.3 CTBE–The Brazilian Bioethanol Science and Technology Laboratory

The *Brazilian Bioethanol Science and Technology Laboratory* – CTBE (http://ctbe.cnpem.br/) is a public laboratory, and part of the network of national laboratories linked to the *Ministry of Science, Technology and Innovation* – MCTI and managed by the *National Centre for Research in Energy and Materials* – Centro Nacional de Pesquisa em Energia e Materiais (CNPEM). Inaugurated in January 2010, in the city of Campinas, its mission is to contribute to the advance on scientific and technological knowledge regarding the production, use, and conversion of biomasses in energy and materials, through research, development, and innovation (RD&I), as well as personnel training. Therefore, it acts in partnership both with the scientific and technological community and with the industrial sector, with its main lines of research being on the production and manufacturing of biomass, functional biology, biotechnology, and biophysics, sustainability in the production of biomass and bioenergy as well as on the integrated evaluation of biorefineries.

These lines of research involve the entire life cycle of a bioproduct, the production of the biomass in the field and its industrial manufacturing process with other bioproducts, as well as the continuous progress made in genetic improvement of microorganisms and the sustainable integration in biorefineries, seeking efficiency and economic viability.

CTBE, thus, focuses mainly in the sugar–energy sector, with projects in the production of bioproducts derived from sugarcane and its lignocellulosic residues (sugarcane bagasse and straw), such as 2G bioethanol and biobutanol. Generating knowledge and innovative technology is the main focus of the private sector, with several industrial partners having established cooperation projects with CTBE.

CTBE relies on an infrastructure consisting of six research and development laboratories, multiuser equipments and a virtual sugarcane biorefinery Virtual Sugarcane Biorefinery (VSB), with particular attention to the pilot plant for process development Planta Piloto para Desenvolvimento de Processos (PPDP), for verifying and demonstrating, at an industrial scale (300 L), the robustness and stability of the technological routes for the production of ethanol and other compounds derived from biomass. This plant is equipped to validate technoeconomic and sustainability indicators of processes in bench scale, minimizing risks of scaling, therefore predicting bottlenecks and optimizing the processes and (capital and operational) investment estimations. The pilot plant is divided into six independent and multipurpose units, representing individual operations required for the production of bioethanol or any other bioproduct derived from cane biomass:

1. Physical treatment of sugarcane bagasse;
2. Physicochemical treatment of sugarcane bagasse;
3. Bioprocesses for the production of microorganisms and metabolites;
4. Enzymatic hydrolysis of lignocellulosic material;
5. Separation and purification;
6. Alcoholic fermentation.

In 2012, CTBE became a strategic partner for Brazilian businesses included in the Joint Plan for Supporting Industrial Technological Innovation in the Sugar-based Energy and Chemical Sectors – PAISS, financed both by the Brazilian Development Bank – BNDES and by Projects Financier – FINEP, such as Lilly (www.lilly. com.br), British Petroleum (www.bp.com) and Rhodia, part of Solvay group (www.rhodia.com.br) (Monteiro, 2015).

CTBE currently works in cooperation with renowned research institutes, universities (USP, UFSCar, UFPE, and UBC, The University of British Columbia), industries (São Martinho, Açucareira-Quatar S/A, Raízen Energia S/A, Petrobras and DSM South America Ltda), and capital markets in areas, such as bioenergy and renewable materials, aimed at creating favorable conditions for technological transfer, disseminating, and accessing the expertise developed in its laboratories.

UBC collaborates with CTBE in the development of studies comparing the various tools for evaluating the life cycle of biofuels. The Quatá S/A sugar mill, part of the Zilor group (www.zilor.com.br), has two different projects with CTBE, one of them involving the validation of production and industrialization scenarios of sugarcane, focused on technologies for collecting straw. The other study, part of the SUCRE project, aims at reducing greenhouse effect emissions (GEE) by significantly increasing the potential of electricity generation exported from sugar mills, using straw as a complement to bagasse. Together with RaízenEnergia S/A (www.raizen.com.br), it is verified whether eliminating farming traffic and precision agriculture may increase productivity and longevity in the sugarcane fields. The partnership with DSM acts on a technological route for developing the production of ethanol from sugarcane bagasse, by using enzymatic and microbial complexes for the consumption of pentose and hexoses obtained from enzymatic hydrolysis, both in a laboratory scale and in the scale of the PPDP from CTBE.

21.3.2.4 National Institute of Technology

The *National Institute of Technology* – Instituto Nacional de Tecnologia (INT) (www.int.gov.br) is one of the other laboratories linked to the *Brazilian Ministry of Science Technology and Innovation* – MCTI and is located in the city of Rio de Janeiro, acting in partnership with the Brazilian productive sector, with innovative technological solutions and specialized services. It is focused on the energy, health, petroleum, and gas fields, as well as defense, renewable energy sources, green chemistry, and social technologies, with 23 highly qualified laboratories, including the *Laboratory of Biocatalysis* – LABIC, from the Division of Catalysis and Chemical Processes, where projects related to the production of 2G ethanol are developed (Monteiro, 2015).

With a team led by Viridiana Santana Ferreira-Leitão, this laboratory operates on the pretreatment of lignocellulosic biomass and on enzymatic technology for cellulose and hemicellulose biomasses originated from sugarcane bagasse and straw, in cooperation with researchers from several Brazilian universities. For the biomass pretreatment, it uses steam treatment assisted by catalysts impregnated in the biomass, aimed at accelerating the process of depolymerization of hemicellulose into its constituent sugars (glucose, fructose, and xylose) (Monteiro, 2015).

21.3.2.5 CETENE—Centre for Strategic Technology of the Northeast

Another national laboratory linked to the *Ministry of Science Technology and Innovation* – MCTI is located in the city of Recife. The *Centre for Strategic Technology of the Northeast* – CETENE (www.cetene.gov.br) was founded in 2005 as a way of supporting the technological development in the Brazilian Northeast, by integrating knowledge, development, and the society.

The center has a biofuels division, which acts on projects for the production of biodiesel, bioethanol, and biogas, with the bioethanol unit aimed at developing its own technology for obtaining 2G bioethanol, via a biochemical route (enzymatic), from sugarcane biomass. For this purpose, it evaluates the optimum pretreatment process of biomass, fermentation trials, the production, purification, and characterization of cellulolytic enzymes of *Trichoderma reesei* and other filamentous fungi, as well as cloning and expression cellulosomal genes in *S. cerevisiae* and bioreactor trials. This unit is integrated with the Bioethanol Network of Pernambuco, proceeding with the studies developed by the universities of the State of Pernambuco: the *Federal University of Pernambuco* – UFPE, the *Federal Rural University of Pernambuco* – UFRPE and the *University of Pernambuco* – UPE, as well as the partnerships established with the *Union of the Sugar* and *Alcohol Industry of Pernambuco* – Sindicato da Indústria do Açúcar e do Álcool de Pernambuco (Sindaçúcar), the *Brazilian Sugarcane* and *Alcohol Technicians Society* – STAB and CTBE, being part of the *National Institute of Science and Technology of Bioethanol* – Instituto Nacional de Ciência e Tecnologia (INCT) of Bioethanol (www.inctdobioetanol.com.br).

21.3.3 Lignocellulosic Ethanol Industrial Plants In Brazil

In Brazil, there are three bioethanol plants that use lignocellulosic material, being present in the cities of São Miguel dos Campos (Alagoas), São Manuel and Piracicaba (both in the State of São Paulo).

The industrial plant of São Miguel dos Campos, named Bioflex 1, is managed by GranBio (www.granbio.com.br), being the first in the southern hemisphere and the biggest in the country, with a capacity for producing 82 million L/year at an investment of approximately R$ 350 million. The Raízen industrial plant (www.raizen.com.br), in the Costa Pinto sugarcane mill, in the city of Piracicaba (part of a joint venture between the Cosan and Shell groups) and the *Center of Sugarcane Technology* – CTC (www.ctcanavieira.com.br/), in the São Manuel sugarcane mill, have approximately the same capacity of 40 million L/year, demanding an investment of R$ 200 million each one (Canalbioenergia, 2014; Novacana, 2013; Quimica, 2015). The industrial plants were developed to use sugarcane straw and bagasse, as a way of increasing the production per cultivated area. At the Raízen plant, production takes place in combination with 1G ethanol plant, with the company running tests until 2017/2018 for adapting the plant to meet the production costs of the 1G ethanol plant (Canalbioenergia, 2015; UNICA, 2015a,b).

In 2015, due to government policies for controlling fuel prices, there was a great drop in investments on the sugarcane sector, with more than a dozen sugar mills having been closed down in the past decade due to low commercial return of the activity (Quimica, 2015). Still, this year was also met with high volatility on oil prices, discouraging the biofuels sector even further (Silva and Bertucco, 2016). The *Brazilian Development Bank* – BNDES points out that lignocellulosic ethanol (2G ethanol) is an alternative capable of eliminating the Brazilian dependency on imported fuel, generating a new cycle of investments on renewable chemistry and with Brazil being established as an international reference in biorefinery. However, none of the industrial plants operate at full capacity, reinforcing the fact that the transition from an experimental (trial) scale to full industrial activity has not yet been accomplished (Valor, 2016).

GranBio uses the physicochemical method of pretreatment via steam explosion, given its advantage of not using chemicals, as well as being highly efficient on destructuring the lignocellulosic matrix. In operation since September 2014, it uses sugarcane straw as feedstock, until then discarded or burned in sugarcane fields, with the company positioning itself as one of the sector's most sustainable companies in the world.

However, over the past two years, Granbio has gone through technical problems, having temporarily ceased operations in 2016, but returning in the end of the same year. One of the problems was linked to at least two fires in the straw storage, generating a feeling of great discomfort in the company, not only due to the loss of feedstock and by having to stall activities to contain the problem, but also on an environmental level, having been fined half a million Reals by the *Brazilian Institute of Environment* – Instituto do Meio Ambiente (IMA) (Valor, 2016). Furthermore, technological difficulties on the steam explosion pretreatment previously mentioned were also verified, given the level of impurities brought from the fields with the straws, with a great amount of sand entering the plant, therefore damaging the steel structures, with the pretreatment process having to be reversed (Novacana, 2016a,b,c,d).

Despite all drawbacks, GranBio predicts investments of around R$ 4 billion in the construction of four cellulosic ethanol plants, two biochemical ethanol plants and biorefineries (Canaonline, 2016). The company has one research center on synthetic biology (BioCelere), in Campinas-SP, focused on the genetic improvement of microorganisms, biomass processing, on the development of fermentation processes and enzymatic hydrolysis. The company also relies on an experimental station (BioVertis) for the development of new biomass sources, having two energy cane cultivars approved by the MAPA, deposited in 2016 and published on the *Official Journal of the Federal Government of Brazil* – Diário Oficial da União (DOU) by the *National Plant Variety Protection Service* – SNPC (Valor, 2017), in partnership with the *Agronomic Institute of Campinas* – IAC and the *Interuniversity Network for the Development of the Sugaralcohol Sector* – Ridesa. Cana-Vertix is a cultivar originated by genetically crossing ancestral and commercial hybrids of sugarcane, with 25% of fibers and only 7% of sucrose, different from the traditional sugarcane, which contains 12% of fibers and 14% of sucrose. This energy cane can be sown in degraded grazing lands, being three times more productive, thus requiring less water and chemical supplies (Monteiro, 2015).

Founded in July 2015, Raízen, only the second company to become interested in the production of 2G ethanol in Brazil, is fruit of a joint venture between the Brazilian Cosan Group and the oil multinational Shell. The production unit is integrated within the Costa Pinto plant, in the city of Piracicaba, in the State of São Paulo, with an installed capacity of 40 million L, using sugarcane straw and sugarcane bagasse as feedstock materials.

The technology employed is part of a biochemical route from the Canadian company Iogen Corporation (www.iogen.ca), which created the Iogen Energy joint venture especially for this project. The process stages include the pretreatment of crude biomass via steam explosion, followed by prehydrolysis with diluted acid and enzymatic hydrolysis for the degradation of cellulose and to form sugar syrups, which will afterward feed the fermentor for bioethanol production. The final stage is distillation, required so that the product complies with the commercial specification of fuel. Iogen Corporation is self-supplied with enzymes for the enzymatic hydrolysis stages and yeast for fermentation, as it maintains a research center in Canada dedicated to the genetic improvement and to the development of more efficient and productive enzymes and yeasts.

The company initially faced some mechanical difficulties with the filtration of lignin, mainly due to the nature of the feedstock process, but is currently satisfied with the performance of the plant and with the growth and consolidation of the process. It is therefore planning to expand the plant, reaching a production of 42 million L until 2018 (Valor, 2016). In the last months of 2016, the Costa Pinto plant produced 211 L of 2G ethanol per tonne of dry biomass; a figure still below its target of 289 L per tonne, given the issues mentioned.

The investment in the Costa Pinto 2G ethanol plant was of R$ 230 million, partially financed by the BNDES, with a contribution of R$ 207.7 million. Raízen predicts the construction of further seven 2G ethanol plants in the country over the next 10 years, reducing CO_2 emissions produced from 1G ethanol in 50%, making this renewable fuel extremely attractive for the sustainability compromises set to energy company. It also intends to export biomass pellets, to be possibly used at production units abroad.

In June 2014, the *Center of Sugarcane Technology* – CTC began operations in its 2G ethanol demonstration plant, using cane bagasse. Inserted in the São Manoel sugarcane mill, located in the municipality of São Manuel in the State of São Paulo, the plant has a manufacturing capacity of 100 tonnes of biomass a day, using a technology developed by its own company (Monteiro, 2015).

At an investment of approximately R$ 80 million, with funds provided by the *Brazilian Development Bank* – BNDES and the Projects Financier – FINEP, as part of the *Joint Plan for Supporting Industrial Technological Innovation in the Sugar-based Energy and Chemical Sectors* – PAISS, the unit wishes to show investors, shareholders, and clients the potential energy developed by the CTC, which can represent an increase of up to 50% on ethanol production, without the need of expanding already sown sugarcane lands.

The CTC process is based on four stages (pretreatment, enzymatic hydrolysis, fermentation, and distillation), being distinguished by integrating the already existent industrial park in the current plant, which reduces the need of further investment.

In 2014, the BNDES approved the funding of R$ 309.6 million for the construction of a third commercial plant of 2G ethanol in the country, named Abengoa Bioenergia São Luiz (www.abengoabioenergy.com), located in Pirassununga, in the State of São Paulo, planned to commence operations before 2018. However, the international plans of Abengoa regarding 2G ethanol were hampered with the sale of the cellulosic ethanol plant located in Hugoton, in the US state of Kansas, following a judicial recovery in the country, forcing its subsidiaries worldwide to sell a great part of their assets.

Novozymes, the main company of enzymes for the production of 2G ethanol, stated that it considers 2020 is still too early for this technology to be consolidated (Novacana, 2016a,b,c,d), with an evident need for modernizing most of the existing plants, besides the technical and economic validation required before a new set of investments is made.

21.4 MACRO AND MICROALGAE

Microalgae are part of the polyphyletic group, which includes a large typology of organisms, as well as the cyanobacteria, even though they may have extremely different characteristics; the first being eukaryotic and the last, prokaryotic. They are able to accumulate, in general, up to 60% of dry weight in specific conditions, mainly under nitrogen limitation; thus, limiting protein synthesis and shifting the energy absorbed to energetic-reserve substances (carbohydrates and lipids) (Silva and Bertucco, 2016). *Spirulina*, *Chlorococcus*, and *Synechococcus* are examples of genera grouped in the cyanobacteria phylum studied for ethanol proposes. The most studied genera of eukaryotic microalgae are *Scenedesmus*, *Chlorella*, *Tetraselmis*, and *Chlamydomonas* (Ho et al., 2013; Markou et al., 2013; Silva and Sforza, 2016; Silva et al., 2017), with an estimated potential of between 100,000 and 200,000 kg of biomass/ha/year (Acién et al., 2012).

In the specific case of macroalgae, the concentration of carbohydrates in biomass (dry weight) can vary between 40% and 85%, consisting of an excellent source of sugars with some potential in fermentation processes (Hong et al., 2014; Wei et al., 2013). Specifically regarding bioethanol, green algae (*Ulva lactuca*, *Ulva pertusa*, *Ulva fasciata*, *Enteromorpha intestinalis*), red algae (*Kappaphycus alvarezii*, *Gelidium amansii*, *Gelidium elegans*, *Gracilaria salicornia*, *Gracilaria verrucosa*, *Eucheuma cottonii*) and brown algae (*Laminaria japonica*, *Laminaria hyperborean*, *Saccharina lattissima*, *Sargassum fulvellum*, *Undaria pinnatifida*, *Alaria crassifolia*) are cited in studies on saccharification and fermentation (Borines et al., 2013; Daroch et al., 2013; Hong et al., 2014; Kim et al., 2014; Kumar et al., 2013; Tan and Lee, 2014; Trivedi et al., 2013; Yanagisawa et al., 2011).

The microalgae have glucose as their principal monomer, in the form of cellulose (cell wall) or starch (energy reserve component), presenting high fermentation levels of the hydrolysed components found in the literature, for instance, 0.163 $g_{ethanol}/g_{biomass}$ for *Arthrospira platensis*, via chemical hydrolysis (Markou et al., 2013), 0.140 $g_{ethanol}/g_{biomass}$ for *Dunaliella tertiolecta*, via the chemoenzymatic process (Lee et al., 2013) and 0.214–0.233 $g_{ethanol}/g_{biomass}$ from *Chlorella vulgaris* FSP-E, via enzymatic and chemical hydrolysis, respectively (Ho et al., 2013), considering a maximum value of 0.511 × 0.5 (stoichiometry based on glucose fermentation to ethanol and CO_2 and an average value of 50% of carbohydrates in biomass), that is, 0.256 $g_{ethanol}/g_{biomass}$.

As microalgae are commonly used for the extraction of hydrocolloids, only remaining the cellulosic component, several studies only focus on the conversion of this fraction, exhibiting low value of conversion of biomass in ethanol, usually ranging from 1.3% to 16% ($g_{ethanol}/g_{algae}$) (Daroch et al., 2013; Wu et al., 2014), with best values only being reached when considering the entire biomass for hydrolysis and after optimizing the enzymatic loading, hydrolysis time, pH, temperature, and solids loading. The production capacity of microalgae is estimated as being more than 700,000 kg of biomass/ha/year), while sugarcane reaches approximately 70,000 (Adams et al., 2009). Therefore, one single conversion of this biomass, which contains between 40% and 85% of carbohydrates, would be considered of great potential to be developed in countries with an unused coastal area. However, the diversity of carbohydrates of microalgae hampers the hydrolysis process, besides not guaranteeing fermentation for most of them from *S. cerevisiae*, main industrial specie.

The use of this biomass type (micro/macroalgae) is still a matter of investigation, having not yet been commercialized. In comparison with the first two production generations of bioethanol, it is still considered an expensive process, due to its low yields (macroalgae)/cultivation cost (macroalgae) and the lack of adapted technology (Cardoso et al., 2011; Harun et al., 2014; Silva and Bertucco, 2016).

21.5 SOME ALTERNATIVES TO IMPROVE THE SUSTAINABILITY AND ENERGY/ENVIRONMENTAL BALANCE OF BRAZILIAN ETHANOL BIOREFINERIES

21.5.1 Biorefinery and Fine Chemicals

Biorefinery is considered a facility that integrates the processes of biomass conversion into biofuels, chemical supplies, materials, food, animal feed, and energy, optimizing resources and minimizing the generation of effluents, and thus leads to increased benefits and profit.

Costa et al. (2016) cites several articles in the literature regarding optimization studies of biorefineries producing biofuels (including bioethanol), only being concerned with issues in the supply chain in a large scale, without considering the possibility of using production surplus, transportation, location, and the size of 2G ethanol plants. In the simulations, different conversion technologies are presented (thermochemical, chemical, and biological), aiming at obtaining high added value of biomass by-products, as well as other fuels.

In the case of Brazil, the total energy used from sugarcane biomass, as well as ethanol and sugar obtained via the conventional form, enables the conversion of plant material using physicochemical, thermochemical, enzymatic processes, or even the biological process, into products that would meet the needs of modern consumption, in a sustainable manner or with a minimum environmental impact (Vaz Junior, 2011). An example of the potential of lignocellulosic materials for obtaining refined chemicals is presented in Fig. 21.1.

Cavalett et al. (2012) emphasize, however, that the analysis of technologies for optimizing industrial processes should not only consider energy savings, but also the economic and environmental impacts, vital for future cane refineries in Brazil, which will support the strategic decision-making process. Aimed at analyzing new technologies and innovations when producing and processing biomass, CTBE developed a computational simulation tool, the VSB. In scenarios evaluated by Cavalett et al. (2012) for 1G ethanol in Brazil, the production plants attached to the sugarcane mills are more flexible, with the process being more cost-effective, in comparison with an autonomous distillery, when directing 50% of cane juice for the production of sugar and 50% for the production of ethanol.

The refining process of this virtual biorefinery, besides evaluating the technology for 2G ethanol, by-products, sugarcane crop management and different strategies for commercializing and using ethanol as fuel, it also contemplates the impacts of the product's final use, with a complete evaluation of the technological sustainability currently being developed, namely of the three

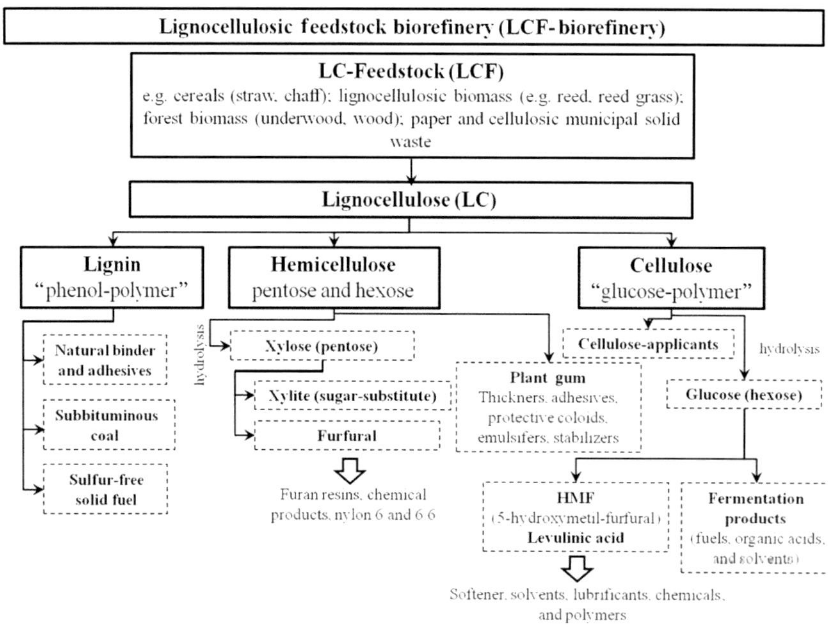

FIGURE 21.1 **Potential use of lignocellulosic residues for obtaining chemical in a biorefinery context.** *Source: Adapted from Kamm, B., Gruber, P.R., Kamm, M., 2006. Biorefineries – Industrial Processes and Products. Wiley-VCH, Weinheim, Germany, ISBN: 3-527-31027-4.*

sustainability axes: economic, environmental, and social. In the economic analysis, it is evaluated the cost of capital investment, calculating the internal rate of return, net present value, production costs, among others. In the environmental analysis, GHG emissions (Greenhouse Gas), energy balance, water consumption, changes in land use, and the several other environmental impacts can be included in the life cycle analysis, such as acidification, eutrophication, human toxicity. Regarding the social analysis, a quantitative evaluation of the social impacts as a result of technological changes in the sugarcane sector, namely automation, increase in plant scale, mechanization of the agricultural sector, among others, is made (Bonomi et al., 2016). Fig. 21.2 illustrates the new methodology for evaluating the sustainability of a virtual biorefinery.

21.5.2 Biomass Combustion

Biomass plays a vital role in the energy, environmental, and socioeconomic context. According to Vaz Junior (2017), the international market of chemical products is estimated in US$ 100 billion a year, with 3% of those related to bioproducts or biomass by-products.

One of the technologies used for exploring biomass is thermochemistry, which includes combustion, gasification, and/or pyrolysis. It is characterized by the use of thermal energy at high temperatures (120–200°C, depending on the type of process carried out) for the chemical reactions to occur, generating various chemical products, as well as electricity, steam, among others. However, as this process requires a considerable amount of energy, it is not of great interest to the chemical industry, except for gasification for generating CO and H_2 (*syngas*), which can be used in organic synthesis routes as precursors of hydrocarbons via the Fischer-Tropsh process (Vaz Junior, 2017).

Combustion, also denominated direct burning, is the only process to be carried out in the presence of oxygen surplus, being characterized as a typical oxidation reaction, generating

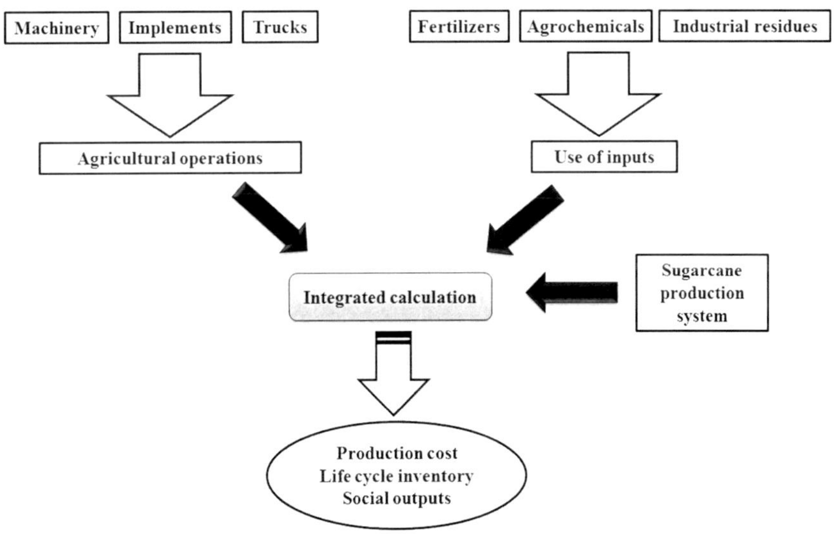

FIGURE 21.2 **Flowchart of the evaluation of the three different sustainability axes of a virtual sugarcane biorefinery.** *Source: Adapted from CTBE (http://ctbe.cnpem.br/en/research/biorefineries-assessment/).*

carbon dioxide (CO_2), steam, and energy (Barreto et al., 2008; Rocha, 2013). When generating thermal energy, the bagasse is burned directly in boilers, with the resulting energy being used for the production of steam, activating the mechanical turbines as well as the turbines used for generating electricity. Alternatively, gases originated from the combustion of bagasse may also be used for activating gas turbines used for generating electricity (Virmond et al., 2013). This is widely used in Brazil, in sugarcane mills, with the industrial plant being energy self-sufficient, and even selling energy surplus in some cases.

Gasification is a process that uses a limited but controlled concentration of oxygen. Pyrolysis, defined as the thermal degradation of organic material in the almost total absence of an oxidising agent, or even in an environment with an oxygen concentration capable of preventing intensive gasification of the organic material, usually takes place at a temperature close to 400°C, until the start of the gasification system, generating (liquids, solids, and gas) products of high-added value (Pedroza et al., 2014; Vieira et al., 2014). The main resulting product of bio-oil, besides being a renewable liquid, may be used in the production of various chemical substances (Guedes et al., 2010). Considered a powerful tool to be used in the several fields of science, the pyrolysis technique has been a great ally for generating energy, proving to be highly efficient in obtaining renewable products, given its increased positive energy and ecological balance (Vieira et al., 2014). In Fig. 21.3 it is possible to visualise the basic aspects of using pyrolysis for the production of chemicals and fuels.

The production of synthesis gas (*syngas*), usually a product of gasification or pyrolysis, undergoes catalytic synthesis or fermentation, thus obtaining hydrocarbons, alcohols, hydrogen, ammonium, natural synthetic gas, and so on, with such processes denominated BTL processes (biomass to liquid) (Lora and Andrade, 2009).

21.5.3 Lignin Utilization

One of the components of the lignocellulosic matrix, lignin, linked to hemicellulose and cellulose, is not formed by sugars, thus it does not contribute to the growth on ethanol yield. Such a component is associated to biomass recalcitrance, hampering its hydrolysis process.

However, it is the second most abundant polymer in the planet, after cellulose (Vishtal and Kraslowski, 2011), lignin exhibits almost the same calorific value as ethanol (27 kJ/g) (Welker et al., 2015); a great capacity not only for generating energy and steam, but also for obtaining various products for the chemical industry (Ludmila et al., 2015; Vishtal and Kraslowski, 2011).

There are several types of lignin, which depend on the source and extraction process, with the most common types being kraft lignin, sodium lignin, lignosulfonate, organosolv lignins, hydrolysis lignins, and ionic liquid lignins (Ludmila et al., 2015; Smolarski, 2012; Vishtal and Kraslowski, 2011), with prices ranging 0–1 US$/kg (Welker et al., 2015; Ludmila et al., 2015). Due to its great variety, its applications are also expectedly diver, with greatest importance to carriers for fertilizers and pesticides, thermoplastic polymer blends, carbon fibers, resins, charcoal, animal feed dispersants, stabilizers for colloidal suspensions, detergents, gels, surfactants, adhesives, concrete additives, varnishes, and dyes used as adsorbents or green solvents (Ludmila et al., 2015; Vishtal and Kraslowski, 2011; Welker et al., 2015).

Following the methods of lignin fragmentation, various added-value products can be obtained, assisting various industrial sectors in reducing the dependence on oil derived products (Li et al., 2016; Welker et al., 2015). Although the extraction of these high added-value products from lignin were considered crucial for the economic viability of integrated biorefineries, its appreciation in chemicals is challenging when considering sustainability and environmental protection, given the difficulty of reaching an efficient bioprocess (Li et al., 2016).

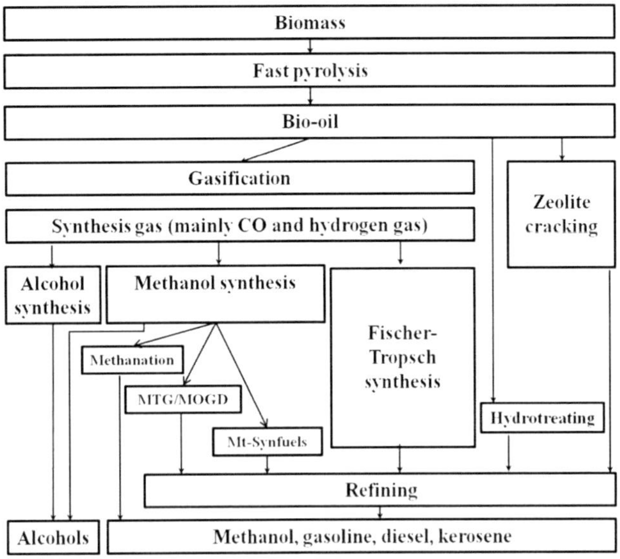

FIGURE 21.3 **Bio-oil to fuels and chemicals.** *MODG*, Methanol to olefins, gasoline, and diesel; *MTG*, methanol to gasoline; *Mt-Synfuels*, process to convert methanol to synthetic transportation fuels (gasoline and diesel). *Source: Adapted from Bridgwater, A.V., 2012. Review of fast pyrolysis of biomass and product upgrading. Biomass Bioenergy 38, 68–94; Virmond, E., Rocha, J.D., Moreira, R.F.P.M., José, H.J., 2013. Valorization of agroindustrial solid residues from biofuel production chains by thermochemical conversion: a review, citing Brazil as a case study. Braz. J. Chem. Eng. 30(2), 197–229.*

The main products obtained from lignin with recognized market value are acetic acid, phenol, methanol, *n*-propanol, vanillin, dimethylfurfural (DMF), dimethyl sulfoxide (DMSO), syringic acid, 4-hydroxybenzaldehyde, vanillic acid, feluric acid, *P*-coumaric acid, benzene, toluene, and xylene, with prices varying 0.1–11 US$/kg (Ludmila et al., 2015). Despite this, lignin is still commonly treated as a residue in the production of energy or in the production of solid coal (Vishtal and Kraslowski, 2011; Welker et al., 2015).

Welker et al. (2015) considers the following key factors for an efficient recovery and destructuring of lignin in the context of a biofinery:

- The development of modified lignins without altering the properties of the cell wall;
- Having a lignin type that can easily be destructured into specific bioproducts;
- Exploring new enzymes that can act more efficiently;
- Developing specialized methods for isolating lignin and for controlling depolymerization according to the characteristics of the biomass, aimed at its industrial application.

In this context, genetic engineering of new crops may represent an interesting tool for reaching those targets (Welker et al., 2015), namely, in the development of a plant with low lignin content (sugar recovery) or high lignin content (fine chemical industry).

21.5.4 Anaerobic Digestion of Vinasse

Sugarcane vinasse is a voluminous effluent generated, product of the distillation process of ethanol (10–15 L of vinasse per liter of ethanol produced) (Koyama et al., 2016). It is a liquid with a high polluting capacity, given its high content of organic matter and low pH, besides having excellent fertilizing capacities (satisfactory NPK contents—nitrogen, phosphorus and potassium nutrients) (Bernal et al., 2017). Therefore, fertigation of sugarcane fields with this effluent is highly diffuse, with productivity only being increased for a limited amount of time, after successive applications of vinasse in the soil. However, when it is carried out without proper planning, it culminates in a physical and chemical deterioration of the soil, hampering the natural microbiota, and threatening water bodies (Bernal et al., 2017; Moraes et al., 2015).

Anaerobic digestion has thus emerged as an alternative waste treatment method to the aerobic digestion processes, mainly due to its higher organic matter degradation rate, with little sludge formed, and the production of energy through the synthesis of biogas, in addition to producing biofertilizers as an ending product (Mohama et al., 2009). It has been widely used to treat municipal sludges and industrial effluents, still with certain limitations in the latter (microbial sensibility to the composition of industrial effluent) (Chen et al., 2008). The organic matter is then removed and basically converted into biogas and sludge originated from the process, indicating better quality of the biofertilizer (Silva and Abud, 2016, 2017), as indicated in Table 21.2.

This process involves gains at an economic/environmental/energetic level for biorefineries, with most studies having demonstrated all the advantages in the use of biofertilizers, opposed to vinasse in natura. The biogas produced (due to great amounts of vinasse) proved viable for electricity production or cogeneration, as well as for vehicular fuel (Moraes et al., 2014). Yet, the process has many challenges to be faced before being consolidated, such as: (1) direct fertigation of cane fields still being an economically viable process; (2) the predominance of empirical approaches in the fundamental anaerobic digestion study of vinasse; (3) unsatisfactory results obtained for the only few plants in industrial scale; and (4) the unappreciation of biogas as an alternative source of energy and of mitigating fossil fuels (Moraes et al., 2015).

Carbon dioxide emissions, one of the GHG, are higher for untreated vinasse, reason why the mitigation provides better values in any scenario that includes anaerobic digestion (Bernal et al., 2017; Moraes et al., 2017). Furthermore, the potential energy can be compared to the demand of many big Brazilian cities or even to the power of some hydroelectrics, as well as helping in the energy and environmental balance of 2G ethanol (Moraes et al., 2014).

21.5.5 CO_2 Sequestration

Almost half of the initial mass of sugar used in fermentation consists of CO_2 (glucose → ethanol + 2 CO_2). This carbon is considered as being of high market value for being used in

TABLE 21.2 Characteristics of Vinasse in Natura and Biodigested

Parameter	Fresh vinasse (in natura)	Biodigested vinasse (biofertilizer)
COD (mg O_2/L)	38911.2	13781.1
Phosphorus (mg P/L in PO_4^{3-})	17.56 ± 0.88	19.90 ± 1.34
a (mg N/L)	711 ± 153	1253.9 ± 213
Total solids (mg/L)	55728 ± 3785	35660 ± 85
Total fixed solids (mg/L)	16921 ± 128	22630 ± 806
Total volatile solids (mg/L)	38807 ± 3657	13030 ± 721

Source: Silva, C.E.F., Abud, A.K.S., 2017. Influence of manure concentration as inoculum in anaerobic digestion of vinasse. Acta Sci. Biol. Sci. 39(2), 173–180. (Manure as inoculum – 0.5%, 30°C and $pH_{initial}$ –7.0).

other processes, such as in the food industry or in chemical synthesis, being stocked as a value-added product and as a way of significantly reducing its environmental impact (CO_2 emission in the atmosphere), besides reducing energy losses and production costs (Kheshgi and Prince, 2005; Xu et al., 2010).

In this context, the development of technologies that may capture and stock this component has become increasingly significant, being an essential step for increasing the economic/environmental sustainability of the process (Mollersten et al., 2003; Rhodes and Keith, 2005; Xu et al., 2010). Among the main applications of CO_2, its application is of significant importance in the food industry and in the supercritical extraction, as well as in the chemical industry, where it can be used to produce organic carbonates or minerals, synthetic fuel, polymers, and as a supercritical solvent in chemical reactions (Xu et al., 2010).

Another possible application would be to use the carbon gas for cultivating microalgae, as the supply/costs of maintenance of this gas at an industrial level can represent up to 50% of the total production costs (Chisti, 2013; Silva et al., 2016). *Dunaliella teriolecta* was cultivated with CO_2 originated from beer fermentation, reaching a growth velocity of 0.58/day and productivity of approximately 0.18 g/L/day (18 klx, 28°C and airflow rate of 1 L/min), which can represent a strong interaction between 1G and 2G plants with 3G ones (Chagas et al., 2015).

21.6 CONCLUSIONS

Given its climate and tropical extension, Brazil is a traditional country for the production of biofuels, namely bioethanol. However, the 1G productive chain is currently saturated, both due to geographical limitations of sowing sugarcane and given the need to produce food. In this context, the last decade saw the development of 2G ethanol, aimed at solving the deficiencies originated from 1G ethanol. Although it is a costly technology, with constraints on industrial development (pretreatment and use of enzymatic complexes), three plants were eventually setup, and with a perspective of increasing this number. For this increase to be reached, the concept of biorefinery needs to be urgently evolved in the Brazilian plants, enabling not only the integration of the various generations of ethanol, but also the incorporation/development of 3G technology. This chapter approaches some of these processes, seen as complements to the production of ethanol in plants, which can be of significant importance for the Brazilian economic and energetic self-sufficiency.

References

Acién, F.G., Fernandez, J.M., Magan, J.J., Molina, E., 2012. Production cost of a real microalgae production plant and strategies to reduce it. Biotechnol. Adv. 30, 1344–1353.

Adams, J.M., Gallagher, J.A., Dannison, I.S., 2009. Fermentation study on *Saccharina latissima* for bioethanol production considering variable pretreatments. J. Appl. Phycol. 21, 569–574.

Agostinho, F., Bertaglia, A.B.B., Almeida, C.M.V.B., Giannetti, B.F., 2015. Influence of cellulose enzyme production on the energetic-environmental performance of lignocellulosic ethanol. Ecol. Model. 315, 46–56.

Barreto, A.C., Freire, M.B.G.S., Nacif, P.G.S., Araújo, Q.R., Freire, F.J., Inácio, E.S.B., 2008. Fracionamento químico e físico do carbono orgânico total em um solo de mata submetido a diferentes usos. Rev. Bras. Ciênc. Solo 32 (4), 1471–1478.

Basso, L.C., Basso, T.O., Rocha, S.N., 2011. Ethanol production in Brazil: the industrial process and its impacts on yeast fermentation. In: Bernardes, M.A.S. (Ed.), Biofuel Production: Recent Developments and Prospects. InTech Books, pp. 85–100. Available from: https://www.intechopen.com/books/biofuel-production-recent-developments-and-prospects/ethanol-production-in-brazil-the-industrial-process-and-its-impact-on-yeast-fermentation.

Belicanta, J., Alchorne, J.A., Silva, M.T., 2016. The Brazilian experience with ethanol fuel: aspects of production, use, quality and distribution logistics. Braz. J. Chem. Eng. 33, 1091–1102.

Bernal, A.P., Santos, I.F.S., Silva, A.P.M., Barros, R.M., Ribeiro, E.M., 2017. Vinasse biogas for energy generation in Brazil: an assessment of economic feasibility, energy potential and avoided CO_2 emissions. J. Clean. Prod. 151, 260–271.

Bonomi, A., Cavalett, O., Cunha, M.P., 2016. Virtual Biorefinery: Optimization Strategy for Renewable Carbon Valorization. Springer.

Borines, M.G., Leon, R.L., Cuello, J.L., 2013. Bioethanol production from the macroalgae *Sargassum* spp. Bioresour. Technol. 138, 22–29.

Bridgwater, A.V., 2012. Review of fast pyrolysis of biomass and product upgrading. Biomass Bioenergy 38, 68–94.

Raízen inicia operação de sua primeira unidade de etanol de segunda geração, 2014. Available from: http://www.canalbioenergia.com.br/raizen-inicia-operacao-de-sua-primeira-unidade-de-etanol-de-segunda-geracao/

Etanol 2G: primeiras plantas comerciais em plena produção. 2015. Available from: https://www.canalbioenergia.com.br/etanol-2g-primeiras-plantas-comerciais-em-plena-producao/

Etanol de milho é viável no Brasil? 2016. Available from: http://www.canaonline.com.br/conteudo/etanol-de-milho-e-viavel-no-brasil.html#.WVj7BaSlzIU

Cardoso, A.S., Vieira, G.E.G., Marques, A.K., 2011. O uso de microalgas para a obtenção de biocombustíveis. R. Bras. Bioci. 9 (4), 542–549.

Cardoso, W.S., Tardin, F.D., Tavares, G.P., Queiroz, P.V., Mota, S.S., Kasuya, C.M., Queiroz, J.H., 2013. Use of sorghum straw (*Sorghum bicolor*) for second generation ethanol production: pretreatment and enzymatic hydrolysis. Quim. Nova 36 (5), 623–627.

Carlin, S.D., Rossetto, R., 2014. Biomassa para a produção de energia. Pesqui. Tecnol. 11 (2), 1–6.

Carvalho, D.M., Perez, A., García, J.C., Colodette, J.L., López, F., Diaz, M.J., 2014. Ethanol-soda pulping of sugarcane bagasse and straw. Cell. Chem. Technol. 48 (3-4), 355–364.

Cavalett, O., Junqueira, T.L., Dias, M.O.S., Jesus, C.D.F., Mantelatto, P.E., Cunha, M.P., Franco, H.C.J., Cardoso, T.F., Maciel Filho, R., Rossell, C.E.V., Bonomi, A., 2012. Environmental and economic assessment of sugarcane first generation biorefineries in Brazil. Clean Technol. Environ. Policy 14, 399–410.

Chagas, A.L., Rios, A.O., Jarenkow, A., Marcilio, N.R., Ayub, M.A.Z., Rech, R., 2015. Production of carotenoids and lipids by *Dunaliella tertuiolecta* using CO_2 from beer fermentation. Process Biochem. 50, 981–988.

Chen, Y., Cheng, J.J., Creamer, K.S., 2008. Inhibition of anaerobic digestion process: a review. Bioresour. Technol. 99, 4044–4064.

Cheng, J.J., Timilsina, G.R., 2011. Status and barriers of advanced biofuel technologies: a review. Renew. Energy 36 (12), 3541–3549.

Cherubin, N., 2013. Novas variedades pedem passagem. RPA News 156, 4–9.

Cherubin, N., 2016. A Cana Energia E Sua Versatilidade. RPA News 185, 20–27.

Chies, V., 2015. Pesquisa investe em capim como fonte de energia. Informativo da Embrapa Agroenergia 59, 4–8. Available from: https://www.embrapa.br/busca-de-noticias/-/noticia/2422024/pesquisa-investe-em-capim-como-fonte-de-energia.

Chisti, Y., 2013. Constraints to commercialization of algal fuels. J. Biotechnol. 167, 201–214.

Collares, D., 2017. Brasil reúne 22 instituições no maior consórcio já feito para pesquisar cana. [8 August, 2017]. Rio de Janeiro, Jornal Dia de Campo. Available from: http://www.diadecampo.com.br/zpublisher/materias/Materia.asp?secao=Not%EDcias&id=34452.

Costa, C.B.B., Potrich, E., Gruz, A.J.C., 2016. Multiobjective optimization of sugarcane biorefinery involving process and environmental aspects. Renew. Energy 96, 1142–1152.

Cunha, F.M., Badino, A.C., Farinas, C.S., 2017. Effect of a novel method for in-house cellulose production on 2G ethanol yields. Biocatal. Agric. Biotechnol. 9, 224–229.

Daroch, M., Geng, S., Wang, G., 2013. Recent advances in liquid biofuel production from algal feedstocks. Appl. Energy 102, 1371–1381.

Dias, M.O.S., Cunha, M.P., Jesus, C.D.F., Rocha, G.J.M., Pradella, J.G.C., Rossell, C.E.V., Maciel Filho, R., Bonomi, A., 2011. Second generation ethanol in Brazil: can it compete with electricity production? Bioresour. Technol. 102 (19), 8964–8971.

Ferreira-Leitão, V., Gottschalk, L.M.F., Ferrara, M.A., Nepomuceno, A.L., Molinari, H.B.C., Bom, E.P.S., 2010. Biomass residues in Brazil: availability and potential uses. Waste Biomass Valorization 1, 65–76.

Grasel, F.S., Stiehl, A.C.R., Bernardi, L.P., Herpich, T.L., Behrens, M.C., Andrade, J.B., Schultz, J., Mangrich, A.S., 2017. Inovação em biorrefinarias I. Produção de etanol de segunda geração a partir de capim-elefante (*Pennisetum purpureum*) e bagaço de cana-de-açúcar (*Saccharum officinarum*). Rev. Virtual Quím. 9 (1), 4–14.

Guedes, C.L.B., Adão, D.C., Quessada, T.P., Borsato, D., Galão, O.F., Di Mauro, E., Pérez, J.M.M., Rocha, J.D., 2010. Avaliação de biocombustível derivado de bio-óleo obtido por pirólise rápida de biomassa lignocelulósica como aditivo para gasolina. Quím. Nova 33 (4), 781–786.

Gupta, A., Verna, J.P., 2015. Sustainable bio-ethanol production from agro-residues: a review. Renew. Sustain. Energy Rev. 41, 550–567.

Harun, R., Yip, J.W.S., Thiruvenkadam, S., Ghani, W.A.W.A., Cherrington, T., Danquah, M.K., 2014. Algal biomass conversion to bioethanol: a step by step assessment. Biotechnol. J. 9, 73–86.

Ho, S., Li, P., Liu, C., Chang, J., 2013. Bioprocess development on microalgae-based CO_2 fixation and bioethanol production using *Scenedesmus obliquus* CNW-N. Bioresour. Technol. 145, 142–149.

Hong, I.K., Jeon, H., Lee, S.B., 2014. Comparison of red, brown and green seaweeds on enzymatic saccharification process. J. Ind. Eng. Chem. 20, 2687–2691.

Kamm, B., Gruber, P.R., Kamm, M., 2006. Biorefineries: Industrial Processes and Products. Wiley-VCH, Weinheim, Germany, ISBN: 3-527-31027-4.

Kheshgi, H., Prince, R.C., 2005. Sequestration of fermentation CO_2 from ethanol production. Energy 30, 1865–1871.

Kim, D., Lee, S., Jeong, G., 2014. Production of reducing sugars from *Enteromorpha intestinalis* by hydrothermal and enzymatic hydrolysis. Bioresour. Technol. 161, 348–353.

Koyama, M.H., Araújo Junior, M.M., Zaiat, M., Ferraz Junior, A.D.N., 2016. Kinetics of thermophilic acidogenesis of typical Brazilian sugarcane vinasse. Energy 116, 1097–1103.

Kumar, S., Gupta, R., Kumar, G., Sahoo, D., Kuhad, R.C., 2013. Bioethanol production from *Gracilaria verrucosa*, a red alga, in a biorefinery approach. Bioresour. Technol. 135, 150–156.

Lee, O.K., Kim, A.L., Seong, D.H., Lee, C.G., Jung, Y.T., Lee, J.W., Lee, E.Y., 2013. Chemeo-enzymatic saccharification and bioethanol fermentation of lipid-extracted residual biomass of the microalga, *Dunaliella tertiolecta*. Bioresour. Technol. 132, 197–201.

Li, S., Liu, S., Colmenares, J.C., Xu, Y., 2016. A sustainable approach for lignin valorization by heterogeneous photocatalysis. Green Chem. 18 (3), 577–607.

Lima A.M., 2011. Estudos recentes e perspectivas da viabilidade técnico-econômica da produção de etanol lignocelulósico Circular Técnica 05. Available from: https://www.infoteca.cnptia.embrapa.br/bitstream/doc/890268/1/CITE05.pdf.

Lopes, M.L., Paulillo, S.C.L., Godoy, A., Cherubin, R.A., Lorenzi, M.S., Giometti, F.H.C., Bernardino, C.D., Amorim Neto, H.B., Amorim, H.V., 2016. Ethanol production in Brazil: a bridge between science and industry. Braz. J. Microbiol. 47, 64–76.

Lora, E.S., Andrade, R.V., 2009. Biomass as energy source in Brazil. Renew. Sustain. Energy Rev. 13 (4), 777–788.

Ludmila, H., Michal, J., Andrea, S., Ales, H., 2015. Lignin, potential products and their market value. Wood Res. 60 (6), 973–986.

Marafon, A.C., Câmara, T.M.M., Santiago, A.D., Rangel, J.H.A., 2012. Cenário energético brasileiro e o potencial do capim elefante como fonte de matéria-prima para a produção de energia renovável. Available from: https://ainfo.cnptia.embrapa.br/digital/bitstream/item/123070/1/06-Cenario-energetico.pdf.

Markou, G., Angelidaki, I., Nerantzis, E., Georgakakis, D., 2013. Bioethanol production by carbohydrate-enriched biomass of *Arthrospira* (*Spirulina*) *platensis*. Energies 6, 3937–3950.

Mohama, S., Acharya, B.K., Madamwar, D., 2009. Distillery spent wash: treatment technologies and potential applications. J. Hazard. Mater. 163, 12–25.

Mollersten, K., Yan, J., Moreira, J.R., 2003. Potential market niche for biomass energy with CO_2 capture and storage: opportunities for energy supply with negative CO_2 emissions. Biomass Bioenergy 25, 273–285.

Monteiro, R.S., 2015. Biocombustíveis de segunda geração – Produto 3. Available from: http://sectordialogues.org/sites/default/files/acoes/documentos/produto3-robson-monteiro1.pdf.

Moraes, B.S., Junqueira, T.L., Pavanello, L.G., Cavalett, O., Mantelatto, P.E., Bonomi, A., Zaiat, M., 2014. Anaerobic digestion of vinasse from sugarcane biorefineries in Brazil from energy, environmental, and economic perspectives: profit or expense? Appl. Energy 113, 825–835.

Moraes, B.S., Zaiat, M., Bonomi, A., 2015. Anaerobic digestion of vinasse from sugarcane ethanol production in Brazil: challenges and perspectives. Renew. Sustain. Energy Rev. 44, 888–903.

Moraes, B.S., Petersen, S.O., Zaiat, M., Sommer, S.G., Triolo, J.M., 2017. Reduction in greenhouse gas emissions from vinasse through anaerobic digestion. Appl. Energy 189, 21–30.

Mosier, N., Wyman, C., Dale, B., Elander, R., Lee, Y.Y., Holtzapple, M., Ladisch, M., 2005. Features of promising technologies for pretreatment of lignocellulosic biomass. Bioresour. Technol. 96 (6), 673–686.

Brasil terá 3 usinas de etanol celulósico até 2015 2013. Available from: https://www.novacana.com/n/etanol/2-geracao-celulose/brasil-usinas-etanol-celulosico-200913/.

O futuro da tecnologia canavieira esta no etanol celulósico, 2016a. Available from: https://www.novacana.com/n/eventos/o-futuro-da-tecnologia-canavieira-esta-no-etanol-celulosico/.

Futuro brilhante do etanol celulósico está ameaçado, 2016b. Available from: https://www.novacana.com/n/etanol/2-geracao-celulose/futuro-brilhante-etanol-celulosico-ameacado-010316/.

Genes-alvos para o melhoramento da cana-energia são identificados, 2016c. Available from: https://www.novacana.com/n/cana/variedades/genes-alvos-melhoramento-cana-energia-identificados-050516/.

Novos incêndios atingem estoque de palha da GranBio em São Miguel dos Campos, 2016d. Available from: https://www.novacana.com/n/industria/usinas/novo-incendio-atinge-estoque-palha-granbio-sao-miguel-campos-040116/.

Ranking das usinas do Brasil por capacidade de moagem. 2017. Available from: https://www.novacana.com/usinas-brasil/ranking/moagem/.

Pedroza, M.M., Sousa, J.F., Vieira, G.E.G., Bezerra, M.B.D., 2014. Characterization of the products from the pyrolysis of sewage sludge in 1 kg/h rotating cylinder reactor. J. Anal. Appl. Pyrol. 105, 108–115.

Biocombustíveis: Etanol celulósico enfrenta crise setorial e petróleo mais barato, 2015. Availabe from: http://www.quimica.com.br/biocombustiveis-etanol-celulosico-enfrenta-crise-setorial-e-petroleo-mais-barato/.

Rhodes, J.S., Keith, D.W., 2005. Engineering economic analysis of biomass IGCC with carbon capture and storage. Biomass Bioenergy 29, 440–450.

Rocha, J.D., 2013. A plataforma termoquímica. Biomassa para Química Verde. Embrapa Agroenergia, Brasília, 139-167.

Rodrigues, D.S., Mendes, T.D., Pacheco, T.F., 2013. Catálise Enzimática Para Desconstrução De Biomassa Lignocelulósica. Microrganismos Na Produção De Biocombustíveis Líquidos. Embrapa, Brasília, 145-187.

Saad, M.B.W., 2010. Avaliação técnica e econômica preliminar da produção de etanol via hidrolise enzimática de bagaço de cana-de-açúcar. EEL SP, USP Digital Library.

Santos, F.A., Queiróz, J.H., Colodette, J.L., Fernandes, A.S., Guimarães, V.M., Rezende, S.T., 2012. Potencial da palha de cana-de-açúcar para a produção de etanol. Quím. Nova 35 (5), 1004–1010.

Silva, C.E.F., Abud, A.K.S., 2016. Anaerobic biodigestion of sugarcane vinasse under mesophilic conditions using manure as inoculum. Rev. Ambient. Agua 11 (4), 763–777.

Silva, C.E.F., Abud, A.K.S., 2017. Influence of manure concentration as inoculum in anaerobic digestion of vinasse. Acta Sci. Biol. Sci. 39 (2), 173–180.

Silva, C.E.F., Bertucco, A., 2016. Bioethanol from microalgae and cyanobacteria: a review and technological outlook. Process Biochem. 51, 1833–1842.

Silva, C.E.F., Sforza, E., 2016. Carbohydrate productivity in continuous reactor under nitrogen limitation: effect of light intensity and residence time on nutrient uptake in *Chlorella vulgaris*. Process Biochem. 51, 2112–2118.

Silva, C.E.F., Gris, B., Sforza, E., La Rocca, N., Bertucco, A., 2016. Effects of sodium bicarbonate on biomass and carbohydrate production in *Synechococcus* PCC 7002. Chem. Eng. Trans. 49, 241–246.

Silva, C.E.F., Sforza, E., Bertucco, A., 2017. Effects of pH and carbon source on *Synechococcus* PCC 7002 cultivation: biomass and carbohydrate production with different strategies for pH control. Appl. Biochem. Biotechnol. 181, 682–698.

Smolarski, N., 2012. High-Value Opportunities for Lignin: Unlocking Its Potential. Frost and Sullivan.

Tan, I.S., Lee, K.T., 2014. Enzymatic hydrolysis and fermentation of seaweed solid wastes for bioethanol production: an optimization study. Energy 78, 53–62.

Tolmasquim, M.T., 2016. Energia Renovável: Hidráulica, Biomassa, Eólica, Solar, Aceânica. Empresa de Pesquisa Energética (EPE), Rio de Janeiro, Available from http://www.epe.gov.br/Documents/Energia%20Renov%C3%A1vel%20-%20Online%2016maio2016.pdf.

Trivedi, N., Gupta, V., Reddy, C.R.K., Jha, B., 2013. Enzymatic hydrolysis and production of bioethanol from common macrophytic green alga *Ulva fasciata* Deme. Bioresour. Technol. 150, 106–112.

Tsunatu, D.Y., Atiku, K.G., Samuel, T.T., Hamidu, B.I., Dahutu, D.I., 2017. Production of bioethanol from rice straw using yeast extracts peptone dextrose. Nigerian J. Technol. 36 (1), 296–301.

Associadas da UNICA mais próximas de produzir etanol celulósico, 2015a. Available from: http://www.unica.com.br/noticia/2981091792031156797/associadas-da-unica-mais-proximas-de-produzir-etanol-celulosico/.

UNICA celebra acordo para o clima na COP21, 2015b. Available from: http://www.unica.com.br/noticia/22757742920320153336/unica-celebra-acordo-para-o-clima-na-cop21/.

2016. COP21: o papel estratégico do setor sucroenergético para que o Brasil atinja suas metas. Available from: http://www.unica.com.br/palavra-do-presidente/41597945920325548352/cop21-por-cento3A-o-papel-estrategico-do-setor-sucroenergetico-para-que-o-brasil-atinja-suas-metas/.

GranBio paralisa usina em Alagoas, 2016. Available from: http://www.valor.com.br/agro/4512288/granbio-paralisa-usina-em-alagoas.

GranBio recebe proteção de duas cultivares da cana energia, 2017. Available from: http://www.valor.com.br/agro/4824584/granbio-recebe-protecao-de-duas-cultivares-de-cana-energia.

Vaz Junior, S., 2011. Biorrefinarias: Cenários e Perspectivas. Embrapa Agroenergia, Brazil.

Vaz Junior, S., 2017. Rumo a uma química renovável brasileira a partir da biomassa vegetal. Rev. Virtual Quím. 9 (1), 238–247.

Vieira, G.E.G., Nunes, A.P., Teixeira, L.F., Colen, A.G.N., 2014. Biomassa: uma visão dos processos de pirólise. Rev. Liberato 15 (4), 105–212.

Virmond, E., Rocha, J.D., Moreira, R.F.P.M., José, H.J., 2013. Valorization of agroindustrial solid residues from biofuel production chains by thermochemical conversion: a review, citing Brazil as a case study. Braz. J. Chem. Eng. 30 (2), 197–229.

Vishtal, A., Kraslowski, A., 2011. Challenges in industrial applications of technical lignins. Bioresources 6 (3), 3547–3568.

Vital, A., 2015. Sorgo biomassa apresenta grande potencial para geração de energia e é alternativa para rotação com a cana-de-açúcar. Rev. Canavieiros 105, 44–49.

Wei, N., Quaterman, J., Jin, Y., 2013. Marine macroalgae An untapped resource for producing fuels and chemicals. Trends Biotechnol. 31 (2), 70–77.

Welker, C.M., Balasubramanian, V.K., Petti, C., Rai, K.M., DeBolt, S., Mendu, V., 2015. Engineering plant biomass lignin content and composition for biofuels and bioproducts. Energies 8, 7654–7676.
Worldwatch Institute, 2007. Biofuels for Transport: Global Potential and Implications for Sustainable Energy and Agriculture. Earthscan, Sterling, VA, London.
Wu, F., Wu, J., Liao, Y., Wang, M., Shih, I., 2014. Sequential acid and enzymatic hydrolysis in situ and bioethanol production from *Gracilaria* biomass. Bioresour. Technol. 156, 123–131.
Xu, Y., Isom, L., Hanna, M.A., 2010. Adding value to carbon dioxide from ethanol fermentations. Bioresour. Technol. 101, 3311–3319.
Yanagisawa, M., Nakamura, K., Ariga, O., Nakasaki, K., 2011. Production of high concentrations of bioethanol from seaweeds that contain easily hydrolysable polysaccharides. Process Biochem. 46, 2111–2116.

Further Readings

Instituto Brasileiro de Geografia e Estatística, 2017. Systematic Survey of Agricultural Production. Available from: http://www.ibge.gov.br/home/estatistica/indicadores/agropecuaria/lspa/lspa_201704_6.shtm.
Rodrigues, R.M., 2012. Cidades brasileiras do passado ao presente, 3 Moderna, São Paulo, SP.
Vaz, S.M., 2011. O setor sucroalcooleiro e a sustentabilidade ambiental. RCA Revista Científica da Ajes: Available from: http://www.revista.ajes.edu.br/index.php/RCA/article/view/81/pdf.

Index

A

AA9 proteins, 261
Accessory enzymes, 254
  ligninases, 254
  loosenins, 254
  lytic polysaccharide monooxygenases, 255
  pectinases, 254
Acetaldehyde, 152, 176, 189, 199, 303
Acetic acid, 91, 92, 189, 194, 196, 219, 303, 325, 329, 437
Acetone, 196, 221, 255, 388
Acetylornithine aminotransferase (ARG9), 374
  mutation in, 374
ADP glucose pyrophosphorylase (ADPGase), 17
Aerobic digestion processes, 438
Aerobic fermentation, 7
Aeromonas hydrophila, 175
Agar agar, 240
Agarose, 368
Agent-based model (ABM), 393
Agro-Ecological Zones (AEZ)
  for quantification of agricultural land, 405
Agronomic Institute of Campinas (IAC), 419, 427
Alcohol separation, 323
Alcohols product distribution, 329
Alcohol synthesis, 323
Algae, bioethanol production, 361
  biofuels, 363
    biodiesel, 363
    biohydrogen, 363
  feedstocks, 362
  macroalgae, 363
  microalgae, 362
  third-generation feedstock, 361
Algae store, 368
Algae sugar composition, for bioethanol production, 369
Algal carbohydrates, 367
Algal polysaccharides, 367
  bioethanol production, 367–369
Aliphatic acid, 300, 322
Ammonia, 219–220
Ammonia fiber explosion (AFEX), 219, 278
Amylases, 249, 351
Anaerobic digestion (AD) process, 346, 438
Anaerobic fermentation, 7
Analytic hierarchy process (AHP), 336
Animal feed generation, 51
Antidetonating additive tetraethyl lead [$Pb(C_2H_5)_4$], 418
ANUCLIM methodology, 406
Arabinose fermentation, 305
Arabinoxylan (AX), 33, 191
Area under curve (AUC), 406, 407
Aromatics, 300
Arthrospira (Spirulina) platensis, 370
Arundodonax, 427
Aryl-alcohol dehydrogenase, 270
Aryl-alcohol oxidase, 270
Ascomycota, 296
Asia-Pacific integrated model (AIM), 393
Aspen Hysys, 324
Aspen Plus, 324
  software, 325
Aspergillus aculeatus, 105, 351
Aspergillus awamori, 127, 222
Aspergillus ficuum, 154
Aspergillus niger, 155, 222
Aspergilus aculeatus, 178
AUC. *See* Area under curve (AUC)
Automobiles, variable fuel engines, 407
AX. *See* Arabinoxylan (AX)
Azadirachta indica L., 239
Azeotropic, 323

B

Bacillus licheniformis, 205, 270, 351
Bacillus subtilis, 306
Bagasse, biorefinery approach, 32, 81, 187, 320
  alternative performance, structure of, 337
  biojet fuel routes, sustainability metrics, 336
  composition, 325
  cost data, 335
    by-product prices, 334
    production, economic criterion, 337
  ethanol production routes, 332
  feasibility analysis, 331
    economic evaluation, 332–334
    environmental sustainability metrics, 334–336
    mass/energy efficiency, 331
    multicriteria decision analysis, 336–338
  motivation/background, 319–320
  normalized weights using ranking method, 338
  OC, breakdown of, 335
  PFD for
    bioethanol production, 328
    syngas fermentation, 330
    thermochemical ethanol production, 330, 338
  pretreatment, reactions involvement, 326
  process design/modeling, 325
    biochemical production, 325–326
    thermochemical production, 327–329
    via syngas fermentation, 329–331
  route selection, process simulation as tool, 324
  technological status, 321–323
  TOPSIS results, 338
Barley, 406
  area with edaphoclimatic potential, 413
Basidiomycetes, 270

Basidiomycota, 296
Beta vulgaris L., 45, 61, 81
Beverages, 48
BHA. *See* Butylated hydroxyl anisole (BHA)
Bioalcohols, 64, 81
Biobased economy, 3
Biobutanol, 64, 157, 165, 233, 364
Biodiesel, 54, 81, 101, 165, 213, 233, 239, 302, 349, 354, 362, 363, 386, 389, 430
Bioenergy Technologies Office (BETO), 334
Bioethanol, 165, 233, 402. *See also* Ethanol production, fuel production
- based feedstocks and their cell wall compositions, 5
- crops, biotechnology, 17
- as feedstock for the chemical industry, 189
- Network of Pernambuco, 430
- production
  - in Brazil, 417
  - global scenario, 4–6
  - from rice/wheat straw, 215
- for sustainable fuel and society, 188

Biofuel, 165, 213, 334, 345
- from algae, 362
- climate as competitive advantage, 405
- for developing countries, 403–404
- environmental or economic sustainability of, 386
- feedstock, 403
  - agricultural, 404
- generation, 386
- production
  - in Brazil, 417
  - mitigating potentially adverse impacts, 402
- production from food crops, 386
- projects, complexity of
  - innovations for analysis, 404
- soil and climatic and topographic variables, 408

Biogas, 21, 64, 73, 102, 109, 138, 165, 273, 355, 389, 424, 438
Biohydrogen, 64, 165, 362, 363
Biological pretreatment, 35, 171, 222, 268, 295
Biomass, 323
- combustion, 435–436
- crystallinity, 171
- cultivation, 344
- effects, different types of pretreatments on, 36
- fermentation products, biochemical conversion route, 322
- generations aimed to bioethanol production, 344
- pretreatment methods
  - advantages and disadvantages, 296
- thermochemical route, 323
- yield, 8

Biomass-to-energy conversion technologies, 187
Biomass-to-ethanol generations, 344
Bio-oil, to fuels, 437
Biopolymers, 260
BioProcess Simulator, 324
Biorefinery, 433. *See also* Brazilian ethanol biorefineries
- approach, 15, 354
  - CBP technology, 16–17
- bioethanol production, 343
- flow chart of bioethanol production
  - from microalgae biomass, 346
- from forest biomass
  - with other feedstocks, comparison of, 241
  - woody, 234
- heavy metals, removal of, 353
- hydrolysis, 350
  - alternatives processes, 351–353
  - *Arthrospira platensis*, 351
  - *Chlamydomonas reinhardtii*, 351
  - *C. vulgaris*, 351
  - enzymatic, 351
  - lipids utilization, 353–354
  - *Scenedesmus bijugatus*, 350
  - *Scenedesmus obliquus*, 350
  - sulfuric acid, 350
  - *Synechococcus* sp. PCC 7002, 351
  - traditional methods, 350
- nitrogen starvation/limitation
  - carbohydrate accumulation, mechanism of, 347
- nutrients recovery, 354
  - CO_2 recycling, 355–356
  - future, 356
  - microalgae, potential bioethanol productivity, 356
  - vinasse, anaerobic digestion of, 354–355
- water salinity reverse osmosis and dilution, 349

Biorefining technologies, bagasse exploiting, 320
Biorenewable energy platforms, 233
BioScope, 393
Biotechnology Design Simulator (BDS), 324
Biotechnology, of bioethanol crops, 17
Blending, of bioethanol with gasoline E2-E20, 165
Bovine serum albumin (BSA), 259, 276
Brazil, bioethanol
- biofuel production, 417
- fourth-generation ethanol, 418
- 2G/3G technologies, 417, 418
- global energetic matrix, 417
- global leaders in fuel ethanol production, 45
- sugarcane exploitation, 418
  - historical background/geographic impositions, 418–421

Brazilian Bioethanol Science and Technology Laboratory (CTBE), 428
Brazilian Development Bank (BNDES), 426, 430
Brazilian ethanol biorefineries
- biomass combustion, 435–436
- CO_2 sequestration, 438–439
- fine chemicals, 433–434
- lignin utilization, 436–437
- sustainability and energy balance of, 433
- vinasse, anaerobic digestion of, 437–438

Brazilian Institute of Environment (IMA), 431
Brazilian Institute of Geography and Statistic (IBGE), 418, 423
Brazilian National Council for Scientific and Technological Development (CNPq), 426
Brazilian Sugarcane Industry Association, 421
Brazil, lignocellulosic ethanol, 421
- bagasse, 423
- elephant grass, 424–425
- industrial plants, 430–432
- macro/microalgae, 432–433
- potential biomasses, 423
- research centres
  - CENPES, 427–428
  - Centre for Strategic Technology of the Northeast (CETENE), 430
  - CTBE, 428–429

Embrapa, 425–427
National Institute of Technology (INT), 429
sorghum, 425
sugarcane straw, 423
wheat/rice straw, 424
Broth distillation, 323
Brown seaweed *(Phaeophyceae)*, 363
Butanol, 157
acetone-butanol-ethanol yield, 157
industrial-scale fermentation, 157
by solventogenic clostridia, 157
JA as feedstock for production, 157
from renewable biomass, 157
zinc micronutrient enhance production, 157
Butylated hydroxyl anisole (BHA), 240

C

Caffeic acid 3-O-methyltransferase (COMT) gene, 17
Calcium alginate, 240, 260
matrix, 260
Calcium carbonate, 64
Caloramator boliviensis, 105
Campinas-SP, 431
Canal irrigation, 95
CanaVialis, 419, 420
Candida glabrata (Cgrd1), 178
Candida shehatae, 178, 223, 237, 308
Carbohydrate accumulation, 347
Carbohydrate content, bioethanol production, 370
genetic engineering, 371–376
genetic material, for higher algal ethanol, 373
agitation with glass beads method, 375
agrobacterium-based transformation of microalgae, 376
algal strains, selection of, 373
biolistic delivery, 375
chloroplast transformation technique, 374
electroporation, 375
genetically modified algal strains, stability of, 376
nuclear transformation techniques, 374
protoplasts, generation of, 375
genetic modifications, role of, 372
medium components/environmental parameters, optimization of, 370–371
nutrients, 370
temperature, 370
metabolism/storage in algae, 372
Carbohydrate, in weed species, 299
Carbohydrate-rich biomass, 417
Carbon dioxide (CO_2), 64, 187, 213
emissions, 31, 432, 438
Carbon emissions, 31
Carbon footprint, 102
Carbon neutrality, 394
Carboxymethylcellulose (CMC) derivates, 260
Cassava, 4, 11, 406
area with edaphoclimatic potential, 412
based ethanol
China, 107–108
Thailand, 108–109
Vietnam, 109–110
as energy crop, 102
carbon footprint, 102
eutrophication potential, 104
water footprint, 103
roots
bioethanol production from, 104–105
new developments, 105–106
processing steps, to ethanol production process, 105
starch, bioethanol production, 393
wastes and peels, bioethanol from, 106–107
CBP. *See* Consodilated bioprocessing (CBP)
Cell immobilization, 240
Cellobiose, 255
Cellobiose dehydrogenase (CDH), 270
Cellulases, 90, 166, 249, 256, 261
Cellulases-xylanases, synergistic effect of, 261
Cellulose, 33, 190, 191
alcohols, 15
biofuel producers, 424
cellulose binding domain (CBD), 259
deactivators, 255
ethanol production, 402
fibers, 190
in glucose, conversion, 422
hydrolysis, 325
to C_6 sugars, 321
reactions involvement, 326
partial chemical structures, 191
Cell walls
architecture, 34
cellulose, 33
classes of linkages, 34
hemicelluloses, 33
microfibrils, 33
mixed linkage β-glucan, 34
pectic polysaccharides, 33
polymers, 33
species taxonomically closer, 35
from sugarcane, 32–35
xyloglucans, 33
Center of Sugarcane Technology (CTC), 419, 430, 432
Centre for Strategic Technology of the Northeast (CETENE), 430
CEPCI. *See* Chemical engineering plant cost index (CEPCI)
Cereal grains, 11
Ceres, 425
Ceriporia lacerata, 171, 237
Ceriporiopsis subvermispora, 171, 198, 222, 269
Cetane number, 293
Chemical engineering plant cost index (CEPCI), 332
Chemical treatments, 188, 278
Chlamydomanos species, 364
Chlamydomonas reinhardtii, 372
Chlorella sorokiniana, 352, 354
Chlorella vulgaris, 349
FSP-E, 433
Chrysolaminarin, 347
Circulated fluidized bed (CFB) reactor, 327
gasifier, 327
cLCA. *See* Consequential life cycle assessment (cLCA)
Clostridium acetobutylicum, 175
Clostridium ljungdahlii, 329
Clostridium phytofermentans, 16, 205
Clostridium tetani, 175
Clostridium thermocellum, 16, 225
Clostridium thermohydrosulfurium, 175
Clostridium thermosaccharolyticum, 175
Clostridium thermosulfurogenes, 175
Clostridum acetobutylicum L7, 157
Cold-cook (BPX) technology, 56
Conodont dactylon, 13
Consequential life cycle assessment (cLCA), 394
Consodilated bioprocessing (CBP), 249, 296, 322
Conventional fossil fuels, 392
Coriolus versicolor, 280

Corn, 45, 46, 406
advantages of using, 48
based ethanol, 46
animal feeds, as beneficial by-product, 54
microorganisms used in fermentation processes for, 53–54
butanol, 388
distribution of corn usage in USA, 46
global corn production and yield from 2015-17, 47
grain milling, 51
major corn producing countries, 46
processing for ethanol production, 50
stover, 11
technological aspects, ethanol production from, 50–54
Costa Pinto plant, 431
P-Coumaric acid, 191
P-Coumaryl alcohol, 193
Covalent linkages, 35
biological pretreatments, 38
chemical pretreatments, 38
Crop land, 405
Cyanidioschyzon merolae, 372
Cyanobacteria, 4, 346
biomass concentration, 348
cultivation, 348
enzymatic hydrolysis, 351
genetically modified, 344, 345
produce glycogen, 347
Cyathus stercolerus, 171, 237

D

Daedalea flavida, 280
DDGS. *See* Distiller dried grains (DDGS)
Degradation, of lignocelluloses, 235
Degree of polymerization (DP), 151, 169, 172
Delignification, 196, 236
efficiency, 237
of forest biomass, 236
microorganisms, 171
oxygen, 236
Depolymerization, 193
Detoxification, 91, 220, 284, 299, 322
Dextrin, 294
Diffusion coefficient, 166
Dilute acid hydrolysis, 322
Distillation, 8, 10, 51, 55, 74, 85, 95, 128, 138, 215, 235, 238, 296, 323, 326, 419
Distiller dried grains (DDGS), 48, 392
Distilleries, 55, 402, 419
DMT. *See* Dry metric tons (DMT)
DNA transformation system, 373
Docosahexaenoic acid (DHA), 362
Dosages, of enzymes, 256
DP. *See* Degree of polymerization (DP)
Dry metric tons (DMT), 366
Dry-milling process, 52
flow chart showing various steps, 52
Dunaliella bardawil, 371
Dunaliella salina, 366
Dunaliella teriolecta, 356, 439
DuPont, 423
Dynamic life cycle assessment (DLCA), 393

E

Economic impacts, of bioethanol, 18–19
Edaphoclimatic zoning model
characteristics of, 407
Eichhornia crassipes, 298
Electricity imports, 331
Elephant grass, 423, 424
Embrapa Agronergia, 425, 427
Endo-xylanase, 90
Energy cane, 420
Energy consumption, 256
Energy Independence and Security Act (EISA), 386
Entner-Doudoroff (ED) pathway, 304
Environmental impacts, 20
Enzymatic hydrolysis, 256, 351, 422
of starch, 250, 251
Enzymatic synergy, 261
Enzyme immobilization, 259, 260
Enzyme inhibition, 255
Enzymes recycling, 259
Enzyme technology, basics of, 250
amylases, 250–252
bioethanol production, 251
cellulases, 252–253
first-generation bioethanol, 256–257
hemicellulases, 253–254
second-generation bioethanol, 257–258
Epoxyethane, 189
Equipment cost (EC), 334
Erwiniachrys anthemi, 306
Erwinia chrysanthemi EC 16, 72
Escherichia coli, 17, 178, 286, 301, 308
as engineered microorganisms, 226
*glg*16 gene, 17
KO11, 72
ribulose-5-p 4-epimerase, 205
Ethanol, as bioenergy resources, 215
Ethanol, as biofuel, 344
Ethanol biorefinery, 346
from microalgae, economic/energetic analysis, 356
reuse of carbon dioxide, 355
Ethanol blended, with petroleum, 293
Ethanol fermentation, 70–72
reaction data, 327
Ethanol production, 8
fuel production, 401
barley area with edaphoclimatic potential, 413
biofuel climate, as competitive advantage, 405
biofuels, for developing countries, 403–404
biofuels projects, complexity of innovations for analysis, 404
cassava area with edaphoclimatic potential, 412
edaphoclimatic zoning model, characteristics of, 407
land suitability, 405
layers of soil and climatic, and topographic variables, 408
maize area with edaphoclimatic potential, 410
methodological approach/data, 405–406
methyl tertiary butyl ether (MTBE), 401
Mexican crop lands, competitive advantages, 406–413
sorghum area with edaphoclimatic potential, 411
sugar beet area with edaphoclimatic potential, 411
sugarcane area with edaphoclimatic potential, 410
value chain, 403
ethanol fuel, 402
wheat area with edaphoclimatic potential, 412
recent advances, 55, 56
Ethyl acetate, 189
Ethylene glycol, 189
1-Ethyl-3-methylimidazolium acetate, 222
Eukaryotic microalgae, 432
Eutrophication, 20, 21, 101, 104, 141, 389
potential, 104
Expansin-like proteins, 249

F

Fatty acid methyl esters (FAME), 363
profile, 354
Fatty acids, 239
Federal Rural University of Pernambuco (UFRPE), 430
Federal University of Pernambuco (UFPE), 430
Federal University of Rio de Janeiro (UFRJ), 427
Fenton reaction, 270
FER. *See* Fossil fuel energy ratio (FER)
Fermentation, 166, 350, 422
batch reactor, 326
inhibitors, 299
effects, overcome via, 299
enzymes to catalyze detoxification of, 299
furan derivatives, 299
organic acids, 299
phenolic compounds, 299
Ferulic acid, 191
First-generation (1G) feedstock, 4
industrial process of bioethanol production from, 8
Flash hydrolysis (FH) process, 354
Food crops, bioethanol production, 388
Food, feed, fiber, and fuels (4Fs), 407
Food security, 294
food costs, 402
Food *vs.* fuel debate, 17–18, 56
Forest biomass
biofuel production from, 235
and global biofuel economy, 243
technoeconomic feasibility of bioethanol production from, 242
Forest resources, for harnessing biofuels
bioprospecting of, 234
Forestry lignocellulosic biomass derived, 319
Forest/woody biomass
to bioethanol, processing routes, 235–236
distillation and bioethanol recovery, 238
fermentation, 237
hydrolysis, 237
pretreatment, 236
technoeconomic feasibility, bioethanol production from, 242
FORTRAN statements, 327
Fossil energies like oil, 31
Fossil fuel energy ratio (FER), 334
Fossil fuels, 165, 187, 233, 243, 267, 319
reserves, 3, 319
Fucoidan, 369
Fuel crisis, 361
Fuel ethanol, production of, 45
Fungal ligninolytic enzymes, 268
Fungal pretreatment
agricultural residues as potential substrate to, 271–272
crop residues, fermentation of, 284–285
alternative systems for saccharification and, 284
development of, 272
carried out at lab scale, 273
solid state fermentation (SSF), 272
disadvantage, 268
experimental implementation, showing stages, 274
optimization of parameters, 273
aeration, 275
moisture, 275
mycelium content, 275
particle size, 276
temperature, 275
time, 273
for producing ethanol, 268
strategies to improve action of, 276–284
alternative strategies, 281
consolidated bioprocessing, 281
decentralized pretreatment, 281
integration for obtaining different by-products, 282
coculture of WRF, 280
combination of fungal and physicochemical pretreatments, 277–278
ligninolytic fungi, screening of, 276
nutrients and inducers of ligninolytic activity, supplementation of, 280
in vitro application of fungal ligninolytic enzymes, 282–284
Furan derivative, 300
Furfural, 68, 172, 220
Fusarium equiseti TWRF-10, 277
Fusarium oxysporum, 93, 225

G

Gallic acids, 255
Gasification, 234, 323, 436
Gasoline, 73, 97, 165, 189, 402
GCAM. *See* Global change assessment model (GCAM)
GDP, 56, 107
Gelatinization, 85
Genetic Algorithms technique, 393
Genetically engineered microorganisms, 301
actinomycetes, ethanol production using, 302
Bacillus subtilis, 306
Escherichia coli, 301–302
Klebsiella oxytoca, 306
Sacch. cerevisiae, 307–308
for pentose fermentation, 308–309
xylose utilization pathway, 309
thermophilic bacteria, 302–304
Z. mobilis, 304–305
Genetically modified (GM) algae, 373, 376
Genetically modified organism (GMO), 373
Geographic information systems (GIS), 393, 404
integrated climate model, 393
Gibbs free energy, 327
GIS. *See* Geographic information systems (GIS)
Global biofuel economy, 243
Global biosphere management model (GLOBIOM), 393
Global change assessment model (GCAM), 393
Global energy, 233
Global fuel ethanol
production in 2017, 48
production scenario, 45
Global industrialization, 294
Global oil, use in planet, 31
Global warming, 3, 50, 165
Global warming potential (GWP), 389, 391
Global wheat production
and usage, 49
and yield from 2015 to 2017, 49
GLOBIOM. *See* Global biosphere management model (GLOBIOM)
Glucoamylase enzymes, 85
δ-Gluconolactone, 255
Glucose, in ethanol, conversion, 423
Glucose 6-phosphate dehydrogenase, 204
Glucuronic acid, 33
Glucurono-arabino-xylan matrix, 190

Glutaraldehyde, 260
Glycomic code, 32
Glycosidic bond, 71
Glyoxal oxidase, 270
GMO. *See* Genetically modified organism (GMO)
GND1 gene, 204
Grain alcohol, 361
Grain-based feedstocks, 4
Grain processing, recent advances, 55, 56
Grains' competition, 4
GranBio, 420, 431
Gravitational sedimentation, 348
Green algae, 433
Greenhouse gas (GHG)
  efficiencies, 72
  emissions, 3, 101, 103, 108–110, 165, 187, 267, 293, 319, 385, 405, 421
    biogas, 389
    carbon dioxide, 438
    COP21 Paris Agreement, 101
    fertilizer production, 390
    global, 320
    renewable jet fuel, 389
Green seaweed *(Chlorophyceae)*, 363
GRE3 gene, 203
GWP. *See* Global warming potential (GWP)

H

Haematococcus pluvialis, 366
Heat of vaporization, 188
Heavy metals, removal of, 353
Helianthus tuberosus L., 149
Hemicellulases, 16, 90, 166, 174, 253, 285
Hemicellulose, 4, 10, 12, 32, 63, 67, 69, 72, 90, 166, 169, 171, 172, 174, 187, 190, 191, 193, 198, 219, 223, 235, 268, 285, 295, 298, 322, 326, 421
Hemicellulosic polysaccharides, 33
Higher heating values (HHV), 354
High-fructose syrup (corn syrup), 48
HMF. *See* Hydroxymethyl furfural (HMF)
Holocellulose, 187
Hordeum vulgare L., 406
Horizontal gene transfer (HGT), 376
HTL. *See* Hydrothermal liquefaction (HTL)
Hydrochloric acid, 294
Hydroenergy, 243
Hydrogen peroxide, 270
Hydrolysates, detoxification, 322
Hydrolysis, 350. *See also* Enzymatic hydrolysis
  alternatives processes, 351–353
  Arthrospira platensis, 351
  cellulose to C_6 sugars, 321, 322
  Chlamydomonas reinhardtii, 351
  C. vulgaris, 351
  dilute acid, 322
  enzymatic, 351
  lipids utilization, 353–354
  Scenedesmus bijugatus, 350
  Scenedesmus obliquus, 350
  sulfuric acid, 350
  Synechococcus sp. PCC 7002, 351
  traditional methods, 350
Hydrothermal liquefaction (HTL), 351
Hydrothermal pretreatments, 188
4-Hydroxybenzoic acids, 255
Hydroxycinnamates, 190
Hydroxy-cinnamic acids, 255
Hydroxymethyl furfural (HMF), 68, 172, 199, 351

I

Immobilization, 249
Inhibitors of fermentation process, 300
Institute of Sugar and Alcohol (IAA), 418
Intergovernmental Panel on Climate Change (IPCC), 393
Inter-Ministerial Council for Sugar and Alcohol (CIMA), 418
Internal combustion engine, 188
International Energy Agency (IEA Bioenergy), 17, 343, 366
Interuniversity Network for the Development of the Sugar-alcohol Sector, 419
Inulin
  characteristics, 151
  during flowering period, 151
  solubility, 151
  structure, 152
  yield, 151
Iogen Corporation, 431
Ipomoea batatas L., 115
Irpex lacteus, 222, 274

J

Jatropha curcas L., 239
Jerusalem artichoke, 4, 149, 150
  bioethanol production from tubers, 152
  consolidated bioprocessing (CBP), 156–157
  separate hydrolysis and fermentation (SHF), 152–155
  simultaneous saccharification and fermentation (SSF), 155
 characteristics, 149, 150
  of inulin and its yields from, 151
 ethanologenic microorganisms, 150, 153
 factors affecting fructoologosaccharides (FOS), 151
 as feedstock for biobutanol production, 157
 inulin content, 151
 production of bioethanol from inulin, 150
 simplified process of producing ethanol from, 157
 technoeconomic feasibility, 158
 total sugar levels, 151
 used for food and animal feeds, 149

K

Kappaphycus alvarezii, 15
Klebsiella oxytoca, 306
  P2, 72
Kluveromyces marxianus, 150
Kluyveromyces cicerisporus, 156
Kluyveromyces fragilis, 155, 156
Kluyveromyces marxianus, 156, 175, 177, 237

L

Laboratory of Biocatalysis (LABIC), 429
Laboratory of Bioprocess Development (Ladebio), 427
Laccase pretreatment, of cellulose, 262
Laminaria digitata, 368
Lantana camara, 298
LCA. *See* Life cycle assessment (LCA)
LCB. *See* Lignocellulosic biomass (LCB)
LCC. *See* Life cycle costing (LCC)
Leuconostoc dextranicum, 92
LHV. *See* Low heating value (LHV)
LHW. *See* Liquid hot water (LHW)
Life cycle assessment (LCA), 20, 21, 386, 391
  biofuel LCA, 394
  biofuel production, 386
  consequential life cycle assessment (cLCA), 394

economics of, 392
environmental benefits, 391–392
feedstock cultivation stage, 390, 391
from food crops, 388
grain crops, 388–389
herbaceous crops, 390
oil seeds, 389
tuber crops, 389
GHG emissions, 394
global energy consumption, 385
land, 390
methodology, 387–388
production costs of, 392
stages of, 387
sustainability improvement, 393–394
water, 390
Life cycle costing (LCC), 395
Lignin, 187, 190, 193, 219, 259
partial chemical structure, 193
structural units, 193
Ligninases, 249
Lignin decay capabilities, in white-rot fungi, 268
Ligninolytic fungi, 269
pretreatment of crop residues by, 269–270
Lignocellulose based feedstock, 16, 199
Lignocellulose degradation, 270
by white rot fungi, 269
Lignocellulosic bioethanol production, 166
carbohydrate metabolism, 176
challenges, 177–178
cofermentation, 177
genetic engineering, 178
sensitivity to products and inhibitors, 177
conventional microorganisms, 168
cost of enzymes for saccharification, 166
economic assessment, 178–179
ethanol production from various sugars and biomass sources, 168
feasibility at industrial scale, 172
feedstocks, major fractions, 169
fermentation, 174
Clostridium species, 175
Kluyveromyces species, 175
metabolic pathway, 175–176
Pichia species, 175
process configurations, 176–177
Thermoanaerobacterium species, 175
Thermoanaerobacter species, 175
thermotolerant/thermophilic ethanologens, 175
hydrolysis of carbohydrates, 166
key steps for biochemical conversion, 166
main aim of pretreatment, 166
pretreatment, 169
along with equipments required, 173
biological, 171
chemical, 171
parameters for, 171
accessible surface area, 172
biomass crystallinity, 171
degree of polymerization (DP), 172
lignin and hemicelluloses contents, 172
physical, 169
physicochemical, 171
saccharification, 174
cellulose hydrolysis, 174
aerobic microorganisms, 174
anaerobic microorganisms, 174
hemicellulose hydrolysis, 174
schematic representation, 167
Lignocellulosic biomass (LCB), 32, 166, 189, 345
heterogenity in carbohydrate composition, 298
as source of sugar for fermentation, 295
Lignocellulosic complex, 191
Lignocellulosic ethanol
in Brazil, 421
conversion of, 421
costs of, 421
production of, 421
Lignocellulosic materials, 45, 70
Lignocellulosic residues
potential use of, 434
Lignocellulosics/macroalgae, recalcitrance of, 344
Linoleic acids, 239
Liquefaction, 9, 50, 51, 52, 56, 86, 88, 125, 126, 351
Liquid hot water (LHW), 322
Low heating value (LHV), 334
Lytic polysaccharide monooxygenases (LPMO), 261

M

Macroalgae, 14
bioethanol production, 366–367
Madhuca indica, 238
Madhuca latifolia, 238
Madhuca longifolia, 238
Magnesium ammonium phosphate ($MgNH_4PO_4$), 352
MAgPIE. *See* Model of agricultural production and its impact on the environment (MAgPIE)
Maleic acid, 220
Maltose, 294
Manihot esculenta Crantz, 81, 101, 406
Manihot glaziovii, 105
Mannanases, 270
Mannitol, 368
Mannosidases, 270
Marine algae, biomass, conversion of, 378
MaxEnt-based technology, 405
MaxEnt executes log-log (clog-clog) transformation, 406
Maximum Entropy Modeling (MaxEnt), 405, 407
MaxEnt software, 406
Mechanical pulverization, 188
Melanoidins, 70
Methyl tertiary butyl ether (MTBE), 401
Michaelis-Menten constant, 260
Microalgae, 14, 346, 347, 364, 432, 433
bioethanol production, 364
biofuel production, challenge of, 364
biomass
for bioethanol production, 365
cultivation, 346
flow chart of bioethanol production, 346
hydrolysis, 353
production, 348
carbohydrate production, 346, 347
cultivation, 348
carbon gas, 439
nitrogen and phosphorous supplied, 350
lipid production, 346
Microfibrils, 33, 368
Micromonas pusilla, 372
Microwave irradiation, 219
Millettia pinnata L., 239
Ministry of Agriculture, Livestock and Supply (MAPA), 418, 425
Ministry of Science, Technology and Innovation (MCTI), 428
Mixed alcohols synthesis (MAS) reactor, 327

Mixed-integer optimization model, 393
Mn(II) supplementation, 280
Model of agricultural production and its impact on the environment (MAgPIE), 393
Mohua (*Madhuca* spp.), 238
  biodiesel from seeds, 239
  bioethanol crop, 238
  biorefinery products, advance technology for production, 239
    cell immobilization, 240
    solid state fermentation (SSF), 239
    submerged fermentation (SmF), 239
  consolidated bioprocessing/engineered microorganism, 241
  economics Mohua biofuel, 240
  thermooxidative stability of (Mohua) oil, 240
  transesterification of (Mohua) oil, 240, 241
Mohua oil methyl ester (MOME), 240
Molasses, 5, 7, 45, 63, 66, 94, 97, 238, 386, 390, 419
MOME. *See* Mohua oil methyl ester (MOME)
Monoethanolamine (MEA), 327
Monolignols, 193
Monte Carlo simulation, 94
MTBE. *See* Methyl tertiary butyl ether (MTBE)
MYB4-OX switchgrass, 17

N

Nannochloropsis gaditana, 354
Nanoporous materials, 260
National Alcohol Programme, 418
National Renewable Energy Laboratory (NREL), 343
National Research Centre for Sorghum (NRCS), 96
Natural gas, 187, 233, 413
Neurospora crassa, 93, 225
NexSteppe, 425
Nitrogen-based fertilizers, 363
Nitrogen starvation/limitation
  mechanism of carbohydrate accumulation, 347
N_2O emissions, 73
Noncompetitive inhibitors, 255
Noncovalent linkages
  effect of pretreatments, 39–40
Nonfermentable substances, 66
Nonfood ethanol, 4
Nonrenewable fuels, 3
Nonrenewable resources, 187
Nonwood biomass, 11
Novozymes, 423, 432
Nutrient recycling, 348, 349, 354
  CO_2 recycling, 355–356
  future, 356
  microalgae, potential bioethanol productivity, 356
  vinasse, anaerobic digestion of, 354–355

O

OC. *See* Operating costs (OC)
Octane number, 165, 188, 293
Offset fossil fuel, 385
Oil-palm tree, 12
Operating costs (OC), 333
Ostreococcus lucimarinus, 372
Ostreococcus tauri, 372
Oxidative systems, 268
Oxygenic photosynthesis, 346
Ozone depletion potential, 391
Ozonolysis, 322

P

Pachysolen tannophilus, 91, 223, 237, 285, 308
Palmitic acids, 239
Panicum virgatum, 12, 425
Paris Agreement, 101
Particle size, 276, 277
Particulate material, 31
PCraster Land Use Change model (PLUC model), 393
Pectin, 33, 35, 63, 64, 69, 105, 187, 269, 303
Pectinases, 69, 90, 135, 249, 351
Penicillium chrysogenum, 275
Pennisetum purpureum, 13
Pentose phosphate pathway (PPP), 199
Peroxides ionic liquid (IL), 219
Phaeodactylum tricornutum, 354, 372
Phanerochaete chrysosporium, 171, 222, 237
Phanerocheate chrysogenum, 280
Phlebia radiata, 280
Phlebia subserialis, 222
Phosphorous, 233
Photobioreactor (PBR), 376
Photofermentation, 345
Photosynthesis, 187, 346
  efficiency, 81
Pichia kudriavzevii, 308
Pichia pastoris, 105
Pichia stipitis, 178, 200, 223, 237
Pilot plant for process development (PPDP), 428
Plant cell walls, 32
Pleurotus eryngii, 222
Pleurotus ostreatus, 222, 237, 296
Pleurotus ostreaus, 171
Pleurotus sajor-kaju, 280
Poaceae, 190
Policy issues, 19–20
Poly (ethylene terephthalate), 189
Polyalkylene glycols, 261
Polylactic acid, 48
  polymers, 48
Polymers in cell wall, sugarcane, 33
Polyphenols, 71
Polysaccharides, 270
Polyvinylchloride, 189
Poria subvermispora, 276
PPDP. *See* Pilot plant for process development (PPDP)
Pressure swing adsorption (PSA), 95
Process flow diagram (PFD)
  for bioethanol, 326
Process simulation, route selection, 324
Process Systems Enterprise Limited (PSE), 324
Production cost, bioethanol, 392
Production technologies, 20–21
ProSimPlus, 324
Prototheca wickerhami, 376
PSA. *See* Pressure swing adsorption (PSA)
*Ps*XR mutant, 201
PvMYB4 overexpression, 17
Pycnoporus cinnarrbarinus, 171

Q

Quinone reductase (QR), 270

R

Recalcitrance, 32
Receiver operating characteristic (ROC), 407
Recombinant enzymes, 258–259
Recycling, 15, 66, 105, 202, 203, 258, 260, 343, 348, 419, 423
Red algae, 433
Redlich-Kwong-Soave cubic equation of state with Boston-Mathias alpha function (RKS-BM), 325
Red seaweed *(Rhodophyceae)*, 363

Renewable feedstocks, 6
first-generation feedstock, 7
starch-containing feedstock, 7
sugar-containing feedstock, 7
second-generation feedstock, 9
cassava bagasse, 11
cereal straws, 11
corn stover, 11
native plants, 12–13
natural nonwoody plant fibers, 13
nonwoody biomass, 11
oil-palm biomass, 12
sugarcane baggase, 11
sweet potato residues, 11
woody biomass, 10
third-generation feedstock, 14
Renewable fuel
bioethanol, prospects of, 378
Renewable Fuels Standards (RFS) program, 50
Rhamonogalacturonan I (RG I), 33
Rhodium-based catalysts, 323
Rice straw, 214, 424
acid pretreatment, 220
alkali pretreatment, 219
ammonia treatment, 219
availability, 214
biological pretreatment, 222
combined pretreatment, 222
consolidated bioprocessing (CBP), 225
modification of genes, 227
modification of strains, 227
engineered microorganisms employed, 226
enzymatic hydrolysis, 223
ethanol production, 223, 224
fermentation, 223–225
separate hydrolysis and fermentation, 225
simultaneous saccharification and cofermentation, 225
simultaneous saccharification and fermentation, 225
IL pretreatment, 221
microwave pretreatment, 219
organosolv pretreatment, 221
physical pretreatment, 216
high-energy irradiation, 216–219
milling, 216
potential of bioethanol production, 215
pretreatment, 215
conditions, 217
with oxidizing agent, 220
straw collection, processing, and transportation, 214
sustainable straw extraction, 215
technoeconomical feasibility of bioethanol production, 227
Ridesa, 420
ROC. *See* Receiver operating characteristic (ROC)
RSTOICH reactor, 327
RYIELD reactor, 327

S

Saccharification, 51, 55, 85, 166, 259
Saccharomyces cerevisiae, 7, 16, 52, 53, 72, 83, 94, 150, 168, 188, 198, 223, 237, 285, 326, 345
deletion of putative suppressor genes, 206
endeavors of evolution in xylose-fermenting, 199–201
evolution, toward second-generation bioethanol production
from lignocellulosic biomass, 198–199
improved flux in oxidative PPP, 204
increasing NADPH pool, 204
innovating integration of recombinant and evolutionary strategies, 205
metabolic pathways in recombinant, 204
optimization of XR/XDH ratio, 203
overcoming xylitol accumulation, 202, 203
overhaul of coenzyme specificity of XR and XDH, 201–203
vs, *Zymomonas mobilis*, comparative analysis, 54
xylitol excretion, 201
Saccharum officinale L., 61
Saccharum officinarum L., 81, 187
Saccharum spontaneum, 298
Scheffersomyces (Pichia) stipites, 91, 205, 308
*Sc*XK gene, 200
Seaweeds, bioethanol production, 366, 367
Second-generation ethanol plant, 334
Second-generation (2G) feedstock, 4
industrial process of bioethanol production from, 10
Separated hydrolysis and fermentation (SHF), 126, 249, 351
Sequential hydrothermal liquefaction (SHTL), 352
Shelled corn
chemical composition, 48
SHF. *See* Separated hydrolysis and fermentation (SHF)
SHTL. *See* Sequential hydrothermal liquefaction (SHTL)
Simulation of nitrogen in cropping systems model ($SIMS_{NIC}$ model), 393
Simultaneous saccharification and fermentation (SSF), 53, 93, 249
Skeletonema costatum, 371
SLR. *See* Solid-liquid ratio (SLR)
SmF. *See* Submerged fermentation (SmF)
SMS. *See* Spent mushroom substrate (SMS)
Socioeconomical advantages, 56
Sodium, phosphorous, and potassium (NPK) levels of soil, 63
Soil conservation, 420
Solid-liquid ratio (SLR), 65
Solid state fermentation (SSF), 239, 272
Solid substrate fermentation, 92
advantages, 92
ethanol productivity, depend on, 92
influence of various process parameters, 92
Sorghum, 81, 425
area with edaphoclimatic potential, 411
biomass, 423
Sorghum bicolor [L.] Moench, 81, 406
Sorghum fermentation
biotechnology, 91
cell immobilization, 93–94
simultaneous saccharification and fermentation (SSF), 93
solid substrate fermentation, 92
use of adjuncts, 94
economics of biofuel production and, 95, 96
comparison of particulars per liter, 96
technoeconomic feasibility, 94
and real-time applications, 94–95
Sorghum grains, 85–86
fermentation and distillation, 87
gelatinization, 87
germination and sprouting, 88
grain crushing and milling, 86
liquefaction, 87

Sorghum grains (*cont.*)
physicomechanical and chemical technologies, 88
decortication, 88
particle size reduction, 88
steam-flaking, 89
processes for conversion into bioethanol, 86
saccharification, 87
techniques to improve ethanol yield, 87–88
use of exogenous enzymes, 88
Spent mushroom substrate (SMS), 282
SSF. *See* Solid state fermentation (SSF)
Starch-based feedstocks, 7
Starch-rich feedstocks, 294
Starchy materials, 70
Steam explosion, 188, 222
physicochemical process of, 426
Stearic acids, 239
Stovers, 81
Straw, 11, 32, 109, 418, 420, 431
Submerged fermentation (SmF), 239
Sucrose-containing bioethanol feedstocks, 7
Sugar beet, 45, 61, 81, 406
area with edaphoclimatic potential, 411
downstream processing, 65–66
extraction of juice, 65
global production, 61–62
largest producing countries, 62
as raw material, 63–64
roots, 61
pretreatment, 65
sugary juices
for bioethanol, 64–65
composition and characteristics, 66–67
Sugar beet ethanol, 72
economics, 72–73
life-cycle assessments, 72–73
Sugar beet pulp, 67
bioethanol production process from, 68
hydrolysis, 69–70
conditions, 70
pretreatment, 67–68
acid pretreatment, 68–69
liquid hot water, 69
mechanical comminution, 68
ultrasound, 69
Sugarcane, 32, 45, 406
area with edaphoclimatic potential, 410
Australia, 420
Barbados, 420
Brazil, 418, 419
Cuba, 420
fuel source for boilers, 420
India, 420
production of ethanol, 419
Sugarcane bagasse, 11, 189
chemical composition, 190
conversion technologies, 193
chemical pretreatment, 194
acid hydrolysis, 194
alkaline pretreatments, 196
hydrochloric acid, 195
nitric acid, 195
phosphoric acid, 195
sulfuric acid, 195
microwave assisting chemical pretreatments, 197–198
organosolv pretreatment, 196–197
oxidation bleaching, 196
physical pretreatment, 194
ultrasonic assisting chemical pretreatments, 197
Sugarcane-based bioethanol
GHG emissions, 390
Sugarcane biomass, 434
Sugarcane biorefinery
flowchart of, 435
Sugarcane cell wall architecture, 34
Sugarcane endo enzymes, 39
Sugarcane molasses. *See* Molasses
Sugarcane straw, 421
Sugarcane vinasse, anaerobic digestion, 437–438
Sugar crops, 4, 61
Sugar-energy sector, 426
Sugar-rich feedstock, 45
Sugars fermentation, 354
Sulfur, 233
Supercritical CO_2, 35
Surfactants, 261
Sustainability, 31
bioethanol from food crops, 390
renewable sources, 5
Sweetener, 48
Sweet potato, 115
as bioenergy crop, 143
critical aspects, 143
chemical composition, 121–124
coproducts from sweet potato stillage recovery, 128
current uses, 116–118
general characteristics of crops, 115–116
harvest period, 115
industrial, 118–121
life cycle analysis, 140–142
processes for conversion into bioethanol, 125
ethanol recovery, 128
fermentation, 127–128
gelatinization, 126
liquefaction, 126
pretreatment, 125–126
saccharification, 127–128
technoeconomic feasibility, 136–139
cost of yeast, 139
energy cost, 138
fixed costs, 138
use of residual solids for animal feed, 136
variables, effect of, 134
enzymes, and temperature, 134–135
solid to liquid ratio, 134–135
wastewater treatment, 134
world production, 116–118
Sweet potato residues (SPRs), 11
Sweet sorghum, 4, 406
as biofuel crop, 82
in Brazil, 425
net returns involved in cost of cultivation, 96
processes for conversion into bioethanol, 82
dehydration, 85
distillation, 85
fermentation, 83–85
juice extraction, 83
stalk juice, 83
Sweet sorghum bagasse (SSB), 89, 280
enzyme hydrolysis, 90
ethanol fermentation from, 91
pretreatments for, 90
alkali pretreatment, 90
dilute acid hydrolysis, 90
steam explosion, 90
processes for conversion into bioethanol, 89
Switchgrass, 425
Synergy ratios, 261
Synthesis gas (*syngas*), 234
energy, 323
fermentation process, 329
production of, 436
Synthetic organic polymers, 260

T

Table cornstarch, 48
Tannic acids, 255
Taproot, 63
Technique for order preference by similarity to an ideal solution (TOPSIS) method, 336
Tetraselmis suecica, 370
Thalassiosira pseudonana, 372
Thalassiosira weissflogii, 376
Thermal hydrolysis, 86
Thermanaero-bacterium sp., 16
Thermoanaerobacter ethanolicus, 16, 175
Thermoanaerobacterium saccharolyticum, 175
Thermoanaerobium brockii, 202
Thermophilic bacteria, 302–304
Thermophilic/thermotolerant microorganisms, 166
Third-generation (3G) feedstock, 4
  industrial process of bioethanol production, 14
Tinea versicolor, 280
Titanium dioxide (TiO_2), 197
Ton of oil equivalent (tOe), 385
Total capital investment (TCI), 332, 334
  estimation methodology, 333
Transesterification reaction, 241
Trichoderma reesei, 178, 225, 430
Triethylamine, 189
Trilicvm aestivum L., 45, 406
Tuber crops, bioethanol production, 393

U

Ultrafiltration, 136, 259
Ulva fasciata, 15
Ulvan, 369
Unbleached hardwood kraft pulp (UHKP), 281
United States
  global leaders in fuel ethanol production, 45
    prefer corn, 50
  Renewable Fuels Standards (RFS) program, 50
University of Pernambuco (UPE), 430
Urban pollution, 3

V

Very high gravity (VHG) technology, 84
Vignis, 420
Vinasse
  AD of, 355
  characteristics of, 438
  produced from sugarcane, 354
Violuric acid, 282
Virtual sugarcane biorefinery (VSB), 428
Volvox carteri, 372

W

Water footprint, 103
Water/nutrient recycle steps
  and harvesting, 348–349
Water recycling, 348, 349
Water salinity, reverse osmosis and dilution, 349
Water stress index, 104
Wet metric tons (WMT), 366
Wet-milling process, 52
  flow chart showing various steps, 52
Wheat, 45, 406
  area with edaphoclimatic potential, 412
  based ethanol, 46
    technological aspects, 55
  bioethanol production, 389
  straw. *See* Wheat straw
Wheat straw, 214
  acid pretreatment, 220
  alkali pretreatment, 219
    ammonia treatment, 219
  availability, 214
  biological pretreatment, 222
  combined pretreatment, 222
  consolidated bioprocessing (CBP), 225
    modification of genes, 227
    modification of strains, 227
  engineered microorganisms employed, 226
  enzymatic hydrolysis, 223
  ethanol production, 223, 224
  fermentation, 223–225
    separate hydrolysis and fermentation, 225
    simultaneous saccharification, 225
      and cofermentation, 225
  IL pretreatment, 221
  microwave pretreatment, 219
  organosolv pretreatment, 221
  physical pretreatment, 216
    high-energy irradiation, 216–219
    milling, 216
  potential of bioethanol production, 215
  pretreatment, 215
    conditions, 217
    with oxidizing agent, 220
  straw collection, processing, and transportation, 214
  sustainable straw extraction, 215
  technoeconomical feasibility of bioethanol production, 227
White-rot fungi (WRF), 268
Wood biomass, 10
Wood-rotting basidiomycetes
  degradation of lignocellulose by, 270
World bioethanol production, 6
World-Clim Website, 406
World fuel ethanol production, 46, 294
  and challenges, 294
World's leading ethanol producers, 294

X

XR/XDH/XK ratios, 203
Xylan, 33, 34, 83, 172, 196, 220, 268, 278, 322
Xylanases, 69, 72, 249, 261
Xylitol, 219
Xylitol dehydrogenase (XDH), 199
Xyloglucan (XG), 33, 191
  binding, 34
Xylose, 14, 53, 72, 90, 107, 175, 188, 191, 195, 200, 202, 205, 223, 258, 308, 325, 429
Xylose reductase (XR), 199
β-Xylosidases, 270

Z

Zea mays L., 46, 81, 406
Zinc micronutrient, 157
ZWF1 gene, 204
Zymomonas mobilis, 53, 54, 127, 150, 155, 168, 223, 237, 240, 304
  MTCC 92, 7

FSC
www.fsc.org
MIX
Papier | Fördert gute Waldnutzung
FSC® C083411

Kolibri 360 GmbH
Ferdinand-Jühlke-Straße 7
99095 Erfurt, Deutschland
produktsicherheit@kolibri360.de